CHAPTER 3 Systems of Numeration

3000 B.C. Egyptian system, an additive system. Babylonian system, a place-value system (without zero).

2000 B.C. Multiplication by duplation and mediation.

230 B.C. Eratosthenes developed the prime number sieve.

A.D. 1000 Beginning of development of the symbols we use for numbers, 1–9. Hindus introduced the number 0 before A.D. 800.

CHAPTER 4 Real Number System

3000 B.C. Egyptians and Babylonians had a knowledge of counting and unit fractions.

600 B.C. Pythagoras and followers took first steps in the development of number theory—friendly numbers, perfect numbers, irrational numbers, prime numbers.

75 B.C. Heron—approximations of square roots.

1491 Calandra first printed process of long division.

1525 Rudolff first used $\sqrt{}$ symbol for square root.

1636 Fermat is often credited with founding modern number theory.

1820 Gauss, considered one of the three greatest mathematicians (with Archimedes and Newton), made many contributions to number theory. Proved the Fundamental Theorem of Arithmetic and contributed to many other branches of mathematics.

1872 Dedekind—irrational numbers.

1874 Cantor made many contributions in the area of real numbers.

1889 Peano's axioms for natural numbers.

1910 *Principia Mathematica*—logical structure of arithmetic.

CHAPTER 5 Algebra

3000 B.C. Egyptians and Babylonians had some knowledge of algebra, solved algebraic problems by trial and error rather than formalized procedures.

1850 B.C. Problems of algebraic nature in Moscow Papyrus.

1650 B.C. Rhind Papyrus used legs walking to right to represent +, legs walking to left to represent −. The word "aha" (heap) was used as the unknown quantity.

A.D. 250 Diophantus wrote *Arithmetica*, an algebraic treatment of number theory. First to take steps toward algebraic notation. Solved first-, second-, and third-degree equations.

820 Arab mathematician al-Khowarizmi wrote *Al-jabr*, an influential book on algebra and Hindu numerals.

1202 Fibonacci wrote *Liber abaci*, a book on arithmetic and elementary algebra that included advanced Hindu Arabic notation, a method of calculating with integers and fractions, solution of equations by algebraic processes, and the Fibonacci sequence, as well as some geometry and trigonometry.

1545 Tartaglia solved cubic equation.

Cardino wrote *Ara magna*, which included negative roots of an equation, imaginary numbers.

1580 Viète wrote *In Artem* and *Canon*. Did much for the development of symbolic algebra. Used vowels to represent unknown quantities.

1630 Girard—algebra, geometry, and trigonometry. First to use abbreviations sin, cos, tan.

Descartes is credited with using the beginning letters of the alphabet for constants and the last letters for variables.

3000 B.C. Bottomry contracts, earliest type of insurance, made in Babylon.

2000 B.C. Recorded evidence that interest was charged by Babylonians.

1661 First bank notes in Europe. Development of checks and bank loans.

1711 London Assurance Corporation and Royal Exchange Assurance Corporation—beginning of property and liability insurance.

1720 DeMoivre developed actuarial mathematics.

1752 First U.S. insurance company, "Philadelphia Contributionship."

1769 Lloyds of London was founded.

1935 Federal Deposit Insurance Corporation (FDIC) was developed.

3RD EDITION

ALLEN R. ANGEL
MONROE COMMUNITY COLLEGE

STUART R. PORTER
MONROE COMMUNITY COLLEGE

A SURVEY OF MATHEMATICS

WITH APPLICATIONS

▲▼ ADDISON-WESLEY PUBLISHING COMPANY

Reading, Massachusetts • Menlo Park, California • New York
Don Mills, Ontario • Wokingham, England • Amsterdam • Bonn
Sydney • Singapore • Tokyo • Madrid • San Juan

Sponsoring Editor: David Pallai
Production Supervisor: Jack Casteel
Text Designer: Vanessa Pineiro, Pineiro Design Associates
Copy Editor: Barbara Willette
Illustrator: Illustrated Arts
Art Consultant: Loretta Bailey
Manufacturing Supervisor: Roy Logan
Cover Design: Marshall Henrichs

We should like to acknowledge the following sources for the chapter-opening photographs.

Chapter 1, Allen R. Angel; Chapter 2, Ira Kirchenbaum/Stock Boston; Chapter 3, Ontario Science Center; Chapter 4, © Eugene Gordon 1982/Photo Researchers, Inc.; Chapter 5, Joan Liftin/Archives Pictures; Chapter 6, Marshall Henrichs; Chapter 7, Gerrard Fritz/Monkmeyer Press Photos; Chapter 8, Allen R. Angel; Chapter 9, Marshall Henrichs; Chapter 10, Marshall Henrichs; Chapter 11, Allen Photographers; Chapter 12, Marshall Henrichs; Chapter 13, Courtesy of International Business Machines Corporation.

Reprinted with corrections, June 1989

Library of Congress Cataloging-in-Publication Data
Angel, Allen R., 1942–
   A survey of mathematics with applications.
   Includes index.
   1. Mathematics—1961–   . I. Porter, Stuart R.,
1932–   . II. Title.
QA39.2.A54 1989   510   88–6334
ISBN 0-201-13696-1

DEFGHIJ–HA–89

Preface

The title of this text is a key to its contents and the type of course for which it is intended. Prospective readers are the nonmathematics, nonscience majors, including students majoring in the liberal arts, elementary education, the social sciences, and business. Many of these students take one or two semesters of mathematics to fulfill a graduation requirement, but some elect mathematics without being required to do so.

Our goals as authors were manifold, but the primary one was to write a text that students could read, understand, and enjoy. We hope that as students read through the text, they will begin to understand and appreciate mathematics and possibly decide to venture on to another mathematics course.

This text meets the needs of those states, including California, Florida, New Jersey and Texas, that now require students to obtain a minimum competency in mathematics as a prerequisite for graduation or transfer.

Features of the Text

- Clear explanations of topics.
- An abundance of detailed worked-out examples.
- Chapters that are, for the most part, independent so that the instructors can teach the topics in any order they wish.
- An ample supply of graded exercises at the end of each section.
- Practical applications are used wherever possible to reinforce the material and motivate the student.
- DID YOU KNOWs provide historical information, and special facts and ideas relating to the material.
- Problem-solving problems at the end of many exercise sets reinforce the problem-solving techniques presented in Chapter 1.
- Research exercises at the end of many exercise sets can be assigned to embellish the material.
- Answers to odd-numbered exercises, all Review Exercises, and all chapters tests are provided at the end of the text.
- The text is accurate. A substantial number of well-qualified mathematics teachers have carefully read through the manuscript, galleys, and pages, to ensure accuracy. Other professionals checked the answers to the exercises.

- Functional use of two color. Important definitions and procedures are highlighted in a second color.
- Review Exercises.
- Chapter Summaries.
- Chapter Tests.

Changes to the 3rd Edition

- A chapter on problem solving has been added.
- A second chapter on algebra has been added.
- Topics in the probability chapter have been rearranged and a section on finding probabilities using counting techniques has been added.
- Material on the computer has been combined into one chapter, and a section on LOGO has been added to the computer chapter.
- A section on the metric system has been integrated into the geometry chapter.
- The logic chapter has been reorganized.
- A section on Infinite Sets has been added.
- More examples and more exercises have been added to most sections.
- Problem solving problems have been added to many exercise sets.
- Research questions have been added to many exercise sets.
- Each chapter now ends with a chapter test to prepare students for their actual test.
- The exercise sets have been renumbered for greater clarity.

Acknowledgments

We would like to thank the many students and faculty members from all over the country who offered suggestions for improving the third edition.

We would also like to thank our colleagues at Monroe Community College for their valued suggestions. In particular we wish to thank: Robert Berry, Larry Clar, Patricia Burgess, Gary Egan, Hubert Haefner, Annette Leopard, Marilyn Semrau, and Mike Zwick. We would like to thank our students for their input.

We would like to thank our wives, Kathy and Joyce, and our children, Robert and Steven, and Todd, Teri, Lisa and Brian. Without their support and great sacrifice, this book could not have become a reality.

It is our pleasure to acknowledge the assistance given us by the staff of the Addison-Wesley Publishing Company. In particular we appreciate the advice and encouragement of our executive editor, Dave Pallai, and our production editor, Jack Casteel.

We also wish to give our thanks to all the conscientious reviewers and supplement authors that follow for their invaluable suggestions.

Instructors Manual and Student's Solution Manual — Susan Shrader, Caldwell College, Caldwell, NJ

CLAST *Supplement* — Julie Monte and Lea Pruet, Daytona Beach Junior College, Daytona Beach, FL

Software Testing Package — John Fraleigh, University of Rhode Island

REVIEWERS

Robert C. Bueker
Western Kentucky University

Carl Carlson
Moorhead State College

Ruth Ediden
Morgan State University

Raymond Flagg
McPherson College

Penelope Fowler
Tennessee Wesleyan College

Gilberto Garza
El Paso Community College

John Hornsby
University of New Orleans

David Lehmann
Southwest Missouri State University

Julie Monte
Daytona Beach Community College

Wing Park
College of Lake County

Bettye Parnham
Daytona Beach Community College

Gerald Schultz
Southern Connecticut State University

Minnie Shuler
Chipola Junior College

Steve Sworder
Saddleback College

Alvin D. Tinsley
Central Missouri State University

Shirley Thompson
Morehouse College

The authors would also like to thank the following people for providing additional information and suggestions for this text.

Donald Bennet
Murray State University

R. C. Bueker
Western Kentucky University

James Carney
Lorain County Community College

Deborah DeVecchi
Roger Williams College

Ben Divers, Jr.
Ferrum College

Penelope L. Fowler
Tennessee Wesleyan College

Jim Hunter
Hopkinsville Community College

Hilbert Johs
Wayne State College

Norman Ladd
College of the Redwoods

Lois G. Leonard
Erie Community College – South Campus

Powell Livesay
Elizabethtown Community College

Hubert J. Ludwig
Ball State University

Sr. Agnes Marie Marusak
Our Lady of the Lake University

Adele J. Miller
Cabrillo College

Fredric N. Misner
Ulster County Community College

Joyce Myster
Morgan State University

Wayne Rhea
Oklahoma Baptist University

Carolyn C. Smith
Tallahassee Community College

Raymond F. Smith
Whittier College

William Tomhaue
Concordia College

Michael Trover
Madisonville Community College

Shirley Wakin
University of New Haven

Sr. Armella Weibel
Alverno College

Norun R. Yamanu
Hawaii Community College

Supplements

Complete Solutions Manual   This supplement contains the worked-out solutions for *all* exercises in the text.

Even-Numbered Exercises Answer Book   Contains the answers to even-numbered exercises.

Printed Test Bank   At least three alternate tests per chapter are included in this valuable supplement. Instructors can use these items as actual tests, or as a reference for creating tests with or without the computer.

A–W Test Edit   This is a computerized test bank containing over 1,000 multiple-choice items. This program also allows the instructor to edit existing test items and/or enter new items. Tests may also be created with both multiple-choice and/or open-ended questions. Departmental software free upon adoption.

Student Solution Manual   This manual contains the worked-out solutions to all odd-numbered exercises in the text.

Guide to CLAST Mathematical Competency *(State of Florida)*   This manual contains all of the necessary material to help students prepare for the computational portion of the CLAST test. It includes worked examples and practice for CLAST skills, as well as practice tests. Prepared by Julie Monte, and Lea Pruet, Daytona Beach Community College.

To the Student

Our primary purpose in writing this text was to provide material that you could read, understand and enjoy. To this end we have used straightforward language and tried to relate mathematical concepts to everyday experiences.

The concepts, definitions and formulas that deserve special attention have been either boxed or set in boldface type. The exercises are graded so that the more difficult problems appear at the end of the exercise set. The starred (*) problems are more challenging and the problem solving problems are the most challenging.

Each chapter has a summary, review exercises, and a chapter test. The summary reviews the important concepts covered in the chapter. The review exercises provide additional practice on the material. The chapter test provides you with an opportunity to check your understanding of the material.

The answers to the odd-numbered exercises, all review exercises, and the chapter tests appear in the Answer section in the back of the text. However, you should use the answers only to check your work.

It is difficult to learn mathematics without becoming involved. To be successful, we suggest you read the text carefully and work each assignment in detail.

We welcome your suggestions and your comments. Good luck in your adventure in mathematics.

Contents

CHAPTER 5

THE REAL NUMBER SYSTEM

CHAPTER 6

ALGEBRA

CHAPTER 7

SYSTEMS OF LINEAR EQUATIONS AND INEQUALITIES

CHAPTER 1

Problem
Solving

1.1   INDUCTIVE REASONING

Every day of our lives we are faced with the challenge of solving problems. If we have a technique for solving problems, we can use it to arrive at a solution more easily. Techniques for arriving at conclusions include basing decisions on similar situations and studying patterns.

So that we can provide some examples of problem solving, let us first review a few facts about certain numbers. The **natural numbers** or **counting numbers** are the numbers 1, 2, 3, 4, 5, 6, 7, 8,.... The three dots mean that 8 is not the last number, but the numbers continue in the same manner (in this case, one by one). A word that we sometimes use is "divisible." If $a \div b$ has a remainder of zero, then a **is divisible by** b. The counting numbers that are divisible by 2 are 2, 4, 6, 8,.... These numbers are called *even counting numbers*. The numbers that are not divisible by 2 are 1, 3, 5, 7, 9,.... These numbers are called *odd counting numbers*.

■ **Example 1** _____

If two odd counting numbers are multiplied together, will the product always be an odd counting number?

Solution: To answer this question, we will examine the product of several pairs of odd numbers.

$$
\begin{array}{lll}
1 \times 3 = 3 & 3 \times 5 \ = 15 & 5 \times 7 \ = 35 \\
1 \times 5 = 5 & 3 \times 7 \ = 21 & 5 \times 9 \ = 45 \\
1 \times 7 = 7 & 3 \times 9 \ = 27 & 5 \times 11 = 55 \\
1 \times 9 = 9 & 3 \times 11 = 33 & 5 \times 13 = 65
\end{array}
$$

Examining the products, we see that none are divisible by 2. We might conclude from these examples that the product of any two odd numbers is an odd number.   ❖

■ **Example 2** _____

If we multiply an odd and an even counting number, will the product be an odd or an even counting number?

Solution: Let us look at a few examples.

$$
\begin{array}{lll}
1 \times 2 = 2 & 3 \times 2 = 6 & 5 \times 2 = 10 \\
1 \times 4 = 4 & 3 \times 4 = 12 & 5 \times 4 = 20 \\
1 \times 6 = 6 & 3 \times 6 = 18 & 5 \times 6 = 30 \\
1 \times 8 = 8 & 3 \times 8 = 24 & 5 \times 8 = 40
\end{array}
$$

"A great discovery solves a great problem, but there is a grain of discovery in the solution of any problem."

George Polya

Examining the products, we see that all of them are divisible by 2. Therefore we might conclude that the product of an odd and an even number is an even number. ❖

In Examples 1 and 2 we cannot conclude that the results are true for all counting numbers. However, from the patterns developed, we could make predictions. This type of reasoning process, arriving at a general conclusion from specific observations or examples, is called inductive reasoning.

> **Inductive reasoning** is the process of reasoning to a general conclusion through observations of specific cases.

Examples 3 and 4 illustrate how we arrive at a conclusion using inductive reasoning.

Example 3

Fingerprints are accepted as evidence in a court of law because, in millions of comparisons, no two people have been found to have exactly the same fingerprints. If we conclude after examining all the specific cases that no two people can have exactly the same fingerprints, we have reasoned inductively. Is it possible that, sometime in the future, two people will be found who do have exactly the same fingerprints? ❖

Inductive reasoning is often used by mathematicians and scientists in predicting answers to complicated problems. For this reason, inductive reasoning is sometimes called the **scientific method.** When a scientist or mathematician makes a prediction that they believe to be true based on specific observations, it is called a **conjecture.** After looking at the products in Example 1 we might wish to conjecture that the product of two odd counting numbers will be an odd counting number.

Example 4

What can you conclude about a number if the sum of the digits of the number is divisible by 3?

Solution: Let us look at some numbers, the sum of whose digits are divisible by 3.

Number	Sum of the Digits	Sum of the Digits Divided by 3	Number Divided by 3
114	$1 + 1 + 4 = 6$	$6 \div 3 = 2$	$114 \div 3 = 38$
234	$2 + 3 + 4 = 9$	$9 \div 3 = 3$	$234 \div 3 = 78$
7020	$7 + 0 + 2 + 0 = 9$	$9 \div 3 = 3$	$7020 \div 3 = 2340$
2943	$2 + 9 + 4 + 3 = 18$	$18 \div 3 = 6$	$2943 \div 3 = 981$
9873	$9 + 8 + 7 + 3 = 27$	$27 \div 3 = 9$	$9873 \div 3 = 3291$

In each of the examples we found that the sum of the digits is divisible by 3, and the number itself is divisible by 3. From these specific examples we might be tempted to generalize or conjecture that "if the sum of the digits of a number is divisible by 3, then the number itself is divisible by 3."   ❖

The result reached by the inductive reasoning process might be correct for the specific cases studied but might not be correct for all cases.

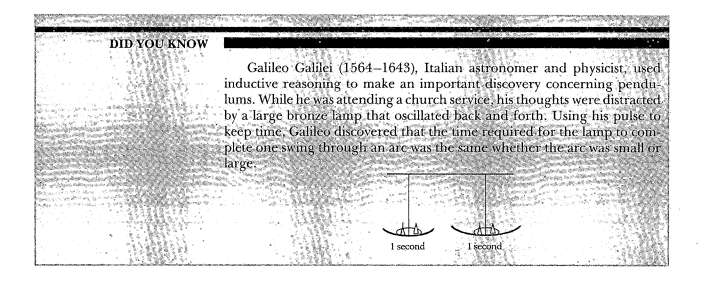

DID YOU KNOW

Galileo Galilei (1564–1643), Italian astronomer and physicist, used inductive reasoning to make an important discovery concerning pendulums. While he was attending a church service, his thoughts were distracted by a large bronze lamp that oscillated back and forth. Using his pulse to keep time, Galileo discovered that the time required for the lamp to complete one swing through an arc was the same whether the arc was small or large.

1 second   1 second

History has shown us that not all conclusions arrived at by inductive reasoning are correct. For example, Aristotle (384–322 B.C.) reasoned inductively that heavier objects fall at a faster rate than lighter objects. About 2000 years later, Galileo dropped two pieces of metal — one ten times heavier than the other — from the Leaning Tower of Pisa in Italy. He found that both hit the ground at exactly the same time, so they must have traveled at the same rate.

Whenever you arrive at a general conclusion using inductive reasoning, you should test it with several special cases to see whether the conclusion appears to be correct. If any of the special cases show that the conclusion is not correct, we call this special case a **counterexample.** Galileo's counterexample disproved Aristotle's conjecture. If a counterexample cannot be found, the conjecture is neither proven nor disproven.

■ **Example 5** ───────────────────────────────────

Pick a number, multiply the number by 4, add 6 to the product, divide the sum by 2, and subtract 3 from the quotient. The answer is ───── .

Solution: Let us go through this one together.

Pick a number:	say, 5
Multiply the number by 4:	$4 \times 5 = 20$
Add 6 to the product:	$20 + 6 = 26$
Divide the sum by 2:	$26 \div 2 = 13$
Subtract 3 from the quotient:	$13 - 3 = 10$

Note that you started with the number 5 and finished with the number 10. Now try this problem with the numbers 2, 3, 4, 6, 7, and 8. Each time, record the number with which you started and the number with which you finished. Using inductive reasoning, can you make a conjecture about the relationship between the starting number and the final number? ❖

The following set of exercises contains some easy and some challenging problems. Don't be disappointed if you cannot find a solution to all the problems on the first try. Don't give up. Try again.

"Your problem may be modest; but if it challenges your own curiosities and brings into play your inventive faculties, and if you solve it by your own means, you may experience the tension and enjoy the triumph of discovery."

George Polya

EXERCISES 1.1

Use inductive reasoning to predict the next line in the pattern for each of the following.

1. $(1 \times 9) + 2 = 11$
$(12 \times 9) + 3 = 111$
$(123 \times 9) + 4 = 1111$

2. $1 = 1$
$1 + 2 = 3$
$1 + 2 + 3 = 6$
$1 + 2 + 3 + 4 = 10$
$1 + 2 + 3 + 4 + 5 = 15$

3.
1	$= 1$
$1 + 3$	$= 4$
$1 + 3 + 5$	$= 9$
$1 + 3 + 5 + 7$	$= 16$

4.

$$\begin{array}{cccc} 1 & 1 & 1 & 1 \\ \underline{+\ 10} & 10 & 10 & 10 \\ 11 & \underline{+\ 100} & 100 & 100 \\ & 111 & \underline{+\ 1000} & 1000 \\ & & 1111 & \underline{+\ 10000} \\ & & & ? \end{array}$$

5.

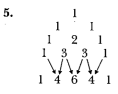

Use inductive reasoning to predict the next three numbers in the pattern (or sequence).

6. $1, 3, 5, 7, \ldots$
7. $12, 10, 8, 6, \ldots$
8. $5, 3, 1, -1, -3, \ldots$
9. $1, -1, 1, -1, 1, \ldots$
10. $1, -3, 9, -27, \ldots$
11. $1, 1/2, 1/4, 1/8, \ldots$
12. $1, 4, 9, 16, 25, \ldots$
13. $1, 1, 2, 3, 5, 8, 13, 21, \ldots$

Look for a pattern in the first three products and use it to find the fourth product.

14.
$$
\begin{array}{r} 1 \\ \times\ 1 \\ \hline 1 \end{array}
\qquad
\begin{array}{r} 11 \\ \times\ 11 \\ \hline 121 \end{array}
\qquad
\begin{array}{r} 111 \\ \times\ 111 \\ \hline 12321 \end{array}
\qquad
\begin{array}{r} 1111 \\ \times\ 1111 \\ \hline ? \end{array}
$$

15.
$$
\begin{array}{r} 9 \\ \times\ 9 \\ \hline 81 \end{array}
\qquad
\begin{array}{r} 99 \\ \times\ 9 \\ \hline 891 \end{array}
\qquad
\begin{array}{r} 999 \\ \times\ 9 \\ \hline 8991 \end{array}
\qquad
\begin{array}{r} 9999 \\ \times\ 9 \\ \hline ? \end{array}
$$

16.
$$
\begin{array}{r} 9 \\ \times\ 9 \\ \hline 81 \end{array}
\qquad
\begin{array}{r} 909 \\ \times\ 9 \\ \hline 8181 \end{array}
\qquad
\begin{array}{r} 90909 \\ \times\ 9 \\ \hline 818181 \end{array}
\qquad
\begin{array}{r} 9090909 \\ \times\ 9 \\ \hline ? \end{array}
$$

17. Study the entries that follow and use the pattern that is exhibited to complete the last two rows:

$1 + 3 = 4$ or 2^2
$1 + 3 + 5 = 9$ or 3^2
$1 + 3 + 5 + 7 = 16$ or 4^2
$1 + 3 + 5 + 7 + 9 = ?$
$1 + 3 + 5 + 7 + 9 + 11 = ?$

18. Consider the following products. What patterns can you detect?

$1 \times 9 = 9$ $4 \times 9 = 36$ $7 \times 9 = 63$
$2 \times 9 = 18$ $5 \times 9 = 45$ $8 \times 9 = 72$
$3 \times 9 = 27$ $6 \times 9 = 54$ $9 \times 9 = 81$

19. Consider the number 142,857 and its first four multiples:

$$
\begin{array}{r} 142857 \\ \times\ \ 1 \\ \hline 142857 \end{array}
\qquad
\begin{array}{r} 142857 \\ \times\ \ 2 \\ \hline 285714 \end{array}
\qquad
\begin{array}{r} 142857 \\ \times\ \ 3 \\ \hline 428571 \end{array}
\qquad
\begin{array}{r} 142857 \\ \times\ \ 4 \\ \hline 571428 \end{array}
$$

a) Observe the digits in the product and use inductive reasoning to make a conjecture about the digits that will appear in the product $142{,}857 \times 5$.
b) Multiply 142,857 by 5 to see whether your conjecture appears to be correct.
c) Can you make a more general conjecture about the digits in the product of a multiplication problem where 142,857 is multiplied by a one-digit positive number?
d) Multiply 142,857 by the digits 6 through 8 and see whether your conjecture appears to be correct.

20. The ancient Greeks labeled certain numbers as triangular numbers. The numbers 1, 3, 6, 10, 15, 21, and so on are triangular numbers.

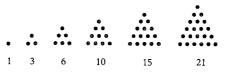

1   3   6   10   15   21

a) Can you determine the next two triangular numbers?
b) Describe a procedure to determine the next five triangular numbers without drawing the figures.
c) Is 72 a triangular number? Explain how you determined your answer.

21. Just as there are triangular numbers, there are also square numbers. The numbers 1, 4, 9, 16, 25, and so on are square numbers.

1   4   9   16   25

a) Determine the next three square numbers.
b) Describe a procedure to determine the next five square numbers without drawing the figures.
c) Is 72 a square number? Explain how you determined your answer.

22. Pick a number, multiply the number by 2, add 4 to the product, divide the sum by 2, and subtract 2 from the quotient.

a) What is the relationship between the number you started with and the final number?
b) Arbitrarily select some different numbers and repeat the given process, recording the original number and the result.

c) Can you make a conjecture about the relationship between the original number and the final number?

23. Pick a number, add 5 to the number, divide the sum by 5, subtract 1 from the quotient, and multiply the difference by 5.

a) What is the relationship between the number you started with and the final number?

b) Arbitrarily select some different numbers and repeat the given process, recording the original number and the result.

c) Can you make a conjecture about the relationship between the original number and the final number?

24. Pick any number and add 1 to it. Find the sum of the new number and the original number. Add 9 to the sum. Divide the sum by 2 and subtract the original number from the quotient.

a) What is the relationship between the number you started with and the final number?

b) Arbitrarily select some different numbers and repeat the given process, recording the original number and the result.

c) Can you make a conjecture about the relationship between the original number and the final number?

For Exercises 25–28, find a counterexample to show that each of the statements is incorrect.

25. The product of any two counting numbers is divisible by 2.

26. Every counting number greater than 5 is the sum of either two or three consecutive counting numbers. For example, 9 = 4 + 5 and 17 = 9 + 8.

27. The product of a number multiplied by itself is even.

28. The sum of any two odd numbers is divisible by 3.

29. a) Construct a quadrilateral (a four-sided figure) and measure the four interior angles with a protractor. What is the sum of their angle measures?

b) Construct three other quadrilaterals, measure the angles, and record the sums. Are your answers the same?

c) Make a conjecture about the sum of the measures of the four interior angles of a quadrilateral.

30. a) Calculate the squares of 15, 25, 35, and 45 and record the results. (Note that the square of 15 is 15 × 15 = 225.) Examine the products and see whether you can establish a pattern in relation to the numbers being multiplied.

b) Make a conjecture about how one can mentally calculate the square of any number whose unit digit is a 5.

c) Calculate the squares of 55, 75, 95, and 105 using your conjecture.

31. a) Select one- and two-digit numbers and multiply each by 99. Record your results.

b) Find the sum of the digits in each of your products in part (a).

c) Make a conjecture about the sum of the digits when a one- or two-digit number is multiplied by 99.

32. From 1, 2, 3, 4, 5, 6, 7, 8, 9, we have selected four numbers at random. They are 2, 3, 7, and 8. Arrange these digits to form the largest and the smallest four-digit numbers. Now subtract the smallest from the largest.

$$\begin{array}{r} 8732 \\ -2378 \\ \hline 6354 \end{array}$$

Now take the result (difference) and arrange the digits to form the largest and smallest number. Again subtract the smaller from the larger.

$$\begin{array}{r} 6543 \\ -3456 \\ \hline 3087 \end{array}$$

Take this result and go through the same process until you arrive at the result 6174.

$$\begin{array}{r} 8730 \\ -0378 \\ \hline 8352 \end{array}$$

$$\begin{array}{r} 8532 \\ -2358 \\ \hline 6174 \end{array}$$

Depending on the original choice of digits, the number of steps varies until one arrives at 6174. This is known as *Kapreher's constant*, named after the man who first discovered and proved the result. (Note that if the digits are 7, 7, 7, and 6, the first subtraction results in 7776 − 6777 = 999. We must think of this result as 0999 and then write 9990 − 0999 = 8991.) Now select any four digits (0, 1, 2, 3, 4, 5, 6, 7, 8, or 9), not all numbers being the same, and show that the process leads to Kapreher's constant.

PROBLEM SOLVING

Solving mathematical puzzles and real-life mathematical problems can be an enjoyable experience. You should work as many exercises in this chapter as possible. By doing so, you will see a variety of different problem-solving techniques. However, before one jumps into solving problems it is essential that a method or technique for solving problems be developed. The process of solving problems was studied and developed by George Polya (1887–1985). He was educated in Europe and taught there before coming to the United States and accepting a position at Stanford University. One of his more widely read books is *How to Solve It*. In this book, Polya outlines four steps in problem solving.

We will use Polya's four points as guidelines for problem solving. Remember that the following are guidelines. They do not always provide us with exact steps to obtain a solution to a problem.

George Polya
Courtesy of World Wide Photos

GUIDELINES FOR PROBLEM SOLVING

A. Understand the problem.
   1. Read the problem carefully. This is the obvious thing to do and one of the most important parts of solving a problem, but it is often slighted. It is essential that you read the problem at least twice.
      a) The first reading of the problem is to obtain a general overview. Do not stop and try to set up a procedure or jot down facts at this time.
      b) Read through a second time (and perhaps a third and a fourth), slowly and completely. As you do, think ahead to the following steps.
   2. Can you identify the unknown? That is, can you determine the thing you want to find? Do you know what you are looking for?
   3. Can you identify the condition or conditions that must be satisfied?
   4. Make a sketch to illustrate the problem whenever possible. Label the sketch, identifying the given information.
   5. Make a list of the given facts. Can you identify the facts that are pertinent to the problem?
   6. Is there sufficient information to solve the problem? Or is the information redundant or contradictory?

B. Devise a plan.
   1. Have you seen the problem before? Or have you seen a similar problem?
   2. How does it relate to other problems you have solved? Can you use the information from the other problems?

3. Can you write a basic equation or formula that relates the unknown and the known quantities? A particular problem might require more than one equation or formula to solve the problem. (We will discuss this procedure in Chapter 6, Algebra.)
4. Look for patterns or relationships that exist in the problem.
5. Is there a simpler related problem you can solve that will help you solve the given problem?
6. Can you make an educated guess?
7. If you know the answer, can you work backwards?
8. Try to restate the problem in a different form as a means of gaining additional information.

C. Carry out the plan.
Use the plan developed in part B to solve the problem. It is important to check all work at each step.

D. Examine the results.
1. Can you check the solution? If at all possible, go back to the original statement of the problem and check the result.
2. Is the answer reasonable? If it is impossible to perform a check, estimate the solution and compare the results.
3. Could you arrive all the same solution using a different method?
4. Can you use the result, or the method, to solve other problems?
5. Can you translate the results into a real-life application?

Example 1

Teri and Tod, who have just purchased a new Ford Taurus, are driving from Youngstown, Ohio, to Rochester, New York. They leave on a bright, sunny Friday at 8:00 A.M. with their son Andrew and Maybe, the dog. Their good friends Liz and Gerry, on the same Friday at 8:00 A.M., leave Syracuse, New York, for Chicago, Illinois, in their three-year-old Chrysler. In Syracuse it is cloudy and beginning to snow. Both parties stop for a break at the same service area on the New York State Thruway and meet in the restaurant. Who is closer to Buffalo, New York — Teri and Tod or Liz and Gerry?

Solution: If we read the question carefully and understand what is asked, we can see that none of the information in the entire paragraph pertains to the question. Sitting in the same restaurant, both parties are the same distance from Buffalo. ❖

In Example 1 we illustrated the importance of reading the problem carefully and being able to determine what, if any, information is pertinent to the question.

Now we will look at some examples of problem solving where a plan is devised.

■ Example 2 ———————————————————————————

The Joneses decide to plant blueberry bushes for fruit and as a hedge. The instructions on the package state that the minimum distance between plants should be four feet, thus implying but not stating that each bush must have two feet on either side for normal growth. For the most effective pollination the Joneses must plant two bushes of each variety. They have selected four varieties, so they must plant eight bushes. What is the total distance required for normal growth? Disregard the thickness of the bushes.

Solution: If the bushes are planted in a straight line, the problem is to determine the total distance required for normal growth of the eight plants. To get a better understanding of the problem, make a diagram.

$$2'\rightarrow\triangle\leftarrow4'\rightarrow\triangle\leftarrow4'\rightarrow\triangle\leftarrow4'\rightarrow\triangle\leftarrow4'\rightarrow\triangle\leftarrow4'\rightarrow\triangle\leftarrow4'\rightarrow\triangle\leftarrow4'\rightarrow\triangle\leftarrow2'$$

We see from the sketch, that with eight plants there are seven spaces between the plants. Thus the total distance is 4 × 7 or 28 feet from the first bush to the last. However, the total distance required for the bushes is 32 feet, since 2 feet is required on each side of the end plants for normal growth.   ❖

In Example 2 we identified the unknown, used a sketch to help understand the problem, and found the solution.

■ Example 3 ———————————————————————————

On a trip to visit a friend, Kathryn remembered to bring her child's medicine but forgot the measuring spoon. The child is supposed to have 140 milliliters (abbreviation m*l*) of medicine twice a day. Mia, the friend, said, "No problem, I have two identical 80 m*l* containers, so I can measure exactly 140 m*l* of medicine." Why is Mia so confident? How can she measure exactly 140 m*l* with two 80 m*l* containers?

Solution: The problem is how to measure exactly 140 m*l* with the two 80 m*l* containers shown in Fig. 1.1(a). One half of 80 m*l* is 40 m*l*, and one half of 40 m*l* is 20 m*l*. The sum of 80, 40, and 20 is 140. Now the problem is how to measure 40 m*l* and 20 m*l*. Mia fills both 80 m*l* containers and empties one into a glass as shown in Fig. 1.1(b). Then she takes the second 80 m*l* container and pours half of the medicine into the first container.

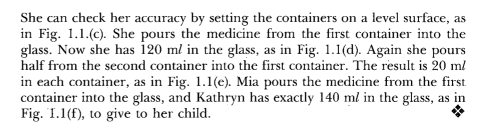

80 ml   80 ml   Glass
(a)

80 ml   Glass
(b)

40 ml   40 ml   Glass
(c)

40 ml   Glass
(d)

20 ml   20 ml   Glass
(e)

20 ml   Glass
(f)

FIGURE 1.1

"You turn the problem over and over in your mind; try to turn it so it appears simpler. . . . Is the problem as simply, as clearly, as suggestively expressed as possible? Could you restate the problem?"

George Polya

She can check her accuracy by setting the containers on a level surface, as in Fig. 1.1.(c). She pours the medicine from the first container into the glass. Now she has 120 ml in the glass, as in Fig. 1.1(d). Again she pours half from the second container into the first container. The result is 20 ml in each container, as in Fig. 1.1(e). Mia pours the medicine from the first container into the glass, and Kathryn has exactly 140 ml in the glass, as in Fig. 1.1(f), to give to her child. ❖

Example 4

A farmer had a mixture of horses and chickens. He told his grandson Adam that he counted 11 heads and 30 legs in the barnyard. His question to Adam was, "How many horses and how many chickens are in the barnyard?"

Solution: The problem is to determine the number of chickens and the number of horses. We know that each chicken and each horse has exactly one head; but a chicken has two legs, and a horse has four legs. One way to obtain a solution is to construct a table. The table for this problem will consist of four columns: number of heads, number of chickens, number of horses, and number of legs.

Number of Chickens	Number of Horses	Number of Heads	Number of Legs
0	11	11	$(0 \times 2) + (11 \times 4) = 44$
1	10	11	$(1 \times 2) + (10 \times 4) = 42$
2	9	11	$(2 \times 2) + (9 \times 4) = 40$
3	8	11	$(3 \times 2) + (8 \times 4) = 38$
4	7	11	$(4 \times 2) + (7 \times 4) = 36$
5	6	11	$(5 \times 2) + (6 \times 4) = 34$
6	5	11	$(6 \times 2) + (5 \times 4) = 32$
7	4	11	$(7 \times 2) + (4 \times 4) = 30$
8	3	11	$(8 \times 2) + (3 \times 4) = 28$
9	2	11	$(9 \times 2) + (2 \times 4) = 26$
10	1	11	$(10 \times 2) + (1 \times 4) = 24$
11	0	11	$(11 \times 2) + (0 \times 4) = 22$

We can see by studying the table that there are exactly 11 heads and 30 legs if there are seven chickens and four horses in the barnyard. ❖

If the information can be organized into a table or a chart, as in Example 4, you may wish to construct one. This is an excellent method for seeing a pattern develop. With a pattern you can analyze the information and arrive at a solution without randomly guessing.

■ **Example 5** _____

The odometer of a motor home showed 14,941 miles. The driver said that this number was *palindromic;* that is, it reads the same backward as forward. "Look at this," the driver said to the passengers, "it will be a long time before this happens again." But after another day's drive, the odometer showed five new palindromic numbers. Can you find the numbers?

Solution: The problem is to find five palindromic numbers larger than 14,941. The numbers must be of the form △ ☐ ◇ ☐ △. We know that the number we would put in place of the triangles must be a 1. Why? Because to replace the triangles with a 2, the motor home would have had to travel over 5,000 miles in a day. This is practically impossible. Now we must replace the squares and the diamond with numbers. Since the number of miles increased, the next number that we could replace the squares with is 5. The number now looks like this: 15 ◇ 51. The diamond could be replaced with any number and the number formed would be palindromic. If the diamond is replaced with a 0, the result, 15051, is the smallest palindromic number that is larger than 14941. The remaining four numbers desired are easily found. All that needs to be done is to replace the diamond with 1, 2, 3, and 4. Thus the next five palindromic numbers are 15051, 15151, 15251, 15351, and 15451.   ❖

■ **Example 6** _____

A magic square is a square array of numbers such that the numbers in the rows, columns, and diagonals have the same sum. Using the digits 1, 2, 3, 4, 5, 6, 7, 8, and 9, construct a magic square.

Solution: The first step is to create a figure with nine cells as in Fig. 1.2(a). The challenge is to place the nine numbers in the cells so that the same sum is obtained for each row, column, and diagonal. It can be seen from Fig. 1.2(a) that the sum is found by adding three numbers. Common sense will tell us that 7, 8, and 9 cannot be in the same row, column, or diagonal. We need some small and large numbers in the same row, column, and diagonal. To help see a relationship, list the numbers in order:

$$1, 2, 3, 4, 5, 6, 7, 8, 9.$$

Note that the middle number is 5 and the smallest and largest numbers are 1 and 9, respectively. The sum of 1, 5, and 9 is 15. If the sum of 2 and 8 is added to 5 the sum is 15. Likewise 3, 5, 7 and 4, 5, 6 have sums of 15. We see that in each group of three numbers the sum is 15 and 5 is a member of the group.

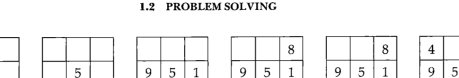

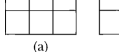

FIGURE 1.2

Since 5 is the middle number in the list of numbers, place 5 in the center square as in Fig. 1.2(b). Place 9 and 1 to the right and left of 5 as in Fig. 1.2(c). Now we will try to place the 2 and the 8. The 8 cannot be placed next to 9, since 8 + 9 = 17, which is greater than 15. To place the smaller number 2 next to the larger number 9, we place the 2 in the lower left hand cell and the 8 in the upper right hand cell as in Fig. 1.2(d). The sum of 8 and 1 is 9. To arrive at a sum of 15, place 6 in the lower right hand cell as in Fig. 1.2(e). The sum of 9 and 2 is 11. To arrive at a sum of 15, place 4 in the upper left hand cell as in Fig. 1.2(f). Now the diagonals 2, 5, 8 and 4, 5, 6 have sums of 15. The numbers that remain to be placed in the empty cells are 3 and 7. With a simple arithmetic check we can see that 3 goes in the top middle cell and 7 in the bottom middle cell as in Fig. 1.2(g). A check shows that the sum in all the rows, columns, and diagonals is 15. ❖

The solution to Example 6 is not unique. There are other arrangements of the nine numbers in the cells that will produce a magic square. Also, there are other techniques of arriving at a solution for a magic square. In fact, the above process will not work if you have an even number of squares — for example, 16 instead of 9. Magic squares are not limited to the operation of addition or to the set of counting numbers.

Estimation

An important concept in problem solving is that of **estimating** an answer to a problem. This is especially important today with the frequent use of the calculator. Some of the questions one might ask are: Does the answer make sense? Is the answer realistic? Is the result possible? For example, suppose you purchase four three-way light bulbs at $2.95 each and the clerk charges you $16.95. Is this amount correct? If you round the cost per item to $3, then the cost of 4 items would be 4 × $3 or $12. If the sales tax rate is 10% (which is a high estimate), then the sales tax is $1.20. The total cost would be approximately $12 + 1.20 = $13.20, and so the quoted price is likely to be too high. The $13.20 can be calculated mentally very quickly. You can then inform the clerk that the cost is too high and should be recalculated.

Here is another example of estimating an answer. You have a meal at a restaurant, and the bill before taxes is $15.75. If you wish to leave a 15% tip, how much should you tip the waitress? To do the calculations mentally, we do the following. Find 10% of the pretax bill and then find 5% of the pretax bill. The sum of these amounts is the amount of the tip. We first find 10% of the bill:

$$10\% \text{ of } \$15.75 = \$1.575 \quad \text{or} \quad \$1.58.$$

Since 5% is one half of 10%, we can find 5% of $15.75 by finding one half of $1.58:

$$1/2 \text{ of } \$1.58 = \$0.79$$

Thus the 15 percent tip is

$$\$1.58 + \$0.79 = \$2.37$$

If we round off to the nearest ten cents, the estimate for the tip is

$$\$1.60 + \$0.80 = \$2.40$$

The worked-out examples in the previous section and in this section will provide you with some insights into the process of problem solving. To give you additional practice, we have included **problem-solving** exercises at the end of many exercise sets. We suggest you try as many of these as possible. In chapter 6, Algebra, we use problem-solving techniques to solve real-life problems. Someone once said, "The only way to become good at what one is doing is by doing it." The following exercise set contains many problems. Some may be done very quickly, while others might take many attempts. It might be necessary to restudy the guidelines and the examples several times to help develop a method for finding a solution to a problem. If you do not succeed on the first try, try again.

EXERCISES 1.2

1. If there are 12 one-cent stamps in a dozen, how many two-cent stamps are in a dozen?
2. Allen had seven oranges and ate all but three. How many were left?
3. A baker discovered that the baking time for a special cake was 1 hour and 10 minutes when she wore her tall baker's hat, but only 70 minutes when she wore her short baker's hat. What is the difference in time?
4. A train leaves Washington for Boston traveling at the rate of 100 miles per hour. A turtle is traveling along the tracks from Boston to Washington traveling at the rate of one mile every five hours. When the train and the turtle meet, which one is nearer to Boston?
5. There are six ice cream bars in a box. How can you divide them among six children so that each gets an ice cream bar but one ice cream bar stays in the box?
6. How much dirt is in a hole 3 feet wide, 6 feet long, and 6 feet deep?
7. A nursery had nine dogwood trees. All but six have been sold. How many does the nursery have left to sell?
8. A girl went to a lawyer for help. The girl was the lawyer's daughter, but the lawyer was not the girl's father. How could this happen?

9. A cone and ice cream together cost $1.10. If the ice cream costs a dollar more than the cone, how much does the cone cost?

10. How long will it take to cut 10 feet of plastic pipe into five 2-foot lengths if it takes 3 minutes per cut?

11. Twelve square posts, 6 inches on a side, are placed in a straight line at 4-foot intervals to construct a fence. How far is it from the first post to the last post?

12. A student started out poorly on his first mathematics test. However, he doubled his score on each of the next two tests. The third test grade was 96. What was the grade on the first test?

13. Fill in the three blanks using the symbols $+$, $-$, $\times$, and $\div$ to make a true statement of equality:

$$7 \underline{\quad} 7 \underline{\quad} (7 \underline{\quad} 7) = 13$$

14. While visiting a friend's home, I saw kittens and children playing in the backyard. Counting heads, I got 18. Counting feet, I got 60. How many kittens and how many children were in the backyard?

15. Suppose you have only an eight-minute hourglass (timer) and a five-minute hourglass. How would you time the cooking of your eggs that require exactly 3 minutes to cook?

16. How can you measure 6 quarts of maple syrup from a tank if you have two 4-quart containers and an empty container?

17. A 24-by-24-foot carpet is partitioned into 4-by-4-foot squares. How many squares will there be?

18. The odometer of a motor home showed 37,473 miles. The driver said that this number was a palindromic number and informed the passengers that it would be a long time before such a number would occur again. But after 2 hours of driving the odometer showed a new palindromic number. How far had they traveled in the motor home in 2 hours. (*Hint:* See Example 5.)

19. Using a balance scale and only the four weights 1 gram, 3 grams, 9 grams, and 27 grams, explain how you could show that an object had the following weights.
a) 5 grams　　　　**b)** 16 grams
(*Hint:* Weights must be added to both sides of the balance scale.)

20. A woman purchased a dress that cost $45 and gave the merchant a $100 bill. After the woman had gone with her dress and her change, the merchant took the $100 bill to the bank. The bank clerk informed him that the $100 bill was counterfeit. What was the total loss to the merchant?

21. A bug is caught in a glass cylinder that is 10 inches tall. After trying to escape, the bug realizes that the only exit is at the top. On day one, the bug climbs 3 inches up from the bottom of the glass cylinder; but at night while resting, it slides back 2 inches. Knowing that there is no other way out, it continues this pattern 3 inches up during the day and down 2 at night. How long does it take the bug to get out of the glass cylinder?

22. The Sunday morning chef is stuck with a pan that holds only two slices of bread for French toast. It takes 30 seconds to brown one side of a piece of toast. How can the chef brown both sides of three slices in $1\frac{1}{2}$ instead of 2 minutes?

(*Hint:* Partially finish two slices and then start the third slice.)

23. Create a magic square using the numbers 1, 3, 5, 7, 9, 11, 13, 15, and 17. The sum of the numbers in every column, row, and diagonal is 27.

24. Create a magic square using the numbers 2, 4, 6, 8, 10, 12, 14, 16, and 18. The sum of the numbers in every column, row, and diagonal is 30.

25. Examine the following 3-by-3 magic squares and find the sum of the four corner entries of each magic square. Conjecture how the sum may be found using a key number in the magic square.

6	5	10
11	7	3
4	9	8

3	2	7
8	4	0
1	6	5

10	9	14
15	11	7
8	13	12

26. For a 3-by-3 magic square, conjecture how the sum of the numbers in any particular row, column, or diagonal may be determined by using a key value in the magic square.

27. For a 3-by-3 magic square, conjecture how the sum of all the numbers in the square may be determined using a key value in the magic square.

28. A plane leaves every 2 hours to fly from Toronto to London, and a plane leaves London for Toronto at the same times. The trip from London to Toronto takes 8 hours. How many planes flying from London to Toronto will a plane leaving Toronto pass in the air on the flight to London. (*Hint:* Draw a sketch.)

29. Assume that the rate of inflation is 6% for the next two years. What will be the cost of goods two years from now, adjusted for inflation, if the goods cost $450.00 today?

30. Ruth's Friendly Car Sales purchased a used car for $1000.00. The sales price was determined by

adding 20% to the purchase price. After six months the car had not been sold. The owner then reduced the sales price by 20% and sold the car. Did the car dealer make a profit or take a loss? How much?

31. The advertisement for a clothing store states that the marked price of each item will be reduced 25%. Peter buys a sport jacket marked $125.00 and a pair of trousers marked $36.00. What is his total cost before sales tax?

32. Consider a domino with six dots as illustrated in Fig. 1.3. Two different ways of connecting the three dots on the left with the three dots on the right are shown. In how many ways can the three dots on the left be connected with the three dots on the right?

FIGURE 1.3

33. In how many ways can four people be placed in a straight line?

34. Assume that any two letters represent a "word." For example, *pz* and *zp* are two different words, but *pp* and *zz* are not words. How many different words can be made using the English alphabet?

35. (a) By drawing straight lines across the face of a clock, can you divide it into three regions so that the sums of the numbers in the three regions are the same?
    (b) Can you do the same for four regions?
    (c) Can you do the same for six regions?
    (*Hint:* Draw parallel lines.)

36. A rancher died, and his will stated that his 17 horses should be distributed to his three sons in the following manner: 1/2 of his horses to the oldest son, 1/3 of his horses to the middle son, and 1/9 of his horses to the youngest son. The sons were good with fractions and calculated that their shares should be 8 1/2 horses, 5 2/3 horses, and 1 8/9 horses. These results caused a disagreement among the sons, since none of them wanted a fractional part of a horse. A neighbor came along and donated his horse to make a total of 18 horses. Then he gave 1/2 of the horses to the oldest son, 1/3 of the horses to the middle son, and 1/9 of the horses to the youngest son. How many horses did each son receive? Seeing that the sons were satisfied with his solution, the neighbor jumped on his horse and rode away. Explain why his solution was possible.

37. A woman has to take a wolf, a goat, and some cabbage across a river. Her boat has room enough for the woman (who must row) plus either the wolf or the goat or the cabbage. If she takes the cabbage with her, the wolf will eat the goat. If she takes the wolf, the goat will eat the cabbage. When the woman is present, the wolf will not eat the goat and the goat will not eat the cabbage. Nevertheless, the woman transports the wolf, goat, and cabbage across the river. How is this possible?
    (*Hint:* The woman must make a number of trips back and forth across the water.)

38. A friend asks you for change for a dollar. You reach into your pocket and find that you have a number of coins, but not the exact change for a dollar (and no silver dollars). What is the largest possible amount of money you can have (in coins) in your pocket and still not have exact change for a dollar?

39. Fill in the five blanks using the symbols $+$, $-$, and $\times$ to make a true statement of equality:

    (7 _____ 7) _____ 7 _____ (7 _____ 7 _____ 7) = 77

40. If a chicken and a half lay an egg and a half in a day and a half, how many eggs will 33 chickens lay in 11 days?
    (*Hint:* How many eggs can a chicken lay in a day?)

41. Given the digits 1 through 7, using each digit only once, construct an addition problem whose sum is 100.
    (*Hint:* At least two of the addends will be two-digit numbers. There will be at least four addends.)

42. Without raising your pencil from the paper, draw five straight line segments connecting all 12 of the displayed dots. Is it possible to end where you started?

        .   .   .   .

        .   .   .   .

        .   .   .   .

    (*Hint:* Extend a horizontal line one unit to the left of the upper dot.)

43. Three different-sized coins are placed in a pile (the smallest on top and the largest on the bottom) on the first position of a three-position board as shown in Fig. 1.4.

FIGURE 1.4

a) If a larger coin may never be placed upon a smaller coin, determine the minimum number of moves required to move the three coins from the first to the third position of the board.
b) Determine the minimum number of moves required if four coins are used.
44. A high school graduate has received two offers: Company A has offered her a starting salary of $12,000 a year with a $600 raise every 6 months. Company B has offered her $12,000 a year with a $2,400 raise every 12 months. Which offer will provide the most income?
45. Larry drives 40 miles to work each morning. The first 20 miles are in light rural traffic, and he averages 50 miles per hour. The second 20 miles are in heavy city traffic, and he averages 25 miles per hour. What is the average speed for the whole trip? (*Hint:* Find the time needed to travel one mile at each speed.)

In Exercises 46–50: (a) Use approximate numbers and mental arithmetic to get a rough estimate of the answer. (b) Use a calculator to find the actual value to two decimal places.

46. $7.3(8.5 + 7.32)$

47. $\dfrac{8.415 - 3.728}{4.016}$

48. $6.732 - \dfrac{4.651}{2.465}$

49. $6.3\left(\dfrac{4}{9} - \dfrac{3}{7}\right)$

50. A painter gets $8.75 per hour with time and a half for any time over 40 hours per week. What are her wages when she works 52 hours in one week?

SUMMARY

It is possible to arrive at a conclusion by looking at patterns. The process of arriving at a general conclusion by looking at specific cases is called inductive reasoning. A conjecture is a statement believed to be true based upon inductive reasoning. If a specific case shows that a conjecture is not correct, we call this special case a counterexample.

The four steps for problem solving developed by Polya are: (1) Understand the problem. (2) Devise a plan. (3) Carry out the plan. (4) Examine the results. Study the examples in Section 1.2. They will help you in developing your problem-solving techniques.

REVIEW EXERCISES

1. What is the largest number you can write using only three digits?
2. What is the largest number you can write using only three distinct digits?
3. What is the largest number you can write using only three odd digits?
4. What is the largest number you can write using only three distinct odd digits?

Use inductive reasoning to predict the next number in the pattern.

5. $3, 5, 7, 9, \ldots$
6. $1, 4, 9, 16, \ldots$
7. $4, -8, 16, -32, \ldots$
8. $5, 7, 10, 14, 19, \ldots$
9. $25, 24, 22, 19, 15, \ldots$
10. $1, 1, 2, 3, 5, 8, 13, \ldots$
11. Given that $24 \div 6 = 4$ (24 is the dividend, 6 is the divisor, and 4 is the quotient), how is the quotient affected if
    a) the divisor is increased?
    b) the divisor is decreased?

12. **a)** Find the sum of the first three, four, five, and ten counting numbers.
   b) Use these results to develop a procedure that can be used to find the sum of the first n counting numbers. (*Hint:* The sum will equal the product of n and a second number expressed in terms of n, all divided by 2.)
   c) Use the procedure to find the sum of the first 20 counting numbers.
   d) Use the procedure to find the sum of the first 50 counting numbers.

13. Complete the magic square:

8	1	6
3		7
4	9	

14. Complete the magic square:

21	7		18
10		15	
14	12	11	17
9	19		

15. Create a magic square using the numbers 13, 15, 17, 19, 21, 23, 25, 27, 63.
16. Create a magic square using the operation of addition for the numbers 20, 40, 60, 80, 100, 120, 140, 160, and 180.
17. **a)** Consider the number 9876, reverse the order of the digits, and find the difference. Reverse the order of the digits in the difference and add this number to the difference. Record your result.
   b) Make up several four-digit numbers with digits in descending order and follow the procedure in part (a). Record your results in each case.
   c) Write a conjecture that describes your results for the procedure followed in part (a).
18. Three men check into a single room in a hotel and pay \$10 apiece. The room should have cost \$25 instead of \$30, and so the bellhop is sent to the room to give \$5 back to the three gentlemen. The men take back \$1 each, and the bellhop is given \$2 for his trouble. Now each of the men paid \$9 for a total of \$27, and the bellhop received \$2. What happened to the missing dollar?
19. Choose a number between 1 and 20. Add 5 to the number. Multiply the sum by 6. Subtract 12 from the product. Divide the difference by 2. Divide the quotient by 3. Subtract the number you started with from the quotient. What is your answer? Try this process with a different number. Can you make a conjecture as to what your final answer will always be?
20. **a)** Take any three-digit number with digits A, B, C. Write the digits twice again with the order of the digits changed as below:

$$ABC$$
$$BCA$$
$$CAB$$

Find the sum of the three three-digit numbers, and divide by the sum of $A + B + C$. What is the quotient?

b) Try this for any three-digit number. What is the quotient?

c) Will the answer found in part (a) be the quotient for any three-digit number? Explain why or why not.

21. If you divide a specific number by 23, you obtain a quotient of 15 and a remainder of 19. Find the whole number.

22. A day-care supervisor was taking some children to an amusement park. She wanted each child to have the same number of tickets without any tickets left over. She was going to have either six or nine children with her. What was the smallest number of tickets she should have taken with her?

23. A faucet is leaking 1 quart of water every 1 3/4 hours. How many quarts of water will it leak in 4 1/2 hours?

24. Would your paycheck be higher at the beginning of the third year if you received a 15 percent raise for the first year followed by a 15 percent cut for the second year or if there were no change in salary for the entire two-year period?

25. The banks will allow a person to take out a mortgage if the monthly payments are not greater than 25 percent of the person's take-home pay. What are the maximum payments you can make if your gross salary is $1800 a month and the payroll deductions are 30 percent of the gross salary?

26. Draw a triangle and a line such that
    a) their intersection is a single point.
    b) their intersection is precisely two points.
    c) their intersection consists of more than two points.

27. Arrange six toothpicks to form four triangles so that each edge of the figure is one toothpick long. (*Hint:* Not a two-dimensional figure.)

28. Four women in a room have an average weight of 130 pounds. A fifth woman who weighs 180 pounds enters the room. Find the average weight of all five women.

29. A colony of microbes doubles in size every second. A single microbe is placed in a jar, and in an hour the jar is full. When was the jar half full?

30. A red cup contains a pint of water, and a blue cup contains a pint of wine. A teaspoon of water is taken from the red cup and placed in the blue cup, and the contents of the blue cup are mixed thoroughly. A teaspoon of the mixture in the blue cup is returned to the red cup. Is there more wine in the red cup or water in the blue cup?

31. Find the sum of the first 500 counting numbers. (*Hint:* Group in pairs.)

32. Which numbers, when divided by themselves, become larger than when multiplied by themselves?

33. Write the number 24 using only odd digits and the operation of addition.

34. A man who has a garden 10 meters square (10 meters by 10 meters) wishes to know how many posts will be required to enclose his land. If the posts are placed exactly 1 meter apart how many are needed? Disregard the thickness of the posts.

35. A duck before two ducks, a duck behind two ducks, and a duck in the middle. How many ducks are there?

36. Can you place the letters $A, B, C, D,$ and E in the chart on the next page so that no letter appears twice in any row, in any column, or in any diagonal? (For example, if A is in the second row of the first column, then A cannot be placed in the first,

second, or third row of the second column.)

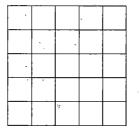

37. You have thirteen coins, all of which look alike. Twelve coins weigh exactly the same, but the other one is heavier. You have a pan balance. Tell how to find the odd coin in just three weighings.

38. What is the smallest nonnegative number that is divisible by 3, 17, and 510? (*Hint:* The answer can be calculated mentally.)

39. Often, the name of a puzzle provides a clue to its solution that beginners will overlook. Below is an *alphamagic square* invented by Lee Swallows, an electrical engineer from the Netherlands. Every row, column, and diagonal will add up to the same number (162). But taking a clue from the word "alphamagic," can you discover another property that makes this an interesting magic square? (*Hint:* Count the number of letters in each number, when written out, and construct a magic square using those numbers.)

44	61	57
67	54	41
51	47	64

40. In a rectangular room, how can you place ten small tables so that there are three against each wall?

41. In how many ways can
   a) two people stand in a line?
   b) three people stand in a line?
   c) four people stand in a line?
   d) five people stand in a line?
   e) Using the results from parts (a) through (d), can you make a conjecture about the number of ways in which n persons can stand in a line?

42. Using each of the digits 1 through 8 place one digit in each box below so that no two consecutive digits are adjacent to each other in a horizontal, vertical, or diagonal direction. (*Hint:* Fill in center positions first.)

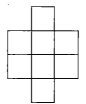

CHAPTER TEST

Use inductive reasoning to determine the next three numbers in the pattern.

1. 0, 4, 8, 12, . . .

2. $1, \dfrac{1}{3}, \dfrac{1}{9}, \dfrac{1}{27}, \ldots$

3. A farmer had 14 horses. All but three have been sold. How many are left?

4. How long will it take a carpenter to cut a 4 in. by 4 in. by 8 ft piece of treated lumber into four equal pieces if each cut takes 2 1/2 minutes?

5. A worker gets $11.25 per hour with time and a half for any time over 44 hours per week. What are his wages when he works a 49-hour week?

6. Create a magic square using the numbers 5, 10, 15, 20, 25, 30, 35, 40, and 45. The sum of the numbers in every row, column, and diagonal is 75.

7. A square is 2 inches on a side. A second square is 4 inches on a side. How many times larger in area is the second square than the first square?

8. Clara drove from her home to the beach that is 30 miles from her house. The first 15 miles she drove at 60 miles per hour, and the next 15 miles she drove at 30 miles per hour. Would the trip take more, less, or the same time if she traveled the entire 30 miles at a steady 45 miles per hour?

9. Given the six numbers 2, 6, 8, 9, 11, and 13, pick five that, when multiplied, give 11,232.

10. The period of time the clock takes to strike seven times at 7 o'clock is seven seconds. Assume that the duration of the strike and the time between strikes is the same. How long does it take the same clock to strike twelve times at 12 o'clock? (*Hint:* Make a chart showing times of strikes and times of breaks between strikes.)

CHAPTER 2

Sets

2.1 SETS

History

Two questions that are frequently asked by students are: (1) Why should we study sets? (2) Where can we use the information gained from studying sets? There are many answers to the first question. The primary one is that the concept and language of sets are used throughout the study of mathematics from kindergarten through college. The second question will be answered, we hope, as the student works through the exercises at the end of each section.

Georg Cantor (1845–1918), born in St. Petersburg, Russia, is recognized as one of the pioneers in the study of the theory of sets. Cantor's creative work in mathematics was nearly lost when his father insisted that he become an engineer rather than a mathematician. When Cantor's father recognized that his son was extremely unhappy while preparing for a career in engineering, he relented and allowed Georg to choose his own career. At the age of 29, Cantor married; while on his honeymoon, he met the mathematician J. W. Richard Dedekind (1831–1916), who eventually became one of his best friends. Cantor and Dedekind both had an interest in infinite sets. Cantor recognized that not all infinite sets are alike. The work he did in this area is now regarded as a masterpiece. It was ridiculed, however, by many prominent mathematicians of his time. A leading critic, Leopold Kronecker, prevented Cantor from gaining a position at the prestigious University of Berlin. Eventually, Cantor was given the recognition due him, but not until late in life. By then, the criticism he had received had taken its toll. He had several nervous breakdowns and spent his last days in a mental hospital.

Georg Cantor
Courtesy of The Granger Collection

A set is a collection of definite, distinguishable objects of perception or thought conceived as a whole.
Georg Cantor

Concepts of Sets

The concept of a set is one that we encounter in many different ways every day of our lives. Intuitively, a set is a collection of objects, which are called **elements** or members of the set. The United States Senate is a collection of 100 people. The Senate as a body is a set, and the individual senators are the members or elements of the set.

A set is **well-defined** if its contents can be clearly determined. The set of justices serving on the United States Supreme Court is a well-defined set, since its contents can be clearly determined. The set of the three best cars is not a well-defined set, since interpretation of the word "best" requires a subjective judgment. Thus the set of the three best cars is not clearly defined. In this text we will use only well-defined sets.

There are three methods commonly used to indicate a set: (1) by description, (2) by roster form, and (3) by set-builder notation.

One method of indicating a set is by **description,** as illustrated in Example 1.

■ **Example 1** _____

a) The set of horses that won the Kentucky Derby in the last five years.
b) The set of natural numbers.
c) The set of natural numbers less than 4. ❖

Listing the elements of a set inside a pair of braces { } is called **roster form.** The braces are an essential part of the notation, since they identify the contents as a set. For example, {1, 2, 3} is notation for the set whose elements are 1, 2, and 3, but (1, 2, 3) and [1, 2, 3] are not used to denote this set. Capital letters will generally be used to name sets.

■ **Example 2** _____

Write the following in roster form.

a) The set of natural numbers.
b) The set of natural numbers less than or equal to 80.
c) The set of natural numbers less than 4.

Solution

a) The set of natural numbers (or counting numbers) is designated by the letter N:

$$N = \{1, 2, 3, 4, 5, \ldots\}.$$

The three dots after the 5, called an **ellipsis,** indicate that the elements in the set continue in the same manner.

b) $A = \{1, 2, 3, 4, \ldots, 80\}$. The 80 after the ellipsis indicates that the set continues in the same manner until the number 80. Note that we arbitrarily labeled this set A.
c) $B = \{1, 2, 3\}$. ❖

■ **Example 3** _____

Write the following sets in roster form.

a) The set of natural numbers between 5 and 8.
b) The set of natural numbers between 5 and 8 inclusive.

Solution

a) $A = \{6, 7\}$

b) $B = \{5, 6, 7, 8\}$. Note that the word "inclusive" indicates that the values of 5 and 8 are included in the set. ❖

The symbol $\in$ is used to indicate membership in a set. The fact that 3 is an element of set B is indicated by the notation $3 \in B$. To indicate that 6 is not an element of set B, we write $6 \notin B$.

The **set-builder notation** (sometimes called set-generator notation) of indicating a set is often used to shorten a verbal description of a set. Set-builder notation is frequently used in the study of algebra, but it does not need to be restricted to that area. The following example illustrates the form of set-builder notation.

$$D \quad = \quad \{ \quad x \quad | \quad \text{Condition(s)} \}$$

Set D	is	the set of	all elements x	such that	the condition(s) x must meet in order to be a member of the set.

▪ Example 4

Given set $B = \{x \mid x \in N \text{ and } x < 4\}$, write set B in roster form.

Solution

$$B \quad = \quad \{ \quad x \quad | \quad x \in N \quad \text{and} \quad x < 4\}$$

Set B	is	the set of	all elements x	such that	x is a member of the natural numbers	and	x is less than 4.

In roster form, $B = \{1, 2, 3\}$. ❖

▪ Example 5

Given set $D = \{x \mid x \text{ is a vowel in the English alphabet}\}$, write set D in roster form.

Solution: This statement is read: Set D is the set of all elements x such that x is a vowel in the English alphabet.

In roster form, $D = \{a, e, i, o, u\}$. ❖

■ **Example 6** _____

Given set $C = \{x \mid x + 3 = 7\}$, write set C in roster form.

Solution: Set C is the set of all elements x such that $x + 3 = 7$. When the x is replaced with a 4, the statement $x + 3 = 7$ is a true statement.

In roster form, $C = \{4\}$. ❖

■ **Example 7** _____

Given set $A = \{x \mid 3 \leq x < 9 \text{ and } x \in N\}$, write set A in roster form.

Solution: To write set A in roster form, we must find all the natural numbers that satisfy the conditions $3 \leq x < 9$. To satisfy these conditions, x must be greater than or equal to 3 *and* less than 9. Thus the solution in roster form is $A = \{3, 4, 5, 6, 7, 8\}$. ❖

If the number of elements in a given set can be expressed as a natural number or if the set contains no elements, the set is said to be **finite.** The set $B = \{2, 4, 6, 8, 10\}$ is finite because the number of elements in the set is 5, and 5 is a natural number. The set of the first hundred natural numbers, $\{1, 2, 3, 4, 5, 6, \ldots, 100\}$, is also a finite set. A set that is not finite is said to be **infinite.** The set of natural numbers is one example of an infinite set. Infinite sets are discussed in more detail in optional Section 2.7.

Another important concept of set theory is equality.

> Set A is **equal to** set B, symbolized by $A = B$, if and only if they contain the same elements.

Set $A = \{1, 2, 3\}$ and set $B = \{3, 1, 2\}$ are equal, since they satisfy the definition above. If two sets are equal, both must contain the same number of elements. The number of elements in a set is called its **cardinal number.**

> The **cardinal number** of set A, symbolized by $n(A)$, is the number of elements in set A.

Both set A and set B have a cardinal number of 3; that is, $n(A) = 3$, and $n(B) = 3$. We can say that set A and set B both have a *cardinality* of 3.

Two sets are said to be **equivalent** if they contain the same number of elements. Any sets that are equal must also be equivalent. However, not all sets that are equivalent are equal. The sets $D = \{a, b, c\}$ and $E = \{*, ?, \triangle\}$ are equivalent, since both have the same cardinal number, 3. They are not equal, however, since their elements differ.

Two sets that are equivalent can be placed in a **one-to-one correspondence.** Sets A and B can be placed in a one-to-one correspondence if every element of set A can be matched with exactly one element of set B and every element of set B can be matched with exactly one element of set A.

Consider sets A and B:

$$A = \{a, b, c\}, \qquad B = \{\#, @, ?\}.$$

Two different one-to-one correspondences for sets A and B follow.

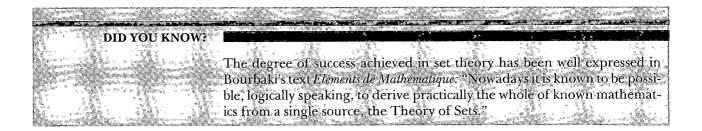

Other one-to-one correspondences between sets A and B are possible. (See Exercise 62.)

EXERCISES 2.1

Determine whether each set is well defined.

1. The set of well-dressed college students.
2. The set of people who are serving this year as governors of the states in the United States.
3. The set of the ten largest populated cities in the world.
4. The set of the five best vocalists living today.
5. The set of students in this class that write with their left hand.
6. The set of states that border Kansas.

Determine whether each of the following sets is finite or infinite.

7. $\{1, 2, 3, 4, 5, \ldots\}$.
8. The set of fractions between 1 and 2.
9. The set of multiples of 2.
10. The set of numbers greater than 7.
11. The set of even numbers greater than 7.
12. The set of integers between 1 and 1,000,000.
13. The set of grains of sand on the beach at Atlantic City, New Jersey, at any instant.
14. The set of alligators in Everglades National Park.

Express each set in roster form.

15. The set of Great Lakes in the United States.
16. The set of living persons who have been President of the United States.
17. The set of natural numbers 5 through 350.
18. $Q = \{x \mid x \in N \text{ and } x \text{ is odd}\}$.
19. $A = \{x \mid x + 5 = 9\}$.
20. The set of countries that border the United States.
21. The set of colors of the rainbow.
22. $B = \{x \mid x > 2 \text{ and } x \in N\}$ (> means "is greater than").
23. $C = \{x \mid x \geq 2 \text{ and } x \in N\}$ ($\geq$ means "is greater than or equal to").
24. The set of even natural numbers.

Express each set in set-builder notation.

25. $B = \{1, 2, 3, 4, 5, 6, 7\}$.
26. $E = \{2, 4, 6, 8, 10, \ldots\}$.
27. $C = \{7, 8, 9, 10, \ldots\}$.
28. D is the set of natural numbers less than 7.
29. E is the set of natural numbers between 1 and 4.
30. $A = \{3, 4, 5, 6, 7\}$.
31. G is the set of natural numbers greater than or equal to 7.
32. $F = \{1, 2, 3, 4, \ldots, 75\}$.

Write a description of each set.

33. $N = \{1, 2, 3, 4, \ldots\}$.
34. $B = \{1, 4, 9, 16, 25, 36, \ldots\}$.
35. $V = \{\text{China, France, USSR, United Kingdom, USA}\}$.
36. $L = \{z\}$.
37. $C = \{w, x, y, z\}$.
38. $A = \{3, 4, 5, 6, 7\}$.
39. $J = \{x \mid x > 1 \text{ and } x < 7 \text{ and } x \in N\}$.
40. $K = \{1, 3, 5, 7, 9, 11, 13, 15, \ldots\}$.

State whether each of the following is true or false.

41. $A \in \{a, e, i, o, u\}$.
42. $a \notin \{a, e, i, o, u\}$.
43. $a \in \{a, e, i, o, u\}$.
44. $5 \in \{x \mid x \text{ is an even number}\}$.
45. $3 \notin \{2, 4, 6, 8, 10\}$.
46. Apple $\in \{\text{grapes, oranges, Ford, Chevrolet, dog}\}$.
47. $3 \notin \{x \mid x \text{ is an odd natural number}\}$.
48. $W \in \{x \mid x \text{ is a vowel in the English alphabet}\}$.

Given sets $A = \{1, 2, 3\}$, $B = \{6, 11, 13, 15, 17\}$, $C = \{\ \}$, and $D = \{\bigcirc, \square, *, \triangle, !, :\}$, find

49. $n(A)$, **50.** $n(B)$, **51.** $n(C)$, **52.** $n(D)$.

Determine whether the following pairs of sets are equal, equivalent, both, or neither.

53. $A = \{1, 2, 3\}$, $B = \{3, 2, 1\}$.
54. $A = \{1, 2, 3\}$, $B = \{3, 4, 5\}$.
55. $A = \{a, b, c\}$, $B = \{1, 2, 3, 4\}$.
56. $A = \{\bigcirc, \triangle, \square\}$, $B = \{?, \triangle, 3\}$.
57. A is the set of state capitals in the United States, B is the set of states in the United States.
58. A is the set of students in this class, B is the set of female students in this class.
59. Set-builder notation is often more versatile and efficient than listing a set in roster form. This is illustrated with the following two sets:

$$A = \{x \mid x > 2, x \in N\},$$
$$B = \{x \mid x > 2\}.$$

a) Write a description of set A and set B.
b) Explain the difference between set A and set B. (*Hint:* Is $4\frac{1}{2} \in A$? Is $4\frac{1}{2} \in B$?)
c) Write set A in roster form.
d) Write set B in roster form.
60. Repeat Exercise 59 using the following sets.

$$A = \{x \mid 2 < x \leq 5, x \in N\},$$
$$B = \{x \mid 2 < x \leq 5\}.$$

61. Consider the following two sets:

$$A = \{x \mid 4 < x < 5, x \in N\},$$
$$B = \{x \mid 4 < x < 5\}.$$

a) Is set A a finite or infinite set? What is its cardinal number?
b) Is set B a finite or an infinite set?
62. a) If set A has one element and set B has one element, how many different one-to-one correspondences can be made?
b) If each of sets A and B contains two elements, how many different one-to-one correspondences can be made?
c) Answer the above question for sets A and B with three elements and with four elements.
*d) Write a general formula for the number of distinct one-to-one correspondences for sets A and B each containing n elements.

SUBSETS

There are some sets that do not contain any elements. This type of set, called an empty set or null set, is symbolized by { } or ∅. Please note that {∅} is not an empty set, since this set contains the element ∅.

> A set that contains no elements is called an **empty** or **null set.** The empty set is symbolized by { } or ∅.

■ **Example 1** —————————————————————————————

Find the set of natural numbers that satisfies the equation $x + 2 = 0$.

Solution: The values that satisfy the equation are those that make the equation a true statement. Only the number -2 satisfies this equation. Since -2 is not a natural number, the solution set of this problem is { } or ∅.

The shell is empty without its oarsmen.
Joe Schuyler Stock Boston Photo

Before we can understand the set operations discussed in the next section, we must understand the concept of the universal set. The universal set is established to meet the specific needs of a particular problem. For example, if we wish to discuss Lassie, the universal set may be the set of all dogs, the set of all collies, the set of all animals, the set of all living things, or some other inclusive set. To avoid confusion, we define a universal set for every problem.

A **universal set,** symbolized by U, is a set that contains all the elements for any specific discussion.

Each of us categorizes items daily. For example, when sending out our bills, we may place the bills in two piles:

1. those to be mailed out immediately;
2. those that need not be mailed immediately.

In doing so, we have subdivided the total set of bills into two smaller sets. Each of the smaller sets is a subset of the original set of bills.

Set A is a **subset** of set B, symbolized by $A \subseteq B$, if and only if all the elements of set A are also elements of set B.

Example 2

Determine whether set A is a subset of set B.

a) $A = \{a, e, f\}$, $B = \{a, b, c, d, e, f, g\}$.
b) $A = \{1, 2, 3, 4, 5\}$, $B = \{1, 2, 3\}$.
c) $A = \{x \mid x$ is a student at Texas A. & M.$\}$,
    $B = \{x \mid x$ is a student at Broome County College$\}$.
d) $A = \{*, \triangle, \square\}$, $B = \{*, \triangle, \square\}$.

Solution

a) Since all the elements of set A are contained in set B, $A \subseteq B$.
b) Since the elements 4 and 5 are in set A but not in set B, $A \nsubseteq B$
    (A is not a subset of B). In this example, however, all the elements of set B are contained in set A; therefore $B \subseteq A$.
c) Since there are students in set A who are not in set B, $A \nsubseteq B$.
d) Since all the elements of set A are contained in set B, $A \subseteq B$. Note that set A = set B. ❖

Set A is a **proper subset** of set B, symbolized by $A \subset B$, if and only if all the elements of set A are elements of set B and set $A \neq$ set B (that is, set B must contain at least one element not in set A).

■ **Example 3**

Determine whether set A is a proper subset of set B.

a) A = {Ford, Dodge, Buick},
    B = {Ford, Chevrolet, Dodge, Plymouth, Chrysler, Buick}.

b) Set A is the set of people in the United States who live north of the Mason-Dixon Line.
    Set B is the set of people living in the United States.

c) A = {a, b, c, d},
    B = {a, b, c, d}.

Solution

a) All the elements of set A are contained in set B, and set B contains at least one element not in set A; thus $A \subset B$.

b) All the elements of set A are contained in set B, and set B has at least one element not in set A; thus $A \subset B$.

c) Since set A = set B, $A \not\subset B$. (However, $A \subseteq B$.)   ❖

Every set is a subset of itself but is not a proper subset of itself (if A is a set, then $A \subseteq A$, but $A \not\subset A$). To show that set A is not a subset of set B, it is necessary to find at least one element of set A that is not an element of set B. For example, if A = {1, 2, 3} and B = {2, 3, 4, 5}, then $A \not\subseteq B$ because $1 \in A$ and $1 \notin B$.

Let A = { } and B = {1, 2, 3, 4}. Is $A \subseteq B$? In order to show $A \not\subseteq B$, you must find at least one element of set A that is not an element of set B. Since this cannot be done, it must be true that $A \subseteq B$. Using the same reasoning, we can show that the empty set is a subset of every set, including itself.

■ **Example 4**

Determine whether the following are true or false.

a) {3} $\in$ {3, 4, 5}

b) {3} $\in$ {{3}, {4}, {5}}

c) {3} $\subseteq$ {3, 4, 5}

d) 3 $\subseteq$ {3, 4, 5}

Solution

a) The statement $3 \in$ {3, 4, 5} is a true statement, since 3 is a member of the set {3, 4, 5}. However, {3} $\in$ {3, 4, 5} is a false statement, since {3} is a set and the set {3} is not an element of the set {3, 4, 5}.

b) The statement $\{3\} \in \{\{3\}, \{4\}, \{5\}\}$ is true because $\{3\}$ is an element in the set. The elements of the set $\{\{3\}, \{4\}, \{5\}\}$ are themselves sets.

c) The statement $\{3\} \subseteq \{3, 4, 5\}$ is true because every element of the first set is an element of the second set.

d) The statement $3 \subseteq \{3, 4, 5\}$ is false, since 3 is an element and not a set.

❖

■ **Example 5**_____

Determine the number of subsets of each of the following sets, and list all their subsets.

a) $A = \{a\}$ **b)** $B = \{1, 2\}$ **c)** $C = \{\square, 0, ?\}$

Set	Subsets	Number of subsets
a) $\{a\}$	$\{a\}, \{\ \}$	$2 = 2 = 2^1$
b) $\{1, 2\}$	$\{1, 2\}, \{1\}, \{2\}, \{\ \}$	$4 = 2 \times 2 = 2^2$
c) $\{\square, 0, ?\}$	$\{\square, 0, ?\}, \{\square, 0\}, \{\square, ?\}$ $\{0, ?\}, \{\square\}, \{0\}, \{?\}, \{\ \}$	$8 = 2 \times 2 \times 2 = 2^3$

❖

Example 5 shows that a set with one element contains two subsets. A set with two elements contains four subsets. A set with three elements contains eight subsets. How many subsets will a set with four elements contain?

If we continue this table with larger and larger sets, we can develop a general formula for finding the number of subsets of any given set.

> The **number of subsets** of a finite set A is 2^n, where n is the number of elements in set A.

Thus a set with four elements will contain 2^4, or $2 \times 2 \times 2 \times 2$, or 16 subsets.

*EXERCISES 2.2*_____

Determine whether the following are true or false.

1. $2 \subseteq \{2, 3, 4, 5, 6\}$ **2.** $\{\ \} \in \{1, 2, 3\}$ **7.** $\{\ \} \subset \{0\}$ **8.** $\{3\} \in \{3, 4, 5\}$

3. $\{\ \} \subseteq \{1, 2, 3\}$ **4.** $? \notin \{0, \triangle, \square\}$ **9.** $\{3\} \in \{\{3\}, \{4\}, \{5\}\}$ **10.** $\{3\} \subset \{3, 4, 5\}$

5. $\{\ \} = \{0\}$ **6.** $\{1, 2, 3\} \subseteq \{1, 2, 3\}$ **11.** $\{3\} \subseteq \{3, 4, 5\}$ **12.** $\{1, 2, 3\} \subset \{1, 2, 3\}$

Determine whether $A = B$, $A \subseteq B$, $B \subseteq A$, $A \subset B$, $B \subset A$, or none of these. (There may be more than one answer.)

13. $A = \{1, 2, 3, 4\}$, $B = \{3, 4, 6, 7\}$.
14. $A = \{x \mid x \in N \text{ and } x < 7\}$,
     $B = \{1, 2, 3, 4, 5, 6\}$.
15. $A = \{\text{apple, peach, cherry, pear}\}$,
     $B = \{\text{apple, peach}\}$.
16. $A = \{1, 2, 3\}$, $B = \{2, 3\}$.
17. $A = \{x \mid x \text{ is a capital city of a state in the United States}\}$,
     $B = \{x \mid x \text{ is a state in the United States}\}$.
18. $A = \{x \mid x \text{ is an odd number less than 9}\}$,
     $B = \{2, 3, 5\}$.
19. $A = \{x \mid 3 < x < 7, x \in N\}$.
     Set B is the set of natural numbers between 3 and 7.
20. Set A is the set of stars in the Milky Way.
     Set B is the set of stars in the universe.

List all the subsets of the given sets.

21. $A = \{$ $\}$
22. $B = \{\square, \bigcirc\}$
23. $C = \{\text{red, green, yellow}\}$

24. $D = \emptyset$
25. Given set $A = \{*, ?, \#, 1\}$:
  a) List all the subsets of set A.
  b) State which of the subsets in part (a) are not proper subsets.
26. A set contains six elements.
  a) How many subsets does it contain?
  b) How many proper subsets does it contain?
  c) Write a formula for determining the number of proper subsets for a set with n elements.

If the statement is true for all sets A and B, write the word "true." If it is not true for all sets A and B, write the word "false." Assume that $A \neq \emptyset$, $U \neq \emptyset$, and $A \neq U$.

27. If $A \subseteq B$, then $A \subset B$.
28. If $A \subset B$, then $A \subseteq B$.
29. $A \subset A$ **30.** $A \subseteq A$
31. $\emptyset \subseteq A$ **32.** $\emptyset \subset A$
33. $\emptyset \subset \emptyset$ **34.** $A \subseteq U$
35. $U \subseteq \emptyset$ **36.** $\emptyset \subset U$
37. $\emptyset \subseteq \emptyset$ **38.** $A \subset U$

PROBLEM SOLVING

39. Bob is purchasing a new clothes washer. The machine that he is selecting can be purchased with no extra features or with as many as five extra features. For the appliance store to display machines with all the possible combinations of features, how many washers must it have on display?

40. The executive committee of the student senate consists of Ashton, Bailey, Katz, and Snyder. Each member of the executive committee has exactly one vote (no abstentions), and a simple majority of the committee is needed to pass or reject a motion. If a motion is neither passed nor rejected, then it is considered blocked and will be considered again. Determine the number of specific ways the members can vote so that a motion can be passed, rejected, or blocked.

2.3   SET OPERATIONS

The operations of arithmetic are $+$, $-$, $\times$, and $\div$. When we see these symbols, we know what procedure to follow to determine the answer. Some of the operations in set theory are $'$, $\cup$, and $\cap$. They represent complement, union, and intersection, respectively.

The **complement** of set A, symbolized by A', is the set of all the elements in the universal set that are not in set A.

☐ **Example 1** _____

Given $U = \{1, 2, 3, 4, 5, 6, 7, 8, 9, 10\}$,
$\qquad A = \{1, 2, 4, 6\}$,
$\qquad B = \{1, 3, 6, 7, 9\}$,
$\qquad C = \{\ \ \}$.

Find   **a)** A',   **b)** B',   **c)** C'.

Solution

a) The elements in U that are not in set A are 3, 5, 7, 8, 9, 10. Thus $A' = \{3, 5, 7, 8, 9, 10\}$.

b) The elements in U that are not in set B are 2, 4, 5, 8, 10. Thus $B' = \{2, 4, 5, 8, 10\}$.

c) Since there are no elements in $\{\ \ \}$, $C' = U$.   ❖

The next operation is union. The word "union" means to unite or join together, and that is exactly what the operation union does.

> The **union** of sets A and B, symbolized by $A \cup B$, is the set containing all the elements that are members of set A or of set B (or of both sets).

The union of set A and set B is the set that contains all the elements from set A and all the elements from set B. If an element is common to both sets, it is listed only once in the union of sets.

The family reunion is the union of all the subsets of the universal set.

Photo by Marshall Henrichs

Example 2

Given $U = \{1, 2, 3, 4, 5, 6, 7, 8, 9, 10\}$,

$A = \{1, 2, 4, 6\}$,
$B = \{1, 3, 6, 7, 9\}$,
$C = \{\ \ \}$.

Find **a)** $A \cup B$,   **b)** $A \cup C$,   **c)** $A' \cup B$,   **d)** $(A \cup B)'$.

Solution

a) $A \cup B = \{1, 2, 4, 6\} \cup \{1, 3, 6, 7, 9\} = \{1, 2, 3, 4, 6, 7, 9\}$

b) $A \cup C = \{1, 2, 4, 6\} \cup \{\ \ \} = \{1, 2, 4, 6\}$

c) Before we can determine $A' \cup B$, it is necessary to determine A'.

$$A' = \{3, 5, 7, 8, 9, 10\}$$
$$A' \cup B = \{3, 5, 7, 8, 9, 10\} \cup \{1, 3, 6, 7, 9\}$$
$$= \{1, 3, 5, 6, 7, 8, 9, 10\}$$

d) In mathematics it is customary to evaluate the information within parentheses first. Thus we find $(A \cup B)'$ by first determining $A \cup B$ and then finding the complement of $A \cup B$.

$$A \cup B = \{1, 2, 3, 4, 6, 7, 9\} \text{ from part (a)}$$
$$(A \cup B)' = \{1, 2, 3, 4, 6, 7, 9\}' = \{5, 8, 10\}$$

FIGURE 2.1 Intersection.
Photo by Allen R. Angel.

The term "intersection" brings to mind the area common to two crossing streets (see Fig. 2.1).

The **intersection** of sets A and B, symbolized by $A \cap B$, is the set containing all the elements that are common to both set A and set B.

Example 3

Given $U = \{1, 2, 3, 4, 5, 6, 7, 8, 9, 10\}$,
$\quad A = \{1, 2, 4, 6\}$,
$\quad B = \{1, 3, 6, 7, 9\}$,
$\quad C = \{ \ \}$.

Find **a)** $A \cap B$,   **b)** $A \cap C$,   **c)** $A' \cap B$,   **d)** $(A \cap B)'$.

Solution

a) $A \cap B = \{1, 2, 4, 6\} \cap \{1, 3, 6, 7, 9\} = \{1, 6\}$. The elements common to both set A and set B are 1 and 6.

b) $A \cap C = \{1, 2, 4, 6\} \cap \{ \ \} = \{ \ \}$. Since there are no elements common to both set A and set C, sets A and C are disjoint sets. Two sets that have no elements in common are called **disjoint sets.**

c) $\quad A' = \{3, 5, 7, 8, 9, 10\}$
$A' \cap B = \{3, 5, 7, 8, 9, 10\} \cap \{1, 3, 6, 7, 9\}$
$\quad\quad\quad = \{3, 7, 9\}$

d) To find $(A \cap B)'$, we must first determine $A \cap B$.

$\quad A \cap B = \{1, 6\}$ from part (a)
$(A \cap B)' = \{1, 6\}' = \{2, 3, 4, 5, 7, 8, 9, 10\}$

Example 4

Given $U = \{a, b, c, d, e, f, g\}$,
$\quad A = \{a, b, e, g\}$,
$\quad B = \{a, c, d, e\}$,
$\quad C = \{b, e, f\}$.

Find

a) $A \cup B$,   **b)** $A \cup C$,   **c)** $(A \cup B) \cap (A \cup C)$,

d) $(A \cup B) \cap C'$,   **e)** $A' \cap B'$.

This photo is the union of two disjoint sets. Name two disjoint sets that appear here. There is more than one correct answer.
Photo by Marshall Henrichs

Solution

a) $A \cup B = \{a, b, e, g\} \cup \{a, c, d, e\} = \{a, b, c, d, e, g\}$

b) $A \cup C = \{a, b, e, g\} \cup \{b, e, f\} = \{a, b, e, f, g\}$

c) $(A \cup B) \cap (A \cup C) = \{a, b, c, d, e, g\} \cap \{a, b, e, f, g\}$
$$= \{a, b, e, g\}$$

d) $(A \cup B) = \{a, b, c, d, e, g\}$
$$C' = \{a, c, d, g\}$$
$(A \cup B) \cap C' = \{a, b, c, d, e, g\} \cap \{a, c, d, g\}$
$$= \{a, c, d, g\}$$

e) $A' \cap B' = \{c, d, f\} \cap \{b, f, g\} = \{f\}$

The words "and" and "or" are very important in many areas of mathematics. You will see these words used in a number of different chapters in this book, including the probability chapter. The word *"or"* is generally interpreted to mean *"union,"* while *"and"* is generally interpreted to mean *"intersection."* Suppose $A = \{1, 2, 3, 5, 6, 8\}$ and $B = \{1, 3, 4, 7, 9, 10\}$. Then the elements that belong to set A *or* set B are 1, 2, 3, 4, 5, 6, 7, 8, 9, and 10 (the union of the sets). The elements that belong to set A *and* set B are 1 and 3 (the intersection of the sets).

EXERCISES 2.3

1. Let U represent the set of all married people in the state of Pennsylvania in 1988. Let A represent the set of all people in the state of Pennsylvania in 1988 who have been married less than 25 years. Describe A'.
2. Let U represent the set of all dogs. Let A represent the set of all dogs that are registered wirehaired fox terriers. Describe A'.

Given that U is the set of college graduates with a degree,
  A is the set of college graduates with an A.A. degree,
  B is the set of college graduates with a B.S. degree,
  C is the set of college graduates with a B.A. degree.

Describe the following in words.

3. $A \cap B$ 4. $A \cup B$ 5. $A' \cup C$
6. $A' \cup B$ 7. $A \cap B \cap C$ 8. $A' \cap B'$

Given that U is the set of cities in the United States,
  A is the set of cities in the United States with a population over 100,000,
  B is the set of cities in the United States with a subway system,
  C is the set of cities in the United States with a philharmonic orchestra.

Describe the following in words.

9. $A \cup C$ 10. $B \cap C$ 11. $B' \cap C$
12. $A \cup B \cup C$ 13. $A' \cup B'$ 14. $A \cap B \cap C$

Given $U = \{1, 2, 3, 4, 5, 6, 7, 8\}$,
  $A = \{1, 2, 4, 5, 8\}$,
  $B = \{2, 3, 4, 6\}$.

Find the following.

15. $A \cup B$ 16. $A \cap B$
17. B' 18. $A \cup B'$
19. $(A \cup B)'$ 20. $A' \cap B'$
21. $(A \cup B)' \cap B$ 22. $(A \cup B) \cap (A \cup B)'$
23. $(B \cup A)' \cap (B' \cup A')$ 24. $A' \cup (A \cap B)$

Given $U = \{0, \square, \triangle, \#, \alpha, \beta\}$,
  $A = \{0, \square, \triangle, \#\}$,
  $B = \{\#, \alpha, \beta\}$
  $C = \{0, \square, \triangle\}$.

Find the following.

25. $A \cup B$ 26. $A \cap B$
27. $A' \cup B$ 28. $(B \cup C)'$
29. $A \cap B'$ 30. $A \cap C'$
31. $(B \cap C)'$ 32. $(A \cup B) \cap C$
33. $(C \cap B) \cup A$ 34. $(C \cup A) \cap B$
35. $(A' \cup C) \cap B$ 36. $(A \cap B') \cup C$
37. $(A \cup B)' \cap C$ 38. $(A \cap C)' \cap B$

Given $U = \{a, b, c, d, e, f, g, h, i, j, k\}$,
  $A = \{a, c, f, g, i\}$,
  $B = \{b, c, d, f, g\}$,
  $C = \{a, b, f, i, j\}$.

Find the following.

39. A' 40. $B \cup C$
41. $A \cap C$ 42. $A' \cup B$
43. $(A \cap C) \cup B$ 44. $(A \cap C)'$
45. $A \cup (C \cap B)'$ 46. $A' \cap (B \cap C)$
47. $(C \cap B) \cap (A' \cap B)$ 48. $A \cup (C' \cup B')$

Given $U = \{0, 1, 2, 3, 4, 5, \ldots\}$,
  $A = \{1, 2, 3, 4, \ldots\}$,
  $B = \{4, 8, 12, 16, \ldots\}$,
  $C = \{2, 4, 6, 8, \ldots\}$.

Find the following.

49. $A \cup B$ 50. $A \cap B$
51. $B \cap C$ 52. $B \cup C$
53. $A \cap C$ 54. $A' \cap C$
55. $B' \cap C$ 56. $(B \cup C)' \cup C$
57. $(A \cap C) \cap B'$ 58. $U' \cap (A \cup B)$

For each of the following, determine whether the answer is $\emptyset$, A, or U. (Assume that $A \neq \emptyset$, $A \neq U$.)

59. $A \cup A'$ 60. $A \cap A'$
61. $A \cup \emptyset$ 62. $A' \cup U$
63. $A \cap \emptyset$ 64. $A \cup U$
65. $A \cap U$ 66. $A \cap U'$

Another set operation is the **difference of two sets.** The difference of two sets A and B, symbolized $A - B$, is defined as

$$A - B = \{x \mid x \in A \text{ and } x \notin B\}$$

$A - B$ is the set of elements that belong to set A but not to set B. For example, if $U = \{1, 2, 3, 4, 5, 6, 7, 8, 9, 10\}$, $A = \{2, 4, 5, 9, 10\}$, and $B = \{1, 3, 4, 5, 6, 7\}$, then $A - B = \{2, 9, 10\}$ and $B - A = \{1, 3, 6, 7\}$.

For Exercises 67 through 70, let $U = \{a, b, c, d, e, f, g, h, i, j, k\}$, $A = \{b, c, e, f, g, h\}$, and $B = \{a, b, c, g, i\}$. Find the following.

67. $A - B$ **68.** $B - A$
69. $A' - B$ **70.** $A - B'$

For Exercises 71 through 76, let $U = \{1, 2, 3, 4, 5, 6, 7, 8, 9, 10, 11, 12, 13, 14, 15\}$, $A = \{2, 4, 5, 7, 9, 11, 13\}$, and $B = \{1, 2, 4, 5, 6, 7, 8, 9, 11\}$. Find the following.

71. $A - B$ **72.** $B - A$ **73.** $(A - B)'$
74. $A - B'$ **75.** $(B - A)'$ **76.** $A \cap (A - B)$

Another set operation is the **Cartesian product.** The Cartesian product of set A and set B is symbolized $A \times B$ and read "A cross B." The Cartesian product[†] of set A and set B is the set of all possible ordered pairs of the form (a, b) where $a \in A$ and $b \in B$. For example, if $A = \{$George, Paul, John, Ringo$\}$ and $B = \{1, 2, 5\}$, then:

$A \times B = \{$(George, 1), (George, 2), (George, 5), (Paul, 1), (Paul, 2), (Paul, 5), (John, 1), (John, 2), (John, 5), (Ringo, 1), (Ringo, 2), (Ringo, 5)$\}$

Let set $A = \{a, b, c\}$ and set $B = \{1, 2\}$.

77. Find $A \times B$.   **78.** Find $B \times A$.
79. Does $A \times B = B \times A$?   **80.** Find $n(A \times B)$.
81. Find $n(B \times A)$.
82. Does $n(A \times B) = n(B \times A)$?

83. If set $A = \{1, 2, 3, 4\}$, find $A \times A$.
84. If set A has m elements and set B has n elements, how many elements will be in $A \times B$?

PROBLEM SOLVING

Determine the relationship between set A and set B if

85. $A \cap B = B$.   **86.** $A \cup B = B$.
87. $A \cap B = \emptyset$.   **88.** $A \cup B = A$.
89. $A \cap B = A$.

Describe the conditions under which each statement would be true.

90. $B - A = B$ **91.** $A - B = A$
92. $A - \emptyset = A$ **93.** $B - \emptyset = B$
94. $B - B' = B$
95. $B - B = \emptyset$
96. $B - A = \emptyset$ **97.** $B - U = \emptyset$
98. $U - B = B'$

---

† Ordered pairs are discussed in Section 6.5.

 2.4 VENN DIAGRAMS

A useful technique for picturing set relations is the Venn diagram, named for the English mathematician John Venn (1834–1923). Venn invented the diagrams and used them to illustrate ideas in his text on symbolic logic, published in 1881.

In a Venn diagram the universal set, U, is usually represented by the set of points inside a rectangle. A subset, A, of the universal set can be represented by the set of points inside a circle within the rectangle (see Fig. 2.2).

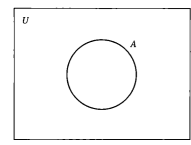

FIGURE 2.2

Two sets may be represented in a Venn diagram in any of four different ways (see Fig. 2.3). If the two sets A and B are disjoint, they may be illustrated as in Fig. 2.3(a). If one set is a proper subset of the other set — that is, $A \subset B$ — they may be illustrated as in Fig. 2.3(b). Two sets containing exactly the same elements — that is, $A = B$ — are shown in Fig. 2.3(c). Two sets with some elements in common are shown in Fig. 2.3(d). The Venn diagram in Fig. 2.3(d) is regarded as the most general form of the diagram. It can be used to show all the possible cases.

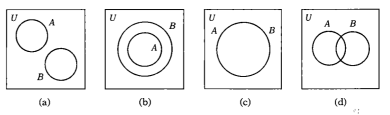

(a)   (b)   (c)   (d)

FIGURE 2.3

Let us examine the Venn diagram in Fig. 2.3(d) more carefully. For convenience in discussion we will call the regions I, II, III, and IV, as shown in Fig. 2.4.

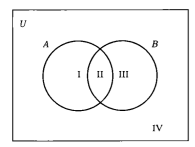

FIGURE 2.4

- **CASE 1: Disjoint Sets**
  When set A and set B are disjoint, they have no elements in common. Therefore region II of the diagram is empty.

- **CASE 2: Subsets**
  When $A \subset B$, every element of set A is also an element of set B. Thus there can be no elements in region I. On the other hand, if $B \subset A$, then region III is empty.

- **CASE 3: Equal Sets**
  When set A = set B, all the elements of set A are elements of set B, and all the elements of set B are elements of set A. Thus regions I and III are empty.

- **CASE 4: Overlapping Sets**
  When sets A and B have elements in common, those elements are in region II. The elements that belong to set A but not to set B are in region I. The elements that belong to set B but not to set A are in region III.

In each of the four cases, any element not belonging to set A or set B is placed in region IV.

■ **Example 1** ——————————————————————————————

Given the Venn diagram in Fig. 2.5, determine the following sets.

a) U **b)** A **c)** B' **d)** $A \cup B$ **e)** $A \cap B$ **f)** $(A \cup B)'$

Solution

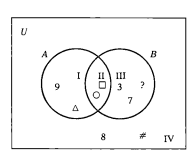

FIGURE 2.5

a) The universal set consists of all the elements within the rectangle. Thus $U = \{9, \triangle, \square, \bigcirc, 3, 7, ?, \#, 8\}$.

b) Set A consists of the elements in regions I and II. $A = \{9, \triangle, \square, \bigcirc\}$.

c) B' consists of the elements outside set B, or the elements in regions I and IV. $B' = \{9, \triangle, \#, 8\}$.

d) $A \cup B$ consists of the elements that belong to set A or set B (regions I, II, and III). $A \cup B = \{9, \triangle, \square, \bigcirc, 3, 7, ?\}$.

e) $A \cap B$ consists of elements that belong to both set A and set B (region II). Thus $A \cap B = \{\square, \bigcirc\}$.

f)
$(A \cup B)' = \{\#, 8\}$.   ❖

The number of elements in a set can be determined by examining a Venn diagram. For example, in Fig. 2.5 we see that $n(A) = 4$, $n(B) = 5$, $n(A \cup B) = 7$, and $n(A \cap B) = 2$. There is a relationship between $n(A \cup B)$ and $n(A)$, $n(B)$, and $n(A \cap B)$, as illustrated below.

$$\boxed{n(A \cup B) = n(A) + n(B) - n(A \cap B)}$$

For Example 1,

$$n(A \cup B) = n(A) + n(B) - n(A \cap B)$$
$$7 = 4 + 5 - 2$$
$$7 = 7, \quad \text{true.}$$

■ **Example 2** ——————————————————————————————

Draw a Venn diagram illustrating the following sets.

$$U = \{a, \#, *, ?, x, L, 6, w\}$$
$$A = \{\#, ?, L, 6\}$$
$$B = \{a, \#, x, w, 6\}$$

Solution: First determine the intersection of sets A and B, that is, $A \cap B = \{\#, 6\}$. Place these elements in region II of Fig. 2.6. Now place the elements in set A that have not been placed in region II, that is, $?$ and L, in region I. Complete region III by determining the elements in set B that

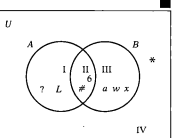

FIGURE 2.6

are not in region II. Thus a, w, and x go in region III. Finally, place out-side both circles those elements in U that are not in either set. The only element in the universal set not in set A or set B is *. Place the * in region IV. ❖

The idea of expressing sets by means of Venn diagrams can be expanded to three or more sets. For three sets, A, B, and C, the diagram has eight regions (see Fig. 2.7).

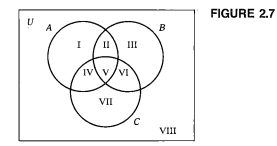

FIGURE 2.7

■ **Example 3** _____

Construct a Venn diagram illustrating the following sets.

$$U = \{1, 2, 3, 4, 5, 6, 7, 8, 9, 10, 11, 12, 13, 14, 15\}$$
$$A = \{1, 2, 3, 4, 7, 9, 11\}$$
$$B = \{2, 3, 4, 5, 10, 12, 14\}$$
$$C = \{1, 2, 4, 8, 9\}$$

Solution: First find the intersection of all three sets. The elements common to all three sets are 2 and 4. These numbers go in region V (see Fig. 2.8). Next complete region II by determining the intersection of A and B, $A \cap B = \{2, 3, 4\}$. $A \cap B$ consists of regions II and V. Since 2 and 4 have already been placed in region V, the 3 must be placed in region II. In a similar manner, complete regions IV and VI. The only elements of set A

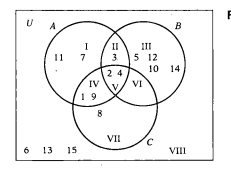

FIGURE 2.8

that have not previously been placed in regions II, IV, and V are 7 and 11. These two elements are only in set A. Therefore we place 7 and 11 in region I. Complete regions III and VII in a similar manner. Finally, determine whether there are any elements of the universal set that have not been included in any of the circles. If so, place them in region VIII. The 6, 13, and 15 must be placed in region VIII.   ❖

Venn diagrams can be used to illustrate and simplify the explanations of many different concepts. One example follows.

■ Example 4 _____

Antigens, made of protein and carbohydrates found in the blood, are substances that produce antibodies to fight infections. Blood may contain antigen A, antigen B, antigen Rh, or any combination of the three. Blood is classified according to the antigens it contains. Blood that contains the Rh antigen is referred to as positive, and blood that lacks this antigen is called negative. Thus if a person's blood contains antigens A, B, and Rh, it is called AB-positive blood. However, if it contains only antigens A and B, it is AB-negative blood. Blood that lacks both the A and the B antigen is called type O. Blood with only the Rh antigen is called O positive, and blood without any of the three antigens is called O negative. Use a Venn diagram to illustrate and name the various types of blood.

Solution: Construct a Venn diagram with three sets, A, B, and Rh, as shown in Fig. 2.9.

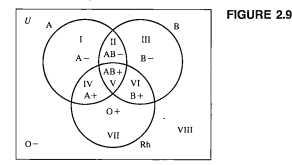

FIGURE 2.9

Since any blood containing the Rh antigen is positive, all blood within the Rh circle is positive, and all blood outside the Rh circle is negative. The intersection of all three sets, region V, is AB+. Region II contains only antigens A and B and is therefore AB−. Region I is A−, since it contains only antigen A. Region III is B−, region IV is A+, and region VI is B+. Region VII is O+, containing only the Rh antigen. Region VIII, which lacks all three antigens, is O−.   ❖

EXERCISES 2.4

1. Construct a Venn diagram illustrating the following sets.

$$U = \{1, 2, 3, 4, 5, 6, 7, 8, 9, 10\}$$
$$A = \{1, 3, 4, 6, 8\}$$
$$B = \{2, 3, 5, 6, 9\}$$

2. Construct a Venn diagram illustrating the following sets.

$$U = \{a, b, c, d, e, f, g, h, i, j, k\}$$
$$A = \{b, c, e, g, h, k\}$$
$$B = \{a, b, d, g, i\}$$
$$C = \{a, b, c, d, g, k\}$$

Listed below are the top five female and top five male tennis players as of March 1988 as ranked by the Association of Tennis Professionals. Their citizenship and whether they play right- or left-handed are also given. Indicate the region in Fig. 2.10 in which each of the tennis players would be placed.

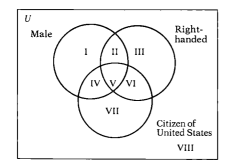

FIGURE 2.10

Female	Citizenship	Plays
3. Stefi Graf	West Germany	Right
4. Martina Navratilova	United States	Left
5. Chris Evert	United States	Right
6. Pam Shriver	United States	Right
7. Hana Manlikova	Czechoslovakia	Right

Male	Citizenship	Plays
8. Ivan Lendl	Czechoslovakia	Right
9. Stefan Edberg	Sweden	Right
10. Mats Wilander	Sweden	Right
11. Jimmy Connors	United States	Left
12. Boris Becker	West Germany	Right

Indicate the region in Fig. 2.11 in which each of the figures in Exercises 13 through 24 would be placed.

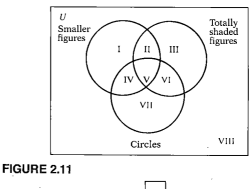

FIGURE 2.11

13. ▲ 14. ◯ 15. ☐

16. ⬤ 17. ⬡ 18. ■

19. ⬣ 20. ◖⬤◗ 21. ⬛

22. △ 23. ⬤ 24. ◯

Use the Venn diagram in Fig. 2.12 to list the sets in Exercises 25 through 32 in roster form.

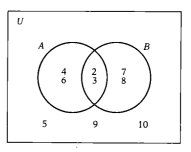

FIGURE 2.12

25. A
26. B
27. U
28. $A \cup B$
29. $A \cap B$
30. $A \cap B'$
31. $A' \cap B$
32. $(A \cup B)'$

Use the Venn diagram in Fig. 2.13 to list the sets in Exercises 33 through 46 in roster form.

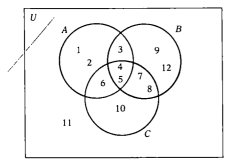

FIGURE 2.13

33. A	**34.** B	**35.** C
36. U	**37.** $A \cap B$	**38.** $A \cap C$
39. $(B \cap C)'$	**40.** $A \cap B \cap C$	**41.** $A \cup B$
42. $B \cup C$	**43.** $(A \cup C)'$	**44.** $A \cup B \cup C$
45. A'	**46.** $(A \cup B \cup C)'$	

47. Given sets U, A, and B, construct a Venn diagram and place the elements in the proper regions.

U = {Sunday, Monday, Tuesday, Wednesday, Thursday, Friday, Saturday}
A = {Monday, Tuesday, Thursday}
B = {Saturday, Sunday, Thursday}

48. Given the sets U, A, and B, construct a Venn diagram and place the elements in the proper regions.

U = {Joyce, Bob, Kathy, Allen, Lisa, Steven, Tod, Stuart}
A = {Joyce, Kathy, Lisa}
B = {Bob, Kathy, Steven, Tod, Lisa}

49. Given sets U, A, B, and C, construct a Venn diagram and place the elements in the proper regions.

U = {Sunday, Monday, Tuesday, Wednesday, Thursday, Friday, Saturday}
A = {Monday, Tuesday, Wednesday, Thursday}
B = {Sunday, Monday, Tuesday, Wednesday, Thursday, Friday}
C = {Tuesday, Sunday}

50. Given the sets U, A, B, and C, construct a Venn diagram and place the elements in the proper regions.

U = {trout, bass, salmon, shark, whale, carp, catfish, whitefish, bluefish, eel}

A = {salmon, shark, whale, eel, whitefish, bluefish}
B = {trout, salmon, eel, bass, carp}
C = {catfish, eel, carp}

51. A hematology text gives the following information on percentages of the different types of blood worldwide.

Type	Positive Blood, %	Negative Blood, %
A	37	6
O	32	6.5
B	11	2
AB	5	0.5

Construct a Venn diagram similar to the one in Example 4 and place the proper percent in each of the eight regions.

52. The Venn diagram given in Fig. 2.14 shows a technique of labeling the regions to indicate membership of elements in a particular region. Define each of the four regions with a set statement. (*Hint:* $A \cap B'$ defines region I.)

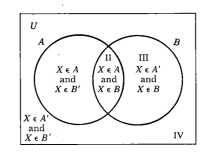

FIGURE 2.14

53. Consider the formula

$$n(A \cup B) = n(A) + n(B) - n(A \cap B).$$

a) Show that this relationship holds for $A = \{a, b, c, d\}$ and $B = \{b, d, e, f, g, h\}$.
b) Make up your own sets A and B each consisting of at least six elements. Using these sets, show that the relationship holds.
c) Using a Venn diagram, can you explain why the relationship holds for any two sets A and B?

54. Given sets U, A, B, and C, construct a Venn diagram and place the elements in the proper regions.

$U = \{1, 2, 3, 4, 5, 6, 7, 8, 9, 10, 11, 12\}$

Set A is the set of even natural numbers less than or equal to 12.
Set B is the set of odd natural numbers less than or equal to 12.
Set C is the set of natural numbers less than or equal to 12 that are multiples of 3. (A multiple of 3 is any number that is divisible by 3.)

55. Given sets U, A, B, and C, construct a Venn diagram and place the elements in the proper regions.

$U = \{1, 2, 3, 4, 5, 6, 7, 8, 9, 10, 11, 12\}$

Set A is the set of even natural numbers less than or equal to 12.
Set B is the set of natural numbers less than or equal to 12 that are multiples of 3.
Set C is the set of natural numbers that are factors of 12. (A factor of 12 is a number that divides 12.)

PROBLEM SOLVING _____

Draw a Venn diagram as in Fig. 2.3 that illustrates the relationship between set A and set B (assume $A \neq B$) if.

56. $A \subset B$,
57. $A \cup B = B$,
58. $A \cap B = B$,
59. $A \cap B = \emptyset$.
60. a) Draw and label as in Exercise 52 a Venn diagram for three sets A, B, and C of a universal set, U.

b) Using the language of sets, name each of the eight regions with a set statement.
*61. a) Construct a Venn diagram illustrating four sets, A, B, C, and D. (*Hint:* Four circles cannot be used, and you should end up with 16 *distinct* regions.) Have fun!
b) Label each region with a set statement (see Exercises 52 and 60). Check all 16 regions to make sure that *each is distinct.*

2.5 VERIFICATION OF SET STATEMENTS

Let $U = \{1, 2, 3, 4, 5\}$, $A = \{1, 2, 3\}$, and $B = \{3, 4\}$. Does $(A \cup B)' = A' \cap B'$? To answer this question, do the following.

Find $(A \cup B)'$.	Find $A' \cap B'$.
$A \cup B = \{1, 2, 3, 4\}$	$A' = \{4, 5\}$
$(A \cup B)' = \{5\}$	$B' = \{1, 2, 5\}$
	$A' \cap B' = \{5\}$

In this specific case, $(A \cup B)' = A' \cap B'$, since both are equal to $\{5\}$. Will $(A \cup B)' = A' \cap B'$ for all sets A and B?

This single case or even a few specific cases cannot verify that the set statements are equal for all sets A and B that we select. To determine whether set statements are equal *for any two sets selected,* we will use Venn diagrams.

Two **set statements are equal** when they represent the same set of elements for any given sets. Two methods can be used to determine whether set statements are equal. One method uses regions, and the other uses shading.

To determine whether set statements are equal using regions, construct a Venn diagram illustrating the sets. Use the diagram to determine the region(s) that correspond to the set statements. If both set statements are represented by the same regions, then the set statements are equal. If the regions are not identical, then the set statements are not equal. Example 1 illustrates this method.

■ **Example 1** _____

Determine whether $(A \cup B)' = A' \cap B'$ for all sets A and B, using the technique of regions.

Solution: Draw a Venn diagram with two sets A and B, as in Fig. 2.15. Label the regions as indicated.

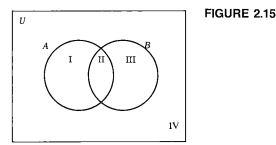

FIGURE 2.15

Find $(A \cup B)'$.   Find $A' \cap B'$.

Set	Corresponding Regions
A	I, II
B	II, III
$A \cup B$	I, II, III
$(A \cup B)'$	IV

Set	Corresponding Regions
A'	III, IV
B'	I, IV
$A' \cap B'$	IV

Both set statements are represented by the same region, IV, of the Venn diagram. Thus $(A \cup B)' = A' \cap B'$ for all sets A and B. ❖

To determine whether two set statements are equal using shading, construct two identical Venn diagrams illustrating the sets. On one of the diagrams, shade the regions that represent one of the set statements. On the other diagram, shade the regions that represent the other set statement. If both Venn diagrams have the same regions shaded, then the set statements are equal. If the diagrams do not have exactly the same regions shaded, then the set statements are not equal.

There is no specific method for shading. Thus the areas we decided to shade horizontally could have been shaded vertically, and vice versa.

Example 2 ———————————————————————

Determine whether $A' \cup B' = (A \cap B)'$ for all sets A and B, using the technique of shading.

Solution: Construct two Venn diagrams, one to represent $A' \cup B'$ and a second to represent $(A \cap B)'$. If the shaded regions representing the sets are identical in both diagrams, then the statements are equal for all sets A and B.

To determine $A' \cup B'$, draw vertical lines to indicate A' and horizontal lines to indicate B' (Fig. 2.16). Since union means joining together, the set $A' \cup B'$ is indicated by the regions that are shaded vertically, horizontally, or in both directions. For emphasis the solution of $A' \cup B'$ is shaded in Fig. 2.17.

To determine $(A \cap B)'$, shade $A \cap B$ horizontally, as in Fig. 2.18. Then shade $(A \cap B)'$ vertically. For emphasis the solution of $(A \cap B)'$ is shaded in Fig. 2.19.

It is now apparent that $A' \cup B' = (A \cap B)'$, since Figs. 2.17 and 2.19 have identical shaded regions. Thus $A' \cup B' = (A \cap B)'$ for all sets A and all sets B.

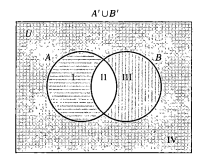

FIGURE 2.16

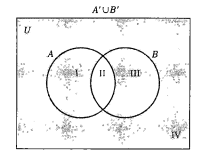

FIGURE 2.17

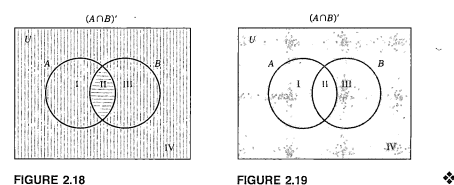

FIGURE 2.18

FIGURE 2.19

Venn diagrams can also be used to determine whether set statements containing three sets are equal. Example 3 illustrates such a problem using the regions technique, and Example 4 illustrates a similar problem solved by shading.

■ **Example 3** _____

Determine whether $A \cap (B \cup C) = (A \cap B) \cup (A \cap C)$ for all sets, $A, B,$ and C by using the technique of regions in Venn diagrams.

Solution: Since the statements include three sets, $A, B,$ and C, three circles must be used. The Venn diagram illustrating the eight regions is shown in Fig. 2.20.

First we will find the regions that correspond to $A \cap (B \cup C)$, then we will find the regions that correspond to $(A \cap B) \cup (A \cap C)$. If both answers are the same, the statements are equal.

Find $A \cap (B \cup C)$.

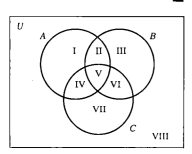

FIGURE 2.20

Set	Corresponding Regions
A	I, II, IV, V
$B \cup C$	II, III, IV, V, VI, VII
$A \cap (B \cup C)$	II, IV, V

The regions that correspond to $A \cap (B \cup C)$ are II, IV, and V.
Now find $(A \cap B) \cup (A \cap C)$.

Set	Corresponding Regions
$A \cap B$	II, V
$A \cap C$	IV, V
$(A \cap B) \cup (A \cap C)$	II, IV, V

The regions that correspond to $(A \cap B) \cup (A \cap C)$ are II, IV, and V.

The work above shows that both statements are represented by the same regions, namely, II, IV, and V, and therefore $A \cap (B \cup C) = (A \cap B) \cup (A \cap C)$ for all sets A, B, and C.   ❖

■ Example 4

Determine whether $(A \cap B) \cup C = A \cap (B \cup C)$, using the technique of shading.

Solution: First find $(A \cap B) \cup C$. Shade $A \cap B$ horizontally (see Fig. 2.21). Shade set C vertically. The answer is the union of the two shaded areas (see Fig. 2.22).

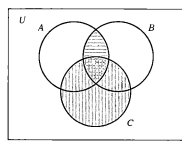

FIGURE 2.21

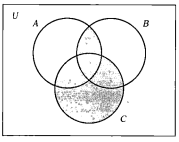

FIGURE 2.22

Now find $A \cap (B \cup C)$. Shade $B \cup C$ horizontally (see Fig. 2.23). Then shade set A vertically. The answer is the intersection of the two shaded areas (see Fig. 2.24). Since the shaded areas in Figs. 2.22 and 2.24 are not identical, we can conclude that $(A \cap B) \cup C \neq A \cap (B \cup C)$ for all sets A, B, and C.

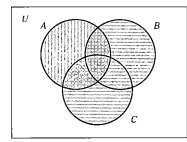

FIGURE 2.23

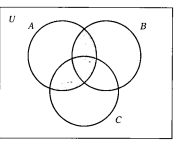

FIGURE 2.24   ❖

DID YOU KNOW?

The practical applications of set theory to the physical world arose in an indirect way. Georg Cantor, F. L. O. Frége, and Ernst Zermelo were pioneers in the area of set theory. Their accomplishments not only affected the foundations of mathematics, but led to a new attitude in mathematics. Their abstract approach to sets led to a more abstract approach to every branch of mathematics and more general approaches to classical problems in geometry, algebra, and analysis. Such new areas as topology, functional analysis, and combinational analysis developed. Each of these areas in turn has been rich in practical applications. The giant step began with the beginning of abstract set theory.

EXERCISES 2.5

Using Venn diagrams, determine whether the following statements are equal for all sets A and B. Use the technique specified by your instructor.

1. $(A \cup B)'$, $A' \cap B'$
2. $(A \cup B)'$, $A' \cap B$
3. $A' \cup B'$, $A \cap B$
4. $(A \cup B)'$, $(A \cap B)'$
5. $A' \cup B'$, $(A \cup B)'$
6. $A \cap B'$, $A' \cup B$
7. $(A \cap B')'$, $A' \cup B$
8. $A' \cap B'$, $(A' \cap B')'$

Using Venn diagrams, determine whether the following statements are equal for all sets A, B, and C. Use the technique specified by your instructor.

9. $A \cup (B \cap C)$, $(A \cup B) \cap C$
10. $A \cup (B \cap C)$, $(B \cap C) \cup A$
11. $A \cap (B \cup C)$, $(B \cup C) \cap A$
12. $A' \cup (B \cap C)$, $A \cap (B \cup C)'$
13. $A \cap (B \cup C)$, $(A \cap B) \cup (A \cap C)$
14. $A \cup (B \cap C)$, $(A \cup B) \cap (A \cup C)$
15. $A \cap (B \cup C)'$, $A \cap (B' \cap C')$
16. $(A \cup B) \cap (B \cup C)$, $B \cup (A \cap C)$
17. $(C \cap B)' \cup (A \cap B)'$, $A \cap (B \cap C)$
18. $(A \cup B)' \cap C$, $(A' \cup C) \cap (B' \cup C)$

19. Let $U = \{1, 2, 3, 4, 5, 6, 7, 8, 9, 10\}$,
    $A = \{1, 2, 3, 4\}$,
    $B = \{3, 6, 7\}$,
    $C = \{6, 7, 9\}$.
    a) Show that $(A \cup B) \cap C = (A \cap C) \cup (B \cap C)$ for these specific sets.
    b) Make up your own sets A, B, and C. Verify that $(A \cup B) \cap C = (A \cap C) \cup (B \cap C)$ for your sets A, B, and C.
    c) Using Venn diagrams, verify that $(A \cup B) \cap C = (A \cap C) \cup (B \cap C)$ for all sets A, B, and C.
20. Let $U = \{a, b, c, d, e, f, g, h, i\}$,
    $A = \{a, c, d, e, f\}$,
    $B = \{c, d\}$,
    $C = \{a, b, c, d, e\}$.
    a) Determine whether $(A \cup C)' \cap B = (A \cap C)' \cap B$ for the specific sets given.
    b) Make up your own specific sets, A, B, and C. Determine whether $(A \cup C)' \cap B = (A \cap C)' \cap B$ for your sets.
    c) Determine by the method of your choice whether $(A \cup C)' \cap B = (A \cap C)' \cap B$ for all sets A, B, and C.

2.6 APPLICATION OF SETS

We can use the information we have learned about set operations and Venn diagrams to solve some practical problems. Examples of these problems are given below.

■ Example 1

At the ABC Sports Shop, 160 people were selected at random and asked to complete the questionnaire below.

   Check the appropriate box or boxes.
   □ **1.** I enjoy cross-country skiing.
   □ **2.** I enjoy downhill skiing.

It was found that 80 people checked box 1, indicating that they enjoy cross-country skiing, 70 people checked box 2, and 20 checked both boxes, indicating that they enjoy both cross-country and downhill skiing.

a) How many people did not check a box?

b) How many people enjoy cross-country skiing but do not enjoy downhill skiing?

c) How many people enjoy downhill skiing but do not enjoy cross-country skiing?

d) How many people enjoy downhill or cross-country skiing?

Solution: The problem provides us with the following information.

1. The number of people surveyed is 160: $n(U) = 160$.

2. The number of people who enjoy cross-country skiing is 80: $n(C) = 80$.

3. The number of people who enjoy downhill skiing is 70: $n(D) = 70$.

4. The number of people who enjoy both cross-country and downhill skiing is 20: $n(C \cap D) = 20$.

We can illustrate this information with a Venn diagram, as in Fig. 2.25. From previous sections we know that $C \cap D$ corresponds to region II. Since $n(C \cap D) = 20$, we write 20 in region II. Set C consists of regions I and II. From statement 2 we know that set C contains 80 people. Therefore region I contains $80 - 20$, or 60 people, and we write the number 60 in region I. Set D consists of regions II and III. Since $n(D) = 70$, there must be a total of 70 in these two regions. Region II contains 20, leaving 50 for region III. The total number of people who enjoy skiing is $n(C \cup D) = 60 + 20 + 50 = 130$. The number of people in region IV is the

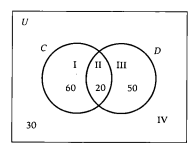

FIGURE 2.25

difference between $n(U)$ and $n(C \cup D)$. There are $160 - 130$, or 30 people in region IV.

a) How many people who were interviewed did not check a box? The people who did not check a box are those members of set U who are not contained in set C or set D. This is region IV in the diagram. The number of people in region IV is 30.

b) How many people enjoy cross-country skiing but do not enjoy downhill skiing? The 60 people in region I are those who enjoy only cross-country skiing.

c) How many people enjoy downhill skiing but do not enjoy cross-country skiing? The 50 people in region III are in this category.

d) How many people enjoy downhill skiing or cross-country skiing? The people in regions I, II, or III enjoy cross-country or downhill skiing. The sum of the numbers in the three regions is 130.   ❖

Similar problems can be done with three sets, as illustrated in the following example.

■ Example 2

Of the first 39 presidents of the United States, 8 held cabinet posts, 13 served as vice-president, 15 served in the United States Senate, 2 served in cabinet posts and as vice-president, 4 served in cabinet posts and in the United States Senate, 6 served in the United States Senate and as vice-president, and 1 served in all three positions. How many presidents served in

a) none of these positions?
b) only the United States Senate?
c) at least one of the three positions?
d) exactly two positions?

Solution: Construct a Venn diagram with three circles: cabinet post, vice-president, and Senate (see Fig. 2.26). Label the eight regions.

Whenever possible, work from the center outward. First find the number of presidents that appear in all three sets. Since one president served in all three positions, place a 1 in region V. Next determine the number to be placed in region II. Two presidents served both in cabinet posts and as vice-president. Regions II and V represent the intersection of presidents who served both in cabinet posts and as vice-president. Therefore the sum of the elements in these two regions must be 2. Since a 1 has previously

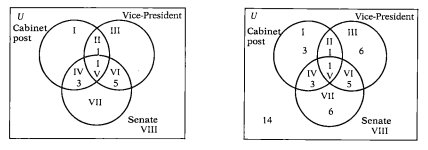

FIGURE 2.26   **FIGURE 2.27**

been placed in region V, this leaves a 1 for region II. Four presidents served in cabinet posts and in the Senate. The sum of regions IV and V must therefore be 4. Since there is already a 1 in region V, place a 3 in region IV. Using similar reasoning, place a 5 in region VI.

Now find the numbers in regions I, III, and VII (Fig. 2.27). A total of 8 presidents served in cabinet posts. The sum of regions II, IV, and V is 5, so there are 3 in region I. The total number of presidents who served as vice-president was 13. The sum of the numbers in regions II, V, and VI is 7, so there must be 6 in region III. Similarly, the number in region VII must be 15 − 9, or 6. So far we have accounted for a total of 25 presidents in regions I through VII. There must be 39 − 25, or 14 presidents who did not serve in any of these capacities. Place the number 14 in region VIII. Now use the completed Venn diagram to determine the answers to questions (a) through (d).

a) Fourteen presidents did not serve in any of the three positions. This number is found in region VIII.

b) Region VII represents the presidents who served only in the Senate. There were six.

c) "At least one" means one or more. Thus the number of presidents who served in at least one of the three positions is found by summing the numbers in regions I through VII. Twenty-five presidents served in at least one of these positions.

d) The presidents represented by regions II, IV, and VI served in exactly two positions. Summing the numbers in the three regions shows that nine presidents served in exactly two positions.   ❖

When you are solving problems of this type, it is a good idea to check your Venn diagram carefully. The most common error made by students is forgetting to subtract the number in region V from the respective values in determining the numbers to place in regions II, IV, and VI.

■ **Example 3** _____

Common ailments of senior citizens are arthritis, arteriosclerosis, and loss of hearing. A survey of 200 senior citizens found that 70 had arthritis, 60 had arteriosclerosis, 80 had loss of hearing, 35 had arthritis and arteriosclerosis, 33 had arthritis and loss of hearing, 31 had arteriosclerosis and loss of hearing, and 15 had all three.

a) How many of the senior citizens in the survey had none of these ailments?

b) How many of the senior citizens had arthritis but neither of the other two ailments?

c) How many had exactly one of these ailments?

d) How many had arteriosclerosis *and* loss of hearing but not arthritis?

e) How many had arteriosclerosis *or* loss of hearing but not arthritis?

f) How many had exactly two of these ailments?

Solution: The solution to this problem can be simplified by constructing a Venn diagram, as explained in Example 2. You should go through the process of developing the Venn diagram yourself. For your convenience the diagram is given in Fig. 2.28. We will use it in explaining the solution of this problem.

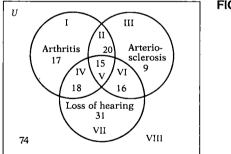

FIGURE 2.28

a) How many had none of these ailments? The 74 senior citizens without any of these ailments are found in region VIII.

b) How many had arthritis but neither of the other two ailments? The 17 people in region I had only arthritis.

c) How many had exactly one of these ailments? Those listed in regions I, III, and VII had only one of the ailments. The sum of the numbers in these regions, 17 + 9 + 31, or 57, is the number of senior citizens with exactly one of the ailments.

d) How many had arteriosclerosis *and* loss of hearing but not arthritis? The 16 people in region VI satisfy these conditions.

e) How many had arteriosclerosis *or* loss of hearing but not arthritis? This is a tough one. The word "or" in this type of problem means one or the other or both. All the people in regions II, III, IV, V, VI, and VII had either arteriosclerosis or loss of hearing or had both. Those in regions II, IV, and V also had arthritis. The senior citizens with either arteriosclerosis or loss of hearing but not arthritis are found in regions III, VI, and VII. The sum of the numbers in these three regions is 9 + 16 + 31, or 56. Fifty-six people satisfy these conditions.

f) How many had exactly two of these ailments? The people in regions II, IV, and VI, a total of 54, had exactly two of the three ailments. ❖

EXERCISES 2.6

1. Interviews of 150 people to determine whether they had money invested in stocks or bonds produced the following results. Of those interviewed, 80 had money invested in stocks, 40 had money invested in bonds, and 23 had money invested in both stocks and bonds.

   a) How many of those interviewed had money invested only in stocks?
   b) How many had money invested only in bonds?
   c) How many had not invested in either stocks or bonds?

2. In the 1988–1989 school year at Lincoln High School, 65 students participated in soccer or basketball, 50 participated in soccer, and 10 participated in both sports. How many students participated in basketball?

3. Two bills brought before the legislature were of great concern to a representative. In the interest of representing her people accurately, she decided to take a sample poll of her constituents. The results showed that out of 200 families in the survey, 55 favored bill number 1, 110 favored bill number 2, and 40 favored both bills.

   a) How many favored only bill number 1?
   b) How many were not in favor of either bill?
   c) Did the majority of those surveyed favor bill number 1? bill number 2?

4. A survey of 50 families showed that 16 had two cars, 18 had a color TV, 24 had a ten-speed bicycle, 4 had two cars and a ten-speed bicycle, 5 had a color TV and a ten-speed bicycle, 8 had two cars and a color TV, and 3 had all three. Find the number of families that had

   a) none of the three,
   b) two cars and a color TV but not a ten-speed bicycle,
   c) only the two cars,
   d) only a ten-speed bicycle,
   e) a color TV or a ten-speed bicycle.

5. In a freshman class at San Diego Community College, 200 students were enrolled in mathematics, 180 in English, 170 in biology, 40 in mathematics and English, 45 in English and biology, 50 in mathematics and biology, and 15 in all three courses. In all, 500 students were surveyed. How many were taking

   a) none of these courses,
   b) only mathematics,
   c) mathematics and English but not biology,
   d) mathematics or English but not biology,
   e) at least one of the courses?

6. Jane questions 35 men to determine whether they like lemon-scented, menthol, or regular shaving cream. Twenty like lemon-scented, 20 like menthol, 19 like regular, 12 like lemon and menthol, 11 like lemon and regular, 10 like regular and menthol, and 8 like all three.

a) How many like only lemon?
b) How many like lemon or regular?
c) How many like only one of the three?
d) How many like lemon and regular but not menthol?
e) How many like lemon or regular but not menthol?
f) How many do not like any of the three?

7. In a recent blood drive the following data on donors were recorded (see Example 4 in Section 2.4): 243 donors had the A antigen, 93 had the B antigen, 28 had the A and the B antigens, 80 had the B and the Rh antigens, 325 had the Rh antigen, 33 donors had none of the antigens, 210 donors had the A and the Rh antigens, and 25 donors had all three antigens. How many donors have
a) A-positive blood?
b) B-negative blood?
c) AB-positive blood?
d) How many donors are represented here?

8. The Democratic Party is in the process of selecting a candidate to run against the incumbent Republican. Three candidates are under consideration. A survey is performed to determine which of the three candidates, A, B, or C, is the best choice. In a survey, 300 randomly selected voters are asked to complete the following questionnaire.
  Check the box or boxes of the candidate(s) you would vote for over the incumbent.

☐ Mr. A   ☐ Mr. B   ☐ Ms. C

When the survey was completed, the results were summarized as follows: 130 checked box A, 110 checked box B, 160 checked box C, 80 checked boxes A and B, 65 checked boxes A and C, 40 checked boxes B and C, 27 checked all three boxes.

a) How many checked only box C?
b) How many did not check any box?
c) How many checked boxes B and C but not A?
d) How many checked box A or B?
e) How many checked box A or B but not C?

9. A statistician was commissioned to determine the newspapers read daily by New York City residents. She randomly selected an appropriate sample and had those surveyed complete the following form.

I read the following newspapers daily.

☐ New York Post
☐ New York Times
☐ New York Daily News
☐ I do not read any of the three papers listed.

When the survey was completed, the results were summarized as follows: 40 checked all three boxes, 69 checked the Times and the Post, 85 checked the Times and the News, 160 checked the Post and the News, 250 checked the News, 270 checked the Post, 350 checked the Times, and 104 checked none of the three.

a) How many read only the New York Times?
b) How many read at least one of the three papers?
c) How many read at most one of the three papers?
d) How many read the Times and the Post but not the News?
e) How many read the Times or the Post but not the News?
f) How many read exactly one of the three papers?

10. José made a survey for the manufacturer of brand X toothpaste to determine how many people preferred wintergreen-flavored, peppermint-flavored, or cherry-flavored toothpaste. One hundred people were asked to complete the following questionnaire.

Check the box or boxes indicating the flavors of brand X toothpaste you would purchase.

A. ☐ Wintergreen-flavored
B. ☐ Peppermint-flavored
C. ☐ Cherry-flavored
D. ☐ I would not purchase brand X.

When the survey was completed, the results were summarized as follows: 44 checked box C, 40 checked box A, 37 checked box B, 16 checked boxes B and C, 15 checked boxes A and B, 11 checked boxes A and C, and 4 checked boxes A, B, and C.

a) How many people in the survey would not purchase brand X toothpaste?
b) How many people would purchase only wintergreen-flavored toothpaste?
c) How many people would purchase cherry- or peppermint-flavored toothpaste?
d) How many people would purchase cherry- or wintergreen- but not peppermint-flavored toothpaste?

11. Molly was hired to interview people to find out what brand of ice cream they liked. She submitted the following report.

Number of people interviewed	100
Number of people who liked Carvel	78

Number of people who liked
Abbott's Frozen Custard 61
Number of people who liked both
Carvel and Abbott's Frozen Custard 40

Every person interviewed liked one or the other or both flavors of ice cream. When Molly turned in her report, she was fired. Why?

12. An immigration agent samples 85 cars going from the United States into Canada. In his report he indicates that 35 of the cars are driven by women, 53 of the cars are driven by U.S. citizens, 43 of the cars have two or more passengers, 27 of the cars are driven by women who are U.S. citizens, 25 of the cars driven by women have two or more passengers, 20 of the cars driven by U.S. citizens have two or more passengers, and 15 of the cars driven by women who are U.S. citizens have two or more passengers. After his supervisor reads the report, she explains to the agent that he made a mistake. Explain how his supervisor knew that the agent's report contained an error.

PROBLEM SOLVING

13. A survey of 500 farmers in a midwestern state showed that 125 farmers grew *only* wheat, 110 farmers grew *only* corn, and 90 farmers grew *only* oats. The number who grew wheat is 200, the number who grew wheat and corn is 60, the number who grew wheat and oats is 50, and the number who grew corn is 180. Find the number who

a) grew at least one of the three,
b) grew all three,
c) did not grow any of the three,
d) grew exactly two of the three.

14. A survey was taken of 400 owners of home computers to determine how many owned the following software: VisiCalc, Home Accountant, and Easy-Writer. The survey showed that 66 owned only Home Accountant and 104 owned only VisiCalc. The number that owned EasyWriter was 150, and the number that owned Home Accountant and EasyWriter was 31. The number that owned Home Accountant and VisiCalc was 33, and the number that owned VisiCalc and EasyWriter was 23. The number that did not own any of the three software packages was 62. Find the number that

a) owned all three software packages,
b) owned exactly one of the software packages,
c) owned exactly two of the software packages,
d) owned both EasyWriter and Home Accountant but not VisiCalc.

*15. Mr. and Mrs. Semrau have a large family. They want to go on a vacation and take all their children. When they discussed where to go on their vacation, they found that of all their children:
a) eight refused to go to the family cottage,
b) seven refused to go to a resort spot on the ocean,
c) nine refused to go on a camping trip,
d) three will go neither to the family cottage nor to the resort spot on the ocean,
e) four will go neither to the resort spot on the ocean nor on the camping trip,
f) six will go neither to the family cottage nor on the camping trip,
g) two will not go to the family cottage, the resort spot on the ocean, or the camping trip,
h) no children will agree to go to all 3 places.
How many children do the Semrau's have?

*16. a) There are five houses.
b) The green house is directly to the right of the ivory house.
c) The Scot has the red house.
d) The dog belongs to the Spaniard.
e) The Slovak drinks tea.
f) The person who eats cheese lives next door to the fox.
g) The Japanese eats fish.
h) Milk is drunk in the middle house.
i) Apples are eaten in the house next to the horse.
j) Ale is drunk in the green house.
k) The Norwegian lives in the first house.
l) The peach eater drinks whiskey.
m) Apples are eaten in the yellow house.
n) The banana eater owns a snail.
o) The Norwegian lives next door to the blue house.

Find:

1) the color of each house,
2) the nationality of the occupant,
3) the type of food eaten in each house,
4) the owner's favorite drink,
5) the owner's pet.
6) Finally, the crucial question is: Does the zebra's owner drink vodka or ale?

2.7 INFINITE SETS (OPTIONAL)

In Section 2.1 we introduced the concept of infinite sets. In this section we discuss this topic in greater depth.

The German mathematician Georg Cantor, 1845–1918, known as the father of set theory, was the first person to explore infinite sets in a systematic fashion. In discussing infinite sets, Cantor made use of the concept of one-to-one correspondence. Two sets A and B can be placed in a one-to-one correspondence if every element of set A can be matched with exactly one element of set B and vice versa. For example, the sets $A = \{*, ?, \sim\}$ and $B = \{a, b, c\}$ can be placed in a one-to-one correspondence as shown below.

$$A = \{*, ?, \sim\}$$
$$\downarrow \ \downarrow \ \downarrow$$
$$B = \{a, b, c\}$$

Note that two finite sets such as A and B that can be placed in a one-to-one correspondence must have the same number of elements (therefore the same cardinality) and must be equivalent sets. Note that $n(A)$ and $n(B)$ both equal 3.

While studying the concept of infinite sets, Cantor introduced the transfinite cardinal number **aleph-null,** symbolized $\aleph_0$. (the first Hebrew letter, *aleph,* with a zero subscript read null). The basic set used in discussing infinite sets is the set of natural or counting numbers, $N = \{1, 2, 3, 4, \ldots\}$. The set of counting numbers is said to have the infinite cardinal number $\aleph_0$. Thus $n(N) = \aleph_0$. If asked the question "How many counting numbers are there?" we would answer that there are $\aleph_0$ of them.

Cantor believed that there were different kinds or different orders of infinity, that is, that some sets of elements are "more infinite" than others. Cantor believed that saying a set of elements is "infinite" was not specific enough. He referred to the set of counting numbers as denumerably infinite.

> Any set that can be placed in a one-to-one correspondence with the set of counting numbers has cardinality $\aleph_0$ and is said to be **denumerably infinite.**

Denumerably infinite sets are considered the smallest type of infinite sets.

Are there more counting numbers than even counting numbers? That is, does the set $\{1, 2, 3, 4, \ldots\}$ contain more elements than the set $\{2, 4, 6, 8, \ldots\}$? Intuition might lead us to answer yes, there are more counting numbers than there are even counting numbers. However, Cantor showed that this is not true. Consider the one-to-one correspondence that follows:

$$\{1, 2, 3, 4, 5, \ldots, \quad n, \ldots\}$$
$$\downarrow \downarrow \downarrow \downarrow \downarrow \qquad \downarrow$$
$$\{2, 4, 6, 8, 10, \ldots, \quad 2n, \ldots\}$$

Note that for every counting number n, its corresponding number is $2n$. Since there is a one-to-one correspondence, the set of counting numbers and the set of even counting numbers are both denumerably infinite, and both have the cardinal number $\aleph_0$.

Using a similar procedure, we can show that the set of odd counting numbers also has cardinality $\aleph_0$. To do this, we need to show a one-to-one correspondence between the counting numbers and odd counting numbers. The one-to-one correspondence follows.

$$\{1, 2, 3, 4, 5, 6, \ldots, \quad n, \ldots\}$$
$$\downarrow \downarrow \downarrow \downarrow \downarrow \downarrow \qquad \downarrow$$
$$\{1, 3, 5, 7, 9, 11, \ldots, \quad 2n - 1, \ldots\}$$

Note that the numbers $1, 3, 5, 7, \ldots$ are all 1 less than the corresponding numbers $2, 4, 6, 8, \ldots$, which we recently denoted as $2n$. Thus any number n in the set of counting numbers will be matched with the number $2n - 1$ in the set of odd counting numbers. Since there is a one-to-one correspondence, the odd counting numbers have cardinality $\aleph_0$. We have shown that both the odd and even counting numbers have cardinality $\aleph_0$. Since the odd counting numbers plus the even counting numbers give the counting numbers, we may reason that

$$\aleph_0 + \aleph_0 = \aleph_0$$

This may seem strange, but it is true.

Consider the set of whole numbers, $\{0, 1, 2, 3, \ldots\}$. We can show that the whole numbers have cardinal number $\aleph_0$ by setting up a one-to-one correspondence with the set of counting numbers as follows.

$$\{1, 2, 3, 4, 5, \ldots, \quad n, \ldots\}$$
$$\downarrow \downarrow \downarrow \downarrow \downarrow \qquad \downarrow$$
$$\{0, 1, 2, 3, 4, \ldots, \quad n - 1, \ldots\}$$

Therefore the whole numbers have cardinality $\aleph_0$. Note that if we add one element, namely 0, to the set of counting numbers, we obtain the whole numbers. Thus we may write

$$\aleph_0 + 1 = \aleph_0$$

■ **Example 1** _____

Show that set A, $A = \{3, 7, 11, 15, \ldots\}$, has cardinal number $\aleph_0$.

Solution: To show that set A has cardinal number $\aleph_0$ we must show a one-to-one correspondence between set A and the set of counting numbers. Begin as follows.

$$N = \{1, 2, 3, \quad 4, \ldots, \quad n, \ldots\}$$
$$\downarrow \downarrow \downarrow \quad \downarrow \qquad \downarrow$$
$$A = \{3, 7, 11, 15, \ldots, \quad , \ldots\}$$

The n in the set of counting numbers is used to represent any counting number. We must determine what expression n will match up with (or map into) in set A. Notice that the numbers in set A differ by 4 units. Thus n will match up with an expression containing $4n$ in set A. If n is matched with $4n$, then the number 1 in the set of counting numbers would match with the number 4, the number 2 in the set of counting numbers would match with the number 8, 3 would match with 12, and so on. Since 4 is one more than the number 3, 8 is one more than the number 7, 12 is one more than the number 11, and so on, 1 must be subtracted from each of the numbers of the form $4n$. If we subtract 1 from $4n$ we obtain $4n - 1$. Thus any number n in the set of counting numbers will match with the number $4n - 1$ in set A. For example, the number 9 in the set of counting numbers will match up with the number $4(9) - 1 = 35$ in set A. The one-to-one correspondence we are seeking is

$$N = \{1, 2, 3, \quad 4, \ldots, \quad n, \ldots\}$$
$$\downarrow \downarrow \downarrow \quad \downarrow \qquad \downarrow$$
$$A = \{3, 7, 11, 15, \ldots, 4n - 1, \ldots\}$$ ❖

Consider the set in Example 1, $\{3, 7, 11, 15, \ldots\}$. Notice that we can place this set in a one-to-one correspondence with a proper subset of itself, as illustrated below:

$$\{3, \quad 7, 11, 15, \ldots, \quad n, \ldots\} \quad \text{Set}$$
$$\downarrow \quad \downarrow \downarrow \downarrow \qquad \downarrow$$
$$\{7, 11, 15, 19, \ldots, n + 4, \ldots\} \quad \text{Proper subset}$$

To obtain the proper subset we removed the first element, 3, from the original set. Note that each element in the proper subset is 4 greater than the corresponding element in the original set.

> An **infinite set** is a set that can be placed in a one-to-one correspondence with a proper subset of itself.

■ Example 2

Show that $\{5, 10, 15, 20, 25, \ldots\}$ is an infinite set.

Solution: To show that the set is infinite, we will set up a one-to-one correspondence between the set and a proper subset of itself. A one-to-one correspondence is illustrated below:

$$\{5, \ 10, \ 15, \ 20, 25, \ldots, \quad n, \ldots\} \qquad \text{Set}$$
$$\downarrow \ \downarrow \ \ \downarrow \ \ \downarrow \ \ \downarrow \qquad \qquad \downarrow$$
$$\{10, \ 15, \ 20, \ 25, \ 30, \ldots, n + 5, \ldots\} \qquad \text{Proper Subset}$$

Since we have illustrated a one-to-one correspondence between the set and a proper subset of itself, the set $\{5, 10, 15, 20, 25, \ldots\}$ is an infinite set. ❖

Note that in Example 2 we showed that $\{5, 10, 15, 20, 25, \ldots\}$ is an infinite set. However, we did not determine its cardinality. We could show that this set has cardinality $\aleph_0$ by placing it in a one-to-one correspondence with the set of counting numbers, as illustrated below:

$$\{1, 2, \ 3, \ 4, \ \ 5, \ldots, \quad n, \ldots\}$$
$$\downarrow \ \downarrow \ \ \downarrow \ \ \downarrow \ \ \downarrow \qquad \quad \downarrow$$
$$\{5, 10, 15, 20, 25, \ldots, 5n, \ldots\}$$

As we mentioned earlier, there are different orders of infinity. Every infinite set can be placed in a proper subset with itself. However, not every infinite set can be placed in a one-to-one correspondence with the set of counting numbers. Cantor showed that the set of integers, the set of rational numbers (fractions of the form p/q, $q \neq 0$) were all infinite sets with cardinal number $\aleph_0$. He also showed that the set of irrational numbers ($\sqrt{2}$ is an example of an irrational number) and the set of real numbers (to be discussed in Chapter 4) was nondenumerably infinite and was a higher order of infinity than $\aleph_0$. The transfinite cardinal number of the real numbers is aleph-one, $\aleph_1$. Since the real numbers do not have cardinality $\aleph_0$ they cannot be placed in a one-to-one correspondence with the set of counting numbers. However, since the real numbers are an infinite set they can be placed in a one-to-one correspondence with a proper subset of itself.

There are also other transfinite cardinal numbers. In fact, there are infinitely many transfinite cardinal numbers, $\aleph_0$, $\aleph_1$, $\aleph_2$, $\aleph_3$, and so on. It has been shown that $\aleph_0$ is the smallest of these "infinite" numbers.

Transfinite cardinal numbers can be added and multiplied, but they cannot be subtracted or divided. The results of adding and multiplying transfinite cardinal numbers are not intuitive. For example,

$$\aleph_0 + 19 = \aleph_0,$$
$$\aleph_0 + \aleph_0 = \aleph_0,$$
$$2 \cdot \aleph_0 = \aleph_0,$$
$$\aleph_0 \cdot \aleph_0 = \aleph_0,$$
$$\aleph_0 \cdot \aleph_1 = \aleph_1.$$

2.7 EXERCISES

For each of the following sets, show that the set is infinite by placing it in a one-to-one correspondence with a proper subset of itself.

1. $\{3, 4, 5, 6, 7, \ldots\}$
2. $\{4, 8, 12, 16, 20, \ldots\}$
3. $\{2, 3, 4, 5, 6, \ldots\}$
4. $\{3, 5, 7, 9, 11, \ldots\}$
5. $\{4, 7, 10, 13, 16, \ldots\}$
6. $\{6, 11, 16, 21, 26, \ldots\}$
7. $\{8, 10, 12, 14, 16, \ldots\}$
8. $\{1, \frac{1}{2}, \frac{1}{3}, \frac{1}{4}, \frac{1}{5}, \ldots\}$
9. $\{1, \frac{1}{3}, \frac{1}{5}, \frac{1}{7}, \frac{1}{9}, \ldots\}$
10. $\{\frac{5}{8}, \frac{6}{8}, \frac{7}{8}, \frac{8}{8}, \frac{9}{8}, \ldots\}$

For each of the following sets show that the set has cardinal number $\aleph_0$ by setting up a one-to-one correspondence between the set of counting numbers and the given set.

11. $\{3, 6, 9, 12, 15, \ldots\}$
12. $\{100, 101, 102, 103, 104, \ldots\}$
13. $\{4, 6, 8, 10, 12, \ldots\}$
14. $\{0, 2, 4, 6, 8, \ldots\}$
15. $\{2, 5, 8, 11, 14, \ldots\}$
16. $\{4, 9, 14, 19, 24, \ldots\}$
17. $\{5, 8, 11, 14, 17, \ldots\}$
18. $\{\frac{1}{2}, \frac{1}{4}, \frac{1}{6}, \frac{1}{8}, \ldots\}$
19. $\{\frac{1}{3}, \frac{1}{4}, \frac{1}{5}, \frac{1}{6}, \frac{1}{7}, \ldots\}$
20. $\{\frac{1}{2}, \frac{2}{3}, \frac{3}{4}, \frac{4}{5}, \frac{5}{6}, \ldots\}$

PROBLEM SOLVING

Show that the following sets have cardinality $\aleph_0$ by setting up a one-to-one correspondence between the set of counting numbers and the given set.

21. $\{1, 4, 9, 25, 36, \ldots\}$
22. $\{2, 4, 8, 16, 32, \ldots\}$

23. $\{3, 9, 27, 81, 243, \ldots\}$

*24. $\{\frac{1}{3}, \frac{1}{6}, \frac{1}{12}, \frac{1}{24}, \frac{1}{48}, \ldots\}$

SUMMARY

There are at least three ways of indicating a set: by description, by roster form, and by set-builder notation. The objects in a set are called its members or elements. Membership in a set may be indicated by the symbol $\in$. The universal set, U, and the empty set, $\emptyset$, are essential in our discussion of sets. The cardinal number of set A, $n(A)$, is the number of elements in set A. The complement of set A, A', is the set of elements in the universal set not in set A. Two sets are equal if they contain the same elements.

    The union of set A and set B, $A \cup B$, is the set of elements in either set A or set B or in both.

    The intersection of set A and set B, $A \cap B$, is the set of elements common to both sets.

    Venn diagrams are used to illustrate sets. They can also be an aid in describing some practical situations.

    The natural numbers are an infinite set that have cardinality aleph-null, $\aleph_0$. Any set that can be placed in a one-to-one correspondence with a proper subset of itself is an infinite set. An infinite set that can be placed in a one-to-one correspondence with the set of counting numbers has cardinality $\aleph_0$.

REVIEW EXERCISES ———————————————————

Answer true or false.

1. _____ The set of students who have blond hair is a well-defined set.
2. _____ The set of presidents who served in the House of Representatives is a well-defined set.
3. _____ $4 \in \{2, 3, 7, 14, 24\}$
4. _____ $\{\ \} \subset \emptyset$
5. _____ $\{1, 3, 5, 7, \ldots \}$ and $\{2, 4, 6, 8, \ldots\}$ are disjoint sets.
6. _____ $\{x \mid x$ is an odd number$\}$ is an example of a set in roster form.
7. _____ $\{1, 3, 7\} = \{2, 4, 7\}$
8. _____ $\{1, 3, 7\}$ is equivalent to $\{2, 4, 7\}$.
9. _____ If $A = \{1, 3, 7, 11, 14, 15, 17\}$, then $n(A) = 7$.
10. _____ $A = \{2, 4, 6, 8, \ldots \}$ is a finite set.
11. _____ $\{4\} \in \{2, 4, 6\}$
12. _____ If $A = \{1, 2, 3\}$ and $B = \{a, b, c, d\}$ then $n(A \times B) = 7$.

Express each set, using roster form.

13. $A = \{x \mid x$ is an even natural number and $x < 13\}$.
14. The set of natural numbers greater than 4.

Express each set, using set-builder notation.

15. $\{3, 4, 5, 6, 7, \ldots \}$
16. The set of odd natural numbers between 4 and 20.

Express each set, using a definition.

17. $\{x \mid 4 < x < 11, x \in N\}$
18. $\{$Alabama, Alaska, Arizona, Arkansas, California, Colorado, $\ldots$, Wyoming$\}$

Let $U = \{1, 2, 3, 4, 5, 6, 7, 8, 9\}$,
 $A = \{1, 2, 3, 4\}$,
 $B = \{5, 6, 7, 8\}$,
 $C = \{3, 4, 5, 6\}$.

Find the following.

19. $A \cap B$
20. $A \cup B'$
21. $A \cap C$
22. $(A \cup B)' \cup C$
23. $A - C$
24. The number of subsets of set B
25. Given the sets below, construct a Venn diagram and place the elements in the proper regions.

$$U = \{1, 2, 3, 4, 5, 6, 7, 8, 9, 10\}$$
$$A = \{1, 2, 3, 4, 5\}$$
$$B = \{3, 4, 5, 6, 7\}$$

Use Fig. 2.29 to find the following.

26. $A \cup B$ **27.** $A \cap B'$

28. $A \cup B \cup C$ **29.** $A \cap B \cap C$

30. $(A \cup B) \cap C$ **31.** $(A \cap B) \cup C$

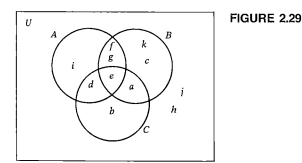

FIGURE 2.29

Construct a Venn diagram to determine whether the following statements are true.

32. $(A' \cup B')' = A \cap B$

33. $(A \cup B') \cup (A \cup C') = A \cup (B \cap C)'$

34. A coffee company was willing to pay \$1 to each person interviewed about his or her likes and dislikes of types of coffee. Of the people interviewed, 200 liked ground coffee, 270 liked instant coffee, 70 liked both, and 50 did not like coffee at all. What was the total amount of money the company had to pay?

A snack shop is conducting a survey to determine the preferences of customers. The results show that 200 people like ice cream, 190 people like hot dogs, 210 people like coffee, 100 people like ice cream and hot dogs, 150 people like hot dogs and coffee, 110 people like ice cream and coffee, 70 people like all three, and 5 people like none of these foods.

35. Find the number of people who completed the survey.

36. Find the number of people who like only hot dogs.

37. Find the number of people who like hot dogs and ice cream but do not like coffee.

38. Find the number of people who like hot dogs or coffee but do not like ice cream.

In Exercises 39 and 40, show that the sets are infinite by placing each set in a one-to-one correspondence with a proper subset of itself.

39. $\{2, 4, 6, 8, 10, \ldots\}$

40. $\{3, 5, 7, 9, 11, \ldots\}$

In Exercises 41 and 42, show that each set has cardinal number $\aleph_0$ by setting up a one-to-one correspondence between the set of counting numbers and the given set.

41. $\{5, 8, 11, 14, 17, \ldots\}$

42. $\{4, 9, 14, 19, 24, \ldots\}$

CHAPTER TEST

State whether each is true or false.

1. $\{a, b, c, d\} = \{b, c, a, d\}$
2. $\{1, 2, 3\}$ is equivalent to $\{3, 2, 1\}$
3. $\{1, 2, 3\} \subset \{1, 2, 3, 4\}$
4. $6 \subseteq \{4, 5, 6\}$
5. $9 \notin \{1, 2, 3, 4\}$
6. $5 \in \{x \mid x > 4, x \in N\}$
7. If $A \cap B = \emptyset$, then A and B are disjoint sets.
8. $\{1, 2, 3\}$ has six subsets.
9. For any set A, $\emptyset \subsetneq A$.
10. For any set A, $A \cup A' = U$.

Write each of the following sets in set-builder notation.

11. $\{4, 5, 6, 7, 8\}$
12. $\{8, 9, 10, 11, 12, 13, \ldots\}$

Given $U = \{1, 2, 3, 4, 5, 6, 7, 8, 9, 10\}$, $A = \{1, 2, 3, 4, 6, 8\}$, $B = \{2, 4, 5, 6, 7\}$, and $C = \{1, 2, 4, 5, 9, 10\}$, find the following.

13. $A \cup B$
14. $A' \cup C$
15. $(A \cup B)' \cap C'$
16. $n(A \cap C)$
17. Given $U = \{a, b, c, d, e, f, g, h, i, j\}$, $A = \{a, c, e, g, i\}$, $B = \{a, b, c, d, e, g\}$, and $C = \{b, c, g, h\}$, draw a Venn diagram illustrating the relationship between sets U, A, B, and C.
18. Use a Venn diagram to determine whether $(A \cup B)' \cap C = A \cap (B \cap C)'$ for all sets A, B, and C.

In a survey of 105 students at a local college the following information was obtained: 48 were taking a business course, 62 were taking a mathematics course, 50 were taking a psychology course, 32 were taking a business and mathematics course, 25 were taking a business and psychology course, 27 were taking a mathematics and psychology course, and 15 were taking all three courses.

19. Construct a Venn diagram, and record the number of students in each region.
20. How many students were not taking any of the three courses?
21. How many students were taking only mathematics?
22. How many students were taking mathematics and psychology?
23. How many students were taking mathematics or psychology?

24.[†] Show that the set is infinite by setting up a one-to-one correspondence between the set and a proper subset of itself.

$$\{4, 5, 6, 7, 8, 9, \ldots\}$$

25.[†] Show that the set has cardinal number $\aleph_0$ by setting up a one-to-one correspondence between the set of counting numbers and the given set.

$$\{1, 3, 5, 7, 9, 11, \ldots\}$$

[†]From optional Section 2.7.

CHAPTER 3

Logic

3.1   STATEMENTS AND LOGICAL CONNECTIVES

History

George Boole
*Courtesy of International Business
Machines Corporation*

"When I use a word," Humpty
Dumpty said, in rather a scornful
tone, "it means just what I choose it
to mean — neither more nor less."
Lewis Carroll

Would you like to develop your process of reasoning so that you can be more effective in influencing others in conversations? Would you like to be more convincing in arguments? These have been personal goals for centuries. Human beings have always been trying to improve their reasoning processes in arriving at logical conclusions.

Before we proceed, let us gain a historical perspective of logic. The ancient Greeks were the first people to analyze the way humans think to arrive at a solution. The results of the Greeks' efforts were organized by Aristotle (384–322 B.C.) in a work called *Organon*. Aristotle has therefore been called the "Father of Logic." His type of logic, referred to as Aristotelian syllogistic reasoning, has been taught and studied for more than 2000 years.

The first new developments in logic after Aristotle were the work of the German mathematician Gottfried Wilhelm Leibniz (1646–1716). As the first serious student of symbolic logic, Leibniz stimulated interest in the subject because of his deep conviction that all mathematical and scientific concepts could be derived from logic.

George Boole (1815–1864), a self-educated English mathematician, is considered by many to be the founder of "symbolic logic" because of his impressive work in this area. Among Boole's publications are *The Mathematical Analysis of Logic* (1847) and *An Investigation of the Law of Thought* (1854).

Since the development of symbolic logic, many mathematicians have studied the subject. Among them was the Reverend Charles Lutwidge Dodgson (1832–1898), who is more commonly known as Lewis Carroll, the author of *Alice's Adventures in Wonderland*. His goal in writing that book was to make logic more interesting and understandable for his students.

Alfred North Whitehead (1861–1947) and Bertrand Russell (1872–1970) were influenced by the thinking of Leibniz. Together they wrote a detailed development of arithmetic, starting with only undefined concepts and the assumptions of logic. The results of their work appeared in three large volumes called *Principia Mathematica* (1910–1913).

"Why do we study logic?" is a question often asked by students. Some reasons why we believe students should be introduced to logic are as follows.

1. Logic is different from topics in mathematics such as arithmetic, algebra, and geometry. It can be interesting and fun even if you do not have a strong background in mathematics.

2. The study of logic often gives students a better understanding of key words and phrases used in the English language. This understanding can help you communicate more clearly in any situation.

3. The study of logic prepares you to study other areas of mathematics. The specialized mathematical meanings of words such as "or" and "and," for example, apply to all areas of mathematics. If you glance through the text, you will see these words used in a mathematical context in many of the chapters. If, for example, you preview the probability chapter, you will see formulas for P(*A* or *B*) and P(*A* and *B*). An understanding of "or," "and," and other words discussed in this chapter may be helpful in achieving success in other chapters of this text and other areas of mathematics.

4. The study of logic can help you analyze everyday situations. Consider these two statements.

   If you buy a ticket, then you can go into the movie theater.
   You can go into the movie theater if and only if you buy a ticket.

   Do these two statements have the same meaning? If not, how do they differ?
       Consider this statement.

   > If you don't pay, then you don't play.

   Can you determine which, if any, of the following statements have the same meaning?

   a) If you pay, then you play.
   b) If you don't play, then you don't pay.
   c) You can play if you pay.
   d) If you play, then you pay.
   e) You can play only if you pay.

   By the end of this chapter you should be able to answer questions like these, in which the precise meaning of words is important.

5. Logic may improve your ability to reason effectively. The study of logic can help you determine whether an argument is valid or is a fallacy. In Section 3.5 we will illustrate how Euler diagrams can be helpful in determining whether an argument is valid. In this chapter we will study different forms of arguments and use logic to determine whether these arguments are valid or are fallacies.

6. Logic will help you better understand the decision-making process of the computer.

Statements and Logical Connectives

In logic a sentence that can be assigned a truth value (either true or false) is called a **statement.** Here are some examples of statements.

1. The earth is round.
2. Steve is 6 feet tall.
3. Gail is president of the United States.

In each case we can say that the sentence is either true or false. In statement 2 we can measure Steve and then say yes he is or no he is not 6 feet tall. Who is president of the United States? Is Gail president?

Here are some examples of sentences that are not statements.

1. Open the door.
2. Sue is the most sincere person in the room.
3. Is this class interesting?

The first expression is a command, to which we cannot assign a truth value. In the second, the words "most sincere" are subjective, unless they have been clearly defined. Who is to say who is the most sincere? "Is this class interesting?" is a question and as such cannot be labeled either true or false.

The three statements presented earlier are examples of **simple statements** because they convey one idea. In everyday life we encounter compound sentences—sentences that convey more than one idea. The common connectives used to form such **compound statements** in logic are "and," "or," "not," "if–then," and "if and only if."

In logic we represent each simple statement with a small letter and each connective with a symbol. It is customary to use the letters p, q, r, and s to represent simple statements. However, other letters may also be used.

The discussion that follows explains how the connectives are used in writing compound statements in **symbolic form,** that is, in symbols rather than words.

And Statements

The first connective that we will discuss is the **conjunction.** The conjunction is symbolized by $\wedge$, and is read "and."

Consider the compound statement that uses the conjunction "and" ($\wedge$): Boots is a cat and Frisky is a dog. Let p and q represent the simple statements.

p: Boots is a cat.
q: Frisky is a dog.

Now we write the compound statement using the letters in place of the simple statements:

p and q.

Finally, we replace the word "and" with the conjunction symbol $\wedge$. The statement in symbolic form is

$p \wedge q$.

We read the symbols $p \wedge q$ as "Boots is a cat and Frisky is a dog."

∎ **Example 1** ———————————————————————————

s: Gene is a student.
j: Gene has a jeep.

Write the following statements in symbolic form.

a) Gene is a student and Gene has a jeep.

b) Gene has a jeep and Gene is a student.

Solution: **a)** $s \wedge j$ **b)** $j \wedge s$ ❖

Generally, the conjunction is expressed as "and." However, it may sometimes appear as *"but," "however,"* or *"nevertheless."*

Not Statements

Now we will introduce the **negation.** The negation is symbolized by $\sim$ and read "not."

Occasionally, it is necessary to change a statement to its opposite meaning. We then use the negation of a statement. For example, the negation of the statement "Steve is at home" is "Steve is not at home." Thus if p represents the simple statement "Steve is at home," $\sim p$ represents the compound statement "Steve is not at home."

■ Example 2

Write in symbolic form the statement "Today is Friday and it is not snowing."

Solution: Let p: Today is Friday.
              q: It is snowing.

We write: p and $\sim q$.
Final form: $p \wedge \sim q$ or $p \wedge (\sim q)$.   ❖

Or Statements

Now we introduce the **disjunction.** The disjunction is symbolized by $\vee$ and read "or."

■ Example 3

Let p: Joyce will go to the movie.
    q: Joyce will eat popcorn.

Write the following statements in symbolic form.

a) Joyce will go to the movie or Joyce will eat popcorn.
b) Joyce will eat popcorn or Joyce will go to the movie.
c) Joyce will not eat popcorn or Joyce will go to the movie.

Solution: a) $p \vee q$ b) $q \vee p$ c) $\sim q \vee p$ ❖

Consider the following problems.

$$2 \times 3 \times 4 = (2 \times 3) \times 4 = 2 \times (3 \times 4) = 24$$
$$2 + 3 + 4 = (2 + 3) + 4 = 2 + (3 + 4) = 9$$

Notice that in multiplying or adding three numbers the parentheses may be placed around any two adjacent numbers without changing the answer. This illustrates the associative properties of multiplication and addition, respectively. The same principle applies to the logical operations of conjunction, $\wedge$, and disjunction, $\vee$.

$$p \wedge q \wedge r = (p \wedge q) \wedge r = p \wedge (q \wedge r)$$
$$p \vee q \vee r = (p \vee q) \vee r = p \vee (q \vee r)$$

Now consider the expression

$$2 + 3 \times 4$$

What is the solution to this calculation? Since there are two different operations and no parentheses are used, there may be some confusion about the answer. Some students may answer 20, while others may answer 14. When parentheses are added around two of the numbers, there is no confusion about which operation is to be performed first.

$$(2 + 3) \times 4 = 5 \times 4 \qquad 2 + (3 \times 4) = 2 + 12$$
$$= 20 \qquad\qquad\qquad = 14$$

When a compound statement in logic contains two or more different connectives, how do we know which connectives to group together? Consider this statement:

John is a horse (h) or Mary is a pilot (p) and Lassie is a dog (d).

Translating the statement into symbolic form, we have

$$h \vee p \wedge d.$$

Does this statement mean $(h \vee p) \wedge d$ or $h \vee (p \wedge d)$? Without punctuation in the statement, we do not know which interpretation is correct.

When a compound statement contains more than one connective, commas can be used to indicate the form of the statement. In translating a statement into symbolic form the commas will indicate which simple statements are grouped together. When written symbolically, parentheses are used to show this grouping.

If the statement were written "John is a horse or Mary is a pilot, and Lassie is a dog," the translation of the statement would be $(h \vee p) \wedge d$. Notice the location of the comma. The simple statements "John is a horse" and "Mary is a pilot" are on the same side of the comma and are thus grouped together. If the statement were written "John is a horse, or Mary is a pilot and Lassie is a dog," the translation of the statement would be $h \vee (p \wedge d)$. Notice the location of the comma. The simple statements "Mary is a pilot" and "Lassie is a dog" are on the same side of the comma and are thus grouped together.

■ **Example 4** _____

p: The crowd is shouting.
q: The team is winning.
r: The batter hit the ball.

Write the following statements in symbolic form.

a) The crowd is shouting, and the team is winning or the batter hit the ball.

b) The crowd is shouting and the team is winning, or the batter hit the ball.

Solution

a) The comma tells us to group the statement "The team is winning," (q), with the statement "The batter hit the ball," (r). Note that both these statements are on the same side of the comma. The statement in symbolic form is $p \land (q \lor r)$. In mathematics we always evaluate the information within parentheses first. Since the conjunction, $\land$, is outside the parentheses and is evaluated last, this statement is considered a conjunction.

b) The comma tells us to group the statement "The crowd is shouting," (p), with the statement "The team is winning," (q). Note that both these statements are on the same side of the comma. The statement in symbolic form is $(p \land q) \lor r$. Since the disjunction, $\lor$, is outside the parentheses, this statement is considered a disjunction. ❖

■ **Example 5** _____

p: Karen is a doctor.
q: Karen is a car owner.
r: Karen is a cab driver.

Write the following compound statements in symbolic form.

a) Karen is a doctor or Karen is a car owner, and Karen is not a cab driver.

b) Karen is not a doctor, and Karen is a car owner or Karen is a cab driver.

Solution: **a)** $(p \lor q) \land \sim r$ **b)** $\sim p \land (q \lor r)$ ❖

An important point to be stressed at this time is that a negation has the effect of negating only the statement that directly follows it. To negate a compound statement, we must use parentheses. When a negation symbol is placed in front of a statement in parentheses, it negates the entire statement in parentheses. The negation symbol in this case is read, "It is not true that" or "It is false that."

■ **Example 6**

Let p: The spring is a source of water.
　　q: The birds are singing.

Write the following symbolic statements in words.

a)　$\sim p \wedge q$　　　b)　$\sim p \wedge \sim q$　　　c)　$\sim (p \wedge q)$

Solution

"Mathematicians are like Frenchmen: whatever you say to them they translate into their own language and forthwith it is something entirely different."

Goethe

a)　The spring is not a source of water and the birds are singing. This statement could also be written "The spring is not a source of water but the birds are singing," since *but* may be used in place of *and.*

b)　The spring is not a source of water and the birds are not singing.

c)　It is false that the spring is a source of water and the birds are singing. ❖

Parts (a) and (b) of Example 6 are conjunctions, since they can be written $(\sim p) \wedge q$ and $(\sim p) \wedge (\sim q)$, respectively. Part (c) is a negation, since it negates the entire statement. The differences in the three statements will be discussed in Section 3.3.

Occasionally, we come across a **neither-nor** statement. For example, "John is *neither* handsome *nor* rich." This means that John is not handsome *and* John is not rich. If p represents "John is handsome" and q represents "John is rich," this statement is symbolized $\sim p \wedge \sim q$.

If–Then Statements

Now we introduce the conditional statement. The **conditional statement** is symbolized by $\rightarrow$, and read "if–then." It is extremely important in mathematics. The statement $p \rightarrow q$ is read "if p then q." The conditional statement consists of two parts; the part that precedes the arrow is the **premise,** or **hypothesis,** and the part that follows the arrow is the **conclusion.** In the statement $p \rightarrow q$, the p is the premise and the q is the conclusion. An example of a conditional statement is "If the computer can talk, then I can play the game."

■ **Example 7**

Let p: The horse is black.
　　q: The barn is red.

Write the following in symbolic form.

a)   If the horse is black, then the barn is red.

b)   If the horse is not black, then the barn is not red.

c)   It is false that if the horse is black, then the barn is red.

Solution: a) $p \rightarrow q$ b) $\sim p \rightarrow \sim q$ c) $\sim(p \rightarrow q)$

Example 8

Let p: Sally is a calculus student.
　　q: Tina is an engineering student.
　　r: Paul is a nursing student.

Write the following symbolic statements in words.

a) $(q \rightarrow \sim p) \lor r$ b) $q \rightarrow (\sim p \lor r)$

Conditional response: If the seal
jumps through the hoop, then it will
be rewarded.
Photo by Marshall Henrichs

Solution

a)   If Tina is an engineering student then Sally is not a calculus student, or Paul is a nursing student.

b)   If Tina is an engineering student, then Sally is not a calculus student or Paul is a nursing student. ❖

Note that in Example 8 the parentheses tell us where to place the commas.

If and Only If Statements

The last connective we will discuss is the biconditional. The **biconditional,** symbolized by ↔, is read "if and only if." The phrase "if and only if" is sometimes abbreviated as "iff." The statement $p \leftrightarrow q$ is read "p if and only if q." We will show in Section 3.3 that $p \leftrightarrow q$ is equivalent to $(p \rightarrow q) \wedge (q \rightarrow p)$. An example of a biconditional statement is "The movie is good if and only if Robert Redford plays the leading role."

▧ Example 9

Let p: The water is pure.
  q: The milk is white.

Write the following symbolic statements in words.

a) $q \leftrightarrow \sim p$ **b)** $\sim(p \leftrightarrow \sim q)$

Solution

a)   The milk is white if and only if the water is not pure.

b)   It is false that the water is pure if and only if the milk is not white. ❖

Table 3.1 summarizes the connectives discussed in this section.

TABLE 3.1			
Formal Name	**Symbol**	**How Read**	**How Used**
Conjunction	$\wedge$	and	$p \wedge q$
Disjunction	$\vee$	or	$p \vee q$
Negation	$\sim$	not	$\sim p$
Conditional	$\rightarrow$	if–then	$p \rightarrow q$
Biconditional	$\leftrightarrow$	if and only if	$p \leftrightarrow q$

We discussed the use of the comma to indicate the proper grouping of statements in symbolic form. However, when parentheses are not used, a problem might occur in translating from the symbolic form to the statement form. For example, how could the statement $p \rightarrow q \wedge r$ be classified? Is the statement a conditional or a conjunction? Where is the punctuation placed in the statement? In arithmetic, when parentheses are not used, we perform multiplication before addition. Thus to evaluate $2 + 3 \times 4$, we first multiply 3×4 and then add 2 to the product. That is,

$$2 + (3 \times 4) = 2 + 12$$
$$= 14$$

This is called the **order or priority of operations.** In logic the manner in which statements without parentheses are grouped is determined by the **dominance of the connectives.** The least dominant is the negation, and the most dominant is the biconditional. The order of dominance is given in Table 3.2.

TABLE 3.2
Dominance of Connectives
1. Biconditional, $\leftrightarrow$
2. Conditional, $\rightarrow$
3. Conjunction, $\wedge$; disjunction, $\vee$
4. Negation, $\sim$

As is indicated in Table 3.2, the conjunction and disjunction have the same level of dominance. Thus to determine whether the symbolic statement $p \wedge q \vee r$ is a conjunction or a disjunction, it is necessary to have grouping symbols (parentheses). When evaluating a symbolic statement that does not contain parentheses, we evaluate the least dominant connective first and the most dominant connective last. For example,

$\sim p \vee q$	means $(\sim p) \vee q$
$p \rightarrow q \vee r$	means $p \rightarrow (q \vee r)$
$p \wedge q \rightarrow r$	means $(p \wedge q) \rightarrow r$
$p \rightarrow q \leftrightarrow r$	means $(p \rightarrow q) \leftrightarrow r$
$p \vee r \leftrightarrow r \rightarrow \sim p$	means $(p \vee r) \leftrightarrow (r \rightarrow \sim p)$
$p \rightarrow r \leftrightarrow s \wedge p$	means $(p \rightarrow r) \leftrightarrow (s \wedge p)$

■ **Example 1** _____

Indicate whether each statement is a negation, conjunction, disjunction, conditional, or biconditional.

a) $p \rightarrow q \vee r$ b) $\sim p \wedge q \leftrightarrow r \vee p$

Solution

a) The statement $p \to q \lor r$ is a conditional statement. The order of dominance given in Table 3.2 provides this information. By using the dominance of connectives the statement can be written with parentheses as $p \to (q \lor r)$.

b) The statement $\sim p \land q \leftrightarrow r \lor p$ is a biconditional, since the biconditional is the most dominant connective. By using the dominance of connectives the statement could be written with parentheses as $(\sim p \land q) \leftrightarrow (r \lor p)$. ❖

EXERCISES 3.1

Determine whether the following are statements.

1. This is an interesting class.
2. Today is Wednesday.
3. Shut the door.
4. Do you want to go to the movies?
5. $7 + 4 = 15$.
6. Strawberries are red.

Indicate whether each statement below is a simple or compound statement. If it is a compound statement, indicate whether it is a negation, conjunction, disjunction, conditional, or biconditional.

7. Today is Monday and the chicken is red.
8. Charlie is not a horse.
9. The bandit will fight if and only if he is cornered.
10. If $2 + 3 = 7$, then the maid is guilty.
11. The house is on Church Street or on Green Street.
12. The pencil is neither blue nor green.
13. The paper is interesting but too long.
14. Boots the cat is not sleeping or she is trying to catch her tail.
15. It is false that today is Monday or today is Tuesday.
16. If today is Monday then tomorrow is Tuesday, or the tea is warm.
17. If today is Monday, then tomorrow is Tuesday or the tea is warm.
18. It is false that if the computer is green, then the cow gives chocolate milk.

Write each of the following compound statements in symbolic form.

Let p: The fish are biting.
  q: Dick is catching fish.

19. The fish are biting and Dick is catching fish.
20. The fish are not biting.
21. The fish are biting or Dick is not catching fish.
22. Dick is catching fish if and only if the fish are not biting.
23. If the fish are biting, then Dick is catching fish.
24. If Dick is catching fish, then the fish are not biting.
25. Neither are the fish biting nor is Dick catching fish.
26. It is false that the fish are biting and Dick is catching fish.

Write each compound statement in words.

Let p: Joan is a football official.
  q: Charles is a tennis player.

27. $\sim p$	28. $\sim q$
29. $p \land q$	30. $q \lor p$
31. $\sim p \to q$	32. $\sim p \leftrightarrow \sim q$
33. $\sim(q \lor p)$	34. $\sim q \land \sim p$
35. $\sim p \lor \sim q$	36. $\sim p \land q$
37. $q \to \sim p$	38. $\sim(p \land q)$

Write each of the following statements in symbolic form.

Let p: The rose is red.
  q: The stem is green.
  r: Today is Saturday.

39. The rose is red and the stem is green, or today is Saturday.
40. If the rose is red or the stem is green, then today is Saturday.

41. If the rose is red, then the stem is green or today is not Saturday.
42. Today is Saturday if and only if the rose is not red.
43. The rose is not red if and only if the stem is not green.
44. If today is Saturday and the stem is green, then the rose is red.
45. Today is Saturday if and only if the stem is green, and the rose is red.
46. It is false that if today is Saturday, then the stem is not green.
47. Today is Saturday or the stem is not green, if and only if the rose is red.
48. If the stem is green, then the rose is red if and only if today is Saturday.

Using the same statements for p, q, and r as in Exercises 39–48, write each of the following symbolic statements in words.

49. $(p \lor q) \land r$
50. $(\sim p \lor q) \land \sim r$
51. $(q \to p) \lor r$
52. $\sim p \land (q \lor r)$
53. $\sim r \to (q \land p)$
54. $(q \land r) \to p$
55. $(r \to q) \land p$
56. $\sim p \to (q \lor r)$
57. $(q \leftrightarrow p) \land r$
58. $q \to (p \leftrightarrow r)$

Read the Did You Know on page 80, then place parentheses in the following statements according to the dominance of connectives.

59. $\sim p \to q \land \sim r$
60. $p \land r \to p$
61. $\sim p \leftrightarrow q \lor r$
62. $q \leftrightarrow p \land r$
63. $q \to p \leftrightarrow r$
64. $\sim p \lor q \to \sim r$

Translate the following statements into symbolic form. Indicate the letters you use to represent each simple statement.

65. The Constitution requires that the president be at least 35 years old, a natural-born citizen of the United States, and a resident of the country for 14 years.
66. "You can fool all of the people some of the time and you can fool some of the people all of the time, but you can't fool all of the people all of the time." (Abraham Lincoln)
67. "If you want to kill an idea in the world today, then get a committee working on it." (C. F. Kettering)
68. "Happiness is beneficial for the body but it is grief that develops the powers of the mind." (Marcel Proust)
69. "I did not know the dignity of their birth, but I do know the glory of their death." (Douglas MacArthur)

Using the letters indicated, translate the following statements into symbolic form.

*70. If the brown witch is in the pumpkin patch (w) or the black cat is on the barn roof (c) then the ghost is milking the cow (g), if and only if the scarecrow is scaring the crows away (s) and the lion is keeping the mice away (l). (*Hint:* See the Did You Know.)
*71. If the plumber is in the basement (p) and the carpenter is in the attic (c), then the dog is barking at the moon (d) or the electrician is stringing wire (e). (*Hint:* See the Did You Know.)
*72. Neither is Dick at work (d) nor is Mary at home (m), or Jane is teaching (j) if and only if Barb (b) and Carol (c) are at the shore. (*Hint:* See the Did You Know.)

3.2 TRUTH TABLES

A truth table is a device used to determine the truth value of a compound statement, that is, to determine whether it is true or false. Consider the compound statement $(p \lor q) \to {\sim}p$. Since both p and q are statements that are either true or false, the compound statement must also be either true or false. The truth tables for the common connectives, which are given in the following discussion, *must be understood* by the student. It is essential that you be able to reproduce Table 3.3, negation (${\sim}$); Table 3.5, conjunction ($\land$); Table 3.7, disjunction ($\lor$); Table 3.10, conditional ($\to$); and Table 3.15, biconditional ($\leftrightarrow$).

Negation

The first truth table is for **negation.** If p is a true statement, then the negation of p, "not p," is a false statement. If p is a false statement, then "not p" is a true statement. These relationships are summarized in Table 3.3.

If a compound statement is composed of two simple statements, p and q, then the truth table must contain the four distinct cases given in Table 3.4.

Conjunction

The conjunction $p \land q$ is true only in the case in which p and q are both true. If either p or q is false, or if both are false, then the conjunction is false.

To illustrate that the conjunction of a true statement with a false statement is a false statement, consider the following situation.

You have recently purchased a new house and are planning a housewarming party for next Saturday. You have just purchased a new carpet and furniture, and you explain to the salesperson that they must be delivered before the party on Saturday. He promises you that the carpet will be delivered Thursday *and* the furniture will be delivered Friday.

On Thursday the carpet is delivered and installed. However, because the furniture doesn't arrive on Friday, you have to call off your party. Has the salesperson kept his promise? Even though the carpet was delivered on Thursday as promised (making one of the simple statements in the conjunction true), the entire statement made by the salesperson (the conjunction) is false. We can therefore see that the conjunction of a true statement with a false statement is a false statement.

Carefully study Table 3.5. The conjunction is true only in case 1, in which both p and q are true.

TABLE 3.3

Negation

p	${\sim}p$
T	F
F	T

TABLE 3.4

	p	q
Case 1	T	T
Case 2	T	F
Case 3	F	T
Case 4	F	F

TABLE 3.5

Conjunction

p	q	$p \land q$
T	T	T
T	F	F
F	T	F
F	F	F

■ Example 1

Construct a truth table for $p \wedge \sim q$.

Solution: Since there are two statements, p and q, construct a truth table with four cases (see Table 3.6a). Then write the truth values under the p in the compound statement and label this column 1 (see Table 3.6b). These truth values are copied directly from the p column on the left. Write the corresponding truth values under the q (from the q column on the left) in the compound statement and call this column 2. Now find $\sim q$ by negating the truth values in column 2 and call this column 3. Using the conjunction table, Table 3.5, and the entries in columns 1 and 3, you arrive at the following results—row 1: $T \wedge F$ is F; row 2: $T \wedge T$ is T; row 3: $F \wedge F$ is F; and row 4: $F \wedge T$ is F. The answer is always the last column completed. Columns 1, 2, and 3 are only aids in arriving at the answer in column 4.

TABLE 3.6(a)

	p	q	$p \wedge \sim q$
Case 1	T	T	
Case 2	T	F	
Case 3	F	T	
Case 4	F	F	

TABLE 3.6(b)

p	q	p	$\wedge$	$\sim$	q
T	T	T	F	F	T
T	F	T	T	T	F
F	T	F	F	F	T
F	F	F	F	T	F
		1	4	3	2

❖

Remember that the negation symbol negates only the statement that directly follows it. The statement $p \wedge \sim q$ actually means $p \wedge (\sim q)$. Many statements can be expressed more clearly by using parentheses. The operations within the parentheses are performed first. For this reason the negation column had to be completed before the conjunction column. In the future, instead of listing a column for q and a separate column for its negation, we will make one column for $\sim q$. The truth values in this column will be the opposite values of those in the q column on the left.

Disjunction

The inclusive **disjunction** $p \vee q$ is true in all cases except case 4, in which p and q are both false (see Table 3.7).

To illustrate that the disjunction of a true statement with a false statement is a true statement, consider the following situation.

You bring your car into an audio store to have an in-dash stereo radio and cassette player installed. You explain to the salesperson that you are leaving on an extended vacation on Saturday. The salesperson promises that the car will be ready Thursday *or* the car will be ready Friday.

TABLE 3.7
Disjunction

p	q	$p \vee q$
T	T	T
T	F	T
F	T	T
F	F	F

The car is not ready Thursday but the work is completed Friday morning. Has the salesperson kept her promise? The salesperson's statement was true. We can therefore see that the disjunction of a true statement with a false statement is a true statement.

■ **Example 2** _____

Construct a truth table for $\sim(q \lor \sim p)$.

Solution: First construct the standard truth table listing the four cases. Then determine the truth values for the statement within the parentheses. The order to be followed is indicated by the numbers below the columns (see Table 3.8). In column 1, copy the values from the q column on the left. Under $\sim p$, column 2, write the negation of the p column on the left. Next complete the "or" column, column 3, using columns 1 and 2 and the truth table for the disjunction. The "or" column is false only when both statements are false, as in case 2. Finally negate the values in the "or" column, and place these negated values in column 4. By examining the truth table you can see that the compound statement $\sim(q \lor \sim p)$ is true only in case 2, that is, when p is true and q is false. ❖

TABLE 3.8

p	q	$\sim$	$(q$	$\lor \sim$	$p)$
T	T	F	T	T	F
T	F	T	F	F	F
F	T	F	T	T	T
F	F	F	F	T	T
		4	1	3	2

■ **Example 3** _____

Construct a truth table for the statement $(p \lor q) \land \sim p$.

Solution: Again set up the standard truth table (see Table 3.9). Then use the "or" table to determine the truth values for the statement within parentheses, $p \lor q$. Record the truth value for $p \lor q$ in column 1. Then negate p and call this column 2. Next find the conjunction of columns 1 and 2. The answer is in column 3. Take note that this statement is true only in case 3, where p is false and q is true. ❖

TABLE 3.9

p	q	$(p \lor q)$	$\land$	$\sim p$
T	T	T	F	F
T	F	T	F	F
F	T	T	T	T
F	F	F	F	T
		1	3	2

Conditional

The fourth basic truth table is the **conditional** table. In Section 3.1 we mentioned that the statement or statements preceding the conditional symbol are called the hypothesis, or premise. The statement or statements following the conditional symbol are called the conclusion. For example, consider $(p \lor q) \rightarrow [\sim(q \land r)]$. In this statement, $(p \lor q)$ is the premise and $[\sim(q \land r)]$ is the conclusion.

The truth table for the conditional statement is given in Table 3.10. The conditional statement is false only when the premise is true and the conclusion is false, as indicated in case 2 of the table.

TABLE 3.10

Conditional

p	q	$p \rightarrow q$
T	T	T
T	F	F
F	T	T
F	F	T

Many students find this truth table more difficult to accept than the previous truth tables. The example below will help you understand why the conditional is false only when the premise is true and the conclusion is false.

We will attempt to illustrate the results of this table with the help of the statement "If you get an A, then I will buy you a car." We will assume that this statement is true except when I have actually broken my promise to you.

Let *p:* You get an A.
  q: I will buy you a car.

Translated into symbolic form, the statement above becomes $p \rightarrow q$. Let's examine the four cases.

- **CASE 1 (T, T)**
  You get an A, and I buy a car for you. I have met my commitment, and the statement is true.

- **CASE 2 (T, F)**
  You get an A, and I do not buy a car for you. I have broken my promise, and the statement is false.

  What happens if you don't get an A? If you don't get an A, I no longer have a commitment to you, and therefore I cannot break my promise.

- **CASE 3 (F, T)**
  You do not get an A, and I buy you a car. I have not broken my promise, and therefore the statement is true.

- **CASE 4 (F, F)**
  You do not get an A, and I don't buy you a car. I have not broken my promise, and therefore the statement is true.

■ **Example 4** _____

Construct a truth table for the statement $p \rightarrow \sim q$.

Solution: Fill out the truth table by placing the appropriate values under p, column 1, and under $\sim q$, column 2 (see Table 3.11). Then, using the information given in the conditional truth table, determine the solution, column 3. In row 1 the premise, p, is true and the conclusion, $\sim q$, is false. Row 1 is T → F, which is F. Row 2 is T → T, which is T. Row 3 is F → F, which is T. Row 4 is F → T, which is T. ❖

TABLE 3.11

p	q	p	→	~q
T	T	T	F	F
T	F	T	T	T
F	T	F	T	F
F	F	F	T	T
		1	3	2

TABLE 3.12

	p	q	r
Case 1	T	T	T
Case 2	T	T	F
Case 3	T	F	T
Case 4	T	F	F
Case 5	F	T	T
Case 6	F	T	F
Case 7	F	F	T
Case 8	F	F	F

To construct a truth table of a statement that consists of three simple statements, such as $(p \wedge q) \wedge r$, we must construct a truth table with eight distinct cases. A truth table with eight distinct cases is illustrated in Table 3.12.

One way to construct this table is to write four Ts and four Fs in the column under p. Under the second statement, q, we alternate two Ts and two Fs. Under the third statement, r, we alternate T, F, T, F, and so on. This technique ensures that each case is unique and no cases are omitted. All truth tables with three statements may be set up in this manner.

Example 5

Construct a truth table for the statement $p \rightarrow (q \wedge \sim r)$.

Solution: Since this statement is composed of three simple statements, there are eight distinct cases (see Table 3.13). Fill out the truth table by placing values under q, column 1, and $\sim r$, column 2. Then take the conjunction of columns 1 and 2 to obtain column 3. Place the values of p in column 4. To obtain the answer, column 5, use columns 3 and 4 and your knowledge of the conditional table. Column 4 represents the truth value of the premise, and column 3 represents the truth value of the conclusion. Remember that the conditional is false only when the premise is true and the conclusion is false, as in cases 1, 3, and 4.

TABLE 3.13

p	q	r	p	→	(q	∧	~r)
T	T	T	T	F	T	F	F
T	T	F	T	T	T	T	T
T	F	T	T	F	F	F	F
T	F	F	T	F	F	F	T
F	T	T	F	T	T	F	F
F	T	F	F	T	T	T	T
F	F	T	F	T	F	F	F
F	F	F	F	T	F	F	T
			4	5	1	3	2

The number of distinct cases in a truth table can be determined from the formula 2^n, where n is the number of distinct simple statements. For example, the compound statement $(p \wedge q) \rightarrow (r \vee \sim s)$ has four simple statements, p, q, r, and s. Its truth table has 2^4, or 16 distinct cases.

Biconditional

The **biconditional statement**, $p \leftrightarrow q$, means $p \rightarrow q$ and $q \rightarrow p$. Symbolically, we can write this as $(p \rightarrow q) \wedge (q \rightarrow p)$. To determine the truth table for

$p \leftrightarrow q$, we will construct the truth table for $(p \rightarrow q) \wedge (q \rightarrow p)$ (see Table 3.14).

TABLE 3.14

p	q	(p	→	q)	∧	(q	→	p)
T	T	T	T	T	T	T	T	T
T	F	T	F	F	F	F	T	T
F	T	F	T	T	F	T	F	F
F	F	F	T	F	T	F	T	F
		1	3	2	7	4	6	5

The compound statement "p if and only if q" is true only when both p and q have the same truth value, that is, when both are true or both are false. The **biconditional** truth table is given in Table 3.15.

TABLE 3.15

Biconditional

p	q	$p \leftrightarrow q$
T	T	T
T	F	F
F	T	F
F	F	T

■ **Example 6** _____

Construct a truth table for the statement $p \leftrightarrow (q \rightarrow \sim r)$.

Solution: Since there are three letters, there must be 2^3, or eight cases. The parentheses indicate that the answer must be under the biconditional, see Table 3.16.

TABLE 3.16

p	q	r	p	↔	(q	→	~r)
T	T	T	T	F	T	F	F
T	T	F	T	T	T	T	T
T	F	T	T	T	F	T	F
T	F	F	T	T	F	T	T
F	T	T	F	T	T	F	F
F	T	F	F	F	T	T	T
F	F	T	F	F	F	T	F
F	F	F	F	F	F	T	T
			4	5	1	3	2

❖

■ **Example 7** _____

Construct a truth table for $\sim r \rightarrow (p \leftrightarrow \sim q)$.

Solution: See Table 3.17.

TABLE 3.17								
p	**q**	**r**	**~r**	**→**	**(p**	**↔**	**~q)**	
T	T	T	F	T	T	F	F	
T	T	F	T	F	T	F	F	
T	F	T	F	T	T	T	T	
T	F	F	T	T	T	T	T	
F	T	T	F	T	F	T	F	
F	T	F	T	T	F	T	F	
F	F	T	F	T	F	F	T	
F	F	F	T	F	F	F	T	
			1	5	2	4	3	

❖

To find the truth value of a compound statement for a specific case, it is not necessary to construct an entire truth table.

■ **Example 8** _____

Determine the truth value of the statement $(q \lor r) \to (\sim p \land r)$ when p is true, q is false, and r is true.

Solution: We could do this example by constructing a truth table and observing the answer to case 3, which is T F T. However, it is much easier to evaluate the single case by substituting the truth value for each simple statement, as illustrated below.

$$(q \lor r) \to (\sim p \land r)$$
$$(F \lor T) \to (F \land T)$$
$$T \to F$$
$$F$$

For this specific case the statement is false.   ❖

■ **Example 9** _____

Determine the truth value of $q \leftrightarrow [(p \lor r) \land \sim s]$ when p is true and q, r, and s are false.

$$q \leftrightarrow [(p \lor r) \land \sim s]$$
$$F \leftrightarrow [(T \lor F) \land T]$$
$$F \leftrightarrow \quad [T \land T]$$
$$F \leftrightarrow \qquad T$$
$$F$$

For this particular case the statement is false. A truth table for this statement would contain 2^4, or 16, cases.   ❖

■ **Example 10** _____

Use your knowledge of the truth tables to determine the truth values of the following compound statements.

a) $2 + 3 = 5$ and $1 + 1 = 3$.

b) $2 + 3 = 5$ or $1 + 1 = 3$.

c)   3 is greater than or equal to 2 $(3 \geq 2)$.

d)   If $3 + 3 = 8$, then $5 + 2 = 11$.

e)   If George Washington was the first president of the United States, then Denver is in Colorado.

Solution

a)   Let p be $2 + 3 = 5$ and let q be $1 + 1 = 3$. Thus $2 + 3 = 5$ and $1 + 1 = 3$ can be expressed as $p \wedge q$. Furthermore, p is a true statement and q is a false statement.

$$p \wedge q$$
$$T \wedge F$$
$$F$$

Therefore "$2 + 3 = 5$ and $1 + 1 = 3$" is a false statement.

b)   Let p and q be as in part (a). Then $2 + 3 = 5$ or $1 + 1 = 3$ can be expressed as $p \vee q$.

$$p \vee q$$
$$T \vee F$$
$$F$$

Therefore "$2 + 3 = 5$ or $1 + 1 = 3$" is a true statement.

> "There are things which seem incredible to most men who have not studied mathematics."
>
> Pascal

c)   Let p be 3 is greater than 2, and let q be 3 is equal to 2. Thus 3 is greater than or equal to 2 can be expressed as $p \vee q$. Further, p is a true statement, and q is a false statement.

$$p \vee q$$
$$T \vee F$$
$$T$$

Therefore "3 is greater than or equal to 2" is a true statement.

d)   Let p be $3 + 3 = 8$, and let q be $5 + 2 = 11$. Thus if $3 + 3 = 8$, then $5 + 2 = 11$ can be written $p \rightarrow q$.

$$p \rightarrow q$$
$$F \rightarrow F$$
$$T$$

Therefore, "if $3 + 3 = 8$, then $5 + 2 = 11$" is a true statement.

e) If George Washington was the first president of the United States, then Denver is in Colorado.

   p: George Washington was the first president of the United States.
   q: Denver is in Colorado.

$$p \rightarrow q$$
$$T \rightarrow T$$
$$T$$

Thus the statement is true.   ❖

▮ Example 11 _____

Determine all the values of x that make the following statements true.

a) $x + 4 = 7$ or $4 + 6 = 10$.
b) If $x + 4 = 6$, then $5 + 3 = 9$.
c) $6 + 3 = 9$ and $x + 5 = 12$.
d) $4 + 3 = 9$ and $x + 5 = 8$.
e) $4 + 3 = 5$ if and only if $x + 3 = 9$.

Solution

a) The "or" statement is true if either part is true. Since $4 + 6 = 10$ is true, the statement will be true for any x.
b) The conditional statement $p \rightarrow q$ is false only when p is true and q is false. Since the conclusion is false, the premise must also be false to make the statement true. Thus x can be any number but 2.
c) The conjunction is true only when both p and q are true. The value $x = 7$ is the only one that makes the statement true.
d) The conjunction is true only when both $4 + 3 = 9$ and $x + 5 = 8$ are true. Since $4 + 3 = 9$ is false, there are no values of x that will make the statement true.
e) The biconditional will be true when p and q have the same truth values. Since $4 + 3 = 5$ is a false statement, the statement $x + 3 = 9$ must also be false in order for the biconditional to be true. The statement $x + 3 = 9$ will be false for all values of x except $x = 6$. Thus the biconditional is true for all values of x except 6.   ❖

▮ A **self-contradiction** is a statement that is false in every case of the truth table.

Example 12

Construct a truth table for the statement $(p \leftrightarrow q) \wedge (p \leftrightarrow \sim q)$.

Solution: See Table 3.18. In this example, the answer is false in each case (see column 5). This is an example of a self-contradiction or a **logically false statement.**

p	q	$(p \leftrightarrow q)$	$\wedge$	$(p$	$\leftrightarrow$	$\sim q)$
T	T	T	F	T	F	F
T	F	F	F	T	T	T
F	T	F	F	F	T	F
F	F	T	F	F	F	T
		1	5	2	4	3

TABLE 3.18

❖

> A **tautology** is a statement that is true in every case of the truth table.

Example 13

Construct a truth table for the statement $(p \wedge q) \rightarrow (p \vee r)$.

Solution: The truth table is Table 3.19, and the answer is given in column 3. In this example, the answer is true in every case. This is an example of a tautology or a **logically true statement.**

TABLE 3.19

p	q	r	$(p \wedge q)$	$\rightarrow$	$(p \vee r)$
T	T	T	T	T	T
T	T	F	T	T	T
T	F	T	F	T	T
T	F	F	F	T	T
F	T	T	F	T	T
F	T	F	F	T	F
F	F	T	F	T	T
F	F	F	F	T	F
			1	3	2

❖

The statement $(p \wedge q) \rightarrow (p \vee r)$ is a special kind of conditional statement. The conditional statement is true in every case of the truth table and is a tautology. Such conditional statements are called implications. In Example 13 we can say that $p \wedge q$ implies $p \vee r$.

∎ An **implication** is a conditional statement that is a tautology.

In any implication the premise of the conditional statement implies the conclusion. This means that *if* the premise is true, *then* the conclusion must also be true. That is, the conclusion will be true whenever the premise is true.

∎ **Example 14** ——————————————————————

Determine whether the conditional statement $[(p \wedge q) \wedge p] \rightarrow q$ is an implication.

Solution: If the truth table of the conditional statement is a tautology, the conditional statement is an implication. Since the truth table is a tautology, (see Table 3.20), the conditional statement is an implication. The premise $[(p \wedge q) \wedge p]$ implies the conclusion q. Note that the premise is true only in case 1 and the conclusion is also true in case 1.

TABLE 3.20

p	q	$[(p \wedge q)$	$\wedge$	$p]$	$\rightarrow$	q
T	T	T	T	T	T	T
T	F	F	F	T	T	F
F	T	F	F	F	T	T
F	F	F	F	F	T	F
		1	3	2	5	4

❖

EXERCISES 3.2 ————————————————————

Construct a truth table for each of the following statements.

1. $p \wedge \sim p$
2. $p \vee \sim p$
3. $q \vee \sim p$
4. $p \wedge \sim q$
5. $\sim(p \wedge \sim q)$
6. $\sim p \vee \sim q$
7. $\sim p \rightarrow q$
8. $\sim q \rightarrow \sim p$
9. $\sim(q \leftrightarrow p)$
10. $\sim(p \rightarrow q)$
11. $\sim p \leftrightarrow q$
12. $p \rightarrow \sim q$
13. $\sim(p \vee \sim q)$
14. $\sim(\sim p \leftrightarrow \sim q)$

Construct a truth table for each of the following statements.

15. $(p \wedge q) \vee \sim q$
16. $(p \vee \sim q) \wedge p$
17. $q \vee (p \vee \sim p)$
18. $(p \vee q) \rightarrow (p \wedge q)$
19. $q \rightarrow (p \rightarrow \sim q)$
20. $(q \vee \sim p) \leftrightarrow (p \wedge q)$
21. $p \leftrightarrow (q \vee \sim p)$

22. $(\sim p \wedge q) \vee (q \wedge \sim p)$
23. $q \rightarrow (q \vee \sim p)$
24. $\sim[\sim q \rightarrow (p \rightarrow q)]$

Construct a truth table for each of the following statements.

25. $p \wedge (q \wedge r)$
26. $p \wedge (q \vee r)$
27. $r \vee (p \wedge q)$
28. $\sim r \vee (\sim p \wedge q)$
29. $r \rightarrow (\sim p \vee q)$
30. $p \rightarrow (q \wedge r)$
31. $r \wedge (\sim p \rightarrow q)$
32. $p \leftrightarrow (q \wedge r)$
33. $(p \vee \sim r) \leftrightarrow q$
34. $(p \vee r) \rightarrow (q \wedge r)$
35. $(\sim p \vee \sim q) \rightarrow r$
36. $[p \wedge (q \vee \sim r)] \leftrightarrow \sim p$
37. $(p \rightarrow q) \leftrightarrow (\sim q \rightarrow \sim r)$

Define the following words.

38. self-contradiction
39. tautology
40. implication

Determine whether each of the following statements is a tautology, a self-contradiction, or neither.

41. $(a \vee b) \vee \sim a$
42. $(a \wedge b) \wedge \sim a$
43. $\sim a \rightarrow b$
44. $(\sim a \rightarrow b) \vee \sim a$
45. $(\sim a \vee \sim b) \rightarrow a$
46. $(a \vee \sim b) \vee b$
47. $(p \wedge q) \wedge (\sim q \wedge r)$
48. $[(p \wedge q) \wedge \sim r] \leftrightarrow [(p \vee q) \wedge r]$

Determine whether each of the following statements is an implication.

49. $(p \vee q) \rightarrow q$
50. $q \rightarrow (p \vee q)$
51. $(p \wedge q) \rightarrow (q \wedge p)$
52. $(p \vee q) \rightarrow (p \vee \sim q)$
53. $[(p \rightarrow q) \wedge (q \rightarrow p)] \rightarrow (p \leftrightarrow q)$
54. $[(p \vee q) \wedge r] \rightarrow (p \vee q)$

If p is true, q is false, and r is true, find the truth value of the following statements.

55. $(\sim p \wedge r) \wedge q$
56. $\sim p \vee (q \wedge r)$
57. $(\sim q \wedge \sim p) \wedge \sim r$

58. $\sim p \rightarrow (q \wedge \sim r)$
59. $p \leftrightarrow (\sim q \wedge r)$
60. $(q \wedge \sim p) \rightarrow \sim r$
61. $(p \wedge \sim q) \wedge r$
62. $\sim[p \rightarrow (q \wedge r)]$
63. $(p \vee r) \leftrightarrow (p \wedge \sim q)$
64. $(\sim p \vee q) \rightarrow \sim r$
65. $(\sim p \leftrightarrow r) \vee (\sim q \leftrightarrow r)$
66. $(r \rightarrow \sim p) \wedge (q \rightarrow \sim r)$

Determine the truth value of the following statements.

67. $2 + 3 = 7$ or $6 + 5 = 11$.
68. If Russia is a member of the United Nations, then Egypt is in Europe and Brazil is in North America.
69. 2 and 3 are prime numbers, if and only if $6 + 4 = 13$. (A prime number is a number that has exactly two divisors, itself and 1.)
70. There are five Great Lakes in the United States or the country of Brazil is in Africa.
71. If Mexico is south of the United States and Canada is north of the United States, then $3^2 = 9$.
72. 17 and 19 are primes, but 2 is the only even prime number.
73. If $4 + 3 = 6$, then $3 + 1 = 5$.
74. If the earth is flat then fish have fins.
75. Abraham Lincoln was the sixteenth President of the United States, if and only if the Queen of England lives in Spain and Thomas Jefferson designed the first televison set.
76. If the Panama Canal connects the Dead Sea and the Indian Ocean, then the Mississippi River flows through New York City or the Amazon River is in Mexico.

Determine all the values of x that make the following true.

77. $x + 3 = 9$ and $4 + 3 = 8$.
78. $x + 3 = 9$ or $4 + 3 = 7$.
79. If $4 + 1 = 5$, then $x + 3 = 7$.
80. If $x + 4 = 7$, then $9 = 2$.
81. $x + 3 = 9$ if and only if $6 = 5 + 1$.
82. $9 + 2 = 13$ if and only if $x + 7 = 12$.
83. If $4 + 1 = 6$, then $x + 3 = 9$.
84. $3 + 7 = 11$ if $x + 2 = 3$.
85. If $x + 4 = 11$, then $4 + 3 = 7$ or $6 + 5 = 11$.
86. $x + 9 = 14$ and $7 + 8 = 11$, if and only if $9 + 6 = 15$.
87. $x + 7 = 9$, and $6 + 8 = 14$ if and only if $9 + 6 = 15$.
88. If $x + 3 = 7$, then $6 + 7 = 17$ or $9 + 8 = 21$.

Another connective is the **exclusive or,** which is symbolized by $\veebar$. The exclusive or is true when either p or q (but not both) is true. The truth values for the exclusive or are given in Table 3.21.

Construct a truth table for each of the following statements.

89. $(p \wedge q) \veebar \sim p$

90. $q \rightarrow \sim(r \veebar p)$

91. $p \vee (q \veebar r)$

TABLE 3.21		
p	q	$p \veebar q$
T	T	F
T	F	T
F	T	T
F	F	F

PROBLEM SOLVING ———————————

92. In a certain city some people are labeled As and others Bs, but they all look alike. The As always tell the truth, but the Bs always lie. A visitor to the city meets three people and asks them whether they are As or Bs. The first mutters something inaudible. The second, pointing to the first, says, "He says that he is an A." The third, pointing to the second, says, "He lies." Knowing beforehand that only one of the three is a B, the visitor concludes what each person is. Can you determine which are the As and which is the B?

93. On page 69 is a photograph of a logic game at the Ontario Science Centre. There are twelve balls on top of the game board, numbered from left to right, with ball 1 on the extreme left and ball 12 on the extreme right. On the platform in front of the players are twelve buttons, one corresponding to each of the balls. When six buttons are pushed, the six respective balls are released. When one or two balls reach an "and" or "or" gate, a single ball may, or may not pass through the gate. The object of the game is to select a proper combination of six buttons that will allow one ball to reach the bottom. Using your knowledge of "and" and "or," select a combination of six buttons that will result in a win.

3.3 **EQUIVALENT STATEMENTS AND DEMORGAN'S LAWS**

Two statements are **equivalent** (symbolized $\Leftrightarrow$) if both statements have exactly the same truth values in every case of the truth table.

To determine whether two statements are equivalent, (a) construct a truth table for each statement and (b) compare the answer columns of the two truth tables. If the answers are identical, then the statements are equivalent. If the answers are not identical, then the statements are not equivalent.

∎ **Example 1** ——————————————————————————————

Show that the following two statements are equivalent.

$$[p \vee (q \vee r)] \qquad [(p \vee q) \vee r]$$

Solution: Construct a truth table for each statement (see Table 3.22). Since the truth tables shown in Table 3.22 have the same answer (column 3 for both tables), the statements are equivalent. Thus we can write

$$[p \lor (q \lor r)] \Leftrightarrow [(p \lor q) \lor r].$$

TABLE 3.22

p	q	r	[p	∨	(q∨r)]	[(p∨q)	∨	r]
T	T	T	T	T	T	T	T	T
T	T	F	T	T	T	T	T	F
T	F	T	T	T	T	T	T	T
T	F	F	T	T	F	T	T	F
F	T	T	F	T	T	T	T	T
F	T	F	F	T	T	T	T	F
F	F	T	F	T	T	F	T	T
F	F	F	F	F	F	F	F	F
			1	3	2	1	3	2

❖

The statement $p \lor q \lor r$ may be evaluated either as $(p \lor q) \lor r$ or as $p \lor (q \lor r)$. Similarly, a statement of the form $p \land q \land r$ may be evaluated as $(p \land q) \land r$ or as $p \land (q \land r)$. These statements are the associative property of disjunction and conjunction, respectively. (You may recall the associative property of addition: $(a + b) + c = a + (b + c)$.)

DeMorgan's Laws

Table 3.23(a) shows that $\sim(p \land q)$ and $\sim p \lor \sim q$ have identical truth values. Thus $\sim(p \land q)$ is equivalent to $\sim p \lor \sim q$. Table 3.23(b) shows that $\sim(p \lor q)$ and $\sim p \land \sim q$ have identical truth tables. Thus $\sim(p \lor q)$ is equivalent to $\sim p \land \sim q$.

TABLE 3.23(a)

p	q	~(p∧q)	~p∨~q
T	T	F	F
T	F	T	T
F	T	T	T
F	F	T	T

TABLE 3.23(b)

p	q	~(p∨q)	~p∧~q
T	T	F	F
T	F	F	F
F	T	F	F
F	F	T	T

These equivalent statements are:

$$\sim(p \land q) \Leftrightarrow \sim p \lor \sim q, \qquad \sim(p \lor q) \Leftrightarrow \sim p \land \sim q.$$

These statements, known as **DeMorgan's laws,** were developed by the English mathematician Augustus DeMorgan. Augustus, the son of a member of the East India Company, was born in India and was blind in one eye at birth. This handicap did not deter him ·from becoming a recognized mathematician in the areas of algebra, sets, and logic.

Agustus DeMorgan
Courtesy of The Granger Collection

DID YOU KNOW?

There are strong similarities between the topics of sets and logic. We can see them by examining DeMorgan's laws for sets and logic.

DeMorgan's Laws: Set Theory	DeMorgan's Laws: Logic
$(A \cap B)' = A' \cup B'$	$\sim(p \wedge q) \Leftrightarrow \sim p \vee \sim q$
$(A \cup B)' = A' \cap B'$	$\sim(p \vee q) \Leftrightarrow \sim p \wedge \sim q$

The complement in set theory, ', is similar to the negation, $\sim$, in logic. The intersection, $\cap$, is similar to the conjunction, $\wedge$, and the union, $\cup$, is similar to the disjunction, $\vee$. If we were to interchange the set symbols with the logic symbols, DeMorgan's laws would remain, but in a different form.

Both ' and $\sim$ can be interpreted as "not."
Both $\cap$ and $\wedge$ can be interpreted as "and."
Both $\cup$ and $\vee$ can be interpreted as "or."

■ **Example 2** _____

Using DeMorgan's laws, write an equivalent statement for "The car is too long or the garage is too short."

Solution: The first step is to write the statement in symbolic form.

Let p: The car is too long.
$\quad q$: The garage is too short.   ❖

The statement in symbolic form is $p \vee q$. To create an equivalent statement using DeMorgan's laws, we do the following: (1) Negate the complete statement, $\sim(p \vee q)$. (2) Negate each statement inside the parentheses, $\sim(\sim p \vee \sim q)$. (3) Change the $\vee$ to an $\wedge$. Thus $p \vee q \Leftrightarrow \sim(\sim p \wedge \sim q)$. The equivalent statement reads: "It is false that the car is not too long and the garage is not too short."

The procedure used in Example 3 can be used to determine a statement equivalent to the original statement using DeMorgan's laws. However, the equivalent statement might be more complicated than the orginal statement.

■ **Example 3** _____

Using DeMorgan's laws, write a statement equivalent to the statement "Today is not a holiday and the children are at home."

Solution

Let p: Today is a holiday.
   q: The children are at home.

The statement in symbolic form is $\sim p \wedge q$. To create an equivalent statement using DeMorgan's laws, we do the following: (1) Negate the complete statement, $\sim(\sim p \wedge q)$. (2) Negate each statement inside the parentheses, $\sim[\sim(\sim p) \wedge \sim q]$. Using truth tables, we can show that $\sim(\sim p) \Leftrightarrow p$. Thus we can write the statement as $\sim(p \wedge \sim q)$. (3) Change the $\wedge$ to $\vee$. Thus $\sim p \wedge q \Leftrightarrow \sim(p \vee \sim q)$. The equivalent statement reads: "It is false that today is a holiday or the children are not at home."   ❖

   It will be left to the student to show that $p \rightarrow q$ and $\sim p \vee q$ have identical truth tables and thus $p \rightarrow q \Leftrightarrow \sim p \vee q$. With these equivalent statements we can write a conditional statement as a disjunction or a disjunction as a conditional statement. Examining the equivalent statements $p \rightarrow q \Leftrightarrow \sim p \vee q$, we see that to change from a conditional statement to a disjunction, we negate the first simple statement, change the connective from $\rightarrow$ to $\vee$, and keep the second simple statement the same. To change from a disjunction to a conditional statement, negate the first simple statement, change from $\vee$ to $\rightarrow$ and keep the second simple statement the same.

■ **Example 4** _____

Write a conditional statement that is equivalent to $p \vee \sim q$.

Solution: Using the procedure discussed in the previous paragraph, we see that $p \vee \sim q$ becomes $\sim p \rightarrow \sim q$ when written as a conditional statement.
                                                                                 ❖

Quantifiers

The words **all, no,** and **some** are referred to as **quantifiers.** In mathematics it is sometimes necessary to negate statements involving quantifiers. If a statement is true, its negation will be false; and if a statement is false, its negation will be true.
   Consider statements involving the quantifier "all." For example, consider the statement "All teachers are rich." This statement is false. Its negation must therefore be true. We may be tempted to write its negation as "No teachers are rich." This statement is also false and is therefore not the negation of "All teachers are rich." The correct way to write the negation of "All teachers are rich" is "Not all teachers are rich" or "At least one

teacher is not rich" or "Some teachers are not rich." Note that the word "some" means "at least one."

Consider statements involving the quantifier "no." For example, consider "No birds can swim." This statement is false, and therefore its negation must be true. We may be tempted to write the negation of "No birds can swim" as "All birds can swim." Since "All birds can swim" is also false, it cannot be the negation of "No birds can swim." The negation of "No birds can swim" is "Some birds can swim" or "At least one bird can swim." Note that each of these statements is true.

Now consider statements involving the quantifier "some." Consider, for example, "Some students have a driver's license." This is a true statement, so its negation must be false. The statement "Some students have a driver's license" means "There is at least one student that has a driver's license." The negation of this statement is "No student has a driver's license." Note that the negation is a false statement.

Consider the statement "Some students do not ride motorcycles." This statement is true. What is its negation? We can rewrite this statement as "There is at least one student that does not ride a motorcycle." The negation of this statement is "All students ride motorcycles." Note that the negation is a false statement.

The negation of statements that use quantifiers is summarized here.

Statement of Form	Negation of Form
All are.	Some are not.
None are.	Some are.
Some are.	None are.
Some are not.	All are.

■ Example 5 ———————————————————————————

Write the negation of each statement.

a)   Some dogs have long tails.

b)   All books have pictures.

c)   Some houses are not homes.

d)   No horse is a donkey.

Solution

a)   Since "some" means "at least one," the statement "Some dogs have long tails" is the same as "At least one dog has a long tail." Since this is a true statement, its negation must be false. The negation is "No dogs have long tails."

b) The statement "All books have pictures" is false. Its negation must therefore be true. We may be tempted to write its negation as "No books have pictures." This statement is also false and is not the negation of "All books have pictures." The correct negation may be written as "Not all books have pictures" or "At least one book does not have pictures."

c) The statement "Some houses are not homes" means that at least one house is not a home. The negation is "All houses are homes."

d) The negation of "No horse is a donkey" is "Some horses are donkeys" or "At least one horse is a donkey."   ❖

EXERCISES 3.3

1. Two statements are connected with the biconditional, and the truth table is constructed. If this truth table is a tautology, what must be true about the two statements? Explain your answer.

Determine whether the following statements are equivalent.

2.	$\sim p \vee \sim q$	$\sim(p \wedge q)$
3.	$\sim(p \vee q)$	$\sim p \vee \sim q)$
4.	$p \rightarrow q$	$\sim p \vee q$
5.	$\sim(p \rightarrow q)$	$p \wedge q$
6.	$p \rightarrow q$	$q \rightarrow \sim p$
7.	$q \rightarrow p$	$\sim p \rightarrow \sim q$
8.	$\sim(p \wedge \sim q)$	$\sim p \vee q$
9.	$p \vee (q \wedge r)$	$(p \vee q) \wedge r$
10.	$\sim p \rightarrow (q \wedge r)$	$p \vee (q \wedge r)$
11.	$p \vee (q \wedge r)$	$(p \vee q) \wedge (p \vee r)$
12.	$(q \vee \sim r) \rightarrow p$	$[\sim(q \vee r)] \rightarrow p$

13. Show that $[(p \rightarrow q) \wedge (q \rightarrow p)] \Leftrightarrow (p \leftrightarrow q)$.

Write the negation of each of the following statements.

14. All grass is green.
15. Some plants are green.
16. No dogs have five legs.
17. Some birds do not fly.
18. No flowers are black.
19. All snow is white.
20. Some students will get an A.
21. Some instruments do not have strings.
22. Not all sheep are white.

Use DeMorgan's laws to write an equivalent statement for each of the following statements.

23. Today is Saturday or I am in school.
24. The car is not small and the gas mileage is not good.
25. The Philadelphia Eagles will play in the Super Bowl or they did not win enough games.
26. It is false that I passed the final exam and I did not study.
27. The test was not hard or I did not study.
28. The figure is not a rectangle or it is a square.
29. It is false that the computer knows the answers or the operator strikes the wrong key.
30. The donkey is not smart or the owner is a poor trainer.

Using the fact that $p \rightarrow q$ is equivalent to $\sim p \vee q$, rewrite the following statements in an alternative form.

31. If a man is healthy, then he is happy.
32. The dog is an animal or the cat is a fish.
33. If $x + 3 = 4$, then $3 + 1 = 4$.
34. $6 + 4 = 10$ or $1 + 6 = 7$.
35. The White House is in Washington or the U.S. Capitol is not in Chicago.
36. The toothpaste does not contain flouride or it does not contain sodium.
37. If the grass is not green, then it did not rain.
38. If the water is not warm, then I will go fishing.

PROBLEM SOLVING

39. Four men, one of whom was known to have committed a certain crime, made the following statements when questioned by the police.

Archie: "Dave did it."
Dave: "Tom did it."
Gus: "I didn't do it."
Tom: "Dave lied when he said I did it."

a) If only one of the four statements is true, who is the guilty man?
b) If only one of the four statements is false, who is the guilty man?

40. There are three students, and each student will be assigned a grade of A, B, or C. No two students will be given the same grade. If Karen gets a grade of A in this course, Charlie will not get a grade of B. If Lynn gets a grade of A or B, Karen will get a grade of C. If Karen gets a grade of B or C, Lynn will get an A. List all the ways the grades may be assigned to the three students.

 THE CONDITIONAL

There are many different ways in which the conditional statement may be written. Consider the following statement:

A person living in Philadelphia is a sufficient condition for a person living in Pennsylvania.

This statement says that a person who lives in Philadelphia also lives in Pennsylvania. Note that a person may live in Pennsylvania and not live in Philadelphia. For example, a person may live in Pittsburgh, Pennsylvania. The above statement is another way of stating that "if a person lives in Philadelphia, then that person lives in Pennsylvania." If we use the simple statements

p: A person lives in Philadelphia.
q: A person lives in Pennsylvania.

then the statement "A person living in Philadelphia is a sufficient condition for a person living in Pennsylvania" is symbolically written as $p \to q$.

"p is a sufficient condition for q" is symbolically written as $p \to q$.

Consider the statement "The apple being red is a sufficient condition for the apple being ripe." This tells us that a red apple is a ripe apple. Note that an apple may be ripe and not be red. For example, there might be some yellow or green apples that are ripe. All this statement tells us is that if the apple is red, then the apple is ripe. If we use the simple statements

p: The apple is red.
q: The apple is ripe.

then the statement "the apple being red is a sufficient condition for the apple being ripe" is symbolically written as $p \to q$.

Now let us examine the following statement, which provides us with a different form of the conditional.

A person living in Pennsylvania is a necessary condition for a person living in Philadelphia.

This statement indicates that for a person to live in Philadelphia a person must live in Pennsylvania. Note that a person may live in Pennsylvania and not live in Philadelphia. For example, a person may live in Pittsburgh, Pennsylvania. The above statement indicates that "if a person is living in Philadelphia, then a person is living in Pennsylvania." If we use the simple statements

p: A person lives in Philadelphia.
q: A person lives in Pennsylvania.

then the statement "A person living in Pennsylvania is a necessary condition for a person living in Philadelphia" is symbolically written as $p \to q$.

"q is a necessary condition for p" is symbolically written as $p \to q$.

Let's consider one more example: "The apple being ripe is a necessary condition for the apple being red." This statement indicates that if the apple is red, then the apple is ripe. Note that it is possible for apples to be ripe and not be red. For example, there might be some green or yellow apples that are ripe. If we use the simple statements

p: The apple is red.
q: The apple is ripe.

then the statement "The apple being ripe is a necessary condition for the apple being red" means "if the apple is red then the apple is ripe." This is symbolically written as $p \to q$.

Another form of the conditional statement is

A person lives in Pennsylvania if a person lives in Philadelphia.

This statement may be equivalently written as "If a person lives in Philadelphia, then a person lives in Pennsylvania." If we use the simple statements

p: A person lives in Philadelphia.
q: A person lives in Pennsylvania.

then the statement "A person lives in Pennsylvania if a person lives in Philadelphia" is symbolically written as $p \to q$.

"q if p" is symbolically written as $p \to q$.

Let's examine one additional form of the conditional:

A person lives in Philadelphia only if a person lives in Pennsylvania.

This statement indicated that for a person to live in Philadelphia that person must live in Pennsylvania. Therefore this statement means that if a person lives in Philadelphia, then a person lives in Pennsylvania. If we use the simple statements

    p: A person lives in Philadelphia.
    q: A person lives in Pennsylvania.

then the statement "A person lives in Philadelphia only if a person lives in Pennsylvania" is symbolically written as $p \to q$.

"p only if q" is symbolically written as $p \to q$.

The various forms of the conditional discussed are as follows.

Statement	Symbolic Form
If p then q	$p \to q$
p only if q	$p \to q$
q if p	$p \to q$
p is a sufficient condition for q	$p \to q$
q is a necessary condition for p	$p \to q$

 Example 1

Let p: The mouse is fat.
  q: The mouse ate the cat.

Write each of the following in symbolic form.

a) If the mouse is fat, then the mouse ate the cat.

b) The mouse is fat only if the mouse ate the cat.

c) The mouse is fat if the mouse ate the cat.

d) The mouse is fat is a sufficient condition for the mouse to have eaten the cat.

e) The mouse is fat is a necessary condition for the mouse to have eaten the cat.

Solution

a) $p \to q$

b) $p \to q$

c) $q \to p$

d) $p \to q$

e) $q \to p$ ❖

The statements $p \leftrightarrow q$ and $(p \to q) \land (q \to p)$ are equivalent statements. The statement $p \to q$ may be interpreted as "p is a sufficient condition for q," and the statement $q \to p$ may be interpreted as "p is a necessary condition for q." Therefore $p \leftrightarrow q$ may be interpreted as "p is a necessary and a sufficient condition for q." Thus p if and only if q (p iff q) can be equivalently stated as "p is a necessary and sufficient condition for q."

Variations of the Conditional

The statement $\sim q \to \sim p$ is a variation of the conditional called the **contrapositive of the conditional,** or simply the **contrapositive.** *The contrapositive is equivalent to the conditional, $(p \to q) \Leftrightarrow (\sim q \to \sim p)$.* Table 3.24(a) shows that the conditional statement and the contrapositive statement are equivalent.

TABLE 3.24(a)

		Conditional	Contrapositive
p	**q**	**$p \to q$**	**$\sim q \to \sim p$**
T	T	T	T
T	F	F	F
F	T	T	T
F	F	T	T

Other variations of the conditional are the **converse of the conditional,** $q \to p$, and the **inverse of the conditional,** $\sim p \to \sim q$. *The converse and the inverse are equivalent to each other, but neither is equivalent to the conditional: $(q \to p) \Leftrightarrow (\sim p \to \sim q)$.* Table 3.24(b) shows that the converse is equivalent to the inverse.

TABLE 3.24(b)

		Converse	Inverse
p	**q**	**$q \to p$**	**$\sim p \to \sim q$**
T	T	T	T
T	F	T	T
F	T	F	F
F	F	T	T

■ Example 2

Illustrate the following statement in the various forms of the conditional. "If today is Sunday, then tomorrow is Monday."

Solution: To write the *converse* of a conditional statement, switch the order of the premise and the conclusion. (See Table 3.25.) To write the *inverse,* negate both the premise and the conclusion. To write the *contrapositive,* switch the order of the premise and the conclusion and then negate both of them.

TABLE 3.25			
Name	**Symbolized**	**Read**	**Example**
Conditional	$p \rightarrow q$	If p, then q.	If today is Sunday, then tomorrow is Monday.
Converse of the conditional	$q \rightarrow p$	If q, then p.	If tomorrow is Monday, then today is Sunday.
Inverse of the conditional	$\sim p \rightarrow \sim q$	If not p, then not q.	If today is not Sunday, then tomorrow is not Monday.
Contrapositive of the conditional	$\sim q \rightarrow \sim p$	If not q, then not p.	If tomorrow is not Monday, then today is not Sunday.

❖

▮ **Example 3** ────────────────────

For p and q as follows, write the conditional statement, the converse, the inverse, and the contrapositive. Determine which of these statements are true.

p: The number is divisible by 9.
q: The number is divisible by 3.

Solution: *Conditional statement:* $(p \rightarrow q)$
If the number is divisible by 9, then the number is divisible by 3. This statement is true. A number divisible by 9 must also be divisible by 3, since 3 is a divisor of 9.

Converse of the conditional: $(q \rightarrow p)$
If the number is divisible by 3, then the number is divisible by 9. This statement is false. *Example:* 6 is divisible by 3, but 6 is not divisible by 9.

Inverse of the conditional: $(\sim p \rightarrow \sim q)$
If the number is not divisible by 9, then the number is not divisible by 3. This statement is false. *Example:* 6 is not divisible by 9, but 6 is divisible by 3.

Contrapositive of the conditional: $(\sim q \rightarrow \sim p)$
If the number is not divisible by 3, then the number is not divisible by 9. The statement is true, since any number that is divisible by 9 must be divisible by 3. ❖

■ **Example 4** ───────────────────────────────

Consider this statement.

> If Sue's car is a Ford, then Sue's car is not a Dodge.

Let p: Sue's car is a Ford.
    q: Sue's car is a Dodge.

The conditional statement written in symbolic form is $p \rightarrow \sim q$. Write the converse, inverse, and contrapositive of the conditional statement in both symbolic and statement form.

Solution: *Converse of the conditional:* $\sim q \rightarrow p$
If Sue's car is not a Dodge, then Sue's car is a Ford.

Inverse of the conditional: $\sim p \rightarrow \sim(\sim q)$ *is equivalent to* $\sim p \rightarrow q$
If Sue's car is not a Ford, then Sue's car is a Dodge.

Contrapositive of the conditional: $\sim(\sim q) \rightarrow \sim p$ *is equivalent to* $q \rightarrow \sim p$
If Sue's car is a Dodge, then Sue's car is not a Ford.   ❖

How many times have you asked someone to restate a sentence? In so doing, you are asking for an equivalent way of conveying the original thought. In many instances the second statement may be understood more easily. For this reason the variations of the conditional are very important in mathematics. Consider the statement "If a^2 is not a whole number, then a is not a whole number." Is this statement true? By writing the contrapositive, "If a is a whole number, then a^2 is a whole number," you will find it easier to see that the statement is true.

EXERCISES 3.4

Write each of the following statements in the if–then form.

1. The mouse is gray only if the cat is red.
2. I will water ski only if the water temperature is 70 degrees.
3. Owning a motorbike is a sufficient condition for buying a helmet.
4. Typing homework assignments is a necessary condition for buying a Macintosh.
5. Watering the garden is a necessary condition for planting seeds.
6. Owning an automobile is a sufficient condition for buying insurance.
7. I can go fishing if I have a license.
8. The fish will bite if they see an appetizing meal.
9. The mouse will kiss the cat only if the cat loves the mouse.
10. Being a lawyer is a sufficient condition for going to school.
11. The Eagles will win the Super Bowl if they win all their games.
12. Driving fast is a necessary condition for driving on the Autobahn in Germany.

13. Being in favor of waste management is a sufficient condition for being elected.
14. The weather is nice only if I watch the Buffalo Bills play.
15. The leaves turn red if the pumpkins turn orange.
16. Splitting the wood is a necessary condition for hitting the wedge.

For each of the following conditionals, write the converse, the inverse, and the contrapositive.

17. If today is Friday, then I am happy.
18. If I am not in school, then I am happy.
19. If the test was hard, then I did not study.
20. If the car is not small, then the gas mileage is not good.
21. If the key does not fit, then the door will not open.
22. If I pass the final exam, then I will pass the course.
23. If there is no lightning, then we can go swimming.
*24. If a person's IQ is 145 or higher, then the person is considered a genius. (Use DeMorgan's law.)
*25. If the figure is a rectangle, then the opposite sides are parallel and the angles are right angles.
*26. If I buy a new car or a used car, then I must buy gasoline.

Write the contrapositive of the following statements.

27. If the cattle are fed the proper diet, then the meat will be tender.
28. If the sum of the digits is divisible by 9, then the number is divisible by 9.

29. If you are going faster than 55 miles per hour, then you are speeding.
30. If two parallel lines are cut by a transversal, then the alternate interior angles are equal.
31. If the penguin is not a bird, then the penguin cannot fly.
32. If a plane figure has three angles, then it is a triangle.
33. If the figure has four equal sides, then it is a rhombus.
34. If the apple tree is sprayed regularly, then it will bear fruit.
35. If the dog eats carrots, then the cat is gray or the mouse is green.
36. If a and b are odd counting numbers, then the product of a and b is an odd counting number.

Write the contrapositive of the following statements. Use the contrapositive to determine whether the conditional statement is true or false.

37. If 2 does not divide the counting number, then 2 does not divide the units digit of the counting number.
38. If $\frac{m \cdot a}{m \cdot b} \neq \frac{a}{b}$, then m is not a counting number.
39. If two lines do not intersect in at least one point, then the two lines are parallel.
40. If $1/n$ is not a natural number, then n is not a natural number.

3.5 EULER CIRCLES AND SYLLOGISTIC ARGUMENTS

There are two kinds of reasoning processes, inductive and deductive. Inductive reasoning was discussed in Chapter 1. Now we will discuss deductive reasoning.

To help explain deductive reasoning, consider the following two statements.

> All German shepherds are dogs.
> All dogs bark.

Can you determine a statement that must necessarily be true when the two previous statements are true? If you reasoned that all German shepherds bark, you reasoned correctly. This is an example of deductive reasoning.

The two statements from which the **conclusion** is drawn are called **premises.**

> **Deductive reasoning** is the process of using logic to reason to a necessary and inescapable conclusion from a given set of premises.

The premises and conclusion together form an **argument.** An example of an argument is

All German shepherds are dogs.

All dogs bark.

∴ All German shepherds bark.

The three dots ∴ are read "therefore." This is an example of a valid argument because the conclusion can be deduced (or necessarily follows) from the given set of premises. An argument in which the conclusion does not necessarily follow from the given set of premises is said to be an invalid argument or a **fallacy.**

When we say an argument is valid, it does not mean that the conclusion is true. It means only that the conclusion necessarily follows from the premises.

> An argument is said to be **valid** when its conclusion necessarily follows, or can be deduced, from a given set of premises.

■ Example 1

Determine whether the following argument is valid or is a fallacy.

All animals that have a tail can fly.

A cow has a tail.

∴ A cow can fly.

Solution: The argument is valid because the conclusion—"A cow can fly"—necessarily follows from the given set of premises. Notice that the argument is valid even though the conclusion is a false statement. ❖

If an argument is valid and all of the premises are true statements, then the conclusion must also be a true statement.

Often it is not easy to determine whether an argument is valid or is a fallacy. Two methods will be discussed in this text that may help in determining whether an argument is valid. One method used with *syllogistic logic arguments* is Euler circles. This technique is discussed in this section. The other method used with *symbolic logic arguments* is truth tables. This technique is discussed in Section 3.6.

Syllogistic Logic

Syllogistic logic was developed by Aristotle around 350 B.C. Aristotle considered the relationships among the four types of statements that follow.

All _____ are _____.

No _____ are _____.

Some _____ are _____.

Some _____ are not _____.

An example of such statements follows. *All doctors are tall. No doctors are tall. Some doctors are tall. Some doctors are not tall.* Since Aristotle's time, other types of statements have been added to the study of syllogistic logic, two of which are

_____ is a _____.

_____ is not a _____.

An example of such statements follows. *Maria is a doctor. Maria is not a doctor.*

Symbolic logic considers the relationships among different forms of statements and will be discussed in detail in Section 3.6.

Although the two logics deal with different types of statements, they have a common use. They can both be used to determine the validity of an argument, or a **syllogism.**

Before we can give an example of a syllogism, let us review some of the Venn diagrams discussed earlier with respect to Aristotle's four statements. (See Figs. 3.1, 3.2, 3.3, and 3.4.)

Statement	Venn Diagram Illustrating Statement
All *A*s are *B*s.	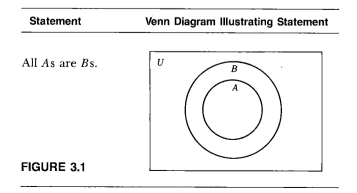

FIGURE 3.1

"All As are Bs" can be represented by the if–then statement "If it is an A, then it is a B."

Statement	Venn Diagram Illustrating Statement

No As are Bs.

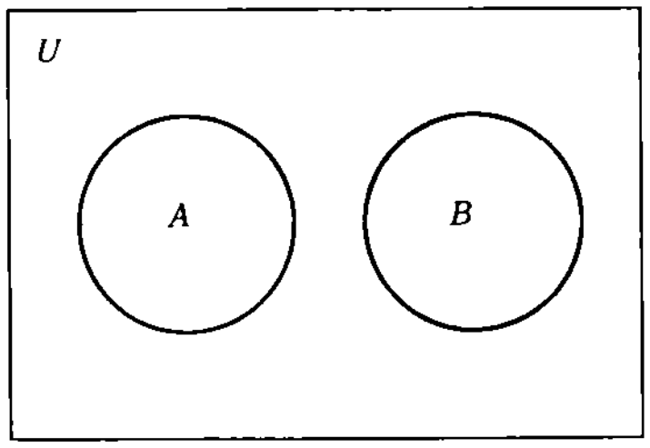

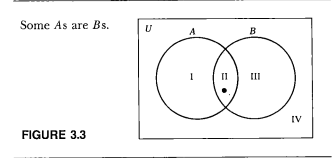

FIGURE 3.2

"No As are Bs" can be represented by the if–then statement "If it is an A, then it is not a B."

Some As are Bs.

FIGURE 3.3

"Some As are Bs" means there is at least one A that is also a B. Therefore Region II must contain at least one element.

Some As are not Bs.

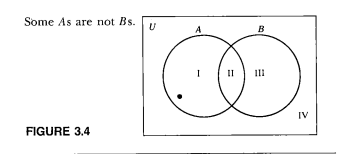

FIGURE 3.4

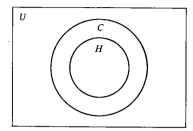

FIGURE 3.5

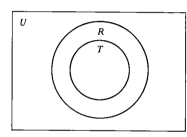

FIGURE 3.6

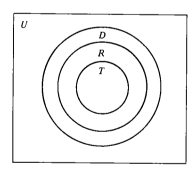

FIGURE 3.7

"Some As are not Bs" means there is at least one A that is not a B. Therefore Region I must contain at least one element. "Some As are not Bs" does not imply that some Bs are As or that some Bs are not As. Regions II and III may or may not contain any elements.

We used As and Bs in our examples, but anything could have been used in their place. For example, "All horses are cows" would be diagrammed in Fig. 3.5. Note that "All horses are cows" may be interpreted as "If it is a horse, then it is a cow."

Leonhard Euler used diagrams as an aid in determining the validity of syllogisms. Diagrams used for this purpose are often called **Euler** (pronounced "oiler") **diagrams.**

Consider the following syllogism.

> All teachers are rich.
> All rich people are doctors.
> ∴ All teachers are doctors.

The first two statements are the premises. The statement following the symbol for "therefore" is called the conclusion. Is this argument valid or is it a fallacy? We will use an Euler diagram as an aid in determining the answer.

The statement "All teachers are rich" can be represented as in Fig. 3.6. The statement "All rich people are doctors" is illustrated in Fig. 3.7. Notice that the set of teachers must be within the set of doctors. The argument is therefore valid. In this example the conclusion necessarily follows from the set of premises. This is an example of deductive reasoning. The use of logic leads us to the inescapable conclusion that all teachers are doctors.

Notice that this argument is valid even though the conclusion, "All teachers are doctors," is obviously a false statement. Similarly, an argument can be a fallacy even if the conclusion is a true statement. When we determine the validity of an argument, we are determining whether the conclusion must follow from the premises. When we say an argument is valid, we are saying that *if* the premises are *all* true statements, *then* the conclusion must also be true.

It is important to understand that it is the form of the argument that determines its validity, not the particular statements. For example, consider the syllogism

> All earth people have two heads.
> All people with two heads can fly.
> ∴ All earth people can fly.

The form of this argument is the same as that of the previous valid argument. Therefore this argument is also valid.

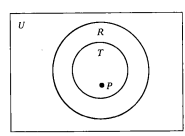

FIGURE 3.8

■ **Example 2** _____

Determine whether the following syllogism is valid or is a fallacy.

> All teachers are rich.
> Pete is a teacher. _____
> ∴ Pete is rich.

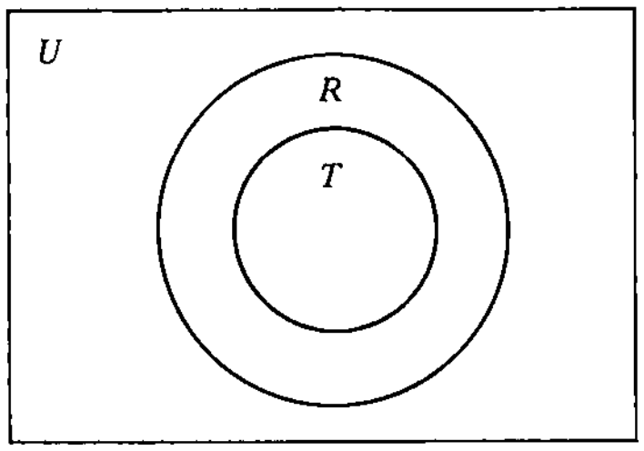

Solution: The statement "All teachers are rich" is illustrated in Fig. 3.8. The second premise, "Pete is a teacher," tells us that Pete must be placed in the inner circle (see Fig. 3.9). The Euler diagram illustrates that we must accept the statement "Pete is rich" as true (when we accept the premises as true). Therefore the argument is valid. ❖

■ **Example 3** _____

Determine whether the following syllogism is valid or is a fallacy.

> All football players are strong.
> Mary is strong. _____
> ∴ Mary is a football player.

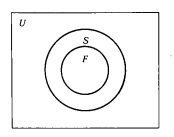

FIGURE 3.9

Solution: The statement "All football players are strong" is illustrated in Fig. 3.10. The next premise, "Mary is strong," tells us that Mary must be placed in the set of strong people. Two possible diagrams are shown in Fig. 3.11(a) and 3.11(b). In Fig. 3.11(a) and Fig. 3.11(b), both premises are satisfied. By examining Fig. 3.11(a), however, we see that in this case Mary is not a football player. Therefore the statement "Mary is a football player" *does not necessarily follow* from the set of premises. Thus the argument is a fallacy.

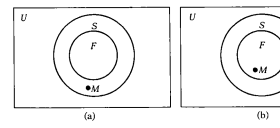

FIGURE 3.11

(a)   (b)

❖

FIGURE 3.10

■ **Example 4** _____

Determine whether the following syllogism is valid or is a fallacy.

> No airplane pilots eat spinach.
> Janet does not eat spinach. _____
> ∴ Janet is an airplane pilot.

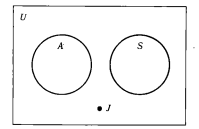

FIGURE 3.12

Solution: The diagram in Fig. 3.12 satisfies the two given premises and also shows that Janet is not an airplane pilot. Therefore the argument is invalid, or a fallacy. ❖

Note that in Example 4, if we placed Janet in circle *A*, the argument would appear to be valid. However, *whenever testing the validity of an argument, we always try to show that the argument is invalid.* If there is any way of showing that the conclusion does not necessarily follow from the premises, then the argument is invalid.

■ **Example 5** _____

Determine whether the following syllogism is valid or invalid.

> All *A*s are *B*s.
> Some *B*s are *C*s.
> ∴ Some *A*s are *C*s

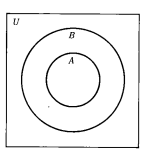

FIGURE 3.13

Solution: The statement "All *A*s are *B*s" is illustrated in Fig. 3.13. The statement "Some *B*s are *C*s" means that there is at least one *B* that is a *C*. We can illustrate this set of premises in many ways. Three illustrations are given in Fig. 3.14.

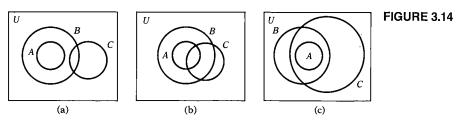

FIGURE 3.14

(a)   (b)   (c)

In all three illustrations we see that (1) all *A*s are *B*s and (2) some *B*s are *C*s. The conclusion is "Some *A*s are *C*s." Since at least one of the illustrations (Fig. 3.14a) shows that the conclusion does not necessarily follow from the given premises, the argument is invalid. (That is, it is possible that both premises may be true and yet the conclusion is not.) ❖

■ **Example 6** _____

Determine whether the following syllogism is valid or invalid.

> Some cats are dogs.
> Some dogs are horses.
> ∴ Some cats are horses.

Solution: Some (but not all) methods of illustrating this syllogism are shown in Fig. 3.15. Note that all three illustrations show that (1) some cats are dogs and (2) some dogs are horses. The conclusion is "Some cats are horses." If *any one* of these diagrams shows that this is not necessarily so, then the conclusion does not necessarily follow from the premises, and the argument is invalid. Since Fig. 3.15(b) does not show that some cats are horses, the argument is invalid.

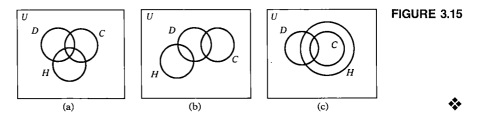

FIGURE 3.15

(a)   (b)   (c)

❖

Let us look at one more example.

■ Example 7

Determine whether the following syllogism is valid or invalid.

> No neurosurgeons are football players.
> All college graduates are football players.
> ∴ No neurosurgeons are college graduates.

Solution: The first premise tells us that the neurosurgeons and football players are disjoint sets, as illustrated in Fig. 3.16. The second premise tells us that the set of college graduates is a subset of the set of football players. This is illustrated in Fig. 3.17.

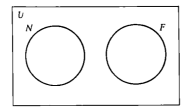   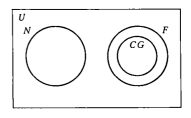

FIGURE 3.16   **FIGURE 3.17**

The set of college graduates and neurosurgeons cannot be made to intersect without violating a premise. Thus no neurosurgeons can be college graduates, and the syllogism is valid. Note that we did not say that this conclusion was true—only that the argument was valid.   ❖

EXERCISE 3.5

Use Euler circles to determine whether each syllogism is valid or is a fallacy.

1. All men love quiche.
   John is a man.
   ∴ John loves quiche.

2. No men love quiche.
   John loves quiche.
   ∴ John is not a man.

3. All frogs can swim.
   All things that can swim can fly.
   ∴ All frogs can fly.

4. All As are Bs.
   All Bs are Cs.
   ∴ All As are Cs.

5. All healthy people are happy.
   Ms. Y. is happy.
   ∴ Ms. Y is healthy.

6. All physicists are intelligent.
   Madame Curie was intelligent.
   ∴ Madame Curie was a physicist.

7. All movie stars have agents.
   Robert Redford is a movie star.
   ∴ Robert Redford has an agent.

8. All numbers that are divisible by 2 are even.
   The number 7 is divisible by 2.
   ∴ The number 7 is even.

9. Some monkeys are hairy.
   Cheeta is a monkey.
   ∴ Cheeta is hairy.

10. Some monkeys are not hairy.
    Cheeta is a monkey.
    ∴ Cheeta is not hairy.

11. Some medical researchers have made important discoveries.
    Some people who have made important discoveries are famous.
    ∴ Some medical researchers are famous.

12. All horses have four legs.
    No animals with four legs can fly.
    ∴ No horses can fly.

13. No prime numbers are divisible by 2.
    The number 10 is not divisible by 2.
    ∴ The number 10 is a prime number.

14. Some cars can float.
    All things that float are light.
    ∴ Some cars are light.

15. Some people love mathematics.
    All people who love mathematics love physics.
    ∴ Some people love physics.

16. Some dogs have two legs.
    All people have two legs.
    ∴ Some people are dogs.

17. No xs are ys.
    No ys are zs.
    ∴ No xs are zs.

18. All pilots can fly.
    All astronauts can fly.
    ∴ Some pilots are astronauts.

19. Some tall people wear glasses.
    Maria wears glasses.
    ∴ Maria is not tall.

20. All rainy days are cloudy.
    Today it is cloudy.
    ∴ Today is a rainy day.

21. All sweet things taste good.
    All things that taste good are fattening.
    All things that are fattening put on pounds.
    ∴ All sweet things put on pounds.

*22. If Jim turns right or left at the stop sign, he will run out of gas before he reaches the service station. He has already gone too far past the service station to return to it before he runs out of gas. He does not see a service station ahead. Which of the following statements is the only one that can possibly be deduced?

   a) He may run out of gas.
   b) He will run out of gas.
   c) He should not have taken this route.
   d) He is lost.
   e) He should turn right at the stop sign.
   f) He should turn left at the stop sign.

3.6 SYMBOLIC ARGUMENTS

In Sections 3.2–3.4 we presented the basic concepts and definitions of symbolic logic. We will now use those basic ideas to determine whether *"symbolic"* arguments are valid. Symbolic arguments differ from syllogistic arguments in the forms of the statements used. Symbolic arguments use the connectives and, or, not, if–then, and if and only if. The validity of syllogistic arguments was determined by using Euler circles. The validity of symbolic arguments is determined by using truth tables. Consider the following statements.

> If Sue goes to the movie, then she will eat popcorn.
> Sue goes to the movie.

If you accept these two statements as true, then using deductive reasoning, you would conclude:

> Sue will eat popcorn.

These three statements constitute what is called a **symbolic argument.** The first two statements are called *premises,* and the third statement is called the *conclusion.* The argument is written:

> If Sue goes to the movie, then she will eat popcorn.
> Sue goes to the movie.
> _____
> Therefore Sue will eat popcorn.

To write the argument in symbolic form, let

$$p: \text{Sue goes to the movie.}$$
$$q: \text{Sue will eat popcorn.}$$

Symbolically, the argument is written

$$p \rightarrow q$$
$$\underline{p \qquad}$$
$$\therefore q$$

Any argument will be either valid or a fallacy (invalid). As with syllogistic logic an argument is valid when its conclusion necessarily follows, or can be deduced, from a given set of premises.

Before we determine whether the argument above is valid or invalid, let us look at the general procedure used to determine the validity of a symbolic argument.

In the argument

$$p_1$$
$$\underline{p_2 \qquad}$$
$$\therefore c$$

p_1 and p_2 are the premises, and c is the conclusion. This argument will be valid if $(p_1 \wedge p_2) \rightarrow c$ is a tautology. For $(p_1 \wedge p_2) \rightarrow c$ to be a tautology the conclusion, c, must be true whenever the conjunction of the premises, $p_1 \wedge p_2$, is true.

An argument in symbolic logic will be valid if the conclusion is true whenever the conjunction of the premises is true.

Now let us determine whether

$$p \rightarrow q$$
$$\underline{p \qquad\qquad}$$
$$\therefore q$$

is a valid argument.

We write the argument in the following form.

$$[(p \rightarrow q) \wedge p] \rightarrow q$$

The part inside the square brackets is the conjunction of the premises, and the part after the $\rightarrow$ is the conclusion. If this statement is a tautology, then the argument is valid. The truth table is true in every case (see Table 3.26). Therefore the statement is a tautology, and the argument is valid. The letters p and q could represent any statements. Once we have demonstrated that an argument in a particular form is valid, all other arguments with exactly the same form will also be valid. In fact, many of these forms are so common they have been assigned names. The argument form just discussed is called the **law of detachment** (or *modus ponens*).

Consider the following argument.

If bees are sweet, then honey is bitter.
Bees are sweet.
———————————————
∴ Honey is bitter.

Translated into symbolic form, it becomes

b: Bees are sweet.
h: Honey is bitter.

$$b \rightarrow h$$
$$\underline{b \qquad}$$
$$\therefore h$$

TABLE 3.26		
p	q	$[(p \rightarrow q) \wedge p] \rightarrow q$
T	T	T T T T T
T	F	F F T F F
F	T	T F F F T
F	F	T F F F F
		1 3 2 5 4

This is exactly the same form as the law of detachment, and therefore it is a valid argument. Notice that the previous argument is valid even though the conclusion, "Honey is bitter," is a false statement. When we say that an argument is valid, we mean that the conclusion necessarily follows from the premises and not that the premises or the conclusion themselves are true statements. It is also possible to have an invalid argument in which the conclusion is a true statement.

The following is an example of the **law of contraposition** (or *modus tollens*).

> If Tim drives the truck, then the shipment will be delivered.
> The shipment will not be delivered.
> ∴ Tim does not drive the truck.

To determine whether the argument is valid, we first write the argument in symbolic form.

> *p:* Tim drives the truck.
> *q:* The shipment will be delivered.

$$p \rightarrow q$$
$$\frac{\sim q}{\therefore \sim p}$$

Now write the argument in the form $[(p \rightarrow q) \wedge \sim q] \rightarrow \sim p$ and construct a truth table (Table 3.27). Since the answer, column 5, has all Ts, the argument is valid.

TABLE 3.27						
p	*q*	[(*p*→*q*)	∧	~*q*]	→	~*p*
T	T	T	F	F	T	F
T	F	F	F	T	T	F
F	T	T	F	F	T	T
F	F	T	T	T	T	T
		1	3	2	5	4

Example 1

Use a truth table to determine whether the following argument is valid or invalid.

> The car is red or the car is new.
> The car is new.
> ∴ The car is red.

Solution: Let *p:* The car is red.
                      q: The car is new.
In symbolic form the argument is

$$p \vee q$$
$$\frac{q}{\therefore p}$$

TABLE 3.28				
p	*q*	[(*p* ∨ *q*)	∧ *q*] → *p*	
T	T	T	T T T T	
T	F	T	F F T T	
F	T	T	T T F F	
F	F	F	F F T F	
		1	3 2 5 4	

Write the argument in the form $[(p \vee q) \wedge q] \to p$. Next construct a truth table, as shown in Table 3.28. The answer to the truth table, column 5, is not true in *every case*. Therefore the statement is not a tautology, and the argument is invalid, or is a fallacy.   ❖

◼︎ **Example 2** _____

Use a truth table to determine whether the following argument is valid or is a fallacy.

If Albert Einstein was a physicist, then Albert Einstein was intelligent. Albert Einstein was intelligent.
_____

∴ Albert Einstein was a physicist.

Solution: Let p: Albert Einstein was a physicist.
  q: Albert Einstein was intelligent.

In symbolic form the argument is
$$p \to q$$
$$\underline{q \quad\quad\quad}$$
$$\therefore p$$

Now write the argument in the form $[(p \to q) \wedge q] \to p$ and construct the truth table (Table 3.29). Since the answer (column 5) is not true in every case, the argument is a fallacy.   ❖

p	**q**	**[(p →q)**	**∧ q]**	**→**	**p**
T	T	T	T	T	T T T
T	F	F	F	F	T T
F	T	T	T	T F	F
F	F	T	F	F T	F
		1	3	2	5 4

◼︎ **Example 3** _____

Use a truth table to determine whether the following argument is valid or invalid.

If I eat pizza with peppers, then I can sleep.
If I can sleep, then I can dream.
_____

∴ If I eat pizza with peppers, then I can dream.

Solution: Let p: I eat pizza with peppers.
  q: I can sleep.
  r: I can dream.

In symbolic form the argument is
$$p \to q$$
$$\underline{q \to r \quad}$$
$$\therefore p \to r$$

Write the argument in the form

$$[(p \rightarrow q) \land (q \rightarrow r)] \rightarrow (p \rightarrow r)$$

Then construct the truth table (Table 3.30). The answer, column 5, is true in every case. Thus the statement is a tautology, and the argument is valid. This argument form is called the **law of syllogism.**

TABLE 3.30							
p	q	r	[(p→q)	∧ (q→r)]	→ (p→r)		
T	T	T	T	T	T	T	T
T	T	F	T	F	F	T	F
T	F	T	F	F	T	T	T
T	F	F	F	F	T	T	F
F	T	T	T	T	T	T	T
F	T	F	T	F	F	T	T
F	F	T	T	T	T	T	T
F	F	F	T	T	T	T	T
			1	3	2	5	4

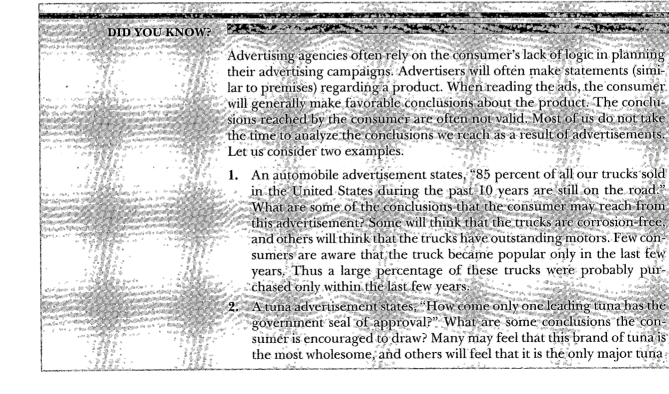

that meets the government's specifications. How many consumers know that there is no mandatory inspection of seafood? The National Marine Fisheries Service, an agency of the Department of Commerce, runs a voluntary inspection program, under which packers pay a fee in return for federal inspection. A few years ago almost all tuna packers participated in this program, but one by one they have dropped out—possibly because of the high fee charged by the government.

EXERCISES 3.6

Determine whether the following symbolic arguments are valid or invalid.

1. $p \to q$
p
∴ q

2. $p \to q$
$\sim q$
∴ $\sim p$

3. $p \lor q$
$\sim p$
∴ $\sim q$

4. $p \lor q$
$\sim q$
∴ p

5. $p \to q$
q
∴ p

6. $\sim p \to q$
$\sim q$
∴ $\sim p$

7. $q \land (\sim p)$
$\sim p$
∴ q

8. $q \land p$
q
∴ $\sim p$

9. $p \to q$
$q \to r$
∴ $p \to r$

10. $p \leftrightarrow q$
$q \to r$
∴ $\sim r \to \sim p$

11. $p \to q$
$q \land r$
∴ $p \lor r$

12. $r \leftrightarrow p$
$\sim p \land q$
∴ $p \land r$

13. $p \lor q$
$r \land p$
∴ q

14. $p \to q$
$q \lor r$
$r \lor p$
∴ p

*15. $p \to q$
$r \to s$
$p \lor r$
∴ $q \lor s$

*16. $p \to q$
$q \to r$
$r \to s$
∴ $p \to s$

Translate the following arguments into symbolic form. Then use a truth table to determine whether the arguments are valid or invalid.

17. If the weather is bad, then the party is canceled.
The weather is bad.
∴ The party is canceled.

18. If the story is interesting, then the people will listen.
The people will listen.
∴ The story is interesting.

19. I am a teacher or I am a preacher.
I am not a preacher.
∴ I am a teacher.

20. The dog is a collie and the cat is a siamese.
If the cat is siamese, then the bird is a parrot.
∴ If the bird is not a parrot, then the dog is not a collie.

21. The painting is a Rembrandt or the painting is a Picasso.
If the painting is a Picasso, then it is not a Rembrandt.
∴ The painting is not a Picasso.

22. If you go in the water, then I will go in the water.
I will not go in the water.
∴ You will not go in the water.

23. The watch is not expensive.
If the watch is expensive, then the watch cost a lot of money.
∴ The watch cost a lot of money.

24. It is snowing and I am going skiing.
If I am going skiing, then I will wear a coat.
∴ If it is snowing, then I will wear a coat.

25. The brick is red or the brick is made of sand.
If the brick is not red, then it is Larry's house.
∴ The brick is made of sand or it is Larry's house.

26. If you study, then you will pass the tests.
If you pass the tests, then you will pass the course.
∴ If you study, then you will pass the course.

27. If George deals the cards, then Jim will win the hand.
If Jim wins the hand, then George did not deal the cards or Charlie played cards.
∴ Jim will win the hand.

*28. If Sue wins the contest or robs a bank, then she will be rich.
If Sue is rich, then she will stop working.

∴ If Sue does not stop working, she did not win the contest.

Translate each argument into symbolic form, using the suggested notation. Test the validity of each argument by making a truth table.

29. The truck is red or the house is blue. The house is not blue. Therefore the truck is not red. (t, h)

30. If all truck drivers are frustrated little boys, then I am a truck driver. I am not a truck driver. Therefore not all truck drivers are frustrated little boys. (f, t)

31. If all the frustrated little boys are truck drivers, then I am in trouble. All frustrated little boys are truck drivers. Therefore I am in trouble. (f, t)

32. If switch A is closed, then switch B is open. Switch B is not open. Therefore switch A is not closed. (a, b)

33. The music is soft and soothing. The music is not soothing or it is soft. Therefore the music is not soft. (s, o)

34. The computer program is hard and the result is short. The result is not short or the computer program is not hard. Therefore the result is not short. (h, s)

35. The birds are singing and the cat is listening. The cat is not listening or the birds are singing. Therefore the birds are not singing. (b, c)

36. If the car is new, then the car has air conditioning. The car is not new and the car has air conditioning. Therefore the car is not new. (c, a)

*37. If switch A is open, then switch B is open. If switch B is open, then switch C is open. Switch A is open. Therefore switch C is open. (a, b, c)

*38. Money is the root of all evil or the salvation of the poor. If money is the salvation of the poor, then it is not the root of all evil. Money is the root of all evil. Therefore money is not the salvation of the poor. (m, p, e)

Supply a conclusion that you believe will make the argument valid for each of the following arguments. Use a truth table to determine whether the argument is valid.

39. If you drink orange juice, then you will be healthy.
You drink orange juice.
Therefore . . .

40. If you do not read a lot, then you will not gain knowledge.
You do not read a lot.
Therefore . . .

41. If you like to eat, then you enjoy meat.
If you enjoy meat, then you are an outdoorsperson.
Therefore . . .

42. If you like the job, then you are adventurous.
If you are adventurous, then you will become an entrepreneur.
Therefore . . .

*43. If the fire will burn, then the logs are dry.
If the fire will not burn, then you will have to dry the logs.
Therefore . . .

Verify that each of the following argument forms is valid.

44. $a \rightarrow b$
a
∴ b

45. $a \rightarrow b$
$\sim b$
∴ $\sim a$

46. $a \rightarrow b$
$b \rightarrow c$
∴ $a \rightarrow c$

47. $a \rightarrow b$
$b \rightarrow c$
∴ $\sim c \rightarrow \sim a$

48. $a \lor b$
$\sim a$
∴ b

49. $a \lor b$
$\sim b$
∴ a

50. In a certain company the positions of president, vice-president, and treasurer are held by Davis, Egan, and Leopard, not necessarily in that order. We know the following:
a) The treasurer is an only child and earns the least money.
b) Leopard, who married Davis's brother, earns more than the vice-president.
c) Assume that the president is the highest paid and the treasurer is the lowest paid.
Match the individual with the position.

51. Solve the following puzzle. Good luck!
The Joneses have four cats. The parents are Tiger and Boots, and the kittens are Sam and Sue. Each cat insists on eating out of its own bowl. To complicate matters, each cat will eat only its own brand of cat food. The colors of the bowls are red, yellow, green, and blue. The different types of cat food are Puss 'n' Boots, Friskies, Nine Lives, and Meow Mix. Tiger will eat Meow Mix if and only if it is in a yellow bowl. If Boots is to eat her food, then it must be in a yellow bowl. Mrs. Jones knows that

the label on the can containing Sam's food is the same color as his bowl. The name of the food that Boots will eat contains her name. Meow Mix and Nine Lives are packaged in a paper bag. The color of Sue's bowl is green if and only if she eats Meow Mix. None of the other cats' eyes are the same color as their bowls. The label on the Friskies can is red. Match each cat with its food and the bowl of the correct color.

PROBLEM SOLVING

52. Explain why the following method may be used to show that an argument is valid. Determine all cases in which the conclusion is false, and show that in each case at least one premise is false.

53. Statements in logic can be translated into set statements—for example, $p \wedge q$ is similar to $P \cap Q$; $p \vee q$ is similar to $P \cup Q$; $p \rightarrow q$ is similar to $P' \cup Q$. Euler circles can also be used to show that arguments similar to those discussed in this section are valid or invalid. Use Euler circles to show that the argument is invalid.

$$p \rightarrow q$$
$$\underline{p \vee q}$$
$$\therefore \ \sim p$$

*54. The following set of premises is taken from a logic textbook written by Lewis Carroll, author of *Alice in Wonderland.*

No kitten that loves fish is unteachable.
No kitten without a tail will play with a gorilla.
Kittens with whiskers always love fish.
No teachable kitten has green eyes.
No kittens have tails unless they have whiskers.

One and only one deduction can be drawn from this set of statements. *Hint:* You will have to reword and reorganize the statements. For example, the fourth statement may be rewritten "Green-eyed kittens cannot be taught." What conclusions can be deduced from these five premises?

RESEARCH ACTIVITIES

55. Show how logic is used in advertising. Discuss several advertisements, and show how logic is used to persuade the reader.

56. Show how logic is used in politics. Collect articles on a particular political campaign, and show how logic is used by the candidates to gain votes.

3.7 SWITCHING CIRCUITS

FIGURE 3.18

Switch	Light bulb
T	on (switch closed)
F	off (switch open)

TABLE 3.31

A common application of logic is switching circuits. To understand the basic concepts, let us examine a few simple circuits that are common in most homes. The typical lamp has a cord, which is plugged into a wall receptacle. Somewhere between the bulb in the lamp and the receptacle is a switch to turn the lamp on and off. When the switch is closed, the current flows, and the bulb lights up. When the switch is open (not making contact), no current can flow, and the bulb will not light. The basic configuration of a switch is shown in Fig. 3.18.

Electric circuits can be expressed as logical statements. We represent switches as letters, using T to represent a closed switch (or current flow) and F to represent an open switch (or no current flow). This relationship is indicated in Table 3.31.

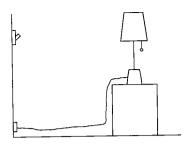

FIGURE 3.19

Occasionally, we have a wall switch connected to a wall receptacle and a lamp plugged into the wall receptacle (Fig. 3.19). We then say that the wall switch and the switch on the lamp are in series, meaning that for the bulb in the lamp to light, both switches must be on at the same time (Fig. 3.20). On the other hand, the bulb will not light if either switch is off or if both switches are off. In either of these conditions the electricity will not flow through the circuit. In a **series circuit** the current can take only one path. If any switch in the path is open, the current cannot flow.

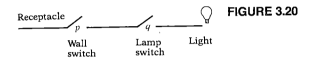

Receptacle / _____ / _____ ◯ **FIGURE 3.20**
p q

Wall   Lamp   Light
switch   switch

To illustrate this symbolically, let p represent the wall switch and q the lamp switch. The letter T will be used to represent both a closed switch and the bulb lighting. The letter F will represent an open switch and the bulb not lighting. Thus we have four possible cases.

- **CASE 1.** Both switches are closed, that is, p is T and q is T. The light is on, T.

- **CASE 2.** Switch p is closed and switch q is open, that is, p is T and q is F. The light is off, F.

- **CASE 3.** Switch p is open and switch q is closed, that is, p is F and q is T. The light is off, F.

- **CASE 4.** Both switches are open, that is, p is F and q is F. The light is off, F.

TABLE 3.32			
p	q	Light	$p \wedge q$
T	T	on (T)	T
T	F	off (F)	F
F	T	off (F)	F
F	F	off (F)	F

Table 3.32 summarizes the results.

The on–off results are the same as the truth table for conjunction $p \wedge q$ if we think of "on" as true and "off" as false. Switches in series will always be indicated with a conjunction, $\wedge$.

Another type of electric circuit used in the home is the **parallel circuit,** in which there are two or more paths that the current can take. If the current can pass through either path or both (see Fig. 3.21), the light will go on. The letter T will be used to represent both a closed switch and the bulb lighting. The letter F will represent an open switch and the bulb not lighting. Thus we have four possible cases.

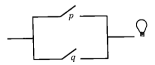

FIGURE 3.21

● **CASE 1.** Both switches are closed, that is, p is T and q is T. The light is on, T.

● **CASE 2.** Switch p is closed and switch q is open, that is, p is T and q is F. The light is on, T.

● **CASE 3.** Switch p is open and switch q is closed, that is, p is F and q is T. The light is on, T.

● **CASE 4.** Both switches are open, that is, p is F and q is F. The light is off, F.

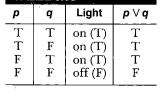

TABLE 3.33

p	q	Light	$p \lor q$
T	T	on (T)	T
T	F	on (T)	T
F	T	on (T)	T
F	F	off (F)	F

Table 3.33 summarizes the results. The on–off results are the same as the $p \lor q$ truth table if we think of "on" as true and "off" as false. Switches in parallel will always be indicated with a disjunction, $\lor$.

We can now combine the basic ideas we have discussed and analyze more complex circuits.

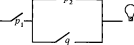

FIGURE 3.22

■ **Example 1** ─────────────────────────────

a) Write a symbolic statement that represents the circuit shown in Fig. 3.22.

b) Construct a truth table to determine when the light will be on.

Solution

a) Switch p_1 is in series with the parallel branch containing switches p_2 and q. The current must flow through p_1; therefore it is in series. After flowing through p_1, the current may flow through the branch containing p_2, the branch containing q, or both branches. Thus p_2 and q are in parallel. The circuit in symbolic form is $p_1 \land (p_2 \lor q)$. Note that the parentheses are important. Without them the statement could be interpreted as $(p_1 \land p_2) \lor q$. The diagram for this statement is illustrated in Fig. 3.23. Since the statements are not equivalent, the circuits cannot produce the same results.

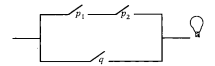

FIGURE 3.23

In Fig. 3.22 we have two switches labeled with the letter p, that is, p_1 and p_2. We used these subscripts for the sake of explanation. Since they are both represented by the letter p, they will either both open together or both close together. Fig. 3.22 can be represented by the statement $p_1 \wedge (p_2 \vee q)$. At this point we will use p to represent both switches p_1 and p_2. Thus our statement becomes $p \wedge (p \vee q)$.

b) The truth table for the statement (Table 3.34) indicates that the light will be on only in the cases in which p is true or when switch p is closed. The results show that we could remove switch q and one of the p switches without changing the electrical properties of the circuit. Fig. 3.22 shows that only one switch (p_1) is needed to control the current flow.   ❖

TABLE 3.34

p	q	$p \wedge (p \vee q)$
T	T	T
T	F	T
F	T	F
F	F	F

■ **Example 2** _____

a) Write a symbolic statement that represents the circuit in Fig. 3.24.

b) Construct a truth table to determine when the light will be on.

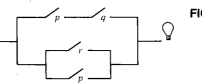

FIGURE 3.24

TABLE 3.35

p	q	r	$(p \wedge q) \vee (p \vee r)$
T	T	T	T
T	T	F	T
T	F	T	T
T	F	F	T
F	T	T	T
F	T	F	F
F	F	T	T
F	F	F	F

Solution

a) The upper branch of the circuit contains two switches in series. We can represent that branch with the statement $p \wedge q$. The lower branch of the circuit has two switches in parallel. We can represent that branch with the statement $r \vee p$. The upper branch is in parallel with the lower branch. Putting the two branches together, we get the statement $(p \wedge q) \vee (r \vee p)$.

b) The truth table for the statement (Table 3.35) shows that cases 6 and 8 are false. Thus the light will be off in these cases and on in all the other cases. In examining the truth values we see that the statement is false only in the cases in which both p and r are false, or when both switches are open. By examining the diagram or truth table, we can see that the current will flow if switch p is closed or switch r is closed. In fact, the top branch of the circuit could be eliminated without affecting the electrical properties of the circuit.   ❖

We now look at the problem in reverse. Suppose that we are given the statement $(p \wedge q) \vee r$ and are asked to construct a circuit corresponding to it.

Remember that ∧ indicates a series branch and ∨ indicates a parallel branch. Working first within the parentheses, we see that switches p and q are in series. This series branch is in parallel with switch r, as indicated in Fig. 3.25.

Occasionally, it is necessary to have one switch closed and another switch open at the same time. This situation can be represented by using p for one of the switches and $\bar{p}$ for the other switch. The switch labeled $\bar{p}$ corresponds to $\sim p$ in a logic statement. For example, in a series circuit, $p \wedge \sim p$ would be represented by Fig. 3.26. In this case the light would never go on; the switches would counteract each other. When switch p is closed, switch $\bar{p}$ is open; and when p is open, $\bar{p}$ is closed.

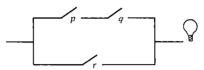

FIGURE 3.25

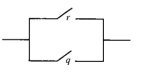

FIGURE 3.26

■ **Example 3** ────────────────────────────────

Draw a switching circuit that represents $[(p \wedge \sim q) \vee (r \vee q)] \wedge s$.

Solution: Switches p and $\bar{q}$ are in series as represented in Fig. 3.27. Switches r and q are in parallel as represented in Fig. 3.28.

FIGURE 3.27

FIGURE 3.28

The two branches are in parallel with each other. The parallel branches are represented in Fig. 3.29. Finally, switch s is in series with the entire circuit, as illustrated in Fig. 3.30.

FIGURE 3.29

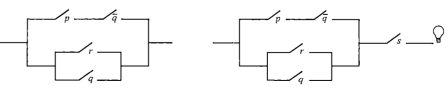

FIGURE 3.30 ❖

EXERCISES 3.7

For each of the following diagrams of circuits, write the corresponding logic statement.

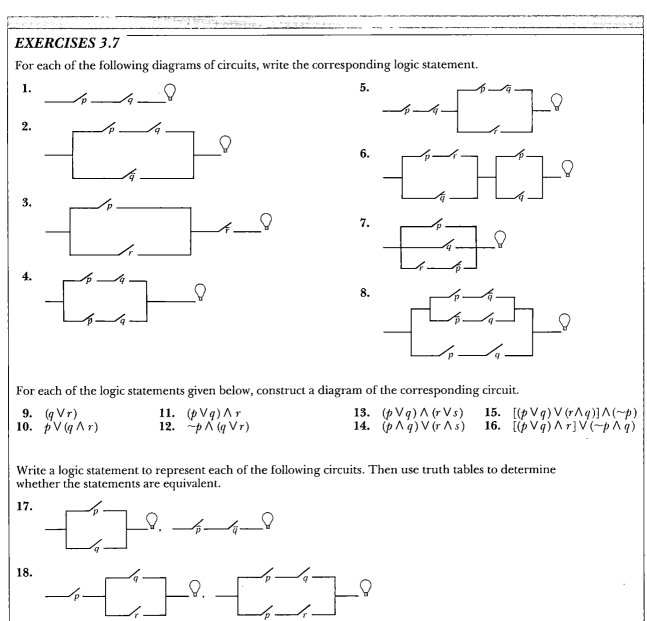

For each of the logic statements given below, construct a diagram of the corresponding circuit.

9. $(q \lor r)$

10. $p \lor (q \land r)$

11. $(p \lor q) \land r$

12. $\sim p \land (q \lor r)$

13. $(p \lor q) \land (r \lor s)$

14. $(p \land q) \lor (r \land s)$

15. $[(p \lor q) \lor (r \land q)] \land (\sim p)$

16. $[(p \lor q) \land r] \lor (\sim p \land q)$

Write a logic statement to represent each of the following circuits. Then use truth tables to determine whether the statements are equivalent.

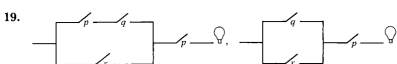

20.

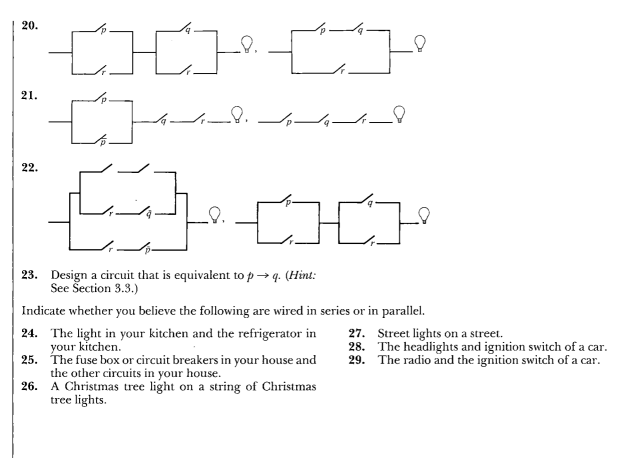

21.

22.

23. Design a circuit that is equivalent to $p \to q$. (*Hint:* See Section 3.3.)

Indicate whether you believe the following are wired in series or in parallel.

24. The light in your kitchen and the refrigerator in your kitchen.

25. The fuse box or circuit breakers in your house and the other circuits in your house.

26. A Christmas tree light on a string of Christmas tree lights.

27. Street lights on a street.

28. The headlights and ignition switch of a car.

29. The radio and the ignition switch of a car.

PROBLEM SOLVING

30. There are three people who must decide whether or not Candidate A will be hired for a position. This will be done by a secret vote. Design a switching circuit for determining a majority vote.

RESEARCH ACTIVITIES

31. How does the computer make use of logic? Visit the computer center at your school, and ask the person in charge for specific references that can assist you in answering this question.

32. Digital computers use gates that work like switches to perform calculations. Information is fed into the gates and information leaves the gates, according to the type of gate. The three basic gates used in computers are the NOT gate, the AND gate, and the OR gate. Other gates used are combinations of these three gates. Do research on the three types of gates. (a) Explain how each gate works. (b) Explain the relationship between each gate and the corresponding logic connectives *not, and,* and *or.* (c) Illustrate how two or more gates can be combined to form a more complex gate.

DID YOU KNOW?

George Boole was an English logician and mathematician to whom all computer technicians and programmers are indebted for his development of symbolic logic, specifically of the binary logic operators "and" and "or." His *Treatise on Differential Equations* (1859) contained a clear and complete account of the general symbolic method. The contents of that book eventually led him to make other, more important discoveries. Boole did not regard logic as a branch of mathematics but pointed out an analogy between the symbols of algebra and those used in logic. Boole's formalism showed the way to "mechanize" logic by operating on only the digits of 0 and 1. This "mechanized" logic became known as **Boolean algebra** and eventually led to binary switching, the basis of the modern-day computer.

Students often wonder why it is necessary to study abstract mathematics. One reason is that present-day abstract mathematical concepts often lead to important and useful applications at a later time. Who would have believed a hundred years ago that the abstract concepts of Boolean algebra would lead to the development of the computer, which affects each of us daily?

SUMMARY

To understand the symbolic logic developed by Boole and others, it is necessary to be able to distinguish between simple and compound statements. The student should be able to recognize the logical connectives: $\sim$, $\land$, $\lor$, $\rightarrow$, and $\leftrightarrow$. The basic truth tables for these connectives must be learned.

p	$\sim q$
T	F
F	T

p	q	$p \land q$	$p \lor q$	$p \rightarrow q$	$p \leftrightarrow q$
T	T	T	T	T	T
T	F	F	T	F	F
F	T	F	T	T	F
F	F	F	F	T	T

Using the basic truth tables, you should be able to construct more complex truth tables.

If all the cases of a truth table are true, then the statement is a tautology. If every case of the table is false, then the statement is a self-contradiction. A conditional statement that is a tautology is called an implication.

With the help of DeMorgan's laws it is possible to create equivalent statements:

$$\sim(p \wedge q) \Leftrightarrow \sim p \vee \sim q$$
$$\sim(p \vee q) \Leftrightarrow \sim p \wedge \sim q$$

The various forms of the conditional are as follows.

Statement	Symbolic Form
If p then q	$p \rightarrow q$
p only if q	$p \rightarrow q$
q if p	$p \rightarrow q$
p is a sufficient condition for q	$p \rightarrow q$
q is a necessary condition for p	$p \rightarrow q$

The variations of the conditional statement are the inverse, the converse, and the contrapositive. The conditional is equivalent to the contrapositive, and the inverse is equivalent to the converse.

Euler circles can be used to determine whether syllogistic arguments are valid or invalid. Truth tables can be used to determine whether symbolic arguments are valid or invalid. An application of logic is switching circuits.

REVIEW EXERCISES

Write the following statements in words. Let

p: The calculator is correct.
q: I entered the correct number.
r: I touched the wrong keys.

1. $p \wedge q$
2. $\sim p \vee r$
3. $p \rightarrow (q \wedge \sim r)$
4. $p \leftrightarrow q$
5. $(p \vee \sim q) \wedge \sim r$
6. $\sim p \leftrightarrow (r \wedge \sim q)$

Using the translations for p, q, and r given above, write the following statements in symbolic form.

7. I entered the correct number or I did not touch the wrong keys.
8. The calculator is correct and I touched the wrong keys.
9. If I touched the wrong keys, then the calculator is not correct or I entered the correct number.
10. I entered the correct number if and only if the calculator is correct, and I did not touch the wrong keys.
11. I touched the wrong keys and the calculator is correct, or I entered the correct number.
12. It is false that I entered the correct number or I did not touch the wrong keys.

Construct a truth table for each of the following statements.

13. $(p \vee \sim q) \wedge p$
14. $\sim p \wedge \sim q$
15. $\sim p \rightarrow (q \wedge \sim p)$
16. $\sim p \leftrightarrow q$
17. $p \wedge (\sim q \vee r)$
18. $p \rightarrow (q \wedge \sim r)$
19. $(p \vee q) \leftrightarrow (p \vee r)$
20. $(p \wedge q) \rightarrow \sim r$

Find the truth value of the following statements.

21. If the moon is square, then Texas is in the United States.
22. George Boole was a mathematician or Charles Dodgson wrote children's stories.
23. If Aristotle is known for his automobile designs, then Ronald Reagan was a dance teacher.
24. $3 + 7 = 11$ or $6 + 5 = 11$, and $7 \cdot 6 = 42$.
25. French is a language, if and only if $2 + 2 = 7$ or $3 + 5 = 8$.

Find all the values of x that make the following statements true.

26. $x + 3 = 9$ or $6 + 1 = 7$.
27. If $3 = 2 + 2$, then $x + 3 = 8$.
28. $8 + 1 = 6$ if and only if $x + 3 = 10$.
29. $x + 3 = 9$ and $6 + 2 = 10$.
30. $2 + 3 = 7$ or $(x + 5 = 7$ and $3 + 4 = 7)$.

Find the truth values of the following statements when p is T, q is F, r is F, and s is T.

31. $(p \vee q) \rightarrow (\sim r \wedge s)$
32. $(q \rightarrow \sim s) \vee (p \wedge q)$
33. $\sim r \leftrightarrow [(p \vee q) \leftrightarrow \sim s]$
34. $\sim[(q \wedge s) \rightarrow (\sim p \vee r)]$
35. $[\sim(q \wedge s)] \rightarrow \sim(\sim p \vee r)$

Determine whether the following statements are equivalent.

36. $\sim p \rightarrow \sim q$ $\quad p \vee \sim q$
37. $\sim p \vee \sim q$ $\quad \sim p \leftrightarrow q$
38. $\sim p \vee (q \wedge r)$ $\quad (\sim p \vee q) \wedge (\sim p \vee r)$
39. $(\sim q \rightarrow p) \wedge p$ $\quad \sim(\sim p \leftrightarrow q) \vee p$

Write the negation of each of the following statements.

40. All wood is hard.
41. No birds are yellow.
42. Some apples are not red.
43. Some squirrels fly.
44. Not all people wear glasses.

Write an equivalent statement for each of the following.

45. The pencil is not yellow or the desk is orange.
46. The Flyers score goals and the Flyers win games.
47. It is not true that New Orleans is in California or Chicago is not a city.
48. If the water is blue, then the fish will not bite.
49. Write the converse, inverse, and contrapositive of the conditional statement "If I study, then I will get a good grade."

Determine whether the following syllogistic arguments are valid.

50. All dogs bark.
Phoenix is not a dog.

∴ Phoenix does not bark.
51. All cats are green.
Charlie is green.

∴ Charlie is a cat.

Determine whether the following symbolic arguments are valid.

52. He went to the movie or the beach.
He went to the movie.

∴ He did not go to the beach.
53. The book is either black or red.
The book is not black.

∴ The book is red.
*54. Barb will join the group or Barb will not join the group. If Barb does join the group, it will be because she is not in town. Barb is in town. Therefore Barb will join the group.
55. Construct the corresponding circuit diagram for $(p \wedge \sim q) \vee (p \wedge r)$.

56.   Write the corresponding logical statement for the following circuit diagram.

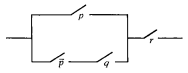

CHAPTER TEST

Write each of the statements in Problems 1–3 in symbolic form, using the following simple statements.

$p:$ The chair is blue.
$q:$ The sofa is red.
$r:$ The teapot is black.

1. The chair is blue and the sofa is red, or the teapot is black.
2. If the sofa is not red, then the teapot is black.
3. It is false that the chair is blue if and only if the sofa is red.

Use p, q, and r as above and write an English sentence for the following symbolic statements.

4. $p \rightarrow (q \lor r)$
5. $\sim(p \leftrightarrow q)$

Construct a truth table for the given statements:

6. $\sim(p \lor q)$

7. $[\sim(p \land r)] \rightarrow \sim q$
8. $q \lor (r \leftrightarrow \sim p)$

Find the truth values for the following statements.

9. $3 + 15 = 18$ or $6 - 11 = 13$
10. If $3 + 5 = 8$ then $6 + 11 \neq 22$.

Given that p is false, q is true, r is false, and s is false, find the truth value of the following statements.

11. $[\sim(p \land r)] \lor (\sim q \rightarrow \sim s)$
12. $(s \land r) \leftrightarrow [\sim(p \lor \sim q)]$

Find all values of x that make the following statements true.

13. $x + 5 = 14$ or $6 + 9 = 15$
14. If $3 + 4 = 7$, then ($x + 5 = 11$ if and only if $2 + 4 = 6$)

Given the conditional statement "If the mouse is gray, then the dog is green."

15. Write the converse of the conditional.
16. Write the contrapositive of the conditional.
17. Write the inverse of the conditional.

Write each of the following statements in the if–then form.

18. The chipmunk is eating dinner only if the sun is shining.
19. The bird having a red head is a sufficient condition for the bird being a woodpecker.
20. Use Euler circles to determine whether the following syllogistic argument is valid or invalid.

    All numbers divisible by 3 are odd.
    The number 14 is divisible by 3.

    ∴ The number 14 is odd.

21. Use truth tables to determine whether the following symbolic argument is valid or invalid.

    If you work hard, then you will get a raise.
    If you get a raise, then you will be able to afford the new computer.

    ∴ If you work hard, then you will be able to afford the new computer.

22. Determine whether the following statements are equivalent.

    $(q \lor \sim r) \rightarrow p$ $[\sim(q \lor r)] \rightarrow p$

Write the negation of the following statements.

23. All jokes are funny.
24. Some test questions are easy.

25. Construct the corresponding circuit diagram for $(p \land r) \lor (\sim q \land r)$.

CHAPTER 4

Systems of
Numeration

SYSTEMS OF NUMERATION

Just as the first attempts to write were made long after the development of speech, the first representation of numbers by symbols came long after people had learned to count. Probably the earliest method of recording the number of animals in a herd was a tally system using physical objects, such as scratch marks in the soil, scratches on a stone, notches on a stick, or knots on a vine.

In ancient times such a tally system adequately served the limited need for recording whatever was counted. As civilization developed, however, more efficient and accurate methods of calculating and keeping records were needed. Because scratch marks and notches were impractical and inefficient, societies developed symbols to replace them. For example, the Egyptians used the symbol ∩ and the Babylonians used the symbol ⊲ to represent the number we symbolize by 10.

A **number** is a quantity, and it answers the question "How many?" A **numeral** is the symbol used to represent the number. One thinks a number but writes a numeral. Symbols such as ∩, ⊲, and 10 are numerals. The distinction between number and numeral will be made here only if it is helpful to the discussion.

In language, relatively few letters are used to construct a large number of words. Similarly, in arithmetic a small variety of numerals can be used to represent all numbers. One of the greatest accomplishments of humankind has been the development of systems of numeration, whereby all numbers are "created" from a few symbols. Without such systems, arithmetic would not have developed.

> A **system of numeration** consists of a set of numerals and a scheme or rule for combining the numerals to represent numbers.

Very few people give any thought to the development of our present-day *Hindu-Arabic (decimal)* system of numeration. In this section we will discuss four types of numeration systems—additive (or repetitive), multiplicative, ciphered, and place-value systems. You will be introduced to several numeration systems used by different cultures. It is not necessary for you to memorize all the symbols, but you should understand the principles behind each system. By the end of this chapter, we hope you will see the advantages of our present-day place-value system over the other types of systems and will gain a better respect for and understanding of the Hindu-Arabic system of numeration.

Additive Systems

The additive system of numeration is one of the oldest and most primitive types of numeration systems. One of the first additive systems, dating back to about 3000 B.C., was the Egyptian hieroglyphic system. There were symbols for the powers of $10 - 10^0$ or 1; 10^1 or 10; 10^2 or 10×10; 10^3 or $10 \times 10 \times 10$; and so on. The Egyptian hieroglyphic numerals and the equivalent Hindu-Arabic numerals are illustrated in Table 4.1.

TABLE 4.1

Egyptian Hieroglyphics

Hindu-Arabic Numerals	Egyptian Numerals	Description
1	|	Staff (vertical stroke)
10	∩	Heel bone (arch)
100	୨	Scroll (coiled rope)
1,000	℥	Lotus flower
10,000	⟍	·Pointing finger
100,000	∝◁	Tadpole (or whale)
1,000,000	⚡	Astonished person

The Egyptian system is based on the additive principle. The number represented by a particular set of numerals is simply the sum of the values of the numerals. Thus to write the number 600 in Egyptian hieroglyphics, one writes the numeral for 100 six times.

■ **Example 1** ————————————————————————————

Write the following numeral as a Hindu-Arabic numeral.

$$\text{⟍⟍ ୨୨୨ }|$$

Solution

$$10{,}000 + 10{,}000 + 100 + 100 + 100 + 1 = 20{,}301 \qquad ❖$$

■ **Example 2** ————————————————————————

Write 43,628 as an Egyptian numeral.

Solution

$$43,628 = 40,000 + 3,000 + 600 + 20 + 8$$

〰〰〰〰 ⌡⌡⌡ 999999 ∩∩ |||||||| ❖

In this system the order of the symbols is not important. For example, ∝⌒|| 99∩ and ||99∝∩ both represent 100,212.

Addition and subtraction were easily accomplished; symbols were simply combined or taken away. Multiplication and division were difficult; they were performed by a process called duplation and mediation, which we will discuss in Section 4.5. The Egyptians had no symbol for zero, but they did have an understanding of fractions. The symbol ⌒ was used to take the reciprocal of a number; thus ⌒ℛ meant 1/3, and ⌒∩⃒ was 1/11. In this system as many as 45 symbols could be required to express a single number from 1 to 100,000.

A second example of an additive system, the Roman numeration system, was developed later than the Egyptian system, but it still dates back to the time of the ancient Romans. Roman numerals were used in most European countries until the eighteenth century, and they are still commonly seen on buildings, on clocks, and in books. Roman numerals were selected letters of the Roman alphabet. They are illustrated in Table 4.2.

TABLE 4.2						
Roman numerals	I	V	X	L	C	D	M
Hindu-Arabic Numerals	1	5	10	50	100	500	1000

The Roman system has two advantages over the Egyptian system. The first is that it uses the subtraction principle as well as the addition principle. Starting from the left, we add each numeral unless its value is smaller than the value of the numeral to its right, in which case we subtract it from that numeral. Only the numbers 1, 10, 100, . . . can be subtracted, and only from the next two higher numbers. For example, C (100) can be subtracted only from D (500) or M (1000). The symbol DC represents 500 + 100, or 600, and CD represents 500 − 100, or 400. Similarly, MC represents 1000 + 100, or 1100, and CM represents 1000 − 100, or 900.

DID YOU KNOW?

Most of our knowledge of the Egyptian numeration system comes from two sources. One is the Moscow Papyrus, written about 1850 B.C. The other is the Rhind Papyrus (Fig. 4.1). While Henry Rhind was on vacation in Egypt in 1858, he accidentally discovered the papyrus in a shop in Luxor and purchased it. When he died, the document was transferred to the British Museum. The roll, which originally measured 18 feet by 13 inches, included several "books," but a number of fragments were missing when it was found. Many of the fragments were later discovered in the possession of the New York Historical Society. The papyrus is a collection of 85 problems copied by the scribe Ahmes. Its introduction explains that the papyrus is a copy of earlier writing from the time of Nemaet Re. This information places the writing of the earlier works in the latter half of the nineteenth century B.C.

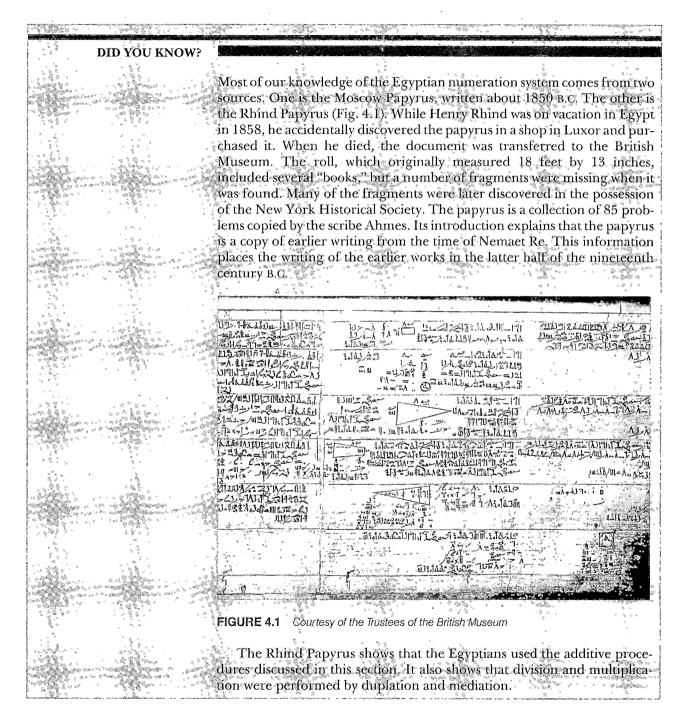

FIGURE 4.1 *Courtesy of the Trustees of the British Museum*

The Rhind Papyrus shows that the Egyptians used the additive procedures discussed in this section. It also shows that division and multiplication were performed by duplation and mediation.

Example 3

Write CLXII as a Hindu-Arabic numeral.

Solution: Since each numeral is larger than the one on its right, no subtraction is necessary.

$$CLXII = 100 + 50 + 10 + 1 + 1 = 162.$$

Example 4

Write DCXLVI as a Hindu-Arabic numeral.

Solution: Checking from left to right, we see that X (10) has a smaller value than L (50). Therefore XL represents $50 - 10$, or 40.

$$DCXLVI = 500 + 100 + (50 - 10) + 5 + 1 = 646$$

Example 5

Write 289 as a Roman numeral.

Solution

$$289 = 200 + 80 + 9 = 100 + 100 + 50 + 10 + 10 + 10 + 9$$

(Nine is not broken down to $5 + 1 + 1 + 1 + 1$ because it will be treated as $10 - 1$.)

$$289 = CCLXXXIX$$

*Photo by
Allen R. Angel*

A second advantage of the Roman numeration system over the Egyptian system is that it makes use of the multiplication principle for numbers over 1000. Placing a bar above the symbol or a group of symbols has the effect of multiplying by the number 1000. Thus $\overline{V} = 5 \times 1000 = 5000$, $\overline{X} = 10 \times 1,000 = 10,000$, and $\overline{CD} = 400 \times 1000 = 400,000$. This greatly reduces the number of symbols needed to write large numbers. Still, as many as 20 symbols may be required in this system to express a single number between 1 and 100,000.

Multiplicative Systems

Multiplicative numeration systems are more similar than the additive systems to our Hindu-Arabic system. The number 642 in a multiplicative system might be written (6) (100) (4) (10) (2) or

$$
\begin{array}{c}
6 \\
100 \\
4 \\
10 \\
2
\end{array}
$$

Note that no addition signs are needed to represent the number. From this illustration, try to formulate a rule explaining how multiplicative systems work.

The principal example of a multiplicative system is the traditional Chinese system, the numerals of which are given in Table 4.3.

TABLE 4.3

Traditional Chinese numerals	一	二	三	四	五	六	七	八	九	十	百	千
Hindu-Arabic Numerals	1	2	3	4	5	6	7	8	9	10	100	1000

Chinese numerals are always written vertically. The number on top will be a number from 1 to 9 inclusive. This number is to be multiplied by the power of 10 below it. The number 20 is written

$$
\begin{array}{c}
二 \\
十
\end{array}
$$

which means $2 \times 10 = 20$. The number 400 is written

$$
\begin{array}{c}
四 \\
百
\end{array}
$$

which means 4×100.

■ **Example 6** _____

Write 428 as a Chinese numeral.

Solution

$$428 = \begin{cases} 400 = & \begin{cases} 4 & 四 \\ 100 & 百 \end{cases} \\ \\ 20 = & \begin{cases} 2 & 二 \\ 10 & 十 \end{cases} \\ \\ 8 = & \quad 8 \quad 八 \end{cases}$$

❖

Have you noticed that there is no symbol for zero in this system? Why is a symbol for zero not needed?

The symbols used in the modern National Chinese system and the mercantile Chinese system are both different from the traditional system. Both of these systems of numeration are positional-value systems rather than multiplicative systems, and both use the symbol 0 for zero.

Ciphered Systems

Ciphered numeration systems require the memorization of many different symbols but have the advantage that numbers can be written in a compact form. The ciphered numeration system that we will discuss is the Ionic Greek (Table 4.4). Others include the Hebrew, Coptic, Hindu, Brahmin, Syrian, Egyptian heiratic, and early Arabic. The Ionic Greek used letters of their alphabets for numerals. The Ionic Greek numeration system was developed around 3000 B.C.

TABLE 4.4
Ionic Greek Numerals

1	α	alpha	60	ξ	xi
2	β	beta	70	o	omicron
3	γ	gamma	80	π	pi
4	δ	delta	90	Q	koph (obsolete)
5	ε	epsilon	100	ρ	rho
6	Ϛ	vau (obsolete)	200	σ	sigma
7	ζ	zeta	300	τ	tau
8	η	eta	400	υ	upsilon
9	θ	theta	500	φ	phi
10	ι	iota	600	χ	chi
20	κ	kappa	700	ψ	psi
30	λ	lambda	800	ω	omega
40	μ	mu	900	ϡ	sampi (obsolete)
50	ν	nu			

The Greek alphabet contains 24 letters. Since 27 symbols were needed, three had to be added. Instead of creating new symbols, the Greeks borrowed the symbols ⌡,Ϙ,ϖ from the Phoenician alphabet.

When a prime (′) is placed above a number, it multiplies that number by 1000. For example,

$$\beta' = 2 \times 1000 = 2000,$$
$$\sigma' = 200 \times 1000 = 200{,}000.$$

The number $24 = 20 + 4$; when it is written as a Greek numeral, the plus sign is omitted:

$$24 = \kappa\delta.$$

■ **Example 7** _____

Write $\chi\,\nu\,\gamma$ as a Hindu-Arabic numeral.

Solution: $\chi = 600$, $\nu = 50$, and $\gamma = 3$. Adding these numbers gives 653. ❖

■ **Example 8** _____

Write 8652 as an Ionic Greek numeral.

Solution:
$$
\begin{aligned}
8652 &= 8000 + 600 + 50 + 2 \\
&= (8 \times 1000) + 600 + 50 + 2 \\
&= \eta' \qquad\qquad \chi \quad \nu \quad \beta \\
&= \eta'\chi\nu\beta
\end{aligned}
$$
❖

EXERCISES 4.1

Write each of the following as a Hindu-Arabic numeral.

1. ◌◌∩|

2. ∩∩◌|

3. ⌁⌁◌◌◌∩|∩||

4. ⟩⟩⟩∩||||

5. ⊠⟩⟩⟩⌁⌁⌁◌◌∩||||
   ⊠⟩⟩

6. ⚥⚥⊠⊠◌◌∩∩∩|
   ⊠◌◌

Write each of the following as an Egyptian numeral.

7. 426
8. 218
9. 1234

10. 1492
11. 162,451
12. 1,342,567

Write each of the following as a Hindu-Arabic numeral.

13. LXIV
14. XLIX
15. CXVII
16. DCLXIV
17. MDCCCXCII
18. MCMLXVI
19. MCMXCIX
20. MMCDXLIV
21. $\overline{\text{V}}$MMDCII
22. $\overline{\text{IV}}$DLV
23. $\overline{\text{X}}$DCLXXIX
24. $\overline{\text{IX}}$DCCCXLII

Write each of the following as a Roman numeral.

25. 38
26. 95
27. 148
28. 397
29. 1978
30. 2435
31. 2606
32. 9436
33. 4321
34. 15,954

Write each of the following as a Hindu-Arabic numeral.

35. 五
十
七

36. 二
百
八
十

37. 一
千
九
百
七
十
六

38. 三
千
一
十
七

39. 二
千
六
百
五

40. 三
千
四
百
八
十
七

Write each of the following as a traditional Chinese numeral.

41. 42
42. 156
43. 248
44. 2483
45. 9007
46. 1999

Write each of the following as a Hindu-Arabic numeral.

47. τξδ
48. χοη
49. μ′β′φε
50. ρ′ν′ωιγ
51. θ′ζ
52. α′τϜΟθ

Write each of the following as an Ionic Greek numeral.

53. 3
54. 78
55. 123
56. 2076
57. 35,412
58. 102,685
59. Construct your own additive numeration system. You can use the subtraction and multiplication properties if you wish. Write 1988 in your system.

60. Construct your own multiplicative numeration system. Write 1988 in your system.

61. Construct your own ciphered numeration system. Write 1988 in your system.

List the advantages and disadvantages of a ciphered system of numeration compared with each of the following systems.

62. An additive system

63. A multiplicative system

64. The Hindu-Arabic system

Write each of the following as numerals in the indicated systems of numeration.

65. ∩ ∩ ❘ into Hindu-Arabic, Roman, Chinese, and Greek.

66. MCMXXXVI into Hindu-Arabic, Greek, and Chinese.

67.
五
百
二
十
六
into Hindu-Arabic, Egyptian, Roman, and Greek.

68. υκβ into Hindu-Arabic, Egyptian, Roman, and Chinese.

4.2 PLACE-VALUE OR POSITIONAL VALUE NUMERATION SYSTEMS

Today the most common type of numeration system is the place-value system. The Hindu-Arabic numeration system, which we use, is an example of this type. In place-value systems the value of the symbol depends on its position in the representation of the number. For example, the 2 in 20 represents 2 tens, and the 2 in 200 represents 2 hundreds. A true positional-value system requires a base and a set of symbols—a symbol for zero and one for each counting number less than the base. In addition, a rule is required that explains how the symbols are combined to represent numbers. Although any number can be written in any base, the most common positional system is the base 10 system (the decimal number system).

In the Hindu-Arabic system the symbols 0, 1, 2, 3, 4, 5, 6, 7, 8, and 9 are called **digits.** The positional values in the Hindu-Arabic system are . . . , $(10)^5$, $(10)^4$, $(10)^3$, $(10)^2$, 10, 1. In the Hindu-Arabic system the first digit to the left of the decimal point is multiplied by 1. The second digit to the left of the decimal point is multiplied by the base 10. The third digit to the left is multiplied by the base squared 10^2 or 100. The fourth digit to the left is multiplied by the based cubed, 10^3 or 1000, and so on. In general, the digit that is n places to the left of the decimal point is multiplied by 10^{n-1}. Therefore the digit eight places to the left of the decimal point is multiplied by 10^7. Using the place value rule, we can write a number in **expanded form.** The number 1234. written in expanded form is

$$(1 \times 10^3) + (2 \times 10^2) + (3 \times 10^1) + (4 \times 1)$$
$$(1 \times 1000) + (2 \times 100) + (3 \times 10) + 4$$

TABLE 4.5

Babylonian numerals	∀	◁
Hindu-Arabic Numerals	1	10

When a whole number is written without a decimal point, it is assumed that the decimal point is to the right of the number. Thus 1234. may be written 1234 (the decimal point is assumed).

The oldest known numeration system that resembled a place-value system was developed by the Babylonians around 2500 B.C. Their system resembled a place-value system with a base of 60, a sexagesimal system. It was not a true place-value system because it lacked a symbol for zero. The lack of a symbol for zero led to a great deal of ambiguity and confusion. Table 4.5 gives the Babylonian numerals.

The positional values in the Babylonian system are

$$\dots (60)^3, (60)^2, 60, 1.$$

In a Babylonian numeral a wider gap was left between the characters to distinguish between the various place values. In reading from right to left the sum of the first group of numerals is multiplied by 1. The sum of the second group is multiplied by 60. The sum of the third group is multiplied by $(60)^2$, and so on.

■ **Example 1** _____

Write the following as a Hindu-Arabic numeral.

◁◁ ◁◁∀∀∀∀

Solution

$$\underbrace{◁◁}_{60's} \qquad \underbrace{◁◁∀∀∀∀}_{units}$$

$$\underbrace{10 + 10}_{60} \qquad \underbrace{10 + 10 + 1 + 1 + 1 + 1}_{units}$$

$$(20 \times 60) + (24 \times 1)$$

$$1200 + 24 = 1224 \qquad ❖$$

The Babylonians used the symbol ▷∀ to indicate subtraction. The numeral

◁▷∀∀∀ represents 10 − 2, or 8. The numeral

◁◁◁ ∀∀∀∀∀ ▷∀◁∀∀

represents 35 − 12, or 23 in decimal notation.

■   **Example 2** _____

Write ∀   ◁∀   ◁◁ ∀̷ ∀∀   as a Hindu-Arabic numeral.

Solution: There are three groups of numerals. The place value of these groups, from right to left, is 1, 60, and 60^2. The numeral in the group on the right has a value of $20 - 2$, or 18. The numeral in the center has a value of $10 + 1$, or 11. The numeral on the left represents 1.

$$(1 \times 60^2) + (11 \times 60) + (18 \times 1)$$
$$= (1 \times 3600) + (11 \times 60) + (18 \times 1)$$
$$= 3600 + 660 + 18$$
$$= 4278$$ ❖

To explain the procedure used to convert from a decimal number to a Babylonian number, we will consider a length of time, 9820 seconds. How do we change this to a more useful measurement? This amount of time may be more understandable if it is converted to hours, minutes, and seconds. Since there are 3600 seconds in an hour (60 seconds to a minute and 60 minutes to an hour), we can find the number of hours by dividing 9820 by 60^2, or 3600.

$$\begin{array}{r} 2 \leftarrow \text{hours} \\ 3600 \overline{\smash{)}9820} \\ \underline{7200} \\ 2620 \leftarrow \text{remaining seconds} \end{array}$$

Now we can determine the number of minutes by dividing the remaining seconds by 60, the number of seconds in a minute.

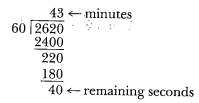

$$\begin{array}{r} 43 \leftarrow \text{minutes} \\ 60 \overline{\smash{)}2620} \\ \underline{2400} \\ 220 \\ \underline{180} \\ 40 \leftarrow \text{remaining seconds} \end{array}$$

Since the remaining number of seconds, 40, is less than the number of seconds in a minute, our task is complete.

9820 seconds = 2 hours, 43 minutes, and 40 seconds

The same procedure is used to convert a decimal (base 10) number to a Babylonian number or any number in a different base.

■ **Example 3** ————————————————————————————

Write 1602 as a Babylonian numeral.

Solution: The Babylonian numeration system has positional values of

$$\ldots, 60^3, 60^2, 60, 1,$$

which can be expressed as

$$\ldots, 216{,}000, 3600, 60, 1.$$

The largest positional value less than or equal to 1602 is 60. To determine how many groups of 60 there are in 1602, divide 1602 by 60:

$$
\begin{array}{r}
26 \\
60\,\overline{\smash)\,1602} \\
\underline{120} \\
402 \\
\underline{360} \\
42
\end{array}
$$

There are 26 groups of 60 and 42 units remaining. Since the remainder, 42, is less than the base, 60, no further division is necessary. The remainder represents the number of units when the number is written in expanded form. Therefore $1602 = (26 \times 60) + (42 \times 1)$. When written as a Babylonian numeral, 1602 is

⟨⟨ ∀ ∀ ∀ ∀ ∀   ⟨⟨⟨⟨ ∀ ∀   ❖

■ **Example 4** ————————————————————————————

Write 6270 as a Babylonian numeral.

Solution: The largest positional value less than or equal to 6270 is 60^2 or 3600. Divide 6270 by 3600:

$$
\begin{array}{r}
1 \\
3600\,\overline{\smash)\,6270} \\
\underline{3600} \\
2670
\end{array}
$$

There is one group of 3600 in 6270. Next divide 2670 by 60 to determine the number of groups of 60 in 2670:

$$
\begin{array}{r}
44 \\
60\,\overline{\smash)\,2670} \\
\underline{240} \\
270 \\
\underline{240} \\
30
\end{array}
$$

There are 44 groups of 60 and 30 units remaining.

$$6270 = (1 \times 60^2) + (44 \times 60) + (30 \times 1)$$

Thus 6270 written as a Babylonian numeral is

$$\nabla \quad \triangleleft\triangleleft\triangleleft\triangleleft \nabla\nabla\nabla\nabla \quad \triangleleft\triangleleft\triangleleft \qquad \clubsuit$$

Another place-value system is the Mayan numeration system. This system was developed, complete with zero, at an early but unknown date. The numbers in this system are written vertically rather than horizontally, with the units position on the bottom. In the Mayan system the number in the bottom row is to be multiplied by 1. The number in the second row from the bottom is to be multiplied by 20. The number in the third row is to be multiplied by 18×20, or 360. You probably expected the third row to be multiplied by 20^2 rather than 18×20. It is believed that the Mayans used 18×20 so that their numeration system would conform to their calendar of 360 days. The positional values above 18×20 are 18×20^2, 18×20^3, and so on. The digits $0, 1, 2, 3, \ldots, 19$ of the Mayan system are formed by a simple grouping of dots and lines, as shown in Table 4.6.

TABLE 4.6

Mayan Numerals

0	1	2	3	4	5	6	7	8	9
⬭	•	••	•••	••••	—	•⎯	••⎯	•••⎯	••••⎯

10	11	12	13	14	15	16	17	18	19
═	•═	••═	•••═	••••═	≡	•≡	••≡	•••≡	••••≡

■ **Example 5** _____

Write $\begin{array}{c}\overline{\bullet\bullet}\\ \bullet\bullet\\ \bullet\bullet\bullet\\ \equiv\end{array}$ as a Hindu-Arabic numeral.

Solution: In the Mayan numeration system the first three positional values are: 18×20
20
1

$$\begin{array}{lll}\overline{\bullet\bullet} = & 7 \times (18 \times 20) = & 2520 \\ \bullet\bullet = & 2 \times 20 = & 40 \\ \underline{\bullet\bullet\bullet} = & 13 \times 1 = & \underline{13} \\ & & 2573\end{array}$$

$\clubsuit$

■ **Example 6**

Write ⋮ (• • • over • (with bar) over • • • •) as a Hindu-Arabic numeral.

Solution:

$$\bullet\bullet\bullet \quad = \quad 8 \times (18 \times 20) = 2880$$

$$\underline{\bullet} \quad = \quad 11 \times 20 \qquad = \quad 220$$

$$\bullet\bullet\bullet\bullet \quad = \quad 4 \times 1 \qquad = \quad \underline{\quad 4\quad}$$

$$3104$$

❖

■ **Example 7**

Write 1023 as a Mayan numeral.

Solution: To convert from a Hindu-Arabic to a Mayan numeral, we use a procedure similar to the one used to convert to a Babylonian numeral. The Mayan positional values are 18×20, 20, 1. The greatest positional value less than or equal to 1023 is 18×20, or 360. Divide 1023 by 360:

$$\begin{array}{r} 2 \\ 360\,\overline{\smash)1023} \\ \underline{720} \\ 303 \end{array}$$

There are two groups of 18×20 in 1023. Next divide the remainder, 303, by 20:

$$\begin{array}{r} 15 \\ 20\,\overline{\smash)303} \\ \underline{20} \\ 103 \\ \underline{100} \\ 3 \end{array}$$

There are 15 groups of 20 with 3 units remaining.

$$1023 = 2 \times (18 \times 20) + (15 \times 20) + (3 \times 1)$$

1023 written as a Mayan numeral is

$$\left\{ \begin{array}{l} 2 \times (18 \times 20) \\ 15 \times 20 \\ 3 \times 1 \end{array} \right\} = \begin{array}{c} \bullet\bullet \\ \equiv \\ \bullet\bullet\bullet \end{array}$$

❖

DID YOU KNOW?

The Hindus are credited with the invention of zero. There is evidence that zero may have been known before the birth of Christ, but no inscription has been found with such a symbol before the ninth century. The symbols used in our system originated in India and were widely used by the Arabs, since they traded regularly with the Hindus—thus the name Hindu-Arabic. However, it was not until the middle of the fifteenth century that the Hindu-Arabic numerals were in the form we know today. The introduction of the Hindu-Arabic numerals and the positional system of numeration revolutionized arithmetic and eventually mathematics. Place-value numeration systems made addition, subtraction, multiplication, and division much easier to learn and very practical to use. Merchants and traders no longer had to depend on the counting board or abacus. A new generation of mathematicians, who computed with the Hindu-Arabic system rather than with pebbles or beads on a wire, were known as the "algorists."

Example 8

Write 3082 as a Mayan numeral.

Solution: The highest positional value less than or equal to 3082 is 18×20, or 360. Divide 3082 by 360:

$$
\begin{array}{r}
8 \\
360 \overline{\smash{)}3082} \\
\underline{2880} \\
202
\end{array}
$$

$$
\begin{array}{r}
10 \\
20 \overline{\smash{)}202} \\
\underline{20} \\
2 \\
\underline{0} \\
2
\end{array}
$$

There are 8 groups of 18×20, 10 groups of 20, and 2 units remaining.

$$
3082 = \left\{ \begin{array}{c} 8 \times (18 \times 20) \\ 10 \times 20 \\ 2 \times 1 \end{array} \right\} =
$$
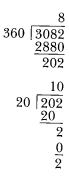

❖

DID YOU KNOW?

For the past 600 years, no significant changes have taken place in the Hindu-Arabic system of numeration. Is there perhaps a simpler, better, more efficient system of numeration than the present one? Have we reached such a level of perfection that there is no longer any possibility of improvement? Where would our numeration system be today if our ancestors had been complacent and not tried to improve the numeration systems of their day?

EXERCISES 4.2

Write the Hindu-Arabic numeral in expanded form.

1. 48
2. 96
3. 942
4. 765
5. 3,452
6. 1,265
7. 64,521
8. 32,687
9. 245,672
10. 148,562

Write each of the following as a Hindu-Arabic numeral.

11. ◁◁◁ ∀∀

12. ◁◁◁◁ ▷∀∀∀∀

13. ◁∀∀ ∀∀∀∀

14. ◁∀ ◁◁▷∀∀

15. ∀ ◁◁◁∀∀ ◁▷∀∀

16. ◁ ◁◁▷∀∀∀ ∀∀

Write each of the following as a Babylonian numeral.

17. 70   18. 120
19. 121   20. 512
21. 3878   22. 3000

Write each of the following as a Hindu-Arabic numeral.

23. •••
    ═══

24. ═══
    ───

25. ••••
    ═══
    ◯
    •

26. ••
    ──
    ••
    ──
    ••

27. •
    ═══
    ••
    ◯

28. ••••
    ═══
    ───

Write each of the following as a Mayan numeral.

29. 15
30. 227
31. 300
32. 406
33. 3060
34. 1978
35. The Babylonians lacked a symbol for zero. Why did this fact lead to confusion?
36. What are the advantages and disadvantages of a place-value system compared with (a) additive numeration systems, (b) multiplicative numeration systems, (c) ciphered numeration systems?

RESEARCH ACTIVITIES

40. Investigate and write a report on the development of the Hindu-Arabic system of numeration. Start with the earliest records of this system in India. Possible sources: *World of Mathematics* by J. R.

37. Create your own place-value system. Write 1988 in your system.

Write each of the following as numerals in the indicated systems of numeration.

38. ◁ ◁ ◁ ▽ ▽ ▽ into Hindu-Arabic and into Mayan.

39. ▬ into Hindu-Arabic and Babylonian.
∙∙
▬
∙∙∙∙

Newman; *A History of Mathematics* by Carl B. Boyer.; *A Introduction to the History of Mathematics* by Howard Eves.

4.3 OTHER BASES

The positional values in the Hindu-Arabic numeration system from right to left are 1, 10, 10^2, 10^3, 10^4, ... The positional values in the Babylonian numeration system from right to left are 1, 60, 60^2, 60^3, 60^4, ... The numbers 10 and 60 are called **bases** in the Hindu-Arabic and Babylonian systems, respectively. In general, any positional-value numeration system with base b has positional values from right to left of 1, b, b^2, b^3, b^4, ... The positional values from right to left can also be represented as b^0, b^1, b^2, b^3, b^4, . . . , because any number (except 0) to the zero power equals 1—that is, $b^0 = 1$—and any number to the first power equals the number itself—that is, $b^1 = b$.

$$
\begin{array}{llll}
\ldots & b^3 & b^2 & b^1 & b^0 \\
\ldots & b^3 & b^2 & b & 1
\end{array} \Big\} \text{ base } b \text{ positional values}
$$

$$
\begin{array}{lllll}
\ldots & 10^3 & 10^2 & 10 & 1 & \text{base 10 positional values} \\
\ldots & 8^3 & 8^2 & 8 & 1 & \text{base 8 positional values} \\
\ldots & 5^3 & 5^2 & 5 & 1 & \text{base 5 positional values} \\
\ldots & 2^3 & 2^2 & 2 & 1 & \text{base 2 positional values}
\end{array}
$$

As was indicated earlier, the Mayan numeration system is based on the number 20. However, it is not a true base 20 positional-value system. Why?

Photo by Marshall Henrichs

Human beings have ten fingers. This fact is believed to be responsible for the almost universal acceptance of base 10 numeration systems. However, there are still some positional-value numeration systems that use bases other than 10. Some societies still use base 2 numeration systems. They include many tribes in Australia and New Guinea, some African Pygmies, and various tribes in South America. Most computers perform their calculations using binary, or base 2, notation. Some tribes of Tierra del Fuego and the South American continent still use number systems with bases 3 and 4. Base 5 number systems are very old, but the only base 5 system in pure form at present seems to be the one used in Saraveca, a South American Arawakan language. Elsewhere, base 5 systems are combined with base 10 or base 20 systems. The pure base 6 system occurs only sparsely in Northwest Africa. Base 6 occurs in other systems in combination with base 12, the *duodecimal system*. We continue to see remains of other base systems in many countries. For example, there are 12 inches in a foot, 12 months in a year. Base 12 is also evident in the dozen, the 24-hour day, and the gross (12 × 12). In English the word "score" means 20, and other traces are found in pre-English Celtic, Gaelic, Danish, and Welsh. Remains of base 60 are found in measurements of time (60 seconds to a minute, 60 minutes to an hour) and angles (60 seconds to one minute, 60 minutes to one degree).

In recent years the base 2, or binary, number system has become very important. The mode of operation of the computer is basically that of a binary system. The computer uses a two-digit "alphabet," which consists of the numerals 0 and 1. Every number and word can be represented by a combination of those two numerals. A single numeral such as 0 or 1 is called a **bit.** Other bases that computers make use of are base 8 and base 16. A group of four bits is called a **nibble,** a group of eight or sixteen bits is called a **byte.** A byte may represent a word or symbol. For example in the ASCII code (American Standard Code for Information Interchange), used by most computers, the byte 01000001 represents capital A, 01100001 represents a lowercase a, 00110000 represents the number 0, and 00110001 represents the number 1.

A place-value system with base b must have b distinct symbols, one for zero and one for each number less than the base. A base 6 system must have symbols for the numbers 0, 1, 2, 3, 4, 5. A base 12 system must have symbols for 0, 1, 2, 3, ... , 11. A number in a base other than 10 will be indicated by a subscript to the right of the number. Thus 123_5 represents a number in base 5. The value of 123_5 is not the same as the value of 123_{10}. A number in base 10 may be written without a subscript, but for clarity in certain problems we will use the subscript 10 to indicate a number in base 10. It is important to remember that there is no single symbol for the base number itself. For example, in base 5 the only symbols are those for the numbers 0, 1, 2, 3, and 4, and all other numbers are constructed from these five.

To change a number in a base other than 10 to a base 10 number, we follow the same procedure we used in Section 4.2 to change the Babylonian and Mayan numbers to base 10 numbers. Multiply each digit in the number by its respective positional value. Then find the sum of the products.

■ **Example 1** _____

Convert 234_6 to base 10.

Solution: In base 6 the positional values are . . . 6^3, 6^2, 6, 1. In expanded form,

$$234_6 = (2 \times 6^2) + (3 \times 6) + (4 \times 1)$$
$$= (2 \times 36) + (3 \times 6) + (4 \times 1)$$
$$= \quad 72 \quad + \quad 18 \quad + \quad 4$$
$$= 94 \qquad \qquad ❖$$

■ **Example 2** _____

Convert 3615_8 to base 10.

Solution

$$3615_8 = (3 \times 8^3) \quad + \quad (6 \times 8^2) + (1 \times 8) + (5 \times 1)$$
$$= (3 \times 512) + (6 \times 64) + (1 \times 8) + (5 \times 1)$$
$$= \quad 1536 \quad + \quad 384 \quad + \quad 8 \quad + \quad 5$$
$$= 1933 \qquad \qquad ❖$$

A base 12 system must have 12 distinct symbols. This text will use the symbols 0, 1, 2, 3, 4, 5, 6, 7, 8, 9, T, E, where T represents ten and E represents eleven. Why will the numerals 10 and 11 have different meanings than we are used to in the base 12 system?

DID YOU KNOW?

Even today, with all our knowledge, it would be rash to think that we have settled on our final counting system.

Some aborigines counted by twos using their hands. Later, increasing in sophistication, we learned to count by tens, using all of our fingers. Today, computers and robots, which simulate human intelligence, count by twos, the binary system. Will we return to the binary system in order to communicate with robots and computers in the future?

Will the binary system with its elemental simplicity be the universal language of mathematics? Perhaps the first comprehensible message from outer space will be a "beep-beep, beep-beep," a method of communication using the binary system.

Example 3

Convert $12T6_{12}$ to base 10.

Solution

$$
\begin{aligned}
12T6_{12} &= (1 \times 12^3) &+ (2 \times 12^2) &+ (T \times 12) &+ (6 \times 1) \\
&= (1 \times 1728) &+ (2 \times 144) &+ (10 \times 12) &+ (6 \times 1) \\
&= 1728 &+ 288 &+ 120 &+ 6 \\
&= 2142
\end{aligned}
$$

Example 4

Convert 101101_2 to base 10.

Solution

$$
\begin{aligned}
&(1 \times 2^5) + (0 \times 2^4) + (1 \times 2^3) + (1 \times 2^2) + (0 \times 2) + (1 \times 1) \\
&= 32 + 0 + 8 + 4 + 0 + 1 \\
&= 45
\end{aligned}
$$

To change from a base 10 number to a number in a different base, we will use the procedure explained in Section 4.2. Divide the number by the highest power of the base less than or equal to the given number. Record this quotient. Then divide the remainder by the next smaller power of the base and record this quotient. Repeat this procedure until the remainder is a number less than the base. The answer is the set of quotients listed from left to right, with the remainder on the far right. This procedure is illustrated in the following examples.

■ **Example 5** ─────────────────────────────

Convert 406 to base 8.

Solution: The positional values in the base 8 system are . . . , 8^3, 8^2, 8, 1, or . . . , 512, 64, 8, 1. The highest power of 8 that is less than or equal to 406 is 8^2, or 64. Divide 406 by 64.

$$\begin{array}{r} 6 \leftarrow \text{first digit in answer} \\ 64\overline{)406} \\ \underline{384} \\ 22 \end{array}$$

Therefore there are six groups of 8^2 in 406. Next divide the remainder, 22, by 8.

$$\begin{array}{r} 2 \leftarrow \text{second digit in answer} \\ 8\overline{)22} \\ \underline{16} \\ 6 \leftarrow \text{third digit in answer} \end{array}$$

There are two groups of 8 in 22 and 6 units remaining. Since the remainder, 6, is less than the base, 8, no further division is required.

$$\begin{aligned} 406 &= (6 \times 64) + (2 \times 8) + (6 \times 1) \\ &= (6 \times 8^2) + (2 \times 8) + (6 \times 1) \\ &= 626_8 \end{aligned}$$ ❖

■ **Example 6** ─────────────────────────────

Convert 273 to base 3.

Solution: The place values in 3 are . . . , 3^6, 3^5, 3^4, 3^3, 3^2, 3, 1, or . . . , 729, 243, 81, 27, 9, 3, 1. The highest power of the base that is less than 273 is 3^5, or 243. Successive divisions by the powers of the base give

$$\begin{array}{ccccc} 1 & 0 & 1 & 0 & 1 \\ 243\overline{)273} & 81\overline{)30} & 27\overline{)30} & 9\overline{)3} & 3\overline{)3.} \\ \underline{243} & \underline{00} & \underline{27} & \underline{0} & \underline{3} \\ 30 & 30 & 3 & 3 & 0 \end{array}$$

To obtain the answer read the quotients from left to right followed by the remainder in the last division.

Since the remainder, 0, is less than the base, 3, no further division is necessary. The number 273 can be represented as one group of 243, no group of 81, one group of 27, no group of 9, one group of 3, and no units.

$$273 = (1 \times 243) + (0 \times 81) + (1 \times 27) + (0 \times 9) + (1 \times 3) + (0 \times 1)$$
$$= (1 \times 3^5) + (0 \times 3^4) + (1 \times 3^3) + (0 \times 3^2) + (1 \times 3) + (0 \times 1)$$
$$= 101010_3 \qquad \clubsuit$$

■ Example 7

Convert 558 to base 12.

Solution: The place values are . . . , 12^3, 12^2, 12^1, 1, or . . . , 1728, 144, 12, 1.

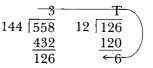

(Remember that T is used to represent ten in base 12.)

$$558 = (3 \times 12^2) + (T \times 12) + (6 \times 1) = 3T6_{12} \qquad \clubsuit$$

EXERCISES 4.3

1. Why is the Mayan positional-value numeration system not a true positional system?

Convert each of the following numbers to base 10.

2. 40_8
3. 12_5
4. 101_2
5. 1011_2
6. 1101_2
7. 67_{12}
8. 20221_3
9. 674_9
10. 654_7
11. 20432_5
12. 101111_2
13. 3001_4
14. $123E_{12}$
15. 123_8
16. 1023_8
17. 10047_8
18. 84721_9

Convert each base 10 number to the base indicated.

19. 8 to base 2
20. 16 to base 2
21. 23 to base 2
22. 46 to base 5
23. 406 to base 8
24. 809 to base 4
25. 1695 to base 12
26. 100 to base 3
27. 230 to base 6
28. 64 to base 2
29. 286 to base 12
30. 1234 to base 5
31. 1011 to base 2
32. 1589 to base 7
33. 2408 to base 8
34. 13469 to base 8

Assume that a base 16 positional-value system uses the numerals 0, 1, 2, 3, 4, 5, 6, 7, 8, 9, A, B, C, D, E, F where A through F represent 10 through 15 respectively. Convert each of the following to base 10.

35. 734_{16}
36. 285_{16}
37. $6D3B_{16}$
38. $24FE_{16}$

Convert the following to base 16.

39. 307
40. 349
41. 5478
42. 34721

Convert 1992 to each of the following bases.

43. 2
44. 3
45. 5
46. 7
47. 8
48. 16

If any of the following numerals is written incorrectly, explain why.

49. 4063_5
50. 1203_3
51. 674_8
52. 1206_{12}

There is an alternative method for changing a number in base 10 to a different base. This method will be used to convert 328 to base 5. Dividing 328 by 5 gives a quotient of 65 and a remainder of 3. Write the quotient below the dividend and the remainder on the right, as illustrated below.

$$5\underline{|328} \qquad \text{remainder}$$
$$65 \qquad\qquad 3$$

Continue this process of division by 5.

$$
\begin{array}{r|l}
5\ \underline{|328} & \text{remainder} \\
5\ \underline{|\ 65} & 3 \uparrow \\
5\ \underline{|\ 13} & 0 \\
5\ \underline{|\ \ 2} & 3 \\
0 & 2
\end{array}
$$
(Since the dividend, 2, is smaller than the divisor, 5, the quotient is 0 and the remainder is 2.)

Note that the division continues until the quotient is zero. The answer is read from the bottom number to the top number in the remainder column. Thus 328 = 2303_5.

53. Convert 683 to base 5 by the method described in Examples 5 through 7.
54. Convert 683 to base 5 by the method described above.
55. Convert 763 to base 8 by the method described in Examples 5 through 7.
56. Convert 763 to base 8 by the method described above.
57. Convert 342_5 to a base 3 number by first changing 342_5 to a base 10 number, then changing the base 10 number to a base 3 number.
58. Convert 342_5 to base 2.

PROBLEM SOLVING _____

59. The American Standard Code for Information Interchange (ASCII), used by most computers, uses the last seven positions of an eight-digit byte to represent letters, numbers, punctuation marks, and so on. How many different orderings of 0s and 1s (or how many different symbols) can be made using the last seven positions of an eight-digit byte.
60. Find b if $111_b = 43$.

RESEARCH ACTIVITY _____

61. Investigate and write a report on how digital computers use the binary number system.

4.4   COMPUTATIONS IN OTHER BASES

Addition of numbers in a base other than 10 is simplified by means of an addition table. The base 5 addition table is given in Table 4.7.

At first glance, this table may appear to contain many errors, but it is correct because in base 5 the only digits are 0, 1, 2, 3, and 4.

In base 10, $3 + 4 = 7$. When changed to a base 5 number, 7 becomes $1(5) + 2(1)$. Thus $3 + 4 = 12_5$. Table 4.7 shows that $3 + 4 = 12_5$ (circled). In base 10, $4 + 4 = 8$. When changed to a base 5 number, 8 becomes $1(5) + 3(1)$. Thus $4 + 4 = 13_5$, as illustrated in the table. The other sums are determined in the same manner.

■ **Example 1** _____

Add 32_5.
    33_5

TABLE 4.7

+	0	1	2	3	4
0	0	1	2	3	4
1	1	2	3	4	10
2	2	3	4	10	11
3	3	4	10	11	⑫
4	4	10	11	12	13

Solution: First determine that $2 + 3$ is 10 from Table 4.7. When the sum consists of two digits, record the digit on the right and carry the digit on the left because it represents groups of 5. In this case we record the 0 and carry the 1, which represents one group of 5.

$$\begin{array}{cc} {}^1 3 & 2_5 \\ 3 & 3_5 \\ \hline & 0_5 \end{array}$$

Add the numbers in the second column, $(1 + 3) + 3 = 4 + 3 = 12_5$. The 2 represents groups of 5, and the 1 represents groups of 25, or 5^2. Record the 12.

$$\begin{array}{ccc} {}^1 3 & & 2_5 \\ & 3 & 3_5 \\ \hline 1 & 2 & 0_5 \end{array}$$

The sum is 120_5.   ❖

■ **Example 2** _____

Add 1234_5.
    2042_5

Solution

$$\begin{array}{cccc} 1 & {}^1 2 & {}^1 3 & 4_5 \\ 2 & 0 & 4 & 2_5 \\ \hline 3 & 3 & 3 & 1_5 \end{array}$$

                                                                    ❖

After some practice you should be able to add in any base without the use of a table. Let's try a few more.

■ **Example 3** _____

Add 1022_3.
  $\underline{2121_3}$

Solution: In solving this problem we will make the necessary conversions by using mental arithmetic. $2 + 1 = 3_{10} = 10_3$. Record the 0 and carry the 1.

$$\begin{array}{ccccc} 1 & 0 & {}^12 & 2_3 \\ 2 & 1 & 2 & 1_3 \\ \hline & & & 0_3 \end{array}$$

$1 + 2 + 2 = 5_{10} = 12_3$. Record the 2 and carry the 1.

$$\begin{array}{ccccc} 1 & {}^10 & {}^12 & 2_3 \\ 2 & 1 & 2 & 1_3 \\ \hline & & 2 & 0 \end{array}$$

$1 + 0 + 1 = 2_{10} = 2_3$. Record the 2.

$$\begin{array}{ccccc} 1 & {}^10 & {}^12 & 2_3 \\ 2 & 1 & 2 & 1_3 \\ \hline & 2 & 2 & 0 \end{array}$$

$1 + 2 = 3_{10} = 10_3$. Record the 10.

$$\begin{array}{c} 1022_3 \\ \underline{2121_3} \\ 10220_3 \end{array}$$

❖

■ **Example 4** _____

Add 1101_2.
  $\underline{111_2}$

Solution: Any sum of 2 will be written as 10_2, and any sum of 3 is 11_2.

$$\begin{array}{ccccc} {}^11 & {}^11 & {}^10 & 1_2 \\ & 1 & 1 & 1_2 \\ \hline 1 & 0 & 1 & 0 & 0_2 \end{array}$$

❖

■ **Example 5** _____

Add 444_5.
$\quad 244_5$
$\quad 143_5$
$\quad \underline{214_5}$

Solution: Adding the digits in the right-hand column, $4 + 4 + 3 + 4 = 15_{10} = 30_5$. Record the 0 and carry the 3. Adding the 3 with the digits in the next column we obtain $3 + 4 + 4 + 4 + 1 = 16_{10} = 31_5$. Record the 1 and carry the 3. Adding the 3 with the digits in the left-hand column gives $3 + 4 + 2 + 1 + 2 = 12_{10} = 22_5$. Record both digits. The sum of these four numbers is 2210_5.

$$
\begin{array}{ccc}
{}^34 & {}^34 & 4_5 \\
2 & 4 & 4_5 \\
1 & 4 & 3_5 \\
\underline{2} & \underline{1} & \underline{4_5} \\
2 \quad 2 & 1 & 0_5
\end{array}
$$

❖

Subtraction can also be performed in other bases. It is important to remember that when you "borrow," you borrow the amount of the base. In the following two examples, we will perform the subtraction in base 10 when convenient and convert the results to the given base.

■ **Example 6** _____

Subtract 3032_5.
$\quad\quad\ \ - \ 1004_5$

Solution: Since 4 is greater than 2, we must borrow one group of 5 (10_5) from the preceding column. This gives a sum of $5 + 2$, or 7 in base 10. Now subtract 4 from 7; the difference is 3. We complete the problem in the usual manner. The 3 in the second column becomes a 2, $2 - 0 = 2$, $0 - 0 = 0$, and $3 - 1 = 2$.

$$
\begin{array}{r}
3032_5 \\
- \ 1004_5 \\
\hline
2023_5
\end{array}
$$

❖

■ **Example 7** _____

Subtract 468_{12}.
$\quad\quad\ \ - \ 295_{12}$

Solution: $8 - 5 = 3$. Next we must subtract 9 from 6. Since 9 is greater than 6, borrowing is necessary. We must borrow one group of 12 from the preceding column. We then have a sum of $12 + 6 = 18$ in base 10. Now we subtract 9 from 18, and the difference is 9. The 4 in the left column becomes a 3, and $3 - 2 = 1$.

$$
\begin{array}{r}
468_{12} \\
- \ 295_{12} \\
\hline
193_{12}
\end{array}
$$

DID YOU KNOW?

The most common type of computer is the digital computer. Digital computers operate with a binary number system that uses only two digits, 0 and 1. To perform any arithmetic operation, the computer translates the base 10 number into a binary number, performs the operation in binary arithmetic, and then translates the answer back into a base 10 number. The answer, a base 10 number, is then displayed on the screen or printed.

There are many digital computers that perform only one operation—addition. Addition in binary arithmetic is performed by a computer in much the same way as addition was presented in this section. Subtraction on these computers is performed through an addition process. One method sometimes used by computers to perform a subtraction is the "end–around carry method." To understand this method, you must understand what is meant by the **complement of a number.** The nines complement of a number is found by subtracting the number from 9 if the number has one digit, from 99 if the number has two digits, and so on. For example the nines complement of 4 is $9 - 4$, or 5, and the nines complement of 253 is $999 - 253$, or 746.

We will show how to subtract by addition in base 10 using the end–around carry method. Then we will show how a similar procedure is used in base 2. This process will work when a smaller number is being subtracted from a larger number.

To use the end-around carry method (base 10):

1. Determine the nines complement of the subtrahend (the number being subtracted).

2. Add the nines complement found in Step 1 to the minuend (the top number). Do not record the carry (it will be a 1) from the sum of the leftmost digits.

3. Add the carry that was not recorded in Step 2 to the sum found in Step 2 to obtain the final answer.

■ **Example 1**

Use the nines complement to find   842.
                                    − 365

Solution: First find the nines complement of 365 by subtracting 365 from 999:

$$\begin{array}{r} 999 \\ -\ 365 \\ \hline 634 \end{array}$$

Now rewrite as follows.

		Check:
	842	842
	+ 634	− 365
	① 476	477
	└→ +1	
	477	

The difference is 477. This answer was obtained using addition by the end–around carry method.   ◆

■ **Example 2**

Use the nines complement to find   3458.
                                    − 73

Solution:

3458	3458
− 73 means	− 0073

Rewrite as

3458
+ 9926 ← nines complement of 0073
① 3384
└→ +1
3385 ← answer

A computer subtracts in much the same way in base 2. The *ones complement* of a binary number is found by subtracting the number from 1 if the number being subtracted contains one digit, from 11 if the number being subtracted contains two digits, and so on.

The ones complement of 101_2 is

$$
\begin{array}{r}
111_2 \\
- \ 101_2 \\
\hline
010_2 \leftarrow \text{ones complement of } 101_2.
\end{array}
$$

The ones complement of 1011_2 is

$$
\begin{array}{r}
1111_2 \\
- \ 1011_2 \\
\hline
0100_2 \leftarrow \text{ones complement of } 1011_2.
\end{array}
$$

Note that the ones complement of a binary number will have all the 1s replaced by 0s and all the 0s replaced by 1s. The ones complement of 11010_2 is 00101_2.

To subtract using the end–around carry method (base 2):

1. Determine the ones complement of the subtrahend.

2. Add the ones complement found in Step 1 to the minuend. Add but do not record the carry (it will be a 1).

3. Add the carry that was not recorded in Step 2 to the sum found in Step 2 to obtain the final answer.

■ **Example 3**

Use the ones complement to find

$$
\begin{array}{r}
1011_2 \\
- \ \ 101_2
\end{array}
$$

Solution

$$
\begin{array}{r}
1011_2 \\
- \ \ 101_2
\end{array}
\quad \text{means} \quad
\begin{array}{r}
1011_2 \\
- \ 0101_2
\end{array}
$$

Rewrite as

$$
\begin{array}{r}
1011_2 \\
+ \ 1010_2 \leftarrow \text{ones complement of } 0101_2 \\
\hline
① \ 0101 \\
\llcorner\!\!\rightarrow + 1 \qquad \text{carry} \\
\hline
110_2 \leftarrow \text{answer}
\end{array}
$$

■ **Example 4**

Use the ones complement to find

$$
\begin{array}{r}
1011_2 \\
- \ \ 11_2
\end{array}
$$

Solution

$$1011_2 \qquad\qquad 1011_2$$
$$-\;\; 11_2 \quad\text{means}\quad -\; 0011_2\,.$$

Rewrite as

$$1011_2$$
$$+\,1100_2 \quad\leftarrow \text{ones complement of } 0011_2$$
$$①\,0111$$
$$\rightarrow +1 \qquad \text{carry}$$
$$1000_2 \leftarrow \text{answer} \qquad\qquad\qquad ❖$$

Since multiplication is no more than repeated addition, and division is no more than repeated subtraction, a computer can perform all arithmetic operations using only addition in base 2.

Multiplication and division in bases other than 10 can also be performed. Initially, the use of a multiplication table may be helpful. However, after a short while you should be able to multiply without the use of a table. Table 4.8 is the multiplication table for base 5.

To construct this table, we can multiply in base 10 and then convert the products to base 5. For example, $4 \times 3 = 12_{10} = 2(5) + 2(1) = 22_5$. This result has been circled in Table 4.8.

TABLE 4.8

×	0	1	2	3	4
0	0	0	0	0	0
1	0	1	2	3	4
2	0	2	4	11	13
3	0	3	11	14	22
4	0	4	13	㉒	31

■ **Example 8** _____

Multiply 13_5.
$$\times\;\; 3_5$$

Solution: Multiply as you would in base 10, but use the base 5 multiplication table to find the products. When the product consists of two digits, record the right digit and carry the left digit. Multiplying $3 \times 3 = 14_5$. Record the 4 and carry the 1.

$$\begin{array}{r} {}^{1}13_5 \\ \times\;\; 3_5 \\ \hline 4 \end{array}$$

$(3 \times 1) + 1 = 4_5$. Record the 4.

$$\begin{array}{r} {}^{1}13_5 \\ \times\ \ 3_5 \\ \hline 44_5 \end{array}$$

The product is 44_5. ❖

It is a tedious task to construct a multiplication table, especially when the base is large. To multiply in a given base without the use of a table, multiply in base 10 and convert the products to the appropriate base number before recording them. This procedure is illustrated in Example 9.

■ **Example 9** ───────────────────────────────────────

Multiply $\quad 43_7$.
$$\underline{\times\ 25_7}$$

Solution: $5 \times 3 = 15_{10} = 2(7) + 1(1) = 21_7$. Record the 1 and carry the 2.

$$\begin{array}{r} {}^{2}43_7 \\ \times\ \ 25_7 \\ \hline 1 \end{array}$$

$(5 \times 4) + 2 = 20 + 2 = 22_{10} = 3(7) + 1(1) = 31_7$. Record the 31.

$$\begin{array}{r} {}^{2}43_7 \\ \times\ \ 25_7 \\ \hline 311 \end{array}$$

$2 \times 3 = 6$. Record the 6.

$$\begin{array}{r} {}^{2}43_7 \\ \times\ \ 25_7 \\ \hline 311 \\ 6 \end{array}$$

$2 \times 4 = 8_{10} = 1(7) + 1(1) = 11_7$. Record the 11. Now add in base 7 to determine the answer. Remember, in base 7 there are no digits greater than 6.

$$\begin{array}{r} {}^{2}43_7 \\ 25_7 \\ \hline 311 \\ 116 \\ \hline 1501_7 \end{array}$$
❖

Division is more complicated than the other operations. A detailed example of a division in base 5 is illustrated below. The same procedure is used for division in any other base.

■ **Example 10** ———————————————————————————

Divide $2_5 \overline{)143_5}$.

Solution: Using Table 4.8, the multiplication table for base 5, we list the multiples of the divisor, 2.

$$2 \times 1 = 2_5$$
$$2 \times 2 = 4_5$$
$$2 \times 3 = 11_5$$
$$2 \times 4 = 13_5$$

Since $2 \times 4 = 13_5$, which is less than 14_5, 2_5 goes into 14_5 four times:

$$
\begin{array}{r}
4 \\
2_5 \overline{)143_5} \\
\underline{13} \\
1
\end{array}
$$

Now bring down the 3 as when dividing in base 10:

$$
\begin{array}{r}
4 \\
2_5 \overline{)143_5} \\
\underline{13} \\
13
\end{array}
$$

We see that $2 \times 4 = 13_5$. Use this information to complete the problem:

$$
\begin{array}{r}
44_5 \\
2_5 \overline{)143} \\
\underline{13} \\
13 \\
\underline{13} \\
0
\end{array}
$$

Thus $143_5 \div 2_5 = 44_5$ with a remainder of zero.

A division problem can be checked by multiplication. If the division was performed correctly, (quotient $\times$ divisor) + remainder = dividend.

$$(44_5 \times 2_5) + 0_5 = 143_5$$

$$
\begin{array}{r}
44_5 \\
\times\ 2_5 \\
\hline
143_5 \text{ (check)}
\end{array}
$$

❖

Example 11 ——————————————————————————————

Divide $4_6 \overline{\smash{\big)}\,2430_6}$.

Solution: The multiples of 4 in base 6 are

$$4 \times 1 = 4_6$$
$$4 \times 2 = 12_6$$
$$4 \times 3 = 20_6$$
$$4 \times 4 = 24_6$$
$$4 \times 5 = 32_6.$$

$$
\begin{array}{r}
404_6 \\
4_6 \overline{\smash{\big)}\,2430_6} \\
\underline{24} \\
03 \\
\underline{00} \\
30 \\
\underline{24} \\
2
\end{array}
$$

Thus the quotient is 404_6, with a remainder of 2_6.

Be careful when subtracting! When you borrow, remember that you borrow 10_6, which is the same as 6 in base 10. *Check:* Does $(404_6 \times 4_6) + 2_6 = 2430_6$?

$$
\begin{array}{r}
404_6 \\
\times 4_6 \\
\hline
2424_6 + 2_6 = 2430_6 \quad \textit{(check)}
\end{array}
$$

❖

EXERCISES 4.4 ——————————————

Add the following in the indicated base.

1. 13_4
   103_4

2. 20_7
   65_7

3. 4023_5
   2334_5

4. 101_2
   11_2

5. 467_{12}
   238_{12}

6. 222_3
   22_3

7. 1012_3
   1011_3

8. 470_{12}
   347_{12}

9. 14631_7
6040_7

†10. $43A_{16}$
496_{16}

Subtract the following in the indicated base.

11. 203_4
$- 103_4$

12. 463_7
$- 124_7$

13. 2334_5
$- 1243_5$

14. 1011_2
$- 101_2$

15. 463_{12}
$- 13T_{12}$

16. 1221_3
$- 202_3$

17. 1001_2
$- 110_2$

18. 1453_{12}
$- 245_{12}$

19. 4223_7
$- 304_7$

†20. $4E7_{16}$
$- 189_{16}$

Multiply the following in the indicated base.

21. 42_5
$\times \ 3_5$

22. 123_5
$\times \ 4_5$

23. 423_7
$\times \ 6_7$

24. 101_2
$\times \ 11_2$

25. 302_4
$\times \ 23_4$

26. 124_{12}
$\times \ 6_{12}$

27. 234_9
$\times \ 23_9$

28. $6T3_{12}$
$\times \ 24_{12}$

29. 111_2
$\times \ 111_2$

Divide the following in the indicated base.

30. $1_2 \overline{)110_2}$
31. $5_6 \overline{)342_6}$
32. $3_5 \overline{)143_5}$
33. $6_8 \overline{)466_8}$
34. $2_4 \overline{)312_4}$
35. $6_{12} \overline{)431_{12}}$
36. $3_7 \overline{)2101_7}$
*37. $14_5 \overline{)301_5}$
*38. $20_4 \overline{)223_4}$

Subtract the following using the end–around carry method.

39. 101_2
$- \ 11_2$

40. 1011_2
$- \ 110_2$

41. 10110_2
$- \ 111_2$

42. 101101_2
$- \ 1101_2$

Subtract the following using the end–around carry method. Use the fours complement since the numbers are in base 5.

43. 423_5
$- 322_5$

---

†See Exercises 35–38 in Exercises 4.3.

44. 231_5
       $-\ \ 13_5$

46. 43210_5
       $-\ \ \ 123_5$

45. 3401_5
       $-\ \ 223_5$

RESEARCH ACTIVITY

47. Investigate and write a report on the use of the duodecimal (base 12) system as a system of numeration. You might wish to contact the Duodecimal Society. The address is Duodecimal society, Nassau Community College, Garden City, NY 11530.

4.5 EARLY COMPUTATIONAL METHODS

Early civilizations used various methods for multiplying and dividing and for finding areas and volumes. Multiplication was performed by *duplation and mediation,* by the *galley method,* and with *Napier rods.* We will explain each method in turn.

Duplation and Mediation

■ **Example 1**

Multiply 17×30 using duplation and mediation.

Solution: Write $17 - 30$. The $-$ is not a minus sign, it is just used to separate 17 and 30. Divide the number on the left, 17, by 2 and drop the remainder. Place the quotient, 8, under the 17. Double the number on the right, 30, obtaining 60, and place it under the 30. You will then have the following paired lines:

$$17 - 30$$
$$8 - 60$$

Continue this process, taking one-half the number in the left-hand column, disregarding the remainder, and doubling the number in the right-hand column, as shown below. When a 1 appears in the left-hand column, stop.

$$17 - 30$$
$$8 - 60$$
$$4 - 120$$
$$2 - 240$$
$$1 - 480$$

Cross out all the numbers in the left-hand column that are even and the corresponding numbers in the right-hand column.

$$17 - 30$$
$$\cancel{8} - \cancel{60}$$
$$\cancel{4} - \cancel{120}$$
$$\cancel{2} - \cancel{240}$$
$$1 - 480$$

Now add the remaining numbers in the right-hand column, obtaining $30 + 480 = 510$, which is the product you want. If you check, you will find that $17 \times 30 = 510$. ❖

The Galley Method

The galley method (sometimes referred to as the Gelosia method) was developed much later than duplation and mediation. To multiply 312×75 using the galley method, you must construct a rectangle consisting of three columns (one for each digit of 312) and two rows (one for each digit of 75), as shown in Fig. 4.2.

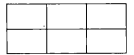

FIGURE 4.2

Place the digits 3, 1, 2 above the boxes and the digits 7, 5 on the right of the boxes, as shown in Fig. 4.3.

Place a diagonal in each box (Fig. 4.4).

Complete each box by multiplying the number on top of the box by the number to the right of the box (Fig. 4.5). Place the units below the diagonal and the tens above.

Add the numbers along the diagonals, as shown in Fig. 4.6, starting with the bottom right diagonal. If the sum in a diagonal is 10 or greater, record the units digit and carry the tens digit to the next diagonal to the left. The result is shown in Fig. 4.7.

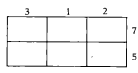

FIGURE 4.3

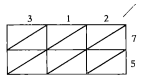

FIGURE 4.4

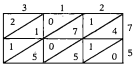

FIGURE 4.5

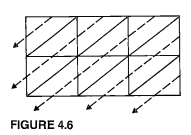

FIGURE 4.6

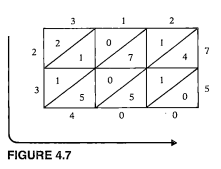

FIGURE 4.7

The answer is read down the left-hand column and along the bottom, as shown by the arrow in Fig. 4.7. The answer is 23,400.

Napier Rods

The third method was developed from the galley method by John Napier in the seventeenth century. His method of multiplication, known as Napier rods, proved to be one of the forerunners of the modern-day computer. Napier developed a collection of separate rods, or strips, as illustrated in Fig. 4.8.

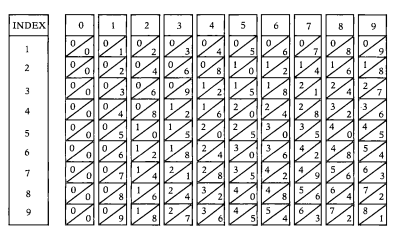

FIGURE 4.8

To multiply 8 × 365, line up the rods, as shown in Fig. 4.9. Opposite the 8 on the index are the following boxes:

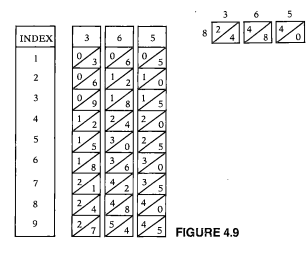

FIGURE 4.9

Add along the diagonals, as in the galley method:

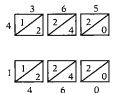

Thus 8 × 365 = 2920. To multiply numbers containing more than one digit, proceed as in the next example.

■ Example 2

Multiply 48 × 365, using Napier rods.

Solution

$$48 \times 365 = (40 + 8) \times 365$$

We can write (40 + 8) × 365 = (40 × 365) + (8 × 365). To find 40 × 365, determine 4 × 365, and multiply the product by 10. To evaluate 4 × 365, set up Napier rods and then evaluate along the diagonals, as indicated:

Therefore 4 × 365 = 1460. Then 40 × 365 = 10 × 1460 = 14,600.

$$48 \times 365 = (40 \times 365) + (8 \times 365)$$
$$= 14,600 + 2920$$
$$= 17,520$$

(8 × 365 = 2920 from the preceding calculation) ❖

EXERCISES 4.5

Multiply, using duplation and mediation.

1. 18 × 30
2. 35 × 23
3. 8 × 145
4. 120 × 90
5. 43 × 221
6. 96 × 53
7. 75 × 82
8. 49 × 124

Multiply, using the galley method.

9. 5 × 365
10. 6 × 365
11. 56 × 365
12. 7 × 12
13. 75 × 12
14. 17 × 256
15. 376 × 431
16. 525 × 698

Multiply, using Napier rods.

17. 4 × 28
18. 7 × 28
19. 47 × 28
20. 7 × 125
21. 5 × 125
22. 75 × 125
23. 8 × 2345
24. 7 × 3456

SUMMARY

A numeral is a symbol used to represent a number. A number is a concept or idea that addresses the question "How many?" A numeration system is a set of numerals and rules that allows only a few symbols to represent any number. We have discussed four basic types of numeration systems: additive, multiplicative, ciphered, and place-value systems. Some numeration systems are a combination of two or more of these four types.

The numerals in additive numeration systems are added together to give the number. For example, if * represents 1, ? represents 10, and □ represents 100, the number □ □ □????** is equal to 342.

In a multiplicative numeration system the numbers are multiplied by powers of the base. The products are then summed to give the number. For example, if * represents 10, □ represents 100, and △ represents 1000, then 2 △ 4 □ 5 * 3 is equal to 2453.

Ciphered numeration systems contain numerals for each number up to $b - 1$, where b represents the base. There are also separate symbols for the products of these numbers times powers of the base. In a ciphered numeration system, for example, if * represents 300, # represents 60, and ? represents 4, the number 364 would be represented by * #?.

In a place-value system, each digit is multiplied by its corresponding positional value to determine the value of a number. In our own base 10 place-value system, 346 means $(3 \times 100) + (4 \times 10) + (6 \times 1)$ or $(3 \times 10^2) + (4 \times 10) + (6 \times 1)$.

Numbers can be converted from base 10 to any given base and from any given base to base 10. The operations of addition, subtraction, multiplication, and division can be performed in other bases.

Ancient societies used various methods to multiply, including duplation and mediation, the galley method, and Napier rods.

REVIEW EXERCISES

Assume an additive numeration system in which $a = 1, b = 10, c = 100$, and $d = 1000$. Find the value of the following numerals.

1. *ddca*	**2.** *bdccda*	**3.** *ddbbaa*
4. *cbdadaaa*	**5.** *dddccbaaaa*	**6.** *ccbaddac*

Assume the same additive numeration system as in Exercise 1. Write the following in terms of *a, b, c,* and *d.*

7. 43	**8.** 167	**9.** 389
10. 1978	**11.** 6004	**12.** 3421

Assume a multiplicative numeration system in which $a = 1, b = 2, c = 3, d = 4,$ $e = 5, f = 6, g = 7, h = 8, i = 9, x = 10, y = 100,$ and $z = 1000.$ Find the value of the following numerals.

13. *ixc*
14. *bxg*
15. *fydxh*
16. *bygxc*
17. *gzdyhxb*
18. *izeygxd*

Assume the same multiplicative numeration system as in Exercises 13–18. Write the following Hindu-Arabic numerals in that system.

19. 43		**20.** 167		**21.** 489	
22. 1978		**23.** 6004		**24.** 2001	

A ciphered numeration system is indicated below.

Decimal	1	2	3	4	5	6	7	8	9
Units	a	b	c	d	e	f	g	h	i
Tens	j	k	l	m	n	o	p	q	r
Hundreds	s	t	u	v	w	x	y	z	A
Thousands	B	C	D	E	F	G	H	I	J
Ten thousands	K	L	M	N	O	P	Q	R	S

Convert the following to Hindu-Arabic numerals.

25. *rc*		**26.** *vg*		**27.** *Itrh*	
28. *NGzqc*		**29.** *Qqb*		**30.** *Pwki*	

Using the table used for Exercises 25–30, write the following as a numeral in the ciphered numeration system.

31. 43		**32.** 167		**33.** 489	
34. 1978		**35.** 23,685		**36.** 75,496	

Convert 1462 to a numeral in each of the following numeration systems.

37. Egyptian		**38.** Roman		**39.** Chinese	
40. Ionic Greek		**41.** Babylonian		**42.** Mayan	

Convert each of the following to a Hindu-Arabic numeral.

43.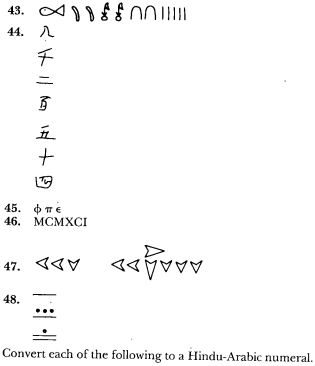

44.

45. $\phi\,\pi\,\epsilon$

46. MCMXCI

47.

48.

Convert each of the following to a Hindu-Arabic numeral.

49. 46_7
50. 101_2
51. 304_5
52. 2663_8
53. $T0E_{12}$
54. 20220_3

Convert 463 to a numeral in the base indicated.

55. base 5
56. base 6
57. base 2
58. base 4
59. base 12
60. base 8

Add in the base indicated.

61. 43_8
$\underline{74_8}$

62. 10110_2
$\underline{1101_2}$

63. TE_{12}
$\underline{46_{12}}$

64. 234_7
$\underline{456_7}$

65. 4023_5
$\underline{4023_5}$

66. 1407_8
$\underline{7014_8}$

Subtract in the base indicated.

67. 4032_7
$- \quad 121_7$

68. 1001_2
$- \quad 101_2$

69. $4TE_{12}$
$- \quad E7_{12}$

70. 4321_5
$- \quad 442_5$

71. 2473_8
$- \quad 567_8$

72. 2021_3
$- \quad 222_3$

Multiply in the base indicated.

73. 23_5
$\times \quad 4_5$

74. 23_4
$\times 22_4$

75. 126_{12}
$\times \quad 42_{12}$

76. 202_3
$\times \quad 22_3$

77. 1011_2
$\times \quad 101_2$

78. 476_8
$\times \quad 23_8$

Divide in the base indicated.

79. $1_2 \overline{)1011_2}$

80. $2_4 \overline{)230_4}$

81. $3_5 \overline{)140_5}$

82. $4_6 \overline{)3020_6}$

*83. $12_4 \overline{)3202_4}$

*84. $14_7 \overline{)213_7}$

85. Multiply 142×24, using the duplation and mediation method.
86. Multiply 142×24, using the galley method.
87. Multiply 142×24, using Napier rods.

CHAPTER TEST

4

1. Explain the difference between a numeral and a number.

Convert each of the following to a Hindu-Arabic numeral.

2. $99\cap\cap\cap$

3. MCMXCIX

4. ◁ ◁ ◁ ▽ ▽ ▽   ◁ ▽

5.
入
千
入
百
三
十
五

Convert each of the following to a numeral in the system of numeration indicated:

6. 325 to Egyptian   7. 1492 to Mayan
8. 4263 to Babylonian   9. 785 to Ionic Greek

10. Describe briefly how you may distinguish between an Additive system, Multiplicative system, Ciphered system, and Place-value system of numeration.

Convert to base 10.

11. 34_5 12. 73_8
13. 842_9 14. 10101_2

Convert to the base indicated.

15. 35 to base 2   16. 93 to base 5
17. 1347 to base 12   18. 3642 to base 8

Perform the indicated operations.

19. 121_3
      $+\ 212_3$

20. 576_8
          $- 347_8$

21. 45_7
          $\times 36_7$

Multiply using duplation and mediation.

22. 15×17

Multiply using the galley method.

23. 187×26

The Real Number System

5.1 PRIME AND COMPOSITE NUMBERS

Multiplication Notation

In algebra, letters called **variables** are often used to represent numbers. A letter that is commonly used for a variable is the letter x. Variables will be discussed further in Chapter 6. Because the multiplication symbol, $\times$, may be confused with the variable x, other symbols for multiplication are often used. A dot placed between two numbers (or between a number and a variable) is used to indicate multiplication. When two numbers are placed next to one another, and one or both numbers are in parentheses, multiplication is also indicated. For example,

$$\begin{array}{lll} 2 \cdot 5 & \text{means} & \text{"2 times 5"} \\ -4 \cdot 9 & \text{means} & \text{"} -4 \text{ times 9"} \\ 3(6) & \text{means} & \text{"3 times 6"} \\ (-2)(8) & \text{means} & \text{"} -2 \text{ times 8"} \end{array}$$

The above notation will be used in this and other chapters.

Prime and Composite Numbers

The numbers we use to count with are called the **counting numbers** or **natural numbers.** Since we begin counting with the number 1, the set of natural numbers begins with 1. The set of natural numbers is frequently denoted by N.

$$N = \{1, 2, 3, 4, \ldots\}$$

Consider the natural number 12. What pairs of numbers can you multiply to equal 12?

$$\begin{array}{c} 12 \cdot 1 = 12 \\ 6 \cdot 2 = 12 \\ 4 \cdot 3 = 12 \\ \uparrow \quad \uparrow \\ \text{Factors} \end{array}$$

The numbers 1, 2, 3, 4, 6, and 12 are factors of 12. Note that each of these numbers divides 12 without a remainder.

> A number a is said to be a **factor** of a number b if a divides b without a remainder.

The factors of 36 are 1, 2, 3, 4, 6, 9, 12, 18, and 36.

TABLE 5.1

Rules of Divisibility

Divisible by	Test	Example
2	The number ends in 0, 2, 4, 6, or 8. (The number is even.)	660,024 is divisible by 2, since the number ends in 4.
3	The sum of the digits of the number is divisible by 3.	660,024 is divisible by 3, since the sum of the digits, 6 + 6 + 0 + 0 + 2 + 4 = 18, and 18 is divisible by 3.
4	The last two digits of the number are a number that is divisible by 4.	660,024 is divisible by 4, since the last two digits are 24, which is
5	The number ends in 0 or 5.	750,265 is divisible by 5, since the number ends in 5.
6	The number is divisible by both 2 and 3.	660,024 is divisible by 6, since it is divisible by both 2 and 3.
8	The last three digits of the number are a number that is divisible by 8.	934,824 is divisible by 8, since the last three digits are 824, which is divisible by 8.
9	The sum of the digits of the number is divisible by 9.	231,921 is divisible by 9, since the sum of the digits, 18, is divisible by 9.
10	The number ends in 0.	465,290 is divisible by 10, since the number ends in 0.
12	The number is divisible by both 3 and 4.	660,024 is divisible by 12, since the number is divisible by both 3 and 4.

Every natural number greater than 1 can be classified as either a prime number or a composite number. A **prime number** is a natural number that has only two factors (or divisors)—itself and one. The number 5 is a prime number since it is divisible only by the factors 1 and 5.

The first eight prime numbers are 2, 3, 5, 7, 11, 13, 17, and 19. The number 2 is the only even number that is prime. All other even numbers have at least three divisors—1, 2, and the number itself.

A **composite number** is a natural number that can be divided by a natural number other than itself and 1. Any natural number greater than 1 that is not prime is composite. The first eight composite numbers are 4, 6, 8, 9, 10, 12, 14, and 15. The number 1 is neither prime nor composite; it is called a **unit.**

The rules of divisibility given in Table 5.1 may be helpful in determining whether a number is a prime number or a composite number. If *a* divides *b,*

"Numerals, the written signs for numbers, are one of 72 items occurring in every human culture known to ethnography."
Anthropologist George P. Murdock

without remainder, then we say that b is **divisible** by a. For example, since 2 divides 6, 6 is divisible by 2. Note that if b is divisible by a, then a must be a factor of b.

Note that we have not listed rules for divisibility for the numbers 7 and 11. There are rules for these numbers, but the rules are complex and difficult to remember. To check divisibility by 7 or 11 it is simpler just to perform the division. We do, however, give the rules for divisibility by 7 and 11 in the exercise set.

■ **Example 1** _____

Determine whether the number 374,832 is divisible by

a) 2   **b)** 3   **c)** 4   **d)** 5   **e)** 6   **f)** 8   **g)** 9   **h)** 10   **i)** 12

Solution

a) Since the number is even, it is divisible by 2.

b) Since the sum of the digits, 27, is divisible by 3, the number is divisible by 3.

c) Since the last two digits, 32, are divisible by 4, the number is divisible by 4.

d) Since the last digit is not a 0 or a 5, the number is not divisible by 5.

e) Since the number is divisible by both 2 and 3, the number is divisible by 6.

f) Since the last three digits, 832, are divisible by 8, the number is divisible by 8.

g) Since the sum of the digits, 27, is divisible by 9, the number is divisible by 9.

h) Since the number does not end in 0, the number is not divisible by 10.

i) Since the number is divisible by both 3 and 4, the number is divisible by 12.

The number 374,832 is divisible by 2, 3, 4, 6, 8, 9, and 12.   ❖

How many prime numbers are less than or equal to 50? The **Sieve of Eratosthenes,** named for the Greek mathematician who first used it more than 2000 years ago, is one method that can be used to answer that question.

To find the prime numbers less than or equal to 50 using this method, list the first 50 counting numbers (Fig. 5.1). Cross out 1. Circle 2, the first prime number. Then cross out every multiple of 2, that is, 4, 6, 8, . . . , 50 (all even numbers). Circle the next prime number, 3. Next cross out all multiples of 3 that are not already crossed out. Continue this process until you reach the prime number p, such that $p \cdot p$, or p^2, is greater than the last number listed. Since $7 \cdot 7$, or 49, is less than 50, cross out the multiples of 7. Since $11 \cdot 11$, or 121, is greater than 50, you need not cross out the multiples of 11. At this point, circle all the remaining numbers. The 15 circled numbers are the prime numbers less than 50.

1 ② ③ 4 ⑤ 6 ⑦ 8 9 10
⑪ 12 ⑬ 14 15 16 ⑰ 18 ⑲ 20
21 22 ㉓ 24 25 26 27 28 ㉙ 30
㉛ 32 33 34 35 36 ㊲ 38 39 40
㊶ 42 ㊸ 44 45 46 ㊼ 48 49 50

FIGURE 5.1

Example 2

Determine whether the number 137 is a prime number.

Solution: Using the information we gained in the discussion on the Sieve of Eratosthenes, we look for the largest prime number whose square is less than 137. We know that $11 \cdot 11 = 121$ and $13 \cdot 13 = 169$. Thus 11 is the largest prime number whose square is less than 137. We then divide 137 by all the prime numbers less than 13. Since no prime number less than 13 divides it, 137 is a prime number. ❖

DID YOU KNOW?

Systems analysts at Chevron Geoscience Company in Houston, Texas, discovered in September 1985, the largest prime number known at this writing using a Cray supercomputer. This prime contains 65,050 digits. But even larger primes lurk beyond. In fact, more than 2500 years ago the Greek mathematician Euclid proved that there are an infinite number of primes. But the primes do not occur in regular intervals, and even today it is impossible to determine where one will turn up. The work required to find a prime number increases with the length of the number, especially if the test entails trying every possible divisor.

If every possible divisor had to be tried to test a number this size, it would require all the computers in the world working trillions of years. But the analysts at Chevron were able to succeed in a reasonable time because they did not select numbers arbitrarily. They chose them from a series of numbers named for Marin Mersenne, a seventeenth-century French monk who studied them. The first few **Mersenne primes** are 3, 7, 31, and 127. Each number is obtained by raising the number 2 to a successively higher prime power and subtracting 1 from the results. For example,

$$2^2 - 1 = 2 \cdot 2 - 1 = 4 - 1 = 3,$$
$$2^3 - 1 = 2 \cdot 2 \cdot 2 - 1 = 8 - 1 = 7,$$
$$2^5 - 1 = 2 \cdot 2 \cdot 2 \cdot 2 \cdot 2 - 1 = 32 - 1 = 31,$$
$$2^7 - 1 = 128 - 1 = 127.$$

Mersenne found that numbers of this form, called **Mersenne numbers,** were often prime numbers. To find the highest prime, each Mersenne number had to be tested, which required enormous patience and a high-speed computer. In 1979 David Slowinski, a senior systems analyst, searched the first 50,000 Mersenne numbers and found the 27th Mersenne prime. It took the Cray supercomputer 200 hours working at a speed of 80 million multiplications and 80 million additions per second to find this 13,395-digit prime. The present largest known prime number, $2^{216,091} - 1$, is the 30th Mersenne prime.

Carl Friedrich Gauss proved the fundamental theorem of arithmetic.

The **fundamental theorem of arithmetic*** states that every composite number can be expressed as a **unique** product of primes.

The order of the factors is immaterial. Thus $2^3 \cdot 3 \cdot 5$ and $3 \cdot 2^3 \cdot 5$ represent the same product of prime factors. For example, let's consider the number 360:

$$360 = 2 \cdot 2 \cdot 2 \cdot 3 \cdot 3 \cdot 5$$
$$= 2^3 \cdot 3^2 \cdot 5$$

The fundamental theorem says that there is no other product of primes that equals 360.

A number of techniques can be used to break a number down into primes. We will illustrate two different procedures.

Carl Friedrich Gauss

---

*A theorem is a statement that has been proven true.

Method 1: Branching

Example 3

Write 300 as a product of primes.

Solution: Select any two numbers whose product is 300. Two possibilities are 10 · 30 and 15 · 20, but there are many other choices. For the first example, consider 15 · 20. Find any two numbers whose product is 15 and any two numbers whose product is 20. Branch the products to their respective numbers, as shown in Fig. 5.2. Continue this process until the numbers in the last row are all prime numbers:

$$300 = 2 \cdot 2 \cdot 3 \cdot 5 \cdot 5$$
$$= 2^2 \cdot 3 \cdot 5^2$$

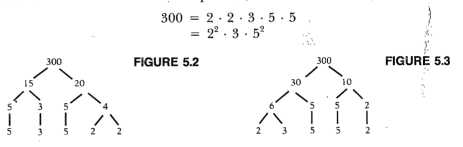

FIGURE 5.2   FIGURE 5.3

Thus the **prime factorization** of 300 is $2^2 \cdot 3 \cdot 5^2$. If we had selected 10 · 30, the results might be as illustrated in Fig. 5.3.

$$300 = 2^2 \cdot 3 \cdot 5^2$$

Notice that the final results are the same, regardless of the factors selected for the first step. ❖

Method 2: Division

To obtain the prime factorization of a number by this method, divide the given number by the smallest prime number by which it is divisible. Place the quotient under the given number. Then divide the quotient by the smallest prime number by which it is divisible.

Again record the quotient. Repeat this process until the quotient is a prime number. The prime factorization is the product of all the prime divisors and the prime (or last) quotient. This procedure is illustrated in Example 4.

Example 4

Write 300 as a product of prime numbers.

Solution: Since 300 is an even number, the smallest prime number that divides it is 2. Divide 300 by 2. Place the quotient, 150, below the 300. Repeat this process of dividing each quotient by the smallest prime number that divides it:

$$
\begin{array}{r|r}
2 & 300 \\
2 & 150 \\
3 & 75 \\
5 & 25 \\
\hline
& 5
\end{array}
$$

Since we obtained a quotient of 5, which is prime, we stop. The prime factorization is

$$300 = 2 \cdot 2 \cdot 3 \cdot 5 \cdot 5 = 2^2 \cdot 3 \cdot 5^2 \qquad \qquad ❖$$

Notice that although we used two different methods, we obtained the same answer in Examples 3 and 4.

> The **greatest common divisor (gcd)** of a set of natural numbers is the largest natural number that divides (without remainder) every number in that set.

The gcd is used to reduce fractions, as will be illustrated in Section 5.3.
   What is the gcd of 12 and 18? One way to determine this is to make a list of the divisors (or factors) of 12 and 18, as illustrated below.

$$\text{Divisors of 12: } \{\mathbf{1, 2, 3,} 4, \mathbf{6,} 12\}$$
$$\text{Divisors of 18: } \{\mathbf{1, 2, 3, 6,} 9, 18\}$$

The common divisors are 1, 2, 3, and 6. The greatest common divisor of 12 and 18 is 6. Two numbers with a gcd of 1 are said to be **relatively prime.** Thus numbers 9 and 14 are relatively prime, since the gcd is 1.
   If the numbers are large, this method of finding the gcd is not practical. The gcd can be found more efficiently by using prime factorization. To find the greatest common divisor of two or more numbers, follow these steps.

1. Determine the prime factorization of each number.
2. Find each prime factor with the smallest exponent that is *common to all* of the prime factorizations.
3. Find the product of the factors found in step 2.

Example 5 illustrates this procedure.

■ **Example 5** —————————————————————————————

Find the gcd of 24 and 60.

Solution: Fig. 5.4 illustrates one method of finding the prime factors of 24 and 60.

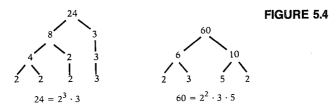

FIGURE 5.4

$24 = 2^3 \cdot 3$ $60 = 2^2 \cdot 3 \cdot 5$

1. The prime factorizations of 24 and 60 are $2^3 \cdot 3$ and $2^2 \cdot 3 \cdot 5$, respectively.
2. The prime factors with the smallest exponents common to 24 and 60 are 2^2 and 3.
3. The product of the factors found in step 2 is $2^2 \cdot 3 = 4 \cdot 3 = 12$. The gcd of 24 and 60 is 12. Twelve is the largest number that divides both 24 and 60. ❖

■ **Example 6** —————————————————————————————

Find the gcd of 36 and 150.

Solution: Fig. 5.5 illustrates one method of finding the prime factors of 36 and 150.

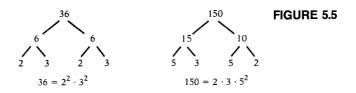

FIGURE 5.5

$36 = 2^2 \cdot 3^2$ $150 = 2 \cdot 3 \cdot 5^2$

1. The prime factors of 36 and 150 are $2^2 \cdot 3^2$ and $2 \cdot 3 \cdot 5^2$, respectively.
2. The prime factors with the smallest exponents common to 36 and 150 are 2 and 3.
3. The product of the factors found in step 2 is $2 \cdot 3 = 6$. The gcd of 36 and 150 is 6. ❖

Prime factorization can also be used to find the least common multiple.

> The **least common multiple (lcm)** of a set of natural numbers is the smallest natural number that is divisible (without remainder) by each element of the set.

The lcm can be helpful in adding or subtracting fractions, as will be illustrated in Section 5.3. What is the least common multiple of 12 and 18? One way to determine the lcm is to make a list of the multiples of each number, as illustrated below.

Multiples of 12:   {12, 24, **36,** 48, 60, **72,** 84, 96, **108,** 120, 132, **144,** . . .}
Multiples of 18:   {18, **36,** 54, **72,** 90, **108,** 126, **144,** 162, . . .}

Some common multiples of 12 and 18 are 36, 72, 108, and 144. The least common multiple is the smallest of these common multiples, so the lcm of 12 and 18 is 36. Note that 36 is the smallest number that is divisible by both 12 and 18.

This method of finding the lcm is often a tedious task. A more efficient method of finding the lcm is to use prime factorization.

To find the least common multiple of two or more numbers by using prime factorization, follow these steps.

1. Determine the prime factorization of each number.
2. List each prime factor with the greatest exponent that *appears in any* of the prime factorizations.
3. Find the product of the factors found in Step 2.

Example 7 illustrates this procedure.

Example 7

Find the lcm of 24 and 60.

Solution

1. Find the prime factors of each number. In Example 5 we determined that

$$24 = 2^3 \cdot 3 \text{ and } 60 = 2^2 \cdot 3 \cdot 5.$$

2. List each prime factor with the greatest exponent that appears in either of the prime factorizations: 2^3, 3, 5.
3. Find the product of the factors found in (2):

$$2^3 \cdot 3 \cdot 5 = 8 \cdot 3 \cdot 5 = 120$$

Thus 120 is the lcm of 24 and 60. It is the smallest number that is divisible by both 24 and 60.   ❖

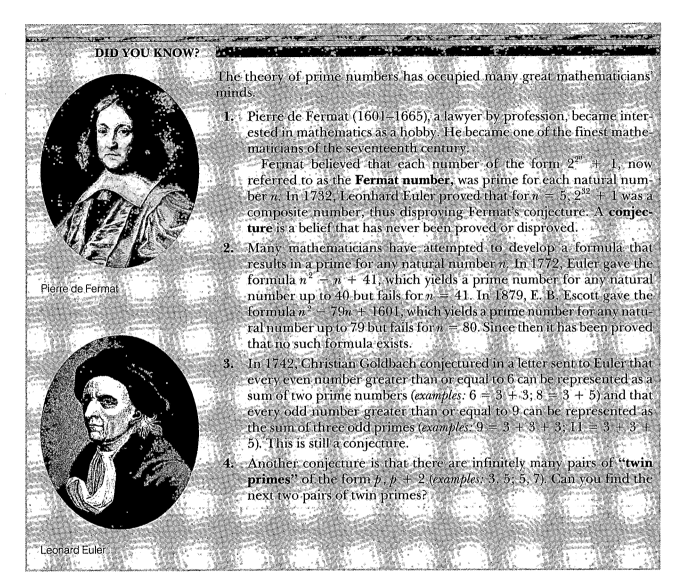

DID YOU KNOW?

The theory of prime numbers has occupied many great mathematicians' minds.

1. Pierre de Fermat (1601–1665), a lawyer by profession, became interested in mathematics as a hobby. He became one of the finest mathematicians of the seventeenth century.

    Fermat believed that each number of the form $2^{2^m} + 1$, now referred to as the **Fermat number,** was prime for each natural number n. In 1732, Leonhard Euler proved that for $n = 5$, $2^{32} + 1$ was a composite number, thus disproving Fermat's conjecture. A **conjecture** is a belief that has never been proved or disproved.

2. Many mathematicians have attempted to develop a formula that results in a prime for any natural number n. In 1772, Euler gave the formula $n^2 - n + 41$, which yields a prime number for any natural number up to 40 but fails for $n = 41$. In 1879, E. B. Escott gave the formula $n^2 - 79n + 1601$, which yields a prime number for any natural number up to 79 but fails for $n = 80$. Since then it has been proved that no such formula exists.

3. In 1742, Christian Goldbach conjectured in a letter sent to Euler that every even number greater than or equal to 6 can be represented as a sum of two prime numbers (*examples:* $6 = 3 + 3$; $8 = 3 + 5$) and that every odd number greater than or equal to 9 can be represented as the sum of three odd primes (*examples:* $9 = 3 + 3 + 3$; $11 = 3 + 3 + 5$). This is still a conjecture.

4. Another conjecture is that there are infinitely many pairs of "twin primes" of the form p, $p + 2$ (*examples:* 3, 5; 5, 7). Can you find the next two pairs of twin primes?

Pierre de Fermat

Leonard Euler

■ **Example 8**

Find the lcm of 36 and 150.

Solution

1. Find the prime factors of each number. In Example 6 we determined that

$$36 = 2^2 \cdot 3^2 \quad \text{and} \quad 150 = 2 \cdot 3 \cdot 5^2$$

2. List each prime factor with the greatest exponent that appears in either of the prime factorizations: 2^2, 3^2, 5^2.

3. Find the product of the factors found in (2):

$$2^2 \cdot 3^2 \cdot 5^2 = 4 \cdot 9 \cdot 25 = 900$$

Thus 900 is the least common multiple of 36 and 150. It is the smallest number divisible by both 36 and 150. ❖

EXERCISES 5.1

1. Define a prime number.
2. Define a composite number.

In Exercises 3–8, determine whether or not the number is divisible by 2, 3, 4, 5, 6, 8, 9, 10, and 12.

3. 243,864
4. 4,068,123
5. 140,505
6. 198,765,432
7. 1,200,684
8. 929,844
9. Find a number that is divisible by 2, 3, 4, 5, and 6.
10. Find a number that is divisible by 2, 3, 4, 6, and 12.
11. Use the Sieve of Eratosthenes technique to find the prime numbers up to and including 100.

Find the prime factorization of the following.

12.	36	**13.**	26	**14.**	212	**15.**	180
16.	312	**17.**	400	**18.**	1026	**19.**	2520
20.	485	**21.**	999	**22.**	1038	**23.**	1168
24.	1492						

Find the greatest common divisor of the following numbers.

25.	15 and 150	**26.**	30 and 48
27.	52 and 78	**28.**	180 and 450
29.	90 and 600	**30.**	180 and 250
31.	96 and 212	**32.**	240 and 260
33.	24, 40, and 100	**34.**	24, 36, and 72

Find the least common multiple of the following numbers.

35.	15 and 150	**36.**	30 and 48
37.	52 and 78	**38.**	180 and 450
39.	90 and 600	**40.**	60 and 90
41.	250 and 52	**42.**	198 and 240
43.	24, 36, and 72	**44.**	12, 75, and 140

45. Find the next two sets of twin primes that follow 5, 7.
46. Show that Goldbach's conjecture is true for the even numbers 6 through 12 and the odd numbers 9 through 15.

Another method that can be used to find the greatest common divisor is known as the **Euclidean algorithm.** We will illustrate this procedure by finding the gcd of 60 and 220.

First divide 220 by 60. Disregard the quotient 3. Then divide 60 by the remainder 40. Continue this process of dividing the divisors by the remainders until you obtain a remainder of 0. The divisor in the last division, in which the remainder is 0, is the gcd.

$$\begin{array}{c} 3 \\ 60 \overline{)220} \\ \underline{180} \\ 40 \end{array} \qquad \begin{array}{c} 1 \\ 40 \overline{)60} \\ \underline{40} \\ 20 \end{array} \qquad \begin{array}{c} 2 \\ 20 \overline{)40} \\ \underline{40} \\ 0 \end{array}$$

Since 40/20 had a remainder of 0, 20 is the gcd.

In Exercises 47–54, use the Euclidean algorithm to find the gcd.

47.	16, 80	**48.**	35, 75
49.	20, 150	**50.**	18, 112
51.	96, 112	**52.**	150, 180
53.	210, 560	**54.**	120, 380

The rule for divisibility by 7 follows. If the number without its unit digit, minus twice the unit digit of the original number, is divisible by 7, then the original number is divisible by 7. For example, consider the number 2996. Its unit digit is 6. Begin by dropping the unit digit. Then subtract twice the unit digit as follows:

$$299 - 2(6) = 299 - 12 = 287$$

Now divide, $287 \div 7 = 41$. Since 7 divides 287, the original number 2996 is divisible by 7.

In Exercises 55–58, use the above procedure to determine whether the number is divisible by 7.

55. 3661   **56.** 4375
57. 39,837   **58.** 56,678

The rule for divisibility by 11 is as follows. Find the sum of the digits in the odd positions (from the right) and the sum of the digits in the even positions. Then find the difference between the sums. If the difference is 0 or a number divisible by 11, then the original number is divisible by 11. For example, to determine whether 136,059 is divisible by 11, we do the following:

$$9 + 0 + 3 = 12 \qquad 5 + 6 + 1 = 12$$

Now subtract: $12 - 12 = 0$. Since the difference is 0, the number 136,059 is divisible by 11.

In Exercises 59–62 use the above procedure to determine whether or not the number is divisible by 11.

59. 61,855   **60.** 134,431
61. 6,185,854   **62.** 672,541

PROBLEM SOLVING

63. A number in which each digit except 0 appears exactly three times is divisible by 3. For example, 888,444,555 and 714,714,714 are both divisible by 3. Explain why this must be true.

64. Explain why the rule for divisibility by 3 works. (*Hint:* Start with *abc*, which represents a three-digit number and write it in expanded form.)

65. Consider the first eight prime numbers greater than 3. The numbers are 5, 7, 11, 13, 17, 19, 23, and 29.
  a) Determine which of these prime numbers differs by 1 from a multiple of the number 6.
  b) Use inductive reasoning and the results obtained in part (a) to make a conjecture regarding prime numbers.
  c) Select a few more prime numbers and determine whether your conjecture appears to be correct.

66. A **palindromic** number is a number that reads the same forwards and backwards.
  a) Check the following palindromic numbers for divisibility by 11 (see Exercises 59–62):
    i) 3553   **ii)** 6116   **iii)** 9229
  b) Explain why a four-digit palindromic number must be divisible by 11.
  c) Is every five-digit palindromic number divisible by 11? Give an example to support your answer.
  d) Is every six-digit palindromic number divisible by 11?
  e) Do you believe that every palindromic number that contains an even number of digits is divisible by 11? Explain your answer.

5.2 THE INTEGERS

In Section 5.1 we introduced the natural or counting numbers:

$$N = \{1, 2, 3, 4, \ldots\}$$

These are the numbers used when counting. Another important set of numbers, the whole numbers, was developed to answer the question "How many?"

$$\text{Whole numbers} = \{0, 1, 2, 3, 4, \ldots\}$$

Note that the set of whole numbers contains the number 0, while the set of counting numbers does not. If a farmer were asked how many chickens he owned, his answer would be a whole number. If the farmer had no chickens, he or she could answer 0. Although we use the number 0 daily and take it for granted, the number 0 as we know it was not used and accepted in our society until the sixteenth century.

DID YOU KNOW?

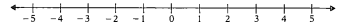

Some years ago a group of French mathematicians who worked under the collective pseudonym of "Monsieur Nicholas Bourbaki" embarked on the development of an encyclopedic description of all mathematics. They found that they had devoted 200 pages simply to introduce the difficulties of that innocent-looking concept, the number "one."

Another important set of numbers is the set of **integers:**

$$\text{Integers} = \{\ldots -4, -3, -2, -1, 0, 1, 2, 3, \ldots\}.$$

Negative integers   Positive integers

The set of integers consists of the negative integers, zero, and the positive integers. The term "positive integers" is yet another name for the natural numbers or counting numbers.

An understanding of addition, subtraction, multiplication, and division of the integers is essential before one can study algebra (Chapter 6). To aid in our explanation of addition and subtraction of integers, we will introduce the real number line (Fig. 5.6).

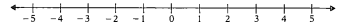

FIGURE 5.6

To construct the real number line, arbitrarily select a point for zero to serve as the starting point. Place the positive integers to the right of 0, equally spaced from one another. Place the negative integers to the left of 0, using the same spacing. The real number line contains the integers and the other real numbers that are not integers. Some examples of real numbers that are not integers are indicated in Fig. 5.7. These other types of numbers will be discussed in the next two sections.

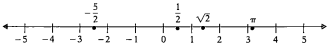

FIGURE 5.7

The arrows at the ends of the real number line indicate that the line continues indefinitely in both directions. Note that for any natural number, n, on the number line, the opposite of that number, $-n$, is also on the number line. This real number line was drawn horizontally, but it could just as well have been drawn vertically. In fact, in the next chapter you will see that the axes of a graph are the union of two number lines, one horizontal and the other vertical.

The number line can be used to determine the order of the integers. The number 3 is greater than 2, written $3 > 2$. Observe that 3 is to the right of 2 on the number line. Similarly, we can see that $0 > -1$ because 0 is to the right of -1 on the number line.

Instead of stating that 3 is greater than 2, we could have stated that 2 is less than 3, written $2 < 3$. Observe that 2 is to the left of 3 on the number line. Similarly, we can see that $-1 < 0$ because -1 is to the left of 0 on the number line. The point of the inequality symbol is always placed next to the smaller of the two numbers.

■ **Example 1** _____

Insert either $>$ or $<$ in the space between the paired numbers to make the statement correct.

a) -4 2   **b)** -2 -4 **c)** -5 -3 **d)** 0 -2

Solution

a) $-4 < 2$, since -4 is to the left of 2.
b) $-2 > -4$, since -2 is to the right of -4.
c) $-5 < -3$, since -5 is to the left of -3.
d) $0 > -2$, since 0 is to the right of -2.   ❖

Addition of Integers

Addition of integers can be represented geometrically with a number line. To accomplish this, begin at 0 on the number line. Represent the first addend by an arrow starting at 0. The arrow will be drawn to the right if the addend is positive. If the addend is negative, the arrow will be drawn to the left. From the tip of the first arrow, draw a second arrow to represent the second addend. The second arrow is drawn to the right or left, as explained above. The sum of the two integers is found at the tip of the second arrow.

■ **Example 2** _____

Evaluate $2 + (-4)$, using the number line.

Solution

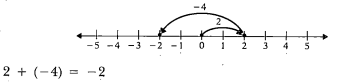

$$2 + (-4) = -2$$ ❖

■ **Example 3** ——————————————————

Evaluate $-2 + (-5)$, using the number line.

Solution

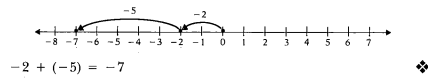

$$-2 + (-5) = -7$$ ❖

■ **Example 4** ——————————————————

Evaluate $-5 + 2$, using the number line.

Solution

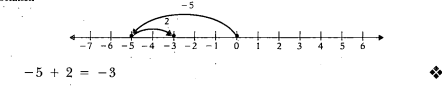

$$-5 + 2 = -3$$ ❖

■ **Example 5** ——————————————————

Evaluate $4 + (-4)$, using the number line.

Solution

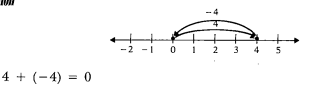

$$4 + (-4) = 0$$ ❖

The number -4 is said to be the additive inverse of 4, and 4 is the additive inverse of -4, because their sum is 0. In general the **additive inverse** of the number n is $-n$, since $n + (-n) = 0$. Inverses will be discussed more formally in Chapter 9, Mathematical Systems.

Subtraction of Integers

Any subtraction problem can be rewritten as an addition problem. To do so, we use the following definition of subtraction.

Subtraction
$a - b = a - (+b) = a + (-b)$

For example, $3 - 5$ actually means that we are subtracting a positive 5 from a positive 3. Thus $3 - 5$ can be written $3 - (+5)$. To change from a subtraction problem to an addition problem, we must (1) change the sign of the number being subtracted *and* (2) change the subtraction sign to an addition sign. Thus $3 - (+5) = 3 + (-5)$. Now we can add $3 + (-5)$ on the number line to obtain the answer (Fig. 5.8).

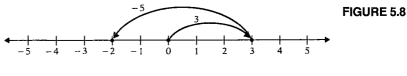

FIGURE 5.8

Thus $3 - 5 = 3 - (+5) = 3 + (-5) = -2$.

■ **Example 6** _____

Evaluate $-3 - (-2)$, using the number line.

Solution: From -3 we are subtracting the number -2. To change this subtraction problem to an addition problem, change the sign of the number being subtracted to $+2$ and change the subtraction sign to an addition sign.

$$-3 - (-2) = -3 + (+2) = -3 + 2$$

We can now add $-3 + 2$ on the number line to obtain the answer -1.

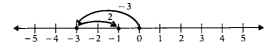

Thus $-3 - (-2) = -3 + 2 = -1$. ❖

■ **Example 7** _____

Evaluate $-3 - 5$, using the number line.

Solution: $-3 - 5 = -3 + (-5)$. Now add on the number line to get the answer, -8.

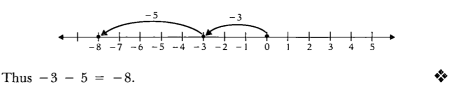

Thus $-3 - 5 = -8$.   ❖

■ **Example 8** _____

Evaluate $4 - (-3)$.

Solution: $4 - (-3) = 4 + (+3) = 4 + 3 = 7$ ❖

Multiplication of Integers

The multiplication property of zero is important in our discusssion of multiplication of integers. It indicates that the product of 0 and any number is 0.

> **Multiplication Property of Zero**
> $$a \cdot 0 = 0 \cdot a = 0$$

For the positive integers we often define multiplication as repeated addition. Thus $2 \cdot 3$ means $2 + 2 + 2$. We will develop the rules for multiplication of integers intuitively, using number patterns. The four possible cases are:

1. Positive integer $\times$ positive integer,
2. Positive integer $\times$ negative integer,
3. Negative integer $\times$ positive integer,
4. Negative integer $\times$ negative integer.

- **CASE 1: Positive integer $\times$ positive integer**
  The product of two positive integers can be defined as repeated addition. This sum will always be positive. Thus a positive integer times a positive integer is a positive integer.

- **CASE 2: Positive integer × negative integer**
  Consider the following patterns.

$$4(3) = 12$$
$$4(2) = 8$$
$$4(1) = 4$$

Note that each time the second term is reduced by 1, the product is reduced by 4. Continuing the process gives

$$4(0) = 0.$$

What comes next?

$$4(-1) = -4$$
$$4(-2) = -8$$

The pattern indicates that a positive integer times a negative integer is a negative integer.

We can also illustrate that a positive integer times a negative integer is a negative integer by using the number line. The expression $3(-2)$ means $(-2) + (-2) + (-2)$. Adding $(-2 + (-2) + (-2)$ on the number line, we obtain a sum of -6 (Fig. 5.9).

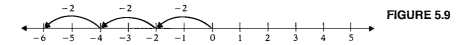

FIGURE 5.9

- **CASE 3: Negative integer × positive integer**
  A procedure similar to that used in Case 2 will indicate that a negative integer times a positive integer is a negative integer.

- **CASE 4: Negative integer × negative integer**
  We have illustrated that a positive integer times a negative integer is a negative integer. We will make use of this fact in the following pattern.

$$4(-3) = -12$$
$$3(-3) = -9$$
$$2(-3) = -6$$
$$1(-3) = -3$$

In this pattern, each time the first term is decreased by 1, the product is increased by 3. Continuing this process gives

$$0(-3) = 0,$$
$$(-1)(-3) = 3,$$
$$(-2)(-3) = 6.$$

This pattern illustrates that a negative integer times a negative integer is a positive integer.

The examples were restricted to integers. However, the rules can be used for any numbers. Let's summarize the rules.

1. The product of two numbers with *like* signs (positive × positive or negative × negative) is a *positive* number.

2. The product of two numbers with *unlike* signs (positive × negative or negative × positive) is a *negative* number.

◼ Example 9

Evaluate $4(-7)$.

Solution: Since the numbers have unlike signs, the product is negative:

$$4(-7) = -28$$ ❖

◼ Example 10

Evaluate $(-3)(2)$.

Solution: Since the numbers have unlike signs, the product is negative:

$$(-3)(2) = -6$$ ❖

◼ Example 11

Evaluate $(-4)(-5)$.

Solution: Since the numbers have like signs, both negative, the product is positive:

$$(-4)(-5) = 20$$ ❖

In Example 6 we showed that $-3 - (-2) = -1$, using the subtraction principle. We can now get the same result by an alternative method. Note that $-3 - (-2)$ is the same as $-3 - 1(-2)$, which, in turn, may be written as $-3 + (-1)(-2)$. Since a negative number times a negative number is a positive number, $(-1)(-2)$ is $+2$. Thus $-3 - (-2)$ can be written as $-3 + 2$, which equals -1.

Division of Integers

You may already realize that there is a relationship between multiplication and division.

$$6 \div 2 = 3 \quad \text{means that} \quad 3 \cdot 2 = 6.$$

$$\frac{20}{10} = 2 \quad \text{means that} \quad 2 \cdot 10 = 20.$$

$$\frac{24}{8} = 3 \quad \text{means that} \quad 3 \cdot 8 = 24.$$

From the examples above we can see that division is the reverse process of multiplication.

> For any a, b, and z, where $b \neq 0$, $a \div b = z$
> means that $z \cdot b = a$.

The reason for specifying that b must not equal zero is that division by 0 is not permitted. The following two cases will illustrate why.

● **CASE 1: What is the quotient when 3 is divided by 0?**
Let us designate the quotient by the letter z.

$$\frac{3}{0} = z$$

From our definition above, $3 \div 0 = z$ means that $z \cdot 0 = 3$. Since there is no number z such that $z \cdot 0 = 3$, the solution cannot be found. For this reason we say the quotient of 3/0 is *undefined*. Any number of the form $\frac{a}{0}$, $a \neq 0$, is said to be undefined.

● **CASE 2: What is the quotient when 0 is divided by 0?**
Let us designate the quotient by the letter z.

$$\frac{0}{0} = z$$

From the definition just given, $0 \div 0 = z$ means that $z \cdot 0 = 0$. But $z \cdot 0 = 0$ is true for any value of z selected, so there is no single solution. Therefore we say the quotient of 0/0 is *indeterminate*.

Now let us state the rules for division of integers. As with multiplication, there are four cases.

- **CASE 1: A positive integer divided by a positive integer is a positive integer.**

$$\frac{6}{2} = 3 \quad \text{since} \quad 3(2) = 6$$

- **CASE 2: A positive integer divided by a negative integer is a negative integer.**

$$\frac{6}{-2} = -3 \quad \text{since} \quad (-3)(-2) = 6$$

- **CASE 3: A negative integer divided by a positive integer is a negative integer.**

$$\frac{-6}{2} = -3 \quad \text{since} \quad (-3)(2) = -6$$

- **CASE 4: A negative integer divided by a negative integer is a positive integer.**

$$\frac{-6}{-2} = 3 \quad \text{since} \quad 3(-2) = -6$$

The examples were restricted to integers. However the rules can be used for any numbers. The rules for division of integers are summarized below.

1. The quotient of two numbers with *like* signs (positive ÷ positive or negative ÷ negative) is a *positive* number.
2. The quotient of two numbers with *unlike* signs (positive ÷ negative or negative ÷ positive) is a *negative* number.

Example 12

Evaluate $\dfrac{20}{-5}$.

Solution: Since the numbers have unlike signs, the quotient is negative:

$$\frac{20}{-5} = -4$$

■ **Example 13** _____

Evaluate $\dfrac{-16}{4}$.

Solution: Since the numbers have unlike signs, the quotient is negative:

$$\frac{-16}{4} = -4$$ ❖

■ **Example 14** _____

Evaluate $\dfrac{-25}{-5}$.

Solution: Since the numbers have like signs (both negative), the quotient is positive:

$$\frac{-25}{-5} = 5$$ ❖

DID YOU KNOW?

The ancient Greeks are considered by many to have been the first true mathematicians. They were the first to study mathematics, not because of its applications, but because of its beauty. The Greeks believed that all nature, all harmony, and everything of beauty could be explained by using ratios of whole numbers (called rational numbers).

Because of this belief the Greeks studied numbers and the relationships between them. The Greeks' belief was reinforced when they discovered that the sounds of plucked strings could be quite pleasing if the strings plucked were in a ratio of 1 to 2 (an octave), 2 to 3 (a fifth), 3 to 4 (a fourth), and so on. In other words, the secret of harmony lies in the ratios of whole numbers, such as 1/2, 2/3, and 3/4.

Nowadays the study of relationships between numbers is called **number theory.**

The Greeks discussed many types of numbers, including prime, composite, and perfect numbers. Other numbers discussed by the Greeks include triangular and square numbers. Examples of these numbers were given in Exercises 1.1.

In their study of the theory of numbers the Greeks also developed and studied the concept of number sequences, which are discussed later in this chapter.

EXERCISES 5.2

Use the number line to evaluate the following.

1. $4 + 3$
2. $4 + (-5)$
3. $(-4) + (2)$
4. $(-3) + (-3)$
5. $7 + (-4)$
6. $(4 + 3) + (-2)$
7. $[(-3) + (-2)] + 5$
8. $[6 + (-4)] + (-3)$
9. $[(-7) + (-3)] + 8$
10. $(2 - 11) + (-7)$
11. $[6 + (-4)] + (-7)$
12. $(11 - 13) + (-9)$

Use the number line to evaluate the following.

13. $3 - 4$
14. $-2 - 6$
15. $6 - 3$
16. $4 - (-2)$
17. $-4 - (-2)$
18. $3 - (-3)$
19. $3 - 8$
20. $3 - (-10)$
21. $-5 - 8$
22. $-7 - 12$
23. $[4 + (-2)] - 3$
24. $3 - (3 + 4)$

Evaluate the following.

25. $4 \cdot 6$
26. $2(-4)$
27. $(-3)(-3)$
28. $(-3)(4)$
29. $[(-3)(-2)] \cdot 3$
30. $[(-2)(4)](-3)$
31. $(3 \cdot 3)(-2)$
32. $[(-1)(-3)](-3)$
33. $[(-6)(-4)](-5)$
34. $(-4)[8(-6)]$
35. $[(-2)(-2)] \cdot [(-6)(9)]$
36. $\{[(-5)(7)](9)\}(-3)$

Evaluate the following.

37. $6 \div 2$
38. $-6 \div 3$
39. $15 \div (-3)$
40. $\dfrac{-20}{-4}$
41. $\dfrac{36}{-4}$
42. $\dfrac{-100}{20}$
43. $\dfrac{24}{-3}$
44. $\dfrac{-22}{-11}$
45. $\dfrac{-90}{-5}$
46. $-120 \div 20$
47. $\dfrac{153}{-3}$
48. $(-800) \div (-8)$

Evaluate the following.

49. $(4 + 4) \div 2$
50. $[4(-3)] + 2$
51. $[4(-3)] - 2$
52. $[(-4)(-3)] - 2$
53. $(3 - 5)(2)$
54. $[12 \div (-4)](-3)$
55. $[4 + (-10)] \div 2$
56. $(3 - 8) \div (-5)$
57. $[(-25)(-2)] \div (6 - 8)$
58. $[18(-3)] \div (-9)$

59. Explain why $\dfrac{a}{b}$ is the same as $\dfrac{-a}{-b}$.

60. Explain why $-\dfrac{a}{b}$ is the same as $\dfrac{a}{-b}$.

61. Mount Everest, the highest point on the earth, is 29,028 feet above sea level. The Marianas Trench, the lowest point on the earth, is 36,198 feet below sea level. Find the vertical height difference between Mount Everest and the Marianas Trench. Consider depth below sea level as negative.

62. A helicopter drops a package from a height of 842 feet above sea level. The package lands in the ocean and settles at a point 927 feet below sea level. What was the vertical distance the package traveled?

63. On the first play a football team loses 23 yards. On the second play it gains 18 yards. Find the total gain or loss for the two plays.

64. The balance in your checkbook is $27. What is the balance if you write a check for $52?

65. A submarine is at a depth of 253 feet below sea level. If the submarine dives another 628 feet, find its depth. Consider depth below sea level as negative.

The ancient Greeks defined a number whose **proper factors** (factors other than the number itself) add up to the number itself as a **perfect number.** For example, 6 is a perfect number because its proper factors are 1, 2, and 3, and $1 + 2 + 3 = 6$. The number 8 is not perfect because its proper factors are 1, 2, and 4, and $1 + 2 + 4 \neq 8$. Determine which, if any, of the following numbers are perfect.

66. 12
67. 28
68. 30
69. 36

PROBLEM SOLVING

70. Find the quotient:

$$\frac{-1 + 2 - 3 + 4 - 5 + \ldots - 99 + 100}{1 - 2 + 3 - 4 + 5 - \ldots + 99 - 100}$$

71. Triangular numbers and square numbers were introduced in Exercises 1.1. There are also **pentagonal numbers,** which were also studied by the Greeks. Four pentagonal numbers are 1, 5, 12, and 22.

    1   5   12   22

a) Determine the next three pentagonal numbers.
b) Describe a procedure to determine the next five pentagonal numbers without drawing the figures.
c) Is 72 a pentagonal number? Explain how you determined your answer.

72. Place the appropriate plus or minus signs between each digit so that the total will equal 1.

    0 1 2 3 4 5 6 7 8 9 = 1

5.3   THE RATIONAL NUMBERS

> The set of **rational numbers,** denoted by Q, is the set of all numbers of the form p/q, where p and q are integers and $q \neq 0$.

"When you can measure what you are talking about and express it in numbers, you know something about it."
Lord Kelvin

The definition above tells us that any number that can be expressed as the quotient of two integers (denominator not 0) is a rational number. The following numbers are examples of rational numbers:

$$\frac{3}{5}, \quad \frac{-2}{7}, \quad \frac{12}{5}, \quad 2, \quad 0$$

The integers 2 and 0 are rational numbers because each can be expressed as the quotient of two integers: $2 = 2/1$ and $0 = 0/1$. In fact, every integer p is a rational number, since it can be written in the form of $p/1$.

Note the following important property of the rational numbers.

> Every **rational number** when expressed as a decimal will be either a terminating or a repeating decimal.

■ **Example 1**

Show that the following rational numbers are terminating decimals.

a) 1/2   **b)** 12/5   **c)** 3/8

Solution: To express the rational number in decimal form, divide the numerator by the denominator.

$$
\begin{array}{r}
0.5 \\
2\,\overline{)1.0} \\
\underline{1\,0} \\
0
\end{array}
\qquad
\begin{array}{r}
2.4 \\
5\,\overline{)12.0} \\
\underline{10} \\
2\,0 \\
\underline{2\,0} \\
0
\end{array}
\qquad
\begin{array}{r}
0.375 \\
8\,\overline{)3.000} \\
\underline{2.4} \\
60 \\
\underline{56} \\
40 \\
\underline{40} \\
0
\end{array}
$$

a)   **b)**   **c)**

Each is a terminating decimal, since at some point the remainder is 0 and the division process terminates. Zeros may be placed to the right of the last digit after the decimal without affecting the value of a terminating decimal. Hence 0.5 = 0.50 = 0.500.   ❖

■ **Example 2**

Show that the following rational numbers are repeating decimals.

a) 1/3   **b)** 23/99   **c)** 2/7

Solution

$$
\begin{array}{r}
0.333\ldots \\
3\,\overline{)1.000} \\
\underline{9} \\
10 \\
\underline{9} \\
10 \\
\underline{9} \\
1
\end{array}
\qquad
\begin{array}{r}
0.2323\ldots \\
99\,\overline{)23.0000} \\
\underline{19\,8} \\
3\,20 \\
\underline{2\,97} \\
230 \\
\underline{198} \\
320 \\
\underline{297} \\
23
\end{array}
\qquad
\begin{array}{r}
0.285714285714\ldots \\
7\,\overline{)2.00000} \\
\underline{1\,4} \\
60 \\
\underline{56} \\
40 \\
\underline{35} \\
50 \\
\underline{49} \\
10 \\
\underline{7} \\
30 \\
\underline{28} \\
2
\end{array}
$$

a)   **b)**   **c)**

The divisions above will never have a 0 remainder. The quotients will repeat a number or set of numbers indefinitely—hence the term "repeating decimal."   ❖

To show that a number or set of numbers repeats, we place a bar above that numbers that repeat.

$$1/3 = 0.333\ldots = 0.\overline{3}$$
$$23/99 = 0.2323\ldots = 0.\overline{23}$$
$$2/7 = 0.285714285714\ldots = 0.\overline{285714}$$

Note that when a fraction is converted to a decimal number, the maximum number of digits that can repeat is $n - 1$, where n is the denominator of the fraction. For example, when 2/7 is converted to a decimal number, the maximum number of digits that can repeat is $7 - 1$ or 6.

We can convert a quotient of integers into a terminating or repeating decimal by dividing the numerator by the denominator. We can also convert a terminating or repeating decimal into a quotient of integers. The following examples indicate how this is done.

■ **Example 3** _____

Convert the following terminating decimals to a quotient of integers.

a) 0.4   **b)** 0.62   **c)** 0.062

Solution: The positional values to the right of the decimal point are illustrated in Fig. 5.10. When converting a terminating decimal to a quotient of integers, we observe the last digit to the right of the decimal point. The position of this digit will indicate the denominator of the quotient of integers.

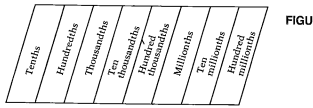

FIGURE 5.10

a) 0.4 = 4/10, since the 4 is in the tenths position.

b) 0.62 = 62/100, since the digit on the right, 2, is in the hundredths position.

c) 0.062 = 62/1000, since the digit on the right, 2, is in the thousandths position.   ❖

Converting repeating decimals to a quotient of integers is more difficult. To do so, we must "create" another repeating decimal with the same repeating digits so that when one repeating decimal is subtracted from the other repeating decimal, the difference will be a whole number. To create a number with the same repeating decimal, multiply the original repeating decimal by 10 if one digit repeats, by 100 if two digits repeat, by 1000 if three digits repeat, and so on. Examples 4, 5, and 6 illustrate this procedure.

■ **Example 4** _____

Convert $0.\overline{3}$ to a quotient of integers.

Solution: $0.\overline{3} = 0.3\overline{3} = 0.33\overline{3}$, and so on.

Let the original repeating decimal be called n; thus $n = 0.\overline{3}$. Since one digit repeats, we multiply both sides of the equation by 10. This gives $10n = 3.\overline{3}$. Then subtract, as indicated below.

$$
\begin{array}{ll}
\begin{aligned}
10n &= 3.\overline{3} \\
n &= 0.\overline{3} \\
\hline
9n &= 3.0
\end{aligned}
\quad \text{or} \quad
\begin{aligned}
10n &= 3.3333\ldots \\
n &= 0.3333\ldots \\
\hline
9n &= 3.0
\end{aligned}
\end{array}
$$

Note that $10n - n = 9n$ and $3.\overline{3} - 0.\overline{3} = 3.0$.
Next, solve for n by dividing both sides by 9:

$$\frac{9n}{9} = \frac{3.0}{9}$$

$$n = \frac{3}{9} = \frac{1}{3}$$

Therefore $0.\overline{3} = 1/3$.   ❖

■ **Example 5** _____

Convert $0.\overline{83}$ to a quotient of integers.

Solution: Let $n = 0.\overline{83}$. Since two digits repeat, multiply both sides of the equation by 100. Thus $100n = 83.\overline{83}$. Now subtract n from $100n$:

$$
\begin{aligned}
100n &= 83.\overline{83} \\
n &= 0.\overline{83} \\
\hline
99n &= 83
\end{aligned}
$$

Finally, divide both sides by 99:

$$\frac{99n}{99} = \frac{83}{99}$$
$$n = \frac{83}{99}$$

Therefore $0.\overline{83} = \frac{83}{99}$.   ❖

■ **Example 6** _____

Convert $12.14\overline{2}$ to a quotient of integers.

Solution: This problem is different from the two preceding examples in that the repeating digit, 2, is not directly to the right of the decimal point. When this situation arises, it is helpful to move the decimal point to the right until the repeating terms are directly to its right. For each place the decimal point is moved, the number is multiplied by 10. In this example the decimal point must be moved two places to the right. Thus the number must be multiplied by 100.

$$n = 12.14\overline{2}$$
$$100n = 100 \times 12.14\overline{2} = 1214.\overline{2}$$

Now proceed as in the previous two examples. Since one digit repeats, we multiply both sides by 10:

$$100n = 1214.\overline{2}$$
$$10 \times 100n = 10 \times 1214.\overline{2}$$
$$1000n = 12142.\overline{2}$$

Now subtract $100n$ from $1000n$ so that the repeating term will drop out:

$$1000n = 12142.\overline{2}$$
$$\underline{100n = 1214.\overline{2}}$$
$$900n = 10928$$

Thus $n = 10928/900$ or $2732/225$.   ❖

A set of numbers is said to be **dense** if between any two distinct members of the set there exists a third distinct member of the set. The set of integers is not dense, since between any two integers there is not another integer. For example, between 1 and 2 there are no other integers. The set of rational numbers is dense because between any two distinct rational numbers there exists a third distinct rational number.

What is $0.\overline{9}$ when expressed as a quotient of integers? We will use the procedure we used earlier in this section to answer this question.

Let $n = 0.\overline{9}$.

$$
\begin{aligned}
10n &= 9.\overline{9} \\
n &= 0.\overline{9} \\
\hline
9n &= 9 \\
n &= 9/9 = 1
\end{aligned}
$$

We have just illustrated that $0.\overline{9} = 1$, a statement that many of you probably question. However, we can show that $0.\overline{9} = 1$, using the denseness property. We stated earlier that the rational numbers are dense. Thus if $0.\overline{9}$ and 1 are two different rational numbers, then we can find a rational number between the two.

$$
\frac{0.\overline{9} + 1}{2} = \frac{1.\overline{9}}{2} = 0.\overline{9}
$$

Since the average of the two numbers, $0.\overline{9}$, is one of the original numbers, the two numbers must be equal.

■ Example 7

Determine a rational number between the following pairs of numbers.

a) 0.243 and 0.244

b) 0.0007 and 0.0008

Solution

a) The number 0.243 can be written as 0.2430, and 0.244 can be written as 0.2440. There are many numbers between these two. Some of them are 0.2431, 0.2435, and 0.243912. How many numbers are there between these two numbers?

b) 0.0007 can be written as 0.00070 and 0.0008 can be written as 0.00080. One number between them is 0.00075.   ❖

It is also possible to find a rational number between any two fractions. One way to do so is to add the fractions and divide the sum by 2. In order to do this we will explain how to add, subtract, multiply, and divide fractions.

Multiplication and Division of Fractions

The top number in a fraction is called the **numerator,** and the bottom number is the **denominator.** We find the product of two fractions by multiplying the numerators together and multiplying the denominators together.

Multiplication of Fractions

$$\frac{a}{b} \cdot \frac{c}{d} = \frac{a \cdot c}{b \cdot d} = \frac{ac}{bd} \qquad b \neq 0, \quad d \neq 0$$

Example 8

Evaluate the following.

a) $\dfrac{2}{3} \cdot \dfrac{5}{7}$ b) $\left(\dfrac{-2}{3}\right)\left(\dfrac{-4}{9}\right)$ c) $\left(2\dfrac{3}{4}\right)\left(1\dfrac{5}{8}\right)$

Solution

a) $\dfrac{2}{3} \cdot \dfrac{5}{7} = \dfrac{2 \cdot 5}{3 \cdot 7} = \dfrac{10}{21}$

b) $\left(\dfrac{-2}{3}\right)\left(\dfrac{-4}{9}\right) = \dfrac{(-2)(-4)}{(3)(9)} = \dfrac{8}{27}$

c) First we change each of the mixed numbers to fractions, as illustrated below.

$$2\frac{3}{4} = \frac{(2 \cdot 4) + 3}{4} = \frac{11}{4}$$

$$1\frac{5}{8} = \frac{(1 \cdot 8) + 5}{8} = \frac{13}{8}$$

Now multiply the fractions together:

$$\left(\frac{11}{4}\right)\left(\frac{13}{8}\right) = \frac{143}{32}$$

Note that 143/32 may be written as a mixed number by dividing the numerator by the denominator, as illustrated below.

$$32 \overline{\smash)143} \begin{array}{r} 4 \\ \underline{128} \\ 15 \end{array}$$

$$\frac{143}{32} = 4\frac{15}{32} \begin{array}{l} \leftarrow \text{remainder} \\ \leftarrow \text{divisor} \end{array}$$

$$\uparrow$$
quotient

The **reciprocal** of a given number is 1 divided by the given number. The product of a number and its reciprocal must equal one. Examples of some numbers and their reciprocals follow.

Number	Reciprocal	Product
3 ·	$\frac{1}{3}$ =	1
$\frac{1}{2}$ ·	2 =	1
$\frac{3}{5}$ ·	$\frac{5}{3}$ =	1
-6 ·	$-\frac{1}{6}$ =	1

To find the quotient of two fractions, multiply the first fraction by the reciprocal of the second fraction.

Division of Fractions

$$\frac{a}{b} \div \frac{c}{d} = \frac{a}{b} \cdot \frac{d}{c} = \frac{ad}{bc}, \quad b \neq 0, \quad d \neq 0, \quad c \neq 0$$

Note that $\dfrac{\frac{a}{b}}{\frac{c}{d}}$ means the same as $\dfrac{a}{b} \div \dfrac{c}{d}$. Thus $\dfrac{\frac{a}{b}}{\frac{c}{d}}$ is evaluated as $\dfrac{a}{b} \cdot \dfrac{d}{c}$.

■ **Example 9** _____

Evaluate the following.

a) $\dfrac{2}{3} \div \dfrac{5}{7}$ b) $\dfrac{\frac{-3}{5}}{\frac{5}{7}}$ c) $\dfrac{\frac{2}{3}}{5}$

Solution

a) $\dfrac{2}{3} \div \dfrac{5}{7} = \dfrac{2}{3} \cdot \dfrac{7}{5} = \dfrac{2 \cdot 7}{3 \cdot 5} = \dfrac{14}{15}$

b) $\dfrac{\frac{-3}{5}}{\frac{5}{7}} = \dfrac{-3}{5} \cdot \dfrac{7}{5} = \dfrac{-3 \cdot 7}{5 \cdot 5} = \dfrac{-21}{25}$

c) $\dfrac{\frac{2}{3}}{5}$ is the same as $\dfrac{\frac{2}{3}}{\frac{5}{1}}$ and $\dfrac{\frac{2}{3}}{\frac{5}{1}} = \dfrac{2}{3} \cdot \dfrac{1}{5} = \dfrac{2 \cdot 1}{3 \cdot 5} = \dfrac{2}{15}$

❖

Addition and Subtraction of Fractions

To add or subtract fractions, it is necessary for the fractions to have a common denominator. A common denominator is another name for a common multiple. The least common multiple (lcm), which was discussed in Section 5.1, will be the lowest common denominator.

To add or subtract two fractions with a common denominator, we add or subtract their numerators and retain the common denominator.

$$\frac{a}{c} + \frac{b}{c} = \frac{a + b}{c}, \quad c \neq 0 \qquad \frac{a}{c} - \frac{b}{c} = \frac{a - b}{c}, \quad c \neq 0$$

■ **Example 10** _____

Evaluate the following.

a) $\dfrac{3}{7} + \dfrac{2}{7}$ b) $\dfrac{5}{9} - \dfrac{1}{9}$

Solution

a) $\dfrac{3}{7} + \dfrac{2}{7} = \dfrac{3 + 2}{7} = \dfrac{5}{7}$

b) $\dfrac{5}{9} - \dfrac{1}{9} = \dfrac{5 - 1}{9} = \dfrac{4}{9}$ ❖

Notice that in Example 10 the denominators of the fractions being added or subtracted were the same. *When adding or subtracting two fractions with unlike denominators, we first rewrite each fraction with a common denominator. Then we add or subtract the fractions.*

To write fractions with a common denominator, we use the fundamental law of rational numbers.

Fundamental Law of Rational Numbers
If a, b, and c are integers, with $b \neq 0$ and $c \neq 0$, then

$$\frac{a}{b} = \frac{a}{b} \cdot \frac{c}{c} = \frac{a \cdot c}{b \cdot c}$$

Examples 11–13 illustrate the procedure for adding and subtracting fractions with unlike denominators.

■ **Example 11** _____

Evaluate $\dfrac{5}{12} - \dfrac{3}{10}$

Solution: Using prime factorization, as discussed in Section 5.1, we find that the lcm of 12 and 10 is 60. We will therefore express each fraction with a denominator of 60. Sixty divided by 12 is 5. Therefore the denominator, 12, must be multiplied by 5 to get 60. If the denominator is multiplied by 5, the numerator must also be multiplied by 5 so that the value of the fraction remains unchanged. Multiplying both numerator and denominator by 5 is the same as multiplying by 1.

We follow the same procedure for the other fraction, $\frac{3}{10}$. Sixty divided by 10 is 6. Therefore we multiply both the denominator, 10, and the numerator, 3, by 6 to obtain an equivalent fraction with a denominator of 60.

$$\frac{5}{12} - \frac{3}{10} = \left(\frac{5}{12} \cdot \boxed{\frac{5}{5}}\right) - \left(\frac{3}{10} \cdot \boxed{\frac{6}{6}}\right)$$

$$= \frac{25}{60} - \frac{18}{60}$$

$$= \frac{7}{60} \qquad \qquad ❖$$

■ **Example 12** _____

Add $\dfrac{1}{36} + \dfrac{1}{150}$.

Solution: In Example 8, Section 5.1, we determined that the lcm of 36 and 150 is 900. We rewrite both fractions using the lcm in the denominator.

$$\frac{1}{36} + \frac{1}{150} = \left(\frac{1}{36} \cdot \boxed{\frac{25}{25}}\right) + \left(\frac{1}{150} \cdot \boxed{\frac{6}{6}}\right)$$

$$= \frac{25}{900} + \frac{6}{900}$$

$$= \frac{31}{900} \qquad \qquad ❖$$

Now that we know how to add and divide fractions, we can find a fraction between any two given fractions by adding the two given fractions and dividing their sum by 2.

∎ **Example 13** _____

Find a rational number between $\frac{1}{4}$ and $\frac{1}{3}$.

Solution

a)   First add $\frac{1}{4}$ and $\frac{1}{3}$:

$$\frac{1}{4} + \frac{1}{3} = \frac{3}{12} + \frac{4}{12} = \frac{7}{12}$$

b)   Next divide the sum by 2:

$$\frac{7}{12} \div 2 = \frac{7}{12} \div \frac{2}{1} = \frac{7}{12} \cdot \frac{1}{2} = \frac{7}{24}$$

Thus $\frac{7}{24}$ is between $\frac{1}{4}$ and $\frac{1}{3}$, written $\frac{1}{4} < \frac{7}{24} < \frac{1}{3}$.   ❖

This procedure will result in a fraction that is exactly halfway between (or the average of) the two given fractions.

Reducing Fractions

A fraction is said to be in its **lowest terms** (or reduced) when the numerator and denominator are relatively prime (that is, have no common divisors other than 1). To reduce a fraction to its lowest terms, divide both the numerator and the denominator by the greatest common divisor. A procedure for finding the greatest common divisor was discussed in Section 5.1.

∎ **Example 14** _____

Reduce $\dfrac{36}{150}$ to its lowest terms.

Solution: In Example 6 of Section 5.1 we determined that the gcd of 36 and 150 is 6. Divide the numerator and the denominator by the gcd, 6.

$$\frac{36 \div 6}{150 \div 6} = \frac{6}{25}$$

Since there are no common divisors of 6 and 25 other than 1, this fraction is in its lowest terms.   ❖

DID YOU KNOW?

Dear Ann Landers: I wonder how many people realize what a trillion is. Let me explain:

If you were to count a trillion $1 bills, one per second, 24 hours a day, it would take you 32,000 years.

Or, to put it differently, it has been figured that with $1 trillion, you could buy a $100,000 house for every family in Kansas, Missouri, Nebraska, Oklahoma, and Iowa.

Then you could put a $10,000 car in the garage of each one of those houses. There would be enough left to build $10 million libraries and $10 million hospitals for 250 cities in those states. There would be enough left over to build $10 million schools for 500 communities.

And there would still be enough left to put in the bank and, from the interest alone, pay 10,000 nurses and teachers, and give a $5,000 bonus for every family in those states.

Worth noting: President Reagan's fanciful Strategic Defense Initiative, the "Star Wars" antimissile scheme, carries a price tag of $3 trillion.

— FRANK A., SOUTH PLAINFIELD, N.J.

EXERCISES 5.3

Determine whether the following are terminating or repeating decimals.

1. 0

2. 41.6

3. $-1.2\overline{3}$

4. 0.3131

5. $0.31\overline{31}$

6. 0.6666

7. 3.14

8. $0.\overline{6}$

9. 0.123123123...

10. 0.456456...

11. $1.\overline{413}$

12. 0.9999...

Express each fraction as a terminating or repeating decimal.

13. $\frac{2}{3}$

14. $\frac{4}{7}$

15. $\frac{7}{9}$

16. $\frac{1}{8}$

17. $\frac{5}{12}$

18. $\frac{22}{7}$

19. $\frac{7}{16}$

20. $\frac{90}{15}$

21. $\frac{124}{35}$

22. $\frac{3}{8}$

23. $\frac{41}{7}$

24. $\frac{1001}{7}$

Express each terminating decimal as a quotient of two integers.

25. 0.4

26. 0.67

27. 0.432

28. 0.6204

29. 2.001

30. 3.201

31. 0.375

32. 0.125

33. 0.8759

34. 0.45623

35. 2.396

36. 1.0015

Express each repeating decimal as a quotient of two integers.

37. $0.\overline{5}$

38. $2.\overline{3}$

39. $0.\overline{47}$

40. $1.\overline{36}$

41. $0.\overline{146}$

42. $2.\overline{101}$

43. $0.4\overline{11}$

44. $1.\overline{123}$

45. $0.1\overline{4}$

46. $0.0\overline{1}$

47. $2.6\overline{3}$

48. $5.3\overline{74}$

Evaluate the following.

49. $\frac{2}{3} \cdot \frac{5}{7}$

50. $\frac{4}{5} \div \frac{3}{8}$

51. $\left(\frac{-2}{5}\right)\left(\frac{-4}{9}\right)$

52. $\frac{-2/7}{4/9}$

53. $\frac{5}{8} \div \frac{8}{5}$

54. $\frac{5}{8} \div \frac{5}{8}$

55. $\left(\frac{2}{3} \cdot \frac{4}{5}\right) \div \frac{1}{2}$

56. $\left(\frac{3}{5} \div \frac{2}{7}\right) \cdot \frac{1}{3}$

57. $\left(\frac{2}{3}\cdot\frac{1}{4}\right)\div\left(\frac{3}{2}\cdot 2\right)$ 58. $\left(\frac{1}{3}\div\frac{3}{5}\right)\cdot\frac{9}{13}$

59. $\left[\left(\frac{-2}{3}\right)\left(\frac{-1}{5}\right)\right]\div\frac{2}{3}$ 60. $\left(\frac{1}{6}\div 2\right)\left(\frac{1}{3}\cdot 4\right)$

61. $\left(\frac{2}{7}\cdot\frac{3}{4}\right)\div\left(\frac{1}{3}\div\frac{1}{4}\right)$ 62. $\left(\frac{2}{3}\div 3\right)\div\left(\frac{1}{2}\cdot\frac{5}{6}\right)$

Reduce each fraction to its lowest terms.

63. $\frac{15}{150}$ 64. $\frac{30}{48}$ 65. $\frac{52}{78}$

66. $\frac{180}{450}$ 67. $\frac{90}{600}$ 68. $\frac{30}{78}$

69. $\frac{30}{450}$ 70. $\frac{125}{280}$ 71. $\frac{96}{212}$

72. $\frac{45}{180}$ 73. $\frac{33}{121}$ 74. $\frac{108}{196}$

Perform the indicated operation, and reduce your answer to lowest terms.

75. $\frac{1}{3}+\frac{1}{4}$ 76. $\frac{1}{5}-\frac{1}{15}$ 77. $\frac{2}{15}+\frac{7}{150}$

78. $\frac{7}{30}+\frac{5}{48}$ 79. $\frac{3}{52}-\frac{1}{78}$ 80. $\frac{9}{35}+\frac{3}{49}$

81. $\frac{15}{48}-\frac{7}{60}$

82. $\frac{1}{24}+\frac{1}{36}+\frac{1}{72}$

83. $\frac{21}{40}-\frac{4}{25}+\frac{7}{90}$

84. $\frac{6}{25}-\frac{3}{100}+\frac{7}{40}$

Alternative methods for adding and subtracting two fractions are given below. These methods may not result in a solution in its lowest terms.

$$\frac{a}{b}+\frac{c}{d}=\frac{ad+bc}{bd} \quad \text{and} \quad \frac{a}{b}-\frac{c}{d}=\frac{ad-bc}{bd}$$

Use the appropriate formula to evaluate the following.

85. $\frac{2}{3}+\frac{3}{4}$ 86. $\frac{5}{7}-\frac{1}{12}$ 87. $\frac{5}{7}+\frac{3}{4}$

88. $\frac{7}{3}-\frac{5}{12}$ 89. $\frac{3}{8}+\frac{5}{12}$ 90. $\left(\frac{2}{3}+\frac{1}{4}\right)-\frac{3}{5}$

Evaluate each of the following.

91. $\left(\frac{1}{5}\cdot\frac{1}{4}\right)+\frac{1}{3}$ 92. $\left(\frac{3}{5}\div\frac{2}{10}\right)-\frac{1}{3}$

93. $\left(\frac{1}{2}+\frac{3}{10}\right)\div\left(\frac{1}{5}+2\right)$

94. $\left(\frac{1}{9}\cdot\frac{3}{5}\right)+\left(\frac{2}{3}\cdot\frac{1}{5}\right)$

95. $\left(3-\frac{4}{9}\right)\div\left(4+\frac{2}{3}\right)$

96. $\left(\frac{2}{5}\div\frac{4}{9}\right)\left(\frac{3}{5}\cdot 6\right)$

Find a rational number between each of the following pairs.

97. 0.45 and 0.46   98. 7.003 and 7.03
99. -1.04 and -1.05 100. 0.4325 and 0.43251
101. 0.9040 and 0.90401   102. 0.312 and 0.313
103. 3.471 and 3.472   104. -0.87641 and
                                -0.87642

Find a rational number between each of the following pairs.

105. $\frac{1}{10}$ and $\frac{1}{100}$ 106. $\frac{1}{2}$ and $\frac{2}{3}$ 107. $\frac{1}{20}$ and $\frac{1}{10}$

108. $\frac{5}{13}$ and $\frac{6}{13}$ 109. $\frac{1}{4}$ and $\frac{1}{5}$ 110. $\frac{1}{3}$ and $\frac{2}{3}$

111. $\frac{2}{7}$ and $\frac{3}{7}$ 112. $\frac{1}{9}$ and $\frac{2}{9}$

113. Are the rational numbers dense? Explain.
114. Are the integers dense? Explain.
115. Find the amount of each ingredient needed to make three servings of rice.

DIRECTIONS
DO NOT BOIL RICE

1. Bring water,* salt, and butter (or margarine) to a boil.

2. Stir in rice. Cover: remove from heat. Let stand 5 minutes. Fluff with fork.

*For softer rice, use 1 tablespoon more water for each serving. Let stand 8 minutes.

Amounts of RICE & WATER
Use equal amounts rice and water. Rice doubles in volume.

TO MAKE	RICE & WATER (equal measures)	SALT	BUTTER OR MARGARINE (if desired)
2 servings	⅔ cup	¼ tsp	1 tsp
4 servings	1⅓ cups	½ tsp	2 tsp

116. If a plumber cuts a $2\frac{3}{16}$ foot length of pipe from an 8 foot length, how much remains?

117. If a carpenter places a $4\frac{5}{16}$ foot length of wood and a $3\frac{3}{8}$ foot length of wood end to end, what is the total length of the two pieces?

118. If a stock drops from $\$15\frac{1}{8}$ to $\$13\frac{3}{4}$, how much did the stock drop?

119. A piece of wood measures $15\frac{3}{8}$ inches. How far from the end should you cut the wood if you wish to cut the length in half?

120. If a piece of wood is $8\frac{3}{4}$ feet long and is to be cut into four equal parts, find the length of each part.

5.4 THE IRRATIONAL NUMBERS AND THE REAL NUMBER SYSTEM

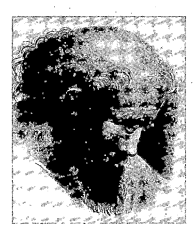

$a^2 + b^2 = c^2$

FIGURE 5.11

Pythagoras

Around 600 B.C. the ancient Greeks discovered a need for a new set of numbers, the irrationals. Pythagoras, a Greek mathematician, discovered that in any right triangle the square of the length of one side (a^2) added to the square of the length of the other side (b^2) equals the square of the length of the hypotenuse, (c^2), as shown in Fig. 5.11. When Pythagoras or a member of his sect tried to evaluate the formula $a^2 + b^2 = c^2$ for $a = 1$ and $b = 1$, the solution was not a rational number.

$$a^2 + b^2 = c^2$$
$$1^2 + 1^2 = c^2$$
$$1 + 1 = c^2$$
$$2 = c^2$$

There is no rational number that when squared—that is, when multiplied by itself—equals 2.

```
 ◄─┼────┼────┼────┼────┼────┼────┼────┼──►
   -3   -2   -1    0    1    2    3    4
```
FIGURE 5.12

In Section 5.2 we introduced the real number line, repeated here as Fig. 5.12. The points on the real number line that are not rational numbers are referred to as irrational numbers. Recall that every rational number is either a terminating or a repeating decimal. Therefore irrational numbers, when represented as decimals, will be nonterminating, nonrepeating decimals.

> An **irrational number** is a real number whose decimal representation is a nonterminating, nonrepeating decimal.

Nonrepeating number patterns can be used to indicate irrational numbers. For example, 6.1011011101111. . . and 0.525225222. . . are both irrational numbers.

The square roots of certain numbers are irrational. The **principal or positive square root** of a number, n, written $\sqrt{n}$, is the positive number that when multiplied by itself gives n. Whenever we mention the term "square root" in this text, we mean the principal square root. For example,

$$\sqrt{25} = 5, \quad \text{since} \quad 5 \cdot 5 = 25$$
$$\sqrt{100} = 10, \quad \text{since} \quad 10 \cdot 10 = 100$$

Both $\sqrt{25}$ and $\sqrt{100}$ are examples of numbers that are rational numbers, since their square roots, 5 and 10, are terminating decimals. What is $\sqrt{2}$ equal to? There is no rational number that when multiplied by itself gives 2. Thus $\sqrt{2}$ is an irrational number. Other irrational numbers include $\sqrt{3}, \sqrt{5}, \sqrt{37}$. Since $\sqrt{2}$ is an irrational number, it cannot be expressed as a terminating or repeating decimal. It can only be approximated by a decimal; that is, $\sqrt{2}$ is *approximately* 1.41. Another important irrational number used to find the area of a circle is pi, symbolized by π. Pi is *approximately* 3.14.

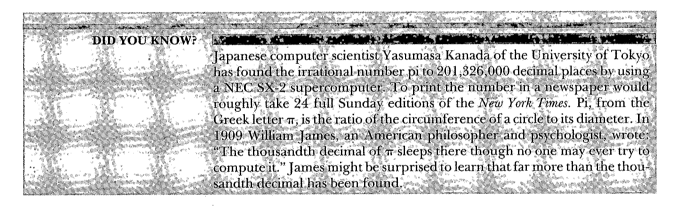

DID YOU KNOW?

Japanese computer scientist Yasumasa Kanada of the University of Tokyo has found the irrational number pi to 201,326,000 decimal places by using a NEC SX-2 supercomputer. To print the number in a newspaper would roughly take 24 full Sunday editions of the *New York Times*. Pi, from the Greek letter π, is the ratio of the circumference of a circle to its diameter. In 1909 William James, an American philosopher and psychologist, wrote: "The thousandth decimal of π sleeps there though no one may ever try to compute it." James might be surprised to learn that far more than the thousandth decimal has been found.

We have discussed procedures for performing the arithmetic operations of addition, subtraction, multiplication, and division with rational numbers. We can perform the same operations with the irrational numbers. Before we can proceed, we must have an understanding of numbers that are called perfect squares. Any number that is the square of a natural number is said to be a **perfect square.**

Natural numbers	1,	2,	3,	4,	5,	6, ...
Squares of the natural numbers	1^2,	2^2,	3^2,	4^2,	5^2,	6^2, ...
or Perfect squares	1,	4,	9,	16,	25,	36, ...

The number inside the **radical sign,** $\sqrt{}$, is called the **radicand.** The number multiplying the radical is called the radical's **coefficient.** For example, in $3\sqrt{5}$ the 3 is the coefficient of the radical and the 5 is the radicand. We can simplify some irrational numbers by determining any perfect square factors in the radicand. After this we can use the following rule:

> **Product Rule for Radicals**
> $$\sqrt{a \cdot b} = \sqrt{a} \cdot \sqrt{b} \quad a \geqslant 0, b \geqslant 0$$

For example, $\sqrt{8} = \sqrt{4 \cdot 2} = \sqrt{4} \cdot \sqrt{2}$
and $\sqrt{50} = \sqrt{25 \cdot 2} = \sqrt{25} \cdot \sqrt{2}$.

After using this rule, we can simplify even further, as illustrated in Example 1.

■ **Example 1** ─────────────────────────────

Simplify $\sqrt{45}$.

Solution: We can recognize that 9 is a perfect square factor of 45, so we write

$$\sqrt{45} = \sqrt{9 \cdot 5} = \sqrt{9} \cdot \sqrt{5}$$
$$= 3\sqrt{5}. \quad \text{Note that } \sqrt{9} = 3.$$

Since 5 has no perfect square factors, $\sqrt{5}$ cannot be simplified further.
❖

■ **Example 2** ─────────────────────────────

Simplify $\sqrt{48}$.

Solution: 16 is a perfect square factor of 48.
$$\sqrt{48} = \sqrt{16 \cdot 3} = \sqrt{16} \cdot \sqrt{3} = 4\sqrt{3} \qquad \text{❖}$$

In this example, if you did not recognize 16 as a perfect square factor of 48, but realized that 4 is a perfect square factor, you could still obtain the correct answer.
$$\sqrt{48} = \sqrt{4 \cdot 12} = \sqrt{4} \cdot \sqrt{12} = 2\sqrt{12}$$

Note that 12 has 4 as a perfect square factor.
$$2\sqrt{12} = 2\sqrt{4 \cdot 3} = 2\sqrt{4} \cdot \sqrt{3}$$
$$= 2 \cdot 2\sqrt{3}$$
$$= 4\sqrt{3}$$

We obtain the same answer by doing a little more work.

Addition and Subtraction of Irrational Numbers

The sum or difference of irrational numbers with the same radicand can be found by using the following rule.

> To add or subtract two or more numbers with the *same* radicand, add or subtract their coefficients. The answer is the sum or difference of the coefficients multiplied by the common radical.

■ **Example 3** ───────────────────────────

Simplify $2\sqrt{7} + 3\sqrt{7}$.

Solution: $2\ \sqrt{7} + 3\ \sqrt{7} = 5\ \sqrt{7}$

add

❖

■ **Example 4** ───────────────────────────

Simplify $3\sqrt{5} - 2\sqrt{5} + \sqrt{5}$.

Solution: This expression can be written

$$3\sqrt{5} - 2\sqrt{5} + 1\sqrt{5} = 2\sqrt{5}.$$

Note that $3 - 2 + 1 = 2$. ❖

■ **Example 5** ───────────────────────────

Simplify $5\sqrt{3} - \sqrt{12}$.

Solution: These radicals cannot be added in their present form since they contain different radicands. When this occurs, determine whether one or more of the radicals can be simplified so that they have the same radicand.

$$5\sqrt{3} - \sqrt{12} = 5\sqrt{3} - \sqrt{4 \cdot 3}$$
$$= 5\sqrt{3} - \sqrt{4} \cdot \sqrt{3}$$
$$= 5\sqrt{3} - 2\sqrt{3} = 3\sqrt{3}$$

❖

■ **Example 6** ───────────────────────────

Simplify $12\sqrt{2} - 2\sqrt{50}$.

Solution: $12\sqrt{2} - 2\sqrt{50}$

$$\begin{aligned} &= 12\sqrt{2} - 2\sqrt{25 \cdot 2} \\ &= 12\sqrt{2} - 2\sqrt{25} \cdot \sqrt{2} \\ &= 12\sqrt{2} - 2 \cdot 5\sqrt{2} \\ &= 12\sqrt{2} - 10\sqrt{2} = 2\sqrt{2} \end{aligned}$$

❖

Multiplication of Irrational Numbers

When multiplying irrational numbers, we again make use of the product rule for radicals. After the radicands are multiplied, simplify the remaining radical when possible.

■ **Example 7** _____

Simplify $\sqrt{2} \cdot \sqrt{8}$.

Solution: $\sqrt{2} \cdot \sqrt{8} = \sqrt{2 \cdot 8} = \sqrt{16} = 4$ ❖

■ **Example 8** _____

Simplify $\sqrt{3} \cdot \sqrt{5}$.

Solution: $\sqrt{3} \cdot \sqrt{5} = \sqrt{3 \cdot 5} = \sqrt{15}$

Since 15 has no perfect square factors, it cannot be simplified.   ❖

■ **Example 9** _____

Simplify $\sqrt{8} \cdot \sqrt{3}$.

Solution: $\sqrt{8} \cdot \sqrt{3} = \sqrt{8 \cdot 3} = \sqrt{24}$

$$\begin{aligned} &= \sqrt{4} \cdot \sqrt{6} \\ &= 2\sqrt{6} \end{aligned}$$

The example could also be done by first simplifying $\sqrt{8}$.

$$\begin{aligned} \sqrt{8} \cdot \sqrt{3} &= \sqrt{4 \cdot 2} \cdot \sqrt{3} \\ &= \sqrt{4} \cdot \sqrt{2} \cdot \sqrt{3} \\ &= 2 \cdot \sqrt{2} \cdot \sqrt{3} \\ &= 2\sqrt{6} \end{aligned}$$

❖

Division of Irrational Numbers

To divide irrational numbers, we make use of the following rule. After performing the division, simplify when possible.

Quotient Rule of Radicals

$$\frac{\sqrt{a}}{\sqrt{b}} = \sqrt{\frac{a}{b}} \quad a \geq 0, b > 0$$

■ **Example 10** _____

Divide $\dfrac{\sqrt{8}}{\sqrt{2}}$.

Solution: Using the above rule, we see that $\dfrac{\sqrt{8}}{\sqrt{2}} = \sqrt{\dfrac{8}{2}} = \sqrt{4} = 2.$ ❖

■ **Example 11** _____

Divide $\dfrac{\sqrt{96}}{\sqrt{2}}$.

Solution:
$$\begin{aligned}
\frac{\sqrt{96}}{\sqrt{2}} &= \sqrt{\frac{96}{2}} = \sqrt{48}\\
&= \sqrt{16 \cdot 3}\\
&= \sqrt{16}\sqrt{3}\\
&= 4\sqrt{3}
\end{aligned}$$
❖

Rationalizing the Denominator

A denominator is **rationalized** when the denominator contains no radical expressions. We rationalize denominators because (without a calculator) it is easier to evaluate the expression when the denominator is a natural number. For example,

if $\sqrt{2} = 1.414$, then $\dfrac{1}{\sqrt{2}} = \dfrac{1}{1.414}$, which is difficult to evaluate. However,

$\dfrac{1}{\sqrt{2}} = \dfrac{\sqrt{2}}{2} = \dfrac{1.414}{2}$, which is easier to evaluate. To rationalize a denominator that contains only a square root, multiply both the numerator and denominator of the fraction by a number that will result in the radicand in the denominator becoming a perfect square. (This is the equivalent of multiplying the fractions by 1, so the value of the fraction does not change.) Then simplify the fractions when possible.

■ Example 12

Rationalize the denominator of $\dfrac{3}{\sqrt{2}}$.

Solution: We need to multiply the numerator and denominator by a number that will make the radicand a perfect square.

$$\frac{3}{\sqrt{2}} = \frac{3}{\sqrt{2}} \cdot \frac{\sqrt{2}}{\sqrt{2}} = \frac{3\sqrt{2}}{\sqrt{4}} = \frac{3\sqrt{2}}{2}$$

Note that the 2s cannot be divided out, since one of the 2s is a radicand and the other is not. ❖

■ Example 13

Rationalize the denominator of $\dfrac{5}{\sqrt{8}}$.

Solution: $\dfrac{5}{\sqrt{8}} = \dfrac{5}{\sqrt{8}} \cdot \dfrac{\sqrt{2}}{\sqrt{2}} = \dfrac{5\sqrt{2}}{\sqrt{16}}$

$$= \frac{5\sqrt{2}}{4}$$

This problem could also have been solved by multiplying both the numerator and denominator by $\sqrt{8}$.

$$\frac{5}{\sqrt{8}} = \frac{5}{\sqrt{8}} \cdot \frac{\sqrt{8}}{\sqrt{8}} = \frac{5\sqrt{8}}{\sqrt{64}} = \frac{5\sqrt{8}}{8}$$
$$= \frac{5\sqrt{4}\sqrt{2}}{8}$$
$$= \frac{5 \cdot 2\sqrt{2}}{8}$$
$$= \frac{5\sqrt{2}}{4}$$ ❖

■ Example 14

Rationalize the denominator of $\dfrac{\sqrt{3}}{\sqrt{6}}$.

Solution: This problem can be solved in a number of ways. We can begin by multiplying both the numerator and denominator by $\sqrt{6}$ to obtain

$$\frac{\sqrt{3}}{\sqrt{6}} = \frac{\sqrt{3}}{\sqrt{6}} \cdot \frac{\sqrt{6}}{\sqrt{6}} = \frac{\sqrt{18}}{6}$$

$$= \frac{\sqrt{9 \cdot 2}}{6} = \frac{\sqrt{9}\sqrt{2}}{6} = \frac{3\sqrt{2}}{6} = \frac{\sqrt{2}}{2}.$$

A second method is to first divide, then rationalize as follows.

$$\frac{\sqrt{3}}{\sqrt{6}} = \sqrt{\frac{3}{6}} = \sqrt{\frac{1}{2}} = \frac{\sqrt{1}}{\sqrt{2}} = \frac{1}{\sqrt{2}}$$

Now rationalize $\dfrac{1}{\sqrt{2}}$.

$$\frac{1}{\sqrt{2}} = \frac{1}{\sqrt{2}} \cdot \frac{\sqrt{2}}{\sqrt{2}} = \frac{\sqrt{2}}{2}$$

Note that the same result is obtained by both methods.   ❖

EXERCISES 5.4

Determine whether the following numbers are rational or irrational.

1. $\sqrt{4}$
2. $\sqrt{3}$
3. 2/5
4. 0.404004000...
5. 3.181881888...
6. π
7. 22/7
8. 3.14
9. 0.68686868...
10. $\sqrt{36}$
11. $\sqrt{12}$
12. 3.14159...
13. $2\sqrt{3}$
14. $\dfrac{\sqrt{3}}{\sqrt{3}}$

Evaluate the following.

15. $\sqrt{36}$
16. $\sqrt{144}$
17. $\sqrt{9}$
18. $-\sqrt{16}$
19. $-\sqrt{169}$
20. $\sqrt{25}$
21. $-\sqrt{81}$
22. $-\sqrt{121}$
23. $\sqrt{196}$
24. $-\sqrt{225}$

Classify each of the following real numbers as belonging to one or more of the following sets: the rational numbers, the integers, the natural numbers, and the irrational numbers.

25. 3
26. -2
27. 5/12
28. $\sqrt{4}$
29. 3.14
30. $\sqrt{5}$
31. $0.31\overline{31}$
32. 0.484884888...
33. π
34. $-2/3$
35. 0.131131113...
36. $8.3143143\overline{14}$
37. 0.56725672...
38. 0.567566756667...

Simplify each of the following radicals.

39. $\sqrt{16}$
40. $\sqrt{27}$
41. $\sqrt{24}$
42. $\sqrt{64}$
43. $\sqrt{81}$
44. $\sqrt{48}$
45. $\sqrt{32}$
46. $\sqrt{98}$
47. $\sqrt{75}$
48. $\sqrt{125}$
49. $\sqrt{72}$
50. $\sqrt{124}$

Perform the indicated operations.

51. $\sqrt{3} + 4\sqrt{3}$
52. $\sqrt{5} - 6\sqrt{5}$
53. $2\sqrt{7} + 13\sqrt{7}$
54. $2\sqrt{2} + 3\sqrt{8}$
55. $2\sqrt{18} + 5\sqrt{2}$
56. $5\sqrt{6} - 3\sqrt{24}$
57. $4\sqrt{12} - 5\sqrt{48} - 3$
58. $13\sqrt{2} + 2\sqrt{18} - 5\sqrt{32}$

59. $15\sqrt{12} - 11\sqrt{75} + 8$

60. $5\sqrt{5} - \sqrt{20} + \sqrt{125}$

Perform the indicated operations. Simplify the answer when possible.

61. $\sqrt{3}\sqrt{2}$

62. $\sqrt{6}\sqrt{3}$

63. $\sqrt{4}\sqrt{8}$

64. $\sqrt{5}\sqrt{10}$

65. $\sqrt{6}\sqrt{12}$

66. $\sqrt{13}\sqrt{26}$

67. $\dfrac{\sqrt{8}}{\sqrt{4}}$

68. $\dfrac{\sqrt{125}}{\sqrt{5}}$

69. $\dfrac{\sqrt{72}}{\sqrt{8}}$

70. $\dfrac{\sqrt{136}}{\sqrt{8}}$

71. $\dfrac{\sqrt{72}}{\sqrt{6}}$

72. $\dfrac{\sqrt{120}}{\sqrt{6}}$

Rationalize the denominator.

73. $\dfrac{3}{\sqrt{2}}$

74. $\dfrac{5}{\sqrt{7}}$

75. $\dfrac{\sqrt{7}}{\sqrt{2}}$

76. $\dfrac{\sqrt{7}}{\sqrt{6}}$

77. $\dfrac{\sqrt{12}}{\sqrt{3}}$

78. $\dfrac{\sqrt{32}}{\sqrt{6}}$

79. $\dfrac{\sqrt{9}}{\sqrt{2}}$

80. $\dfrac{\sqrt{15}}{\sqrt{3}}$

81. $\dfrac{\sqrt{10}}{\sqrt{6}}$

82. $\dfrac{7}{\sqrt{7}}$

PROBLEM SOLVING

83. a) In Section 5.4 we introduced the Pythagorean theorem, $a^2 + b^2 = c^2$. Construct a right triangle, and then construct a square on each side,

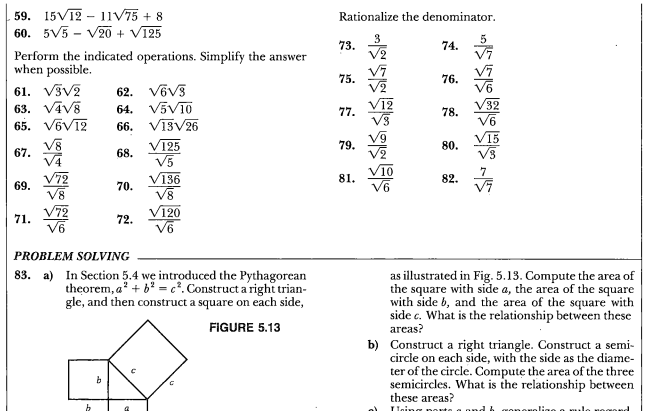

FIGURE 5.13

as illustrated in Fig. 5.13. Compute the area of the square with side a, the area of the square with side b, and the area of the square with side c. What is the relationship between these areas?

b) Construct a right triangle. Construct a semicircle on each side, with the side as the diameter of the circle. Compute the area of the three semicircles. What is the relationship between these areas?

c) Using parts a and b, generalize a rule regarding the areas of similar figures constructed on the sides of a right triangle.

 5.5 # REAL NUMBERS AND THEIR PROPERTIES

Now that we have discussed both the rational and irrational numbers, we can discuss the real numbers and the properties of the real number system. The union of the rational numbers and the irrational numbers is the **real numbers.** The set of real numbers is symbolized by $\mathbb{R}$.

Figure 5.14, which illustrates the relationships among the various parts of the real number system, may help you to better understand that system.

By examining the branches of the tree in Fig. 5.14 we can see that any negative integer is also an integer, a rational number, and a real number. Any positive integer is also a whole number, an integer, a rational number, and a real number. The number 0 is a whole number, an integer, a rational number, and a real number. Fractions are also rational numbers and real numbers. Irrational numbers are also real numbers.

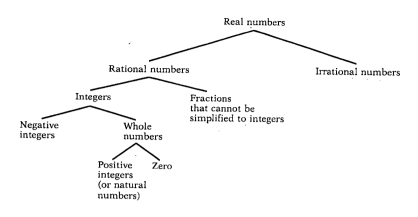

FIGURE 5.14

Properties of the Real Number System

We are now prepared to discuss the properties of the real number system. The first property that we will discuss is closure.

> If an operation is performed on any two elements of a set and the result is an element of the set, we say that the set is **closed** under that given operation.

Is the sum of any two natural numbers a natural number? If we add any two natural numbers, the result will be a natural number. Thus we say that the natural numbers are closed under the operation of addition.

Are the natural numbers closed under the operation of subtraction? The answer is no. For example, $3 - 5 = -2$, which is not a natural number. Therefore the natural numbers are not closed under the operation of subtraction.

Example 1

Determine whether the integers are closed under the operations of (a) multiplication, and (b) division.

Solution

a) If we multiply any two integers, will the product always be an integer? The answer is yes. Thus the integers are closed under the operation of multiplication.

b) If we divide any two integers, will the quotient be an integer? The answer is no. For example, $6 \div 5 = \frac{6}{5}$, which is not an integer. Therefore the integers are not closed under the operation of division.   ❖

The **commutative property of addition** states that the order in which two real numbers are added is immaterial. Thus for any two real numbers a and b, $a + b = b + a$. For example, $2 + 5 = 5 + 2 = 7$.

The **commutative property of multiplication** states that the order in which two real numbers are multiplied is immaterial. Thus for any two real numbers a and b, $a \cdot b = b \cdot a$. For example, $2 \cdot 5 = 5 \cdot 2 = 10$.

The commutative property does not hold for the operations of subtraction or division. For example, $5 - 3 \neq 3 - 5$; $6 \div 2 \neq 2 \div 6$.

The **associative property of addition** states that when adding three or more numbers, parentheses may be placed around any two adjacent numbers. Parentheses are used to indicate the order in which the operations are to be performed. The associative property of addition states that for any three real numbers, a, b, and c, $(a + b) + c = a + (b + c)$. For example,

$$(2 + 3) + 4 = 2 + (3 + 4),$$
$$5 + 4 = 2 + 7,$$
$$9 = 9.$$

The **associative property of multiplication** states that $(a \cdot b) \cdot c = a \cdot (b \cdot c)$ for any real numbers a, b, and c. For example,

$$(2 \cdot 3) \cdot 4 = 2 \cdot (3 \cdot 4),$$
$$6 \cdot 4 = 2 \cdot 12,$$
$$24 = 24.$$

Does the associative property hold for the operations of subtraction and division? Does $(4 - 2) - 3 = 4 - (2 - 3)$? Does $(8 \div 4) \div 2 = 8 \div (4 \div 2)$?

Note the difference between the commutative property and the associative property. The commutative property involves a change in *order*, while the associative property involves a change in *grouping*.

Another property of the real numbers is the **distributive property of multiplication over addition.** For any real numbers a, b, and c, the distributive property states that $a \cdot (b + c) = (a \cdot b) + (a \cdot c)$. For example, let $a = 2$ and $b = 3$ and $c = 4$; then

$$2 \cdot (3 + 4) = (2 \cdot 3) + (2 \cdot 4),$$
$$2 \cdot 7 = 6 + 8,$$
$$14 = 14.$$

This illustrates that you may either add first and then multiply or multiply first and then add. Note that the distributive property involves two operations—addition and multiplication.

Although positive integers were used in the specific examples, any real numbers could have been used. If we assume that x represents a real number, then $x + 4 = 4 + x$ and $(x + 4) + 3 = x + (4 + 3)$ by the commutative and associative properties of addition, respectively. Similar statements could be made for the commutative and associative properties of multiplication. Also, $2(x + 4) = 2 \cdot x + 2 \cdot 4 = 2x + 8$ by the distributive property.

The properties mentioned in this section are summarized in Table 5.2, where a, b, and c are any real numbers.

TABLE 5.2	
Commutative property of addition	$a + b = b + a$
Commutative property of multiplication	$a \cdot b = b \cdot a$
Associative property of addition	$(a + b) + c = a + (b + c)$
Associative property of multiplication	$(a \cdot b) \cdot c = a \cdot (b \cdot c)$
Distributive property of multiplication over addition	$a \cdot (b + c) = (a \cdot b) + (a \cdot c)$

We frequently use these properties without realizing that we are using them. To add $13 + 4 + 6$, we may add the $4 + 6$ first to get 10. To this sum we then add 13 to get 23. Here we have done the equivalent of placing parentheses around the $4 + 6$. We can do so because of the associative property of addition.

To multiply 102×11 in our heads, we might multiply $100 \times 11 = 1100$ and $2 \times 11 = 22$ and add these two products to get 1122. We are permitted to do this because of the distributive property.

$$(102 \times 11) = (100 + 2) \times 11 = (100 \times 11) + (2 \times 11)$$
$$= 1100 + 22 = 1122.$$

■ **Example 2** _____

State the name of the property illustrated.

a) $4 + 11 = 11 + 4$

b) $(3 + 7) + 13 = 3 + (7 + 13)$

c) $(x \cdot y) \cdot z = x \cdot (y \cdot z)$

d) $13(x + z) = 13 \cdot x + 13 \cdot z$

e) $3 + (x + 4) = 3 + (4 + x)$

f) $5(11 \cdot 9) = (11 \cdot 9)5$

Solution

a) Commutative property of addition.

b) Associative property of addition.

c) Associative property of multiplication.

d) Distributive property of multiplication over addition.

e) In this expression the only change made is the order of the x and 4. We have changed $x + 4$ to $4 + x$, or used the commutative property of addition.

f) Note that we changed the order of 5 and $(11 \cdot 9)$. To do this, we used the commutative property of multiplication.   ❖

■ **Example 3** ─────────────────────────────

Use the distributive property to simplify $3(5 + \sqrt{2})$.

Solution: $3(5 + \sqrt{2}) = (3 \cdot 5) + (3\sqrt{2})$
                        $= 15 + 3\sqrt{2}$ ❖

■ **Example 4** ─────────────────────────────

Use the distributive property to simplify $\sqrt{3}(5 + \sqrt{2})$.

Solution: $\sqrt{3}(5 + \sqrt{2}) = (\sqrt{3} \cdot 5) + (\sqrt{3} \cdot \sqrt{2})$
                        $= 5\sqrt{3} + \sqrt{6}$

Note that $\sqrt{3} \cdot 5$ is not equal to $\sqrt{15}$, but is written $5\sqrt{3}$.   ❖

■ **Example 5** ─────────────────────────────

Use the distributive property to multiply $3(x + 2)$. Then simplify the result.

Solution: $3(x + 2) = 3 \cdot x + 3 \cdot 2$
                     $= 3x + 6$ ❖

EXERCISES 5.5

Determine whether the natural numbers are closed under the given operation.

1. addition
2. subtraction
3. multiplication
4. division

Determine whether the integers are closed under the given operation.

5. addition

6. subtraction
7. multiplication
8. division

Determine whether the rational numbers are closed under the given operation.

9. addition
10. subtraction
11. multiplication
12. division

Determine whether the irrational numbers are closed under the given operation.

13. addition
14. subtraction
15. multiplication
16. division

Determine whether the real numbers are closed under the given operation.

17. addition
18. subtraction
19. multiplication
20. division

21. Does $3 + (4 + 5) = 3 + (5 + 4)$ illustrate the commutative property or the associative property? Explain your answer.

22. Give an example to show that the commutative property of addition may be true for the negative integers.

23. Give an example to show that the commutative property of multiplication may be true for the negative integers.

24. Does the commutative property hold for the integers under the operation of subtraction? Give an example to support your answer.

25. Does the commutative property hold for the real numbers under the operation of subtraction? Give an example to support your answer.

26. Does the commutative property hold for the integers under the operation of division? Give an example to support your answer.

27. Does the commutative property hold for the rational numbers under the operation of division? Give an example to support your answer.

28. Give an example to show that the associative property of addition may be true for the negative integers.

29. Give an example to show that the associative property of multiplication may be true for the negative integers.

30. Does the associative property hold for the integers under the operation of subtraction? Give an example to support your answer.

31. Does the associative property hold for the integers under the operation of division? Give an example to support your answer.

32. Does the associative property hold for the real numbers under the operation of subtraction? Give an example to support your answer.

33. Does the associative property hold for the real

numbers under the operation of division? Give an example to support your answer.

34. Does $a + (b \cdot c) = (a + b) \cdot (a + c)$? Give an example to support your answer.

State the name of the property illustrated.

35. $3 + [4 + (-2)] = (3 + 4) + (-2)$
36. $3 \cdot x = x \cdot 3$
37. $(x + 3) + 4 = x + (3 + 4)$
38. $2(3 + 6) = (2 \cdot 3) + (2 \cdot 6)$
39. $(3 \cdot 4) \cdot 5 = 3 \cdot (4 \cdot 5)$
40. $(1/3 + 1/4) + 2 = 1/3 + (1/4 + 2)$
41. $x + y = y + x$
42. $x \cdot y = y \cdot x$
43. $(3 + 4) + (5 + 6) = (5 + 6) + (3 + 4)$
44. $4 + (2 + 3) = (2 + 3) + 4$
45. $2(x + y) = 2x + 2y$
46. $(2 + 3) + 4 = (3 + 2) + 4$
47. $(x + y) + z = z + (x + y)$
48. $(a + b) + c = c + (a + b)$
49. $a \cdot (x + y) = (x + y) \cdot a$
50. $(a + b) + (c + d) = (c + d) + (a + b)$

Use the distributive property to multiply. Then simplify the resulting expression.

51. $4(x + 5)$ **52.** $5(x + 4)$
53. $\sqrt{3}(2 + \sqrt{6})$ **54.** $\sqrt{2}(\sqrt{6} + 3)$
55. $\sqrt{3}(x + \sqrt{3})$ **56.** $x(3 + y)$

Name the property used to go from step to step.

57. $3(x + 2) + 5 = (3x + 3 \cdot 2) + 5$
                $= (3x + 6) + 5$
58. $= 3x + (6 + 5)$
                $= 3x + 11$

Name the property used to go from step to step.

59. $5(x + 3) + 4x = (5 \cdot x + 5 \cdot 3) + 4x$
                 $= (5x + 15) + 4x$
60. $= 5x + (15 + 4x)$
61. $= 5x + (4x + 15)$
62. $= (5x + 4x) + 15$
                 $= 9x + 15$

Name the property used to go from step to step.

63. $5 + 3(x + 2) + 4x = 5 + (3 \cdot x + 3 \cdot 2) + 4x$
                      $= 5 + (3x + 6) + 4x$
64. $= 5 + (6 + 3x) + 4x$
65. $= (5 + 6) + 3x + 4x$
                      $= 11 + 7x$
66. $= 7x + 11$

Determine whether each of the following activities can be used to illustrate the property indicated.

67. Putting on shoes and socks—commutative property.

68. Putting sugar and cream in coffee—commutative property.

69. Brushing your teeth, washing your face, and combing your hair—associative property.

70. Turning on the lamp and reading a book—commutative property.

71. Cracking an egg, pouring out the egg, and cooking an egg—associative property.

72. Removing the gas cap, putting the nozzle in the tank, and turning on the gas—associative property.

73. Writing on the blackboard and erasing the blackboard — commutative property.

74. Sending in a tax return and signing the tax return —commutative property.

75. A coffee machine dropping the cup, dispensing the coffee, and then adding the sugar — associative property.

Can the removing of vest and jacket be used to illustrate the commutative property?   *Photos by Marshall Henrichs*

76. Construct a diagram showing the relationship between the sets of natural numbers, whole numbers, integers, rational numbers, irrational numbers, and real numbers. (*Hint:* Your diagram should show, for example, that the set of natural numbers is a subset of the set of whole numbers, integers, rational numbers, and real numbers.)

5.6   RULES OF EXPONENTS AND SCIENTIFIC NOTATION

We have discussed addition, subtraction, multiplication, and division of real numbers. Other operations that can be performed on real numbers include finding powers and roots.

Exponents

In order to properly understand the priority of operations and certain topics in algebra, you must have an understanding of exponents and roots. In the expression 5^2, the 2 is referred to as the **exponent** and the 5 is referred to as **the base.** We read 5^2 as 5 to the second power, or 5 squared. This means

$$5^2 = \underbrace{5 \cdot 5}$$

2 factors of 5

The number 5 to the third power, or 5 cubed, written 5^3, means

$$5^3 = \underbrace{5 \cdot 5 \cdot 5}$$

3 factors of 5

In general, the number b to the nth power, written b^n, means

$$b^n = \underbrace{b \cdot b \cdot b \cdot \cdots \cdot b}$$

n factors of b

Example 1

Evaluate the following.

a) 3^2 b) $(-3)^2$ c) 4^3 d) 1^{100} e) 8^1

Solution

a) $3^2 = 3 \cdot 3 = 9$
b) $(-3)^2 = (-3)(-3) = 9$
c) $4^3 = 4 \cdot 4 \cdot 4 = 64$
d) $1^{100} = 1$. (The number 1 times itself any number of times equals 1.)
e) $8^1 = 8$. (Any number with an exponent of 1 is the number itself.)

Rules of Exponents

Now that we know how to evaluate powers of numbers we can discuss the rules of exponents. Consider

$$2^2 \cdot 2^3 = \underbrace{2 \cdot 2} \cdot \underbrace{2 \cdot 2 \cdot 2} = 2^5$$

2 factors   3 factors

This example illustrates the product rule for exponents.

Product Rule

$$a^m \cdot a^n = a^{m+n}$$

Therefore by using the product rule, $2^2 \cdot 2^3 = 2^{2+3} = 2^5$.

Example 2

Use the product rule to simplify.

a) $5^2 \cdot 5^6$ b) $8^3 \cdot 8^5$

Solution

a) $5^2 \cdot 5^6 = 5^{2+6} = 5^8$

b) $8^3 \cdot 8^5 = 8^{3+5} = 8^8$

Consider

$$\frac{2^5}{2^2} = \frac{2 \cdot 2 \cdot 2 \cdot \cancel{2} \cdot \cancel{2}}{\cancel{2} \cdot \cancel{2}} = 2 \cdot 2 \cdot 2 = 2^3$$

This example illustrates the quotient rule for exponents.

Quotient Rule for Exponents

$$\frac{a^m}{a^n} = a^{m-n}, \qquad a \neq 0$$

Therefore $\dfrac{2^5}{2^2} = 2^{5-2} = 2^3$.

Example 3

Use the quotient rule to simplify.

a) $\dfrac{5^8}{5^5}$ b) $\dfrac{8^{12}}{8^5}$

Solution

a) $\dfrac{5^8}{5^5} = 5^{8-5} = 5^3$

b) $\dfrac{8^{12}}{8^5} = 8^{12-5} = 8^7$

Consider $2^3 \div 2^3$. By the quotient rule we see that

$$\frac{2^3}{2^3} = 2^{3-3} = 2^0$$

But $\frac{2^3}{2^3} = \frac{8}{8} = 1$. Therefore 2^0 must equal 1. This example illustrates the zero exponent rule.

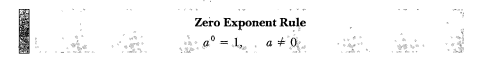

Zero Exponent Rule

$$a^0 = 1, \quad a \neq 0$$

Note that 0^0 is not a real number.

▢ **Example 4** _____

Use the zero exponent rule to simplify.

a) 5^0 **b)** $(-3)^0$

Solution: **a)** $5^0 = 1$ **b)** $(-3)^0 = 1$ ❖

Consider $2^3 \div 2^5$. Using the quotient rule, we find that

$$\frac{2^3}{2^5} = 2^{3-5} = 2^{-2}$$

But $\frac{2^3}{2^5} = \frac{\cancel{2} \cdot \cancel{2} \cdot \cancel{2} \cdot}{\cancel{2} \cdot \cancel{2} \cdot \cancel{2} \cdot 2 \cdot 2} = \frac{1}{2^2}$. Since $\frac{2^3}{2^5}$ equals both 2^{-2} and $\frac{1}{2^2}$, then 2^{-2} must

equal $\frac{1}{2^2}$. This example illustrates the negative exponent rule.

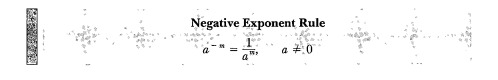

Negative Exponent Rule

$$a^{-m} = \frac{1}{a^m}, \quad a \neq 0$$

▢ **Example 5** _____

Use the negative exponent rule to simplify.

a) 5^{-2} **b)** 8^{-1}

Solution

a) $5^{-2} = \dfrac{1}{5^2} = \dfrac{1}{25}$ **b)** $8^{-1} = \dfrac{1}{8^1} = \dfrac{1}{8}$ ❖

Consider $(2^3)^2$:

$$(2^3)^2 = (2^3)(2^3) = 2^6$$

This example illustrates the power rule for exponents.

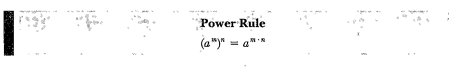

Power Rule

$$(a^m)^n = a^{m \cdot n}$$

Thus $(2^3)^2 = 2^{3 \cdot 2} = 2^6$

 Example 6 _____

Use the power rule to simplify

a) $(3^5)^4$ **b)** $(4^3)^6$

Solution

a) $(3^5)^4 = 3^{5 \cdot 4} = 3^{20}$
b) $(4^3)^6 = 4^{3 \cdot 6} = 4^{18}$ ❖

Summary of the Rules of Exponents

$a^m \cdot a^n = a^{m+n}$	Product rule
$\dfrac{a^m}{a^n} = a^{m-n}, a \neq 0$	Quotient rule
$a^0 = 1, a \neq 0$	Zero exponent rule
$a^{-m} = \dfrac{1}{a^m}; a \neq 0$	Negative exponent rule
$(a^m)^n = a^{m \cdot n}$	Power rule

Scientific Notation

When working with scientific problems, one often deals with very large and very small numbers. For example, the distance from the earth to the sun is about 93,000,000 miles. The wavelength of a yellow color of light is about 0.0000006 meters. Because it is difficult to work with many zeros, scientists often express such numbers with exponents. For example, the number 93,000,000 might be written 9.3×10^7 and the number 0.0000006 might be written 6.0×10^{-7}. Numbers such as 9.3×10^{7} and 6.0×10^{-7} are in a form called **scientific notation.** Each number written in scientific notation is written as a number greater than or equal to 1 and less than 10 ($1 \leq a < 10$) multiplied by some power of 10.

Some examples of numbers in scientific notation are the following.

$$3.7 \times 10^{3} \qquad 2.05 \times 10^{-3}$$
$$5.6 \times 10^{8} \qquad 1.00 \times 10^{-5}$$

Consider the number 68,400:

$$68{,}400 = 6.84 \times 10{,}000$$
$$= 6.84 \times 10^{4} \qquad \text{(Note that } 10{,}000 = 10 \cdot 10 \cdot 10 \cdot 10 = 10^{4}\text{)}$$

Therefore $68{,}400 = 6.84 \times 10^{4}$. Note that to go from 68,400 to 6.84 the decimal point was moved four places to the left. Also note that the exponent on the 10, the 4, is the same as the number of places the decimal point was moved to the left. Here is a simplified procedure for writing a number in scientific notation:

To write a number in scientific notation

1. Move the decimal point in the original number to the right or left until you obtain a number greater than or equal to 1 and less than 10.

2. Count the number of places you have moved the decimal point to obtain the number in Step 1. If the decimal point was moved to the left, the count is to be considered positive. If the decimal point was moved to the right, the count is to be considered negative.

3. Multiply the number obtained in Step 1 by 10 raised to the count found in Step 2. (Note that the count determined in Step 2 is the exponent on the base 10).

■ **Example 7** _____

Write the following numbers using scientific notation.

a) 50,700 **b)** 0.000756
c) 852,000 **d)** 0.0014

Solution

a) 50,700 means 50,700. **b)** $0.000756 = 7.56 \times 10^{-4}$

$\underbrace{50{,}700.}_{\text{4 places to left}} = 5.07 \times 10^{4}$ $\qquad$ 4 places to right

c) $852,000. = 8.52 \times 10^5$ **d)** $0.0014 = 1.4 \times 10^{-3}$

    5 places   3 places
    to left   to right

To convert from a number given in scientific notation to decimal notation:

1. Observe the exponent of the power of 10.

2. **a)** If the exponent is positive, move the decimal point in the number to the right the same number of places as the exponent. It might be necessary to add zeros to the number.

   b) If the exponent is negative, move the decimal point in the number to the left the same number of places as the exponent. It might be necessary to add zeros.

Example 8

Write each number in decimal notation.

a) 5.6×10^4 **b)** 2.14×10^{-3} **c)** 9.75×10^8

Solution

a) Moving the decimal four places to the right gives

$$5.6 \times 10^4 = 56,000$$

Note that $10^4 = 10,000$. Thus $5.6 \times 10^4 = 5.6 \times 10,000 = 56,000$.

b) Move the decimal three places to the left.

$$2.14 \times 10^{-3} = 0.00214$$

c) Move the decimal eight places to the right.

$$9.75 \times 10^8 = 975,000,000$$

We can use the rules of exponents in working with numbers written in scientific notation.

Example 9

Multiply $(4.3 \times 10^6)(2 \times 10^{-4})$. Write the answer in decimal notation.

Solution: $(4.3 \times 10^6)(2 \times 10^{-4}) = (4.3 \times 2)(10^6 \times 10^{-4})$
$$= 8.6 \times 10^2$$
$$= 860$$

Example 10

Divide $\dfrac{0.0000093}{0.003}$. Write the answer in scientific notation:

Solution: First write each number in scientific notation.

$$\frac{0.0000093}{0.003} = \frac{9.3 \times 10^{-6}}{3 \times 10^{-3}} = \left(\frac{9.3}{3}\right)\left(\frac{10^{-6}}{10^{-3}}\right)$$
$$= 3.1 \times 10^{-6 - (-3)}$$
$$= 3.1 \times 10^{-6 + 3}$$
$$= 3.1 \times 10^{-3}$$ ❖

Example 11

Multiply $(42,100,000)(0.008)$. Write the answer in decimal notation.

Solution: Change each number to scientific notation form:

$$(42,100,000)(0.008) = (4.21 \times 10^{7})(8 \times 10^{-3})$$
$$= (4.21 \times 8)(10^{7} \times 10^{-3})$$
$$= 33.68 \times 10^{4}$$
$$= 336,800$$ ❖

Scientific Notation on the Calculator

What will your calculator show when you multiply very large or very small numbers? The answer depends on whether your calculator has the ability to display an answer in scientific notation. On calculators without the ability to express numbers in scientific notation you will probably get an error message because the answer will be too large or too small for the display.

For example, on a calculator without scientific notation:

$\boxed{C}$ 8000000 $\boxed{\times}$ 600000 $\boxed{=}$ Error

If your calculator has the ability to give an answer in scientific notation form, you will probably get the following:

$\boxed{C}$ 8000000 $\boxed{\times}$ 600000 $\boxed{=}$ 4.8   12

This 4.8   12 means 4.8×10^{12}.

On a calculator that uses scientific notation:

$\boxed{C}$ .0000003 $\boxed{\times}$ .004 $\boxed{=}$ 1.2   − 9

This 1.2   − 9 means 1.2×10^{-9}.

EXERCISES 5.6

Evaluate each of the following.

1. 4^2
2. 3^3
3. $(-2)^2$
4. $(-2)^3$
5. 5^0
6. $(1/2)^2$
7. $(5/6)^2$
8. $(-3)^4$
9. $3^2 \cdot 4^3$
10. $6^2/2^2$
11. $3^5/3^2$
12. $3^2 \cdot 3^3$
13. $2^2/2^5$
14. $4^2 \cdot 3^0$
15. $(-5)^0$
16. $(-2)^4$
17. 2^4
18. $2^5 \cdot 4^0$
19. 2^{-2}
20. 2^{-3}
21. $(3^2)^3$
22. $(1^4)^5$
23. $\dfrac{5^7}{5^5}$
24. $4^2 \cdot 4$
25. 6^{-3}
26. 2^{-4}
27. $3^2 \cdot 3^2$
28. $(5^2)^3$
29. $\dfrac{3^5}{3^4}$
30. $2^{-2} \cdot 2$

Express each number in scientific notation.

31. 55,000
32. 4,610,000
33. 900
34. 0.00062
35. 0.053
36. 0.0000561
37. 19,000
38. 5,260,000,000
39. 0.00000186
40. 0.0003
41. 0.00000423
42. 54,000
43. 107
44. 0.02
45. 0.153
46. 416,000

Express each number in decimal notation.

47. 3.1×10^3
48. 1.63×10^{-4}
49. 6×10^7
50. 6.15×10^5
51. 2.13×10^{-5}
52. 2.74×10^{-7}
53. 3.12×10^{-1}
54. 4.6×10^1
55. 9×10^6
56. 7.3×10^4
57. 2.31×10^2
58. 1.04×10^{-2}
59. 3.5×10^4
60. 2.17×10^{-6}
61. 1×10^4
62. 1×10^{-3}

Perform the indicated operation and express each number in decimal notation.

63. $(4 \times 10^2)(3 \times 10^5)$
64. $(2 \times 10^{-3})(3 \times 10^2)$
65. $(5.1 \times 10^1)(3 \times 10^{-4})$
66. $(1.6 \times 10^{-2})(4 \times 10^{-3})$

67. $\dfrac{6.4 \times 10^5}{2 \times 10^3}$
68. $\dfrac{8 \times 10^{-3}}{2 \times 10^1}$
69. $\dfrac{8.4 \times 10^{-6}}{4 \times 10^{-3}}$
70. $\dfrac{25 \times 10^3}{5 \times 10^{-2}}$
71. $\dfrac{4 \times 10^5}{2 \times 10^4}$
72. $\dfrac{16 \times 10^3}{8 \times 10^{-3}}$

Perform the indicated operation by first converting each number to scientific notation. Write the answer in scientific notation.

73. $(700{,}000)(6{,}000{,}000)$
74. $(0.0006)(5{,}000{,}000)$
75. $(0.003)(0.00015)$
76. $(230{,}000)(3000)$
77. $\dfrac{1{,}400{,}000}{700}$
78. $\dfrac{20{,}000}{0.0005}$
79. $\dfrac{0.00004}{200}$
80. $\dfrac{0.0012}{0.000006}$
81. $\dfrac{150{,}000}{0.0005}$

82. List the numbers from smallest to largest.
    $4.8 \times 10^5, 3.2 \times 10^{-1}, 4.6, 8.3 \times 10^{-4}$

83. List the numbers from smallest to largest.
    $9.2 \times 10^{-5}, 8.4 \times 10^3, 1.3 \times 10^{-1}, 6.2 \times 10^4$

84. The distance from the earth to the planet Jupiter is approximately 4.5×10^8 miles. If a spacecraft traveled at a speed of 25,000 miles per hour, how long, in hours, would it take the spacecraft to travel from the earth to Jupiter. Use distance = rate × time.

85. If a computer can do a calculation in 0.000004 seconds, how long, in seconds, would it take the computer to do 8 trillion (8,000,000,000,000) calculations?

*86. The half-life of a radioactive isotope is the time required for half the quantity of the isotope to decompose. The half-life of uranium-238 is 4.5×10^9 years, and the half-life of uranium-234 is 2.5×10^5 years. How many times greater is the half life of uranium-238 than uranium-234?

*87. A treaty between the United States and Canada requires that during the tourist season a minimum of 100,000 cubic feet of water per second flow over Niagara Falls (another 130,000–160,000 cubic feet/sec are diverted for power generation). Find the minimum amount of water that will flow over the falls in a 24-hour period during the tourist season.

PROBLEM SOLVING

88. a) Light travels at a speed of 1.86×10^5 miles per second. A *light year* is the distance that light travels in one year. Determine the number of miles in a light year.

    b) The earth is approximately 93,000,000 miles from the sun. How long does it take light from the sun to reach the earth?

5.7 SEQUENCES

Now that we can recognize the various sets of real numbers and know how to add, subtract, multiply, and divide real numbers, we can discuss sequences. A **sequence** is a list of numbers in which each number except the first is related to the number or numbers that precede it. Three types of sequences that have many uses are the arithmetic sequence, the geometric sequence and the Fibonacci sequence. We will first discuss the arithmetic sequence.

Arithmetic Sequences

A sequence in which each term after the first term differs from the preceding term by a constant amount is called an **arithmetic sequence.** The amount by which each pair of successive terms differs is called the **common difference,** d. The common difference can be found by subtracting any term from the term that directly follows it.

Examples of Arithmetic Sequences	Common Differences
$1, 5, 9, 13, 17, \ldots$	$d = 5 - 1 = 4$
$-7, -5, -3, -1, 1, \ldots$	$d = -5 - (-7) = -5 + 7 = 2$
$\frac{5}{2}, \frac{3}{2}, \frac{1}{2}, -\frac{1}{2}$	$d = \frac{3}{2} - \frac{5}{2} = -\frac{2}{2} = -1$

■ **Example 1** _____

Write the first five terms of the arithmetic sequence with first term 4 and common difference 3.

Solution: The first term is 4. The second term is $4 + 3$ or 7. The third term is $7 + 3$ or 10, and so on. Thus the sequence is 4, 7, 10, 13, 16.   ❖

■ **Example 2** _____

Write the first four terms of the arithmetic sequence with first term 3 and common difference -2.

Solution: 3, 1, -1, -3, -5 ❖

When discussing a sequence, we often represent the first term as a_1, (read "a sub 1"), the second term as a_2, the fifteenth term as a_{15}, and so on.

We use the notation a_n to represent the general or nth term of a sequence. Thus a sequence may be symbolized as

$$a_1, a_2, a_3, a_4, \ldots, a_n, \ldots$$

When we know the first term of an arithmetic sequence and the common difference, we can find the value of any specific term using the following formula.

> **General or nth term of an arithmetic sequence**
>
> $$a_n = a_1 + (n - 1)d$$

Example 3

Find the seventh term of the arithmetic sequence whose first term is 3 and whose common difference is -6.

Solution: We are asked to find the seventh term, or a_7. We obtain the answer by replacing n in the formula with 7, a_1 with 3, and d with -6.

$$a_n = a_1 + (n - 1)d$$
$$a_7 = 3 + (7 - 1)(-6)$$
$$= 3 + (6)(-6)$$
$$= 3 - 36$$
$$= -33$$

The seventh term is -33. As a check we have shown the first seven terms of the sequence:

$$3, -3, -9, -15, -21, -27, -33$$

Example 4

Write an expression for the general or nth term, a_n, for the sequence 6, 9, 12, 15, . . .

Solution: In this sequence, $a_1 = 6$ and $d = 3$. We substitute these values in $a_n = a_1 + (n - 1)d$ to obtain an expression for the nth term, a_n:

$$a_n = a_1 + (n - 1)d$$
$$a_n = 6 + (n - 1)3$$
$$a_n = 6 + 3n - 3$$
$$a_n = 3 + 3n$$

Note that when $n = 1$, the first term is $3 + 3(1) = 6$. When $n = 2$, the second term is $3 + 3(2) = 9$, and so on.

We can find the sum of the first n terms in an arithmetic sequence by using the following formula.

> **Sum of the first n terms in an arithmetic sequence**
> $$s_n = \frac{n(a_1 + a_n)}{2}$$

In the above formula, s_n represents the sum of the first n terms, a_1 is the first term, a_n is the nth term, and n is the number of terms in the sequences from a_1 to a_n.

■ **Example 5** ───

Find the sum of the first 25 natural numbers.

Solution: The sequence we are discussing is

$$1, 2, 3, 4, 5, \ldots, 25$$

In this sequence, $a_1 = 1, a_{25} = 25$, and $n = 25$. The sum of the first 25 terms is

$$s_n = \frac{n(a_1 + a_n)}{2}$$
$$s_{25} = \frac{25(1 + 25)}{2}$$
$$s_{25} = \frac{25(26)}{2} = 325$$

If you add $1 + 2 + 3 + 4 + \ldots + 25$, you will see that the sum of the terms is 325.   ❖

Geometric Sequences

The next type of sequence we will discuss is the geometric sequence. A **geometric sequence** is one in which the ratio of any term to the term that directly precedes it is a constant. This constant is called the **common ratio.** The common ratio, r, can be found by taking any term except the first and dividing that term by the preceding term.

Example of Geometric Sequences	Common Ratio
$2, 4, 8, 16, 32$	$r = 4 \div 2 = 2$
$-3, 6, -12, 24, -48$	$r = 6 \div (-3) = -2$
$\dfrac{2}{3}, \dfrac{2}{9}, \dfrac{2}{27}, \dfrac{2}{81}$	$r = \dfrac{2}{9} \div \dfrac{2}{3} = \left(\dfrac{2}{9}\right)\left(\dfrac{3}{2}\right) = \dfrac{1}{3}$

To construct a geometric sequence when the first term, a_1, and common ratio are known, multiply the first term by the common ratio to get the second term. Then multiply the second term by the common ratio to get the third term, and so on.

■ **Example 6**

Write the first five terms of the geometric sequence with first term 5 and common ratio 1/2.

Solution: The first term is 5. The second term is $5(\frac{1}{2})$ or $\frac{5}{2}$. The third term is $(\frac{5}{2})(\frac{1}{2})$ or $\frac{5}{4}$, and so on. The sequence is

$$5, \frac{5}{2}, \frac{5}{4}, \frac{5}{8}, \frac{5}{16}.$$

Note that each term is found by multiplying the preceding term by 1/2. ❖

When we know the first term of a geometric sequence and the common ratio we can find the value of the general or nth term, a_n, by using the following formula.

> **General or nth term of a geometric sequence**
>
> $$a_n = a_1 r^{n-1}$$

■ **Example 7**

Find the seventh term of the geometric sequence whose first term is -3 and whose common ratio is -2.

Solution: In this sequence, $a_1 = -3$, $r = -2$, and $n = 7$. Substituting the values, we obtain

$$
\begin{aligned}
a_n &= a_1 r^{n-1}\\
a_7 &= -3(-2)^{7-1}\\
a_7 &= -3(-2)^6\\
a_7 &= -3(64)\\
a_7 &= -192
\end{aligned}
$$

As a check we have illustrated the first seven terms of the sequence: $-3, 6, -12, 24, -48, 96, -192$. ❖

■ **Example 8** _____

Write an expression for the general or nth term, a_n, of the sequence 5, 15, 45, 135, ...

Solution: In this sequence, $a_1 = 5$ and $r = 3$. We substitute these values in $a_n = a_1 r^{n-1}$ to obtain an expression for the nth term, a_n:

$$a_n = a_1 r^{n-1}$$
$$a_n = 5(3)^{n-1}$$

Note that when $n = 1$, $a_1 = 5(3)^0 = 5(1) = 5$. When $n = 2$, $a_2 = 5(3)^1 = 15$, and so on. ❖

We can find the sum of the first n terms of a geometric sequence by using the following formula.

> **Sum of the first n terms of a geometric sequence**
>
> $$s_n = \frac{a_1(1 - r^n)}{1 - r}, \quad r \neq 1$$

In this formula, s_n represents the sum of the first n terms, where a_1 is the first term, r is the common ratio, and n is the number of terms.

■ **Example 9** _____

Find the sum of the first five terms in the geometric sequence whose first term is 3 and whose common ratio is 4.

Solution: In this sequence, $a_1 = 3$, $r = 4$, and $n = 5$. Substituting the values, we obtain

$$s_n = \frac{a_1(1 - r^n)}{1 - r}$$
$$s_5 = \frac{3[1 - (4)^5]}{1 - 4}$$
$$s_5 = \frac{3(1 - 1024)}{-3}$$
$$s_5 = \frac{1(-1023)}{-1} = \frac{-1023}{-1} = 1023$$

The sum of the first five terms of the sequence is 1023. The first five terms of the sequence are 3, 12, 48, 192, 768. If you add these five numbers, you will obtain the sum of 1023. ❖

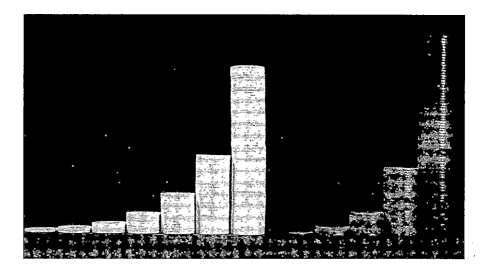

What is the constant multiplier in
the geometric sequence?
Photo by Marshall Henrichs

DID YOU KNOW?

The number 0 followed by the first eight terms of the geometric sequence $3, 6, 12, 24, 48, 96, 192, 384, \ldots, a_n, \ldots$ is known collectively as the **Titius-Bode Law.** These numbers, discovered in 1766 by the two German astronomers, were of great importance to astronomy in the eighteenth and nineteenth centuries. When 4 was added to each term and the result was divided by 10, the new sequence obtained corresponded quite well with the observed mean distance from the sun to the known principal planets of the solar system (in a certain type of astronomical unit). The discovery of Uranus in 1781 with a mean distance in agreement with the Titius-Bode Law stimulated the search for an undiscovered planet at a predicted 2.8 astronomical units. This search led to the discovery of Ceres and other members of the asteroid belt. Although the Titius-Bode Law broke down after the discoveries of Neptune and Pluto, many scientists still believe that in the future some sound physical reason for the Titius-Bode Law will emerge.

■ **Example 10** _____

Determine whether the following sequences are arithmetic or geometric, and find the next two terms.

a) $2, 5, 8, 11, \ldots$ **b)** $2, 6, 18, 54, \ldots$

c) $2, -4, 8, -16, \ldots$ **d)** $7, 3, -1, -5, \ldots$

Solution

a) Each term is 3 more than the preceding term. Therefore this sequence is arithmetic with $d = 3$. The next two terms are 14 and 17.

b) Each term is 3 times the preceding term. Therefore this sequence is geometric with $r = 3$. The next two terms are 162 and 486.

c) Each term is -2 times the preceding term, so the sequence is geometric with $r = -2$. The next two terms are 32 and -64.

d) Each term is 4 less than the preceding term, so the sequence is arithmetic with $d = -4$. The next two terms are -9 and -13.   ❖

The Fibonacci Sequence

We cannot leave the topic of sequences without discussing a very interesting and exciting sequence known as the Fibonacci sequence. To form the Fibonacci sequence, let both the first and second terms be 1. Then add these two terms to get the third term. Next add the second and third terms to get the fourth term, and so on. The **Fibonacci sequence** is

$$1, 1, 2, 3, 5, 8, 13, 21, 34, 55, 89, \ldots$$

This sequence was studied by Leonardo of Pisa (also called Fibonacci) in the early thirteenth century. The sequence became known when Fibonacci, a very distinguished mathematician, published a book called *Liber abaci*. This book performed an invaluable service by spreading throughout Europe the use of Hindu-Arabic numerals in arithmetic operations. However, the book is best remembered for its presentation of a mathematical puzzle involving rabbits. The solution of this puzzle generated the interesting sequence of numbers known as the Fibonacci sequence. This famous problem is given in Exercise 95 at the end of this section.

In the middle of the nineteenth century, mathematicians made a serious study of the sequence and found strong similarities between it and many natural phenomena. Some of the similarities are seen in the reproduction of honeybees and rabbits, the arrangement of bracts (scalelike structures) on a pinecone, the number of scales on a pineapple, the arrangement of leaves on the stem or branch of certain plants, and the arrangement of florets in the head of certain flowers. In addition, the petal count around the head of many plants will constantly yield a Fibonacci number. For example, some daisies will contain 21 petals, and others will contain 34, 55, or 89 petals (therefore people who play the "love me, love me not" game on a daisy will very likely pluck 21, 34, 55, or 89 petals before arriving at an answer). Each day, more and more similarities

Leonardo of Pisa
(also called Fibonacci)
Courtesy of The Granger Collection

between the Fibonacci sequence and natural phenomena are found. The Fibonacci Society is an organization that has been formed to study and publish literature concerning this sequence.

Photo by Marshall Henrichs

Fibonacci Numbers and Divine Proportions

In 1753, while studying the Fibonacci sequence, Robert Simson, a mathematician at the University of Glasgow, noticed that when he took the ratio of any term to the term that immediately precedes it, the value he obtained remained in the vicinity of one specific number. To illustrate this, we will indicate the ratio of various pairs of sequential numbers.

Numbers	Ratio
1, 1	$1/1 = 1$
1, 2	$2/1 = 2$
2, 3	$3/2 = 1.5$
3, 5	$5/3 = 1.66\ldots$
5, 8	$8/5 = 1.6$
8, 13	$13/8 = 1.625$

The ratio of the 50th term to the 49th term is 1.6180. Simson finally proved that the ratio of the $(n + 1)$ term to the nth term as n gets larger and larger is the irrational number $(\sqrt{5} + 1)/2$, which begins $1.61803\ldots$. Simson had no difficulty recognizing this number, since it was well known to mathematicians as the **golden number.** Many years earlier the Bavarian astronomer and mathematician Johannes Kepler had written that for him the golden number symbolized the Creator's intention "to create like from like." The history and influence of the golden number can be traced back many hundreds of years earlier, back at least to the time of the ancient Greeks (about the sixth century B.C.). The ancient Greeks sought to incorporate the unifying principals of beauty and perfection in all aspects of their life including their science, art, architecture, and philosophy. Even such a basic task as dividing items into smaller parts was studied using the philosophical belief in "divine proportions." What probably began with the Greeks as a "feeling of beauty or rightness" has become known as the **divine proportion,** "the smaller part is to the larger part as the larger part is to the whole."

FIGURE 5.15

To illustrate the divine proportion, consider the line segment AB in Fig. 5.15. When the line segment AB is divided by a point C, such that the ratio of the smaller part, CB, to its larger part, AC, is equal to the ratio of the larger part, AC, to the whole, AB, that ratio is the golden number.

$$\frac{BC}{AC} = \frac{AC}{AB} = \frac{\sqrt{5} + 1}{2}$$

When the golden proportion is illustrated geometrically as in Fig. 5.15, it is often referred to as a **golden section.** From the golden section the **golden rectangle** can be formed, as shown in Fig. 5.16.

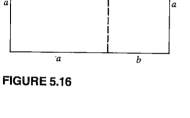

FIGURE 5.16

$$\frac{\text{Length}}{\text{Width}} = \frac{a + b}{a} = \frac{a}{b} = \frac{\sqrt{5} + 1}{2}$$

Note that when a square is cut off at one end of a golden rectangle, as in Fig. 5.17, the remaining rectangle has the same properties as the original golden rectangle ("like from like") and is therefore itself a golden rectangle. It is interesting to note that the curve derived from a succession of diminishing golden rectangles, as shown in Fig. 5.18, is the same as the curve of the spiraled nautilus shown in Fig. 5.19. This same curve is seen on the horns of certain animals, on certain flower heads, pinecones, and

FIGURE 5.17

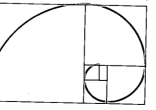

FIGURE 5.18

FIGURE 5.19
Photo by
Susan Van Etten

pineapples. This curve also closely approximates a curve used in mathematics called the logarithmic spiral.

The golden rectangle was a major influence on ancient Greek art and architecture. The main measurements of many buildings of antiquity, including the Parthenon in Athens (Fig. 5.20), are governed by golden sections and rectangles. There is even some evidence that the structure of the great pyramids of Giza, Egypt, constructed over a thousand years before the rise of Greek civilization, is based upon the golden section and golden ratio.

FIGURE 5.20   *Menschenfreund/Taurus Photos*

Many artists of the Renaissance who studied Greek art incorporated golden sections and rectangles into their works. The golden rectangle can be seen in "La Parade" by George Seurat (1859–1891), a French neoimpressionist artist (Figs. 5.21 and 5.22).

FIGURE 5.21
*Metropolitan Museum of
Art, bequest of Stephen C.
Clark, 1960*

FIGURE 5.22

Reprinted with permission from 1977 Yearbook of Science and the Future, ©1976. Encyclopaedia Britannica, Inc., Chicago, Illinois.

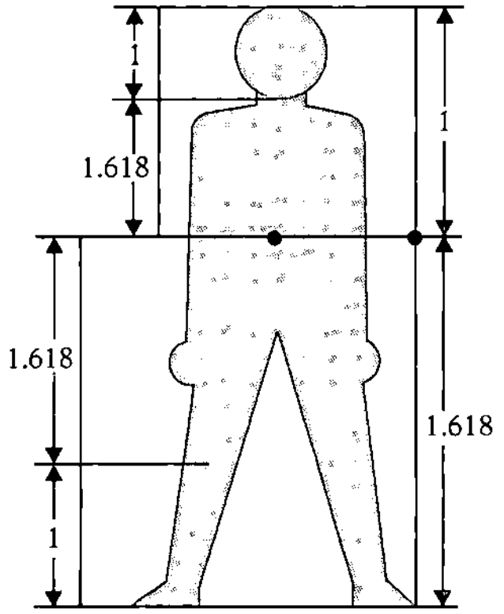

FIGURE 5.23

The Modulor, a scale of proportions developed by the architect Le Corbusier.

The twentieth century architect Le Corbusier developed a scale of proportions that he called the Modulor, which was based on a human body whose height is divided in golden sections starting at the navel (Fig. 5.23).

The visual arts deal with what is pleasing to the eye, whereas musical composition deals with what is pleasing to the ear. While art achieves some of its goals by using division of planes and area, music achieves some of its goals by a similar division of time, using notes of various duration and spacing. It is not surprising therefore that patterns that can be expressed mathematically in terms of Fibonacci relationships have been found in the works of several composers, including Bach and Beethoven. A number of twentieth century works, including Ernst Krenek's *Fibonacci Mobile,* have been deliberately structured by using Fibonacci proportions.

There have been a number of studies that have tried to explain why the Fibonacci series and related items have been linked to so many real-life situations. In the nineteenth century the German physicist and psychologist Gustav Fechner tried to determine which dimensions were most pleasing to the eye. In his study, Fechner, along with the psychologist Wilhelm Wundt, found that most people do unconsciously favor golden dimensions when purchasing greeting cards, mirrors, and other rectangular objects. This discovery has been heavily used by commercial manufacturers in their packaging and labeling designs, by retailers in their store displays, and in other areas of business and advertising.

EXERCISES 5.7

Write the first five terms of the arithmetic sequence for the given first term, a_1, and common difference, d.

1. $a_1 = 4, d = 3$
2. $a_1 = 5, d = 2$
3. $a_1 = -4, d = 5$
4. $a_1 = -3, d = 1$
5. $a_1 = 6, d = -2$
6. $a_1 = 8, d = -5$
7. $a_1 = 24, d = -8$
8. $a_1 = 1/2, d = 2$
9. $a_1 = 1/2, d = 1/2$
10. $a_1 = 5/2, d = -(3/2)$

Find the indicated term for the arithmetic sequence for the given first term, a_1, and common difference, d.

11. Find a_5, $a_1 = 12, d = 6$
12. Find a_9, $a_1 = -8, d = -5$
13. Find a_8, $a_1 = -4, d = 8$
14. Find a_{12}, $a_1 = 9, d = -7$
15. Find a_{10}, $a_1 = -30, d = -12$

16. Find a_{20}, $a_1 = 3/5$, $d = -2$
17. Find a_{15}, $a_1 = -(1/2)$, $d = 5$
18. Find a_{11}, $a_1 = -20$, $d = -(1/2)$

Write an expression for the general or nth term, a_n, for the given arithmetic sequences.

19. $4, 8, 12, 16, \ldots$
20. $5, 2, -1, -4, \ldots$
21. $8, 18, 28, 38, \ldots$
22. $-3, -6, -9, -12, \ldots$
23. $-(5/2), -(3/2), -(1/2), 1/2, \ldots$
24. $-18, -9, -0, 9$
25. $-5, -(5/2), 0, 5/2, \ldots$
26. $0, 4, 8, 12, \ldots$
27. $-3, -(7/2), -4, -(9/2), \ldots$
28. $-100, -95, -90, -85, \ldots$

Find the sum of the terms of the arithmetic sequence.

29. $1, 2, 3, 4, \ldots, 12$
30. $5, 10, 15, \ldots, 45$
31. $20, 18, 16, 14, \ldots, 0$
32. $-1, -4, -7, -10, \ldots, -25$
33. $8, 5, 2, -1, \ldots, -16$
34. $1/2, 5/2, 9/2, 13/2, \ldots, 29/2$
35. $-9, -(17/2), -8, -(15/2), \ldots, -(1/2)$
36. $100, 110, 120, 130, \ldots, 190$
37. $10, 5, 0, -5, \ldots, -20$
38. $3/5, 4/5, 5/5, 6/5, \ldots, 4$

Write the first five terms of the geometric sequence with the given first term, a_1, and common ratio, r.

39. $a_1 = 3, r = 2$
40. $a_1 = 2, r = 3$
41. $a_1 = 5, r = -2$
42. $a_1 = 3, r = 1/2$
43. $a_1 = -2, r = -1$
44. $a_1 = 6, r = -3$
45. $a_1 = -2, r = 1/2$
46. $a_1 = -10, r = -(1/2)$
47. $a_1 = 5, r = 3/5$
48. $a_1 = -4, r = -(2/3)$

Find the indicated term for the geometric sequence for the given first term, a_1, and common ratio, r.

49. Find a_6, $a_1 = 2, r = 3$
50. Find a_4, $a_1 = 5, r = 2$
51. Find a_5, $a_1 = 3, r = 2$
52. Find a_6, $a_1 = -2, r = -2$
53. Find a_4, $a_1 = -8, r = -3$
54. Find a_7, $a_1 = 10, r = -3$

55. Find a_3, $a_1 = 3, r = 1/2$
56. Find a_7, $a_1 = -3, r = -3$
57. Find a_5, $a_1 = 1/2, r = 2$
58. Find a_4, $a_1 = 4, r = 1/3$

Write an expression for the general or nth term, a_n, for the indicated geometric sequence.

59. $2, 4, 8, 16, \ldots$
60. $3, 12, 48, 192, \ldots$
61. $1, 3, 9, 27, \ldots$
62. $-6, 6, -6, 6, \ldots$
63. $-8, -4, -2, -1, \ldots$
64. $1/3, 1, 3, 9, \ldots$
65. $-6, 18, -54, 162, \ldots$
66. $10, 10/3, 10/9, 10/27, \ldots$
67. $-4, -(8/3), -(16/9), -(32/27), \ldots$

Find the sum of the first n terms of the geometric sequence for the given values of a_1 and r.

68. $n = 3, a_1 = 2, r = 5$
69. $n = 4, a_1 = 2, r = 2$
70. $n = 4, a_1 = 5, r = 3$
71. $n = 5, a_1 = 3, r = 5$
72. $n = 6, a_1 = 3, r = 4$
73. $n = 5, a_1 = 4, r = 3$
74. $n = 4, a_1 = -6, r = 2$
75. $n = 5, a_1 = -3, r = -2$
76. Find the sum of the first 50 natural numbers.
77. Find the sum of the first 50 even natural numbers.
78. Find the sum of the first 50 odd natural numbers.
79. Find the sum of the first 20 multiples of 3.
80. Donna is given a starting salary of $20,200 and promised a $1200 raise per year after each of the next eight years. Find her salary during her eight years of work.
81. Each swing of a pendulum is 3 inches shorter than its preceding swing. The first swing is 8 feet.
   a) Find the length of the twelfth swing.
   b) Determine the total distance traveled by the pendulum during the first twelve swings.
82. Each time a ball bounces, the height attained by the ball is 6 inches less than previous height attained. If on the first bounce the ball reaches a height of 6 feet, find the height attained on the eleventh bounce.
83. If you are given $1 on January 1, $2 on January 2, $3 on the 3rd, and so on, how much money will you have accumulated by January 31?
84. A certain substance decomposes and loses 20% of its weight each hour. If there are originally 200 grams of the substance, how much remains after six hours?

85. If your salary increases at a rate of 6% per year, find your salary during your 15th year if your present salary is $20,000.
86. The population in the United States is presently 217.3 million. If the population grows at a rate of 6% per year, find the population in 12 years.
87. When dropped, a ball rebounds to four-fifths of its original height. How high will the ball rebound after the fourth bounce if it is dropped from a height of 30 feet.
88. Explain how to construct the Fibonacci sequence.

89. a) Find the eighth and ninth terms of the Fibonacci sequence.
    b) Divide the ninth term by the eighth term, rounding to the nearest thousandth.
90. A sequence related to the Fibonacci sequence is the **Lucas sequence.** The Lucas sequence is formed in a manner similar to the Fibonacci sequence. The first two numbers of the Lucas sequence are 1 and 3. Write the first eight terms of the Lucas sequence.

PROBLEM SOLVING

91. A geometric sequence has $a_1 = 82$, and $r = 1/2$, find s_6.
92. Determine how many numbers between 7 and 1610 are divisible by 6.
93. Find r and a_1 for the geometric sequence with $a_2 = 24$ and $a_5 = 648$.
94. A ball is dropped from a height of 30 feet. On each bounce it attains a height four-fifths of its original height (or of the previous bounce). Find the total vertical distance traveled by the ball after it has

completed its fifth bounce (therefore has hit the ground six times).
95. The famous Fibonacci puzzle goes like this: A certain man puts a pair of rabbits in a place enclosed on all sides by a wall. How many pairs of rabbits will be born there in the course of one year if it is assumed that every month a pair of rabbits produces another pair and that all new rabbits begin to bear young two months after their own birth?

RESEARCH ACTIVITIES

96. Where does the Fibonacci sequence appear as a natural phenomenon? You might wish to contact the Fibonacci Society.

97. Where has the golden ratio been used in art and architecture? An art teacher or a staff member of an art museum might be able to give you some information and a list of resources.

SUMMARY

The set of natural or counting numbers is {1, 2, 3, 4, . . .}. A natural number greater than 1 that can be divided only by itself and 1 is called a prime number. Any natural number greater than 1 that is not a prime number is a composite number. Every composite number can be represented as a unique product of prime factors. Prime factorization can be used to find the greatest common divisor and the least common multiple of a set of numbers.

The set of whole numbers is {0, 1, 2, 3, 4, 5, . . .}. The set of integers is {. . . , −3, −2, −1, 0, 1, 2, 3, . . .}. An understanding of addition, subtraction, multiplication, and division of integers is essential for students who plan to study algebra.

The set of rational numbers is all the numbers of the form p/q, where p and q are integers and $q \neq 0$. Every rational number when expressed as a decimal will be either a terminating or a repeating decimal.

An irrational number is a nonterminating, nonrepeating decimal. Like radicals may be added and subtracted. Radicals may also be multiplied and divided. To rationalize a denominator means to remove all radicals from the denominator. The union of the set of rational numbers and the set of irrational numbers is the set of real numbers. Some of the properties of the real numbers are the commutative properties of addition and multiplication, the associative properties of addition and multiplication, and the distributive property of multiplication over addition.

The rules of exponents include the following.

$$a^m \cdot a^n = a^{m+n}$$
$$\frac{a^m}{a^n} = a^{m-n}, \qquad a \neq 0$$
$$a^0 = 1, \qquad a \neq 0$$
$$a^{-m} = \frac{1}{a^m}, \qquad a \neq 0$$
$$(a^m)^n = a^{m \cdot n}$$

Scientific notation can be helpful in working problems in which the numbers are very large or very small.

A sequence is a set of numbers in which each number except the first is related to the number or numbers that precede it. Three types of sequences discussed in this chapter are arithmetic sequences, geometric sequences, and the Fibonacci sequence.

REVIEW EXERCISES

In Exercises 1 and 2, determine whether or not the number is divisible by 2, 3, 4, 5, 6, 8, 9, 10, and 12.

  1. 148,632
  2. 400,644

Find the prime factorization of the following numbers.

  3. 192   **4.** 240   **5.** 180   **6.** 1260   **7.** 960

Find the gcd and the lcm of the following numbers.

  8. 32, 20   **9.** 80, 15   **10.** 148, 216
  11. 840, 320   **12.** 60, 40, 96   **13.** 36, 108, 144

Perform the indicated operation, and reduce your answer to lowest terms.

  14. $\frac{2}{3} + \frac{1}{5}$ **15.** $\frac{3}{5} - \frac{2}{4}$

  16. $\frac{7}{18} + \frac{9}{12}$ **17.** $\frac{4}{5} \cdot \frac{7}{9}$

  18. $\frac{5}{9} \div \frac{6}{7}$ **19.** $\left(\frac{4}{5} + \frac{5}{7}\right) \div \frac{4}{5}$

20. $\left(\frac{2}{3} \cdot \frac{1}{7}\right) \div \frac{4}{7}$ **21.** $\left(\frac{1}{5} + \frac{2}{3}\right)\left(\frac{3}{8}\right)$

22. $\left(\frac{1}{5} \cdot \frac{2}{3}\right) + \left(\frac{1}{5} \div \frac{1}{2}\right)$

Use the number line to evaluate the following.

23. $4 + 3$ **24.** $4 + (-2)$
25. $4 - 6$ **26.** $-3 + (-2)$
27. $-4 - 3$ **28.** $-4 - (-5)$
29. $(-1 + 9) - 4$ **30.** $-1 + (9 - 4)$
31. $-1 - (9 - 4)$

Evaluate the following.

32. $(-2)(-6)$ **33.** $-5(3)$ **34.** $6(-4)$

35. $\dfrac{-4}{-2}$ **36.** $\dfrac{8}{-2}$

37. $[8 \div (-4)](-3)$
38. $[(-4)(-3)] \div 2$
39. $[(-30) \div (10)] \div (-1)$

Express each fraction as a terminating or repeating decimal.

40. 5/8   **41.** 8/10   **42.** 9/12
43. 15/4   **44.** 3/7   **45.** 5/12
46. 3/8   **47.** 7/8   **48.** 2/7

Express each decimal as a quotient of two integers.

49. 0.624 **50.** $0.\overline{6}$ **51.** 2.43
52. $1.\overline{84}$ **53.** 12.083 **54.** 0.0042
55. $2.1\overline{5}$ **56.** $2.\overline{34}$ **57.** $5.06\overline{2}$

Find a rational number between the following numbers.

58. 0.0042 and 0.0043   **59.** 3/4 and 3/5
60. 2.406 and 2.407   **61.** 9/12 and 10/12
62. 6/12 and 0.51   **63.** 0.2 and 3/9

Simplify each of the following. Rationalize the denominator when necessary.

64. $\sqrt{12}$ **65** $\sqrt{72}$
66. $\sqrt{2} + 3\sqrt{2}$ **67.** $\sqrt{3} - 4\sqrt{3}$
68. $\sqrt{8} + 6\sqrt{2}$ **69.** $\sqrt{3} - 7\sqrt{27}$
70. $\sqrt{75} + \sqrt{27}$ **71.** $\sqrt{3} \cdot \sqrt{6}$

72. $\sqrt{8} \cdot \sqrt{6}$ **73.** $\dfrac{\sqrt{18}}{\sqrt{2}}$

74. $\dfrac{\sqrt{56}}{\sqrt{2}}$ **75.** $\dfrac{3}{\sqrt{2}}$

76. $\dfrac{\sqrt{3}}{\sqrt{5}}$ **77.** $5(3 + \sqrt{5})$

78. $\sqrt{3}(4 + \sqrt{6})$

State the name of each property illustrated.

79. $x + 2 = 2 + x$
80. $3 \cdot x = x \cdot 3$
81. $(2 + 3) + 4 = 2 + (3 + 4)$
82. $5 \cdot (2 + x) = 5 \cdot 2 + 5 \cdot x$
83. $(6 + 3) + 4 = 4 + (6 + 3)$
84. $(3 + 5) + (4 + 3) = (4 + 3) + (3 + 5)$
85. $(3 \cdot a) \cdot b = 3 \cdot (a \cdot b)$
86. $a \cdot (2 + 3) = (2 + 3) \cdot a$
87. $(x + 3)2 = (x \cdot 2) + (3 \cdot 2)$
88. $x \cdot 2 + 6 = 2 \cdot x + 6$

Determine whether the following sets of numbers are closed under the given operation.

89. Natural numbers, addition
90. Integers, addition
91. Integers, division
92. Real numbers, subtraction
93. Irrational numbers, multiplication
94. Rational numbers, division

Evaluate each of the following.

95. 2^3 **96.** 3^{-2}

97. $\dfrac{5^5}{5^4}$ **98.** $5^2 \cdot 5$

99. 3^0 **100.** 5^{-3}
101. $(2^3)^2$ **102.** $(3^2)^2$

Write each number in scientific notation.

103. 3,200   **104.** 0.0000423
105. 0.00168   **106.** 4,950,000

Express each number in decimal notation.

107. 4.2×10^2 **108.** 3.87×10^{-5}
109. 1.75×10^{-4} **110.** 1×10^5

Perform the indicated operations and express the answer in scientific notation.

111. $(2 \times 10^6)(3.2 \times 10^{-4})$ **112.** $(3 \times 10^2)(4.6 \times 10^2)$
113. $\dfrac{8.4 \times 10^3}{4 \times 10^2}$ **114.** $\dfrac{1.5 \times 10^{-3}}{5 \times 10^{-4}}$

Perform the indicated operation by first converting each number to scientific notation. Write the answer in decimal notation.

115. $(80,000)(420,000)$ **116.** $(75,000)(0.0003)$
117. $\dfrac{9,600,000}{3,000}$ **118.** $\dfrac{0.000002}{0.0000004}$
119. At noon there were 12,000 bacteria in the culture. At 6:00 P.M. there were 300,000 bacteria in the culture. How many times greater is the number of bacteria at 6:00 P.M. than at noon?

Determine whether the following sequences are arithmetic or geometric. Then determine the next two terms of the sequence.

120. $3, 8, 13, 18, \ldots$
121. $-4, 12, -36, 108, \ldots$
122. $0, -4, -8, -12, \ldots$
123. $1, \frac{1}{2}, \frac{1}{4}, \frac{1}{8}, \ldots$
124. $1, 4, 7, 10, 13, \ldots$
125. $2, -2, 2, -2, 2, \ldots$

Write the first five terms of the sequence with the given first term, a_1, and common difference, d, or ratio, r.

126. $a_1 = 4, d = 5$
127. $a_1 = -6, d = -3$
128. $a_1 = -\frac{1}{2}, d = -2$
129. $a_1 = 4, r = 2$
130. $a_1 = 16, r = 1/2$
131. $a_1 = \frac{1}{2}, r = -1/2$

Find the indicated term of the sequence with the given first term, a_1, and common difference, d, or ratio, r.

132. Find a_5, given $a_1 = 6, d = 2$
133. Find a_7, given $a_1 = -6, d = -5$
134. Find a_{10}, given $a_1 = 20, d = 8$
135. Find a_4, given $a_1 = 8, r = 3$
136. Find a_5, given $a_1 = 4, r = \frac{1}{2}$
137. Find a_4, given $a_1 = -6, r = 2$

Find the sum of the arithmetic sequence.

138. $4, 2, 0, -2, \ldots, -18$
139. $-4, -3\frac{3}{4}, -3\frac{1}{2}, -3\frac{1}{4}, \ldots, -2\frac{1}{4}$
140. $100, 94, 88, 82, \ldots, 58$

Find the sum of the first n terms of the geometric series with the given values of a_1 and r.

141. $n = 3, a_1 = 4, r = 3$
142. $n = 4, a_1 = 2, r = 4$
143. $n = 5, a_1 = 3, r = -2$

First determine whether the sequence is arithmetic or geometric, then write an expression for the general or nth term, a_n.

144. $7, 4, 1, -2, \ldots$
145. $0, 5, 10, 15, \ldots$
146. $4, 5/2, 1, -1/2, \ldots$
147. $3, 6, 12, 24, \ldots$
148. $4, -4, 4, -4, \ldots$
149. $5, 5/3, 5/9, 5/27, \ldots$

CHAPTER TEST

1. Which of the numbers 2, 3, 4, 5, 6, 8, 9, 10, and 12 divide 481, 248.
2. Find the prime factorization of 360.
3. Evaluate $[(-8) + (-5)] + 7$
4. Evaluate $-7 - 15$.
5. Evaluate $[(-50)(-2)] \div (8 - 10)$
6. Determine a rational number between 0.435 and 0.436.
7. Determine a rational number between 1/8 and 1/9.
8. Write 3/8 as a terminating or repeating decimal.
9. Express 2.45 as a quotient of two integers.
10. Evaluate $(\frac{5}{16} \div 3) + (\frac{4}{5} \cdot \frac{1}{2})$.
11. Perform the operation and reduce the answer to the lowest terms:

$$17/24 - 3/40$$

12. Simplify $\sqrt{18} + \sqrt{50}$
13. Rationalize $\dfrac{\sqrt{5}}{\sqrt{3}}$
14. Determine whether the integers are closed under the operation of multiplication. Explain your answer.

Name the properties illustrated.

15. $(2 + x) + 3 = 2 + (x + 3)$
16. $3(x + y) = 3x + 3y$

Evaluate

17. $\dfrac{4^5}{4^2}$
18. $2^3 \cdot 2^2$
19. 7^{-2}
20. Perform the operation by first converting the numerator and denominator to scientific notation. Write the answer in scientific notation.

$$\frac{64,000}{0.008}$$

21. Write an expression for the general or nth term, a_n, of the sequence $-4, -8, -12, -16, \ldots$.
22. Find the sum of the terms of the arithmetic sequence $-2, -5, -8, -11, \ldots, -32$.
23. Find a_5 when $a_1 = 3$ and $r = 3$.
24. Find the sum of the first five terms of the sequence when $a_1 = 3$ and $r = 4$.
25. Write an expression for the general or nth term, a_n, of the sequence $3, 6, 12, 24, \ldots$.

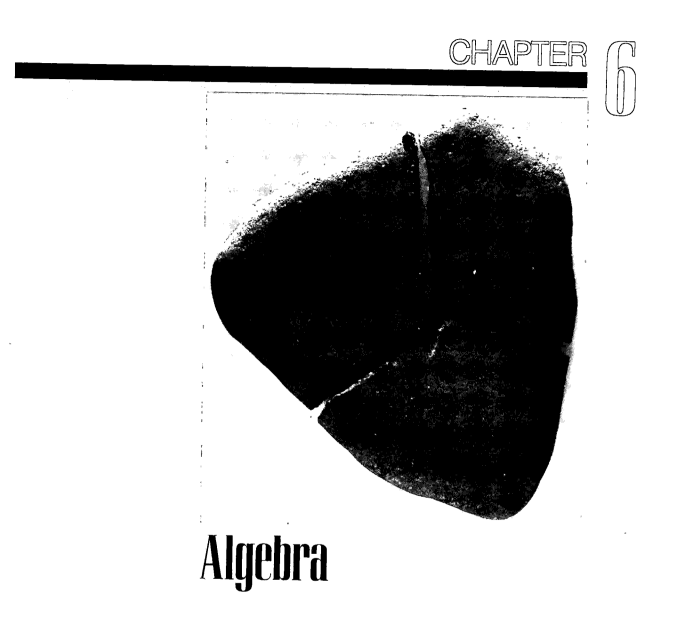

CHAPTER 6

Algebra

PRIORITY OF OPERATIONS

History

René Descartes.
Courtesy of International Business Machines Corporation

Algebra is the study of equations and methods for solving them. The word "algebra" is a Latin variation of the Arabic word *al-jabr* (meaning reunion of broken parts), which was the title of a book written by the Arab mathematician Mohammed ibn-Musa al Khwarizmi around A.D. 825.

The study of algebra began in Babylonia and Egypt around 1700 B.C. Throughout history, many individuals from various countries have contributed to the development of algebra. An outstanding contributor was the French mathematician René Descartes (1596–1650). A story is told that Descartes was lying on his bed watching a spider spin a web on the ceiling. As he watched the spider do its creative work, Descartes tried to recall the various paths the spider had taken. He conceived the idea of having two sets of parallel lines, one set perpendicular to the other set, with the spider web on top of these intersecting lines (Fig. 6.1). On this rectangular array of lines he could sketch curves. Descartes thus invented what is called the rectangular coordinate system. This development led to the combining of algebra and geometry into a single area of study called analytic geometry.

FIGURE 6.1
Descartes's vision
Photo by Marshall Henrichs

Why study algebra? Solutions for many problems in everyday life can be found by using arithmetic, but with a knowledge of algebra the solutions can be found with less effort.

Terminology

Before you can learn algebra, you must become familiar with some of the terminology.

A **variable** is a letter that is used to represent a number. A symbol that represents a specific quantity is called a **constant.** In algebra we often use the letters x and y to represent variables. However, any letter may be used as a variable.

Since the "times" sign might be confused with the variable x, in algebra we often use a dot placed between two numbers or variables to indicate multiplication. Thus $3 \cdot 4$ means 3 times 4, and $x \cdot y$ means x times y. When two letters or a number and a letter are placed next to one another, multiplication is also indicated. Thus $3x$ means 3 times x, and xy means x times y. When a number and a letter are placed next to one another, the number is usually first. When two letters are together, the letters are usually in alphabetical order. Other examples of multiplication problems are the following:

$3 \cdot x$	means	3 times x
$4y$	means	4 times y
$5(x)$	means	5 times x
$(5)(6)$	means	5 times 6

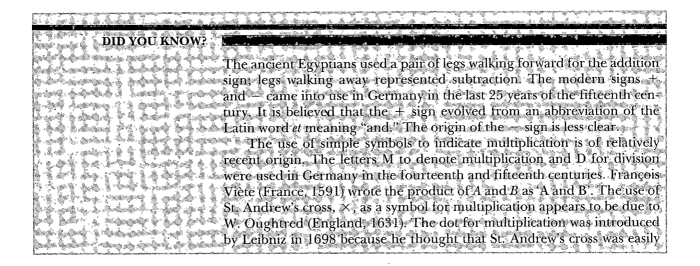

confused with the letter X. Juxtaposition (placing numbers and letters side by side, as in $3x$ or $4xy$) was first used by Descartes in 1637.

The Greeks had no sign for division. The Hindus simply wrote the divisor underneath the dividend. The Arabic author Al-Hassar in the twelfth century used a horizontal line between the dividend and the divisor, as we do today. The sign $\div$ for division was introduced by the Swiss mathematician Rahn in 1659. This sign was not commonly used in Europe but was widely adopted in England and the United States. The symbol $\div$ is most prevalent in English-speaking countries.

An **algebraic expression** (or simply **expression**) is a collection of variables, numbers, parentheses, and operation symbols. Here are some examples of algebraic expressions:

$$x \qquad 7 + 7x \qquad x + 2$$

$$3(2x + 3) \qquad \frac{3x + 1}{2x - 3} \qquad x^2 + 7x + 3$$

To **evaluate an expression** means to find the value of the expression for a given value of the variable.

Two algebraic expressions joined by an equal sign form an **equation.** Here are some examples of equations:

$$x + 2 = 4$$
$$3x + 4 = 1$$
$$x + 3 = 2x$$

The **solution** to an equation is the number or numbers that replace the variable to make the equation a true statement. The solution to the equation $x + 3 = 4$ is $x = 1$. When we find the solution to an equation, we **solve the equation.**

The solution to any equation can be **checked** by substituting the value obtained for the variable in the original equation. To check the answer to the previous equation, we do the following:

CHECK
$$x = 1$$
$$x + 3 = 4$$
$$1 + 3 = 4$$
$$4 = 4 \qquad \text{True}$$

Since the same number is obtained on both sides of the equal sign, the solution is correct.

To evaluate expressions and solve equations, you must have an understanding of exponents. Exponents were discussed in Section 5.5. Recall that in the expression 5^2, the 2 is referred to as the **exponent,** and the 5 is referred to as the **base.** We read 5^2 as "5 to the second power," or "5 squared." This means

$$5^2 = \underbrace{5 \cdot 5.}_{2 \text{ factors of 5}}$$

The number 5 to the third power, or 5 cubed, written 5^3, means

$$5^3 = \underbrace{5 \cdot 5 \cdot 5.}_{3 \text{ factors of 5}}$$

In general, the number b to the nth power, written b^n, means

$$b^n, = \underbrace{b \cdot b \cdot b \cdot \cdots \cdot b.}_{n \text{ factors of } b}$$

An exponent refers only to its base. In the expression -5^2 the base is 5. The expression -5^2 actually means $-1(5)^2$ and has a value of $-1(25)$ $= -25$. In the expression $(-5)^2$, the base is -5. Note that $(-5)^2 = (-5)(-5) = 25$.

$$-5^2 = -(5)^2 = -(5)(5) = -25$$
$$(-5)^2 = (-5)(-5) = 25$$

Priority of Operations

To evaluate an expression or to check an equation, we need to know the **priority,** or **order, of operations.** For example, suppose we wish to evaluate the expression $2 + 3x$ when $x = 4$. Substituting 4 for x, we obtain $2 + 3 \cdot 4$. What is the value of $2 + 3 \cdot 4$? Does it equal 20, or does it equal 14? The correct answer is 14. In mathematics, unless parentheses indicate otherwise, we always perform multiplications before additions. Thus

$$2 + 3 \cdot 4 = 2 + (3 \cdot 4) = 2 + 12 = 14$$

Table 6.1 indicates the order of operations to be followed in evaluating an expression. Note that the evaluation of expressions that contain exponents and roots was discussed in Chapter 5.

TABLE 6.1
Priority of Operations

1. First, perform all operations within parentheses or other grouping symbols (according to the order given below).
2. Next, perform all exponential operations (that is, raising to powers or finding roots).
3. Next, perform all multiplication and division from left to right.
4. Finally, perform all addition and subtraction from left to right.

■ **Example 1** ─────────────────────────────

Evaluate the following, using the priority of operations.

a) $6 + 13 - 11$ b) $15 - 6 \div 2 + 5$

c) $6 + 4^3 \div 2$ d) $8 \div \sqrt{64} + 7$

Solution

a)   Since addition and subtraction have the same priority, we perform the operations from left to right. Shading is used to indicate the order in which the operations are performed.

$$6 + 13 - 11 = 19 - 11$$
$$= 8$$

b) $15 - 6 \div 2 + 5 = 15 - 3 + 5$ Division is performed before
$$= 12 + 5 \qquad\qquad \text{addition or subtraction.}$$
$$= 17$$

c) $6 + 4^3 \div 2 = 6 + 64 \div 2$ Raising to a power is a higher
priority than division or addition.
$$= 6 + 32 \qquad\qquad \text{Division is performed before}$$
$$= 38 \qquad\qquad\quad \text{addition.}$$

d) $8 \div \sqrt{64} + 7 = 8 \div 8 + 7$
$$= 1 + 7$$
$$= 8 \qquad\qquad\qquad\qquad\qquad\qquad\qquad ❖$$

■ **Example 2** ─────────────────────────────

Evaluate the following, using the priority of operations.

a) $-2^2 + (6 - 2)^3 \div 32$

b) $3(4 - 7)^2 + 4 \cdot 8 - 5$

c) $15 \cdot 6 \div 2 \div 5^2$

Solution

a) $-2^2 + (6 - 2)^3 \div 32 = -2^2 + 4^3 \div 32$ Evaluate within ( ).
$$= -4 + 64 \div 32 \qquad \text{Raise to powers.}$$
$$= -4 + 2 \qquad\qquad\quad \text{Divide before}$$
$$= -2 \qquad\qquad\qquad\quad \text{adding.}$$
Notice that $-2^2 = -(2)^2$ or -4.

b) $3(4 - 7)^2 + 4 \cdot 8 - 5 = 3(-3)^2 + 4 \cdot 8 - 5$

$$= 3(9) + 4 \cdot 8 - 5$$
$$= 27 + 32 - 5$$
$$= 59 - 5$$
$$= 54$$

c) $15 \cdot 6 \div 2 \div 5^2 = 15 \cdot 6 \div 2 \div 25$

$$= 90 \div 2 \div 25$$
$$= 45 \div 25$$
$$= 45/25 \text{ or } 9/5 \text{ or } 1.8 \qquad ❖$$

We will now evaluate expressions containing variables using the priority of operations.

■ **Example 3** ─────────────────────────────

Evaluate x^2 when (a) $x = 4$ and (b) $x = -3$.

Solution

a) Substitute 4 for x in the expression x^2:

$$x^2 = (4)^2 = 16$$

b) Substitute -3 for x in the expression x^2:

$$x^2 = (-3)^2 = 9 \qquad ❖$$

■ **Example 4** ─────────────────────────────

Evaluate $-x^2$ when (a) $x = 3$ and (b) $x = -5$.

Solution

a) Substitute 3 for x:

$$-x^2 = -(3)^2 = -(9) = -9$$

b) Substitute -5 for x:

$$-x^2 = -(-5)^2 = -(25) = -25 \qquad ❖$$

■ **Example 5** ─────────────────────────────

Evaluate $4x^2 - 3$ when $x = 3$.

Solution: Substitute 3 for x, then evaluate using the priority of operations:

$$\begin{aligned} 4x^2 - 3 &= 4(3)^2 - 3 \\ &= 4(9) - 3 \\ &= 36 - 3 \\ &= 33 \end{aligned}$$ ❖

■ **Example 6** _____

Evaluate $3x^2 + 4x + 5$ when $x = -2$.

Solution: Substitute -2 for x, then evaluate using the priority of operations.

$$\begin{aligned} 3x^2 + 4x + 5 &= 3(-2)^2 + 4(-2) + 5 \\ &= 3(4) - 8 + 5 \\ &= 12 - 8 + 5 \\ &= 4 + 5 \\ &= 9 \end{aligned}$$ ❖

■ **Example 7** _____

Evaluate $-3x^2 + 4xy - y^2$ when $x = 3$ and $y = 4$.

Solution: Substitute 3 for x and 4 for y, then evaluate using the priority of operations.

$$\begin{aligned} -3x^2 + 4xy - y^2 &= -3(3)^2 + 4(3)(4) - 4^2 \\ &= -3(9) + 12(4) - 16 \\ &= -27 + 48 - 16 \\ &= 21 - 16 \\ &= 5 \end{aligned}$$ ❖

■ **Example 8** _____

Determine whether 3 is a solution to the equation $2x^2 + 4x - 9 = 21$.

Solution: To determine whether 3 is a solution to the equation, substitute 3 for each x in the equation. Then evaluate the left side of the equation using the priority of operations. If this leads to a 21 on the left side of the equal sign, then both sides of the equation have the same value, and 3 is a solution.

$$2x^2 + 4x - 9 = 21$$
$$2(3)^2 + 4(3) - 9 = 21$$
$$2(9) + 12 - 9 = 21$$
$$18 + 12 - 9 = 21$$
$$30 - 9 = 21$$
$$21 = 21 \quad \text{True}$$

Since 3 makes the equation a true statement, 3 is a solution to the equation. ❖

EXERCISES 6.1

Evaluate each of the following, using the priority of operations given in Table 6.1.

1. $3 + 7 \cdot 9$
2. $24 \div 6 \cdot 2$
3. $\sqrt{16} + 9 \div 3 \cdot 3 - 7$
4. $3 - \sqrt[3]{8} + 6^2 - 14$
5. $11 \cdot 4 - 2 + 7^2 - 6 + \sqrt{9}$
6. $9 - 13 + 6 \div 3 - \sqrt{25} + 4^2$
7. $3(11 - 6) + 7\sqrt{16} - 17$
8. $2 + 3(5 - 2 \cdot 3^2) + 11$
9. $-2^2 + 3(4 - 1)^2$
10. $3 - 4 + 5 - 6 - 7 \cdot 3$
11. $(3 \cdot 4 + 7)^2 - 6 \cdot 2 - 3$
12. $6^2 \div \sqrt{4} \cdot (-4)^2 \div 6^2$

Evaluate each of the following for the given value of the variable.

13. $x^2, x = 3$
14. $x^2, x = -6$
15. $-x^2, x = 7$
16. $-x^2, x = -3$
17. $x^3, x = 5$
18. $x^3, x = -2$
19. $4x^2, x = 5$
20. $3x^3, x = -5$
21. $4x^3, x = 0$
22. $-2x^2, x = 3$
23. $-3x^2, x = -4$
24. $-2x^3, x = -5$
25. $-x^3, x = -3$
26. $-4x^3, x = 4$

Evaluate each expression for the given value(s) of the variables(s).

27. $x + 5$ $\quad x = 2$
28. $2x - 4$ $\quad x = \frac{3}{2}$
29. $-3x + 8$ $\quad x = -4$
30. $x^2 + 2x - 4$ $\quad x = 5$
31. $-x^2 - 3x + 8$ $\quad x = -2$
32. $3x^2 + 7x - 11$ $\quad x = -1$
33. $\frac{1}{2}x^2 + 4x - 6$ $\quad x = \frac{1}{3}$
34. $x^3 - 3x^2 + 7x + 11$ $\quad x = 2$
35. $4x^3 - 6x + 5$ $\quad x = \frac{1}{2}$
36. $x^2 + 4xy$ $\quad x = 2, y = 3$
37. $x^2 - 2xy + y^2$ $\quad x = -1, y = -1$
38. $3x^2 + \frac{2}{5}xy - \frac{1}{2}y^2$ $\quad x = 5, y = -2$
39. $4x^2 - 12xy + 9y^2$ $\quad x = 3, y = 2$
40. $(x + 3y)^2$ $\quad x = 4, y = -1$

Determine whether the given values are a solution to the equation.

41. $2x - 4 = 0, x = 2$
42. $3x - 6 = 4, x = 3$
43. $x + 2y = 0, x = 2, y = -1$
44. $3x + 2y = 4, x = 0, y = 2$
45. $x^2 + 3x - 4 = 5, x = 2$
46. $2x^2 - x - 5 = 0, x = 3$
47. $2x^2 - x = 28, x = 4$
48. $y = x^2 + 3x - 5, x = 1, y = -1$
49. $y = -x^2 - 3x + 2, x = 2, y = -8$
50. $x^2 + 3y = 1, x = -2, y = -1$

6.2 LINEAR EQUATIONS IN ONE VARIABLE

In Section 6.1 we stated that two algebraic expressions joined by an equal sign form an equation. The solution to some equations, such as $x + 3 = 4$, can be found easily by trial and error. However, to solve more complex equations, such as $2x + 3 = \frac{5}{2}$ or $3x - 2x + 3 = 5(3x + 2)$, we must understand the meaning of like terms, and learn four basic properties.

The parts that are added or subtracted in an algebraic expression are called **terms.** The expression $4x - 3y - 5$ contains three terms, namely, $4x$, $-3y$, and -5. The $+$ and $-$ signs that break the expression into terms are a part of the terms. When listing the terms of an expression, however, it is not necessary to include the $+$ sign at the beginning of the term.

The numerical part of a term is called its **numerical coefficient** or simply its **coefficient.** In the term $4x$, the 4 is the numerical coefficient. In the term $-4y$, the -4 is the numerical coefficient.

Like terms are terms that have the same variables with the same exponents on the variables. **Unlike terms** have different variables or different exponents on the variables.

Like Terms		Unlike Terms
$3x$, $-4x$	(same variable, x)	$3x$, 2
$4y$, $6y$	(same variable, y)	$3x$, $4y$
5, -6	(both constants)	x, 3
$-6x^2$, $7x^2$	(same variable with same exponent)	$3x^2$, $5x^3$

Very often we will need to **simplify** an expression. To simplify an expression means to combine like terms. We can do this by using the commutative, associative, and distributive properties discussed in Chapter 5. For convenience we list the properties here.

$a(b + c) = ab + ac$	Distributive property
$a + b = b + a$	Commutative property of addition
$ab = ba$	Commutative property of multiplication
$(a + b) + c = a + (b + c)$	Associative property of addition
$(ab)c = a(bc)$	Associative property of multiplication

■ **Example 1** _____

Combine like terms in the expression $4x + 3x$.

Solution: We use the distributive property (in reverse) to combine like terms:

$$4x + 3x = (4 + 3)x \qquad \text{Distributive property}$$
$$= 7x$$

❖

■ **Example 2** _____

Combine like terms in the expression $4x - 3x$.

Solution

$$4x - 3x = (4 - 3)x \qquad \text{Distributive property}$$
$$= x \ (\text{or } 1x)$$

❖

■ **Example 3** _____

Combine like terms in the expression $5a - 7a$.

Solution

$$5a - 7a = (5 - 7)a \qquad \text{Distributive property}$$
$$= -2a$$

❖

■ **Example 4** _____

Combine like terms in the expression $3x + 2x + 1$.

Solution

$$3x + 2x + 1 = (3 + 2)x + 1 \qquad \text{Distributive property}$$
$$= 5x + 1$$

❖

■ **Example 5** _____

Combine like terms in the expression $12 + x + 7$.

Solution

$$12 + x + 7 = x + 12 + 7 \qquad \text{Commutative property of addition}$$
$$= x + (12 + 7) \qquad \text{Associative property of addition}$$
$$= x + 19$$

Thus $12 + x + 7 = x + 19$.

❖

In future examples when we rearrange terms, we might not show each step in the process. The student should realize that the terms can be rearranged by using the properties mentioned on page 270.

■ **Example 6** _____

Combine like terms in the expression $3y + 4x - 3 - 2x$.

Solution

$4x$ and $-2x$ are the only like terms.

$$4x - 2x + 3y - 3 \qquad \text{Rearranging terms}$$
$$2x + 3y - 3 \qquad \text{Combining like terms} \qquad \clubsuit$$

Example 7

Combine like terms.

$$-2x + 3y - 4x + 3 - y + 5$$

Solution

$-2x$ and $-4x$ are like terms.
$3y$ and $-y$ are like terms.
3 and 5 are like terms.
Grouping the like terms together gives

$$\underbrace{-2x - 4x}_{-6x} + \underbrace{3y - y}_{2y} + \underbrace{3 + 5}_{8.}$$

We used the commutative and associative properties to rearrange the terms. The order of the terms in the answer is not critical. Thus $2y - 6x + 8$ is also an acceptable answer. $\clubsuit$

Solving Equations

To solve any equation, we must **isolate the variable.** This means that we must get the variable by itself on one side of the equal sign. The four properties we are about to discuss will be used to isolate the variable. We will give each property followed by an example of how the property is used. The first property we will discuss is the addition property.

> **Addition property.** If $a = b$, then $a + c = b + c$ for all real numbers $a, b,$ and c.

The rule says that the same number can be added to both sides of an equation without changing the solution.

Example 8

Find the solution to the equation $x - 6 = 4$.

Solution: To solve an equation, it is necessary to isolate the variable. In this equation we add 6 to both sides of the equation.

$$x - 6 + 6 = 4 + 6$$

Since -6 and $+6$ have the sum of 0, we can write

$$x + 0 = 10 \quad \text{or simply} \quad x = 10.$$

CHECK
$$x = 10$$
$$x - 6 = 4$$
$$10 - 6 = 4$$
$$4 = 4 \quad \text{True}$$ ❖

To solve this equation, we used the addition property.

> **Subtraction property.** If $a = b$, then $a - c = b - c$ for all real numbers a, b, and c.

The rule states that the same number can be subtracted from both sides of an equation without changing the solution.

Example 9

Find the solution to the equation $x + 5 = 7$.

Solution: To solve the equation, it is necessary to isolate the variable. In this equation we subtract 5 from both sides of the equation.

$$x + 5 - 5 = 7 - 5$$
$$x + 0 = 2$$
$$x = 2$$

Note that we did not subtract 7 from both sides of the equation, since this would not result in getting x on one side of the equals sign by itself. ❖

To solve this equation, we used the subtraction property.

> **Multiplication property.** If $a = b$, then $a \cdot c = b \cdot c$ for all real numbers a, b, and c where $c \neq 0$.

The rule states that both sides of the equation can be multiplied by the same nonzero number without changing the solution.

Example 10

Find the solution to the equation $\frac{x}{3} = 2$.

Solution

To solve this equation, we multiply both sides of the equation by 3:

$$3\left(\frac{x}{3}\right) = 3\,(2)$$

$$\frac{3x}{3} = 6$$

$$1x = 6$$

$$x = 6 \qquad \text{❖}$$

   To solve this equation, we used the multiplication property. Note that $\frac{x}{3}$ is the same as $\frac{1}{3}x$ or $\frac{1}{3}$ times x.

> **Division property.** If $a = b$, then $a/c = b/c$ for all real numbers a, b, and c, $c \neq 0$.

   The rule states that both sides of an equation can be divided by the same number without changing the solution. There is one added restriction in this case: the divisor cannot be equal to zero. Why?

Example 11

Find the solution to the equation $5x = 35$.

Solution: To solve this equation, we divide both sides of the equation by 5:

$$\frac{5x}{5} = \frac{35}{5}$$

$$x = 7 \qquad \text{❖}$$

   To solve this equation, we used the division property.

   An **algorithm** is a general procedure for accomplishing a task. The following is an algorithm for solving **linear (or first degree)** equations. A linear equation is one in which the exponent that appears on any variable is 1. The reason such equations are called linear is explained in Section 6.6.

A general procedure for solving linear equations is:

1. If the equation contains fractions, multiply both sides of the equation by the lowest common denominator (or least common multiple). This step will eliminate all fractions from the equation.

2. Use the distributive property to remove parentheses when necessary.

3. Combine like terms on the same side of the equals sign.

4. Use the addition or subtraction property to collect all terms with a variable on one side of the equals sign and all constants on the other side of the equals sign. It may be necessary to use the addition or subtraction property more than once. This process will eventually result in an equation of the form $ax = b$.

5. Solve for the variable using the division or multiplication property. This will result in an answer in the form $x = c$, where c is some real number.

There are variations on this procedure that can be used to solve equations. Remember that the primary objective in solving any equation is to isolate the variable.

■ **Example 12** _____

Solve the following equations:

a) $x - 3 = 7$ **b)** $x + 5 = 11$ **c)** $\frac{1}{9}x = 10$ **d)** $6x = 72$

Solution: To solve these equations, we must get the variable x by itself on one side of the equals sign.

a)
$$x - 3 = 7$$
$$x - 3 + 3 = 7 + 3 \qquad \text{Add 3 to both sides of the equation.}$$
$$x = 10 \qquad \text{(Addition property)}$$

b)
$$x + 5 = 11$$
$$x + 5 - 5 = 11 - 5 \qquad \text{Subtract 5 from both sides of the equation.}$$
$$x = 6 \qquad \text{(Subtraction property)}$$

c)
$$\frac{1}{9}x = 10$$
$$9(\tfrac{1}{9}x) = 9(10) \qquad \text{Multiply both sides of the equation by 9.}$$
$$x = 90 \qquad \text{(Multiplication property)}$$

d)
$$6x = 72$$
$$\frac{6x}{6} = \frac{72}{6} \qquad \text{Divide both sides of the equation by 6.}$$
$$x = 12 \qquad \text{(Division property)}$$ ❖

One or more of the four properties may be used in solving a linear equation. Consider the following examples.

■ **Example 13** _____

Solve the equation $3x + 4 = 13$. Then check your solution by replacing x in the equation with the value found for x.

Solution

$$3x + 4 = 13$$
$$3x + 4 - 4 = 13 - 4 \qquad \text{Subtract 4 from both sides of the equation.}$$
$$3x = 9 \qquad \text{(Subtraction property)}$$
$$\frac{3x}{3} = \frac{9}{3} \qquad \text{Divide both sides of the equation by 3.}$$
$$x = 3 \qquad \text{(Division property)}$$

Your first reaction might have been to divide both sides of the equation by 3 before subtracting. Would this have given a different result? Let us see what happens.

$$3x + 4 = 13$$
$$\frac{3x + 4}{3} = \frac{13}{3} \qquad \text{Divide both sides of the equation by 3.}$$
$$\qquad \qquad \text{(Division property)}$$
$$\frac{3x}{3} + \frac{4}{3} = \frac{13}{3}$$
$$x + \frac{4}{3} = \frac{13}{3}$$
$$x + \frac{4}{3} - \frac{4}{3} = \frac{13}{3} - \frac{4}{3} \qquad \text{Subtract 4/3 from both sides of the equation.}$$
$$\qquad \qquad \text{(Subtraction property)}$$
$$x = \frac{9}{3}$$
$$x = 3$$

As you can see, you obtain the same result, but you must work with fractions. In general, when you are solving an equation that does not contain fractions or decimals, it is simpler to perform the operations of addition and subtraction before the operations of multiplication and division.

CHECK $3x + 4 = 13$
$$3(3) + 4 = 13 \qquad \text{Replacing } x \text{ in the equation with 3.}$$
$$9 + 4 = 13$$
$$13 = 13, \text{ True}$$

Thus 3 is the solution to $3x + 4 = 13$.   ❖

As you can see from Example 13, there are different ways of solving equations and finding the correct solution. However, some ways are more efficient than others.

■ **Example 14** _____

Solve the equation $2 = 3 + 5(p + 1)$ for p.

Solution: The goal is to isolate the variable p. To accomplish this, we will follow the general procedure for solving equations:

$$2 = 3 + 5(p + 1)$$
$$2 = 3 + 5p + 5 \qquad \text{Distributive property (Step 2)}$$
$$2 = 3 + 5 + 5p \qquad \text{Commutative property}$$
$$2 = 8 + 5p \qquad \text{Combine like terms. (Step 3)}$$
$$2 - 8 = 8 - 8 + 5p \qquad \text{Subtract 8 from both sides of the equation.}$$
$$\text{(Step 4)}$$

$$-6 = 5p$$
$$-\frac{6}{5} = \frac{5p}{5} \qquad \text{Divide both sides of the equation by 5.}$$
$$\text{(Step 5)}$$

$$-\frac{6}{5} = p$$

❖

■ **Example 15** _____

Solve the equation and check your solution: $\dfrac{2x}{3} + \dfrac{1}{3} = \dfrac{3}{4}$.

Solution: Fractions scare some people. Remain calm; if you follow the rules, this example is no more difficult than the others. A technique used to simplify the problem is to multiply each term of the equation by the lowest common denominator, lcd (see Chapter 5). In this example the lcd is 12, since 12 is the smallest number that is divisible by both 3 and 4.

$$12\left(\frac{2x}{3} + \frac{1}{3}\right) = 12\left(\frac{3}{4}\right) \qquad \text{Multiply both sides of the equation by the lcd. (Step 1)}$$
$$12\left(\frac{2x}{3}\right) + 12\left(\frac{1}{3}\right) = 12\left(\frac{3}{4}\right) \qquad \text{Distributive property (Step 2)}$$

$$\overset{4}{\cancel{12}}\left(\frac{2x}{\cancel{3}}\right) + \overset{4}{\cancel{12}}\left(\frac{1}{\cancel{3}}\right) = \overset{3}{\cancel{12}}\left(\frac{3}{\cancel{4}}\right)$$

$$8x + 4 = 9$$
$$8x + 4 - 4 = 9 - 4 \qquad \text{Subtract 4 from both sides of the equation. (Step 4)}$$

$$8x = 5$$

$$\frac{8x}{8} = \frac{5}{8}$$

Divide both sides of the equation by 8. (Step 5)

$$x = \frac{5}{8}$$

CHECK

$$\frac{2x}{3} + \frac{1}{3} = \frac{3}{4}$$

$$\overset{1}{\underset{3}{\cancel{2}}}\left(\frac{5}{\underset{4}{\cancel{8}}}\right) + \frac{1}{3} = \frac{3}{4}$$

Substituting 5/8 in place of x in the equation. Notice that $\frac{2x}{3}$ is the same as $\frac{2}{3}x$ or $\frac{2}{3}$ times x.

$$\frac{5}{12} + \frac{1}{3} = \frac{3}{4}$$

$$\frac{5}{12} + \frac{4}{12} = \frac{3}{4}$$

$$\frac{9}{12} = \frac{3}{4}$$

$$\frac{3}{4} = \frac{3}{4}, \text{ True}$$

The check shows that 5/8 is the solution of the equation. You could have worked the problem without multiplying both sides of the equation by the lcd. Try it! ❖

■ **Example 16** _____

Solve the equation $3x + 4 = 5x + 6$.

Solution: Note that the equation has an x on both sides of the equals sign. In equations of this type you might wonder what to do first. It really does not matter as long as you don't forget your goal of isolating the variable x. Two methods of solving this equation will be illustrated.

METHOD 1

$$3x + 4 = 5x + 6$$

$$3x + 4 - 4 = 5x + 6 - 4 \qquad \text{Subtract 4 from both sides of the equation.}$$

$$3x = 5x + 2$$

$$3x - 5x = 5x - 5x + 2 \qquad \text{Subtract } 5x \text{ from both sides of the equation.}$$

$$-2x = 2$$

$$\frac{-2x}{-2} = \frac{2}{-2} \qquad \text{Divide both sides of the equation by } -2.$$

$$x = -1$$

METHOD 2

$$3x + 4 = 5x + 6$$
$$3x + 4 - 6 = 5x + 6 - 6 \qquad \text{Subtract 6 from both sides of the equation.}$$

$$3x - 2 = 5x$$
$$3x - 3x - 2 = 5x - 3x \qquad \text{Subtract } 3x \text{ from both sides of the equation.}$$

$$-2 = 2x$$
$$\frac{-2}{2} = \frac{2x}{2} \qquad \text{Divide both sides of the equation by 2.}$$

$$-1 = x$$

Note that in Method 1 the variables are collected on the left side of the equals sign, while in Method 2 the variables are collected on the right side of the equals sign. The same answer is obtained by both methods.

CHECK
$$3x + 4 = 5x + 6$$
$$3(-1) + 4 = 5(-1) + 6$$
$$1 = 1, \text{True} \qquad \text{❖}$$

■ **Example 17** _____

Solve the equation and check your solution: $4x - 0.48 = 0.8x + 4$.

Solution: This problem may be solved with the decimals, or you may multiply each term by 100 and eliminate the decimals. We will solve the problem with the decimals:

$$4x - 0.48 = 0.8x + 4$$
$$4x - 0.48 + 0.48 = 0.8x + 4 + 0.48 \qquad \text{Add 0.48 to both sides of the equation.}$$

$$4x = 0.8x + 4.48$$
$$4x - 0.8x = 0.8x - 0.8x + 4.48 \qquad \text{Subtract } 0.8x \text{ from both sides of the equation.}$$

$$3.2x = 4.48$$
$$\frac{3.2x}{3.2} = \frac{4.48}{3.2} \qquad \text{Divide both sides of the equation by 3.2.}$$

$$x = 1.4$$

CHECK
$$4x - 0.48 = 0.8x + 4$$
$$4(1.4) - 0.48 = 0.8(1.4) + 4 \qquad \text{Substitute 1.4 in place of } x \text{ in the equation.}$$
$$5.6 - 0.48 = 1.12 + 4$$
$$5.12 = 5.12, \text{True} \qquad \text{❖}$$

As you can see, decimals are no more difficult than fractions or integers if you follow the rules.

In Chapter 5 we explained that $a - b$ can be expressed as $a + (-b)$. We will use this principle in explaining the next example.

Example 18

Solve $9 = -7 + 8(r - 5)$ for r.

Solution

$$9 = -7 + 8(r - 5)$$
$$9 = -7 + 8[r + (-5)] \qquad \text{Definition of subtraction of integers}$$
$$9 = -7 + 8(r) + 8(-5) \qquad \text{Distributive property}$$
$$9 = -7 + 8r + (-40)$$
$$9 = -7 + (-40) + 8r \qquad \text{Commutative property of addition}$$
$$9 = -47 + 8r \qquad \text{Combine like terms.}$$
$$9 + 47 = -47 + 47 + 8r \qquad \text{Add 47 to both sides of the equation.}$$
$$56 = 8r \qquad \text{Combine like terms.}$$
$$\frac{56}{8} = \frac{8r}{8} \qquad \text{Divide both sides of the equation by 8.}$$
$$7 = r$$

Not every equation has a solution, and some equations have more than one solution. Example 19 illustrates an equation that has no solution, and Example 20 illustrates an equation that has an infinite number of solutions.

Example 19

Solve $2(x - 3) + x = 5x - 2(x + 5)$.

Solution

$$2(x - 3) + x = 5x - 2(x + 5)$$
$$2x - 6 + x = 5x - 2x - 10 \qquad \text{Distributive property}$$
$$3x - 6 = 3x - 10 \qquad \text{Combine like terms.}$$
$$3x - 3x - 6 = 3x - 3x - 10 \qquad \text{Subtract } 3x \text{ from both sides of the equation.}$$
$$-6 = -10 \qquad \text{False}$$

During the process of solving an equation, if you obtain a false statement like $-6 = -10$, or $4 = 0$, then the equation has **no solution.** An equation that has no solution is called an **inconsistent equation.** The equation $2(x - 3) + x = 5x - 2(x + 5)$ is inconsistent and has no solution.

Example 20

Solve $3(x + 2) - 5(x - 3) = -2x + 21$.

Solution

$$3(x + 2) - 5(x - 3) = -2x + 21$$
$$3x + 6 - 5x + 15 = -2x + 21 \qquad \text{Distributive property}$$
$$-2x + 21 = -2x + 21$$

Note that at this point both sides of the equation are the same. Every real number will satisfy this equation. This equation has an infinite number of solutions. An equation of this type, one that is true for all real numbers, is called an **identity.** When solving an equation, if you notice the same expression appears on both sides of the equation, the equation is an identity. The solution to any identity is **all real numbers.** If you continue to solve an equation that is an identity, you will end up with $0 = 0$, as illustrated below:

$$-2x + 21 = -2x + 21$$
$$-2x + 2x + 21 = -2x + 2x + 21 \qquad \text{Add } 2x \text{ to both sides of the equation.}$$

$$21 = 21$$
$$21 - 21 = 21 - 21 \qquad \text{Subtract 21 from both sides of the equation.}$$

$$0 = 0 \qquad \text{True} \qquad ❖$$

DID YOU KNOW?

The earliest work to be regarded as algebra was done by the Greek philosopher Diophantus of Alexandria (A.D. 250). His work was devoted to solving equations, for which he had to invent a suitable notation. He developed rules for generating powers of a number and for multiplication and division of simple quantities.

During the sixth century the ideas of Diophantus were greatly improved on by various Hindu mathematicians, and many deficiencies in the Greek symbolism were corrected. We owe the development of symbolic algebra (the use of symbols to represent numbers) to the sixteenth-century French mathematician François Viète (1540–1608).

The main step in the development of modern algebra was the evolution of a correct understanding of negative quantities, contributed by Albert Girard (1595–1637). Girard's work was later overshadowed by that of his contemporary René Descartes (1596–1660). Many consider Descartes's work to be the starting point of modern-day algebra. In 1707, Sir Isaac Newton (1643–1727) gave this symbolic mathematics the name Universal Arithmetic.

Formulas

A **formula** is an equation that has a real-life application. For example, to find the area of a rectangle, we can use the formula

$$\text{area} = \text{length} \cdot \text{width} \quad \text{or} \quad A = lw.$$

Some formulas contain **subscripts.** Subscripts are numbers (or other variables) placed below and to the right of variables. They are used to help clarify a formula. For example, if there are two different resistances in a circuit, the resistances may be symbolized as R_1 and R_2. The total resistance in the circuit may be symbolized as R_T. Subscripts are read using the word "sub"; for example, R_1 is read "R sub 1."

To **evaluate a formula,** substitute the given values for their respective variables, then evaluate using the priority of operations given in Table 6.1.

■ Example 21

A formula used in the study of electronics is

$$R_T = \frac{R_1 R_2}{R_1 + R_2}$$

Find the value of the total resistance, R_T, when $R_1 = 100$ ohms and $R_2 = 150$ ohms.

Solution: Substitute 100 for R_1 and 150 for R_2, then evaluate:

$$R = \frac{R_1 R_2}{R_1 + R_2} = \frac{100(150)}{100 + 150} = \frac{15{,}000}{250} = 60$$

Thus R_T has a value of 60 ohms.   ❖

Many formulas contain Greek letters. Some Greek letters that are used in formulas are: μ (mu), σ(sigma), Σ (capital sigma), δ (delta), ϵ (epsilon), π (pi), θ (theta), and λ (lambda). Example 22 makes use of Greek letters.

■ Example 22

A formula used in the study of statistics is

$$Z = \frac{\bar{x} - \mu}{\dfrac{\sigma}{\sqrt{n}}}$$

Find the value of Z when $\bar{x}$ (read "x bar") $= 120$, $\mu = 100$, $\sigma = 16$, and $n = 4$.

Solution

$$Z = \frac{\bar{x} - \mu}{\dfrac{\sigma}{\sqrt{n}}} = \frac{120 - 100}{\dfrac{16}{\sqrt{4}}} = \frac{20}{\dfrac{16}{2}} = \frac{20}{8} = 2.5$$

❖

Often in mathematics and science courses, you are given a formula or an equation expressed in terms of one variable and asked to express it in terms of a different variable. To do this, treat each of the variables, except the one you are solving for, as if they were constants. Then solve for the variable desired, using the properties previously discussed.

When we graph equations in Sections 6.6, we will sometimes have to solve the equation for the variable y. Example 23 illustrates how this can be done.

■ **Example 23** _____

Solve the equation $2x + 3y - 6 = 0$ for y.

Solution: We must isolate the term containing the y. Therefore we will move the constant, -6, and the term containing the x to the right-hand side of the equation.

$$2x + 3y - 6 = 0$$

$$2x + 3y - 6 + 6 = 0 + 6 \qquad \text{Add 6 to both sides of the equation.}$$

$$2x + 3y = 6$$

$$-2x + 2x + 3y = -2x + 6 \qquad \text{Subtract } 2x \text{ from both sides of the equation.}$$

$$3y = -2x + 6$$

$$\frac{3y}{3} = \frac{-2x + 6}{3} \qquad \text{Divide both sides of the equation by 3.}$$

$$y = \frac{-2x + 6}{3}$$

$$y = \frac{-2x}{3} + \frac{6}{3}$$

$$y = -\frac{2}{3}x + 2$$

❖

Note that once we have found $y = (-2x + 6)/3$, we have solved the equation for y. We rewrote the equation in the form $y = -\frac{2}{3}x + 2$. This form of the equation is convenient for graphing equations, as will be explained in Section 6.6.

EXERCISES 6.2

Combine like terms.

1. $2x - 3x$
2. $-2x - 3x$
3. $x + 3x - 7$
4. $2 + 2x + 3x$
5. $5x + 2y + 3 + y$
6. $x - 4x + 3$
7. $-3x + 2 - 5x$
8. $3x + 4x - 2 + 5$
9. $2 - 3x - 2x + 1$
10. $3x - 4y + 6x + 7y + 12$

Solve the following equations.

11. $r + 9 = 17$
12. $5x - 6 = 24$
13. $16 = 6x + 4$
14. $11 = 3 - 4y$
15. $12 = 14 + 2x$
16. $15k - 7 = -37$
17. $\frac{1}{2}x + \frac{1}{3} = \frac{2}{3}$
18. $\frac{1}{2}y + \frac{1}{3} = \frac{1}{4}$
19. $0.3x + 4.2 = 3.6$
20. $3x - 0.02 = -0.548$
21. $5t + 6 = 2t + 15$
22. $\frac{x}{4} + 2x = -\frac{2}{3}$
23. $27 - 6x = 4x - 7x$
24. $2r + 8 = 5 + 3r$
25. $\frac{x}{15} = 2 + \frac{x}{5}$
26. $12x - 1.2 = 3x + 1.5$
27. $4(x + 3) = 36 + 4x$
28. $6y + 3(4 + y) = 8$
29. $6(x + 1) = 4x + 2(x + 3)$
30. $\frac{x}{3} + 4 = \frac{2x}{5} - 6$
31. $0.34x - 0.15 = 0.23x + .41$
32. $\frac{2}{3}(x + 5) = \frac{1}{4}(x + 2)$

33. $2x + 3 - 5x = 11 + 3x - 7x - 5$
34. $3x + 4 - 11x = 14 + 7x - 5 + 6x$

Evaluate each of the following formulas for the given value of the variables.

35. $P = 2l + 2w, l = 6, w = 4$
36. $E = IR, I = 0.1, R = 100$
37. $A = \frac{1}{2}bh, b = 20, h = 12$
38. $T = \frac{PV}{K}, P = 100, V = 200, K = 50$
39. $a = F/m, F = 500, m = 50$
40. $V = \frac{1}{2}at^2, a = 32, t = 4$
41. $A = \frac{1}{2}h(b_1 + b_2), h = 12, b_1 = 10, b_2 = 8$
42. $Z = \frac{\bar{x} - \mu}{\sigma}, \bar{x} = 80, \mu = 60, \sigma = 10$
43. $i = prt, p = 2000, r = 0.06, t = 2$
44. $Z = \frac{\bar{x} - \mu}{\frac{\sigma}{\sqrt{n}}}, \bar{x} = 160, \mu = 140, \sigma = 5, n = 25$
45. $R_T = \frac{R_1 R_2}{R_1 + R_2}, R_1 = 50, R_2 = 100$
46. $d = -16t^2 + 80t + 120, t = 2$
47. $A = (a + b + c)/3, a = 5, b = 7, c = 12$
48. $S = \frac{n}{2}(f + l), n = 8, f = 6, l = 15$
49. $F = \frac{9}{5}C + 32, C = 10$
50. $C = \frac{5}{9}(F - 32), F = 98.6$

Solve for y in each of the following equations.

51. $3x + 2y = 5$
52. $2x - 3y = 6$
53. $4x + 7y = 14$
54. $-x + 2y = 16$
55. $2x - 3y + 6 = 0$
56. $3x + 4y = 0$
57. $4x + 3y - 2z = 4$
58. $5x - 3z = 2 + 7y$
59. $2x - 3y + 5z = 0$

PROBLEM SOLVING

60. The total pressure, P, in pounds per square inch, exerted on an object x feet below the sea is given by the formula $P = 14.70 + 0.43x$. The 14.70 represents the weight in pounds of the column of air (from sea level to the top of the atmosphere) standing over a one-inch-by-one-inch square of seawater. The $0.43x$ represents the weight in pounds of a column of water one inch by one inch by x feet (see Fig. 6.2).

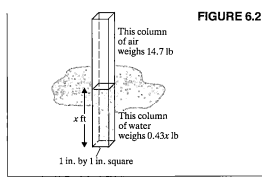

This column
of air
weighs 14.7 lb

x ft

This column
of water
weighs 0.43x lb

1 in. by 1 in. square

FIGURE 6.2

a)   A submarine is built to withstand a total pressure of 148 pounds per square inch. How
deep can that submarine go?

b)   If the pressure gauge in the submarine regis-
ters a total pressure of 128.65 pounds per
square inch, how deep is the submarine?

6.3   APPLICATION OF LINEAR EQUATIONS IN ONE VARIABLE

In Chapter 1 we discussed problem solving. In Step B3 on page 9 we wrote,
"Can you write a basic equation or formula that relates the unknown and the
known quantities? A particular problem might require more than one equation or formula to solve the problem. (We will discuss this procedure further
in Chapter 7.) You might wish to review Section 1.2 briefly at this time.

In this section we will do two things: (1) show how to translate a written
problem into a mathematical equation and (2) show how linear equations can
be used in solving everyday problems.

■ **Example 1** ————————————————————————————————————

Write the following in mathematical terms.

a)   6 less than 7 times a number

b)   11 more than twice a number

c)   11 decreased by 5 times x

Solution:

a)   The first thing we must do is select a letter to represent the number.
Let us arbitrarily select the letter n. Seven times the number is written
as $7n$. From this product we subtract 6. Thus the solution is $7n - 6$.

b)   Again we must select a letter to represent the number. In this example
we will select x. Twice the number is represented as $2x$. To this product
we add 11. The solution is $2x + 11$.

c)   Does 11 decreased by 5 times x mean $(11 - 5)x$ or $11 - 5x$? If we refer
to the priority of operations discussed earlier, we recall that multiplication and division are to be evaluated before addition and subtraction.
Without any punctuation to indicate otherwise, we will follow this

order when writing mathematical expressions. Therefore the correct answer to the question as stated is $11 - 5x$. If the statement had read "11 decreased by 5, times x," the statement would be $(11 - 5)x$. ❖

■ Example 2

Write an equation for each of the following.

a) Six more than a number is 11.

b) A number decreased by 13 is 6 times the number.

Solution

a) First we must recognize that the word "is" indicates an equals sign. Let us select n as the number. We write 6 more than a number as $n + 6$. This expression is placed on one side of the equals sign. The 11 is placed on the other side.

$$n + 6 = 11$$

b) Let n be the number. The number decreased by 13 can be written $n - 13$. Six times the number can be written as $6n$. The $n - 13$ is placed on one side of the equals sign and the $6n$ on the other side of the equals sign. Putting it together, we have the equation

$$n - 13 = 6n.$$ ❖

One of the main reasons for studying algebra is that it can be used to solve everyday problems. We discussed problem solving in Chapter 1. Solving a problem by representing it as an equation and then solving the equation is an extension of problem solving. Below is a general procedure for solving word problems.

To solve a word problem:

1. Read the problem carefully at least twice to make sure you understand it.
2. If possible, draw a sketch to help visualize the problem.
3. Determine which quantity you are being asked to find. Choose a letter to represent this unknown quantity. Write down exactly what this letter represents.
4. Write the word problem as an equation.
5. Solve the equation for the unknown quantity.
6. Answer the question or questions asked.
7. Check the solution.

This general procedure for solving word problems is illustrated in the following examples.

■ **Example 3** _____

A car rental agency charges $210 per week plus $0.18 per mile. How far can you travel, to the nearest mile, on a maximum budget of $400?

Solution: In this problem the unknown quantity is the number of miles you can travel. Let us select m to represent the number of miles you can travel. We must construct an equation using the given information that will allow us to solve for m.

Analysis

$$m = \text{number of miles you can travel,}$$
$$\$0.18m = \text{amount spent for } m \text{ miles traveled at } 18¢ \text{ per mile,}$$
$$\text{rental fee} + \text{mileage charge} = \text{total amount spent,}$$
$$\$210 + \$0.18m = \$400.$$

Equation

$$210 + 0.18m = 400$$
$$210 - 210 + 0.18m = 400 - 210$$
$$0.18m = 190$$
$$\frac{0.18m}{0.18} = \frac{190}{0.18}$$
$$m = 1055.\overline{5}$$
$$m = 1055$$

Answer

Therefore you can travel 1055 miles. (If you go 1056 miles, you have traveled too far.)

CHECK

The check is made with the information given in the original problem.

$$\text{total amount spent} = \text{rental fee} + \text{mileage charge}$$
$$= 210 + 0.18(1055)$$
$$= 210 + 189.90$$
$$= 399.90$$

The check shows that we are correct. Note that if one more mile was traveled, the total cost would exceed $400. ❖

■ **Example 4** _____

Forty hours of overtime must be split among three workers. Twice the number of hours must be assigned to one worked as to each of the other two.

a) Set up an equation that can be used to find the number of hours of over-time to be assigned to each worker.
b) Find the number of hours of overtime that will be assigned to each worker.

Solution: Two workers receive the same amount of overtime, and the third worker receives twice that amount.

Let x = number of hours of overtime for the first worker,
x = number of hours of overtime for the second worker,
$2x$ = number of hours of overtime for the third worker.

$$x + x + 2x = \text{total amount of overtime, 40 hours.}$$
$$x + x + 2x = 40$$
$$4x = 40$$
$$x = 10$$

Thus two workers are assigned 10 hours, and the third worker is assigned 2(10), or 20, hours of overtime.   ❖

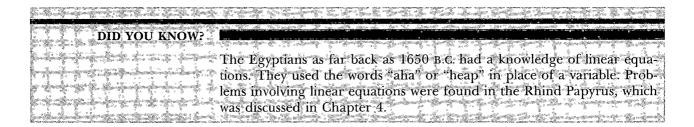

DID YOU KNOW?

The Egyptians as far back as 1650 B.C. had a knowledge of linear equations. They used the words "aha" or "heap" in place of a variable. Problems involving linear equations were found in the Rhind Papyrus, which was discussed in Chapter 4.

■ Example 5

Mrs. Chang wants to build a sandbox for her daughter. She has only 26 feet of wood to build the perimeter of the sandbox. What should be the dimensions of the sandbox if she wants the length to be 3 feet greater than the width?

Solution: Knowing the formula for finding the perimeter of a rectangle is essential. The formula is $P = 2l + 2w$, where P is the perimeter, l is the length, and w is the width. A simple diagram such as Fig. 6.3 is often helpful in solving problems of this type.

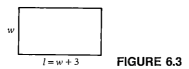

$l = w + 3$ **FIGURE 6.3**

Let w equal the width of the sandbox. Since the length is 3 feet more than the width, $l = w + 3$. The total distance around the sandbox, P, is 26 feet.

Substitute the known quantities in the formula:

$$\begin{aligned} P &= 2w + 2l \\ 26 &= 2w + 2(w + 3) \\ 26 &= 2w + 2w + 6 \\ 26 &= 4w + 6 \\ 20 &= 4w \\ 5 &= w \end{aligned}$$

The width of the sandbox is 5 feet, and the length of the sandbox is $5 + 3$ or 8 feet.   ❖

In shopping and other daily activities we are occasionally asked to solve problems using percents. The word percent means "per hundred." Thus, for example, 7% means 7 per hundred, or 7/100. When 7/100 is converted to a decimal number, we obtain 0.07. Thus 7% = 0.07.[*]

Let us look at one example involving percent.

■ **Example 6** _____

Mr. Dobson is planning to open a hot dog stand in New York City. What should be the price of the hot dog before tax if the total cost of the hot dog, including a 7% sales tax, is to be $1.25?

Solution: We are asked to find the cost of the hot dog before sales tax.

Let x = the cost of the hot dog before sales tax.

Then $0.07x$ = 7% of the cost of the hot dog (the sales tax).

$$\begin{aligned} \text{cost of hot dog before tax} + \text{tax on hot dog} &= 1.25 \\ x + 0.07x &= 1.25 \\ 1.07x &= 1.25 \\ \frac{1.07x}{1.07} &= \frac{1.25}{1.07} \\ x &= \frac{1.25}{1.07} \\ x &= 1.168 \quad \text{or} \quad 1.17 \end{aligned}$$

Thus the cost of the hot dog before tax is $1.17 to the nearest cent.   ❖

_____

*Section 8.3 has a more detailed discussion of percent.

DID YOU KNOW?

The = sign was introduced in England in 1557 by Robert Recorde. By the middle of the nineteenth century, Recorde's symbol had been adopted by several influential writers and was commonly used in England. However, there were several competing symbols for equals, particularly in Europe. The symbols [and ‖ were used by some writers, but the greatest competition came from the symbol ∞ used by Descartes. In the seventeenth century, Descartes's symbol was in common use on the Continent and was even occasionally used in England. Recorde's symbol was used by both Newton and Leibniz, and the influence of these great mathematicians eventually led to the general adoption of the = sign. By the beginning of the eighteenth century, Recorde's symbol had emerged as the standard symbol for equals. Before the development of a symbol for equals, the words "aequales," "aequanton," "facient," and "gleich" were used.

EXERCISES 6.3

Write the following as mathematical expressions.

1. 11 decreased by 7 times x
2. $4r$ increased by 16
3. 5 more than 6 times y
4. 10 decreased by twice n
5. 4 times r decreased by 11
6. 5 more than t
7. 5 decreased by x, times 2
8. 5 decreased by two times x
9. The sum of 6 and x, divided by 7
10. The sum of 6 and x divided by 7
11. 5 less than the quantity 7 times y, increased by 11
12. w more than 11 times z, decreased by twice w

Write an equation and solve.

13. The sum of a number and 11 is 19.
14. A number multiplied by 3 is 18.
15. A number divided by 7 is 3.
16. A number decreased by 6 is 22.
17. Eleven increased by twice a number is 9.
18. Seven less than 4 times a number is 17.
19. Four more than 3 times a number is 16.
20. Twice a number decreased by 11 is the same as 4 more than the number.
21. A number increased by 11 is the same as 1 more than 3 times the number.

22. A number divided by 3 is the same as 4 less than the number.
23. Three more than a number is 5 times the sum of the number and 7.
24. The product of 3 and a number, decreased by 4 is 4 times the number.

Set up an algebraic equation that can be used to solve the problem. Solve the equation and find the desired value(s).

25. The O'Briens have $2\frac{2}{3}$ cups of pancake flour. They need $7\frac{1}{2}$ cups of the flour for a given recipe. How many more cups of flour do they need?
26. Eighty-four hours of overtime must be split among four workers. The two younger workers are assigned the same number of hours. The third worker is assigned twice as much time as each of the younger workers, and the fourth worker is assigned three times as much time as each of the younger workers. How much overtime is assigned to each worker?
27. Carol budgeted $250 for renting a car. How far can she travel if the charges are $90 plus 18¢ per mile?
28. Ms. Riebe would like to build a rectangular sandbox for her children. She has a total of 20 feet of

wood that she will use to build the perimeter of the sandbox. What will be the dimensions of the box if the width of the box is to be $\frac{2}{3}$ of its length?

29. Mr. Umoja worked a 55-hour week last week. He is not sure of his hourly rate, but he knows that he is paid $1\frac{1}{2}$ times his regular hourly rate for all hours over a 40-hour week. His pay last week was $400. What is his hourly rate?

30. In a day, machine A produces twice as many bowls as machine B. Machine C produces 50 more bowls than machine B. If the total production is 6170 bowls, how many bowls does each produce?

31. Stock A is selling at $140 per share. Stock B is selling at $37 per share. An investor has a total of $10,000 to invest. She wishes to purchase 6 times as many shares of Stock A as of Stock B.
    a) How many shares of each stock will she purchase?
    b) How much money will be left over? *Note:* Stock may be purchased only in whole shares.

32. At a 20% off sale, a box of floppy diskettes costs $18. Find the regular price of the floppy diskettes.

33. The Donahues purchased a car. If the total cost including a 5% sales tax was $12,836.25, find the cost of the car before tax.

34. Mr. Brown will need 186 feet of fence to enclose his rectangular yard. If the length of the yard is 9 feet more than the width, what are the dimensions aof his yard?

35. A bookcase is to be built as illustrated below. If the height of the bookcase is to be 2 feet longer than the width of a shelf, and the total amount of wood

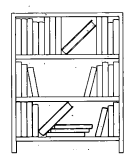

to be used is 32 feet, find the dimensions of the bookcase.

36. The cost of doing the family laundry for a month at a local laundromat is $42. A new washer and dryer cost a total of $632. How many months would it take for the cost of doing the laundry at the laundromat to equal the cost of a new washer and dryer?

37. Telephones are selling for $18.95. Sandy's charge for renting her telephone from the telephone company is $2.95 per month. If Sandy purchases her phone, how long will it take her to recover the cost of the phone?

38. Jim is about to join a racquet club. He has a choice of two plans. Plan A has a flat monthly fee of $22 and no charge for court time. Plan B has no monthly fee, but he must pay $6 per hour for court time. How many hours would Jim have to play to make Plan A the less expensive plan?

39. Ms. Saplin is being hired as a salesperson. She is given a choice of two salary plans. Plan A is $100 a week plus 5% commission on sales. Plan B is a straight 15% commission on sales. How many dollars in sales are necessary per week for Plan B to result in the same salary as Plan A?

40. Florence has been told that with her half-off airfare coupon her airfare from New York to San Diego will be $227.00. The $227.00 includes a 7% tax *on the regular fare*. On the way to the airport, Florence realizes that she has lost her coupon. What will her regular fare be, including tax?

41. The cost of renting a compact car at the Hertz car rental agency is $35 per day plus 8 cents a mile. The cost of renting the same car at the Avis car rental agency is $25 per day plus 12 cents a mile. How far would you have to drive in one day to make the cost of renting from Hertz equal the cost of renting from Avis?

42. The cost of a telephone call from Denver to St. Louis is 75 cents for the first minute and 50 cents for each additional half minute or any part thereof. If Donna lives in St. Louis and her friend lives in Denver, how long can Donna talk on the telephone if she has only $6.25?

PROBLEM SOLVING

43. The cost of a Volkswagen Rabbit with a diesel engine is $8875. The cost of the Rabbit with a standard gasoline engine is $8283. The Rabbit with the diesel engine gets 45 miles per gallon, while the Rabbit with the standard engine gets 34 miles per gallon. If diesel gasoline costs $1.17 per gallon and regular unleaded gasoline costs $1.15 per gallon, how many miles would the owner have to drive the Rabbit with the diesel engine to make up for its higher initial cost?

44. Some states allow a husband and wife to file individual tax returns (on a single form) even though they have filed a joint federal tax return. It is usually to the taxpayers' advantage to do so when both husband and wife work. The smallest amount of tax owed (or the largest refund) will occur when the husband's and wife's taxable incomes are the same.

Mr. Ranieri's 1989 taxable income was $18,200, and Mrs. Ranieri's taxable income for that year was $16,400. The Ranieris' total tax deduction for the year was $3640. This deduction can be divided between Mr. and Mrs. Ranieri any way they wish. How should the $3640 be divided between them to result in each individual's having the same taxable income and therefore the greatest tax refund?

6.4   LINEAR INEQUALITIES

The first sections of this chapter have dealt with equations. There are many occasions when we may encounter statements of inequality. The symbols of inequality are given in Table 6.2.

TABLE 6.2

$a < b$ means a is less than b.
$a \leq b$ means a is less than or equal to b.
$a > b$ means that a is greater than b.
$a \geq b$ means that a is greater than or equal to b.__

Locks are used to raise or lower the water level ($13' \leq x \leq 16'$).
Photo by Marshall Henrichs

A statement of inequality can be used to indicate a set of real numbers. For example, $x < 2$ represents the set of all real numbers less than 2. It is impossible to list all these numbers. Some numbers less than 2 are $-1/2$, -1, 0, -2, 97/163, -1.234.

A method of picturing all real numbers less than 2 is to graph the solution on a number line. The number line was discussed in Chapter 5.

To indicate the solution set of $x < 2$ on the number line, we draw an open circle at 2 and a line to the left of 2 with an arrow at its end (Fig. 6.4). This technique indicates that all the points to the left of 2 are part of the solution. The open circle indicates that the solution does not include the number 2.

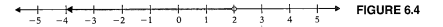 **FIGURE 6.4**

■ Example 1 _____

Graph the solution set of $x \geq -1$, where x is a real number, on the number line.

Solution: The numbers greater than or equal to -1 are all the points on the number line to the right of -1 and -1 itself (Fig. 6.5). The solid dot at -1 shows that -1 is included in the solution.

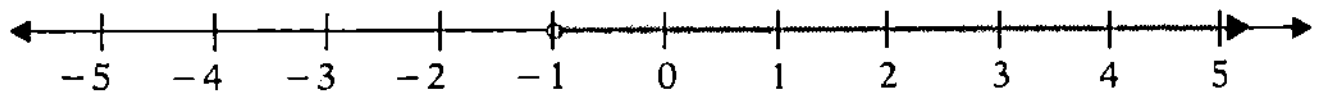

FIGURE 6.5 ❖

The inequality statements $a < b$ and $b > a$ have the same meaning. Note that the inequality symbol points to a (the smaller number) in both cases. Thus one may be written in place of the other. Likewise, $a > b$ and $b < a$ have the same meaning. Note that the inequality symbol points to b (the smaller number) in both cases. We make use of this fact in the following example.

■ Example 2 _____

Graph $-4 \leq x$, where x is a real number, on the number line.

Solution: We can restate $-4 \leq x$ as $x \geq -4$. Both statements have identical solutions. Any number that is greater than or equal to -4 satisfies the inequality $x \geq -4$. Thus the graph includes -4 and all the points to the right of -4 on the number line (Fig. 6.6).

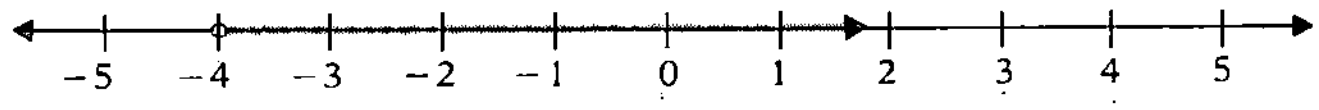

FIGURE 6.6 ❖

We can find the solution to an inequality by adding, subtracting, multiplying, and dividing both sides of the inequality by the same number or expression. We use the same procedures as those discussed in Section 6.2 to isolate the variable, with two important exceptions: (1) *When both sides of an inequality are multiplied by a negative number, the order of the inequality is reversed.* This is sometimes referred to as changing the **sense of the inequality.** (2) *When both sides of an inequality are divided by a negative number, the order or sense of the inequality is reversed.*

Consider the inequality $3 < 8$. If we multiply both sides of the inequality by -1, we must change the sense of the inequality.

$$3 < 8$$
$$-1(3) > -1(8) \qquad \text{Multiply both sides of the inequality by } -1 \text{ and change the sense of the inequality.}$$
$$-3 > -8$$

Note that $3 < 8$ and $-3 > -8$ (or $-8 < -3$) are both true statements.

■ **Example 3** ————————————————————————————

Solve the inequality $-x > 5$.

Solution: To solve this inequality, we must eliminate the negative sign before the x. To do this, we can multiply both sides of the inequality by -1 and change the sense of the inequality:

$$-x > 5$$
$$-1(-x) < -1(5) \qquad \text{Multiply both sides of the inequality by } -1 \text{ and change the sense of the inequality.}$$
$$x < -5 \qquad\qquad\qquad\qquad\qquad\qquad\qquad\qquad ❖$$

Consider the inequality $4 < 8$. If we divide both sides of this inequality by -1, we must change the sense of the inequality:

$$4 < 8$$
$$\frac{4}{-1} > \frac{8}{-1} \qquad \text{Divide both sides of the inequality by } -1 \text{ and change the sense of the inequality.}$$
$$-4 > -8$$

Note that $4 < 8$ and $-4 > -8$ (or $-8 < -4$) are both true statements.

■ **Example 4** _____

Solve the inequality $-2x > 6$.

Solution: To solve this inequality, we must eliminate the -2 before the x. We can do this by dividing both sides of the inequality by -2 and changing the sense of the inequality:

$$-2x > 6$$

$$\frac{-2x}{-2} < \frac{6}{-2} \qquad \text{Divide both sides of the inequality by } -2 \text{ and change the sense of the inequality.}$$

$$x < -3 \qquad\qquad\qquad\qquad\qquad\qquad\qquad ❖$$

Examples 3 and 4 illustrate that when we multiply or divide an inequality by a negative number, the order of the inequality changes.

■ **Example 5** _____

Graph the solution set of $2x - 4 < 6$, where x is a real number, on the number line.

Solution: To find the solution, we must isolate x on one side of the inequality symbol:

$$2x - 4 < 6$$
$$2x - 4 + 4 < 6 + 4 \qquad \text{Add 4 to both sides of the inequality.}$$
$$2x + 0 < 10$$
$$2x < 10$$
$$\frac{2x}{2} < \frac{10}{2} \qquad \text{Divide both sides of the inequality by 2.}$$
$$x < 5 \qquad \text{The order of the inequality does not change, since both sides were divided by a positive number.}$$

Thus the solution to $2x - 4 < 6$ is all real numbers less than 5 (Fig. 6.7).

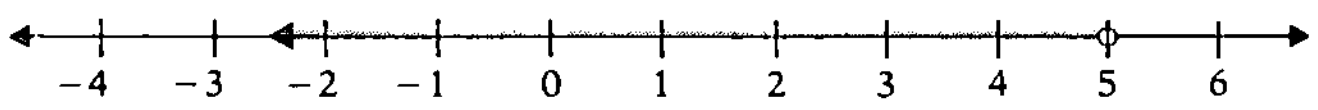

FIGURE 6.7   ❖

■ **Example 6** _____

Graph the solution set of $x + 4 < 7$, where x is an integer, on the number line.

Solution: Following the methods used above, we can see that the solution set is the set of integers less than 3, or $\{ \cdots , -3, -2, -1, 0, 1, 2\}$. To graph the solution, we make large dots at the corresponding points on the number line (see Fig. 6.8). The three smaller dots to the left of -3 indicate that all the integers to the left of -3 are included.

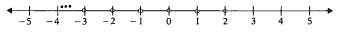

FIGURE 6.8   ❖

An inequality of the form $a < x < b$ is called a **compound inequality.** Consider the compound inequality $-3 < x \leq 2$. This means that $-3 < x$ *and* $x \leq 2$.

■ **Example 7** _____

Graph the solution set of the inequality $-3 < x \leq 2$,

a) where x is an integer,

b) where x is a real number.

Solution

a) The solution set is all of the integers between -3 and 2, including the point 2 but not including the point -3, or $\{-2, -1, 0, 1, 2\}$ (Fig. 6.9).

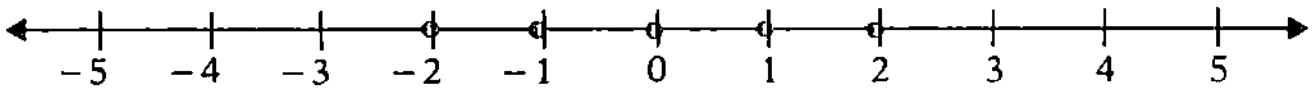

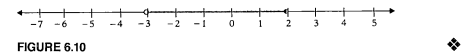

FIGURE 6.9

b) The solution set consists of all the real numbers between -3 and 2, including the 2 but not including the -3. The solution is shown in Fig. 6.10.

FIGURE 6.10   ❖

■ **Example 8** _____

Solve the compound inequality for x.

$$-4 < \frac{x + 3}{2} \leq 5$$

Solution: To solve a compound inequality, we must isolate the x as the middle term. To do this, we use the same principles used to solve inequalities:

$$-4 < \frac{x + 3}{2} \le 5$$

$$2(-4) < 2\left(\frac{x + 3}{2}\right) \le 2(5) \qquad \text{Multiply the three terms by 2.}$$

$$-8 < x + 3 \le 10$$

$$-8 - 3 < x + 3 - 3 \le 10 - 3 \qquad \text{Subtract 3 from all three terms.}$$

$$-11 < x \le 7 \qquad \qquad \qquad \qquad \qquad ❖$$

■ **Example 9** ⎯⎯⎯⎯⎯⎯⎯⎯⎯⎯⎯⎯⎯⎯⎯⎯⎯⎯⎯⎯⎯⎯⎯⎯

A student must have an average that is greater than or equal to 80% but less than 90% on five tests to receive a final grade of B. Bill's grades on the first four tests were 98%, 76%, 86%, and 92%. What range of grades on the fifth test would give him a B in the course?

Solution: The unknown quantity is the range of grades on the fifth test. We first construct an inequality that can be used to find the range of grades on the fifth exam. The average (mean) is found by adding the grades and dividing the sum by the number of exams.
Let x = the fifth grade:

$$\text{Average} = \frac{98 + 76 + 86 + 92 + x}{5}$$

For Bill to obtain a B, his average must be greater than or equal to 80 but less than 90:

$$80 \le \frac{98 + 76 + 86 + 92 + x}{5} < 90$$

$$80 \le \frac{352 + x}{5} < 90$$

$$5(80) \le 5\left(\frac{352 + x}{5}\right) < 5(90) \qquad \text{Multiply the three terms of the inequality by 5.}$$

$$400 \le 352 + x < 450$$

$$400 - 352 \le 352 - 352 + x < 450 - 352 \qquad \text{Subtract 352 from all three terms.}$$

$$48 \le x < 98$$

Thus a grade of 48% up to but not including a grade of 98% will result in a grade of B. ❖

Write an inequality to represent the speed limits on the sign.
Photo by Allen R. Angel

"In mathematics the art of posing problems is easier than that of solving them."
George Cantor

EXERCISES 6.4

Graph the solution set of the inequality, where x is a real number, on the real number line.

1. $x < 4$ 2. $x \geq -3$
3. $x + 4 < 9$ 4. $4x \geq 2$
5. $-3x \leq 15$ 6. $-2x < 6$
7. $\frac{x}{4} < -2$ 8. $\frac{x}{2} > 4$
9. $\frac{x}{3} \leq 4$ 10. $\frac{-x}{5} \geq 2$
11. $-2x - 3 \geq 5$ 12. $3x + 5 > -7 + 3x$
13. $2(x + 6) \leq 15$
14. $-4(x + 2) - 2x > -6x + 2$
15. $3(x + 4) - 2 < 3x + 10$
16. $-2 \leq x \leq 1$
17. $-1 < x + 3 < 8$ 18. $\frac{1}{2} < \frac{x + 4}{2} \leq 4$

Graph the solution set of the inequality, where x is an integer, on the number line.

19. $x \leq 0$ 20. $-5 > x$
21. $3x \geq 15$ 22. $-4x \leq 16$
23. $-3x < 12$ 24. $4 - x > 5$
25. $\frac{x}{5} < -1$ 26. $\frac{x}{3} > 2$
27. $-\frac{x}{6} \geq 3$ 28. $\frac{2x}{3} \leq 4$
29. $-13 < -7y + 1$ 30. $3x + 6 < -2 + 11x$
31. $2(x + 8) \leq 3x + 17$
32. $5(x + 3) < -(6 + x) + 4$
33. $3(x + 4) - 2 \leq 3x + 10$
34. $-2 \leq x \leq 10$
35. $1 > -x > -5$ 36. $-2 < 2x + 3 < 6$
37. $5 < 3x - 4 \leq 10$ 38. $-3 < \frac{x - 3}{2} \leq -2$

39. In Example 9, what range of grades on the fifth test would result in Bill's receiving (a) a grade of B if his grades on the first four tests were 78%, 64%, 88%, and 76%; (b) a grade of C with the same test grades? To obtain a C, he must have an average greater than or equal to 70, but less than 80%.
40. Two employees are told that the sum of their hours of overtime must be less than 21. Overtime will not be paid for any fractional part of an hour. Employee A is assigned three times as many hours of overtime as Employee B. For how many hours of overtime will each employee be paid? Write a statement of inequality and solve it.
41. For a small plane to take off safely, it must have a maximum load of no more than 1500 pounds. The gasoline weighs 400 pounds and the passengers weigh a total of 670 pounds.
    a) Write an inequality that can be used to determine the maximum weight of the luggage that the plane can safely carry.
    b) Determine the maximum weight of the luggage.
42. The janitor must move a large shipment of books from the first floor to the fifth floor. Each box of books weighs 150 lb, and the janitor weighs 180 lb. The sign on the elevator reads "Maximum weight 1200 lb."
    a) Write a statement of inequality to determine the maximum number of boxes of books the janitor can place on the elevator at one time. (The janitor must ride in the elevator with the books.)
    b) Determine the maximum number of boxes that can be moved in one trip.
43. Mr. Green wants to invest $10,000, part at 5% and part at 7%. What is the least he can invest at 7% to earn a minimum return of $616.00?
44. After Mrs. Washington is seated in a restaurant, she realizes that she has only $14.00. If she must pay 7% tax and wishes to leave a 15% tip, what is the price range of meals she can order?
45. For a business to realize a profit, its revenue, R, must be greater than costs, C; that is, a profit will result only if $R > C$ (the company breaks even when $R = C$). A record manufacturer has a weekly cost equation $C = 300 + 1.5x$ and a weekly revenue equation $R = 2x$, where x is the number of records produced and sold in a week. How many records must be sold weekly for the company to make a profit?

PROBLEM SOLVING

46. Joan's five test grades for the semester are 86%, 74%, 68%, 96%, and 72%. Her final exam counts 1/3 of her final grade. What range of grades on her final exam would result in Joan's receiving a final grade of B in the course. (See Example 9.)

6.5 THE RECTANGULAR COORDINATE SYSTEM

Many problems cannot be solved with a single variable. Thus it is helpful if we can work with equations with two variables (for example, $x + 2y = 6$). To illustrate the solution to equations of this type, we must understand the **rectangular coordinate system** (also called the **Cartesian coordinate system**).

The rectangular coordinate system consists of two number lines drawn perpendicular (at right angles) to each other. We call the horizontal line the **x-axis** and the vertical line the **y-axis.** The point of intersection of the x- and y-axis is called the **origin.** Take note that the numbers on the axes to the right and above the origin are positive. The numbers on the axes to the left and below the origin are negative. The axes divide the plane into four parts, which we call the first, second, third, and fourth quadrants, as indicated in Fig. 6.11.

When identifying or plotting a point in the rectangular coordinate system, we always state its value on the x-axis first (called the x-coordinate), followed by its value on the y-axis (called the y-coordinate). Consider the point illustrated in Fig. 6.12. What are its coordinates? To determine the coordinates of the point, project the point down on the x-axis and then project it across to the y-axis, as illustrated in Fig. 6.13. When projected down we see that the point has an x-coordinate of 5. When projected across, it has a y-coordinate of 3. We indicate the location of the point by means of an **ordered pair** of the form (x, y). Note that the x-coordinate is placed first in the ordered pair, followed by the y-coordinate. The ordered pair that represents the point is $(5, 3)$.

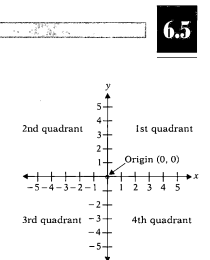

FIGURE 6.11

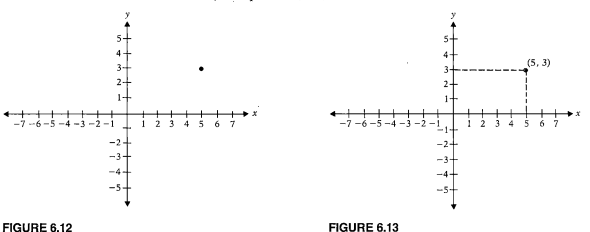

FIGURE 6.12

FIGURE 6.13

The origin is represented by the ordered pair $(0, 0)$. Every point on the plane can be represented by one and only one ordered pair (x, y), and every ordered pair (x, y) represents one and only one point on the plane.

■ **Example 1** _____

Give the coordinates of points A, B, C, and D in Fig. 6.14.

Solution: Point A has an x-coordinate of 3 and y-coordinate of 2; therefore its ordered pair is (3, 2). Similarly, the ordered pairs for points B, C, and D are (−2, 3), (−3, −1), and (2, −4), respectively.   ❖

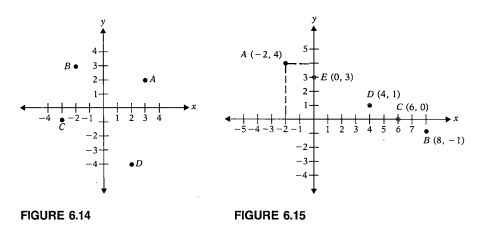

FIGURE 6.14   **FIGURE 6.15**

■ **Example 2** _____

Plot the points $A(-2, 4)$, $B(8, -1)$, $C(6, 0)$, $D(4, 1)$, and $E(0, 3)$.

Solution: Point A has an x-coordinate of −2 and a y-coordinate of 4. Project a vertical line up from −2 on the x-axis and a horizontal line from 4 on the y-axis. The two lines intersect at the point denoted by A (Fig. 6.15). The other points are plotted in a similar manner.   ❖

■ **Example 3** _____

The points A, B, and C are three vertices of a parallelogram. Plot the three points and find the coordinates of the fourth point, D.

$$A(1, 2)   B(2, 4)   C(7, 4)$$

Solution: A parallelogram is a figure that has opposite sides that are of equal length and are parallel. (Parallel lines are two lines in the same plane that do not intersect.) The horizontal distance between points B and C is 5 units. Therefore the horizontal distance between points A and D must. also be 5 units. This problem has two possible solutions, as illustrated in Fig. 6.16 (*a*) and (*b*).

The solutions are the points (6, 2) and (−4, 2).

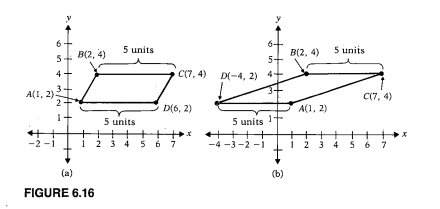

(a) (b)

FIGURE 6.16

❖

EXERCISES 6.5

Plot the following points on the same set of axes.

1. (2, 3)   **2.** (−4, 1)
3. (2, −3)   **4.** (0, 0)
5. (4, 0)   **6.** (0, 4)
7. (0, 7)   **8.** $(3\frac{1}{2}, 4\frac{1}{2})$

Plot the following points on the same set of axes.

9. (1, 3)   **10.** (0, 3)
11. (−2, −3)   **12.** (0, −4)
13. (−3, −1)   **14.** (−3, 0)
15. (4, −1)   **16.** (4.5, 3.5)

17. Write the coordinates for each point on the graph in Fig. 6.17.

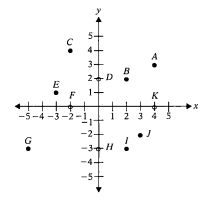

FIGURE 6.17

18. Write the coordinates for each point on the graph in Fig. 6.18

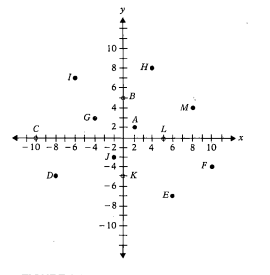

FIGURE 6.18

The points A, B, and C are three vertices (the points where two lines meet) of a rectangle. Plot the three points. Find the coordinates of the fourth point, D, to complete the rectangle. Find the area of the rectangle using the formula $A = lw$.

19. A (2, 3), B (7, 3), C (7, 5)
20. A (−4, 2), B (7, 2), C (7, 8)

The points A, B, and C are three vertices of a parallelogram. Plot the three points. Find the coordinates of the fourth point, D, to complete the parallelogram. *Note:* There are two possible answers for point D.

21. $A(3, 2), B(5, 5), C(9, 5)$
22. $A(1, 6), B(-5, 6), C(-3, -1)$

For what value of b will the line joining the points P and Q be parallel to the indicated axis?

23. $P(-4, 3), Q(b, 1)$; y-axis
24. $P(-5, 2), Q(7, b)$; x-axis
25. $P(3b - 1, 5), Q(8, 4)$; y-axis
26. $P(-6, 2b + 1), Q(2, 7)$; x-axis

6.6 GRAPHING LINEAR EQUATIONS IN TWO VARIABLES

Consider the following equation in two variables, x and y: $y = x + 1$. Every ordered pair that makes the equation a true statement is a solution to the equation (or satisfies the equation). Examples of some ordered pairs that satisfy the equation $y = x + 1$ are (1, 2), (2, 3), (3, 4), (4, 5), (6.5, 7.5), and (-4, -3).

CHECK

(1, 2)	(2, 3)	(3, 4)	(4, 5)
$y = x + 1$	$y = x + 1$	$y = x + 1$	$y = x + 1$
$2 = 1 + 1$	$3 = 2 + 1$	$4 = 3 + 1$	$5 = 4 + 1$
$2 = 2$	$3 = 3$	$4 = 4$	$5 = 5$

$$(6.5, 7.5) \qquad (-4, -3)$$
$$y = x + 1 \qquad y = x + 1$$
$$7.5 = 6.5 + 1 \qquad -3 = -4 + 1$$
$$7.5 = 7.5 \qquad -3 = -3$$

Are there any other points that satisfy the equation? How many other points satisfy the equation? Since we cannot list all of the solutions, we illustrate them by means of a graph. A **graph** is an illustration of all of the points whose coordinates satisfy an equation.

The points (1, 2), (2, 3), (3, 4), (4, 5), (6.5, 7.5), and (-4, -3) are plotted in Fig. 6.19. With a straightedge we can draw one line that contains all these points. We write the equation alongside the line so that when more than one line appears on the graph, the lines can be identified.

This line, when extended indefinitely in either direction, passes through all the points on the plane that satisfy the equation $y = x + 1$. Since there are an infinite number of points that satisfy the equation, the graph in Fig. 6.19 is the only way the solution set can be illustrated. The arrows on the ends of the line indicate that the line extends indefinitely.

All equations of the form $ax + by = c$, $a \neq 0$, $b \neq 0$, will be straight lines when graphed. Thus such equations are called **linear equations in two variables.** To graph linear equations in two variables, we must obtain at least two ordered pairs that satisfy the equation. To obtain these ordered pairs, it is often helpful to solve the equation for y. After solving the equation for y, we substitute values for x in the equation and find the corresponding values for y. Since only two points are needed to draw a line, only two points are needed to graph a linear equation. It is always a good idea to graph a third point as a check point. If no error has been made, all three points will be in a line (collinear).

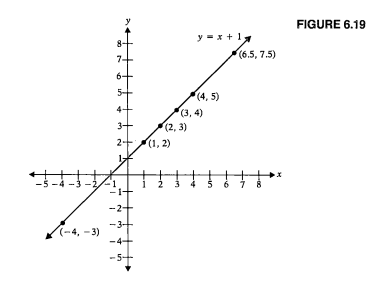

FIGURE 6.19

TABLE 6.3

x	y
0	6
1	9
-2	0

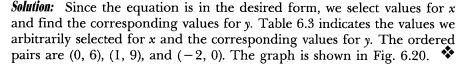

Example 1

Graph $y = 3x + 6$.

Solution: Since the equation is in the desired form, we select values for x and find the corresponding values for y. Table 6.3 indicates the values we arbitrarily selected for x and the corresponding values for y. The ordered pairs are $(0, 6)$, $(1, 9)$, and $(-2, 0)$. The graph is shown in Fig. 6.20. ❖

In example 1 take note of two special points on the graph, $(-2, 0)$ and $(0, 6)$. It is at these points that the line crosses the x-axis and the y-axis, respectively. The numbers -2 and 6 are called the x-**intercept** and the y-**intercept,** respectively. Another method that can be used to graph linear equations is to find the x- and y-intercepts of the graph. *To find the x-intercept, let y equal 0 and solve the equation for x. To find the y-intercept, let x equal 0 and solve the equation for y.* After the intercepts are found, mark them on the graph. Abritrarily select a third point as a check point. If the three points are collinear, draw the straight line through the three points.

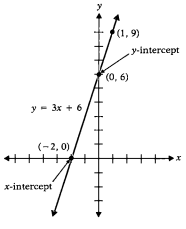

FIGURE 6.20

■ **Example 2**

Graph $2x + 3y = 6$.

Solution: We will graph this equation by finding the x- and y-intercepts. If we were to use the method in Example 1, we would have to solve the equation for y.

To find the x-intercept, set $y = 0$ and solve for x:

$$2x + 3(0) = 6$$
$$2x = 6$$
$$x = 3$$

The x-intercept is 3. Thus (3, 0) is an ordered pair that satisfies the equation. To find the y-intercept, set $x = 0$ and solve for y:

$$2(0) + 3y = 6$$
$$3y = 6$$
$$y = 2$$

The y-intercept is 2. Thus (0, 2) is an ordered pair that satisfies the equation.

As a check point, arbitrarily choose a value for x and solve for y. This ordered pair should be in the same line with the first two ordered pairs. Let us choose $x = 1$ and find the corresponding value for y:

$$2x + 3y = 6$$
$$2(1) + 3y = 6$$
$$2 + 3y = 6$$
$$3y = 4$$
$$y = 4/3$$

The check point is the ordered pair (1, 4/3) or $(1, 1\frac{1}{3})$.

Since all three points are collinear in Fig. 6.21, the points are correct. If the points were not collinear, we would have made a mistake. Now draw a line through the three points to obtain the graph (see Fig. 6.22).

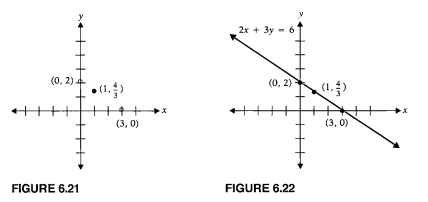

FIGURE 6.21   **FIGURE 6.22**

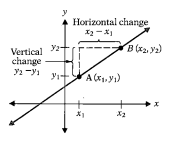

FIGURE 6.23

A third method for graphing linear equations is to make use of the slope and y-intercept. We have already discussed the y-intercept. Now we will discuss the slope of a line.

The slope of a line is a measure of the "steepness" of a line. The **slope of a line** is a ratio of the vertical change to the horizontal change for any two points on the line. Consider Fig. 6.23. Point A has coordinates (x_1, y_1), and point B has coordinates (x_2, y_2). We see that the vertical change between points A and B is $y_2 - y_1$, and the horizontal change between points A and B is $x_2 - x_1$. Thus the slope, which is often symbolized with the letter m, can be found as follows:

$$\textbf{Slope} = \frac{\text{vertical change}}{\text{horizontal change}}$$
$$m = \frac{y_2 - y_1}{x_2 - x_1}$$

The Greek letter delta, Δ, is used to represent the words "the change in." Therefore slope may be defined as

$$m = \frac{\Delta y}{\Delta x}$$

A line may have a positive slope, or a negative slope, or zero slope, as indicated in Fig. 6.24. A line with a positive slope rises from left to right (Fig. 6.24a). A line with a negative slope falls from left to right (Fig. 6.24b). A horizontal line, which neither rises nor falls, has a slope of zero (Fig. 6.24c). Since a vertical line does not have any horizontal change (the x value remains constant), and since we cannot divide by 0, the slope of a vertical line is undefined (Fig. 6.24d).

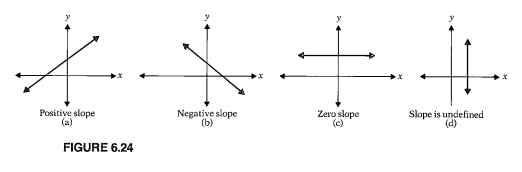

FIGURE 6.24

■ **Example 3** _____

Determine the slope of the line that passes through the points $(-1, -3)$ and $(1, 5)$.

Solution: Let us begin by drawing a sketch, illustrating the points and the line (Fig. 6.25a).

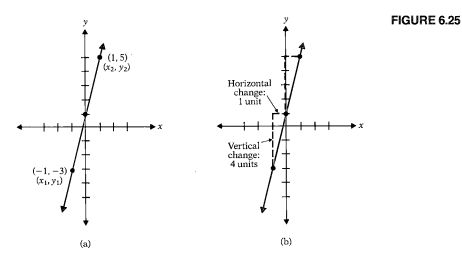

FIGURE 6.25

(a)   (b)

We will let (x_1, y_1) be $(-1, -3)$ and (x_2, y_2) be $(1, 5)$.

$$\text{Slope} = \frac{y_2 - y_1}{x_2 - x_1} = \frac{5 - (-3)}{1 - (-1)} = \frac{5 + 3}{1 + 1} = \frac{8}{2} = \frac{4}{1} \quad \text{or} \quad 4$$

Thus the slope is 4. This means that there is a vertical change of 4 units for each horizontal change of 1 unit (see Fig. 6.25b). Note that the slope is positive, and the line rises from left to right.   ❖

Note in Example 3 that we elected to let (x_1, y_1) be $(-1, -3)$ and (x_2, y_2) be $(1, 5)$. We would have obtained the same results if we let (x_1, y_1) be $(1, 5)$ and (x_2, y_2) be $(-1, -3)$. Try this now and see.

An equation given in the form $y = mx + b$ is said to be in slope-intercept form.

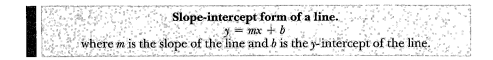

Slope-intercept form of a line.
$$y = mx + b$$
where m is the slope of the line and b is the y-intercept of the line.

Consider the equation $y = 3x + 4$. The graph of this equation is illustrated in Fig. 6.26. By examining the graph we can see that the y-intercept is 4. We can also see that the graph has a positive slope, since it rises from left to right. Since the vertical change is 3 units for every 1 unit of horizontal change, the slope must be 3/1 or 3.

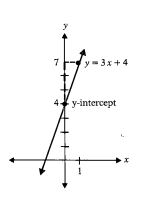

FIGURE 6.26

We could graph this equation by marking the y-intercept at 4 and then moving *up* 3 units and to the *right* 1 unit to get another point. If the slope were -3, we could start at the y-intercept and move *down* 3 units and to the *right* 1 unit.

■ **Example 4**

Graph $y = -2x + 3$ using the slope and y-intercept.

Solution: Plot 3 on the y-axis. Then plot the next point by moving *down* two units and to the *right* 1 unit (see Fig. 6.27). A third point has been plotted in the same way. ❖

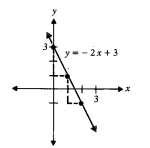

FIGURE 6.27

■ **Example 5**

Graph $y = \frac{5}{3}x - 2$ using the slope and y-intercept.

Solution: Plot -2 on the y-axis. Then plot the next point by moving *up* 5 units and to the *right* 3 units (see Fig. 6.28). ❖

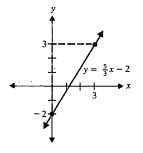

FIGURE 6.28

When an equation is not given in slope-intercept form, we can write the equation in slope-intercept form by solving the equation for y. The following example illustrates this procedure.

■ **Example 6**

a) Write $3x + 4y = 4$ in a slope-intercept form.
b) Graph the equation.

Solution

a) Solve the given equation for y:

$$3x + 4y = 4$$
$$3x - 3x + 4y = -3x + 4$$
$$4y = -3x + 4$$
$$\frac{4y}{4} = \frac{-3x + 4}{4}$$
$$y = -\frac{3x}{4} + \frac{4}{4}$$
$$y = -\frac{3}{4}x + 1$$

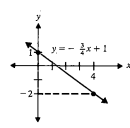

FIGURE 6.29

Thus in slope-intercept form the equation is $y = -\frac{3}{4}x + 1$.

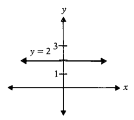

FIGURE 6.30

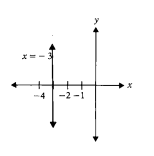

FIGURE 6.31

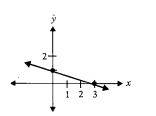

FIGURE 6.32

b)   The y-intercept is 1, and the slope is $-3/4$. Place a dot at 1 on the y-axis then move *down* 3 units and to the *right* 4 units to obtain the second point (see Fig. 6.29). ❖

■ **Example 7** ─────────────────────────────

Determine the equation of the line in Figure 6.30.

Solution: If we determine the slope and the y-intercept of the line, then we can write the equation using slope-intercept form, $y = mx + b$. We see from the graph that the y-intercept is 1; thus $b = 1$. The slope of the line is negative, since the graph falls from left to right. The change in y is one unit for each 3-unit change in x. Thus m, the slope of the line, is $-1/3$:

$$y = mx + b$$
$$y = -\frac{1}{3}x + 1$$

The equation of the line is $y = -\frac{1}{3}x + 1$. ❖

■ **Example 8** ─────────────────────────────

In the Cartesian coordinate system, graph (a) $y = 2$ and (b) $x = -3$.

Solution

a)   For any value of x, the value of y is 2. Therefore the graph will be a horizontal line through $y = 2$ (Fig. 6.31).

b)   For any value of y, the value of x is -3. Therefore the graph will be a vertical line through $x = -3$ (Fig. 6.32).

Note that the graph of $y = 2$ has a slope of 0. The slope of $x = -3$ is undefined. ❖

EXERCISES 6.6

Determine which ordered pairs satisfy the given equation.

1. $2x + 3y = 0$ $(3, -2), (0, 0), (2, -2)$
2. $x - 4y = 6$ $(2, -3), (0, -3/2), (6, 0)$
3. $3x - 2y = -10$ $(0, 5), (-10/3, 0), (2, -3)$
4. $2x = 6y + 3$ $(0, -1/2), (2, -1), (-3/2, 0)$
5. $\frac{x}{2} + 3y = 4$ $(0, 4/3), (8, 0), (10, -2)$
6. $2y = 3x - 5$ $(1, -1); (-3, -7), (2, 5)$

7. $3x - 4y = -4$ $(0, 1), (-4, -2), (2, -1/2)$
8. $\frac{x}{2} + \frac{3y}{4} = 2$ $(0, 8/3), (1, 11/4), (4, 0)$

Graph each equation and state the slope if it exists (see Example 8).

9. $x = 4$
10. $x = -2$
11. $y = 3$

12. $y = -1$

Graph each equation by plotting points as in Example 1.

13. $y = x + 2$
14. $y = x - 3$
15. $y = 2x + 7$
16. $y = 3x - 2$
17. $y = -x + 3$
18. $y = -2x - 1$
19. $y = -2x + 6$
20. $y = 3x - 3$
21. $y = \frac{1}{2}x + 4$
22. $y = \frac{2}{3}x - 1$
23. $y = -\frac{1}{2}x + 3$
24. $y = -\frac{3}{4}x$

Graph each equation using the x- and y-intercepts as in Example 2.

25. $x + y = 4$
26. $x - y = 2$
27. $x + 2y = 8$
28. $x - 2y = 6$
29. $2x - 3y = 6$
30. $3x - 2y = 6$
31. $3y = 2x - 12$
32. $4x = 3y - 12$
33. $y = 6x + 6$
34. $y = -2x - 8$
35. $2x + 4y = 6$
36. $-3x + y = 4$
37. $4x - 8y = -8$
38. $5y = 3x + 10$

Find the slope of the line through the given points. If the slope is undefined, so state.

39. $(4, 2)$ and $(2, 3)$
40. $(3, 5)$ and $(6, -1)$
41. $(-2, -5)$ and $(3, 2)$
42. $(2, -2)$ and $(-3, 5)$
43. $(5, 2)$ and $(-3, 2)$
44. $(-3, -5)$ and $(-1, -2)$
45. $(8, -3)$ and $(-8, 3)$
46. $(2, 6)$ and $(2, -3)$
47. $(-2, 3)$ and $(1, -1)$
48. $(-7, -5)$ and $(5, -6)$

Graph each equation using the slope and y-intercept as in Examples 4 through 6.

49. $y = x + 4$
50. $y = 2x + 1$

51. $y = -x - 2$
52. $y = -2x + 3$
53. $y = -3x + 1$
54. $y = 3x - 5$
55. $y = -\frac{3}{5}x + 3$
56. $y = -x - 2$
57. $y = \frac{4}{7}x - 1$
58. $2x + y = 6$
59. $3x + 2y = 6$
60. $-2x - 2y = 8$
61. $-\frac{1}{2}x - 2y = 4$
62. $3x - 2y + 6 = 0$
63. $3x + 4y - 8 = 0$

Determine the equation of the graph.

64.

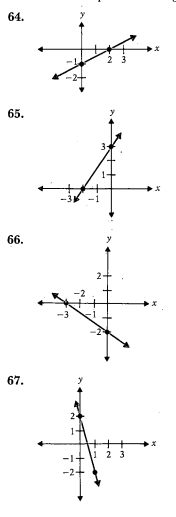

65.

66.

67.

PROBLEM SOLVING

68. **a)** Two lines are parallel when they do not intersect no matter how far they are extended. Explain how you can determine, without graphing the equations, whether two equations will be parallel lines when graphed.

b) Determine whether the graphs of the equations $2x - 3y = 6$ and $4x = 6y + 6$ are parallel lines.

6.7 LINEAR INEQUALITIES IN TWO VARIABLES

In Section 6.4 we introduced linear inequalities in one variable. Now we will introduce linear inequalities in two variables. Some examples of linear inequalities in two variables are: $2x + 3y \leqslant 7, x + 7y \geqslant 5$, and $x - 3y < 6$.

We were able to indicate a solution set of linear inequalities in one variable on a number line. The solution set of a linear inequality in two variables will be indicated on a coordinate plane.

An inequality that is strictly $<$ or $>$ will have as its solution set a **half-plane.** A half-plane is the set of all the points on one side of a line. An inequality that is $\leqslant$ or $\geqslant$ will have as its solution set the set of points that consists of a half-plane and a line. To indicate that the line is part of the solution set, we draw a solid line. To indicate that the line is not part of the solution set, we draw a dashed line.

■ **Example 1** _____

Draw the graph of $x + 3y < 9$.

Solution: To obtain the solution set, we start by graphing $x + 3y = 9$. Since the original inequality is strictly "less than," we draw a dashed line (Fig. 6.33). The dashed line indicates that the points on the line are not part of the solution set.

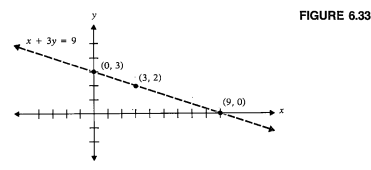

FIGURE 6.33

The line $x + 3y = 9$ divides the plane into three parts—the line itself and two half-planes. The line is the boundary between the two half-planes.

The points in one half-plane will satisfy the inequality $x + 3y < 9$. The points in the other half-plane will satisfy the inequality $x + 3y > 9$.

To determine the solution set of the inequality $x + 3y < 9$, we pick any point on the plane that is not on the line. The simplest point to work with is the origin, $(0, 0)$. Substituting $x = 0$ and $y = 0$ in $x + 3y < 9$ gives

$$x + 3y < 9$$
$$\text{Is } 0 + 3(0) < 9?$$
$$0 + 0 < 9$$
$$0 < 9$$

Since 0 is less than 9, the point $(0, 0)$ is part of the solution set. All the points on the same side of the line $x + 3y = 9$ as the point $(0, 0)$ are members of the solution set. We indicate this by shading the half-plane that contains $(0, 0)$. The graph is shown in Fig. 6.34.

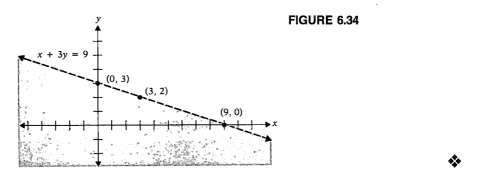

FIGURE 6.34

■ **Example 2** _____

Draw the graph of $4x - 2y \geq 12$.

Solution: We first draw the graph of the equation $4x - 2y = 12$. We use a solid line, since the points on the boundary are included in the solution set. Now we pick a point that is not on the line. Let us take $(0, 0)$ as the test point.

$$4x - 2y \geq 12$$
$$\text{Is } 4(0) - 2(0) \geq 12?$$
$$0 \not\geq 12$$

Since zero is not greater than or equal to 12 $(0 \not\geq 12)$, the solution set is the line and the half-plane that does not contain the point $(0, 0)$. The graph is shown in Fig. 6.35.

If we had arbitrarily selected the point $(3, -5)$, we would have found that $4(3) - 2(-5) \geq 12$, or $22 \geq 12$. Thus the point would be in the half-plane containing the solution set.   ❖

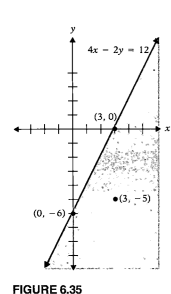

FIGURE 6.35

■ **Example 3**

Draw the graph of $y < x$.

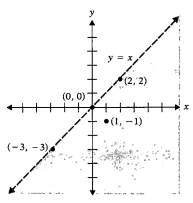

Solution: The inequality is strictly "less than," so the boundary is not part of the solution set. In graphing the equation $y = x$, we draw a dashed line (Fig. 6.36). Since $(0, 0)$ is *on* the line, we cannot use it as a check. Let us pick the point $(1, -1)$.

$$y < x$$
$$-1 < 1$$

Since $-1 < 1$, the solution set is the half-plane containing the point $(1, -1)$.

FIGURE 6.36   ❖

EXERCISES 6.7

Draw the graph for each of the following.

1. $x > 3$
2. $y \leq 2$
3. $y > x + 2$
4. $y \leq x - 4$
5. $y \geq 2x - 6$
6. $y < -2x + 2$
7. $3x - 4y > 12$
8. $x + 2y > 4$
9. $5x - 2y \leq 10$
10. $2y - 5x \geq 10$
11. $3x + 2y < 6$
12. $-x + 2y < 2$
13. $x + y > 0$
14. $x + 2y \leq 0$
15. $3y - 2x \leq 0$
16. $y \geq 2x + 7$
17. $3x + 2y > 12$
18. $y \leq 3x - 4$
19. $\frac{2}{5}x - \frac{1}{2}y \leq 1$
20. $0.1x + 0.3y \leq 0.4$
21. $0.2x + 0.5y \leq 0.3$

22. Ann is allowed to talk a maximum of 15 minutes on the telephone. She has two friends (x and y) with whom she wishes to share the time.
    a) State this problem as an inequality in two variables.
    b) Graph the inequality.
23. Jose has 30 feet of fencing to place around a sandbox he plans to build for his children.
    a) Write an inequality illustrating all possible dimensions of the rectangular sandbox. ($P = 2l + 2w$ is the formula for the perimeter of a rectangle.)
    b) Graph the inequality.

6.8 SOLVING QUADRATIC EQUATIONS USING FACTORING AND USING THE QUADRATIC FORMULA

In Section 6.2 we solved linear, or first degree, equations. In those equations the exponent on all variables was 1. Now we introduce the **quadratic equation.** The **standard form of a quadratic equation** in one variable is

Standard Form of a Quadratic Equation
$ax^2 + bx + c = 0, \quad a \neq 0$

Note that the greatest exponent on x is 2, and the left side of the equation is equal to zero. To solve a quadratic equation means to find the value or values that make the equation true. In this section we will solve quadratic equations by factoring and by the quadratic formula.

Before we discuss factoring, we will introduce the **FOIL method** of multiplying two binomials. A **binomial** is an expression that contains two terms, in which each exponent that appears on a variable is a whole number. Examples of binomials are $x + 3$, $x - 5$, $3x + 5$, and $4x - 2$.

To multiply two binomials, we can use the FOIL method as illustrated below:

$$(a + b)(c + d) = a \cdot c + a \cdot d + b \cdot c + b \cdot d$$

> **F:** multiply the **FIRST** terms in the binomials.
> **O:** multiply the **OUTER** terms in the binomials.
> **I:** multiply the **INNER** terms in the binomials.
> **L:** multiply the **LAST** terms in the binomials.

After we multiply the first, outer, inner, and last terms, we combine all like terms.

Example 1

Multiply $(x + 2)(x + 5)$.

Solution: Using the FOIL method of multiplication, we obtain

$$
\begin{aligned}
(x + 2)(x + 4) &= x \cdot x + (x)(4) + 2 \cdot x + 2 \cdot 4 \\
&= x^2 + 4x + 2x + 8 \\
&= x^2 + 6x + 8
\end{aligned}
$$

Note that $4x$ and $2x$ were combined to get $6x$. ❖

Factoring Trinomials of the Form $ax^2 + bx + c$, $a = 1$

The expression $x^2 + 6x + 8$ is an example of a trinomial. A **trinomial** is an expression containing three terms in which each exponent that appears on a variable is a whole number.

Example 2

Multiply $(3x + 2)(x - 5)$.

Solution

$$(3x + 2)(x - 5) = 3x \cdot x + 3x(-5) + 2 \cdot x + 2(-5)$$
$$= 3x^2 - 15x + 2x - 10$$
$$= 3x^2 - 13x - 10 \qquad ❖$$

In Example 1 we showed that

$$(x + 2)(x + 4) = x^2 + 6x + 8$$

Since the product of $x + 2$ and $x + 4$ is $x^2 + 6x + 8$, we say that $x + 2$ and $x + 4$ are **factors** of $x^2 + 6x + 8$. **To factor an expression** means to write the expression as a product of its factors. For example, to factor $x^2 + 6x + 8$, we would write

$$x^2 + 6x + 8 = (x + 2)(x + 4)$$

Let us look at the factors more closely:

$$2 + 4 = 6$$
$$2 \cdot 4 = 8$$
$$x^2 + 6x + 8 = (x + 2)(x + 4)$$

Note that the sum of the two numbers in the factors is $2 + 4$ or 6. The 6 is the coefficient of the x-term. Also note that the product of the numbers in the two factors is $2 \cdot 4$ or 8. The 8 is the constant. In general, when factoring an expression of the form $x^2 + bx + c$, we need to find two numbers whose product is c and whose sum is b. When we determine the two numbers, the factors will be of the form

$$(x + \square)(x + \square)$$
$$\uparrow \qquad \uparrow$$
one   other
number number

■ **Example 3** _____

Factor $x^2 + 5x + 6$.

Solution: We need to find two numbers whose product is 6 and whose sum is 5. Since the product is $+6$, the two numbers must both be positive or both be negative. Since the coefficient of the x-term is positive, only the positive factors of 6 need to be considered. Can you explain why? We begin by listing the positive numbers whose product is 6.

Factors of 6	Sum of Factors
1(6)	1 + 6 = 7
2(3)	2 + 3 = 5

Since $2 \cdot 3 = 6$ and $2 + 3 = 5$, 2 and 3 are the numbers we are seeking. Thus we write

$$x^2 + 5x + 6 = (x + 2)(x + 3)$$

Note $(x + 3)(x + 2)$ is also an acceptable answer.   ❖

■ **Example 4** _____

Factor $x^2 - x - 12$.

Solution: We must find two numbers whose product is -12 and whose sum is -1. (Remember: $-x$ means $-1x$.) Begin by listing the factors of -12:

Factors of -12	Sum of Factors
$-12(1)$	$-12 + 1 = -11$
$-6(2)$	$-6 + 2 = -4$
$-4(3)$	$-4 + 3 = -1$
$-3(4)$	$-3 + 4 = 1$
$-2(6)$	$-2 + 6 = 4$
$-1(12)$	$-1 + 12 = 11$

We have listed all the factors of -12. The numbers we are looking for are -4 and 3. We listed all factors in this example so that you would see that, for example, $-4(3)$ is a different set of factors than $-3(4)$. Once you find the factors you are seeking, there is no reason to go any further.

$$x^2 - x - 12 = (x + 3)(x - 4)$$ ❖

Factoring Trinomials of the Form $ax^2 + bx + c, a \neq 1$

Now we discuss how to factor an expression of the form $ax^2 + bx + c$ where a, the coefficient of the squared term, is not equal to 1.

Consider the multiplication problem $(2x + 1)(x + 3)$:

$$
\begin{aligned}
(2x + 1)(x + 3) &= 2x \cdot x + 2x \cdot 3 + 1 \cdot x + 1 \cdot 3 \\
&= 2x^2 + 6x + x + 3 \\
&= 2x^2 + 7x + 3
\end{aligned}
$$

Since $(2x + 1)(x + 3) = 2x^2 + 7x + 3$, the factors of $2x^2 + 7x + 3$ are $2x + 1$ and $x + 3$.

Let us study these factors more closely.

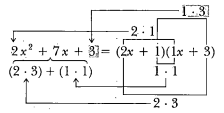

$$2x^2 + 7x + 3 = (2x + 1)(1x + 3)$$

Notice that the product of the coefficient of the first terms in the multiplication of the binomials equals 2, the coefficient of the squared term. The sum of the products of the coefficients of the outer and inner terms equals 7, the coefficient of the x term. The product of the last terms equals 3, the constant.

A procedure to factor expressions of the form $ax^2 + bx + c$, $a \neq 1$ follows.

To factor expressions of the form $ax^2 + bx + c$, $a \neq 1$:

1. Write all pairs of factors of the coefficient of the squared term, a.
2. Write all pairs of factors of the constant, c.
3. Try various combinations of these factors until the correct middle term, bx, is found.

■ **Example 5**

Factor $3x^2 + 17x + 10$.

Solution: The only factors of 3 are 1 and 3. Therefore we write

$$3x^2 + 17x + 10 = (3x \quad)(x \quad)$$

The number 10 has both positive and negative factors. However, since both the constant, 10, and the sum of the products of the outer and inner terms, 17, are positive, the two factors must be positive. Why? The positive factors of 10 are 1(10) and 2(5). Below is a listing of the possible factors:

Possible Factors	Sum of Products of Outer and Inner Terms
$(3x + 1)(x + 10)$	$31x$
$(3x + 10)(x + 1)$	$13x$
$(3x + 2)(x + 5)$	$17x$ ←Correct middle term
$(3x + 5)(x + 2)$	$11x$

Thus $3x^2 + 17x + 10 = (3x + 2)(x + 5)$ ❖

Note that factoring problems of this type may be checked by using the FOIL method of multiplication. We will check the results to Example 5:

$$(3x + 2)(x + 5) = 3x \cdot x + 3x \cdot 5 + 2 \cdot x + 2 \cdot 5$$
$$= 3x^2 + 15x + 2x + 10$$
$$= 3x^2 + 17x + 10$$

Since we obtained the expression we started with, our factoring is correct.

■ **Example 6** ————————————————————————

Factor $6x^2 - 11x - 10$.

Solution: The factors of 6 will be either $6 \cdot 1$ or $2 \cdot 3$. Therefore the factors may be of the form $(6x \quad)(x \quad)$ or $(2x \quad)(3x \quad)$. When there is more than one set of factors for the first term, we generally try the medium-sized factors first. If this does not work, we try the other factors. Thus we write

$$6x^2 - 11x - 10 = (2x \quad)(3x \quad)$$

The factors of -10 are $(-1)(10)$, $(1)(-10)$, $(-2)(5)$, and $(2)(-5)$. Since there are eight different factors of -10, there will be eight different pairs of possible factors to try. Can you list them?

The correct factoring is

$$6x^2 - 11x - 10 = (2x - 5)(3x + 2)$$ ❖

Note that in Example 6 we were fortunate to find that the factors were of the form $(2x \quad)(3x \quad)$. If we had not been able to find the correct factors using these, we would have tried $(6x \quad)(x \quad)$.

Solving Quadratic Equations by Factoring

To solve a quadratic equation by factoring we use the *zero-factor* property:

Zero-Factor Property
If $a \cdot b = 0$, then $a = 0$ or $b = 0$.

The zero-factor property indicated that if the product of two factors is 0, then one (or both) of the factors must have a value of 0.

■ **Example 7** _____

Solve the equation $(x - 2)(x + 4) = 0$.

Solution: Using the zero-factor property, either $x - 2$ or $x + 4$ must equal 0 for the product to equal 0. Thus we set each individual factor equal to 0 and solve each resulting equation for x:

$$(x - 2)(x + 4) = 0$$
$$x - 2 = 0 \quad \text{or} \quad x + 4 = 0$$
$$x = 2 \qquad\qquad x = -4$$

Thus the solutions are 2 and -4.

CHECK

$$x = 2 \qquad\qquad\qquad\qquad x = -4$$
$$(x - 2)(x + 4) = 0 \qquad\qquad (x - 2)(x + 4) = 0$$
$$(2 - 2)(2 + 4) = 0 \qquad\qquad (-4 - 2)(-4 + 4) = 0$$
$$0\,(6) = 0 \qquad\qquad\qquad\quad (-6)(0) = 0$$
$$0 = 0 \quad \text{True} \qquad\qquad 0 = 0 \quad \text{True} \quad ❖$$

To solve a quadratic equation by factoring:
1. Use the addition or subtraction property to make one side of the equation equal to 0.
2. Factor the side of the equation not equal to 0.
3. Use the zero-factor property to solve the equation.

Examples 8 and 9 illustrate this procedure.

■ **Example 8** _____

Solve the equation $x^2 - 8x = -15$.

Solution: First add 15 to both sides of the equation to make the right side of the equation equal to 0:

$$x^2 - 8x = -15$$
$$x^2 - 8x + 15 = -15 + 15$$
$$x^2 - 8x + 15 = 0$$

Now factor the left side of the equation. We need to find two numbers whose product is 15 and whose sum is -8. Since the product of the numbers is positive and the sum of the numbers is negative, the two numbers must both be negative. Can you explain why? The numbers are -3 and

-5. Note that $(-3)(-5) = 15$ and $-3 + (-5) = -8$.

$$x^2 - 8x + 15 = 0$$
$$(x - 3)(x - 5) = 0$$

Now use the zero-factor property to find the solution:

$$x - 3 = 0 \quad \text{or} \quad x - 5 = 0$$
$$x = 3 \qquad\qquad x = 5$$

The solutions are 3 and 5.   ❖

■ Example 9

Solve the equation $2x^2 - 11x + 12 = 0$.

Solution: $2x^2 - 11x + 12$ factors into $(2x - 3)(x - 4)$. Thus we write

$$2x^2 - 11x + 12 = 0$$
$$(2x - 3)(x - 4) = 0$$
$$2x - 3 = 0 \quad \text{or} \quad x - 4 = 0$$
$$2x = 3 \qquad\qquad x = 4$$
$$x = \frac{3}{2}$$

The solutions are $\frac{3}{2}$ and 4.   ❖

Solving Quadratic Equations Using the Quadratic Formula

Not all quadratic equations can be solved by factoring. When a quadratic equation cannot be easily solved by factoring, we can solve the equation with the **quadratic formula.** The quadratic formula can be used to solve any quadratic equation.

Quadratic Formula

For a quadratic equation in standard form, $ax^2 + bx + c = 0, a \neq 0$, the quadratic formula is

$$x = \frac{-b \pm \sqrt{b^2 - 4ac}}{2a}$$

To use the quadratic formula, we first write the quadratic equation in standard form. Then we determine the values for a (the coefficient of the squared term), b (the coefficient of the x term), and c (the constant). Then we substitute the values of a, b, and c into the quadratic formula and evaluate the expression.

■ **Example 10** _____

Solve the equation $x^2 + 6x - 16 = 0$ using the quadratic formula.

Solution: In this equation, $a = 1$, $b = 6$, and $c = -16$. Substituting these values into the quadratic formula gives

$$x = \frac{-b \pm \sqrt{b^2 - 4ac}}{2a} = \frac{-6 \pm \sqrt{6^2 - 4(1)(-16)}}{2(1)}$$

$$= \frac{-6 \pm \sqrt{36 + 64}}{2}$$

$$= \frac{-6 \pm \sqrt{100}}{2}$$

$$= \frac{-6 \pm 10}{2}$$

$$\frac{-6 + 10}{2} = \frac{4}{2} = 2 \qquad \frac{-6 - 10}{2} = \frac{-16}{2} = -8$$

The solutions are 2 and -8.   ❖

Note that Example 10 can also be solved by factoring. We suggest that you solve Example 10 by factoring now.

■ **Example 11** _____

Solve $2x^2 - 9x = -5$ using the quadratic formula.

Solution: Begin by writing the equation in standard form by adding 5 to both sides of the equation:

$$2x^2 - 9x + 5 = 0$$
$$a = 2, \qquad b = -9, \qquad c = 5$$
$$x = \frac{-b \pm \sqrt{b^2 - 4ac}}{2a} = \frac{-(-9) \pm \sqrt{(-9)^2 - 4(2)(5)}}{2(2)}$$

$$= \frac{9 \pm \sqrt{81 - 40}}{4}$$

$$= \frac{9 \pm \sqrt{41}}{4}$$

The solutions are $(-9 + \sqrt{41})/4$ and $(9 - \sqrt{41})/4$.   ❖

Note that the solutions to Example 11 are irrational numbers. It is also possible for a quadratic equation to have no real solution. In solving an equation by the quadratic formula, if the radicand (the expression inside the square root) is a negative number, then the quadratic equation has **no real solution.**

EXERCISES 6.8

Factor the trinomial. If the trinomial cannot be factored, so state.

1. $x^2 + 5x + 6$ **2.** $x^2 + 8x + 12$
3. $x^2 + x - 12$ **4.** $x^2 - x - 12$
5. $x^2 - 6x + 8$ **6.** $x^2 + 2x - 24$
7. $x^2 - 2x + 1$ **8.** $x^2 + 3x - 10$
9. $x^2 - 25$ **10.** $x^2 - 10x + 21$
11. $x^2 - x - 30$ **12.** $x^2 - 7x - 8$
13. $x^2 + 4x - 32$ **14.** $x^2 + 3x - 28$
15. $x^2 - 14x + 40$ **16.** $x^2 + 16x + 63$

Factor the trinomial. If the trinomial cannot be factored, so state.

17. $2x^2 + 3x - 2$ **18.** $2x^2 - 7x - 15$
19. $4x^2 + 4x + 1$ **20.** $3x^2 + 2x - 1$
21. $5x^2 + 12x + 4$ **22.** $2x^2 - 9x + 10$
23. $5x^2 - 7x - 6$ **24.** $4x^2 + 16x + 15$
25. $2x^2 - 17x + 30$ **26.** $5x^2 - 14x + 8$
27. $3x^2 - 14x - 24$ **28.** $6x^2 + 5x + 1$

Solve each equation.

29. $(x - 3)(x + 4) = 0$ **30.** $(2x - 3)(3x + 1) = 0$
31. $(2x + 5)(4x - 3) = 0$ **32.** $(x - 6)(5x - 4) = 0$

Solve each equation using factoring.

33. $x^2 + 3x + 2 = 0$ **34.** $x^2 - 3x + 2 = 0$
35. $x^2 + 2x - 8 = 0$ **36.** $x^2 - 8x + 15 = 0$
37. $x^2 - 7x = -6$ **38.** $x^2 - 15 = 2x$
39. $x^2 = 4x - 3$ **40.** $x^2 - 13x + 40 = 0$
41. $x^2 = 11x - 28$ **42.** $x^2 - 18 = 7x$
43. $x^2 + 12x + 20 = 0$ **44.** $x^2 + 5x - 36 = 0$
45. $2x^2 + x - 3 = 0$ **46.** $3x^2 + 13x = -4$
47. $5x^2 + 11x = -2$ **48.** $2x^2 = -5x + 3$
49. $3x^2 - 4x = -1$ **50.** $5x^2 + 16x + 12 = 0$
51. $4x^2 - 9x + 2 = 0$ **52.** $6x^2 + x - 2 = 0$

Solve each equation using the quadratic formula. If the equation has no real solution, so state.

53. $x^2 - 4x - 21 = 0$ **54.** $x^2 - 9x + 20 = 0$
55. $x^2 + 8x + 15 = 0$ **56.** $x^2 - 4x + 3 = 0$
57. $x^2 - 8x = 9$ **58.** $x^2 = -8x - 15$
59. $x^2 - 2x + 3 = 0$ **60.** $x^2 - 5x = 4$
61. $2x^2 - x - 3 = 0$ **62.** $2x^2 - 7x + 2 = 0$
63. $3x^2 - 2x = -5$ **64.** $2x^2 + x = 5$
65. $4x^2 - 5x - 3 = 0$ **66.** $4x^2 - x - 1 = 0$
67. $3x^2 = 9x - 5$ **68.** $2x^2 + 7x + 5 = 0$
69. $3x^2 - 10x + 7 = 0$ **70.** $4x^2 + 7x - 1 = 0$
*71. $2x^2 + 6x - 3 = 0$ ***72.** $3x^2 - 2x - 6 = 0$

PROBLEM SOLVING

73. The radicand in the quadratic formula, $b^2 - 4ac$, is called the **discriminant.** How many solutions will the quadratic equation have if the discriminant is (a) greater than 0, (b) equal to 0, or (c) less than zero. Explain your answer.

6.9 QUADRATIC FUNCTIONS

In Section 6.6 we discussed the techniques of graphing a linear equation in two variables. The graphs of these equations in every case were straight lines. In this section we will discuss techniques of graphing equations that are not linear, that is, equations whose graphs are not straight lines.

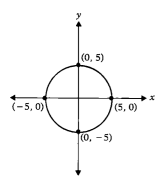

FIGURE 6.37

Algebraic statements of equality in two variables are called relations. A **relation** is a set of ordered pairs. A relation may be illustrated in many different ways, as shown in 1 through 4 below.

1. A set of ordered pairs: $\{(3, 4), (-3, -4), (2, 7), (8, 5), (5, 9)\}$
2. A table of values:

x	1	2	4	1	4
y	1	3	7	5	2

3. A graph (Fig. 6.37):
4. A statement of equality in two variables: $y = x^2$.

The table of values in (2) can be written as a set of ordered pairs, namely, (1, 1), (2, 3), (4, 7), (1, 5), and (4, 2). The graph in (3) is an illustration of an infinite number of ordered pairs. The equation in (4) will generate an infinite number of ordered pairs.

The set of ordered pairs in (1) is different from the set of ordered pairs in (2) in one important way. In (2), each value of x does not have a unique value of y. We can see from the table that when $x = 4$, y can be equal to 2 or 7. In (1), each value of x is paired with a *unique* value of y; when this happens, the relation is called a *function*.

> A **function** is a relation in which no two ordered pairs have the same first coordinate and a different second coordinate.

We can determine whether or not a relation is a function by examining the ordered pairs or by examining the graph of the relation. For any graph, if you can draw a straight line parallel to the y-axis and the line cuts the graph at more than one point, the relation shown is not a function. If the straight line cuts the graph at one and only one point, the graph is a function. (When the line cuts the graph at one and only one point, each value of x has a unique value of y.) This test, which can be used to determine if a relation is a function, is called the **vertical line test.**

For the graph in (3) we can draw a line parallel to the y-axis such that it will cut the curve at two points (see Fig. 6.38). Thus the graph is not a function (it is a relation).

Other examples of relations and functions are illustrated in Fig. 6.39.

The graphs in Figs. 6.39b and Fig. 6.39c show relations that are not functions, since they do not pass the vertical line test. Figure 6.39a shows a relation that is also a function.

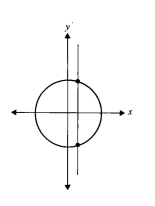

FIGURE 6.38

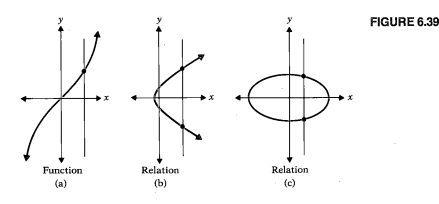

FIGURE 6.39

Function
(a)

Relation
(b)

Relation
(c)

Note that all functions are relations, but not all relations are functions.

Letters such as f, g, and h are often used to name functions. If we are given a function such that for each value of x there is a unique value of y, we may say that y is a function of x and write $y = f(x)$. The notation $f(x)$ is read "f of x." When we are given an equation that is a function, we may replace the y in the equation with $f(x)$, since $f(x)$ represents y. For example, if we are given the function $y = x + 6$, we may write the function as $f(x) = x + 6$.

To evaluate a function for a specific value of x, replace each x in the function with the given value, then evaluate. For example, to evaluate $f(x) = x + 6$ when $x = 2$, we replace each x with a 2:

$$f(x) = x + 6$$
$$f(2) = 2 + 6 = 8$$

Thus $f(2) = 8$. Since $f(x)$ equals y, when $x = 2, y = 8$.

▤ Example 1

For $f(x) = x^2 - 6x + 9$, find (a) $f(5)$ and (b) $f(-2)$.

Solution

a) $f(x) = x^2 - 6x + 9$
If $x = 5, f(5) = 5^2 - 6(5) + 9 = 25 - 30 + 9 = 4$. Thus
$$f(5) = 4.$$

b) $f(x) = x^2 - 6x + 9$
If $x = -2, f(-2) = (-2)^2 - 6(-2) + 9 = 4 + 12 + 9 = 25$. Thus
$$f(-2) = 25.$$ ❖

The graph of every equation of the form $y = ax^2 + bx + c$, $a \neq 0$ will pass the vertical line test and is therefore a function.

> An equation of the form $y = ax^2 + bx + c$ where $a, b,$ and c are real numbers, $a \neq 0$, is called a **quadratic function**.

Since $f(x)$ can be used in place of y, a quadratic function may also be represented as $f(x) = ax^2 + bx + c, a \neq 0$.

 Example 2 _____

The function $y = x^2$ is a quadratic function with $a = 1, b = 0,$ and $c = 0$. ❖

 Example 3 _____

The function $y = 2x^2 - 8$ is a quadratic function with $a = 2, b = 0,$ and $c = -8$. ❖

Example 4 _____

The function $y = 2x^2 - 4x - 6$ is a quadratic function with $a = 2, b = -4,$ and $c = -6$. ❖

Example 5 _____

The function $y = -3x^2 + 12$ is a quadratic function with $a = -3, b = 0,$ and $c = 12$. ❖

For every quadratic function there is a set of ordered pairs that satisfies that particular function. For example, for the function $y = x^2$ we can determine ordered pairs that satisfy the function by substituting values in place of x and obtaining values for y. We will list the x- and y-values in Table 6.4 for convenience.

TABLE 6.4

$y = x^2$		x	y
$x = -3,$	$y = (-3)^2 = 9$	-3	9
$x = -2,$	$y = (-2)^2 = 4$	-2	4
$x = -1,$	$y = (-1)^2 = 1$	-1	1
$x = 0,$	$y = 0^2 = 0$	0	0
$x = 1,$	$y = 1^2 = 1$	1	1
$x = 2,$	$y = 2^2 = 4$	2	4
$x = 3,$	$y = 3^2 = 9$	3	9

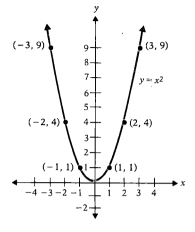

FIGURE 6.40

There are an infinite number of ordered pairs that satisfy the function $y = x^2$. We have found only a few. To show all of the ordered pairs that satisfy the function, we sketch a graph of the function. This can be done by plotting the points in Table 6.4 and connecting the points with a smooth curve as in Fig. 6.40.

The graph of every quadratic function will be a **parabola.** Two parabolas are illustrated in Fig. 6.41.

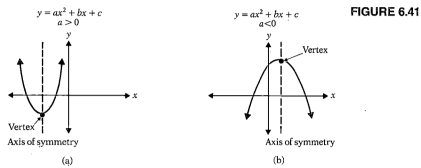

FIGURE 6.41

The parabola in Fig. 6.41a opens upward, and the parabola in Fig. 6.41b opens downward. The sign of the numerical coefficient of the squared term, a, determines whether the parabola opens upward or downward. If $a > 0$, the parabola opens upward; and if $a < 0$, the parabola opens downward.

The **vertex** of a parabola is the lowest point on a parabola that opens upward and the highest point on a parabola that opens downward. Every parabola is **symmetric** about a vertical line through its vertex. This line is called the **axis of symmetry** of the parabola. The x-coordinate of the vertex and the equation of the axis of symmetry can be found by using the following equation.

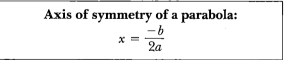

Once the x-coordinate of the vertex has been determined, the y-coordinate can be found by substituting the value found for the x-coordinate into the quadratic equation and evaluating the equation. This procedure is illustrated in Example 6.

■ **Example 6**

Consider the equation $y = 2x^2 - 4x - 6$:

a)   Determine whether the graph will be a parabola that opens upward or downward.

b)   Find the equation of the axis of symmetry of the parabola.

c)   Find the vertex of the parabola.

Solution

a)   Since $a = 2$, which is greater than 0, the parabola opens upward.

b)   To find the axis of symmetry, we use the equation $x = -b/2a$. In $y = 2x^2 - 4x - 6$, $a = 2$, $b = -4$, and $c = -6$:

$$x = \frac{-b}{2a} = \frac{-(-4)}{2(2)} = \frac{4}{4} = 1$$

The equation of the axis of symmetry is $x = 1$.

c)   The x-coordinate of the vertex is 1 (from part b). To find the y-coordinate, we substitute 1 for x in the equation, then evaluate:

$$y = 2x^2 - 4x - 6$$
$$y = 2(1)^2 - 4(1) - 6$$
$$y = 2(1) - 4 - 6$$
$$y = 2 - 4 - 6$$
$$y = -8$$

Therefore the vertex of the parabola is located at the point $(1, -8)$ on the graph.   ❖

■ Example 7

Consider the quadratic equation $y = x^2 - 6x + 8$.

a)   Determine whether the graph will be a parabola that opens upward or downward.

b)   Find the equation of the axis of symmetry of the parabola.

c)   Find the vertex of the parabola.

d)   Graph the parabola.

Solution

a)   Since $a = 1$, which is greater than 0, the parabola opens upward.

b)

$$x = \frac{-b}{2a} = \frac{-(-6)}{2(1)} = \frac{6}{2} = 3$$

The equation of the equation of the axis of symmetry is $x = 3$.

c) The x-coordinate of the vertex is 3 (from part b). To find the y-coordinate, substitute 3 for x in the equation and then evaluate:

$$y = x^2 - 6x + 8$$
$$y = 3^2 - 6(3) + 8$$
$$y = 9 - 18 + 8 = -1$$

The y-coordinate of the vertex is -1. The vertex is therefore $(3, -1)$.

d) To graph the parabola, we will plot the vertex and three points to the left of the vertex (Fig. 6.42a). Using symmetry, we will complete the graph (see Fig. 6.42b).

x	y
3	-1
2	0
1	3
0	8

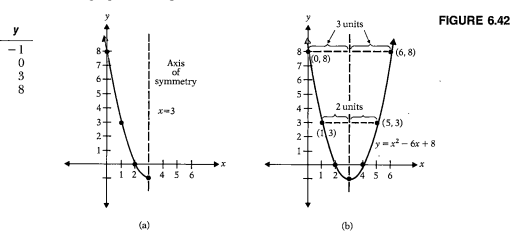

(a)   (b)

FIGURE 6.42

Note the symmetry in Fig. 6.42. Instead of selecting three points to the left of the vertex, we could have selected three points to the right of the vertex and used symmetry to complete the graph.   ❖

If we examine the graph of $y = x^2 - 6x + 8$ (Fig. 6.42), we see that the graph crosses the x-axis at 2 and 4. These values are called the *roots* of the graph. The **roots** are the values of x where the graph crosses the x-axis (or the x-intercepts). Whenever the graph crosses the x-axis, the y-value must be 0. We can find the roots of a parabola algebraically by substituting 0 for y in the quadratic equation and solving the resulting equation by factoring or by the quadratic formula. We will now find the roots of $y = x^2 - 6x + 8$ by factoring:

$$y = x^2 - 6x + 8$$
$$0 = x^2 - 6x + 8$$
$$\text{or} \quad x^2 - 6x + 8 = 0$$
$$(x - 4)(x - 2) = 0$$
$$x - 4 = 0 \quad \text{or} \quad x - 2 = 0$$
$$x = 4 \qquad\qquad x = 2$$

Note that we obtained the x-intercepts or roots of the parabola in Fig. 6.42(b).

When the x-coordinate of the vertex is a fractional value, it is sometimes more difficult to make use of symmetry in drawing the graph. When this occurs, plot a sufficient number of points on both sides of the vertex to get an accurate graph.

■ **Example 8**

Consider the graph $y = -2x^2 + 3x + 4$.

a) Determine whether the graph will be a parabola that opens upward or downward.

b) Find the equation of the axis of symmetry of the parabola.

c) Find the vertex of the parabola.

d) Find the roots of the equation if they exist.

e) Graph the parabola.

Solution

a) Since $a = -2$, which is less than 0, the parabola opens downward.

b)

$$x = \frac{-b}{2a} = \frac{-3}{2(-2)} = \frac{-3}{-4} = \frac{3}{4}$$

The equation of the axis of symmetry is $x = 3/4$.

c) The x-coordinate of the vertex is 3/4:

$$y = -2x^2 + 3x + 4$$
$$y = -2\left(\frac{3}{4}\right)^2 + 3\left(\frac{3}{4}\right) + 4$$
$$y = -2\left(\frac{9}{16}\right) + \frac{9}{4} + 4$$
$$y = -\frac{9}{8} + \frac{9}{4} + 4$$
$$y = -\frac{9}{8} + \frac{18}{8} + \frac{32}{8}$$
$$y = \frac{41}{8} \quad \text{or} \quad 5\frac{1}{8}$$

The vertex is $(\frac{3}{4}, 5\frac{1}{8})$ or $(0.75, 5.125)$.

d) We will find the roots by substituting 0 for y and then using the quadratic formula:

$$y = -2x^2 + 3x + 4$$
$$0 = -2x^2 + 3x + 4$$
$$\text{or} \quad -2x^2 + 3x + 4 = 0$$
$$a = -2, \quad b = 3, \quad c = 4$$

$$x = \frac{-b \pm \sqrt{b^2 - 4ac}}{2a}$$

$$= \frac{-3 \pm \sqrt{3^2 - 4(-2)(4)}}{2(-2)}$$

$$= \frac{-3 \pm \sqrt{9 + 32}}{-4}$$

$$= \frac{-3 \pm \sqrt{41}}{-4} \qquad \sqrt{41} \approx 6.4 \text{ from a calculator}$$
or Appendix B.

$$x \approx \frac{-3 + 6.4}{-4} \approx \frac{3.4}{-4} \approx -0.85 \quad \text{or} \quad x \approx \frac{-3 - 6.4}{-4} \approx \frac{-9.4}{-4} \approx 2.35$$

e) Plot the vertex $(\frac{3}{4}, 5\frac{1}{8})$ and the roots -0.85 and 2.35. Then select three values of x that are less than $\frac{3}{4}$ and three values of x that are greater than $\frac{3}{4}$. After we find the corresponding values of y, we plot the ordered pairs to obtain the graph (see Fig. 6.43).

x	y
-2	-10
-1	-1
0	4
1	5
2	2
3	-5

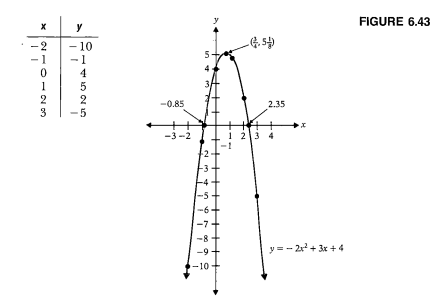

FIGURE 6.43

$y = -2x^2 + 3x + 4$

Note that although the graph is symmetric about $x = \frac{3}{4}$, it is more difficult to make use of symmetry, since the x-coordinate of the vertex is not an integer. ❖

In using the quadratic formula to find the roots of a graph, *if the radicand, $b^2 - 4ac$, is a negative number, then the graph has no real roots. A parabola with no real roots will not intersect the x-axis.*

DID YOU KNOW?

Mathematics is not only a body of knowledge; it is also a language. Often referred to as a universal language, mathematics is so precise that it may be understood by people around the world. In his book *The Ascent of Man*, Jacob Bronowski states, "And the exact fit of the numbers describes the exact laws that bind the universe. In fact, the numbers that compose right-angled triangles have been proposed as messages which we might send out to planets in other star systems as a test for the existence of rational life there." It is almost certain that when and if we come in contact with life forms from outer space, our initial method of communication will be through some type of number pattern.

The vocabulary of the language of mathematics is symbols. You have all used the digits 0 through 9 and the basic operations $+$, $-$, $\times$, and $\div$. There are many other mathematical symbols, some of which you will see in this text. The grammar and usage of the language of mathematics is governed by the rules of logic.

Why has mathematics evolved as a language of symbols? One important reason is that symbolism enables the mathematician to write lengthy expressions in a compact form so that the eye can see the entire statement and the mind can retain it. To describe the algebraic expression $3xy^2 + 2xyz$ would require a long, complex phrase such as "The product of three times a number multiplied by a second number that is multiplied by itself and added to two times the product of the first number, the second number, and the third number." The English philosopher Alfred North Whitehead stated, "By relieving the brain of all unnecessary work a good notation sets the mind free to concentrate on more advanced problems and in effect increases the mental power of the race."

A second reason for symbolic language is its clarity. Many words and phrases in English and other languages are vague or ambiguous. Many words or phrases in one language have no equivalent word or phrase in another language. This is not so with the language of mathematics. In mathematics, each symbol has a precise meaning, so the resulting expressions are clear.

A third reason is that symbolism allows us to consider a large or infinite number of separate cases with a common property. For example, we can use the symbolic representation $ax + b = 0$ to represent all linear equations in one variable.

Mathematics has not always been a language of symbols. The Egyptians, Babylonians, and Greeks knew much of the geometry and algebra we use today, but they wrote their work out in words. Symbolism entered mathematics in the sixteenth and seventeenth centuries, when advances in science required mathematics to become more efficient. In 1545, Gerolamo Cardano published *Ars Magna* ("The Great Skill"), in which he introduced the limited use of symbolism. About 1560, Raphael Bombelli made some attempts at simplified writing of algebraic formulas, but the first systematic effort to introduce an algebraic sign language is generally attributed to François Viète in his *Isagoge in Artem Analyticam* ("Introduction to the Analytic Art").

EXERCISES 6.9

Determine which of the following are functions.

1. $\{(3, 4), (4, 5), (5, 6), (6, 7)\}$

2. $\{(-1, 2), (1, 2), (3, 4), (5, 6)\}$

3.

x	-1	2	5	3	-1	-2
y	3	4	7	1	4	7

4.

x	2	3	3	4	5	-1
y	5	7	9	11	-1	-3

5.

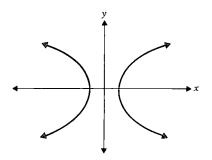

6.

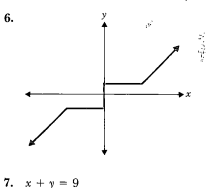

7. $x + y = 9$

8. $y = x^2 + 1$

Determine which of the following are functions.

9. $\{(2, 3), (3, 5), (4, 6), (2, 7)\}$

10. $\{(-1, 3), (-2, 4), (-3, 5), (-4, 7)\}$

11.

x	-1	-2	-3	-7	-8
y	3	5	6	4	3

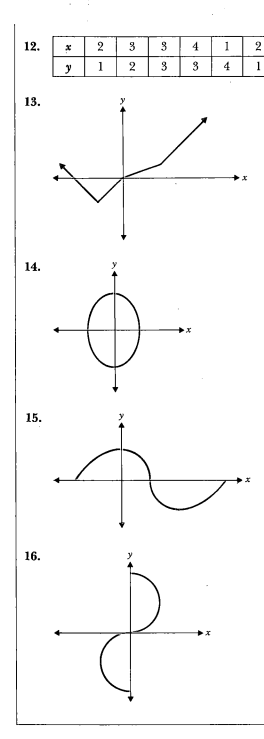

12.

x	2	3	3	4	1	2
y	1	2	3	3	4	1

13.

14.

15.

16.

Evaluate each function for the given value of x.

17. $f(x) = x + 5, x = 3$
18. $f(x) = -2x - 6, x = -2$
19. $f(x) = -5x + 3, x = -1$
20. $f(x) = 4x - 6, x = 0$
21. $f(x) = -3x - 2, x = 4$
22. $f(x) = 6x + 5, x = -3$
23. $f(x) = x^2 + 2x + 4, x = 6$
24. $f(x) = x^2 - 12, x = 6$
25. $f(x) = x^2 - 2x - 8, x = -2$
26. $f(x) = -x^2 + 3x + 7, x = 2$
27. $f(x) = -x^2 + 5x - 12, x = 3$
28. $f(x) = 4x^2 + 3x - 9, x = 2$
29. $f(x) = -5x^2 - 6x - 12, x = -3$
30. $f(x) = -3x^2 + 5x - 9, x = -2$

For each function:

a) Determine whether the parabola will open upward or downward.
b) Find the equation of the axis of symmetry.
c) Find the vertex.
d) Find the roots if they exist.
e) Draw the graph.

31. $y = x^2 + 1$
32. $y = x^2 - 4$
33. $y = x^2 - 25$
34. $y = x^2 + 2$
35. $y = -x^2 - 9$
36. $y = -x^2 - 16$
37. $y = -x^2 + 2$
38. $y = -2x^2 + 8$
39. $y = 2x^2 - 3$
40. $y = -3x^2 - 6$
41. $y = x^2 - 2x - 6$
42. $y = x^2 + 2x - 12$
43. $y = x^2 + 5x + 6$
44. $y = x^2 - 7x - 8$
45. $y = -x^2 + 4x - 6$
46. $y = -x^2 + 8x - 8$
47. $y = 3x^2 - 12x - 2$
48. $y = 2x^2 - 2x - 12$
49. $y = -3x^2 - 5x + 2$
50. $y = -2x^2 - 3x - 9$

SUMMARY

To evaluate algebraic expressions, you must know the priority of operations.

Priority of Operations

1. First, perform all operations within parentheses or other grouping symbols (according to the order given below).
2. Next, perform all exponential operations (that is, raising to powers or finding roots).
3. Next, perform all multiplication and division from left to right.
4. Finally, perform all addition and subtraction from left to right.

An essential ingredient in solving linear equations is an understanding of the four basic rules, which we called the addition, subtraction, multiplication, and division properties. Three other basic concepts that are also needed to help solve equations are the distributive, commutative, and associative properties.

To solve a linear equation, you must isolate the variable on one side of the equation.

To solve a word problem, follow these steps:

1. Read the problem carefully at least twice to make sure you understand it.

2. If possible, draw a sketch of the problem.

3. Choose a letter to represent the unknown quantity, and write down exactly what the letter represents.

4. Write the word problem as an equation.

5. Solve the equation.

6. Answer the question(s) asked.

7. Check the solution.

To solve a linear inequality, you must remember that if you mutliply or divide both sides of the inequality by a negative number, you must also be sure to change the order of the inequality. One method of indicating the solution to a linear inequality in one variable is to graph the solution on a number line.

The solution set of a linear equation in two variables is an infinite set of ordered pairs which can be illustrated as a graph on a coordinate plane. The slope of a line is a ratio of the vertical change to the horizontal change between any two points on the line.

The standard form of a quadratic equation is $ax^2 + bx + c = 0, a \neq 0$. A quadratic equation can be solved by factoring or by the quadratic formula:

$$x = \frac{-b \pm \sqrt{b^2 - 4ac}}{2a}$$

A relation is a set of ordered pairs. A function is a relation in which no two ordered pairs have the same first element and a different second element. If the relation is a function, then each value of x has a unique value of y.

Tha graph of a quadratic equation of the form $y = ax^2 + bx + c, a \neq 0$, is a parabola.

REVIEW EXERCISES

Evaluate each of the following by using the priority of operations given in Table 6.1.

1. $3 - 3 \cdot 6 + 1$
2. $2 - 6 \div 3 + 4$
3. $5 \cdot 7 - 3^2$
4. $\sqrt{64} + 18 \div 3$
5. $-3^2 + 5 \div 2^2$
6. $-2^3 + 18 \div \sqrt{9} + 7$
7. $2 + [5(2 + 3^2) + 5] \div 10 + 2$
8. $4 + 16 \div 8 \cdot 2^2 - 10$

Evaluate each expression for the given value(s) of the variable.

9. $x + 6,$ $x = 3$
10. $3x - 5,$ $x = -4$
11. $x^2,$ $x = 4$
12. $-x^2,$ $x = -3$
13. $-x^2,$ $x = 5$
14. $-3x^3,$ $x = 2$
15. $x^2 - 3x + 1,$ $x = 4$
16. $2x^2 + 9x - 2,$ $x = -1$
17. $5x^3 + 11x - 2,$ $x = \frac{1}{2}$
18. $4x^2 - 2xy + 3y^2,$ $x = 2, y = -1$

Check to see whether the value(s) given for the variable(s) is a solution for the equation.

19. $3x - 12 = 6, x = 6$
20. $2x - 3y = 8, x = 4, y = 0$
21. $y = x^2 - 6x + 3, x = 2, y = 8$
22. $x^2 + 3y = 10, x = 5, y = -5$

Solve the following equations for the given variable.

23. $3s + 7 = 22$
24. $2r + 3 = 5r - 15$
25. $6y + 2(y - 3) = 10$
26. $4(x - 2) = 3 + 5(x + 4)$
27. $\frac{x}{3} + \frac{2}{5} = 4$
28. $3x + 0.03 = 0.41$

Write the following in mathematical terms.

29. 8 decreased by 7 times x
30. 4 more than the sum of 9 and x
31. 7 more than 6 times r
32. 11 less than 8 divided by q

Write an equation that can be used to solve the problem. Solve the equation and find the desired value(s).

33. 12 decreased by 3 times a number is 21.
34. Twice a number decreased by 11 is the same as 4 more than the number.
35. 9 times the sum of a number and 4 is 72.
36. 14 more than 10 times a number is the same as 8 times the sum of the number and 12.

Write the equation and then find the solution.

37. Gina wishes to build a bookcase whose height is 3 feet more than the width. The bookcase is to have two shelves, not including the top and bottom pieces. If she has 24 feet of lumber with which she will build the bookcase, what will be the length and width of the bookcase?
38. Sixty hours of overtime must be split among three workers. One worker must be assigned twice the number of hours assigned to the other two. Find the number of hours of overtime for each worker.
39. In a sales contest the first prize is $300 more than two times the second prize. The total amount of prize money is $1200. Find the amount allocated for first and second prizes.
40. A tiller can be rented for $30 an hour and purchased for $480. How many months would the tiller have to be rented for the rental cost to equal the cost of purchasing a tiller?

Evaluate each of the formulas for the given value of the variables.

41. $A = \frac{1}{2}bh$, $b = 6, h = 7$
42. $A = \frac{1}{2}h(b + d), h = 14, b = 9, d = 11$

Solve for y.

43. $-2x + 4y = 6$
44. $3x - 4y = 12$
45. $2x - \frac{1}{3}y = 4$

Graph the solution set for the set of real numbers.

46. $4 + 5x > -16$ **47.** $3x + 7 \geqslant 5x + 9$
48. $3(x + 9) \leqslant 4x + 11$ **49.** $-2 \leqslant x < 4$

Graph the solution set for the set of integers.

50. $2 + 7x > -12$ **51.** $11 + 3x \leqslant 32$
52. $-5 \leqslant x \leqslant 2$ **53.** $-1 < x \leqslant 7$

Graph the following ordered pairs.

54. $(3, 5)$
56. $(-3, -4)$

55. $(-3, 2)$
57. $(6, -7)$

Points A, B, and C are vertices of a rectangle. Plot the points. Find the coordinates of the fourth point, D, to complete the rectangle. Find the area of the rectangle.

58. $A(-2, 2), B(3, 2), C(-2, -3)$
59. $A(-3, 1), B(-3, -2), C(4, -2)$

For each equation find three ordered pairs that satisfy the equation; then graph the equation.

60. $x + y = 3$
62. $x = y$

61. $2x + 3y = 12$
63. $x = 3$

Graph each equation using the x- and y-intercepts.

64. $x + 2y = 4$
66. $3x - y = 6$

65. $3x - 2y = 6$
67. $2x + 3y = 9$

Find the slope of the line through the given points.

68. $(3, 5), (-1, 6)$
70. $(-1, -4), (5, 3)$

69. $(4, 2), (6 -3)$
71. $(6, 2), (-6, -2)$

Graph each equation by plotting the y-intercept and plotting a second point by making use of the slope.

72. $y = 3x - 2$
74. $y = -\frac{1}{2}x + 4$

73. $y = 2x + 3$
75. $y = -x - 1$

Determine the equation of the graph.

76.   **77.**

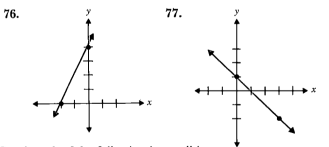

Graph each of the following inequalities.

78. $5x + 7y < 35$
80. $2x - 3y > 12$

79. $3x + 2y \geqslant 12$
81. $-3x - 5y \leqslant 15$

Factor the trinomial. If the trinomial cannot be factored, so state.

82. $x^2 + 9x + 20$
84. $x^2 - 10x + 24$
86. $2x^2 - 7x - 15$

83. $x^2 + x - 6$
85. $x^2 - 11x + 24$
87. $3x^2 + 5x - 2$

Solve each equation using factoring.

90. $x^2 + 3x + 2 = 0$ **91.** $x^2 - 6x = -5$
92. $2x^2 - 9x - 18 = 0$ **93.** $3x^2 = -7x - 2$

Solve each equation using the quadratic formula. If the equation has no real solution, so state.

94. $x^2 - 5x + 1 = 0$ **95.** $x^2 - 3x + 2 = 0$
96. $x^2 + 2x + 6 = 0$ **97.** $x^2 - x - 3 = 0$

Determine whether each of the following is a function.

98.

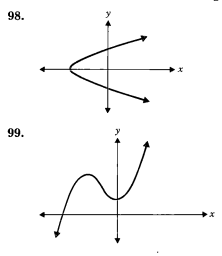

99.

Evaluate $f(x)$ for the given value of x.

100. $f(x) = 3x - 2, x = 2$
101. $f(x) = -4x + 5, x = -3$
102. $f(x) = 2x^2 - 3x + 4, x = 5$
103. $f(x) = -3x^2 - 6x + 8, x = 2$

For each function:
a) Determine whether the parabola will open upward or downward.
b) Find the equation of the axis of symmetry.
c) Find the vertex.
d) Find the roots if they exist.
e) Draw the graph.

104. $y = x^2 - 7$
105. $y = -x^2 + 7$
106. $y = 3x^2 - 24x - 30$

6 *CHAPTER TEST*

1. Evaluate $15 + 10 \div 5 \cdot 3 - 6^2$.
2. Evaluate $-3x^2 + 6x - 4$ when $x = -4$.

Solve the equation.

3. $5x - 4 = 2(4x + 8)$
4. $2(x - 3) + 4x = 3x - 2(x + 5)$

Write an equation to represent the problem, then solve the equation.

5. Four times a number decreased by 5 is 19. Find the number.
6. A salesperson must select between two payment plans. Plan 1 is a $100-a-week salary plus a 6% commission on sales. Plan 2 is a $300-a-week salary plus 4% commission on sales. What must the weekly sales be if the two plans are to result in the same salary?
7. Evaluate $A = L(W - B)$ when $L = 9$, $W = 11$, and $B = 3$.
8. Solve $3x - 2y = 6$ for y.
9. Graph the solution set of $4x - 6 < 2x + 3$ on the real number line.
10. Find the slope of the line through the points $(6, 2)$ and $(-3, 5)$.

Graph the equation.

11. $y = 2x + 3$
12. $2x - 3y = 12$
13. Graph the inequality $2y \geqslant -2x + 4$.
14. Solve the equation $x^2 - x = 12$ by factoring.
15. Solve the equation $2x^2 + 3x - 4 = 0$ using the quadratic formula.
16. Determine whether the following graph is a function. Explain your answer.

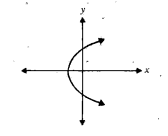

17. Evaluate $f(x) = -4x^2 + 3x - 8$ when $x = 2$.
18. For the equation $y = x^2 - 2x + 4$.
    a) Determine whether the parabola will open upward or downward
    b) Find the equation of the axis of symmetry
    c) Find the vertex
    d) Find the roots if they exist
    e) Draw the graph.

CHAPTER 7

Systems of Linear Equations and Inequalities

7.1   # SYSTEMS OF LINEAR EQUATIONS

In the previous chapter we discussed linear equations in two variables. In algebra it is often necessary to find the common solution to two or more such equations. We refer to the equations in this type of problem as a **system of linear equations** or as **simultaneous linear equations.** A solution to a system of equations is the ordered pair or ordered pairs that satisfy all equations in the system. A system of linear equations may have exactly one solution, no solution, or infinitely many solutions.

The solution to a system of linear equations may be found by a number of different techniques.

In this section we will illustrate how a system of equations may be solved by graphing. In Section 7.2 we will illustrate two algebraic methods, substitution and the addition method, for solving a system of linear equations.

■ **Example 1** ————————————————————————

Determine which of the ordered pairs is a solution to the following system of equations:

$$2x - 5y = 10$$
$$3x + y = 15$$

a) $(0, -2)$ b) $(5, 0)$ c) $(4, 3)$

Solution: For the ordered pair to be a solution to the system it must satisfy all equations in the system.

a)
$$2x - 5y = 10 \qquad\qquad 3x + y = 15$$
$$2(0) - 5(-2) = 10 \qquad\qquad 3(0) + (-2) = 15$$
$$10 = 10 \quad \text{True} \qquad\qquad -2 = 15 \quad \text{False}$$

Since $(0, -2)$ does not satisfy both equations, it is not a solution to the system.

b)
$$2x - 5y = 10 \qquad\qquad 3x + y = 15$$
$$2(5) - 5(0) = 10 \qquad\qquad 3(5) + 0 = 15$$
$$10 = 10 \quad \text{True} \qquad\qquad 15 = 15 \quad \text{True}$$

Since $(5, 0)$ satisfies both equations, it is a solution to the system.

c)
$$2x - 5y = 10 \qquad\qquad 3x + y = 15$$
$$2(4) - 5(3) = 10 \qquad\qquad 3(4) + 3 = 15$$
$$-7 = 10 \quad \text{False} \qquad\qquad 15 = 15 \quad \text{True}$$

Since $(4, 3)$ does not satisfy both equations in the system, it is not a solution to the system.   ❖

To find the solution set to a system of linear equations graphically, we graph both of the equations on the same set of axes. The point or points of intersection of the graphs are the solution to the system of equations. When two equations are graphed, three situations are possible. The two lines may intersect at one point, as in Example 2; or the two lines may be parallel and not intersect, as in Example 3; or the two equations may represent the same line, as in Example 4.

■ Example 2

Find the solution to the following system of linear equations graphically:

$$x + 2y = 4$$
$$2x - 3y = 1$$

Solution: To find the solution, we draw the graph of both equations on the same set of axes. We will start by graphing $x + 2y = 4$ (see Fig. 7.1). Three points that satisfy the equation $x + 2y = 4$ are (4, 0), (0, 2), and (2, 1). Next we graph the equation $2x - 3y = 1$. Three points that satisfy this equation are (1/2, 0), (0, -1/3), and (2, 1). The point of intersection of the two lines is (2, 1).

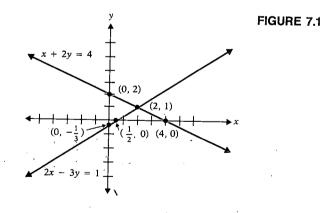

FIGURE 7.1

Thus the solution set is the ordered pair (2, 1). This is the only point that satisfies *both* equations.

CHECK

$x + 2y = 4$	$2x - 3y = 1$
$2 + 2(1) = 4$	$2(2) - 3(1) = 1$
$2 + 2 = 4$	$4 - 3 = 1$
$4 = 4,$ True	$1 = 1,$ True ❖

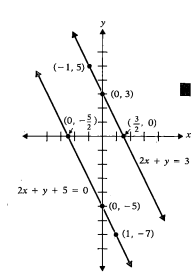

FIGURE 7.2

The system of equations in Example 2 is an example of a **consistent system of equations.** A consistent system of equations is one that has a solution.

■ **Example 3** _____

Find the solution to the following system of equations graphically:

$$2x + y = 3$$
$$2x + y + 5 = 0$$

Solution: Three ordered pairs that satisfy the equation $2x + y = 3$ are $(0, 3)$, $(3/2, 0)$ and $(-1, 5)$. Three ordered pairs that satisfy the equation $2x + y + 5 = 0$ are $(0, -5)$, $(-5/2, 0)$ and $(1, -7)$. The graphs of both equations are given in Fig. 7.2. In that diagram the two lines do not intersect; they are parallel. If this is true, the system has *no solution.*   ❖

The system of equations in Example 3 has no solution. A system of equations that has no solution is called an **inconsistent system.**

■ **Example 4** _____

Find the solution to the following system of equations graphically:

$$x - \frac{1}{2}y = 2$$
$$y = 2x - 4$$

Solution: Three ordered pairs that satisfy the equation $x - \frac{1}{2}y = 2$ are $(1, -2)$, $(2, 0)$, and $(-1, -6)$. Three ordered pairs that satisfy the equation $y = 2x - 4$ are $(0, -4)$, $(-2, -8)$, and $(3, 2)$. Graph the equations on the same set of axes (Fig. 7.3). We see that all six points are on the same line. The two equations represent the same line. Therefore every ordered pair that is a solution for one equation is also a solution for the other equation. Every point on the line satisfies both equations; thus this system has an *infinite number of solutions.*   ❖

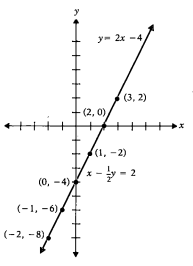

FIGURE 7.3

When a system of equations has an infinite number of solutions, as in Example 4, it is called a **dependent system.** Note that a dependent system is also a consistent system, since it has a solution.

Figure 7.4 summarizes the three possibilities for a system of linear equations.

Consistent   Inconsistent   Dependent   **FIGURE 7.4**

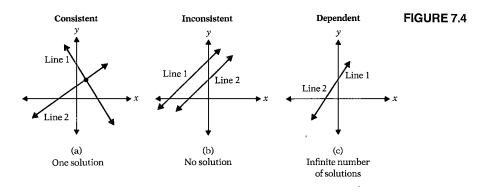

(a)	(b)	(c)
One solution	No solution	Infinite number of solutions

▪ Example 5

A college is considering purchasing one of two types of computer systems. System 1 is a minicomputer that costs \$12,000 with terminals that cost \$2000 each. System 2 is a networking system in which the networking device costs \$4000 and the terminals cost \$3000 each.

a) Write a system of equations to represent the cost of the two types of computer systems each with n terminals.

b) Graph both equations on the same set of axes and determine the number of terminals needed for both systems to have the same cost.

c) If ten terminals are to be used, which system is the least expensive?

Solution: Let n = the number of terminals used. The total cost of each computer system is the initial cost plus the cost of the terminals.

a)   minicomputer system: $c = 12,000 + 2000n$
          network system: $c = 4000 + 3000n$

b)

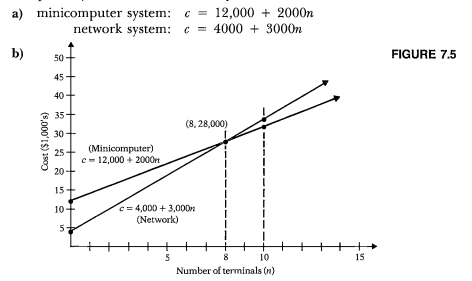

FIGURE 7.5

We have graphed the cost, c, versus the number of terminals, n, for up to15 terminals (see Fig. 7.5). From the graph we see that the lines intersect at the point (8, 28,000). Thus for eight terminals, both systems would have the same cost, $28,000.

c) From the graph we can see that for more than eight terminals the minicomputer is the least expensive system. Thus for ten terminals the minicomputer is the least expensive.   ❖

EXERCISES 7.1

Solve the following systems of equations graphically.

1. $x = 4$
    $y = 5$

2. $x = 6$
    $y = -2$

3. $x = -4$
    $y = 5$

4. $x = -4$
    $y = -5$

Solve each system of equations graphically. If the system does not have a single ordered pair as a solution, state whether the system is inconsistent or dependent.

5. $y = x + 3$
    $y = -x - 5$

6. $x + y = -2$
    $x - y = 0$

7. $2x - y = 6$
    $x + 2y = -2$

8. $x - 3y = 3$
    $3y - 4x = 6$

9. $y - x = 3$
    $x - y = 4$

10. $x + y = 1$
    $x - y = 5$

11. $y = 2x - 4$
    $2x + y = 0$

12. $2x + y = 3$
    $2y = 6 - 4x$

13. $x - 2y = 3$
    $x + 2y = -9$

14. $y = x + 3$
    $y = -1$

15. $x = 1$
    $x + y + 3 = 0$

16. $3x + 2y = 6$
    $6x + 4y = 12$

17. $3x - 2y = 6$
    $2y - 3x = 3$

18. $x = \frac{1}{2}y - 3$
    $y - 2x = 3$

19. $y = \frac{1}{3}x - 2$
    $x - 3y = 6$

20. $2(x + 2) + 2y = 0$
    $3x + 2(y - 1) = 0$

21. a) If the two lines in a system of equations have different slopes, how many solutions will the system have? Explain your answer.
b) If two lines in a system of equations have the same slope but different y-intercepts, how many solutions will the system have? Explain your answer.
c) If two lines in a system of equations have the same slope and the same y-intercept, how many solutions will the system have? Explain your answer.

After answering question 21, determine without graphing whether each system of equations has exactly one solution, no solution, or an infinite number of solutions.

22. $2x + y = 6$
    $-2x + y = 4$

23. $3x + 2y = 4$
    $4y = -6x + 6$

24. $x - y - 7 = 0$
    $4x + 4y = 5$

25. $4x - y = 5$
    $4y - x = 5$

26. $3x - y = 7$
    $y = -2x + 3$

27. $2x - 3y = 6$
    $x - \frac{3}{2}y = 3$

28. $x - 4y = 12$
    $x = 4y + 3$

29. $3x = 6y + 5$
    $y = \frac{1}{2}x - 3$

30. $3y = 6x + 4$
    $-2x + y = 4/3$

31. $x - 2y = 6$
    $-x + 2y = -6$

32. $12x - 5y = 4$
    $3x + 4y = 6$

Two lines are **perpendicular** when they meet at right angles (90° angles). Two lines are perpendicular to each other when their slopes are **negative reciprocals** of each other. The negative reciprocal of 2 is $-1/2$, the negative reciprocal of 3/5 is $-1/(3/5)$ or $-5/3$, and so on. If a represents any real number, except 0, its negative reciprocal is $-1/a$. Note that the product of a number and its negative reciprocal is -1. Determine, by finding the slope of each line, whether the lines will be perpendicular to each other when graphed.

33. $3x - 2y = 5$
    $2x - 3y = 5$

34. $2x + 3y = 8$
    $2x - 3y = 6$

35. $4x + y = 6$
    $4y = x + 12$

36. $6x - 5y = 2$
    $-10x = 2 + 12y$

37. In Example 5, if the minicomputer costs $10,000 with terminals that cost $3000, and the networking device costs $5000 with terminals that cost $4000:
a) Write the system of equations to represent the cost of the two types of computer systems.

b) Graph both equations (for up to and including ten terminals) on the same set of axes.
c) Determine the number of terminals that must be used for both systems to have the same cost.
38. The total cost of printing a book consists of a setup charge and an additional fee for material for each book printed. The ABC Printing Company charges a $1600 setup fee, plus $6 per book it prints. The XYZ Printing Company has a setup fee of $1200, plus $8 per book it prints.
a) Write a system of equations to represent the cost of printing the books with each company.
b) Graph both equations (for up to and including 300 books) on the same set of axes.
c) Determine the number of books that need to be printed for both companies to have the same cost.

d) If 100 books are to be printed, which is the less expensive printer?
39. Ms. Chan, a salesperson, has the option of selecting the method by which her salary will be determined. Option 1 is a straight 5% of her total dollar sales, and option 2 is a weekly salary of $150 plus 2% of her total dollar sales.
a) Write a system of equations to represent her salary by each option.
b) Graph both equations (for up to and including $10,000 in sales) on the same set of axes.
c) Find the total dollar sales needed for her salary from option 1 to equal her salary from option 2.
d) If she expects to average $8000 per week in sales, which option should she choose?

7.2 SOLVING SYSTEMS OF EQUATIONS BY THE SUBSTITUTION AND ADDITION METHODS

In Section 7.1 we solved systems of equations by graphing. In this section we introduce two other methods used to solve systems of linear equations: the substitution method and the addition method. We will first discuss the substitution method.

Substitution Method

To solve a system of equations by substitution, solve one of the equations for one of the variables. After you have solved one equation for one variable, make a substitution in the other equation. This procedure is illustrated in Examples 1, 2, and 3.

■ **Example 1** _____

Use the method of substitution to find the solution set of the following system of equations:

$$x + 2y = 4$$
$$2x - 3y = 1$$

Solution: Solve one of the equations for one of the variables. You may solve for either of the variables in either equation. However, if you solve for a variable with a numerical coefficient of 1, you might avoid working with fractions. For example, in the equation $x + 2y = 4$ the numerical coefficient of the x-term is 1, and the numerical coefficient of the y-term is 2. Let us solve for x in the equation $x + 2y = 4$:

$$x + 2y = 4$$
$$x + 2y - 2y = 4 - 2y$$
$$x = 4 - 2y$$

Now we substitute $4 - 2y$ for x in the *other equation* and solve for y:

$$2x - 3y = 1$$
$$2(4 - 2y) - 3y = 1$$
$$8 - 4y - 3y = 1$$
$$8 - 7y = 1$$
$$8 - 8 - 7y = 1 - 8$$
$$-7y = -7$$
$$\frac{-7y}{-7} = \frac{-7}{-7}$$
$$y = 1$$

Now substitute $y = 1$ in the equation solved for x, and solve for the variable x:

$$x = 4 - 2y$$
$$x = 4 - 2(1)$$
$$x = 2$$

Thus the solution is the ordered pair (2, 1). This answer checks with the solution obtained graphically in Section 7.1, Example 2.   ❖

■ **Example 2** _____

Solve the following system of linear equations by substitution:

$$2x + y = 3$$
$$2x + y + 5 = 0$$

Solution: Solve for y in the first equation:

$$2x + y = 3$$
$$2x - 2x + y = 3 - 2x$$
$$y = 3 - 2x$$

Now substitute $3 - 2x$ in place of y in the second equation:

$$2x + y + 5 = 0$$
$$2x + (3 - 2x) + 5 = 0$$
$$2x + 3 - 2x + 5 = 0$$
$$8 = 0 \qquad \text{False}$$

Since 8 cannot be equal to 0, there is no solution to the system of equations. Thus the system of equations is inconsistent. This answer checks with the solution obtained graphically in Section 7.1, Example 3.   ❖

■ **Example 3** _____

Solve the following system of equations by substitution:

$$x - \frac{1}{2}y = 2$$
$$y = 2x - 4$$

Solution: The second equation is already solved for y, so we will substitute $2x - 4$ for y in the first equation:

$$x - \frac{1}{2}y = 2$$
$$x - \frac{1}{2}(2x - 4) = 2$$
$$x - x + 2 = 2$$
$$2 = 2, \quad \text{true}$$

Since 2 is equal to 2, the system has infinitely many solutions. Thus the system of equations is dependent. This answer checks with the solution obtained in Section 7.1, Example 4.   ❖

If neither of the equations in a system of linear equations has a variable with a coefficient of 1, it is generally easier to solve the system by using the **addition (or elimination) method.**

To solve a system of linear equations by the addition method, we need to obtain two equations whose sum will be a single equation containing only one variable. To achieve this goal, we will rewrite the system of equations as two equations whose coefficients of one of the variables are opposites (or additive inverses) of each other. To obtain the desired equations, it might be necessary to multiply one or both equations in the original system by a number. When an equation is to be multiplied by a number, we will place brackets around the equation and place the number that is to multiply the equation before the brackets. For example, if we write $4[2x + 3y = 6]$ it means that each term in the equation $2x + 3y = 6$ is to be multiplied by 4:

$$4[2x + 3y = 6] \quad \text{gives} \quad 8x + 12y = 24$$

This notation will make our explanations much more efficient and easier for you to follow.

■ **Example 4** _____

Solve the following system of equations by the addition method:

$$x + 2y = 4$$
$$-x + y = 2$$

Solution: If we add the two given equations, the variable x will not appear in the sum, and the sum will contain only one variable, y. Add the two equations to obtain one equation in one variable. Then solve for the remaining variable:

$$\begin{array}{r} x + 2y = 4 \\ \underline{-x + y = 2} \\ 3y = 6 \\ \dfrac{3y}{3} = \dfrac{6}{3} \\ y = 2 \end{array}$$

Now substitute 2 for y in either of the original equations to find the value of x:

$$\begin{array}{r} x + 2y = 4 \\ x + 2(2) = 4 \\ x + 4 = 4 \\ x = 0 \end{array}$$

The solution to the system is $(0, 2)$.   ❖

■ **Example 5** _____

Solve the following system of equations by the addition method:

$$2x + y = 8$$
$$3x + y = 5$$

Solution: If we add the two equations, we will obtain an equation containing both an x-term and a y-term. Since we want our sum to have only one variable, we need to do something before adding. If we multiply either equation by -1 before adding, the variable y will be eliminated. We will choose to multiply the first equation by -1:

$$\begin{array}{lll} -1[2x + y = 8] & \text{gives} & -2x - y = -8 \\ \quad 3x + y = 5 & & \quad 3x + y = 5 \end{array}$$

$$\begin{array}{r} -2x - y = -8 \\ \underline{3x + y = 5} \\ x = -3 \end{array}$$

Now solve for y by substituting -3 for x in either of the original equations:

$$2x + y = 8$$
$$2(-3) + y = 8$$
$$-6 + y = 8$$
$$y = 14$$

The solution is $(-3, 14)$.   ❖

■ Example 6

Solve the following system of equations by the addition method:

$$2x + y = 6$$
$$3x + 3y = 9$$

Solution: We can multiply the top equation by -3 and then add to eliminate the variable y:

$$-3[2x + y = 6] \quad \text{gives} \quad -6x - 3y = -18$$
$$3x + 3y = 9 \qquad\qquad\qquad 3x + 3y = 9$$

$$
\begin{aligned}
-6x - 3y &= -18 \\
\underline{3x + 3y} &= \underline{9} \\
-3x &= -9 \\
x &= 3
\end{aligned}
$$

Now find y:

$$2x + y = 6$$
$$2(3) + y = 6$$
$$6 + y = 6$$
$$y = 0$$

The solution is $(3, 0)$.   ❖

Note that in Example 6 we could have eliminated the variable x by multiplying the top equation by 3 and the bottom equation by -2, then adding. Try this now.

■ Example 7

Solve the following system of equations by the addition method:

$$3x - 4y = 8$$
$$2x + 3y = 9$$

Solution: In this system we cannot eliminate a variable by multiplying only one equation by an integer value and then adding. To eliminate a variable, we will need to multiply each equation by a different number. To eliminate the variable x, we can multiply the top equation by 2 and the bottom by -3 (or the top by -2 and the bottom by 3) and then add the two equations. If we wish, we can instead eliminate the variable y by multiplying the top equation by 3 and the bottom by 4 and then adding the two equations. We will eliminate the variable x:

$$2[3x - 4y = 8] \quad \text{gives} \quad 6x - 8y = 16$$
$$-3[2x + 3y = 9] \qquad\qquad -6x - 9y = -27$$

$$
\begin{array}{r}
6x - 8y = 16 \\
-6x - 9y = -27 \\
\hline
-17y = -11 \\
y = \dfrac{11}{17}
\end{array}
$$

We can now find x by substituting $\frac{11}{17}$ for y in either of the original equations. Although it can be done, it gets pretty messy. Instead, we will solve for x by eliminating the variable y from the two original equations:

$$3[3x - 4y = 8] \quad \text{gives} \quad 9x - 12y = 24$$
$$4[2x + 3y = 9] \qquad\qquad 8x + 12y = 36$$

$$
\begin{array}{r}
9x - 12y = 24 \\
8x + 12y = 36 \\
\hline
17x = 60 \\
x = \dfrac{60}{17}
\end{array}
$$

The solution to the system is $\left(\frac{60}{17}, \frac{11}{17}\right)$.   ❖

When solving a system of linear equations by the addition method, if you obtain the equation $0 = 0$, it indicates that the system is dependent (both equations represent the same line) and there are an infinite number of solutions. When solving, if you obtain an equation such as $0 = 6$, or any other equation that is false, it means that the system is inconsistent (the two equations represent parallel lines), and there is no solution.

■ **Example 8** ─────────────────────────────

Hertz automobile rental agency charges $26 a day plus 15 cents a mile for a specific car. For the same car, Avis charges $18 a day plus 20 cents a mile. How far would you have to drive in one day for the total cost of Hertz to equal the total cost of Avis?

Solution: We are asked to find the number of miles that must be driven for each company to have the same total cost, c. First write a system of equations to represent the total cost for each of the companies in terms of the daily fee and the mileage charge.

Let x = number of miles driven

$$\text{total cost} = \text{daily fee} + \text{mileage charge}$$
$$\text{Hertz:} \quad c = 26 + 0.15x$$
$$\text{Avis:} \quad c = 18 + 0.20x$$

Since we wish to determine when the cost will be the same, we will set the two costs equal to each other (substitution method) and solve the resulting equation:

$26 + 0.15x = 18 + 0.20x$	
$26 - 18 + 0.15x = 18 - 18 + 0.20x$	Subtract 18 from both sides of the equation.
$8 + 0.15x = 0.20x$	
$8 + 0.15x - 0.15x = 0.20x - 0.15x$	Subtract $0.15x$ from both sides of the equation.
$8 = 0.05x$	
$\dfrac{8}{0.05} = \dfrac{0.05x}{0.05}$	Divide both sides of the equation by 0.05.
$160 = x$	

Thus if you travel 160 miles, the two companies' costs will be the same. If you travel less than 160 miles, Avis will be less expensive. If you travel more than 160 miles, Hertz will be less expensive.   ❖

▪ **Example 9** ———————————————————————————

A druggist needs 100 milliliters of a 10% phenobarbital solution. She has only a 5% solution and a 25% solution available. How many milliliters of each solution should she mix to obtain the desired solution?

Solution: First we will set up a system of equations. The unknown quantities are the amount of the 5% solution and the amount of the 25% solution that must be used.

Let x = number of ml of 5% solution
y = number of ml of 25% solution

We know that 100 milliliters of solution are needed. Thus

$$x + y = 100$$

The total amount of pure phenobarbital in a solution is determined by multiplying the percent of phenobarbital by the number of milliliters of solution. The second equation comes from the fact that

total amount of phenobarbital in 5% solution	+	total amount of phenobarbital in 25% solution	=	total amount of phenobarbital in mixture
$0.05x$	+	$0.25y$	=	$0.10(100)$
	or	$0.05x + 0.25y$	=	10

The system of equations is

$$x + y = 100$$
$$0.05x + 0.25y = 10$$

We will solve this system of equations using the addition method:

$$-5[x + y = 100] \qquad \text{gives} \qquad -5x - 5y = -500$$
$$100[0.05x + 0.25y = 10] \qquad\qquad 5x + 25y = 1000$$

$$
\begin{aligned}
-5x - 5y &= -500 \\
\underline{5x + 25y} &= \underline{1000} \\
20y &= 500 \\
\frac{20y}{20} &= \frac{500}{20} \\
y &= 25
\end{aligned}
$$

Now find x:

$$
\begin{aligned}
x + y &= 100 \\
x + 25 &= 100 \\
x &= 75
\end{aligned}
$$

Therefore 75 milliliters of a 5% phenobarbital solution must be mixed with 25 milliliters of a 25% phenobarbital solution to obtain 100 milliliters of a 10% phenobarbital solution. ❖

Example 9 can also be done by using substitution. Try solving Example 9 by substitution now.

EXERCISES 7.2

Solve each system of equations by the substitution method. If the system does not have a single ordered pair as a solution, state whether the system is inconsistent or dependent.

1. $y = x + 5$
   $y = x - 4$

2. $y = 2x + 3$
   $y = x - 2$

3. $x - 2y = 6$
   $2x + y = -3$

4. $3x - y = 3$
   $4y - 3x = 6$

5. $y - x = 4$
   $x - y = 3$

6. $x + y = 1$
   $y - x = 5$

7. $x = 2y - 4$
   $2y + x = 0$

8. $2y + x = 3$
   $2x = 6 - 4y$

9. $y - 2x = 3$
   $y + 2x = -9$

10. $x = y + 3$
    $x = -1$

11. $y = 1$
    $y + x + 3 = 0$

12. $x + 2y = 6$
    $y = 2x + 3$

13. $y = 4x - 3$
    $y = 2x + 8$

14. $x - 2y - 6 = 0$
    $3x - y = 5$

15. $x = 3y + 1$
    $y = 2x - 1$

16. $x + 3y = 8$
    $2x - y - 4 = 0$

17. $4x - y = 5$
    $y = 4x - 3$

18. $x + 2y = 4$
    $y = -\frac{1}{2}x + 2$

Solve each system of equations by the addition method. If the system does not have a single ordered pair as a solution, state whether the system is inconsistent or dependent.

19. $2x - y = 8$
    $2x + y = 4$

20. $3x - y = 4$
    $x + y = 4$

21. $x + 2y = 6$
    $-x + y = 0$

22. $x - 3y = 4$
    $-x + 2y = -5$

23. $2x - y = -4$
    $-3x - y = 6$

24. $x + y = 6$
    $-2x + y = -3$

25. $2x + y = 6$
    $3x + y = 5$

26. $2x + y = 11$
    $x + 3y = 18$

27. $2x - 3y = 4$
    $2x + y + 4 = 0$

28. $2x - y = 7$
    $3x + 2y = 0$

29. $5x - 2y = -4$
    $4y = -3x + 34$

30. $4x - 2y = 6$
    $y = 2x - 3$

31. $4x + y = 6$
    $-8x - 2y = 20$

32. $2x + 3y = 6$
    $5x - 4y = -8$

33. $5x + 4y = 10$
    $-3x - 5y = 7$

34. $3x + 4y = 10$
    $4x + 5y = 14$

In Exercises 35 through 41, write a system of equations that can be used to solve the problem. Then solve the system and determine the solution.

35. National Automobile Rental Agency charges $30 a day plus 15 cents per mile. Budget Automobile Rental Agency charges $36 per day plus 12 cents per mile. How many miles would Maria have to drive in one day for the total cost from Budget to equal the total cost from National?

36. A plane can travel 280 miles per hour with the wind and 240 miles per hour against the wind. Find the speed of the plane in still air and the speed of the wind.

37. Jill can make a weekly salary of $200 plus 5% commission on sales, or a weekly salary consisting of a straight 15% commission on sales. Determine the amount of sales necessary for the 15% straight commission salary to equal the $200 plus 5% commission salary.

38. Ramon wishes to mix 30 pounds of coffee to sell for a total cost of $100. To obtain the mixture, he will mix coffee that sells for $3 per pound with coffee that sells for $5 per pound. How many pounds of each type of coffee should he use?

39. Gina owns a dairy. She has milk that is 5% butterfat and skim milk without butterfat. How much of the 5% milk and how much of the skim milk should she mix to make 100 gallons of milk that is 3.5% butterfat?

40. In chemistry class, Mark has an 80% acid solution and a 50% acid solution. How much of each solution should he mix to get 100 liters of a 75% acid solution?

41. Animals in an experiment are to be kept on a strict diet. Each animal is to receive, among other things, 20 grams of protein and 6 grams of carbohydrates. The scientist has only two food mixes available of the following compositions:

	Protein (%)	Carbohydrates (%)
Mix A	10	6
Mix B	20	2

How many grams of each mix should be used to obtain the right diet for a single animal?

MATRICES

We have discussed solving a system of equations by graphing, using substitution, and using the addition method. In Section 7.4 we will discuss solving a system of linear equations by using matrices. So that you will become familiar with matrices, in this section we explain how to add, subtract, and multiply matrices. We will also explain how a matrix may be multiplied by a real number. Matrix techniques are easily adapted to computers.

A **matrix** is a rectangular array of elements. In this text we will indicate a matrix by using square brackets. Matrices (the plural of matrix) may be used to display information and to solve systems of linear equations.

The following matrix displays information about a survey of 500 registered voters.

	Democrats	Republicans	Conservatives	Liberals	Others
Men	100	93	20	35	4
Women	80	92	21	47	8

The numbers inside the brackets are called **elements** of the matrix. The matrix above contains ten elements. Because it has two rows and five columns, it is referred to as a 2 by 5 (2 × 5) matrix. A matrix that contains the

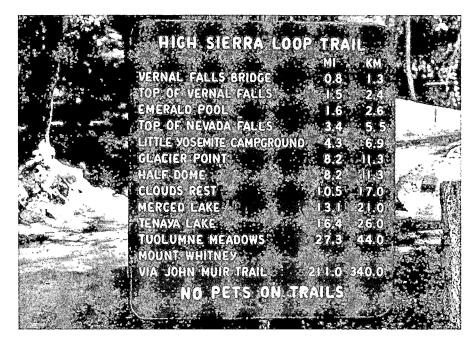

Trails in Yosemite National Park are displayed in matrix form.   *Photo by Allen Angel*

same number of rows and columns is called a **square matrix.** Following are examples of 2 × 2 and 3 × 3 square matrices.

$$\begin{bmatrix} 2 & 3 \\ 5 & 2 \end{bmatrix} \qquad \begin{bmatrix} 4 & 6 & -1 \\ 2 & 3 & 0 \\ 5 & 2 & 1 \end{bmatrix}$$

Two matrices are equal if and only if they have the same elements in the same relative positions.

Example 1 _____

Given $A = B$, find x and y:

$$A = \begin{bmatrix} 1 & 4 \\ 6 & 3 \end{bmatrix}, \qquad B = \begin{bmatrix} x & 4 \\ 6 & y \end{bmatrix}$$

Solution: Since the corresponding elements must be the same, $x = 1$ and $y = 3$. ❖

Addition of Matrices

Two matrices can be added only if they have the same dimensions (same number of rows and same number of columns). To obtain the sum of two matrices with the same dimensions, add the corresponding elements of the two matrices.

Example 2 _____

$A = \begin{bmatrix} 1 & 4 \\ -2 & 6 \end{bmatrix}, \qquad B = \begin{bmatrix} 3 & 8 \\ 6 & 0 \end{bmatrix}$. Find $A + B$.

Solution

$$A + B = \begin{bmatrix} 1 & 4 \\ -2 & 6 \end{bmatrix} + \begin{bmatrix} 3 & 8 \\ 6 & 0 \end{bmatrix}$$

$$= \begin{bmatrix} 1 + 3 & 4 + 8 \\ -2 + 6 & 6 + 0 \end{bmatrix} = \begin{bmatrix} 4 & 12 \\ 4 & 6 \end{bmatrix}$$ ❖

The following example illustrates an application of addition of matrices.

■ **Example 3**

The Ski Swap Corporation owns and operates two stores. The number of pairs of downhill skis and cross-country skis sold in each store for the months of January through June and July through December are illustrated in the matrices below:

$$
\begin{array}{c}
\text{Store } X \\
\begin{array}{cc} \text{CC} & \text{DH} \end{array} \\
\begin{array}{l} \text{Jan.–June} \\ \text{July–Dec.} \end{array}
\begin{bmatrix} 200 & 230 \\ 452 & 500 \end{bmatrix} = A
\end{array}
\qquad
\begin{array}{c}
\text{Store } Y \\
\begin{array}{cc} \text{CC} & \text{DH} \end{array} \\
\begin{bmatrix} 230 & 190 \\ 377 & 502 \end{bmatrix} = B
\end{array}
$$

Find the total number of each type of ski sold by the corporation during each time period.

Solution: To solve the problem, we add the matrices A and B:

$$
\begin{array}{l} \text{Jan.–June} \\ \text{July–Dec.} \end{array}
\begin{array}{cc} \text{CC} & \text{DH} \end{array}
\begin{bmatrix} 200 + 230 & 230 + 190 \\ 452 + 377 & 500 + 502 \end{bmatrix}
=
\begin{array}{cc} \text{CC} & \text{DH} \end{array}
\begin{bmatrix} 430 & 420 \\ 829 & 1002 \end{bmatrix}
$$

We can see from the sum matrix that during the period of January through June, 430 pairs of cross-country skis and 420 pairs of downhill skis were sold by the corporation. During the period of July through December a total of 829 pairs of cross-country skis and 1002 pairs of downhill skis were sold.   ❖

The matrix

$$
I = \begin{bmatrix} 0 & 0 \\ 0 & 0 \end{bmatrix}
$$

is the **additive identity matrix** for 2×2 matrices. We will denote this matrix with the letter I. Note that for any 2×2 matrix A, $A + I = I + A = A$.

Subtraction of Matrices

Only matrices with the same dimensions may be subtracted. To subtract matrices with the same dimensions, subtract each entry in one matrix from the corresponding entry in the other matrix.

■ **Example 4**

Find $A - B$ if

$$
A = \begin{bmatrix} 4 & 6 \\ 5 & -1 \end{bmatrix}, \qquad B = \begin{bmatrix} 3 & -4 \\ 7 & -3 \end{bmatrix}
$$

Solution

$$A - B = \begin{bmatrix} 4 & 6 \\ 5 & -1 \end{bmatrix} - \begin{bmatrix} 3 & -4 \\ 7 & -3 \end{bmatrix}$$

$$= \begin{bmatrix} 4 - 3 & 6 - (-4) \\ 5 - 7 & -1 - (-3) \end{bmatrix} = \begin{bmatrix} 1 & 10 \\ -2 & 2 \end{bmatrix} \quad \diamondsuit$$

Multiplying a Matrix by a Real Number

A matrix may be multiplied by a real number by multiplying each entry in the matrix by the real number. Sometimes when we multiply a matrix by a real number, we call that real number a **scalar.**

■ **Example 5** _____

For matrices A and B, find (a) $3A$ and (b) $3A - 2B$.

$$A = \begin{bmatrix} 4 & 6 \\ -3 & 5 \end{bmatrix}, \qquad B = \begin{bmatrix} -1 & 5 \\ 2 & 6 \end{bmatrix}$$

Solution

a) $3A = 3\begin{bmatrix} 4 & 6 \\ -3 & 5 \end{bmatrix} = \begin{bmatrix} 3(4) & 3(6) \\ 3(-3) & 3(5) \end{bmatrix} = \begin{bmatrix} 12 & 18 \\ -9 & 15 \end{bmatrix}$

b) We found $3A$ in part (a). Now we will find $2B$:

$$2B = 2\begin{bmatrix} -1 & 5 \\ 2 & 6 \end{bmatrix} = \begin{bmatrix} 2(-1) & 2(5) \\ 2(2) & 2(6) \end{bmatrix} = \begin{bmatrix} -2 & 10 \\ 4 & 12 \end{bmatrix}$$

$$3A - 2B = \begin{bmatrix} 12 & 18 \\ -9 & 15 \end{bmatrix} - \begin{bmatrix} -2 & 10 \\ 4 & 12 \end{bmatrix}$$

$$= \begin{bmatrix} 12 - (-2) & 18 - 10 \\ -9 - 4 & 15 - 12 \end{bmatrix} = \begin{bmatrix} 14 & 8 \\ -13 & 3 \end{bmatrix} \quad \diamondsuit$$

Multiplication of Matrices

Multiplication of matrices is slightly more difficult than addition of matrices. Multiplication of matrices is possible only when the number of *columns* of the first matrix, A, is the same as the number of *rows* of the second matrix, B. We will use the notation

$$A$$
$$3 \times 4$$

to indicate that matrix A has three rows and four columns. Suppose matrix A is a 3×4 matrix and matrix B is a 4×5 matrix, then

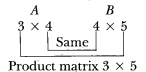

$$\begin{array}{cc} A & B \\ 3 \times 4 & 4 \times 5 \end{array}$$

Same

Product matrix 3×5

This indicates that matrix A has four columns and matrix B has four rows. Therefore we can multiply these two matrices. The product matrix will have the same number of rows as matrix A and the same number of columns as matrix B. Thus the dimensions of the product matrix are 3×5.

Example 6

Determine which of the following pairs of matrices can be multiplied:

a) $A = \begin{bmatrix} 3 & 2 \\ 5 & 7 \end{bmatrix}$, $B = \begin{bmatrix} 0 & 6 \\ 4 & 1 \end{bmatrix}$ b) $A = \begin{bmatrix} 2 & 3 \\ 5 & 6 \end{bmatrix}$, $B = \begin{bmatrix} 2 & 4 & -1 \\ 6 & 8 & 0 \end{bmatrix}$

c) $A = \begin{bmatrix} 2 & 1 & 4 \\ 3 & 2 & 8 \end{bmatrix}$, $B = \begin{bmatrix} 2 & 1 & 3 \\ 1 & 0 & -2 \end{bmatrix}$

Solution

a)
$$\begin{array}{cc} A & B \\ 2 \times 2 & 2 \times 2 \end{array}$$

Same

Since matrix A has two columns and matrix B has two rows, the two matrices can be multiplied. The product is a 2×2 matrix.

b)
$$\begin{array}{cc} A & B \\ 2 \times 2 & 2 \times 3 \end{array}$$

Same

Since matrix A has two columns and matrix B has two rows, the two matrices can be multiplied. The product is a 2×3 matrix.

c)
$$\begin{array}{cc} A & B \\ 2 \times 3 & 2 \times 3 \end{array}$$

Same

Since matrix A has three columns and matrix B has two rows, the two matrices cannot be multiplied.

To explain matrix multiplication, we will use matrices A and B given below:

$$A = \begin{bmatrix} 3 & 2 \\ 5 & 7 \end{bmatrix} \qquad B = \begin{bmatrix} 0 & 6 \\ 4 & 1 \end{bmatrix}$$

Since A contains two rows and B contains two columns, the product matrix will contain two rows and two columns. To multiply two matrices, we use a row-column scheme of multiplying. The numbers in the first row of matrix A are multiplied by the numbers in the first column of matrix B (see Fig. 7.6).

$$A \times B = \begin{bmatrix} 3 & 2 \\ 5 & 7 \end{bmatrix} \begin{bmatrix} 0 & 6 \\ 4 & 1 \end{bmatrix}$$

First row   First column   **FIGURE 7.6**

$$\begin{bmatrix} 3 & 2 \\ 5 & 7 \end{bmatrix} \qquad \begin{bmatrix} 0 & 6 \\ 4 & 1 \end{bmatrix}$$

$(3 \times 0) + (2 \times 4) = 0 + 8$
$= 8$

The 8 is placed in the first-row, first-column position of the product matrix. The other numbers in the product matrix are obtained in a similar way, as illustrated in Fig. 7.7.

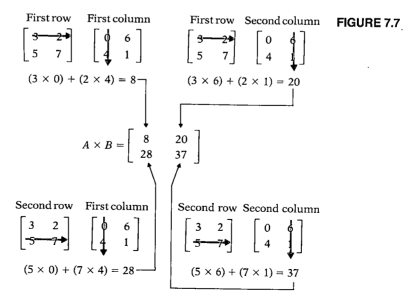

First row   First column   First row   Second column   **FIGURE 7.7**

$$\begin{bmatrix} 3 & 2 \\ 5 & 7 \end{bmatrix} \qquad \begin{bmatrix} 0 & 6 \\ 4 & 1 \end{bmatrix} \qquad\qquad \begin{bmatrix} 3 & 2 \\ 5 & 7 \end{bmatrix} \qquad \begin{bmatrix} 0 & 6 \\ 4 & 1 \end{bmatrix}$$

$(3 \times 0) + (2 \times 4) = 8$ $(3 \times 6) + (2 \times 1) = 20$

$$A \times B = \begin{bmatrix} 8 & 20 \\ 28 & 37 \end{bmatrix}$$

Second row   First column   Second row   Second column

$$\begin{bmatrix} 3 & 2 \\ 5 & 7 \end{bmatrix} \qquad \begin{bmatrix} 0 & 6 \\ 4 & 1 \end{bmatrix} \qquad\qquad \begin{bmatrix} 3 & 2 \\ 5 & 7 \end{bmatrix} \qquad \begin{bmatrix} 0 & 6 \\ 4 & 1 \end{bmatrix}$$

$(5 \times 0) + (7 \times 4) = 28$ $(5 \times 6) + (7 \times 1) = 37$

We can shorten the procedure as follows:

$$A \times B = \begin{bmatrix} 3 & 2 \\ 5 & 7 \end{bmatrix} \begin{bmatrix} 0 & 6 \\ 4 & 1 \end{bmatrix}$$

$$= \begin{bmatrix} 3(0) + 2(4) & 3(6) + 2(1) \\ 5(0) + 7(4) & 5(6) + 7(1) \end{bmatrix}$$

$$= \begin{bmatrix} 8 & 20 \\ 28 & 37 \end{bmatrix}$$

In general, if

$$A = \begin{bmatrix} a & b \\ c & d \end{bmatrix} \quad \text{and} \quad B = \begin{bmatrix} e & f \\ g & h \end{bmatrix},$$

then

$$A \times B = \begin{bmatrix} a & b \\ c & d \end{bmatrix} \begin{bmatrix} e & f \\ g & h \end{bmatrix} = \begin{bmatrix} ae + bg & af + bh \\ ce + dg & cf + dh \end{bmatrix}$$

Let's do one more multiplication.

■ **Example 7** ⎯⎯⎯⎯⎯⎯⎯⎯⎯⎯⎯⎯⎯⎯⎯⎯⎯⎯⎯⎯⎯⎯⎯⎯

Find $A \times B$, given

$$A = \begin{bmatrix} 2 & 3 \\ 5 & 6 \end{bmatrix} \quad \text{and} \quad B = \begin{bmatrix} 2 & 4 & -1 \\ 6 & 8 & 0 \end{bmatrix}.$$

Solution: Matrix A contains two columns and matrix B contains two rows. Thus the matrices can be multiplied. Since A contains two rows and B contains three columns, the product matrix will contain two rows and three columns.

$$A \times B = \begin{bmatrix} 2 & 3 \\ 5 & 6 \end{bmatrix} \begin{bmatrix} 2 & 4 & -1 \\ 6 & 8 & 0 \end{bmatrix}$$

$$= \begin{bmatrix} 2(2) + 3(6) & 2(4) + 3(8) & 2(-1) + 3(0) \\ 5(2) + 6(6) & 5(4) + 6(8) & 5(-1) + 6(0) \end{bmatrix}$$

$$= \begin{bmatrix} 22 & 32 & -2 \\ 46 & 68 & -5 \end{bmatrix} \qquad \qquad ❖$$

We previously discussed the additive identity matrix. **Square matrices** also have a **multiplicative identity matrix.** The multiplicative identity matrices for a 2 × 2 and a 3 × 3 matrix, denoted by I, follow. Note that in any multiplicative identity matrix 1s go diagonally from top left to bottom right, and all other elements in the matrix are 0s.

$$I = \begin{bmatrix} 1 & 0 \\ 0 & 1 \end{bmatrix}, \qquad I = \begin{bmatrix} 1 & 0 & 0 \\ 0 & 1 & 0 \\ 0 & 0 & 1 \end{bmatrix}$$

Note that for any square matrix, A, $A \times I = I \times A = A$.

■ **Example 8** ─────────────────────────────────────

Using the multiplicative identity matrix for a 2 × 2 matrix and matrix A, show that $A \times I = A$.

$$A = \begin{bmatrix} 4 & 3 \\ 2 & 1 \end{bmatrix}$$

Solution: The identity matrix is

$$I = \begin{bmatrix} 1 & 0 \\ 0 & 1 \end{bmatrix}$$

$$A \times I = \begin{bmatrix} 4 & 3 \\ 2 & 1 \end{bmatrix} \begin{bmatrix} 1 & 0 \\ 0 & 1 \end{bmatrix}$$

$$= \begin{bmatrix} 4(1) + 3(0) & 4(0) + 3(1) \\ 2(1) + 1(0) & 2(0) + 1(1) \end{bmatrix}$$

$$= \begin{bmatrix} 4 & 3 \\ 2 & 1 \end{bmatrix} = A$$

❖

The following example illustrates an application of multiplication of matrices.

■ **Example 9** ─────────────────────────────────────

A manufacturer of bathing suits produces three types of bathing suits. On a given day it produces 20 of type X, 30 of type Y, and 50 of type Z. Each suit of type X requires four units of material and one hour of work to produce; each of type Y requires three units of material and two hours of work to produce; each of type Z requires five units of material and three hours to produce. Use multiplication of matrices to determine the total number of units of material and the total number of hours needed for a day's production.

Solution: We will let matrix A represent the number of each type of bathing suit produced:

$$\begin{array}{cccc} \text{Type} & X & Y & Z \\ A = & [20 & 30 & 50]. \end{array}$$

The units of material and time requirements for each type are indicated in matrix B:

$$\begin{array}{ccc} & \text{Material} & \text{Hours} \\ B = & \begin{bmatrix} 4 & 1 \\ 3 & 2 \\ 5 & 3 \end{bmatrix} & \begin{array}{c} \text{Type } X \\ \text{Type } Y \\ \text{Type } Z \end{array} \end{array}$$

The product of A and B, $A \times B$, will give the total number of units of material and the total number of hours of work needed for the day's production.

$$A \times B = [20 \quad 30 \quad 50] \begin{bmatrix} 4 & 1 \\ 3 & 2 \\ 5 & 3 \end{bmatrix}$$

$$= [20(4) + 30(3) + 50(5) \quad 20(1) + 30(2) + 50(3)]$$

$$= [420 \quad 230]$$

Thus a total of 420 units of material and a total of 230 hours are required per day. ❖

DID YOU KNOW?

Matrices might seem abstract, but they play an important role in our everyday life. One of the major problems in applying mathematics to real-life applications lies in developing a procedure of systemizing and working with the great number of variables and large amounts of data that occur in real-life situations. Matrices lend themselves nicely to such situations because they can be used to organize and analyze large amounts of data with a large number of variables. Matrices can also be evaluated by computers, and many tedious calculations can thus be avoided.

At present, some areas of study that use matrices are physics, engineering, economics, business, social science, weather forecasting, game theory, electronics, and many branches of mathematics. In fact, matrix theory is so important that video games, computers, and many other electronic aspects of our civilization would be impossible without it.

Three mathematicians played important roles in the development of matrix theory. They are James Sylvester (1814–1897), William Rowan Hamilton (1805–1865), and Arthur Cayley (1821–1895). Sylvester and Cayley were good friends, and although they did not work together, they cooperated in developing matrix theory. Sylvester was the first to use the term "matrix." Hamilton, a noted physicist, astronomer, and mathematician, also used what was essentially the algebra of matrices under the name of "linear and vector functions." The mathematical concept of "vector space" grew out of Hamilton's work on the algebra of vectors.

EXERCISES 7.3

1. Bring in an article that shows information illustrated in matrix form.
2. Records are kept each day for a month at a local cinema that houses three movie theatres A, B, and C. The daily average Monday through Sunday receipts for the three theatres are as follows. A: \$654, \$785, \$458, \$345, \$1478, \$2109, \$543; B: \$764, \$778, \$568, \$451, \$1024, \$1689, \$853; C: \$567, \$764, \$873, \$407, \$2034, \$2432, \$567. Express this information in the form of a 3 × 7 matrix.

Find $A + B$.

3. $A = \begin{bmatrix} 2 & 3 \\ 6 & 2 \end{bmatrix}$ $B = \begin{bmatrix} -4 & -2 \\ 0 & 3 \end{bmatrix}$

4. $A = \begin{bmatrix} 3 & 0 & 2 \\ 0 & 2 & 2 \end{bmatrix}$ $B = \begin{bmatrix} 5 & 7 & 3 \\ 3 & -2 & 4 \end{bmatrix}$

5. $A = \begin{bmatrix} 4 & 1 \\ 3 & 4 \\ 2 & 7 \end{bmatrix}$ $B = \begin{bmatrix} -2 & 3 \\ 0 & 5 \\ 2 & 1 \end{bmatrix}$

6. $A = \begin{bmatrix} 3 & 0 & 2 \\ 5 & 1 & -3 \\ -1 & 4 & -6 \end{bmatrix}$ $B = \begin{bmatrix} -4 & 2 & -1 \\ 5 & -3 & 6 \\ -2 & 0 & 6 \end{bmatrix}$

Find $A - B$.

7. $A = \begin{bmatrix} 7 & 9 \\ 3 & -1 \end{bmatrix}$ $B = \begin{bmatrix} 4 & -2 \\ -3 & 5 \end{bmatrix}$

8. $A = \begin{bmatrix} 5 & -3 & 6 \\ 2 & 4 & 2 \end{bmatrix}$ $B = \begin{bmatrix} 0 & -4 & -5 \\ 1 & 3 & 8 \end{bmatrix}$

9. $A = \begin{bmatrix} 5 & 3 & -1 \\ 7 & 4 & 2 \\ 6 & -1 & -5 \end{bmatrix}$ $B = \begin{bmatrix} 4 & 3 & 6 \\ -2 & -4 & 9 \\ 0 & -2 & 4 \end{bmatrix}$

10. $A = \begin{bmatrix} -4 & 3 \\ 6 & 2 \\ 1 & -5 \end{bmatrix}$ $B = \begin{bmatrix} -6 & -8 \\ -10 & -11 \\ 3 & -7 \end{bmatrix}$

In Exercises 11 through 16, let

$A = \begin{bmatrix} 1 & 2 \\ 0 & 5 \end{bmatrix}$ $B = \begin{bmatrix} 4 & 7 \\ 0 & 2 \end{bmatrix}$ $C = \begin{bmatrix} -2 & 3 \\ 4 & 0 \end{bmatrix}$

Find the following.

11. $2B$
12. $-3B$
13. $2B + 3C$
14. $2B + 3B$
15. $3B - 2C$
16. $4C - 2A$

Find $A \times B$.

17. $A = \begin{bmatrix} 2 & 3 \\ 6 & 2 \end{bmatrix}$ $B = \begin{bmatrix} 4 & 2 \\ 0 & 3 \end{bmatrix}$

18. $A = \begin{bmatrix} -1 & 1 \\ 0 & 3 \end{bmatrix}$ $B = \begin{bmatrix} 4 & 1 \\ 0 & -1 \end{bmatrix}$

19. $A = \begin{bmatrix} 2 & 3 & -1 \\ 0 & 4 & 6 \end{bmatrix}$ $B = \begin{bmatrix} 2 \\ 4 \\ 1 \end{bmatrix}$

20. $A = \begin{bmatrix} 2 & 1 \\ 1 & 1 \end{bmatrix}$ $B = \begin{bmatrix} 1 & -1 \\ -1 & 2 \end{bmatrix}$

21. $A = \begin{bmatrix} 2 & 5 \\ 1 & 3 \end{bmatrix}$ $B = \begin{bmatrix} 3 & -5 \\ -1 & 2 \end{bmatrix}$

22. $A = \begin{bmatrix} 2 & 3 & 1 \\ -2 & -1 & 0 \\ 4 & 5 & 6 \end{bmatrix}$ $B = \begin{bmatrix} 1 & 0 & 0 \\ 0 & 1 & 0 \\ 0 & 0 & 1 \end{bmatrix}$

If possible, find $A + B$ and $A \times B$. If an operation cannot be performed, explain why.

23. $A = \begin{bmatrix} 0 & 3 \\ 2 & 4 \end{bmatrix}$ $B = \begin{bmatrix} 1 & 2 & 3 \\ 2 & -2 & 3 \end{bmatrix}$

24. $A = \begin{bmatrix} 2 & 3 & 4 \\ 4 & 6 & -2 \end{bmatrix}$ $B = \begin{bmatrix} 2 & -3 & 4 \\ 0 & 4 & 3 \end{bmatrix}$

25. $A = \begin{bmatrix} 4 & 5 & 3 \\ 6 & 2 & 1 \end{bmatrix}$ $B = \begin{bmatrix} 3 & 2 \\ 4 & 6 \\ -2 & 0 \end{bmatrix}$

26. $A = \begin{bmatrix} 1 & 2 \\ 3 & 4 \\ 5 & 6 \end{bmatrix}$ $B = \begin{bmatrix} 1 & 2 \\ 3 & 4 \\ 5 & 6 \end{bmatrix}$

27. $A = \begin{bmatrix} 1 & 2 \\ 3 & 4 \end{bmatrix}$ $B = \begin{bmatrix} 5 \\ 6 \end{bmatrix}$

28. $A = \begin{bmatrix} 5 \\ 6 \end{bmatrix}$ $B = \begin{bmatrix} 1 & 2 \\ 3 & 4 \end{bmatrix}$

Show that the *commutative property of addition,* $A + B = B + A$, holds for matrices A and B.

29. $A = \begin{bmatrix} 2 & 3 \\ -1 & 4 \end{bmatrix}$ $B = \begin{bmatrix} 8 & 5 \\ 6 & 1 \end{bmatrix}$

30. $A = \begin{bmatrix} -4 & 3 \\ 5 & 7 \end{bmatrix}$ $B = \begin{bmatrix} -5 & -8 \\ 0 & -7 \end{bmatrix}$

31. $A = \begin{bmatrix} 0 & -1 \\ 3 & -4 \end{bmatrix}$ $B = \begin{bmatrix} 8 & 1 \\ 3 & -4 \end{bmatrix}$

32. Make up two matrices with the same dimensions, A and B, and show that $A + B = B + A$.

Show that the *associative property of addition*, $(A + B) + C = A + (B + C)$ holds for the matrices given.

33. $A = \begin{bmatrix} 4 & -3 \\ -1 & 5 \end{bmatrix}$ $B = \begin{bmatrix} 5 & 7 \\ 3 & 4 \end{bmatrix}$,

$C = \begin{bmatrix} 0 & -4 \\ 5 & 1 \end{bmatrix}$

34. $A = \begin{bmatrix} -9 & -8 \\ -7 & -6 \end{bmatrix}$ $B = \begin{bmatrix} -5 & -4 \\ -3 & -2 \end{bmatrix}$,

$C = \begin{bmatrix} -1 & 0 \\ 1 & 2 \end{bmatrix}$

35. $A = \begin{bmatrix} 7 & 4 \\ 9 & -36 \end{bmatrix}$ $B = \begin{bmatrix} 5 & 6 \\ -1 & -4 \end{bmatrix}$,

$C = \begin{bmatrix} -7 & -5 \\ -1 & 3 \end{bmatrix}$

36. Make up three matrices with the same dimensions, A, B, and C, and show that $(A + B) + C = A + (B + C)$.

Determine whether the *commutative property of multiplication*, $A \times B = B \times A$, holds for the matrices given.

37. $A = \begin{bmatrix} 1 & 2 \\ 3 & -1 \end{bmatrix}$ $B = \begin{bmatrix} 2 & 0 \\ 1 & 3 \end{bmatrix}$

38. $A = \begin{bmatrix} 1 & -1 \\ 3 & 5 \end{bmatrix}$ $B = \begin{bmatrix} 4 & 2 \\ -1 & 3 \end{bmatrix}$

39. $A = \begin{bmatrix} 4 & 2 \\ 1 & -3 \end{bmatrix}$ $B = \begin{bmatrix} -4 & -2 \\ -1 & 3 \end{bmatrix}$

40. $A = \begin{bmatrix} -3 & 2 \\ 6 & -5 \end{bmatrix}$ $B = \begin{bmatrix} -\frac{5}{3} & -\frac{2}{3} \\ -2 & -1 \end{bmatrix}$

41. $A = \begin{bmatrix} 3 & 2 & 1 \\ 4 & 2 & 0 \\ 0 & -2 & 5 \end{bmatrix}$ $B = \begin{bmatrix} 1 & 0 & 0 \\ 0 & 1 & 0 \\ 0 & 0 & 1 \end{bmatrix}$

42. Make up two matrices with the same dimensions, A and B, and determine whether $A \times B = B \times A$.

Show that the *associative property of multiplication*, $(A \times B) \times C = A \times (B \times C)$, holds for the matrices given.

43. $A = \begin{bmatrix} 1 & 2 \\ 4 & 0 \end{bmatrix}$ $B = \begin{bmatrix} 2 & 3 \\ 1 & 0 \end{bmatrix}$ $C = \begin{bmatrix} 4 & 2 \\ 3 & 1 \end{bmatrix}$

44. $A = \begin{bmatrix} -2 & 3 \\ 0 & 4 \end{bmatrix}$ $B = \begin{bmatrix} 4 & 5 \\ 7 & 2 \end{bmatrix}$ $C = \begin{bmatrix} 3 & 4 \\ -2 & 5 \end{bmatrix}$

45. $A = \begin{bmatrix} 4 & 3 \\ -6 & 2 \end{bmatrix}$ $B = \begin{bmatrix} 1 & 2 \\ 0 & 1 \end{bmatrix}$ $C = \begin{bmatrix} 4 & 3 \\ 0 & -2 \end{bmatrix}$

46. $A = \begin{bmatrix} -1 & -2 \\ -3 & -4 \end{bmatrix}$ $B = \begin{bmatrix} 1 & 0 \\ 0 & 1 \end{bmatrix}$ $C = \begin{bmatrix} 0 & 0 \\ 0 & 0 \end{bmatrix}$

47. $A = \begin{bmatrix} 3 & 4 \\ -1 & -2 \end{bmatrix}$ $B = \begin{bmatrix} 0 & 1 \\ 1 & 0 \end{bmatrix}$ $C = \begin{bmatrix} 2 & 0 \\ 3 & 0 \end{bmatrix}$

48. Make up three matrices with the same dimensions, A, B, and C, and show that $(A \times B) \times C = A \times (B \times C)$.

49. The Original Cookie Factory bakes and sells four types of cookies: chocolate chip, sugar, molasses, and peanut butter. Matrix A shows the number of units of various ingredients used in baking a dozen of each type of cookie:

	Sugar	Flour	Milk	Eggs	
$A =$	2	2	1/2	1	Chocolate chip
	3	2	1	2	Sugar
	0	1	0	3	Molasses
	1/2	1	0	0	Peanut butter

The cost, in cents per cup or per egg, for each ingredient when purchased in large quantities and in small quantities is given in matrix B:

	Large quantities	Small quantities	
$B =$	10	12	sugar
	5	8	flour
	8	8	milk
	4	6	eggs

Use matrix multiplication to find a matrix representing the comparative cost per item for large and small quantities purchased.

In Exercise 49, suppose that a typical day's order consists of 30 dozen chocolate chips cookies, 20 dozen sugar cookies, 8 dozen molasses cookies, and 15 dozen peanut butter cookies.

50. Express these orders as a 1×4 matrix, and use matrix multiplication to determine the amount of each ingredient needed to fill the day's order.

51. Use matrix multiplication to determine the cost under the two purchase options to fill the day's order.

Two matrices whose sum is the additive identity matrix are said to be **additive inverses.** That is, if $A + B = B + A = I$, where I is the additive identity matrix, then A and B are additive inverses. Determine whether A and B are additive inverses.

52. $A = \begin{bmatrix} 6 & 3 \\ 4 & -2 \end{bmatrix}$ $B = \begin{bmatrix} -6 & -3 \\ -2 & 4 \end{bmatrix}$

53. $A = \begin{bmatrix} 4 & 6 & 3 \\ 2 & 3 & -1 \\ -1 & 0 & 6 \end{bmatrix}$ $B = \begin{bmatrix} -4 & -6 & -3 \\ -2 & -3 & 1 \\ 1 & 0 & -6 \end{bmatrix}$

Two matrices whose product is the multiplicative identity matrix are said to be **multiplicative inverses.** That is, if $A \times B = B \times A = I$, where I is the multiplicative identity matrix, then A and B are multiplicative inverses. Determine whether A and B are multiplicative inverses.

54. $A = \begin{bmatrix} 5 & -2 \\ -2 & 1 \end{bmatrix}$ $B = \begin{bmatrix} 1 & 2 \\ 2 & 5 \end{bmatrix}$

55. $A = \begin{bmatrix} 7 & 3 \\ 2 & 1 \end{bmatrix}$ $B = \begin{bmatrix} 1 & -3 \\ -2 & 7 \end{bmatrix}$

PROBLEM SOLVING

Determine whether the following statements are true or false. Give an example to support your answer.

56. $A - B = B - A$

57. For scalar a and matrices B and C, $a(B + C) = aB + aC$.

RESEARCH ACTIVITIES

58. Read Chapter 9, then determine whether the set of all 2×2 matrices under the operation of addition form a commutative group.

7.4 # SOLVING SYSTEMS OF EQUATIONS USING MATRICES

In Section 7.3 we introduced matrices. Now we will discuss the procedure to solve a system of linear equations using matrices.

We will illustrate how to solve a system of two equations and two unknowns. Systems of equations containing three equations and three unknowns (called a third-order system) and higher-order systems can also be solved by using matrices.

To solve a system of equations using matrices, we first write the **augmented matrix.** An augmented matrix is one made up of two smaller matrices—one for the coefficients of the variables in the equations and one for the constants in the equation. To determine the augmented matrix, first write each equation in standard form, $ax + by = c$. For the system of equations

$$a_1x + b_1y = c_1$$
$$a_2x + b_2y = c_2$$

the augmented matrix is written

$$\left[\begin{array}{cc|c} a_1 & b_1 & c_1 \\ a_2 & b_2 & c_2 \end{array}\right]$$

For the system of equations

$$x + 2y = 8$$
$$3x - y = 7$$

the augmented matrix is

$$\left[\begin{array}{cc|c} 1 & 2 & 8 \\ 3 & -1 & 7 \end{array}\right]$$

Note that the bar in the augmented matrix separates the numerical coefficients from the constants. Since the matrix is just a shortened way of writing the system of equations, we can solve the system of equations using matrices in a manner very similar to solving a system of equations using the addition method.

To solve a system of equations using matrices, we use row transformations to obtain new matrices that have the same solution as the original system. We will discuss three row transformation procedures.

Row Transformations

1. Any two rows of a matrix may be interchanged (this is the same as interchanging any two equations in the system of equations).

2. All the numbers in any row may be multiplied by any nonzero real number. (This is the same as multiplying both sides of an equation by any nonzero real number.)

3. All the numbers in any row may be multiplied by any nonzero real number, and these products may be added to the corresponding numbers in any different row of numbers.

To solve a system of equations using matrices, we use row transformations to obtain an augmented matrix whose numbers to the left of the vertical bar are the same as in the *multiplicative identity matrix*. From this type of augmented matrix we can determine the solution to the system of equations. For example, if we get

$$\begin{bmatrix} 1 & 0 & \bigm| & 3 \\ 0 & 1 & \bigm| & -2 \end{bmatrix}$$

it tells us that $1x = 3$ or $x = 3$, and $1y = -2$ or $y = -2$. Thus the solution to the system of equations that yielded this augmented matrix is $(3, -2)$.

Now let's work an example.

■ **Example 1** _____

Solve the following system of equations using matrices:

$$x + 2y = 5$$
$$3x - y = 8$$

Solution: First write the augmented matrix:

$$\begin{bmatrix} 1 & 2 & \bigm| & 5 \\ 3 & -1 & \bigm| & 8 \end{bmatrix}$$

Our goal is to obtain a matrix of the form

$$\begin{bmatrix} 1 & 0 & \bigm| & c_1 \\ 0 & 1 & \bigm| & c_2 \end{bmatrix}$$

where c_1 and c_2 may represent any real numbers. It is generally easier to work by columns. Therefore we will try to get the first column of the augmented matrix to be $\frac{1}{0}$ and the second column to be $\frac{0}{1}$. Since the element in the top left position is already a 1, we must work to change the 3 into a 0. We will use row transformation procedure 3 to change the 3 into a 0. If we multiply the top row of numbers by -3 and add these products to the second row of numbers, the element in the second row, first column will become a 0:

$$\left[\begin{array}{rr|r} 1 & 2 & 5 \\ 3 & -1 & 8 \end{array}\right]$$

The top row of numbers multiplied by -3 gives

$$1(-3) \qquad 2(-3) \qquad 5(-3)$$

Now add these products to their respective numbers in row 2.

$$\left[\begin{array}{cc|c} 1 & 2 & 5 \\ 3 + 1(-3) & -1 + 2(-3) & 8 + 5(-3) \end{array}\right] = \left[\begin{array}{rr|r} 1 & 2 & 5 \\ 0 & -7 & -7 \end{array}\right]$$

Our next step is to obtain a 1 in the second row, second column. At present, -7 is in this position. To change the -7 to a 1, we will use row transformation procedure 2. If we multiply -7 by $-\frac{1}{7}$, the product will be 1. Therefore we will multiply all the numbers in row 2 by $-\frac{1}{7}$. This gives

$$\left[\begin{array}{cc|c} 1 & 2 & 5 \\ 0(-\frac{1}{7}) & -7(-\frac{1}{7}) & -7(-\frac{1}{7}) \end{array}\right] = \left[\begin{array}{cc|c} 1 & 2 & 5 \\ 0 & 1 & 1 \end{array}\right]$$

Our next step is to obtain a 0 in the first row, second column. At present a 2 is in this position. If we multiply row two by -2 and add the numbers to row one, we get a 0 in the desired position:

$$\left[\begin{array}{cc|c} 1 + 0(-2) & 2 + 1(-2) & 5 + 1(-2) \\ 0 & 1 & 1 \end{array}\right] = \left[\begin{array}{cc|c} 1 & 0 & 3 \\ 0 & 1 & 1 \end{array}\right]$$

We now have the desired augmented matrix:

$$\left[\begin{array}{cc|c} 1 & 0 & 3 \\ 0 & 1 & 1 \end{array}\right]$$

With this matrix we see that $1x = 3$ or $x = 3$ and $1y = 1$ or $y = 1$. The solution to the system is (3, 1).

CHECK:

$$
\begin{array}{ll}
x + 2y = 5 & 3x - y = 8 \\
3 + 2(1) = 5 & 3(3) - 1 = 8 \\
5 = 5 \quad \text{True} & 8 = 8 \quad \text{True}
\end{array}
$$

■ **Example 2**

Solve the following system of equations using matrices:

$$2x - 3y = 10$$
$$2x + 2y = 5$$

Solution: First write the augmented matrix:

$$\left[\begin{array}{cc|c} 2 & -3 & 10 \\ 2 & 2 & 5 \end{array}\right]$$

To obtain a 1 in the top row, first column, multiply the top row by $\frac{1}{2}$.

$$\left[\begin{array}{cc|c} 1 & -\frac{3}{2} & 5 \\ 2 & 2 & 5 \end{array}\right]$$

To get a 0 in the second row, first column, multiply the top row by -2, and add the products to the second row:

$$\left[\begin{array}{cc|c} 1 & -\frac{3}{2} & 5 \\ 2 + 1(-2) & 2 + \left(-\frac{3}{2}\right)(-2) & 5 + 5(-2) \end{array}\right] = \left[\begin{array}{cc|c} 1 & -\frac{3}{2} & 5 \\ 0 & 5 & -5 \end{array}\right]$$

To obtain a 1 in the second row, second column, multiply the second row by $\frac{1}{5}$:

$$\left[\begin{array}{cc|c} 1 & -\frac{3}{2} & 5 \\ 0\left(\frac{1}{5}\right) & 5\left(\frac{1}{5}\right) & -5\left(\frac{1}{5}\right) \end{array}\right] = \left[\begin{array}{cc|c} 1 & -\frac{3}{2} & 5 \\ 0 & 1 & -1 \end{array}\right]$$

To obtain a 0 in the first row, second column, multiply the second row by $\frac{3}{2}$, and add the products to the first row:

$$\left[\begin{array}{cc|c} 1 + \frac{3}{2}(0) & -\frac{3}{2} + \frac{3}{2}(1) & 5 + (-1)\left(\frac{3}{2}\right) \\ 0 & 1 & -1 \end{array}\right] = \left[\begin{array}{cc|c} 1 & 0 & \frac{7}{2} \\ 0 & 1 & -1 \end{array}\right]$$

The solution to the system of equations is $\left(\frac{7}{2}, -1\right)$.   ❖

EXERCISES 7.4

Use matrices to solve each system of equations.

1. $x - y = -5$
   $-2x + y = 4$

2. $x - y = -2$
   $x + y = 2$

3. $x + 2y = -4$
   $2x - y = -3$

4. $-x + y = 3$
   $x + y = -5$

5. $x + y = 6$
   $2x - y = 3$

6. $x + 2y = 8$
   $2x - 3y = 2$

7. $x - 3y = 1$
   $2x + y = -5$

8. $2x + 4y = 6$
   $4x - 2y = -8$

9. $2x + 4y = 6$
$3x - y = 2$

10. $3x + 2y = 4$
$5x + 5y = 9$

11. $5x + y = 3$
$-6x - y = -5$

12. $2x - 5y = 10$
$3x + y = 15$

Use matrices to solve each problem.

13. If John buys 2 pounds of chocolate-covered cherries and 3 pounds of chocolate-covered mints, his total cost is $13. If he buys 1 pound of chocolate-covered cherries and 2 pounds of chocolate-covered mints, his total cost is $7. Find the cost of 1 pound of chocolate-covered cherries and 1 pound of chocolate covered mints.

14. The length of a rectangle is 3 feet greater than its width. If its perimeter is 14 feet, find the length and width.

7.5 SYSTEMS OF LINEAR INEQUALITIES

In earlier sections we showed how to find the solution to a system of linear equations in two variables. Now we are going to explore the techniques of finding the solution set to a system of linear inequalities in two variables.

The solution set of a system of linear inequalities is the set of points that satisfy all inequalities in the system. The solution set of a system of linear inequalities may consist of infinitely many ordered pairs. To determine the solution set to a system of linear inequalities, graph each inequality on the same set of axes. The ordered pairs common to all of the inequalities are solution sets to the system.

■ Example 1

Graph the following system of inequalities and indicate the solution set:

$$x + y < 1$$
$$x - y < 5$$

Solution: Our method of attack will be to graph both inequalities on the same set of axes. First draw the graph of $x + y < 1$. When drawing the graph, remember to use a dashed line, since the inequality is "less than" (see Fig. 7.8). If you have forgotten how to graph inequalities, review Section 6.7.

Now, on the same set of axes, we find the half plane determined by the inequality $x - y < 5$ (see Fig. 7.9). The solution set of the system of linear inequalities consists of all the points common to the two shaded half planes. These are the points in the region on the graph containing both color shadings. As we can see in Fig. 7.9 the two lines intersect at $(3, -2)$. This ordered pair can also be found by any of the algebraic methods discussed in Sections 7.2 and 7.3.

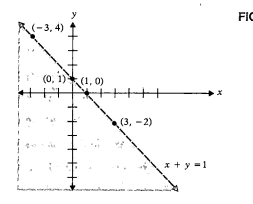

FIGURE 7.8

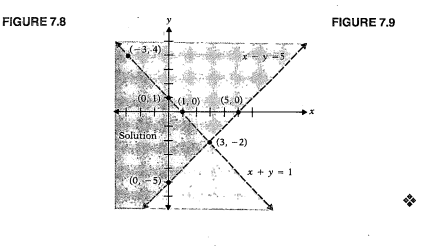

FIGURE 7.9

⊞ **Example 2**

Graph the following system of inequalities and indicate the solution set:

$$2x + 3y \geq 4$$
$$2x - y > -6$$

Solution: Graph the inequality $2x + 3y \geq 4$. Remember to use a solid line, since the inequality is "greater than or equal to" (see Fig. 7.10).

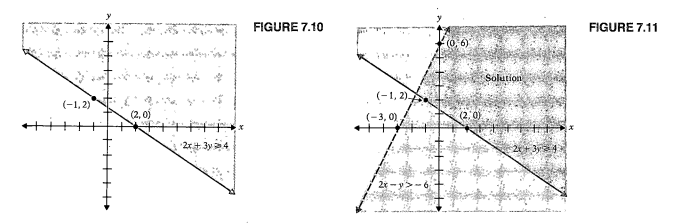

FIGURE 7.10   FIGURE 7.11

On the same set of axes, draw the graph of $2x - y > -6$. Remember to use a dashed line, since the inequality is "greater than" (see Fig. 7.11). The solution is the region of the graph that contains both color shadings and the part of the solid line that satisfies the inequality $2x - y > -6$. Note that the point of intersection of the two inequalities is not a part of the solution.   ❖

■ **Example 3**

Graph the following system of inequalities and indicate the solution set:

$$x \leqslant 5$$
$$y > -1$$

Solution: Graph the inequality $x \leqslant 5$ (see Fig. 7.12).

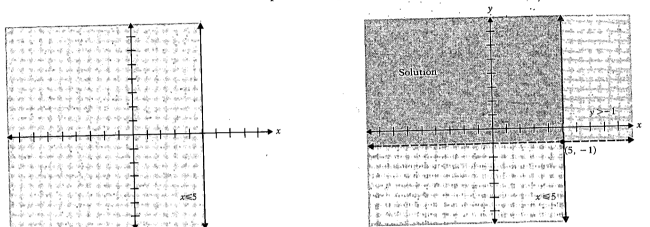

FIGURE 7.12   **FIGURE 7.13**

On the same set of axes, graph $y > -1$ (see Fig. 7.13). The solution is that region of the graph that is shaded in both colors and the part of the solid line that satisfies the inequality $x \leqslant 5$. The point of intersection of the two lines, $(5, -1)$, is not part of the solution, since it does not satisfy the inequality $y > -1$.   ❖

EXERCISES 7.5

In Exercises 1 through 12, graph each system of linear inequalities and indicate the solution set.

1. $y > 3x$
    $y > x + 2$
2. $y \leqslant 2x + 4$
    $y > -2x + 4$
3. $x + y < 3$
    $x - y < 7$
4. $3x + 2y > 6$
    $x + 2y < 4$
5. $x + y \leqslant 3$
    $2x + 3y < 6$
6. $2x - y \leqslant 3$
    $x - y < 3$

7. $x - 3y \leqslant 3$
    $x + 2y \geqslant 4$
8. $x + 2y \geqslant 4$
    $3x - y \geqslant -6$
9. $y \leqslant 2x$
    $x \geqslant 2y$
10. $y \geqslant 3$
    $x + y < 1$
11. $x \geqslant 1$
    $y \leqslant 1$
12. $x \leqslant 0$
    $y \leqslant 0$
13. $x \geqslant 0$
    $y \geqslant 0$
14. $x \leqslant 1$
    $x + 2y < -1$
15. $3x \geqslant 2y + 6$
    $x \leqslant y + 7$
16. $2x - 5y < 7$
    $x > 2y + 1$

LINEAR PROGRAMMING (OPTIONAL)

Many problems in government, business, and industry require the most cost-effective solution. There are many different ways of making a mouse-trap, but some of them are less costly than others. If we can express the relationships in the manufacturing process as linear inequalities, then we have a linear programming problem.

The typical linear programming problem has many variables and is generally so lengthy that it is solved on a computer by a technique called the simplex algorithm. The simplex method was developed in the 1940s by George B. Dantzig. Linear programming is being used to solve problems in the social sciences, health care, land development, nutrition, and so on.

We will not discuss the simplex method in this textbook. We will merely give a brief introduction to how linear programming works. The student can find a detailed explanation in books on finite mathematics.

In a linear programming problem, there are limitations called **constraints.** Each constraint is represented as a linear inequality. The list of constraints is a system of linear inequalities. When the system of inequalities is graphed we often obtain a region bounded on all sides by line segments. The line segments form the boundaries of the region. Since the boundaries of the region form a polygon, the region is called a **polygon region.** The points where two or more boundaries intersect are called the **vertices** of the polygon. The points on the boundary of the region and the points inside the polygonal region are the solution set for the system of inequalities. From all the vertices there is at least one point that, when substituted in an equation of the form $K = Ax + By$, will give us the maximum value for K. A different point from among the vertices will give us a minimum value of K. (K is the value we wish to maximize or minimize.) The linear programming problem is to determine which ordered pair will give us this maximum (or minimum) value. The fundamental principle of linear programming provides us with a rule for finding the maximum and minimum values.

Fundamental principle. If a linear equation of the form $K = Ax + By$ is evaluated at each point in a closed polygonal region, the maximum and minimum values of the equation occur at vertices of the region.

The following example illustrates how the Fundamental Principle is used to solve a linear programming problem.

Example 1

A company makes two types of rocking chairs, a plain chair and a fancy chair. Each rocking chair must be assembled and then finished. The plain chair takes 4 hours to assemble and 4 hours to finish. The fancy chair takes 8 hours to assemble and 12 hours to finish. The company can provide at most 160 worker-hours of assembling and 180 worker-hours of finishing a day. If the profit on the plain chair is $10.00 and the profit on the fancy chair is $18.00, how many rocking chairs of each type should the company make per day to maximize profits? What is the maximum profit?

Solution: From the information given above, we can establish the following facts.

	Assembly Time (hr)	Finishing Time (hr)	Profit ($)
Plain chair	4	4	10.00
Fancy chair	8	12	18.00

$$\text{Let } x = \text{ the number of plain chairs}$$
$$y = \text{ the number of fancy chairs}$$
$$10x = \text{ profit on the plain chairs}$$
$$18y = \text{ profit on the fancy chairs}$$
$$P = \text{ the total profit}$$

The total profit is the sum of the profit on the plain chairs and the profit on the fancy chairs. Thus $P = 10x + 18y$.

The maximum profit, P, is dependent on several conditions, called constraints. The number of chairs manufactured each day cannot be a negative amount. This gives us the constraints $x \geq 0$ and $y \geq 0$. Another constraint is determined by the total number of hours allocated for assembling. Since it takes 4 hours to assemble the plain chair, the total number of hours per day to assemble x plain chairs is $4x$. It takes 8 hours to assemble a fancy chair; therefore the total number of hours needed to make y fancy chairs is $8y$. The maximum number of hours allocated for assembling is 160. Thus the third constraint is $4x + 8y \leq 160$. The final constraint is determined by the number of hours allotted for finishing. It takes 4 hours to finish a plain chair, or $4x$ hours to finish x plain chairs. It takes 12 hours to finish a fancy chair, or $12y$ hours to finish y fancy chairs. The total number of hours allotted for finishing is 180. Therefore the fourth constraint is $4x + 12y \leq 180$. Thus the four constraints are

$$x \geq 0$$
$$y \geq 0$$
$$4x + 8y \leq 160$$
$$4x + 12y \leq 180$$

The list of constraints is a system of linear inequalities in two variables. The solution to the system of inequalities is the set of ordered pairs that satisfies all the constraints. These points are illustrated in Fig. 7.14. Note that the solution to the system consists of the colored region and the solid boundaries. The points (0, 0), (0, 15), (30, 5), and (40, 0) are the points where two of the boundaries intersect. These points can also be found by the substitution method described in Section 7.2.

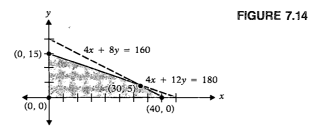

FIGURE 7.14

In this example we wish to maximize the profit. The total profit is given by the formula $P = 10x + 18y$. According to the fundamental principle, the maximum profit will be found at one of the vertices of the polygonal region.

Calculate P for each one of the vertices:
$$P = 10x + 18y$$
$$\text{At } (0, 0), \quad P = 10(0) + 18(0) = 0$$
$$\text{At } (40, 0), \quad P = 10(40) + 18(0) = 400$$
$$\text{At } (30, 5), \quad P = 10(30) + 18(5) = 390$$
$$\text{At } (0, 15), \quad P = 10(0) + 18(15) = 270$$

The maximum profit is at (40, 0), which means that the company should manufacture 40 plain rocking chairs and no fancy rocking chairs. The maximum profit would be $400. The minimum profit would be at (0, 0), when no rocking chairs of either style would be manufactured.

A variation of the original problem could be that the company knows that it cannot sell more than 15 plain rocking chairs per day. With this additional

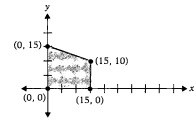

FIGURE 7.15

constraint, we now have the following set of constraints:

$$x \geqslant 0$$
$$x \leqslant 15$$
$$y \geqslant 0$$
$$4x + 12y \leqslant 180$$
$$4x + 8y \leqslant 160$$

The graph of these constraints is shown in Fig. 7.15.

We see that the vertices of the polygonal region are $(0, 0)$, $(0, 15)$, $(15, 10)$, and $(15, 0)$. To determine the maximum profit, we calculate P for each of these vertices:

$$P = 10x + 18y$$

$$\text{At } (0, 0), \quad P = 10(0) + 18(0) = 0$$
$$\text{At } (0, 15), \quad P = 10(0) + 18(15) = 270$$
$$\text{At } (15, 10), \quad P = 10(15) + 18(10) = 330$$
$$\text{At } (15, 0), \quad P = 10(15) + 18(0) = 150$$

With this set of constraints we see that the maximum profit of $330.00 occurs when the company manufactures 15 plain rocking chairs and 10 fancy rocking chairs.

DID YOU KNOW?

"Perhaps the most exciting development—and certainly the one with the greatest potential effect on the many users of mathematics—was a new algorithm (i.e. a step by step problem-solving method) for solving the famous 'linear programming' problem. The problem is solved repeatedly by every major industrial firm as it tries to minimize cost or maximize profit subject to the constraints of labour, capital, and various resources.

"For 30 years the standard method of solving this problem has been the simplex algorithm invented by George Dantzig of Stanford University.

"In 1979 Soviet mathematician L. G. Khachian proposed a new method that is guaranteed to work in a reasonable amount of time and is simple enough to program on a hand held calculator. His method . . . uses a sequence of high-dimensional ellipses to approximate a solution. . . . The new algorithm is able to 'look ahead' much as a mountain climber may descend into a valley in order to reach a slope from which he can then launch his final assault on the summit.

"The importance of this new result is not just that it guarantees a quick solution and is easy to program but also that it behaves in an 'intelligent' way that may presage future breakthroughs in algorithms."

From the 1980 *Encyclopaedia Britannica* Book of the Year

EXERCISES 7.6

In Exercises 1 through 6 a set of constraints and a profit formula are given.

a) Draw the graph of the constraints, and find the points of intersection of the boundaries.
b) Use these points of intersection to determine the maximum and minimum profit.

1.
$x + y \leqslant 5$
$x + 3y \leqslant 9$
$x \geqslant 0$
$y \geqslant 0$
$P = 2x + 4y$

2.
$x + y \leqslant 7$
$x + 2y \leqslant 10$
$x \geqslant 0$
$y \geqslant 0$
$P = 4x + 5y$

3.
$x + y \leqslant 4$
$x + 3y \leqslant 6$
$x \geqslant 0$
$y \geqslant 0$
$P = 6x + 7y$

4.
$x + y \leqslant 50$
$x + 3y \leqslant 90$
$x \geqslant 0$
$y \geqslant 0$
$P = 20x + 40y$

5. $3x + 2y \geqslant 6$
$3x + 4y \leqslant 24$
$x \geqslant 0$
$y \geqslant 0$
$y \leqslant 3$
$P = 1.5x + 3.5y$

6. $x + 2y \leqslant 12$
$4x + 5y \geqslant 20$
$x \geqslant 0$
$y \geqslant 0$
$x \leqslant 8$
$P = 5x + 7y$

7. The P^3 Company manufactures two types of power lawn mowers. Type A is self-propelled, and type B must be pushed by the operator. The company can produce a maximum of 18 mowers per week. It can make a profit of $20 on mower A and a profit of $25 on mower B. The company's planners want to make at least 2 mowers of type A but not more than 5. To keep certain customers happy, they must make at least 2 mowers of type B.

a) List the constraints.
b) List the profit formula.
c) Graph the set of constraints.
d) Find the vertices of the polygonal region.
e) How many of each type should be made to maximize the profit?
f) Find the maximum profit.

8. Richard makes wooden toy cars and trucks. The materials for a car cost $2.00, and the materials for a truck cost $2.25. He can make a total of 15 toys in his spare time in a month. He wants to make at least 2 cars and 2 trucks but no more than 5 trucks. He sells the cars for $6.00 and the trucks for $7.50. How many of each type should he make each month to maximize his profit?

SUMMARY

When you are graphing two linear equations in two unknowns, one of three things may happen: (1) The two lines intersect at a single point (the solution is a single ordered pair), (2) the two lines are parallel (there is no solution), or (3) the two lines coincide (there is an infinite number of solutions). Figure 7.16 illustrates these three possibilities.

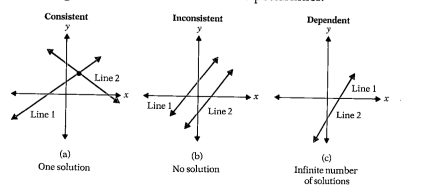

Consistent

(a)
One solution

Inconsistent

(b)
No solution

Dependent

(c)
Infinite number
of solutions

FIGURE 7.16

The solution to a system of equations may also be found algebraically by (1) substitution, (2) the addition method, or (3) using matrices. A system of linear inequalities may be solved by graphing both inequalities on the same set of axes. The solution to a system of inequalities is the set of points that satisfy all inequalities in the system.

We can add, subtract, and multiply matrices if they have the correct dimensions (rows and columns).

Linear programming, an increasingly important area of mathematics, is an application of systems of linear inequalities.

REVIEW EXERCISES

Solve each system of equations graphically. If the system does not have a single ordered pair as a solution, state whether the system is inconsistent or dependent.

1. $x = 5$
   $y = 6$

2. $x + 2y = 4$
   $3x + 4y = 6$

3. $x = 3$
   $x + y = 4$

4. $x + 2y = 5$
   $2x + 4y = 4$

Determine without graphing whether the system of equations has exactly one solution, no solution, or an infinite number of solutions.

5. $2x + 3y = 4$
   $3x + 2y = 5$

6. $2x - 3y = 6$
   $-2x + 3y = -6$

7. $y = 4x - 6$
   $4x - y = 4$

8. $2x - 4y = 8$
   $-2x + y = 6$

Solve each system of equations by the substitution method. If the system does not have a single ordered pair as a solution, state whether the system is inconsistent or dependent.

9. $x + y = 6$
   $x + 2y = 0$

10. $x - 2y = 4$
    $y = 3x - 2$

11. $2x - y = 4$
    $3x - y = 2$

12. $2x + y = 4$
    $2y = -4x + 6$

Solve each system of equations by the addition method. If the system does not have a single ordered pair as a solution, state whether the system is inconsistent or dependent.

13. $x + y = 12$
    $-x + 2y = -3$

14. $x + 3y = 4$
    $x + 2y = 1$

15. $2x - 3y = 1$
    $x + 2y = -3$

16. $3x + 4y = 6$
    $2x - 3y = 4$

17. $3x - 5y = 15$
    $2x - 4y = 0$

18. $3x + y = 6$
    $-6x - 2y = -12$

Use matrices to solve each system of equations.

19. $x + 2y = 4$
$x + y = 2$

20. $-x + y = 4$
$x + 2y = 2$

21. $2x + y = 4$
$4x - y = 2$

22. $-2x - y = 0$
$3x + 2y = 4$

23. A taxi charges a fixed fee plus a mileage fee. If a 10-mile trip costs $22 and a 5-mile trip costs $14.50, find the fixed fee and the mileage fee.

24. A chemist has a 30% acid solution and an 80% acid solution. How much of each should he mix together to obtain five gallons of a 50% solution?

Given $A = \begin{bmatrix} 4 & -2 \\ 3 & 5 \end{bmatrix}$, $B = \begin{bmatrix} -3 & -6 \\ 4 & 2 \end{bmatrix}$, find the following.

25. $A + B$
26. $A - B$
27. $3A$
28. $2A - 3B$
29. $A \times B$
30. $B \times A$

Graph each system of linear inequalities and indicate the solution set.

31. $2x + y < 8$
$y \geq 2x - 1$

32. $y < -2x + 3$
$y \geq 4x - 2$

33. $x + 3y \leq 6$
$2x + 7y \geq 14$

34. $x - y > 5$
$6x + 5y \leq 30$

35. A set of constraints and a profit formula are given below. Graph the constraints and find the points of intersection of the boundaries. Use these points of intersection to determine the maximum profit.

$$x + y \leq 10$$
$$2x + 1.8y \leq 18$$
$$x \geq 0$$
$$y \geq 0$$
$$P = 6x + 5y$$

CHAPTER TEST

1. Solve the system of equations graphically:
$$y = 2x - 4$$
$$2x + y = 4$$

2. Determine without graphing whether the system of equations has exactly one solution, no solution, or an infinite number of solutions:
$$3x - 2y = 6$$
$$2x = 3y + 6$$

Solve the system of equations by the method indicated.

3. $2x + y = 4$
   $x + 2y = -1$
   (substitution)

4. $y = 4x - 6$
   $y = -2x + 18$
   (substitution)

5. $x + y = 4$
   $2x + y = 3$
   (addition)

6. $4x + 3y = 5$
   $2x + 4y = 10$
   (addition)

7. $2x - 3y = 5$
   $5x + 4y = 12$
   (addition)

8. $x - 3y = 6$
   $x + y = 2$
   (matrices)

Given $A = \begin{bmatrix} -6 & 2 \\ 3 & -5 \end{bmatrix}$ $B = \begin{bmatrix} -1 & 5 \\ 4 & -2 \end{bmatrix}$, find the following.

9. $A + B$
10. $3A - B$
11. $A \times B$
12. Graph the system of linear inequalities and indicate the solution set.
$$y \geq -2x + 4$$
$$2x - y < 8$$

13. A grocer plans to mix candy that sells for $1.40 per pound with candy that sells for $2.60 per pound to get 10 pounds of a mixture to sell for $2.00 per pound. How many pounds of each type of candy should she mix to obtain the desired mixture?

CHAPTER 8

Consumer
Mathematics

RATIO AND PROPORTION

There are occasions when the consumer has the opportunity to save money. The material in this chapter will provide you with some tools that can help you make better financial decisions.

A consumer often needs to compare two or more quantities. Ratios are often used to make such comparisons.

> A **ratio** of two quantities is the quotient resulting when the first quantity is divided by the second.

A ratio of 3 to 8 may be written 3 to 8, 3:8, or 3/8. Although a ratio is usually expressed in fractional form, it can also be written as a decimal or a percent. When a ratio is expressed in fraction form, the fraction is reduced to simplest form. For example, $\frac{6}{8}$ would be expressed as $\frac{3}{4}$.

A ratio may also be used to compare more than two quantities. When we compare three quantities a, b, and c, we write the ratio as a:b:c.

Example 1

The length of the Sutherlands' boat is 16 feet. The width of Keuka Lake from a point in front of their home is 2640 feet. Find the ratio of the length of the boat to the width of the lake.

Solution: The ratio of the length of the boat to the width of the lake at that specific point is $\frac{16}{2640}$. In simplified form the ratio is $\frac{1}{165}$ or 1:165.   ❖

The ratio 1:165 in Example 1 tells us that it takes 165 boat lengths end to end to equal the distance from shore to shore in front of the Sutherlands' home.

Shoppers often find themselves comparing the price of two or more different brand names of the same item. The shopper can easily make this comparison by means of unit pricing. The unit price is a ratio, which is found as follows:

$$\text{Unit price} = \frac{\text{cost of the package}}{\text{number of units in the package}}$$

Example 2

At a local supermarket a 37-oz can of Grandma Brown's baked beans was selling for $0.89. The store-brand 40-oz can of baked beans was selling for $0.91. Which brand has the lower cost per ounce?

Solution: To determine which has the lower cost per ounce, we find the unit price of each brand:

$$\text{Grandma Brown's: } 89/37 = 2.41 \text{ cents per ounce}$$
$$\text{Store brand: } 91/40 = 2.28 \text{ cents per ounce}$$

Therefore the store-brand can of beans is 0.13 cents per ounce cheaper. ❖

Ratios may also be used to compare more than two quantities, as illustrated in Example 3.

Example 3

The owner of a candy store has a special ratio for obtaining a specific blend of mixed nuts. The mixture contains peanuts, Brazil nuts, English walnuts, pecans, and cashews in a ratio of 7:4:4:3:2, respectively. If the owner has an order for 30 pounds of the mixture, how many pounds of each kind of nut are required?

Solution: The ratio 7:4:4:3:2 implies that for each 7 lb of peanuts the owner needs 4 lb of Brazil nuts, 4 lb of English walnuts, 3 lb of pecans, and 2 lb of cashews. To make one batch, the total number of pounds is 7 + 4 + 4 + 3 + 2 or 20 lb. Therefore to fill an order for 30 lb of the mixture, he would need $\frac{7}{20}$ of 30 lb which is $10\frac{1}{2}$ lb of peanuts ($\frac{7}{20} \cdot 30 = \frac{210}{20} = 10\frac{1}{2}$), $\frac{4}{20}$ of 30 lb or 6 lb of Brazil nuts and 6 lb of English walnuts, $\frac{3}{20}$ of 30 lb or $4\frac{1}{2}$ lb of pecans, and $\frac{2}{20}$ of 30 lb or 3 lb of cashews. To check whether there is a total of 30 pounds of nuts in this mixture, we add the number of pounds of each kind of nut:

$$10\frac{1}{2} \text{ lb peanuts}$$
$$6 \text{ lb Brazil nuts}$$
$$6 \text{ lb English walnuts}$$
$$4\frac{1}{2} \text{ lb pecans}$$
$$+ \ \underline{3 \text{ lb cashews}}$$
$$30 \text{ lb mixture}$$

❖

A **proportion** is a statement of equality between two ratios.

We know that

$$\frac{1}{4} = \frac{1}{4} \times \frac{5}{5} = \frac{5}{20}$$

Thus the statement $\frac{1}{4} = \frac{5}{20}$ is a proportion. We can also write $1:4 = 5:20$. Both statements are read "1 is to 4 as 5 is to 20."

Consider the proportion $\frac{a}{b} = \frac{c}{d}$. Multiply both sides of the proportion by the common denominator, bd:

$$\frac{a}{\cancel{b}}(\cancel{b}d) = \frac{c}{\cancel{d}}(b\cancel{d})$$

$$ad = bc$$

From this discussion we can conclude the following rule:

$$\boxed{\text{If } \frac{a}{b} = \frac{c}{d}, \text{ then } ad = bc.}$$

For the proportion $\frac{a}{b} = \frac{c}{d}$, a and d are called the **extremes,** and b and c are called the **means.** Thus the product of the means, bc, equals the product of the extremes, ad. This process is sometimes called **cross multiplication.**

Example 4

Solve for n in the proportion $8/10 = n/15$.

Solution: Using the rule, we can write

$$\frac{8}{10} = \frac{n}{15}$$
$$(8)(15) = (10)(n)$$
$$120 = 10n$$
$$n = \frac{120}{10}$$
$$= 12$$

❖

This simple tool is powerful when one realizes that tax bills, water bills, electric bills, gas bills, recipes, the amount of medicine to be given a patient, and many other items can be calculated by using proportions.

When solving practical problems using proportions, you must be careful to set up the proportion carefully. In setting up the proportion, the same type of units must be placed in the same relative positions. For example, the following two proportions are correct:

Correct $\dfrac{45 \text{ miles}}{4 \text{ gallons}} = \dfrac{90 \text{ miles}}{8 \text{ gallons}}$ or $\dfrac{20 \text{ dollars}}{3 \text{ pounds}} = \dfrac{60 \text{ dollars}}{9 \text{ pounds}}$

The following proportion is set up incorrectly since the same type of units are in different relative positions in the proportion:

Wrong
$$\frac{\cancel{45 \text{ miles}}}{\cancel{4 \text{ gallons}}} \cancel{=} \frac{\cancel{8 \text{ gallons}}}{\cancel{90 \text{ miles}}}$$

■ **Example 5** _____

The water rate in Monroe County is $1.04 per 1000 gallons of water used. What is the water bill if 25,000 gallons are used?

Solution: This problem may be solved by setting up a proportion. One proportion that can be used is

$$\frac{\text{Cost of 1000 gallons}}{1000 \text{ gallons}} = \frac{\text{Cost of 25,000 gallons}}{25,000 \text{ gallons}}$$

Since we wish to find the cost for 25,000 gallons of water, we will call this quantity x. The proportion then becomes

$$\frac{1.04}{1000} = \frac{x}{25,000}$$

Now we solve for x:

$$(1.04)(25,000) = 1000x$$
$$1000x = 26,000$$
$$x = \frac{26,000}{1000} = 26$$

The cost of 25,000 gallons of water is $26.00. ❖

■ **Example 6** _____

Insulin comes in 10-cc vials labeled in the number of units of insulin per cubic centimeter (cc) of fluid. A vial of insulin marked U40 means that there are 40 units of insulin per cc of fluid. If a patient needs 25 units of insulin, how many cc of fluid should be drawn up into the syringe from the U40 vial?

Solution: The unknown quantity is the number of cc to be drawn into the syringe. We will call this quantity x. One proportion that can be used to find the number of cc is

$$\frac{40 \text{ units}}{1 \text{ cc}} = \frac{25 \text{ units}}{x \text{ cc}}$$
$$40x = 25(1)$$
$$40x = 25$$
$$x = \frac{25}{40} = 0.625$$

The person putting the insulin in the syringe should draw up 0.625 cc of the fluid.   ❖

EXERCISES 8.1

The student might find it convenient to use a calculator for many of the exercises in this chapter.

There are 20 girls and 15 boys in a class.

1. What is the ratio of boys to girls?
2. What is the ratio of girls to boys?
3. What is the ratio of boys to the entire class?
4. What is the ratio of girls to the entire class?

An experimental drug is given to 97 people who have a certain ailment. Fifty-three of these people are cured by the drug.

5. What is the ratio of those cured to those not cured?
6. What is the ratio of those cured to the total number given the drug?
7. What is the ratio of those not cured to those cured?
8. What is the ratio of those not cured to the total number given the drug?

9. The sum of $27 is to be divided among three people in the ratio of 2:3:4. What amount will each receive?
10. Two truck salespeople sold trucks in the ratio of 7:5. At the end of the year the company offered a $3600 bonus to be divided between them on the basis of their sales. How much should each salesperson get?

Determine which item has the lower unit price in each case.

11. A 27-oz can of pears for 47¢ or a 24-oz can for 44¢.
12. A king-size (128-oz) bottle of detergent for $3.15 or the medium-size (48-oz) bottle for $1.15.
13. A 3-lb canned ham for $6.79 or a 5-lb canned ham for $10.79.
14. A 5-oz package of noodles for 39¢ or a 12-oz package of noodles for 93¢.
15. A 16-oz package of carrots for 49¢ or a 24-oz package for 72¢.
16. Two automobile tires for $129.95 or four tires for $247.50.

17. Brand X powder laundry detergent is packaged in three sizes. There is a 17-oz size for $1.07, a 42-oz size for $2.35, and a 72-oz size for $3.98.
    a) Determine which size has the lowest unit price.
    b) If you have a manufacturer's coupon for 25¢ that can be used for any size package and the store doubles the value of the coupon, determine which size has the lowest unit price.
18. A 25-oz box of brand X raisin bran sells for $2.99, and a 20-oz box of brand Y raisin bran sells for $2.29.
    a) Determine which brand has the lowest unit price.
    b) If you have a 40¢ manufacturer's coupon for brand X and a 25¢ manufacturer's coupon for brand Y, determine which has the lowest unit price.
    c) If the store doubles the value of the manufacturer's coupons, determine which has the lowest unit price.
19. Sam has a store coupon for brand X soda. The coupon gives the following choices: (i) 50¢ off two six-packs of 12-oz cans, (ii) 40¢ off two six-packs of 16-oz bottles, (iii) 15¢ off when buying two 1-liter bottles. At the store, Sam finds the following: a 1-liter container (33.8 oz) for 51¢, a six-pack of 16-oz bottles for $2.49, and a six-pack of 12-oz cans for $1.89.
    a) Which of the three options has the lowest unit price without using the store coupon?
    b) Which of the three options has the lowest unit price using the store coupon?
    c) Which of the three options has the lowest unit price if you have a manufacturer's coupon for 45¢ off on two six-packs of 16-oz bottles? Both the store coupon and the manufacturer's coupon may be used.
20. Sue has a 25¢ manufacturer's coupon for brand X coffee, a 35¢ manufacturer's coupon for brand Y coffee, and a 30¢ manufacturer's coupon for brand Z coffee. The prices of the coffee are: brand X, $2.98 for 16 oz; brand Y, $2.68 for 13 oz; and brand Z, $2.77 for 14 oz.
    a) Determine which has the lowest unit price using the manufacturer's coupon.

b) If you have a 25 store coupon for brand Y coffee that may be used in addition to the manufacturer's coupon, determine which coffee has the lowest unit price.

Solve the following proportions for x.

21. $x/6 = 2/3$ **22.** $15/25 = 3/x$
23. $21/1000 = x/10$ **24.** $4/9 = x/11$
25. $5/x = 9/35$ **26.** $x/7 = 9/11$
27. $4/x = 12.5/13.7$ **28.** $0.45/0.4 = x/2$

The water rate in Berks County is $1.55 per 1000 gallons of water used. What is the water bill if a resident uses

29. 17,000 gallons?
30. 32,500 gallons?
31. How many gallons of water can the customer use if the water bill is not to exceed $30.00?

The property tax rate of the town of Brighton is $43.299 per $1000 of assessed value. What is the tax if the property is assessed at

32. $7500?
33. $17,500?
34. What can the maximum assessed value of a house be if the taxes are not to exceed $1500?

The cost of electricity in Granville, Ohio, is $0.1475 per kilowatt hour. What is the monthly electric bill if a customer uses

35. 855 kilowatt hours?
36. 1675 kilowatt hours?
37. 355 kilowatt hours?
38. Find the maximum amount of electricity that can be used if the total cost of electricity is not to exceed $110 per month.

The cost of gas for heating a house in Reading, Pennsylvania, is $1.62 per cubic foot. What is the monthly gas bill if a customer uses

39. 123 cubic feet?
40. 180 cubic feet?
41. 70 cubic feet?
42. Find the maximum amount of gas that can be used if the total cost for gas is not to exceed $90 per month.

How many cc of insulin would be given for the following doses? (Refer to Example 6.)

43. 12 units of U40   **44.** 35 units of U40

45. 50 units of U80   **46.** 100 units of U80

47. A mason lays 120 blocks in 1 hour and 40 minutes. How long will it take her to lay 450 blocks?
48. The electrical resistance of a wire increases as its length increases. The resistance of 6 feet of wire is 3.75 ohms. What is the resistance in a 14-foot piece of the same wire?
49. A car can travel 24 miles on 1 gallon of gasoline. How far can it travel on 14 gallons of gasoline?
50. A quality control worker can check 15 units in 2.5 minutes. How long will it take her to check 80 units?
51. A blueprint for a shopping mall is in the scale 1:200. Thus 1 foot on the blueprint represents 200 feet of actual length. One part of the mall is to be 340 feet long. How long will this part be on the blueprint?
52. If a 40-lb bag of fertilizer covers 6000 square feet of area, how many pounds of fertilizer are needed to cover an area of 25,000 square feet?
53. The instructions on a bottle of insecticide say: "Use 3 teaspoons of insecticide per gallon of water." If your sprayer has an 8-gallon capacity, how many teaspoons of insecticide would you mix with the 8 gallons of water?
54. A model railroad is made in the scale of 1:87. A boxcar measures 10.4 meters. How large will the boxcar be in the model railroad set?
55. The instructions on the Bisquick package call for the following ingredients to make three quick waffles:

2 packets Bisquick baking mix   1 egg
2 T vegetable oil $1\frac{1}{3}$ c milk

How much of each ingredient should be used if you wish to make eight quick waffles?

56. Leanne decided to make her favorite candy, honey rice flake bars. As she checked the ingredients, she found that she had only 2 tablespoons of honey, not 6 as called for in the recipe. Determine the amount of each of the ingredients that she would use to make the recipe with 2 tablespoons of honey if the whole recipe calls for the following ingredients:

6 T sugar $6\frac{1}{2}$ oz rice flaked cereal
$\frac{1}{2}$ t salt $1\frac{1}{2}$ c seedless raisins or mixed
$\frac{1}{4}$ c water   candied peels
6 T honey

8.2 PERCENT

Maureen and Howard both invested in the stock market, and they are trying to determine which one made the better investment. One way to compare the two investments is to compare the ratios of the amount of money made to the amount invested. The amount of money made is called the interest, or yield, on the investment.

Maureen invested $500 and made $300. Howard invested $400 and made $200. The fact that Maureen made more money than Howard does not necessarily mean that Maureen's was the better investment, since they invested different amounts. The ratios of the amount of interest to the amount invested are 300/500 for Maureen and 200/400 for Howard.

Reducing the two ratios gives us

$$\frac{300}{500} = \frac{3}{5} \times \frac{100}{100} = \frac{3}{5} \quad \text{and} \quad \frac{200}{400} = \frac{1}{2} \times \frac{200}{200} = \frac{1}{2}$$

We can now compare the fractions $\frac{3}{5}$ and $\frac{1}{2}$ by writing each of these fractions with a common denominator. One common denominator is 10, but for our purposes we will make the common denominator 100. This will give us a standard of comparison:

$$\frac{3}{5} = \frac{3}{5} \times \frac{20}{20} = \frac{60}{100} \quad \text{and} \quad \frac{1}{2} = \frac{1}{2} \times \frac{50}{50} = \frac{50}{100}$$

By comparing the two investments, we can see that Maureen did make the better investment. She made $60 for every $100 invested, and Howard made $50 for every $100 invested. Maureen made 60 percent on her investment, and Howard made 50 percent on his investment.

The word "percent" comes from the Latin *per centum* meaning "per hundred." A **percent** is simply a ratio of some number to 100. Thus 15/100 = 15% and $x/100 = x\%$. One way *to change a fraction to a percent* involves finding an equal fraction with a denominator of 100.

■ **Example 1** _____

Change 3/4 to a percent.

Solution: One method of changing 3/4 to a percent is to use a proportion:

$$\frac{3}{4} = \frac{x}{100}$$
$$4x = 300$$
$$x = 75$$

Therefore 3/4 = 75/100, or 75%.

A second method of changing 3/4 to a percent is to divide the numerator by the denominator, multiply the quotient by 100, and add a percent sign.

$$\frac{3}{4} \times 100 = 0.75 \times 100 = 75$$

Now add the percent sign to get 75%. ❖

Example 2

Change 1/8 to a percent.

Solution: We will use a proportion to change 1/8 to a percent:

$$\frac{1}{8} = \frac{x}{100}$$
$$8x = 100$$
$$x = 12\tfrac{1}{2}$$

Therefore $1/8 = (12\tfrac{1}{2})/100$ or $12\tfrac{1}{2}\%$. ❖

A number written as a decimal may be converted to a percent in several different ways. One method is illustrated in Example 3.

Example 3

Change 0.235 to a percent.

Solution

$$0.235 = \frac{0.235}{1} \times \frac{100}{100} = \frac{23.5}{100} \text{ or } 23.5\%$$

Thus $0.235 = 23.5\%$. ❖

A convenient method of converting a number in decimal form to a percent is to multiply the number by 100 and add a percent sign. This is equivalent to moving the decimal point two places to the right and adding a percent sign. For example, 0.235 = 23.5%.
To change from a number expressed as a percent to a number expressed as a decimal, write the number as a fraction with a denominator of 100. Then write the fraction as a decimal. Example 4 illustrates this procedure.

■ **Example 4** _____

a) Change 35% to a decimal.

b) Change $\frac{1}{2}$% to a decimal.

Solution: a)   As previously stated, 35% means 35/100:

$$35\% = \frac{35}{100} = 0.35$$

Thus 35% = 0.35.

b) $\frac{1}{2}\% = \frac{1/2}{100} = \frac{1}{2} \div \frac{100}{1} = \frac{1}{2} \times \frac{1}{100} = \frac{1}{200} = 0.005$

Thus $\frac{1}{2}$% = 0.005.   ❖

From Example 4 we can see that converting a percent to a decimal can be accomplished by dividing the number by 100 and dropping the percent sign. This is equivalent to moving the decimal point two places to the left and dropping the percent sign. Thus for example, 35% = 0.35.

■ **Example 5** _____

Of 673 people surveyed, 418 said they favored the school budget. Find the percent in favor of the school budget.

Solution: The percent in favor is the ratio of the number in favor to the total number surveyed. To find the percent in favor, divide the number in favor by the total. Move the decimal point two places to the right and add a percent sign:

$$\text{Percent in favor} = \frac{418}{673} = 0.6211 = 62.11\%$$

or 62.1% (to the nearest tenth of a percent).   ❖

All answers in this section will be given to the nearest tenth of a percent. When rounding answers off to the nearest tenth of a percent, we carry the division out to four places after the decimal point and then round to the nearest thousandths position.

■ **Example 6** _____

At Metropolitan Community College the grades of those completing Mathematics 100 were as follows: 96 received F, 216 received D, 785 received C, 348 received B, and 185 received A. Find the percent that received each grade.

Solution: The total completing the course is 96 + 216 + 785 + 348 + 185 = 1630.

$$\text{Percent receiving F} = \frac{96}{1630} = 0.0588 = 5.9\%$$

$$\text{Percent receiving D} = \frac{216}{1630} = 0.1325 = 13.3\%$$

$$\text{Percent receiving C} = \frac{785}{1630} = 0.4815 = 48.2\%$$

$$\text{Percent receiving B} = \frac{348}{1630} = 0.2134 = 21.3\%$$

$$\text{Percent receiving A} = \frac{185}{1630} = 0.1134 = \underline{11.3\%}$$

$$100.0\%$$

Note that every possible outcome was considered, so the sum of the percents is 100%. In some problems there may be a slight round-off error. ❖

The percent increase or decrease, or percent change, over a period of time is found by the following formula:

$$\textbf{Percent change} = \frac{\substack{\text{amount in} \\ \text{latest period}} - \substack{\text{amount in} \\ \text{previous period}}}{\text{amount in previous period}} \times \frac{100}{100}$$

If the latest amount is greater than the previous amount, the answer will be positive and will indicate a percent increase. If the latest amount is smaller than the previous amount, the answer will be negative and will indicate a percent decrease.

■ **Example 7** _____

In 1987, Ulster County Ford sold 812 new cars. In 1988 it sold 937 new cars. Find the percent increase or decrease in new car sales from 1987 to 1988.

Solution: The previous period is 1987, and the latest period is 1988.

$$\text{Percent change} = \frac{937 - 812}{812} \times \frac{100}{100}$$

$$= \frac{125}{812} \times \frac{100}{100}$$

$$= 0.1539 \times \frac{100}{100}$$

$$= \frac{15.39}{100}$$

$$= 15.4\% \text{ increase} \qquad ❖$$

■ **Example 8** ——————————————————————

In 1985 Ms. Johnson received a salary of $10,000 and a commission of $8000. In 1988 she received a salary of $15,000 and a commission of $12,000. Find the following.

a)   The change in Ms. Johnson's total income from 1985 to 1988.

b)   The percent change in her total income for these three years.

Solution: a)   Ms. Johnson's total income in 1985 was her salary plus commission, $10,000 + $8,000 = $18,000. Her total income in 1988 was $15,000 + $12,000 = $27,000. The change in her income from 1985 to 1988 was $27,000 − $18,000 = $9,000.

$$
\begin{aligned}
\textbf{b)} \quad \text{Percent change} &= \frac{\$27,000 - \$18,000}{\$18,000} \times \frac{100}{100} \\
&= \frac{\$9,000}{\$18,000} \times \frac{100}{100} \\
&= 0.50 \times \frac{100}{100} \\
&= \frac{50}{100} \\
&= 50\%
\end{aligned}
$$

Ms. Johnson's income rose 50% over the three-year period from 1985 to 1988.   ❖

A similar formula is used to calculate percent markup or markdown:

$$
\textbf{Percent markup} = \frac{\text{selling price} - \text{dealer's cost}}{\text{dealer's cost}} \times \frac{100}{100}
$$

A negative answer indicates a markdown.

■ **Example 9** ——————————————————————

K Mart department stores pay $48.76 for Hart glass fireplace screens. They regularly sell them for $79.88. At a sale they sell them for $69.99. Find the following.

a)   The regular percent markup.

b)   The sale-price percent markup.

c)   The percent decrease of the sale price from the regular price.

Photo by Susan Van Etten

Solution: a) Percent markup on regular price

$$= \frac{\$79.88 - \$48.76}{\$48.76} \times \frac{100}{100}$$

$$= 0.6382 \times \frac{100}{100}$$

$$= \frac{63.82}{100}$$

$$= 63.8\%$$

b) Percent markup on sale price

$$= \frac{\$69.99 - \$48.76}{\$48.76} \times \frac{100}{100}$$

$$= 0.4353 \times \frac{100}{100}$$

$$= \frac{43.53}{100}$$

$$= 43.5\%$$

c) Percent decrease from regular price

$$= \frac{\$69.99 - \$79.88}{\$79.88} \times \frac{100}{100}$$

$$= -0.1238 \times \frac{100}{100}$$

$$= \frac{-12.38}{100}$$

$$= -12.4\%$$

The sale price is 12.4% lower than the regular price.   ❖

In day-to-day affairs we may be asked to solve any one of the following three types of problems.

1. What is 38% of 109?

2. What percent of 40 is 27?

3. Fifteen percent of what number is 12?

To answer these questions we will write each problem as an equation. The word "is" means is equal to, $=$. In each problem we will represent the unknown quantity with the letter x. Therefore the above problems can be represented as

1. 38% of 109 $= x$

2. $x\%$ of 40 $= 27$

3. 15% of $x = 12$

The word "of" in these types of problems indicates multiplication. To solve each problem change the percent to a decimal, and express the problem as an equation, then solve the equation for the variable x. The solutions follow.

1. 38% of $109 = x$

   $0.38(109) = x$ (38% is written as 0.38 in decimal form.)

   $41.42 = x$

Thus 38% of 109 is 41.42.

2. $x\%$ of $40 = 27$

   $(0.01x)40 = 27$ ($x\%$ is written as $0.01x$ in decimal form.)

   $0.40x = 27$

   $\dfrac{0.40x}{0.40} = \dfrac{27}{0.40}$

   $x = \dfrac{27}{0.40}$

   $= 67.5$

Thus 67.5% of 40 is 27.

3. 15% of $x = 12$

   $0.15(x) = 12$

   $\dfrac{0.15x}{0.15} = \dfrac{12}{0.15}$

   $x = 80$

Thus 15% of 80 is 12.

■ Example 10

a) A sofa was purchased for $875. The down payment was 20% of the purchase price. How much was the down payment?

b) Fifty-seven, or 95%, of the students in a first-grade class have a pet. How many students are in the class?

c) Adam scored 48 points on a 60-point test. What percent did he get correct?

Solution: a) We want to find the amount of the down payment. Let $x =$ the down payment.

$$\text{down payment} = 20\% \text{ of purchase price}$$
$$x = 20\% \text{ of } 875$$
$$x = 0.20(875)$$
$$x = 175$$

The down payment is $175.

b) We want to find the number of students in the first-grade class. We know that 57 students represent 95% of the class.

Let x = the number of students in the first grade class

$$95\% \text{ of } x = 57$$
$$0.95x = 57$$
$$\frac{0.95x}{0.95} = \frac{57}{0.95}$$
$$x = 60$$

The number of students in the class is 60.

c) We want to determine the percent that Adam got correct. We need to determine what percent of 60 is 48.

Let x = percent that Adam got correct

$$x\% \text{ of } 60 = 48$$
$$0.01x(60) = 48$$
$$0.60x = 48$$
$$x = \frac{48}{0.60} \quad \text{or} \quad 80$$

Adam scored 80% on the test.   ❖

EXERCISES 8.2

Change the following to percents. Express your answer to the nearest tenth of a percent.

1. $\frac{3}{4}$ 2. $\frac{7}{8}$ 3. $\frac{11}{12}$
4. $\frac{1}{8}$ 5. 0.34752   6. 0.006594
7. 16.38   8. 3.75   9. 200.34

Change the following percents to decimals.

10. 34%   11. 27.3%   12. 3.45%
13. 0.003%   14. 137.5%   15. 3.1416%
16. $\frac{1}{2}$%   17. $\frac{3}{4}$%   18. $\frac{7}{8}$%

19. A family estimates that $5000 of its total income of $24,800 is used to purchase food. Find the percent of the family's income that is spent on food.

20. Of approximately 84.78 million working Americans, 34.2% are blue-collar workers. Find the number of blue-collar workers in the United States.

21. In 1988 there were approximately 3,000,000 state motor vehicle registrations in Alaska. Of these approximately 13,000 were for buses, 1,150,000 were for trucks, and 1,837,000 were for cars. Find the percent of registrations for (a) buses, (b) trucks, (c) cars.

22. The population of the United States rose from approximately 203.2 million in 1970 to approximately 226.5 million in 1980. Find the percent increase from 1970 to 1980.

23. The consumer price index in 1967 was 100.0, in 1975 it was 161.2, and in 1987 it was 330.5. Find the following.
   a) The percent increase in the index from 1967 to 1975
   b) The percent increase from 1975 to 1987

24. Mr. Brown's present salary is $32,000. He is getting an increase of 8% in his salary next year. What will his new salary be?

25. In the United States in 1969 there were 27.1 million people classified as living in poverty. In 1979 there were 25.3 million classified as living in

poverty. Find the percent increase or decrease of people living in poverty from 1969 to 1979.

26. A Kirby vacuum cleaner dealership sold 430 units in 1987 and 407 units in 1988. Find the percent increase or decrease in units sold.

27. If the Kirby dealership in Exercise 26 pays $320 for each unit and sells it for $699, what is the percent markup?

28. The regular price of a model 6211 Zenith color TV is $539.62. During a sale, Hill TV is selling it for $439. Find the percent decrease in the price of this TV.

29. The Smiths had 8 grandchildren in 1987. In 1988 they had 13 grandchildren. Find the percent increase in the number of grandchildren from 1987 to 1988.

30. The cost of a fish dinner to the owner of the Golden Wharf restaurant is $6.95. The fish dinner is sold for $9.75. Find the percent markup.

Write each statement as an equation and solve.

31. What is 18% of 200?
32. Eight is 15% of what number?

33. Sixty-eight is what percent of 193?
34. Fifteen is 120% of what number?
35. What is 17% of 400?
36. What is 3% of 9?
37. Seven is 18% of what number?
38. Twenty-four is 120% of what number?
39. What percent of 70 is 14?
40. What percent of 292 is 73?

41. A computer store is advertising a 10% discount on a particular model computer. After the 10% discount you pay $3425 for the computer. What was the original price of the computer?

42. A football team played 12 games. They lost 1 game.
    a) What percent of the games did they lose?
    b) What percent did they win?

43. A furniture store advertised a table at a 15% discount. The original price was $115.00, and the sale price was $100. Was the sale price consistent with the ad? Explain.

44. Charlie sold a truck and made a profit of $675. His profit was 18% of the sales price. What was the sales price?

8.3　PROMISSORY NOTES AND SIMPLE INTEREST

There may be times when you wish to borrow money from a bank or other lending institution. The amount of money that a bank is willing to lend you is called **credit.**

The amount of credit and the interest rate that you may obtain depend on the assurance you can give the lender that you will be able to repay the loan. Your credit is determined by your business reputation for honesty, by your earning power, and by what you can pledge as security to cover the loan. **Security** is anything of value pledged by the borrower that the lender may sell or keep if the borrower does not repay the loan. Acceptable security may be a business, a mortgage on a property, the title to an automobile, savings accounts, or stocks or bonds. The more marketable the security, the easier it is to obtain the loan, and in some cases marketability may help in getting a cheaper interest rate.

Bankers sometimes grant loans without security, but they require the signature of one or more other persons, called **cosigners,** who guarantee that the loan will be repaid.

For either of the two types of loans—secured loans and cosigner loans—the borrower must sign a promissory note. An example of a promissory note is shown in Fig. 8.1. The promissory note is used more frequently by businesses than by individuals. The most common method of borrowing by individuals is the installment loan or the credit card. Installment loans and credit cards will be discussed in Section 8.5. The promissory note will generally contain the following information:

a) The effective date of the note (when the note was written)
b) The term or maturity date, or due date (the date the money must be repaid)
c) The name of the lender, or creditor (the bank, the person, or the institution from which the money is being borrowed)
d) The principal, or face value, of the note (how much is being borrowed)
e) The signature of the borrower, or debtor
f) The rate of simple interest (if any)

FIGURE 8.1

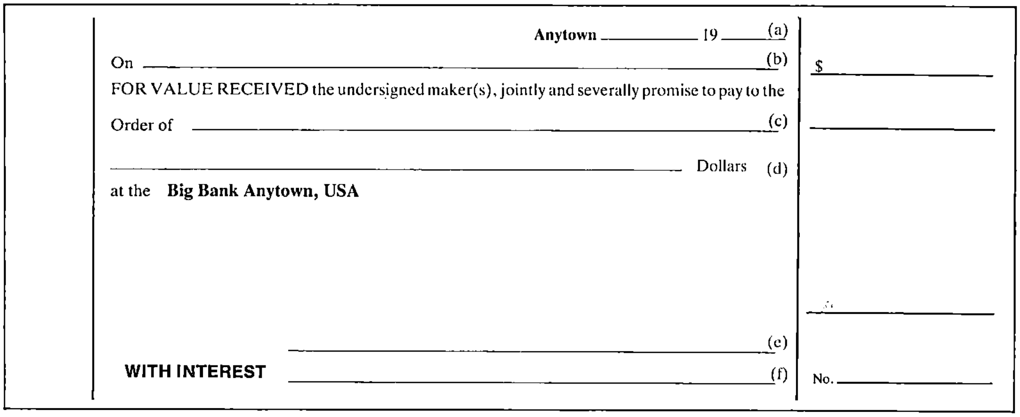
Anytown ________ 19 ____(a)
On ____________________(b) $ ________
FOR VALUE RECEIVED the undersigned maker(s), jointly and severally promise to pay to the
Order of ____________________(c)
____________________ Dollars (d)
at the **Big Bank Anytown, USA**
____________________(e)
WITH INTEREST ____________________(f) No. ________

An understanding of simple interest is essential in the understanding of promissory notes and installment buying.

Interest, *i,* is the money the borrower pays for the use of the lender's money. The formula used to find simple interest is

Simple Interest Formula
Interest = principal × rate × time

In the simple interest formula, the **principal,** *p,* is the amount of money lent or borrowed. The **rate,** *r,* is often called the rate of interest. The rate is generally given as a percent. The **time,** *t,* is the number of days, months, or years for which the money will be lent. Time is expressed in the same period as the rate (that is, in days, months, or years). For example, if the rate is 2% per month, then the time must be expressed in months. The simple interest formula can be expressed as

$$i = prt$$

The rate means annual rate unless otherwise stated. Principal and interest are expressed in dollars.

There are three distinct ways that the simple interest formula may be used, depending on how time is defined. In this text we will discuss two of the methods. The most common—the one that we will use at this time—is called **ordinary interest.** When computing ordinary interest, one considers each month to have 30 days and a year to have 12 months or 360 days. Later in this section we will introduce a second method called the *Banker's Rule* for determining simple interest.

■ **Example 1** _____

Find the simple interest for 3 months on $640 at 6% per year.

Solution: We use the formula $i = prt$. We know that $p = \$640, r = 6\%$ (or converted to a decimal, 0.06), and t in years $= 3/12 = 1/4 = 0.25$. We substitute in the formula

$i = p \times r \times t$
$i = \$640 \times 0.06 \times 0.25$ (convenient for hand calculators)
$\ \ = \$9.60$

The simple interest on $640 at 6% for 3 months is $9.60.   ❖

■ **Example 2** _____

Bonnie borrowed $300 from a bank for two months. The rate of interest charged by the bank is $0.03\overline{3}\%$ per day.

a) Find the interest on the loan.

b) Find the amount that Bonnie will pay the bank at the end of the two months.

Solution

a) To find the interest on the loan, we use the formula $i = prt$. Be careful, because the rate is given in terms of days. Therefore the time *must* be expressed in terms of days. Two months $= 60$ days. Substituting, we obtain

$i = p \times r \times t$
$\ \ = \$300 \times 0.00033\overline{3} \times 60$
$\ \ = \$6$

b) The amount to be repaid is equal to the principal plus interest:

$$A = p + i$$
$$= \$300 + \$6$$
$$= \$306 \qquad \diamondsuit$$

Example 3

A three-month note with a face value of $120 was repaid on the due date with interest at 12%. Find (a) the amount of interest and (b) the amount repaid.

Solution

a) $i = p \times r \times t$

$$= \$120 \times \frac{\cancel{12}}{100} \times \frac{3}{\cancel{12}}$$

$$= \frac{360}{100}$$

$$= \$3.60$$

b) Amount = principal + interest
$$= \$120 + \$3.60$$
$$= \$123.60 \qquad \diamondsuit$$

Interest on a loan that is repaid in a single payment may be collected either in advance (when you borrow the money) or on the day you repay the loan. Suppose that you borrow $500 at 13 percent for one year. The full principal is to be repaid with interest on the date the loan is due (the date of maturity). The bank gives you $500 when you obtain the loan. On the date of maturity one year later you must repay principal, $500, plus interest, $65, or a total of $565.

When interest is collected in advance, the bank is said to have **discounted the note.** The $65 interest charge is often called the **bank discount.** On the day the $500 loan is made, you receive from the bank $500 − $65 = $435. The amount you pay back on the date of maturity is the face value of the note, $500 in this case.

Example 4

For the bank-discounted note described above, the borrower pays $65 interest for the use of $435 for a period of one year.

a) What is the true rate of interest?

b) Is it more or less than the stated rate of 13 percent?

Solution

a) Substituting the appropriate values in the interest formula, $i = prt$, we get $\$65 = \$435 \times r \times 1$. Now solve for r, as explained in Section 6.2:

$$65 = 435r$$
$$\frac{65}{435} = \frac{\cancel{435}r}{\cancel{435}}$$
$$\frac{65}{435} = r$$
$$0.1494 = r$$

Thus the annual rate is about 14.9% for the discounted note.

b) The rate of interest is about 1.9% more for the bank-discounted note. ❖

■ **Example 5** _____

June lent her friend $300. Six months later her friend paid her back the original $300 plus $30.00 interest. What is the annual rate of interest June received?

Solution: Using the formula $i = prt$, we get

$$\$30 = \$300 \times r \times 0.50$$
$$30 = 150r$$
$$\frac{30}{150} = r$$
$$0.2 = r$$

The annual rate of interest paid is 20%. ❖

A promissory note has a date of maturity, at which time the principal and interest are due. It is possible to make partial payments on a promissory note before the date of maturity. The method by which these payments are credited is determined by the United States rule.

The **United States rule** is the result of a Supreme Court decision on the question "How are partial payments on a loan credited to the loan?"

The United States rule works as follows. If a partial payment is made on the loan, interest is computed on the principal from the first day of the loan until the date of the first partial payment. The partial payment is used to pay the interest first; then the rest of the payment is used to pay the principal. The next time a partial payment is made, interest is calculated on the unpaid principal from the date of the previous date of payment. Again the payment goes first to pay the interest, with the rest of the payment used to reduce the principal. An individual can make as many partial payments as he or she wishes; the procedure is repeated for each payment. The balance due on the date of maturity is found by computing interest due since the last partial payment and adding this interest to the unpaid principal.

Before we give an example of the United States rule, it is necessary to understand the Banker's Rule for calculating simple interest. The **Banker's Rule** states that for *time* in the interest formula $i = prt$, a year is 360 days and any fractional part of a year is the exact number of days. Thus for 3 years, $t = 3$; for 45 days $t = 45/360$.

To determine the exact number of days in a period, we can use Table 8.1. We will illustrate how to use Table 8.1 in Example 6.

TABLE 8.1

DAYS IN EACH MONTH

DAY OF MONTH	31 JAN	28 FEB	31 MAR	30 APR	31 MAY	30 JUN	31 JUL	31 AUG	30 SEP	31 OCT	30 NOV	31 DEC
DAY 1	1	32	60	91	121	152	182	213	244	274	305	335
DAY 2	2	33	61	92	122	153	183	214	245	275	306	336
DAY 3	3	34	62	93	123	154	184	215	246	276	307	337
DAY 4	4	35	63	94	124	155	185	216	247	277	308	338
DAY 5	5	36	64	95	125	156	186	217	248	278	309	339
DAY 6	6	37	65	96	126	157	187	218	249	279	310	340
DAY 7	7	38	66	97	127	158	188	219	250	280	311	341
DAY 8	8	39	67	98	128	159	189	220	251	281	312	342
DAY 9	9	40	68	99	129	160	190	221	252	282	313	343
DAY 10	10	41	69	100	130	161	191	222	253	283	314	344
DAY 11	11	42	70	101	131	162	192	223	254	284	315	345
DAY 12	12	43	71	102	132	163	193	224	255	285	316	346
DAY 13	13	44	72	103	133	164	194	225	256	286	317	347
DAY 14	14	45	73	104	134	165	195	226	257	287	318	348
DAY 15	15	46	74	105	135	166	196	227	258	288	319	349
DAY 16	16	47	75	106	136	167	197	228	259	289	320	350
DAY 17	17	48	76	107	137	168	198	229	260	290	321	351
DAY 18	18	49	77	108	138	169	199	230	261	291	322	352
DAY 19	19	50	78	109	139	170	200	231	262	292	323	353
DAY 20	20	51	79	110	140	171	201	232	263	293	324	354
DAY 21	21	52	80	111	141	172	202	233	264	294	325	355
DAY 22	22	53	81	112	142	173	203	234	265	295	326	356
DAY 23	23	54	82	113	143	174	204	235	266	296	327	357
DAY 24	24	55	83	114	144	175	205	236	267	297	328	358
DAY 25	25	56	84	115	145	176	206	237	268	298	329	359
DAY 26	26	57	85	116	146	177	207	238	269	299	330	360
DAY 27	27	58	86	117	147	178	208	239	270	300	331	361
DAY 28	28	59	87	118	148	179	209	240	271	301	332	362
DAY 29	29		88	119	149	180	210	241	272	302	333	363
DAY 30	30		89	120	150	181	211	242	273	303	334	364
DAY 31	31		90		151		212	243		304		365

Add 1 day for Leap Year if February 29 falls between the two dates

■ **Example 6** _____

Using Table 8.1, find (a) the due date of a loan made on March 15 for 180 days and (b) the number of days from April 18 to July 31.

Solution

a) To determine the due date of the note, we do the following: (i) In Table 8.1 we find 15 in the left column (headed "Day of Month"), then move three columns to the right (heading at the top of the column is March) and find the number 74. This tells us that March 15 is the 74th day of the year. Add 74 and 180, since the note will be due 180 days after March 15:

$$\begin{array}{r} 74 \text{ days} \\ + \; 180 \text{ days} \\ \hline 254 \text{ days} \end{array}$$

Thus the due date of the note is the 254th day of the year. Look for 254 in Table 8.1. We find 254 in the column headed September, and the number in the same row as 254 in the left-hand column (headed "Day of Month") is 11. Thus the due date of the note is September 11.

b) To determine the number of days from April 18 to July 31, we do the following: (i) Using Table 8.1, we find that April 18 is the 108th day of the year and July 31 is the 212th day of the year. (ii) Find the difference:

$$\begin{array}{r} 212 \text{ days} \\ - \; 108 \text{ days} \\ \hline 104 \text{ days} \end{array}$$

Thus the number of days from April 18 to July 31 is 104 days.   ❖

■ **Example 7** _____

Find the interest on $300 at 12% for the period March 3 to May 3 using the Banker's Rule.

Solution:
Using Table 8.1, we find that the exact number of days from March 3 to May 3 is 61. Thus the period of time in years would be 61/360. Substituting in the simple interest formula,

$$\begin{aligned} i &= prt \\ &= 300 \times \frac{12}{100} \times \frac{61}{360} \\ &= \$6.10 \end{aligned}$$

Using the Banker's Rule to compute the simple interest on $300.00 at 12% for 61 days, we find that the interest due is $6.10.   ❖

Now we will illustrate how a partial payment is credited by using the United States rule.

 Example 8

On March 3, Robert wrote a 90-day note for $475.00 at an interest rate of 15.5%. On April 1 he made a partial payment of $200.00, and on May 10 he made a partial payment of $150.00.

a) Find the interest and the amount credited to the principal on April 1.

b) Find the interest and the amount credited to the principal on May 10.

c) Find the amount that must be paid on the due date.

Solution

a) First determine the number of days from March 3 to April 1, the date of the first partial payment. The number of days is 29. Now we use the Banker's Rule to compute the simple interest on $475 at 15.5% for 29 days. Substituting in the simple interest formula, we get

$$i = 475 \times \frac{15.5}{100} \times \frac{29}{360}$$
$$= \$5.93$$

The interest of $5.93 that is due on April 1 is deducted first from the payment of $200.00. The remaining amount of the payment is then applied to the principal:

$200.00	Partial payment
− 5.93	Interest
$194.07	Amount to be applied to principal
$475.00	Original principal
− 194.07	Amount to be applied to principal
$280.93	New principal after April 1 payment

b) Using the Banker's Rule, calculate the interest for the period from April 1 to May 10. The number of days from April 1 to May 10 is 39.

$$i = 280.93 \times \frac{15.5}{100} \times \frac{39}{360}$$
$$= \$4.72$$

Now subtract the interest of $4.72 from the partial payment of $150.00 to obtain $145.28. Thus $145.28 is applied against the new principal.

$280.93	New principal
$-$145.28	Amount to be applied to principal
$135.65	Balance due on principal on date of maturity of loan

c) The date of maturity of the loan is June 1. The number of days from May 10 to June 1 is 22. The interest is computed on the remaining balance of $135.65:

$$i = 135.65 \times \frac{15.5}{100} \times \frac{22}{360}$$
$$= \$1.28$$

Therefore the balance due on the maturity date of the loan is the sum of the principal and the interest, $135.65 plus $1.28, or $136.93. *Note:* The sum of the days in the three calculations, $29 + 39 + 22$, equals the total number of days of the note, 90. ❖

EXERCISES 8.3

Find the interest for the following problems, using the simple interest formula. (The rate is an annual rate unless otherwise noted.)

1. $p = \$20, r = 10\%, t = 1$ year
2. $p = \$425, r = 5.5\%, t = 2$ years
3. $p = \$650, r = 14\%, t = 60$ days
4. $p = \$225.75, r = 12\frac{1}{2}\%, t = 6$ months
5. $p = \$587, r = .045\%$ daily rate, $t = 2$ months
6. $p = \$6742.75, r = 6.05\%, t = 90$ days
7. $p = \$3450.72, r = 0.33\overline{3}\%$ daily rate, $t = 24$ days
8. $p = \$4375, r = 1\frac{1}{2}\%$ per month, $t = 6$ months
9. $p = \$8750, r = 1\frac{1}{4}\%$ per month, $t = 1$ year
10. $p = \$15,650, r = 1\frac{3}{4}\%$ per month, $t = 1\frac{1}{2}$ years

Use the simple interest formula to find the missing value.

11. $p = $ _____ $, r = 6\%, t = 60$ days, $i = \$6.00$
12. $p = \$300, r = $ _____ $, t = 2$ months, $i = \$6.00$
13. $p = \$400, r = 6\%, t = $ _____ $, i = \$6.00$
14. $p = $ _____ $, r = 8.5\%, t = 2$ years, $i = \$144.50$
15. $p = \$1375, r = $ _____ $, t = 4$ years, $i = \$357.50$

16. $p = $ _____ $, r = 11\%, t = 3$ months, $i = \$13.20$
17. $p = \$675, r = $ _____ $, t = 4$ months, $i = \$29.25$
18. $p = \$980, r = $ _____ $, t = 9$ months, $i = \$84.53$
19. Joe borrowed $1500 from his bank for 60 days at an interest rate of 9%. He gave the bank stock certificates as security.
    a) How much did he pay for the use of the money?
    b) What is the amount he paid to the bank on the date of maturity?
20. Mrs. Lewis borrowed $1500 from the bank for six months. Her friend Ms. Harris was cosigner of Mrs. Lewis's promissory note. The bank collected $12\frac{1}{2}\%$ interest on the date of maturity.
    a) How much money did Mrs. Lewis receive from the bank on the date she signed the note?
    b) Find the amount she repaid to the bank on the due date of the note.
21. Juan borrowed $2500 for five months from his bank, using United States Government bonds as security. The bank discounted the loan at 13%.
    a) How much did Juan pay the bank for the use of its money?

b) How much did he receive from the bank?
c) What was the true rate of interest he paid?

22. Carol borrowed $3650 from the bank for eight months. The bank discounted the loan at 10%.
  a) How much did Carol pay the bank for the use of its money?
  b) How much did she receive from the bank?
  c) What was the true rate of interest she paid?

23. Enrico wants to borrow $350 for six months from his bank, using his savings account as security. The bank rules are that the maximum amount that he can borrow is 80% of the amount in his savings account. The interest rate is 2% higher than the interest rate being paid on the savings account. The current rate is $5\frac{1}{4}$% on the savings account.
  a) How much money must he have in his account in order to borrow $350?
  b) What is the rate of interest the bank will charge for the loan?

c) Find the amount Enrico must repay in six months.

Find the exact time from the first date to the second date.

24. April 4 to September 23
25. May 19 to October 4
26. June 19 to November 5
27. March 29 to August 12
28. February 15 to June 15
29. January 17 to May 30 (leap year)

Determine the due date, using the exact time.

30. 90 days after May 15
31. 120 days after June 8
32. 120 days after April 15
33. 60 days after September 1
34. 180 days after August 15

In Exercises 35–41 a partial payment is made on the date indicated. Use the United States rule to determine the balance due on the note at the date of maturity.

	Note				Partial Payment(s)	
	Principal	Rate	Effective Date	Maturity Date	Amount	Date
35.	$1,500	14%	July 1	Oct. 15	$300	Aug. 1
36.	$2,400	13%	April 15	July 1	$500	May 1
37.	$3,000	14.5%	May 1	Nov. 1	$1,000	June 1
38.	$7,500	12%	April 15	Oct. 1	$1,000	Aug. 1
39.	$1,000	12.5%	Jan. 1	Feb. 15	$300	Jan. 15
40.	$1,800	15%	Aug. 1	Nov. 1	$500	Sept. 1
					$500	Oct. 1
41.	$2,000	16%	Nov. 15	Jan. 1	$500	Dec. 1
					$800	Dec. 15

42. The Zwick Balloon Company signed a $2500 note with interest at 17% for 180 days on March 1. The company made payments of $750 on May 1 (61 days) and $835 on July 1 (61 days). How much will the company owe on the date of maturity?

43. The Sweet Tooth Restaurant borrowed $3000 on a note dated May 15 with interest of 16%. The maturity date of the loan is September 1 (109 days). They made partial payments of $875 on June 15 (31 days) and $940 on August 1 (47 days). Find the amount due on the maturity date of the loan.

*44. Mr. Harrison borrowed $600 for three months. The banker said he must repay the loan at the rate of $200 per month plus interest. The bank was charging at the rate of 2% above the "prime interest rate." The **prime interest rate** is the rate charged to preferred customers of the bank. The first month the prime rate was 11%, the second month 11.5%, and the third month 12%.
  a) Find the amount he paid the bank at the end of the first month; at the end of the second month; at the end of the third month.
  b) What was the total amount of interest he paid the bank?

RESEARCH ACTIVITIES

45. Visit your local bank and investigate the procedure for obtaining a loan. Determine (1) what type of security the bank requires, (2) what rate of interest the bank will charge for the loan, and (3) when the loan must be repaid.

46. There are three methods of determining the time when calculating simple interest with the simple interest formula. They are ordinary, Banker's Rule, and exact time.

a) Visit a local bank to determine how time as used in the simple interest formula differs for each of the three methods.

b) Compare the results for each method on a loan for $500 at 8% for the period January 2, 1988, to April 2, 1988.

8.4 COMPOUND INTEREST

"Money makes money, and the money that money makes makes more money."
Benjamin Franklin

In Section 8.3 we discussed using other people's money for our benefit. In this section we want to discuss some ways we can put our own money to work for our benefit. An **investment** may be defined as the use of money or capital for income or profit. We can divide investments into two classes: fixed investments and variable investments. In a **fixed investment** the amount invested as principal is guaranteed, and the interest is computed at a fixed rate. By "guaranteed" we mean that the exact amount invested will be paid back together with any accumulated interest. An example of a fixed investment, which we will discuss in this section, is a savings account. Another fixed investment is a government savings bond. In a **variable investment** neither the principal nor the interest is guaranteed. Examples of variable investments are stocks and commercial bonds.

Albert Einstein was once asked to name the greatest discovery of man. His reply was, "Compound interest."

Earlier, we introduced simple interest. Simple interest is calculated once for the period of a loan using the formula $i = prt$. The interest paid on savings accounts at most banks is compound interest. This means that the bank computes the interest periodically (for example, daily or quarterly) and adds this interest to the original principal. The interest for the following period is computed by using the new principal (original principal plus interest). In effect the bank is computing interest on interest, which is called compound interest.

Figure 8.2 shows typical advertisements for savings accounts, stressing the advantages of compound interest and length of term.

Interest that is computed on the principal and any accumulated interest is called **compound interest**.

■ Example 1

Find the amount (A) to which $1000 will have grown at the end of one year if the interest at 12% is compounded quarterly.

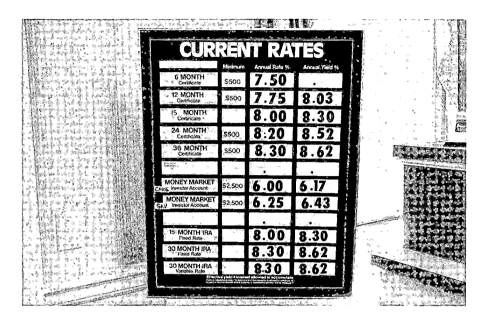

FIGURE 8.2
Photo by Susan Van Etten

Solution: Compute the interest for the first quarter using the simple interest formula. Add this interest to the principal to find the amount at the end of the first quarter:

$$i = prt$$
$$i = \$1000 \times 0.12 \times 0.25 = \$30.00$$
$$A = \$1000 + \$30 = \$1030$$

Find the amount at the end of the second quarter:

$$i = \$1030 \times 0.12 \times 0.25 = \$30.90$$
$$A = \$1030 + \$30.90 = \$1060.90$$

Find the amount at the end of the third quarter:

$$i = \$1060.90 \times 0.12 \times 0.25 = \$31.83$$
$$A = \$1060.90 + \$31.83 = \$1092.73$$

Find the amount at the end of the fourth quarter:

$$i = \$1092.73 \times 0.12 \times 0.25 = \$32.78$$
$$A = \$1092.73 + \$32.78 = \$1125.51$$

This example shows the effect of paying interest on interest, or compounding interest. In one year the amount of $1000 has grown to $1125.51, compared to $1120 that would have been obtained with a simple interest rate of 12%. Thus in one year alone there is a gain of $5.51 with compound

TABLE 8.2
Amount (in $) to Which $1 Will Grow with Compound Interest

Number of Periods (N)	1%	2%	3%	4%	5%	Interest Rate per Period 6%	7%	8%	9%	10%	11%	12%
1	1.010	1.020	1.030	1.040	1.050	1.060	1.070	1.080	1.090	1.100	1.110	1.120
2	1.020	1.040	1.061	1.082	1.103	1.124	1.145	1.166	1.188	1.210	1.232	1.254
3	1.030	1.061	1.093	1.125	1.158	1.191	1.225	1.260	1.295	1.331	1.368	1.405
4	1.041	1.082	1.126	1.170	1.216	1.262	1.311	1.360	1.411	1.464	1.518	1.574
5	1.051	1.104	1.159	1.217	1.276	1.338	1.403	1.469	1.539	1.611	1.685	1.762
6	1.062	1.126	1.194	1.265	1.340	1.419	1.501	1.587	1.677	1.772	1.870	1.974
7	1.072	1.149	1.230	1.316	1.407	1.504	1.606	1.714	1.828	1.949	2.076	2.211
8	1.083	1.172	1.267	1.369	1.477	1.594	1.718	1.851	1.993	2.144	2.305	2.478
9	1.094	1.195	1.305	1.423	1.551	1.689	1.838	1.999	2.172	2.358	2.558	2.773
10	1.105	1.219	1.344	1.480	1.629	1.791	1.967	2.159	2.367	2.594	2.839	3.106
11	1.116	1.243	1.384	1.539	1.710	1.898	2.105	2.332	2.580	2.853	3.152	3.479
12	1.127	1.268	1.426	1.601	1.796	2.012	2.252	2.518	2.813	3.138	3.498	3.896
13	1.138	1.294	1.469	1.665	1.886	2.133	2.410	2.720	3.066	3.452	3.883	4.363
14	1.149	1.319	1.513	1.732	1.980	2.261	2.579	2.937	3.342	3.797	4.310	4.887
15	1.161	1.346	1.558	1.801	2.079	2.397	2.759	3.172	3.642	4.177	4.784	5.474
16	1.173	1.373	1.605	1.873	2.183	2.540	2.952	3.426	3.970	4.595	5.311	6.130
17	1.184	1.400	1.653	1.948	2.292	2.693	3.159	3.700	4.328	5.054	5.895	6.866
18	1.196	1.428	1.702	2.026	2.407	2.854	3.380	3.996	4.717	5.560	6.544	7.690
19	1.208	1.457	1.754	2.107	2.527	3.026	3.617	4.316	5.142	6.116	7.263	8.813
20	1.220	1.486	1.806	2.191	2.653	3.207	3.870	4.661	5.604	6.727	8.062	9.646
21	1.232	1.516	1.860	2.279	2.786	3.400	4.141	5.034	6.109	7.400	8.949	10.804
22	1.245	1.546	1.916	2.370	2.925	3.604	4.430	5.437	6.659	8.140	9.934	12.100
23	1.257	1.578	1.974	2.465	3.072	3.820	4.741	5.871	7.258	8.954	11.026	13.552
24	1.270	1.608	2.033	2.563	3.225	4.049	5.072	6.341	7.911	9.850	12.239	15.179
30	1.348	1.811	2.427	3.243	4.322	5.743	7.612	10.063	13.268	17.449	22.892	29.960
36	1.431	2.040	2.898	4.104	5.792	8.147	11.424	15.968	22.251	30.913	42.818	59.136
42	1.519	2.297	3.461	5.193	7.762	11.557	17.144	25.339	37.318	54.764	80.088	116.723
48	1.612	2.587	4.132	6.571	10.401	16.394	25.729	40.211	62.585	97.017	149.797	230.391

interest. In the first interest period, the interest is computed on the basis of the original principal. During the following interest periods, the interest is computed on the original principal plus the accumulated interest.

The computation of compound interest is a long process, particularly if the interest is compounded four or more times a year. A simple and less time-consuming way to find compound interest is to use a compound-interest table, such as Table 8.2.

■ Example 2

Calculate the interest on $650 at 8% compounded semiannually for three years.

Solution: Since interest is compounded semiannually, there are two periods per year. Hence in three years the number of periods will be 2 × 3, or 6. Eight percent per year is the same as 1/2 × 8%, or 4%, every six months or 4% per period. To solve the problem, we use six periods and a rate of 4% per period. Now we refer to Table 8.2.

1. We find 6 under the first column, "Number of Periods".
2. We go across this row until we find the column headed 4%.
3. We find 1.265; that is, $1.265 is the amount to which $1 will grow in three years at 8% compounded semiannually.
4. We multiply 1.265 by $650. (Why?)

$$1.265 \times \$650 = \$822.25$$

5. $822.25   Amount at end of three years
   − 650.00   Original investment
   ─────────
   $172.25   Amount of compound interest

At the end of three years the original investment of $650 has grown to $822.25. Thus the interest gained is $172.25.   ❖

■ **Example 3** _____

Calculate the amount of interest on $1000 at 12% compounded quarterly for one year.

Solution: Since the interest is compounded quarterly, there are four periods in one year. Since the rate for one year is 12% the rate of each period is $\frac{1}{4} \times 12\% = 3\%$.

Using Table 8.2 for four periods and a rate of 3%, we find 1.126. We now multiply 1.126 by $1000 and obtain $1126.

$1126   Amount at end of one year
− 1000   Original investment
─────────
$ 126   Amount of compound interest   ❖

We can also calculate compound interest by using the compound interest formula.

┌─────────────────────────────────┐
│   **Compound Interest Formula**   │
│ $A = p(1 + \frac{r}{n})^{nt}$ │
└─────────────────────────────────┘

In this formula, A is the amount, p is the principal, r is the annual rate of interest, t is the time in years, and n is the number of periods per year.

■ **Example 4** _____

Using the compound interest formula, find the amount to which $1 will grow when compounded annually at a rate of 3% per year for 6 years.

Solution: Since the interest is compounded annually, $n = 1$.

$$A = p\left(1 + \frac{r}{n}\right)^{nt}$$
$$= 1\left(1 + \frac{0.03}{1}\right)^{(1)(6)}$$
$$= 1(1 + 0.03)^6$$
$$= (1.03)^6$$
$$= 1.194$$

       ❖

     If we refer to Table 8.2, we see that the factor for six periods at a 3% rate is also 1.194.

     The arithmetic in Example 4 can be simplified with a calculator. With a four-function calculator we multiply 1.03 by itself six times.

     Using a calculator with a $\boxed{y^x}$ key, we enter 1.03 and then press the $\boxed{y^x}$ key. We then enter 6 and press the $\boxed{=}$ key. The result is 1.194.

■ **Example 5** _____

Calculate the interest on $650 at 8% compounded semiannually for three years, using the compound interest formula.

Solution: Since interest is compounded semiannually, there are two periods per year. Thus $n = 2$, $r = 0.08$, and $t = 3$. Substituting into the formula, we find the amount, A:

$$A = p\left(1 + \frac{r}{n}\right)^{nt}$$
$$= 650\left(1 + \frac{0.08}{2}\right)^{(2)(3)}$$
$$= 650(1.04)^6$$
$$= 650(1.265)$$
$$= \$822.25$$

     Since the total amount is $822.25 and the original principal is $650, the interest must be $822.25 − $650, or $172.25. These figures check with those obtained in Example 2.   ❖

     In an advertisement for savings accounts and long-term deposits such as IRAs (Individual Retirement Accounts), we see the words **effective annual**

FIGURE 8.3

yield. In the advertisement in Figure 8.3 we see a simple interest rate of 7.25% where the principal is compounded daily and provides the same return as a simple interest rate of 7.52% annually.

When computing the effective annual yield banks use ordinary interest. That is they use 360 for the number of days in a year.

Let us show that a 7.25% interest rate compounded daily yields the same interest as a 7.52% simple interest rate. We will begin by finding the simple interest for a principal of $1 for a period of 1 year.

$$i = prt$$
$$i = 1(.0752)(1) = .0752$$

Thus the amount at the end of 1 year is $1 + $0.0752 = $1.0752.

Now let us find the amount at the end of 1 year when a principal of $1 is invested at 7.25% compounded daily.

$$A = P\left(1 + \frac{0.0725}{360}\right)^{360(1)}$$
$$= 1(1 + 0.000201388)^{360}$$
$$= 1(1.000201388)^{360}$$
$$= 1.0752$$

We see that $1 invested at a rate of 7.25% compounded daily yields the same amount (and the same interest) as $1 invested at a 7.52% simple interest rate. Thus the effective annual yield of 7.25% compounded daily is 7.52%.

There are many different types of savings accounts. Many savings institutions compound interest daily. Some pay interest from the day of deposit to the day of withdrawal, and others pay interest from the first of the month on all deposits made before the tenth of the month. In each of these accounts in which interest is compounded daily, interest is entered into the depositor's account only once each quarter. Some savings banks will not pay any interest on a day-to-day account if the depositor withdraws money to the extent that the balance falls below $5.

Long-term accounts or **certificates of deposit** are special savings accounts in which the depositor agrees to leave money for a specified period of time, and under these conditions the bank agrees to pay a higher rate of interest. If the depositor withdraws the money early, the bank may pay no interest or may pay the same amount that is paid on a regular savings account.

DID YOU KNOW?

The Board of Governors of the Federal Reserve System establishes the maximum rates that banks belonging to the Federal Reserve System may pay on certificates of deposit and savings accounts. The Federal Deposit Insurance Corporation has similar authority with respect to insured commercial banks that are not members of the Federal Reserve System. The Federal Home Loan Bank Association establishes the maximum rates paid by federally insured savings and loan associations.

The frequency of compounding is not restricted. Therefore banks and savings and loan associations try to attract depositors by increasing the frequency of compounding. By compounding more frequently, the bank gives the depositor more interest than under simple rates.

EXERCISES 8.4

In Exercises 1–8, use Table 8.2 to compute the amount and the compound interest paid on each of the following investments.

1. $1000 for 3 years at 6% compounded annually
2. $1200 for 4 years at 8% compounded semiannually
3. $2400 for 2 years at 8% compounded quarterly
4. $800 for 1 year at 16% compounded quarterly
5. $2500 for 3 years at 7% compounded annually
6. $4500 for 4 years at 12% compounded quarterly
7. $3200 for 6 years at 18% compounded semiannually
8. $2800 for 2 years at 12% compounded monthly

In Exercises 9–16, use the formula $A = p(1 + \frac{r}{n})^{nt}$ to compute the amount and the compound interest repaid for each of the following investments.

9. $2000 for 2 years at 6% compounded annually
10. $2000 for 2 years at 6% compounded semiannually
11. $2000 for 2 years at 6% compounded quarterly
12. $2000 for 2 years at 6% compounded monthly
13. $2000 for 2 years at 12% compounded annually
14. $2000 for 2 years at 12% compounded semiannually
15. $2000 for 2 years at 12% compounded quarterly
16. $2000 for 2 years at 12% compounded monthly

17. When Steven was born, his father deposited $1000 in his name in a savings account. The account was paying 8% interest compounded annually.
    a) What was the value of the account after 15 years?
    b) If the money had been invested at 8% compounded semiannually, what would the value of the account have been after 15 years?

18. When Joan was born, her father deposited $2000 in her name at a savings and loan association. At the time that bank was paying 8% interest compounded annually on savings. After 10 years the association changed to an interest rate of 8% compounded semiannually. How much had the $2000 amounted to after 18 years when the money was withdrawn for Joan to use to help pay her college expenses?

19. Charlie borrowed $3000 from his brother Jim. He agreed to repay the money at the end of two years, giving Jim the same amount of interest that he would have received if the money had been invested at 8% compounded quarterly. How much money did Charlie repay his brother?

20. Kathy borrowed $6000 from her daughter Linette. She repaid the money at the end of two years. If Linette had left the money in a bank account that paid 12% compounded monthly, how much interest would she have accumulated?

21. Whitey invested $6000 at 12% compounded quarterly. Three years later he withdrew the full amount and used it for the down payment on a house. How much money did he put down on the house?

22. Compute the interest and the total amount on each of the following principals at 12% compounded monthly for two years.
    a) $100   b) $200   c) $400
    d) Is there a predictable outcome when the principal is doubled?

23. Determine the interest and the amount on $1000 with interest compounded semiannually for ten years at the following rates.

a) 2%   b) 4%   c) 8%
d) Is there a predictable outcome when the rate is doubled?

24. Compute the interest and the amount on $1000 at 6% compounded semiannually for the following time periods.
a) 6 years   b) 12 years   c) 24 years
d) Is there a predictable outcome when the time is doubled?

25. Show that the effective annual yield of a 6.77% interest rate compounded daily is 7.00% as illustrated in the advertisement.

> **Investors**
> **Superfund**
> Money Market Account
>
> **7·00%**
> Annual Yield
>
> **6·77%**
> Annual Percentage Rate

PROBLEM SOLVING

26. Given that the amount is $3633.39, the principal is $2000, and time period is 5 years, find r if compounded monthly.

27. The question often arises how much must be deposited in an account today to have a set amount, A, of dollars x years in the future. The principal, p, that would have to be invested to provide this sum is called the **present value.** The formula for determining the present value for compound interest is

$$p = \frac{A}{(1 + \frac{r}{n})^{nt}}$$

At the birth of a child, what principal should the parents invest in order to provide the child with $30,000 at age 18 if the money earns 8% compounded quarterly?

28. If the cost of a loaf of bread is $1.25 in 1988 and the annual average inflation rate is $3\frac{1}{2}\%$, what will be the cost of a loaf of bread in 1998?

29. Dick borrowed $2000 from Jean. The terms of the loan are as follows: The period of loan is 3 years, and the rate of interest is 8% compounded semiannually. What rate of simple interest would be equivalent to the rate Jean charged Dick?

8.5 INSTALLMENT BUYING

In Section 8.3 we discussed promissory notes and discounted notes. When borrowing money by either of these methods, the borrower normally repays the loan as a single payment at the end of the scheduled time period. There may be circumstances under which it is more convenient for the borrower to repay the loan on a weekly or monthly basis or to use some other convenient time period. One method of doing this is to borrow money on the **installment plan.**

There are two types of installment loans, open-end and fixed payment. An **open-end installment loan** is one in which you can make variable payments each month. Examples of open-end installment loans are charge accounts with department stores, MasterCard, and Visa. A **fixed installment loan** is one in which you pay a fixed amount of money for a set number of payments. Examples of items purchased with fixed-payment installment loans are cars, boats, appliances, and furniture. These loans are generally repaid in 24, 36, 48, or 60 equal monthly payments.

Photo by Marshall Henrichs

Any individual who wishes to borrow money or purchase goods or services on the installment plan is given a credit rating. The lending institution determines whether you are a good "credit risk" by examining your income, assets, and liabilities and by looking at your history of repaying debts.

The advantage of installment buying is that the buyer has the use of an article while still paying for it. If the article is essential, installment buying may serve a real need. A disadvantage is that some people buy more on the installment plan than they can afford. Another disadvantage is the large amount of money (interest) the purchaser pays for the use of someone else's money. The method of determining the interest charged on an installment plan may vary with different lenders.

To provide the borrower with a means of comparing interest charged, Congress passed the Truth in Lending Act in 1969. The law requires that the lending institution tell the borrower two things, the annual percentage rate and the finance charge. The **annual percentage rate (APR)** is the true rate of interest charged for the loan. The APR is calculated by using a complex formula, so we will use the tables provided by the government to determine the APR. The technique of using the tables to find the APR will be illustrated in Example 1. The total **finance charge** is the total amount of money the borrower must pay for the use of money. The finance charge includes the interest plus any additional fees charged. The additional fees may include service charges, credit investigation fees, mandatory insurance premiums, and so on.

The finance charge a consumer pays when purchasing goods or services on the installment plan is the difference between the total installment price and the cash price. The **total installment price** is the sum of all the monthly payments and the down payment, if any.

■ Example 1

Jane is purchasing a new computer and accessories for $3600, including taxes. She is making a $600 down payment and will finance the balance, $3000, through a bank. The banks says that the payments will be $105 per month for 36 months.

a) Determine the finance charge.   b) Determine the APR.

Solution

a) The total installment price = down payment + total monthly installment payments.

$$= 600 + (36 \times 105)$$
$$= 600 + 3780$$
$$= \$4380$$

The finance charge = total installment price − cash price.
$$= 4380 - 3600$$
$$= \$780$$

b) To determine the annual percentage rate, we will use Table 8.3. To find the APR using the table, we must first divide the finance charge by the amount financed and multiply the quotient by 100. This gives the finance charge per $100 of the amount financed:

$$\frac{\text{Finance charge}}{\text{Amount financed}} \times 100 = \frac{780}{3000} \times 100 = 0.26 \times 100 = 26$$

TABLE 8.3

Annual Percentage Rate Table for Monthly Payment Plans

Number of Payments	10.00%	10.25%	10.50%	10.75%	11.00%	11.25%	11.50%	11.75%	12.00%	12.25%	12.50%	12.75%	13.00%	13.25%	13.50%	13.75%
							(Finance Charge per $100 of Amount Financed)									
1	0.83	0.85	0.87	0.90	0.92	0.94	0.96	0.98	1.00	1.02	1.04	1.06	1.08	1.10	1.12	1.15
2	1.25	1.28	1.31	1.35	1.38	1.41	1.44	1.47	1.50	1.53	1.57	1.60	1.63	1.66	1.69	1.72
3	1.67	1.71	1.76	1.80	1.84	1.88	1.92	1.96	2.01	2.05	2.09	2.13	2.17	2.22	2.26	2.30
4	2.09	2.14	2.20	2.25	2.30	2.35	2.41	2.46	2.51	2.57	2.62	2.67	2.72	2.78	2.83	2.88
5	2.51	2.58	2.64	2.70	2.77	2.83	2.89	2.96	3.02	3.08	3.15	3.21	3.27	3.34	3.40	3.46
6	2.94	3.01	3.08	3.16	3.23	3.31	3.38	3.45	3.53	3.60	3.68	3.75	3.83	3.90	3.97	4.05
7	3.36	3.45	3.53	3.62	3.70	3.78	3.87	3.95	4.04	4.12	4.21	4.29	4.38	4.47	4.55	4.64
8	3.79	3.88	3.98	4.07	4.17	4.26	4.36	4.46	4.55	4.65	4.74	4.84	4.94	5.03	5.13	5.22
9	4.21	4.32	4.43	4.53	4.64	4.75	4.85	4.96	5.07	5.17	5.28	5.39	5.49	5.60	5.71	5.82
10	4.64	4.76	4.88	4.99	5.11	5.23	5.35	5.46	5.58	5.70	5.82	5.94	6.05	6.17	6.29	6.41
11	5.07	5.20	5.33	5.45	5.58	5.71	5.84	5.97	6.10	6.23	6.36	6.49	6.62	6.75	6.88	7.01
12	5.50	5.64	5.78	5.92	6.06	6.20	6.34	6.48	6.62	6.76	6.90	7.04	7.18	7.32	7.46	7.60
13	5.93	6.08	6.23	6.38	6.53	6.68	6.84	6.99	7.14	7.29	7.44	7.59	7.75	7.90	8.05	8.20
14	6.36	6.52	6.69	6.85	7.01	7.17	7.34	7.50	7.66	7.82	7.99	8.15	8.31	8.48	8.64	8.81
15	6.80	6.97	7.14	7.32	7.49	7.66	7.84	8.01	8.19	8.36	8.53	8.71	8.88	9.06	9.23	9.41
16	7.23	7.41	7.60	7.78	7.97	8.15	8.34	8.53	8.71	8.90	9.08	9.27	9.46	9.64	9.83	10.02
17	7.67	7.86	8.06	8.25	8.45	8.65	8.84	9.04	9.24	9.44	9.63	9.83	10.03	10.23	10.43	10.63
18	8.10	8.31	8.52	8.73	8.93	9.14	9.35	9.56	9.77	9.98	10.19	10.40	10.61	10.82	11.03	11.24
19	8.54	8.76	8.98	9.20	9.42	9.64	9.86	10.08	10.30	10.52	10.74	10.96	11.18	11.41	11.63	11.85
20	8.98	9.21	9.44	9.67	9.90	10.13	10.37	10.60	10.83	11.06	11.30	11.53	11.76	12.00	12.23	12.46
21	9.42	9.66	9.90	10.15	10.39	10.63	10.88	11.12	11.36	11.61	11.85	12.10	12.34	12.59	12.84	13.08
22	9.86	10.12	10.37	10.62	10.88	11.13	11.39	11.64	11.90	12.16	12.41	12.67	12.93	13.19	13.44	13.70
23	10.30	10.57	10.84	11.10	11.37	11.63	11.90	12.17	12.44	12.71	12.97	13.24	13.51	13.78	14.05	14.32
24	10.75	11.02	11.30	11.58	11.86	12.14	12.42	12.70	12.98	13.26	13.54	13.82	14.10	14.38	14.66	14.95
25	11.19	11.48	11.77	12.06	12.35	12.64	12.93	13.22	13.52	13.81	14.10	14.40	14.69	14.98	15.28	15.57
26	11.64	11.94	12.24	12.54	12.85	13.15	13.45	13.75	14.06	14.36	14.67	14.97	15.28	15.59	15.89	16.20
27	12.09	12.40	12.71	13.03	13.34	13.66	13.97	14.29	14.60	14.92	15.24	15.56	15.87	16.19	16.51	16.83
28	12.53	12.86	13.18	13.51	13.84	14.16	14.49	14.82	15.15	15.48	15.81	16.14	16.47	16.80	17.13	17.46
29	12.98	13.32	13.66	14.00	14.33	14.67	15.01	15.35	15.70	16.04	16.38	16.72	17.07	17.41	17.75	18.10
30	13.43	13.78	14.13	14.48	14.83	15.19	15.54	15.89	16.24	16.60	16.95	17.31	17.66	18.02	18.38	18.74

This tells us that the individual who took out the loan pays $26 interest for each $100 being financed. To use the table, look for 36 in the left-hand column under the heading "Number of Payments." Then move across to the right until you find the value closest to $26. The value that is closest to $26 is $26.12. Moving to the top of this column, we find the value 15.75%. Therefore 15.75% is the annual percentage rate. ❖

■ **Example 2** _____

Allen borrowed $9800 in 1988 to purchase a car. He does not recall the APR of the loan but remembers that there are 48 payments of $255.75. Determine the APR.

TABLE 8.3 (continued)

Number of Payments	Annual Percentage Rate															
	10.00%	10.25%	10.50%	10.75%	11.00%	11.25%	11.50%	11.75%	12.00%	12.25%	12.50%	12.75%	13.00%	13.25%	13.50%	13.75%
						(Finance Charge per $100 of Amount Financed)										
31	13.89	14.25	14.61	14.97	15.33	15.70	16.06	16.43	16.79	17.16	17.53	17.90	18.27	18.63	19.00	19.38
32	14.34	14.71	15.09	15.46	15.84	16.21	16.59	16.97	17.35	17.73	18.11	18.49	18.87	19.25	19.63	20.02
33	14.79	15.18	15.57	15.95	16.34	16.73	17.12	17.51	17.90	18.29	18.69	19.08	19.47	19.87	20.26	20.66
34	15.25	15.65	16.05	16.44	16.85	17.25	17.65	18.05	18.46	18.86	19.27	19.67	20.08	20.49	20.90	21.31
35	15.70	16.11	16.53	16.94	17.35	17.77	18.18	18.60	19.01	19.43	19.85	20.27	20.69	21.11	21.53	21.95
36	16.16	16.58	17.01	17.43	17.86	18.29	18.71	19.14	19.57	20.00	20.43	20.87	21.30	21.73	22.17	22.60
37	16.62	17.06	17.49	17.93	18.37	18.81	19.25	19.69	20.13	20.58	21.02	21.46	21.91	22.36	22.81	23.25
38	17.08	17.53	17.98	18.43	18.88	19.33	19.78	20.24	20.69	21.15	21.61	22.07	22.52	22.99	23.45	23.91
39	17.54	18.00	18.46	18.93	19.39	19.86	20.32	20.79	21.26	21.73	22.20	22.67	23.14	23.61	24.09	24.56
40	18.00	18.48	18.95	19.43	19.90	20.38	20.86	21.34	21.82	22.30	22.79	23.27	23.76	24.25	24.73	25.22
41	18.47	18.95	19.44	19.93	20.42	20.91	21.40	21.89	22.39	22.88	23.38	23.88	24.38	24.88	25.38	25.88
42	18.93	19.43	19.93	20.43	20.93	21.44	21.94	22.45	22.96	23.47	23.98	24.49	25.00	25.51	26.03	26.55
43	19.40	19.91	20.42	20.94	21.45	21.97	22.49	23.01	23.53	24.05	24.57	25.10	25.62	26.15	26.68	27.21
44	19.86	20.39	20.91	21.44	21.97	22.50	23.03	23.57	24.10	24.64	25.17	25.71	26.25	26.79	27.33	27.88
45	20.33	20.87	21.41	21.95	22.49	23.03	23.58	24.12	24.67	25.22	25.77	26.32	26.88	27.43	27.99	28.55
46	20.80	21.35	21.90	22.46	23.01	23.57	24.13	24.69	25.25	25.81	26.37	26.94	27.51	28.08	28.65	29.22
47	21.27	21.83	22.40	22.97	23.53	24.10	24.68	25.25	25.82	26.40	26.98	27.56	28.14	28.72	29.31	29.89
48	21.74	22.32	22.90	23.48	24.06	24.64	25.23	25.81	26.40	26.99	27.58	28.18	28.77	29.37	29.97	30.57
49	22.21	22.80	23.39	23.99	24.58	25.18	25.78	26.38	26.98	27.59	28.19	28.80	29.41	30.02	30.63	31.24
50	22.69	23.29	23.89	24.50	25.11	25.72	26.33	26.95	27.56	28.18	28.80	29.42	30.04	30.67	31.29	31.92
51	23.16	23.78	24.40	25.02	25.64	26.26	26.89	27.52	28.15	28.78	29.41	30.05	30.68	31.32	31.96	32.60
52	23.64	24.27	24.90	25.53	26.17	26.81	27.45	28.09	28.73	29.38	30.02	30.67	31.32	31.98	32.63	33.29
53	24.11	24.76	25.40	26.05	26.70	27.35	28.00	28.66	29.32	29.98	30.64	31.30	31.97	32.63	33.30	33.97
54	24.59	25.25	25.91	26.57	27.23	27.90	28.56	29.23	29.91	30.58	31.25	31.93	32.61	33.29	33.98	34.66
55	25.07	25.74	26.41	27.09	27.77	28.44	29.13	29.81	30.50	31.18	31.87	32.59	33.26	33.95	34.65	35.35
56	25.55	26.23	26.92	27.61	28.30	28.96	29.69	30.39	31.09	31.79	32.49	33.20	33.91	34.62	35.33	36.04
57	26.03	26.73	27.43	28.13	28.84	29.54	30.25	30.97	31.68	32.39	33.11	33.83	34.56	35.28	36.01	36.74
58	26.51	27.23	27.94	28.66	29.37	30.10	30.82	31.55	32.27	33.00	33.74	34.47	35.21	35.95	36.69	37.43
59	27.00	27.72	28.45	29.18	29.91	30.65	31.39	32.13	32.87	33.61	34.36	35.11	35.86	36.62	37.37	38.13
60	27.48	28.22	28.96	29.71	30.45	31.20	31.96	32.71	33.47	34.23	34.99	35.75	36.52	37.29	38.06	38.33

Solution: First determine the finance charge by subtracting the cash price from the total amount paid:

$$\text{Finance charge} = (255.75 \times 48) - 9800$$
$$= 12276 - 9800$$
$$= \$2476$$

Next divide the finance charge by the amount of the loan and multiply this quotient by 100:

$$\frac{2476}{9800} \times 100 = 25.27$$

Now find 48 payments in the left-hand column of Table 8.3. Move to the

TABLE 8.3 (continued)

Number of Payments	14.00%	14.25%	14.50%	14.75%	15.00%	15.25%	15.50%	15.75%	16.00%	16.25%	16.50%	16.75%	17.00%	17.25%	17.50%	17.75%
									Annual Percentage Rate							
							(Finance Charge per $100 of Amount Financed)									
1	1.17	1.19	1.21	1.23	1.25	1.27	1.29	1.31	1.33	1.35	1.37	1.40	1.42	1.44	1.46	1.48
2	1.75	1.78	1.82	1.85	1.88	1.91	1.94	1.97	2.00	2.04	2.07	2.10	2.13	2.16	2.19	2.22
3	2.34	2.38	2.43	2.47	2.51	2.55	2.59	2.64	2.68	2.72	2.76	2.80	2.85	2.89	2.93	2.97
4	2.93	2.99	3.04	3.09	3.14	3.20	3.25	3.30	3.36	3.41	3.46	3.51	3.57	3.62	3.67	3.73
5	3.53	3.59	3.65	3.72	3.78	3.84	3.91	3.97	4.04	4.10	4.16	4.23	4.29	4.35	4.42	4.48
6	4.12	4.20	4.27	4.35	4.42	4.49	4.57	4.64	4.72	4.79	4.87	4.94	5.02	5.09	5.17	5.24
7	4.72	4.81	4.89	4.98	5.06	5.15	5.23	5.32	5.40	5.49	5.58	5.66	5.75	5.83	5.92	6.00
8	5.32	5.42	5.51	5.61	5.71	5.80	5.90	6.00	6.09	6.19	6.29	6.38	6.48	6.58	6.67	6.77
9	5.92	6.03	6.14	6.25	6.35	6.46	6.57	6.68	6.78	6.89	7.00	7.11	7.22	7.32	7.43	7.54
10	6.53	6.65	6.77	6.88	7.00	7.12	7.24	7.36	7.48	7.60	7.72	7.84	7.96	8.08	8.19	8.31
11	7.14	7.27	7.40	7.53	7.66	7.79	7.92	8.05	8.18	8.31	8.44	8.57	8.70	8.83	8.96	9.09
12	7.74	7.89	8.03	8.17	8.31	8.45	8.59	8.74	8.88	9.02	9.16	9.30	9.45	9.59	9.73	9.87
13	8.36	8.51	8.66	8.81	8.97	9.12	9.27	9.43	9.58	9.73	9.89	10.04	10.20	10.35	10.50	10.66
14	8.97	9.13	9.30	9.46	9.63	9.79	9.96	10.12	10.79	10.45	10.62	10.78	10.95	11.11	11.28	11.45
15	9.59	9.76	9.94	10.11	10.29	10.47	10.64	10.82	11.00	11.17	11.35	11.53	11.71	11.88	12.06	12.24
16	10.20	10.39	10.58	10.77	10.95	11.14	11.33	11.52	11.71	11.90	12.09	12.28	12.46	12.65	12.84	13.03
17	10.82	11.02	11.22	11.42	11.62	11.82	12.02	12.22	12.42	12.62	12.83	13.03	13.23	13.43	13.63	13.83
18	11.45	11.66	11.87	12.08	12.29	12.50	12.72	12.93	13.14	13.35	13.57	13.78	13.99	14.21	14.42	14.64
19	12.07	12.30	12.52	12.74	12.97	13.19	13.41	13.64	13.86	14.09	14.31	14.54	14.76	14.99	15.22	15.44
20	12.70	12.93	13.17	13.41	13.64	13.88	14.11	14.35	14.59	14.82	15.06	15.30	15.54	15.77	16.01	16.25
21	13.33	13.58	13.82	14.07	14.32	14.57	14.82	15.06	15.31	15.56	15.81	16.06	16.31	16.56	16.81	17.07
22	13.96	14.22	14.48	14.74	15.00	15.26	15.52	15.78	16.04	16.30	16.57	16.83	17.09	17.36	17.62	17.88
23	14.59	14.87	15.14	15.41	15.68	15.96	16.23	16.50	16.78	17.05	17.32	17.60	17.88	18.15	18.43	18.70
24	15.23	15.51	15.80	16.08	16.37	16.65	16.94	17.22	17.51	17.80	18.09	18.37	18.66	18.95	19.24	19.53
25	15.87	16.17	16.46	16.76	17.06	17.35	17.65	17.95	18.25	18.55	18.85	19.15	19.45	19.75	20.05	20.36
26	16.51	16.82	17.13	17.44	17.75	18.06	18.37	18.68	18.99	19.30	19.62	19.93	20.24	20.56	20.87	21.19
27	17.15	17.47	17.80	18.12	18.44	18.76	19.09	19.41	19.74	20.06	20.39	20.71	21.04	21.37	21.69	22.02
28	17.80	18.13	18.47	18.80	19.14	19.47	19.81	20.15	20.48	20.82	21.16	21.50	21.84	22.18	22.52	22.86
29	18.45	18.79	19.14	19.49	19.83	20.18	20.53	20.88	21.23	21.58	21.94	22.29	22.64	22.99	23.35	23.70
30	19.10	19.45	19.81	20.17	20.54	20.90	21.26	21.62	21.99	22.35	22.72	23.08	23.45	23.81	24.18	24.55

right until you find the value that is closest to 25.27. The closest value to 25.27 is 25.23. Looking at the top of the column, we find the APR to be 11.50%. ❖

Another, perhaps more common, variation of the installment plan is purchasing items on a charge account. This is an example of an open-end installment loan. A typical charge account contract with a bank may have the terms indicated in Table 8.4.

There is no finance or interest charge if you pay the entire new balance for *purchases* by the payment due date on the monthly statement. The bank

TABLE 8.3 (continued)

Number of Payments	14.00%	14.25%	14.50%	14.75%	15.00%	15.25%	15.50%	15.75%	16.00%	16.25%	16.50%	16.75%	17.00%	17.25%	17.50%	17.75%
								Annual Percentage Rate								
31	19.75	20.12	20.49	20.87	21.24	21.61	21.99	22.37	22.74	23.12	23.50	23.88	24.26	24.64	25.02	25.40
32	20.40	20.79	21.17	21.56	21.95	22.33	22.72	23.11	23.50	23.89	24.28	24.68	25.07	25.46	25.86	26.25
33	21.06	21.46	21.85	22.25	22.65	23.06	23.46	23.86	24.26	24.67	25.07	25.48	25.88	26.29	26.70	27.11
34	21.72	22.13	22.54	22.95	23.37	23.78	24.19	24.61	25.03	25.44	25.86	26.28	26.70	27.12	27.54	27.97
35	22.38	22.80	23.23	23.65	24.08	24.51	24.94	25.36	25.79	26.23	26.66	27.09	27.52	27.96	28.39	28.83
36	23.04	23.48	23.92	24.35	24.80	25.24	25.68	26.12	26.57	27.01	27.46	27.90	28.35	28.80	29.25	29.70
37	23.70	24.16	24.61	25.06	25.51	25.97	26.42	26.88	27.34	27.80	28.26	28.72	29.18	29.64	30.10	30.57
38	24.37	24.84	25.30	25.77	26.24	26.70	27.17	27.64	28.11	28.59	29.06	29.53	30.01	30.49	30.96	31.44
39	25.04	25.52	26.00	26.48	26.96	27.44	27.92	28.41	28.89	29.38	29.87	30.36	30.85	31.34	31.83	32.32
40	25.71	26.20	26.70	27.19	27.69	28.18	28.68	29.18	29.68	30.16	30.68	31.18	31.68	32.19	32.69	33.20
41	26.39	26.89	27.40	27.91	38.41	28.92	29.44	29.95	30.46	30.97	31.49	32.01	32.52	33.04	33.56	34.08
42	27.06	27.58	28.10	28.62	29.15	29.67	30.19	30.72	31.25	31.78	32.31	32.84	33.37	33.90	34.44	34.97
43	27.74	28.27	28.81	29.34	29.88	30.42	30.96	31.50	32.04	32.58	33.13	33.67	34.22	34.76	35.31	35.86
44	28.42	28.97	29.52	30.07	30.62	31.17	31.72	32.28	32.83	33.39	33.95	34.51	35.07	35.63	36.19	36.76
45	29.11	29.67	30.23	30.79	31.36	31.92	32.49	33.06	33.63	34.20	34.77	35.35	35.92	36.50	37.08	37.66
46	29.79	30.36	30.94	31.52	32.10	32.68	33.26	33.84	34.43	35.01	35.60	36.19	36.78	37.37	37.96	38.56
47	30.48	31.07	31.66	32.25	32.84	33.44	34.03	34.63	35.23	35.83	36.43	37.04	37.64	38.25	38.86	39.46
48	31.17	31.77	32.37	32.98	33.59	34.20	34.81	35.42	36.03	36.65	37.27	37.88	38.50	39.13	39.75	40.37
49	31.86	32.48	33.09	33.71	34.34	34.96	35.59	36.21	36.84	37.47	38.10	38.74	39.37	40.01	40.65	41.29
50	32.55	33.18	33.82	34.45	35.09	35.73	36.37	37.01	37.65	38.30	38.94	39.59	40.24	40.89	41.55	42.20
51	33.25	33.89	34.54	35.19	35.84	36.49	37.15	37.81	38.46	39.12	39.79	40.45	41.11	41.78	42.45	43.12
52	33.95	34.61	35.27	35.93	36.60	37.27	37.94	38.61	39.28	39.96	40.63	41.31	41.99	42.67	43.36	44.04
53	34.65	35.32	36.00	36.68	37.36	38.04	38.72	39.41	40.10	30.79	41.48	42.17	42.87	43.57	44.27	44.97
54	35.35	36.04	36.73	37.42	38.12	38.82	39.52	40.22	40.92	41.63	42.33	43.04	43.75	44.47	45.18	45.90
55	36.05	36.76	37.46	38.17	38.88	39.60	40.31	41.03	41.74	42.47	43.19	43.91	44.64	45.37	46.10	46.83
56	36.76	37.48	48.20	38.92	39.65	40.38	41.11	41.84	42.57	43.31	44.05	44.79	45.53	46.27	47.02	47.77
57	37.47	38.20	38.94	39.68	40.42	41.16	41.91	42.65	43.40	44.15	44.91	45.66	46.42	47.18	47.94	48.71
58	38.18	38.93	39.68	40.43	41.19	41.95	42.71	43.47	44.23	45.00	45.77	46.54	47.32	48.09	48.87	49.65
59	38.89	39.66	40.42	41.19	41.96	42.74	43.51	44.29	45.07	45.85	46.64	47.42	48.21	49.01	49.80	50.60
60	39.61	40.39	41.17	41.95	42.74	43.53	44.32	45.11	45.91	46.71	47.51	48.31	49.12	49.92	50.73	51.55

Source: Bittinger and Rudolph, *Business Mathematics*, © 1980, Addison-Wesley, Reading, Massachusetts. Pg. 168, Table 6.1. Reprinted with permission.

TABLE 8.4

Type of charge	Periodic rates*		Annual percentage rate*
	Monthly	Daily	
Purchases	1.30%		15.60%
Cash advances		0.04273%	15.60%

*These rates vary with different charge accounts and localities.

will typically add a minimum finance charge of $0.50 if you did not pay the bill in full for the previous month. However, if you *borrow money* through this account, there will be a finance charge from the date you borrowed the money until the date the money is repaid. When you make purchases or borrow money, the minimum monthly payment is sometimes determined by dividing the balance due by 36 and rounding the answer up to the nearest whole dollar, thus ensuring repayment in 36 months. However, if the balance due for any month is less than $360, the minimum monthly payment is $10. These general guidelines may vary with different banks and with different stores. Frequently, you will find that stores such as Sears, J. C. Penney, and others will require that the installment loan be repaid in 24 months rather than 36 months.

Example 3

While on vacation, John Smith charged the following items to his Master-Card account:

Airplane ticket	$285
Hotel room	120
Meals	115
	$520

His bank requires repayment within 36 months and charges the rate given in Table 8.4.

a) Determine the minimum payment he must make the first month.

b) Determine a schedule of payments for paying $90 per month until the loan is repaid.

c) How much would he have to pay each month in order to repay the loan in five months?

Solution: a) To determine the minimum payment we divide the total bill by 36:

$$\frac{\$520}{36} = \$14.44$$

Rounding up to the nearest dollar, we determine that the minimum payment is $15 for the first month.

TABLE 8.5					
Payment Number	Balance Due	Interest on Unpaid Balance	Payment Including Interest	Amount Paid on Balance (Principal)	Unpaid Balance (after Payment)
1	$520.00	*	$90.00	$90.00	$430.00
2	430.00	$5.59	90.00	84.41	345.59

* No interest is paid on the first payment, since it is paid before the due date.

b) To show repayment of the loan by paying $90 per month, we set up the schedule given in Table 8.5.

According to the terms given in Table 8.4, the interest rate is 1.30% per month. Using the simple interest formula, we find that the interest for the second month is $5.59:

$$i = prt$$
$$= \$430 \times 0.013 \times 1$$
$$= \$5.59 \quad \text{(interest for one month)}$$

We find the *amount paid on the balance* by subtracting the interest for the month from the $90 monthly payment: $90 − $5.59 = $84.41. We find the *unpaid balance* by subtracting the amount paid on the balance from the balance due: $430 − $84.41 = $345.59. This procedure is followed each month until the debt is paid, as shown in Table 8.6.

TABLE 8.6					
Payment Number	Balance Due	Interest on Unpaid Balance	Payment Including Interest	Amount Paid on Balance (Principal)	Unpaid Balance (after Payment)
1	$520.00	—	$90.00	$90.00	$430.00
2	430.00	$5.59	90.00	84.41	345.59
3	345.59	4.49	90.00	85.51	260.08
4	260.08	3.38	90.00	86.62	173.46
5	173.46	2.25	90.00	87.75	85.71
6	85.71	1.11	86.82	85.71	—
Total		$16.82	$536.82	$520.00	

The last payment is $85.71 + $1.11 = $86.82. The sum of the amounts in column 3, *"Interest on Unpaid Balance,"* tells us that the interest paid is $16.82. The sum of the amounts in column 5, *"Amount Paid on Balance,"* is equal to the original amount charged, $520. This is true only if no additional purchases are charged during the repayment period. Note that the total *payment including interest* is equal to the sum of the total *interest on the unpaid balance* plus the total *amount paid on the balance.*

c) To repay the loan in five months, we first determine the monthly payments by dividing the amount charged by 5: $520/5 = $104. This amount is due each month in addition to the interest on the balance due (Table 8.7).

TABLE 8.7					
Payment Number	Balance Due	Interest on Unpaid Balance	Payment Including Interest	Amount Paid on Balance (Principal)	Unpaid Balance (after Payment)
1	$520	*	$104.00	$104	$416
2	416	$5.41	109.41	104	312
3	312	4.06	108.06	104	208
4	208	2.70	106.70	104	104
5	104	1.35	105.35	104	—
Total		13.52	533.52	$520	

*No interest.

The interest on the unpaid balance for the second payment (column 3) is found by calculating the simple interest on the unpaid balance after the first payment:

$$i = \$416 \times 0.013 \times 1$$
$$= \$5.41$$

The interest on the unpaid balance is $5.41. The payment including the interest is $109.41 ($104 + 5.41). This procedure is followed for the three remaining monthly payments. The total interest is the sum of the amounts in column 3, *"Interest on Unpaid Balance,"* which is $13.52. The sum of the amounts in column 5, *"Amount Paid on Balance,"* must be $520. The sum of the amounts in column 3, $13.52, plus the sum of the amounts in column 5, $520, must equal the sum of the amounts in column 4, $533.52. The five monthly payments will be $104, $109.41, $108.06, $106.70, and $105.35. ❖

■ **Example 4** ─────────────────────────────────

Jim needed $250 to pay some bills. He borrowed the money from his charge account on September 1, 1988, and repaid it on September 21, 1988. How much interest did Jim pay for the loan? Use the rates given in Table 8.4.

Solution: The agreement says that the interest is calculated for the exact number of days of the loan, starting with the day the money is obtained. The time of the loan in this case is 20 days. The principal is $250, and the rate for borrowing money is 0.04273% per day. Using the simple interest formula and remembering that the rate and time are in terms of days, we find that the interest is $2.14:

$$i = prt$$
$$= 250 \times 0.0004273 \times 20$$
$$= \$2.14$$

The amount due on September 21 was $252.14 ($250 + $2.14).   ❖

DID YOU KNOW?

When you apply for a loan, the decision whether to grant you the loan may be made with a scorecard. Many lending institutions are finding that their decisions on granting credit are far more accurate if the decision is based on numbers rather than on feelings. For example, Sears keeps its losses below 1% by this method. However, the scorecard does not eliminate all bad risks. Furthermore, a credit officer may veto the system's impersonal judgment and make his or her own decision. Records show that in 95% of the cases in which the scoring system has been vetoed by the loan officer, the loans were hard to collect. Some areas that are considered in determining an individual's rating on a credit application scorecard are stability (phone, years on same job, years at same address, home ownership), occupation, income, savings accounts, past credit record, the age of the applicant, and the age of the automobile owned.

Statistics show that an individual with a high income, a large house, and a fancy car may be turned down for a loan, whereas a person with a small income but a better record of paying debts may be granted a loan.

■ **Example 5** ─────────────────────────────

Fran needed an additional $500 to buy a sailboat. She found that the Easy Loan Company charged 10.5% simple interest on the amount borrowed for the duration of the loan. It also required that the loan be repaid in five equal monthly payments. The Friendly Loan Company offered loans of $500 to be repaid in eight monthly payments of $66.50.

a) How much interest would be charged for the loan from the Easy Loan Company?

b) How much interest would be charged for the loan from the Friendly Loan Company?

c) What was the APR on the Easy Loan Company loan?

d) What was the APR on the Friendly Loan Company loan?

Solution: **a)**   To find the interest charged by the Easy Loan Company, we use the simple interest formula:

$$i = prt$$
$$= \$500 \times 0.105 \times \frac{5}{12}$$
$$= \$21.88$$

Thus the interest charged is $21.88 for the loan.

b)   To find the interest charged by the Friendly Loan Company, we must subtract the amount borrowed from the total amount repaid:

$$\text{Interest} = \text{total amount repaid} - \text{amount borrowed}$$
$$= (\$66.50 \times 8) - \$500$$
$$= \$532 - \$500$$
$$= \$32$$

c)   To find the APR on the loan from the Easy Loan Company, divide the interest by the amount financed, multiply by 100, and determine the APR from Table 8.3:

$$\frac{21.88}{500} \times 100 = 4.376$$

The APR found in the table is 17.25%.

d)   To find the APR on the loan from the Friendly Loan Company, divide the interest by the amount financed, multiply by 100, and determine the APR from Table 8.3:

$$\frac{32}{500} \times 100 = 6.4$$

The APR found in the table is 16.75%.

Notice that Fran pays fewer dollars in interest with the loan from the Easy Loan Company ($21.88 versus $32). However, she is paying a higher rate of interest with the Easy Loan Company (17.25% versus 16.75%). The reason the actual interest is less even though the APR is greater is that the time period for repayment with the Easy Loan Company is shorter than that of the Friendly Loan Company.   ❖

DID YOU KNOW?

Not every VISA or MasterCard is a credit card. Although they have the same general appearance, some of these plastic cards are **debit** or **transaction cards.** When you purchase an item with a VISA or MasterCard debit card, the cost of the item is subtracted from your checking (or savings) account. You cannot purchase an item that costs more than the amount of money in your account unless you have a line of credit. When you have a line of credit, you can purchase an item on credit up to the stated amount. With credit cards you do not pay interest on goods purchased if you pay the entire balance on the due date of the payment. When you purchase an item on credit with a debit card, you begin paying interest when the purchase is made.

EXERCISES 8.5

1. Ruth bought a microwave oven on a monthly purchase plan. The oven sold for $460. She paid $100 down and $17.75 a month for 24 months.
   a) What interest did Ruth pay over the period of the loan?
   b) What is the APR?

2. The Dechs can buy an electric refrigerator for a cash price of $710. The installment terms are a down payment of $170 and 18 monthly payments of $33.15 each.
   a) How much interest will the Dechs pay?
   b) What is the APR?

3. Mr. and Mrs. Gerling wish to buy furniture that has a cash price of $450. On the installment plan they must pay one third of the cash price as a down payment and make nine monthly payments of $35.75.
   a) How much interest will the Gerlings pay?
   b) What is the APR?

4. Suppose the Gerlings in Exercise 3 had used a credit card rather than the installment plan and paid $60 on their account each month.
   a) How many months would it have taken them to repay the loan?
   b) What would the total interest on the loan have been?

5. A diamond ring sells for $750 cash. On the installment plan the ring may be purchased with a 25% down payment and the balance paid in 18 equal monthly installments. Compute the monthly payment if the total finance charge is $75.

6. Sara wants to purchase a new television. The purchase price is $890. If she purchases the set today and pays cash, she must take money out of her savings account. Another option is to charge the TV on her credit card, take the set home today, and pay next month. Next month she will have cash and will not have to take any money out of her savings account. Assume the following: (i) she has no other charges on her credit card. (ii) The period of time that she has to pay on her credit card is 30 days. (iii) The simple interest rate on her savings account is $5\frac{1}{4}\%$ compounded annually. How much is she saving by using the credit card instead of taking the money out of her savings account?

7. A certain automobile finance agency requires that the buyer make a down payment of one third of the cash price. The agency subtracts the down payment from the cash price and then adds the cost of disability and life insurance.[*] The resulting sum is the unpaid balance. The total interest on the loan is computed on the unpaid balance at 6% simple interest for the duration of the loan. The Stearns family bought a second-hand car on these installment terms. The cash price of the car was $1440, the insurance was $30, and the balance was to be paid in 15 monthly payments.
   a) What was the down payment?
   b) What was the interest charge?
   c) What was the APR?

---

* Disability insurance will cover the monthly payments if the insured individual becomes disabled and cannot work. The life insurance will pay off the balance due on the car if the insured individual dies.

d) What was the monthly payment including interest and insurance?

8. Penny Miller needs $250. She finds that the ABC Loan Company charges 9.7% simple interest on the amount borrowed for the duration of the loan and requires the loan to be repaid in 10 equal monthly payments. The XYZ Loan Company offers loans of $250 to be repaid in 12 monthly payments of $22.85.
   a) How much interest is charged by the ABC Loan Company?
   b) How much interest is charged by the XYZ Company?
   c) What is the APR on the ABC Loan Company loan?
   d) What is the APR on the XYZ Loan Company loan?

9. On a weekend vacation trip to Florida, Florence and Ray charged two airline tickets at $320.00 each, hotel expenses totaling $190.80, and meals at hotels and restaurants totaling $215.75 on their credit card. When the bill arrived, they found that they could not pay the total amount due. They paid $300 the first month and decided to pay the balance in installments of $200 per month. (Use Table 8.4.)
   a) Set up a table showing the repayment of the loan.
   b) Determine the total amount of interest paid.

10. Last month, Dale Brooks charged automobile repairs amounting to $178.83 on her credit card.
    a) If the amount paid on the principal is to be the same for each payment, how much will she have to pay each month in order to pay the loan off in three months?
    b) How much interest will she pay?

11. John Jones borrowed $375 against his charge account on September 12 and repaid the loan on October 14 (32 days later).

a) How much interest did John pay on the loan?
b) What amount did he pay the bank when he repaid the loan?

12. Amina bought a new Dodge van for $9095, including sales tax. She made a down payment of $1500 and financed the balance through her bank. She repaid the loan at the rate of $172.46 per month for 60 months. What was the APR?

13. Mark and Lori redecorated their bathroom. The total cost of materials was $1975, which was charged on their credit card. When the bill arrived, they found that they could not pay the total amount. They decided to pay $350 the first month and to pay the balance in installments of $300 per month.
    a) Set up a table showing the repayment of the loan.
    b) Determine the amount of interest paid.

14. Sue Shrader bought a new car, but she cannot now remember the original purchase price. Her payments are $379.50 per month for 36 months. She remembers that the car salesperson said that the simple interest rate for the period of the loan was 6%. She also recalls that she was allowed $2500 on her old car. Find the original purchase price.

15. The cash price of a motorboat and trailer is $7500. The package may be purchased for a 10% down payment and 36 monthly payments. For convenience the dealer computes the finance charge at 6% simple interest on the amount financed.
    a) What is the finance charge on the boat and trailer?
    b) How much is the monthly payment?
    c) Determine the total cost of the package on the installment plan.
    d) What annual percentage rate must the dealer disclose to the buyer?

PROBLEM SOLVING

16. The Perkins' credit card bill includes any transactions that occur between the 14th of the current month and the 15th of the preceding month. In May they paid the balance due on their credit card. On June 20 they purchased clothing for $375.84 and charged the total amount. On June 30 they needed cash and borrowed $200 with their credit card. On July 10 they went to the dinner theater and charged $110. On July 20 they received their bill for the period from June 15 to July 14.
    a) What is the balance due on the bill?
    b) On July 20 they repay the $200 borrowed on June 30th. If they make no more charges on their credit card, what will be the total amount due on August 20th?

RESEARCH ACTIVITIES

17. Renting an automobile might be cheaper than buying. Determine whether this statement is true or false for your driving needs. To help you arrive at a conclusion, do the following.
   a) Decide on the automobile you would like to drive, including the accessories.
   b) Determine the cost of the automobile.
   c) Determine the period of time you wish to keep the automobile.
   d) Determine how much of the cost you would finance.
   e) Determine the amount of interest you would lose from the money withdrawn from your savings account.
   f) Determine the cost of insurance.
   g) Estimate the resale value of the car when sold.
   h) Determine the cost of renting the same car for the same period of time.
   i) Determine who pays for the insurance and the repairs on the rental car.
   j) Finally, determine which is less expensive, renting or buying the car.

18. Assume that you are married and have a child. You do not own a washer and dryer and have no money to buy the appliances. Would it be cheaper to borrow money on an installment loan and buy the appliances or to continue going to the local coin-operated laundry for five years until you have saved enough to pay cash for a washer and dryer?

With the aid of parents or friends, establish how many loads of laundry you would be doing each week. Then determine the cost of doing that number of loads at a coin-operated laundry. (Do not forget the cost of transportation to the laundry.) Shop around for a washer and dryer, and determine the total cost on an installment plan. Do not forget to include the cost of gas, electricity, and water. This information can be obtained from a local gas and electric company. With this information you should be able to make a clear decision about whether you should buy now or wait for five years.

BUYING A HOUSE WITH A MORTGAGE

The Great American Dream is to own one's own house. Does the house buyer understand what is involved in making the largest purchase of a lifetime and how it might change one's style of living? The purchaser will normally be committed to 10, 15, 20, 25, or 30 years of mortgage payments. It is therefore essential, before the "Dream House" is selected, that the buyer consider the following questions: Should the property be in the city, suburbs, or country? Should it be an apartment, a condominium, or a house? Is the community appealing from an economic, cultural, and educational point of view? Is the location convenient with respect to relatives, friends, work, shopping, schools, and entertainment? Are the taxes within the buyer's budget? What are the monthly costs of heat, electricity, telephone, and insurance? Will the property have resale value if the owner must move?

Before purchasing a home, the purchaser must ask the all-important question "Can I afford the house?" Great care must be taken to answer this question accurately. If a family buys a house above its means, it will have a difficult time living within its income. When deciding whether or not to purchase a particular house, the purchaser must also consider two more questions that are critical: Do I have enough cash for the down payment? Can I afford the monthly mortgage payments with my current income? These two items—down payment and mortgage payments over time—constitute the buyer's total price of a house.

When the buyer does not have enough money to pay cash for a house, or chooses not to pay cash, he or she usually seeks a mortgage from a bank or other lending institution. Before the bank will approve a mortgage, which is a long-term loan, the bank will require the buyer to have a specified minimum amount for the down payment. The **down payment** is the amount of cash the lending institution insists the buyer must pay to the seller before it will grant the buyer a mortgage. If the buyer has the down payment and meets the other criteria for the mortgage, the lending institution prepares a written agreement stating the terms of the loan. Items such as the repayment schedule, the length of the loan, specification of whether the loan can be assumed by another party, and the penalty if payments are late are included in the agreement. The party borrowing the money accepts the terms of this agreement and gives the lending institution the title or deed of the property as security. This agreement is called the mortgage. When making payments on the loan, we say that we are making a payment on the mortgage.

> **Homeowner's mortgage:** A long-term loan in which the property is pledged as security for payment of the difference between the down payment and the sale price.

Their first house.
Photo by Allen R. Angel

The two most popular types of loans available today through savings banks and commercial banks are **variable-rate loans** and **conventional loans.** The major difference between the two loans is that the interest rate for a conventional loan is fixed for the duration of the loan, while the interest rate for the variable-rate loan may change every 3 months, 6 months; 1 year, or some other predetermined time. We will first discuss the requirements for a loan that are the same in either case and then give examples of both types of loans.

The size of the down payment required depends on who is loaning the money; how old the property is, and whether or not it is easy to borrow money at that particular time. The down payment required by the lending institution can vary from 10% to 50% of the purchase price of the property. A larger down payment is required when money is "tight," that is, when it is difficult to borrow money. Furthermore, most lending institutions tend to require larger down payments on older homes and smaller down payments on newer homes.

Most lending institutions now require the buyer to pay one or more points for their loan at the time of the closing (the final step in the sale process). **One point** amounts to 1% of the mortgage money (the amount being borrowed). By charging points the bank is able to reduce the rate of interest on the mortgage. This has the effect of reducing the size of the monthly payments and enables more people to purchase houses. However, by charging points the stated rate of interest is not the APR for the loan. The APR would be determined by adding the amount paid for points to the total interest paid and then using an APR table.

■ Example 1 _____

The Smiths wish to purchase a house selling for $55,000. Their banker has informed them that the bank is requiring a 20% down payment and requires a payment of two points at the time of closing.

a) What is the required down payment?

b) With the 20% down payment, what is the mortgage on the property?

c) What is the cost of two points on the mortgage determined in part (b)?

Solution

a) The required down payment is 20% of $55,000:

$$0.20 \times \$55,000 = \$11,000$$

b) To find the mortgage on the property, subtract the down payment from the selling price of the house:

$$\$55,000 - \$11,000 = \$44,000$$

c) To find the cost of two points on a mortgage of $44,000, find 2% of $44,000:

$$0.02 \times \$44,000 = \$880$$

When the Smiths sign the mortgage, they must pay the bank $880. The $11,000 down payment is paid to the seller.   ❖

Banks use a formula to determine the maximum monthly payment that they feel is within the purchaser's ability to pay. The banker will first determine the buyer's **adjusted monthly income** by subtracting from the gross monthly income (total income before any deductions) any fixed monthly payments with more than six payments remaining (such as for a car, furniture, TV, or boat). The banker then multiplies the adjusted monthly income by 25%. (This percent may vary in different locations.) In general, this product is the maximum monthly payment the lending institution feels the purchaser can afford to pay for a house. This estimated payment must cover principal, interest, and taxes, and it is sometimes required to cover fire insurance as well. Note that the taxes and insurance are not necessarily paid to the bank; they may be paid directly to the tax collector and the insurance company. Example 2 shows how a bank uses the formula to determine whether or not the buyer qualifies for a mortgage.

■ Example 2 _____

In Example 1 the Smiths' gross monthly income is $2700. They have ten remaining payments of $125 per month on their car and eight remaining payments of $65 per month on new furniture. The taxes on the house they would like to purchase are $90 per month, and the fire insurance is $15 per month.

a) What maximum monthly payment does the bank feel the Smiths can afford?

b) The Smiths would like to get a 20-year, $44,000 conventional mortgage. Do they qualify for this mortgage if the current rate of interest is 12.5%?

Solution

a) To find the maximum monthly payment that the bank feels the Smiths can afford, first find their adjusted monthly income:

$$
\begin{array}{ll}
\$2700 & \text{Gross income} \\
\underline{-190} & \text{Monthly payments (car and furniture)} \\
\$2510 & \text{Adjusted monthly income}
\end{array}
$$

Next find 25% of the adjusted monthly income:

$$0.25 \times \$2510 = \$627.50$$

This is the maximum monthly payment the bank determines that the Smiths can afford with their current income.

b) To determine whether the Smiths can qualify for a 20-year conventional mortgage with their current income, it is necessary to determine the monthly payment on the mortgage. The monthly payment on the mortgage includes principal and interest. All lending institutions and most lawyers have a set of tables that tell the monthly mortgage payment per thousand dollars for a specific number of years at a specific interest rate. Table 8.8 shows selected monthly mortgage payments. With the current rate of interest of 12.5%, a 20-year loan would have a monthly payment of $11.36 per thousand dollars of mortgage (circled in the table).

To determine the Smith's payment, we divide the mortgage by $1000:

$$\frac{\$44,000}{\$1,000} = 44$$

The monthly mortgage payment is found by multiplying the number of thousands of dollars of mortgage, 44, by the value found in Table 8.8:

$$\$11.36 \times 44 = \$499.84$$

TABLE 8.8

Monthly Payment per $1,000 of Mortgage, Including Principal and Interest

Rate %	Number of Years				
	20	25	30	35	40
8	$8.36	$7.72	$7.34	$7.10	$6.95
8.5	8.68	8.05	7.69	7.47	7.33
9	9.00	8.40	8.05	7.84	7.72
9.5	9.33	8.74	8.41	8.22	8.11
10	9.66	9.09	8.70	8.60	8.50
10.5	9.98	9.44	9.15	8.98	8.89
11	10.32	9.80	9.52	9.37	9.28
11.5	10.66	10.16	9.90	9.76	9.68
12	11.01	10.53	10.29	10.16	10.08
12.5	(11.36)	10.90	10.67	10.55	10.49
13	11.72	11.28	11.06	10.95	10.90
13.5	12.07	11.66	11.45	11.35	11.30
14	12.44	12.04	11.85	11.76	11.71
14.5	12.80	12.42	12.25	12.16	12.12
15	13.17	12.81	12.64	12.57	12.53
15.5	13.54	13.20	13.05	12.98	12.94
16	13.91	13.59	13.45	13.38	13.36
16.5	14.29	13.98	13.85	13.79	13.77
17	14.67	14.38	14.26	14.21	14.18

To the $499.84 we add the monthly cost of fire insurance, $15, and real estate taxes, $90, for a total cost of $604.84. Since $604.84 is less than the maximum monthly payment the bank has determined that the Smiths can afford ($627.50), the Smiths will most likely be granted the loan. ❖

To show the effect of increasing the time period on the monthly payments, we have calculated the total monthly payments on the mortgage in Example 2 for the other periods of time. The results are $584.60 for 25 years, $574.48 for 30 years, $569.20 for 35 years, and $566.56 for 40 years. By extending the time period of the mortgage we decrease the amount of the monthly payment. Increasing the number of payments, however, will increase the total cost of the house because of the additional interest paid.

■ **Example 3** _____

The Smiths in Examples 1 and 2 obtain a 20-year $44,000 conventional mortgage at 12.5% on a house selling for $55,000. Their monthly mortgage payment, including principal and interest, is $499.84.

a) Determine the amount the Smiths will pay for their house.

b) How much of the cost will be interest?

c) How much of the first payment on the mortgage is applied to the principal?

Solution

a) To find the total amount the Smiths will pay for the house, we do the following:

1. Multiply the monthly payment times 12 to find the amount paid annually to the bank in mortgage payments (interest and principal).

2. Multiply the annual amount in mortgage payments by the number of years, in this case 20.

3. Find the sum of the mortgage payments for 20 years and the down payment. This sum will be the total cost of the house.

$499.84	Mortgage payment for one month
× 12	
$5998.08	Mortgage payments for one year
× 20	Number of years
$119,961.60	Mortgage payments for 20 years
+ $11,000.00	Down payment
$130,961.60	Total cost of the house

Note: The result might not be the exact cost of the house, since the final payment on the mortgage might be slightly more or less than the regular monthly payment.

b) To determine the amount of interest paid over 20 years, subtract the purchase price of the house from the total cost:

$130,961.60   Total cost

− 55,000.00   Purchase price

$75,961.60   Interest

c) We find the amount of the first payment applied to the principal by subtracting the amount of interest on the first payment from the monthly payment. We can find the interest by using the simple interest formula:

$$i = prt$$
$$= \$44{,}000 \times 0.125 \times \frac{1}{12}$$
$$= \$458.33$$

Now subtract the interest for the first month from the monthly mortgage payment. The difference will be the amount paid on the principal the first month:

$499.84   Monthly mortgage payment

− 458.33   Interest paid for the first month

$ 41.51   Principal paid for the first month

We see that the first payment of $499.84 consists of $458.33 in interest and $41.51 in principal. The $41.51 is applied to reduce the loan. Thus the balance due after the first payment is $44,000 − $41.51, or $43,958.49.   ❖

The principal and the interest can be calculated in this manner for all the payments. This would be a tedious task, however. With a computer it is easy to prepare a list containing the payment number, payment on the interest, payment on the principal, and balance of the loan. Such a list is called a loan **amortization schedule.** Part of an amortization schedule for the Smith's loan in Example 2 is given in Table 8.9.

Now let us consider variable-rate mortgages. Generally, with variable-rate mortgages, the monthly mortgage payment remains the same for a 1-, 2-, or 5-year period even though the interest rate may change every 3 months, 6 months, or some other predetermined period. The interest rate in a variable-rate mortgage may be based on an index that is determined by the

Federal Savings and Loan Insurance Corporation or it may be based on the interest rate of a 3-month, 6-month, or 1-year Treasury bill. When the base is a 3- or 6-month Treasury bill, the actual interest rate charged for the mortgage is determined by adding 3% to $3\frac{1}{2}\%$ (called the add on rate) to the rate of the Treasury bill. Thus if the rate of the Treasury bill is 6% and the add on rate is 3%, then the interest rate charged is 9%.

TABLE 8.9

Amortization Schedule

Annual % rate: 12.5   Monthly payment: $499.84   Loan: $44,000
Term: Years 20   Months 0   Periods 240 months

| Payment | Payment on | | Balance |
Number	Interest	Principal	of Loan
1	$458.33	$ 41.51	$43,958.49
2	457.90	41.93	43,916.55
3	457.46	42.38	43,874.18
4	457.02	42.82	48,831.36
23	447.69	52.15	42,926.03
24	447.15	52.69	42,873.34
119	358.71	141.13	34,294.61
120	357.24	142.60	34,152.00
179	236.96	262.88	22,485.64
180	234.23	265.61	22,220.02
239	10.25	489.59	494.81
240	5.15	494.81	

■ **Example 4** _____

The Jacksons purchased a house for $76,000 with a down payment of $25,000. They obtained a 30-year variable-rate mortgage. The terms of the mortgage are: the interest rate is based on a 6-month Treasury bill, the effective interest rate is 3% above the rate of the Treasury bill on the date of adjustment (3% is the add on rate), the interest rate is adjusted every 6 months, the interest rate will not change more than 1% (up or down) when the interest rate is adjusted, the maximum interest rate that can be charged for the duration of the loan is 16%, there is no lower limit on the interest rate, the initial mortgage interest rate is 9.5%, the monthly payment of interest and principal is adjusted every 5 years.

a) Determine the initial monthly payment for interest and principal.

b) Determine the adjusted interest rate in 6 months if the interest rate on the Treasury bill at that time is 6.25%.

Solution

a) To determine the initial monthly payment for interest and principal, we do the following: divide the amount of the loan, $51,000 ($76,000 − $25,000), by $1,000. The result is 51. Now multiply the number of thousands of dollars of mortgage, 51, by the value found in Table 8.8, with $r = 9.5\%$ for 30 years:

$$\$8.41 \times 51 = \$428.91$$

Thus the initial monthly payment for interest and principal is $428.91. This amount will not change for the first five years of the mortgage.

b) The adjusted interest rate in six months will be

$$\text{treasury bill rate} + \text{add on rate} = \text{adjusted interest rate}$$
$$6.25\% + 3\% = 9.25\%.$$

Since this rate is less than the initial interest rate and the monthly payment remains the same, the additional money paid the bank is applied to the principal.   ❖

In Example 4 the monthly interest and principal payment of $428.91 would pay off the loan in 30 years if the interest remained constant at 9.5%. What happens if the interest rate drops and stays lower than the initial 9.5% for the length of the loan? In this case at the end of each 5-year period the bank may give the payee a choice: (a) keep the same monthly payment and pay off the loan in less than 30 years or (b) reduce the monthly payment and pay off the loan in 30 years. What happens if the interest rates increase above the initial 9.5% rate? In this case part or, if necessary, all of the principal part of the monthly payment would be used to meet the interest obligation. At the end of the 5-year period the bank will increase the monthly payment so that the loan can be repaid in the 30-year time period. There is generally one restriction, that the bank may not increase the monthly payment by more than 25% of the initial monthly payment. If the monthly payment cannot be increased sufficiently to repay the loan in 30 years, then the bank will increase the time period by the number of years required.

Other Types of Mortgages

Conventional mortgages and variable-rate mortgages are not the only methods of financing the purchase of a house. In the following paragraphs are brief descriptions of other methods. Two of these, FHA and VA mortgages, are backed by the government.

FHA mortgage: A house can be purchased with a smaller down payment than with a conventional mortgage if the individual qualifies for a Federal Housing Administration (FHA) loan. He or she applies for the loan through a local bank. The bank provides the money, but the FHA insures the loan. The down payment for an FHA loan is as low as 3 to 5 percent of the purchase price, rather than the standard 10 to 50%. Another advantage is that FHA loans can be assumed at the original rate of interest on the loan. For example, if you purchase a home today that has an 8% FHA mortgage, you, the new buyer, will be able to assume the 8% mortgage regardless of the current interest rates. However, to be able to assume a mortgage, the purchaser must be able to make a down payment equal to the difference between the purchase price and the balance due on the original mortgage.

There are some drawbacks to FHA loans. The maximum loan allowed is $72,000 for a single-family house, except in high-cost cities, where the maximum is set at $90,000. The seller must pay points. If the seller refuses to pay the points, the buyer cannot obtain an FHA mortgage for that property. The borrower must pay an FHA insurance premium as part of the monthly mortgage payment. The insurance premium is calculated at rate of one-half percent of the unpaid balance of the loan on the anniversary date of the loan. Thus even though the insurance premium decreases each year, it adds to the monthly payments.

VA mortgages: The Veterans Administration requires a 25% down payment for all VA-backed loans. Each veteran applying for a loan can receive government backing of the loan up to $27,500. The amount depends on the veteran's benefit package, which is determined from his or her tour of duty in the service. The amount the government backs can be used in lieu of part or all of the down payment. Thus it is possible for a veteran with maximum benefits to purchase a $110,000 house with no down payment (25% of $110,000 = $27,500).

For both VA and FHA mortgages the government sets maximum interest rates the lender can charge. Some other facts about a VA loan are that the loan is always assumable, the seller may be asked to pay points, and there is no monthly insurance premium.

In the mid-1970s, interest rates were very high, and it was difficult to obtain conventional, VA, or FHA mortgages. As a result of the high interest rates and tight money, lending institutions developed **alternative mortgage instruments** known as "AMIers." These new mortgages are designed to be customized for the home buyer's pocketbook. We will look at three of these alternative mortgage instruments.

Growing-equity mortgage (GEM): With a GEM mortgage the interest rate is fixed for the life of the mortgage, but the payments are scheduled to rise from 4% to 10% per year. By increasing the monthly mortgage payment you decrease the time period of the loan. For example, by increasing your monthly mortgage payment you may pay off the loan on your home in 15 years instead of 25 years. With this shorter payback period, the borrower gets a lower interest rate on the original loan.

Graduated payment mortgage (GPM): A GPM mortgage is designed so that for the first five to ten years the size of the mortgage payment is smaller than the payments for the remaining time of the mortgage. After the 5- to 10-year period the mortgage payments remain constant for the duration of the mortgage. Depending on the size of the mortgage, the lender might find that the monthly payments made for the first few years may actually be less than the interest owed on the loan for those few years. The interest not paid during the first few years of the mortgage is then added to the original loan. This type of loan is strictly for the individual who is confident that his or her annual income will be increasing at a rate commensurate with the higher mortgage payments.

Balloon-payment mortgage (BPM): This type of loan is for the person who needs time to find a permanent loan. It typically works this way: The individual pays the interest for three to five years, the period of the loan. At the end of the 3- to 5-year period, the buyer must repay the entire principal unless the lender agrees that the loan may be extended for another period of time.

For further information on any of these mortgages, consult your local banker.

DID YOU KNOW?

The contract to buy a house—or the purchase offer, as it is commonly called—should include the following information.

1. The names of the buyers and sellers
2. The sale price of the house
3. The amount of the down payment
4. The date the buyer is to take possession of the house
5. A legal description of the property
6. A list of any personal property that is to be included in the sale
7. A description of the mortgage terms
8. The date by which the transaction must be completed
9. Easement rights, if any, to the property
10. Specification of closing costs and their payment

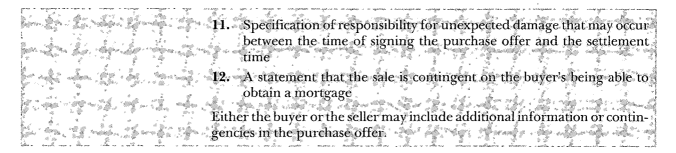

11. Specification of responsibility for unexpected damage that may occur between the time of signing the purchase offer and the settlement time

12. A statement that the sale is contingent on the buyer's being able to obtain a mortgage

Either the buyer or the seller may include additional information or contingencies in the purchase offer.

EXERCISES 8.6

1. Eric is buying a house selling for $45,000. The bank is requiring a minimum down payment of 25%. The current mortgage rate is 12%.
   a) Determine the amount of the required down payment.
   b) Determine the monthly mortgage payment for a 35-year loan with a minimum down payment.

2. Ann is buying a house selling for $110,000. The bank is requiring a minimum down payment of 10%. The current mortgage rate is 10.5%.
   a) Determine the amount of the required down payment.
   b) Determine the monthly mortgage payment for a 40-year loan with a minimum down payment.

3. Barb is buying a house selling for $58,000. The bank is requiring a minimum down payment of 10%. To obtain a 30-year mortgage at 12.5%, she must pay two points at the time of closing.
   a) What is the required down payment?
   b) With the 10% down payment, what is the mortgage on the property?
   c) What is the cost of two points on the mortgage determined in part (b)?

4. The Schwartzes are buying a house that sells for $120,000. The bank is requiring a minimum down payment of 15%. To obtain a 30-year mortgage at 11.5%, they must pay three points at the time of closing.
   a) What is the required down payment?
   b) With the 15% down payment, what is the mortgage on the property?
   c) What is the cost of three points on the mortgage determined in part (b)?

5. Barb's gross monthly income is $2500. She has 12 remaining payments of $95 on furniture and appliances. The taxes and fire insurance on the house are $105 per month.
   a) What maximum monthly payment does the bank feel that Barb can afford?
   b) Barb would like to get a 30-year, $52,200 mortgage. Does she qualify for this mortgage with a 12.5% interest rate?

6. The Dechs' gross monthly income is $6200. They have 15 remaining payments of $290 on a new car. The taxes on the house are $150 and the fire insurance is $45 per month.
   a) What maximum monthly payment does the bank feel that the Dechs can afford?
   b) The Dechs would like to get a 30-year, $102,000 mortgage. Do they qualify for this mortgage with an 11.5% interest rate?

7. Jane obtains a 30-year, $52,000 conventional mortgage at 12.5% on a house selling for $58,000. Her monthly payment, including principal and interest, is $554.84.
   a) Determine the total amount Jane will pay for her house.
   b) How much of the cost will be interest?
   c) How much of the first payment on the mortgage is applied to the principal?

8. The Peppers obtain a 25-year, $102,000 conventional mortgage at 11.5% on a house selling for $120,000. Their monthly mortgage payment, including principal and interest, is $1036.32.
   a) Determine the amount the Peppers will pay for their house.
   b) How much of the cost will be interest?
   c) How much of the first payment on the mortgage is applied to the principal?

9. The Millers purchased a home selling for $95,000 with a 20% down payment. The period of the conventional mortgage is 35 years, and the rate of interest is 14.5%.
    a) Determine the amount of the down payment.
    b) Determine the monthly mortgage payment.
    c) Determine the total cost of the house.
    d) Determine how much of the first payment on the loan is applied to the principal.
10. The Maxwells found a house that is selling for $65,000. The taxes on the house are $1200 per year, and the fire insurance is $175 per year. They are requesting a conventional loan from the local bank. The bank is currently requiring a 25% down payment, and the interest rate is 10.5%. The Maxwells' monthly income is $3750. They have more than 6 monthly payments remaining on a car, a boat, and furniture. The total monthly payments for these items is $420.
    a) Determine the required down payment.
    b) Determine the Maxwells' adjusted monthly income.
    c) Determine the maximum monthly payment the bank feels that they can afford.
    d) Determine the monthly payments of principal and interest for a 20-year loan.
    e) Determine their total monthly payment, including insurance and taxes.
    f) Determine whether the Maxwells qualify for the 20-year loan.
    g) Determine how much of the first payment on the loan is applied to the principal.
11. Kathy would like to buy a condominium selling for $75,000. The taxes on the property are $900 per year, and the fire insurance is $336 per year. Kathy's gross monthly income is $3300. She has 15 monthly payments of $135 remaining on her van. She has $20,000 in her savings account. The bank is requiring 20% down and is charging 13.5% interest.
    a) Determine the required down payment.
    b) Determine the maximum monthly payment that the bank feels Kathy can afford.
    c) Determine the monthly payment of principal and interest for a 35-year loan.
    d) Determine her total monthly payments, including insurance and taxes.
    e) Does Kathy qualify for the loan?
    f) Determine how much of the first payment on the mortgage is applied to the principal.
    g) Determine the total amount she pays for the

condominium with a 35-year conventional loan.
    h) Determine the total interest paid for the 35-year loan.
12. The Christophers are negotiating with two banks for a mortgage to buy a house selling for $75,000. The terms at bank A are a 10% down payment, an interest rate of 13%, a 30-year conventional mortgage, and three points to be paid at the time of closing. The terms of bank B are a 20% down payment, an interest rate of 14.5%, a 25-year conventional mortgage and no points. Which loan should the Christophers select in order for the total cost of the house to be less?
13. The Kellys purchased a house for $65,000 with a down payment of $13,000. They obtained a 30-year variable rate mortgage. The terms of the mortgage are as follows: The interest rate is based on a 3-month Treasury bill, the effective interest rate is 3.25% above the rate of the Treasury bill on the date of adjustment, the interest rate is adjusted every 3 months, the interest rate will not change more than 1% (up or down) when the interest rate is adjusted, the maximum interest rate that can be charged for the duration of the loan is 16%, there is no lower limit on the interest rate, the initial mortgage interest rate is 8.5%, and the monthly payment of interest and principal is adjusted annually.
    a) Determine the initial monthly payment for interest and principal.
    b) Determine the interest rate in 3 months if the interest rate on the Treasury bill at the time is 5.65%.
    c) Determine the interest rate in 6 months if the interest rate on the Treasury bill at the time is 4.85%.
14. The Grahams can afford to pay $950 a month in mortgage payments. If the bank will give them a 25-year conventional mortgage at 12% and requires a 25% down payment, what is (a) the maximum mortgage the bank will grant the Grahams? (b) the highest-price house they can afford?
*15. The Baileys purchased a house for $55,000 with a down payment of $6000. They obtained a 30-year variable-rate mortgage. The terms of the mortgage are as follows: the interest rate is based on a 3-month Treasury bill, the effective interest rate is 3.25% above the rate of the Treasury bill on the date of adjustment, the interest rate is adjusted every 3 months, the interest rate will not change more than 1% (up or down) when the interest rate

is adjusted, the maximum interest rate that can be charged for the duration of the loan is 16%, there is no lower limit on the interest rate, the initial mortgage interest rate is 9.5%, and the monthly payment of interest and principal is adjusted semi-annually.

a) Determine the initial monthly payment for interest and principal.

b) Determine an amortization schedule for months 1–3.

c) Determine the interest rate for months 4–6 if the interest rate on the Treasury bill at the time is 6.13%.

d) Determine an amortization schedule for months 4–6.

e) Determine the interest rate for months 7–9 if the interest rate on the Treasury bill at the time is 6.21%.

RESEARCH ACTIVITIES

16. An important part of buying a house is the closing. The exact procedures for closing differ with individual cases and in different parts of the country. In any closing, however, there are expenses to both parties, the buyer and the seller. To determine what is involved in the closing of a property in your community, contact a lawyer, a real estate agent, or a banker. Explain that you are a student and that your objective is to understand the procedure for closing a real estate purchase and the costs to both the buyer and the seller. Select a specific piece of property that is for sale. List the asking price, and then determine the total closing costs to both the buyer and the seller. Here is a partial list of the most common costs. Consider these in your research.

a) Fee for title search and title insurance
b) Credit report on buyer
c) Fees to the lender for services in granting the loan
d) Fee for property survey
e) Fee for recording of the deed
f) Appraisal fee
g) Lawyer's fee
h) Escrow accounts (taxes, insurance)
i) Mortgage assumption fee

17. Visit local banks and investigate the different types of mortgages that are available. List the advantages and disadvantages of each type. Determine which type you would select and explain your decision.

SUMMARY

Ratios and proportions are often used in comparing quantities. A ratio is a quotient of two numbers. A proportion is a statement of equality between two ratios.

A percent is a ratio of some number to 100. Percents are often used to compare two or more quantities.

A promissory note is a written agreement containing the terms of a loan between two parties. The agreement contains the effective date, face value, rate of simple interest, maturity date, name of the lender, and signature of the borrower. Two types of loans are the security loan and the cosigner loan. The interest charged on a promissory note is simple interest, calculated with the formula $i = prt$. If the lender collects the interest on the effective date of the note, it is called a discounted note.

Investments may be classified as fixed or variable. A fixed investment, such as a savings account, guarantees the return of the original principal plus a fixed amount of interest. Interest that is computed on both principal and accumulated interest is called compound interest.

Installment buying is an accepted way of making purchases. To understand installment buying, it is essential that the consumer understand interest and annual percentage rate (APR). As a safeguard against misquoted rates of interest, it is beneficial to be able to calculate the APR.

The consumer should realize that purchasing items on a charge account may be a form of installment buying. When the balance is not fully paid at the end of each billing cycle, interest is charged on the unpaid balance of the previous month. The miminum payment on a MasterCard or Visa account is determined by dividing the unpaid balance by 36 and rounding up to the nearest dollar.

The home buyer who does not pay cash for the house must have enough cash to make the down payment. In addition the buyer must have sufficient income to pay the monthly mortgage payments. To qualify for a conventional or variable-rate mortgage, the buyer's monthly payments for principal and interest on the mortgage, fire insurance, and taxes should not exceed 25% of his or her adjusted gross monthly income.

REVIEW EXERCISES

1. Two 32-oz bottles of Canada Dry club soda cost a total of $1.39. A six-pack of 16-oz bottles cost $2.29. Which purchase has the lower unit price?
2. Jane's car can travel 230 miles on 12.8 gallons of gas. How many gallons of gas would she need to drive from Denver to Los Angeles, a distance of 1139 miles?
3. A model car is made in the scale of 1:20. If the bumper on the actual car is 36 inches long what is the length of the bumper on the model?
4. What is 13% of 28?
5. Seventy-two is 30% of what number?
6. Mr. Jason purchased a car for $7200. A week later he saw the same car advertised for $6500. What is the percent increase or decrease of the advertised price over the price he paid for the car?

Find the interest for the following problems using the simple interest formula:

7. $p = \$1200, r = 5\%, t = 30$ days
8. $p = \$675, r = 4\%, t = 90$ days
9. $p = \$1275, r = 12\%, t = 120$ days
10. $p = \$450, r = 8\frac{1}{2}\%, t = 3$ years
11. $p = \$5500, r = 11\frac{1}{2}\%, t = 6$ months
12. Bobby borrowed $3500 from the bank for eight months. His mother was cosigner of the promissory note. The rate of interest on the note was $12\frac{1}{2}\%$. How much did Bobby pay the bank on the date of maturity?
13. Sue borrowed $2300 from her bank for 90 days at a simple interest rate of $10\frac{1}{2}\%$.
    a) How much interest did she pay for the use of the money?
    b) How much did she pay the bank on the date of maturity?

14. The Manbecks borrowed $4000 for 18 months from the bank, using stock as security. The bank discounted the loan at $11\frac{1}{2}\%$.
   a) How much interest did the Manbecks pay the bank for the use of the money?
   b) How much did they receive from the bank?
   c) What was the true rate of interest?

15. Carol borrowed $800 for six months from her bank, using her savings account as security. A bank rule limits the amount that can be borrowed in this manner to 85% of the amount in the savings account. The rate of interest is two percent higher than the interest rate being paid on the savings account. The current rate on the savings account is $5\frac{1}{2}\%$.
   a) What rate of interest will the bank charge for the loan?
   b) Find the amount that Carol must repay in six months.
   c) How much money must she have in her account in order to borrow $800?

16. Donald bought a new Dodge Omni for $6640. He was required to make a 25% down payment. He financed the balance with the dealer on a 48-month payment plan. The salesperson told him that the interest rate was 8% simple interest on the principal for the duration of the loan.
   a) Find the amount of the down payment.
   b). Find the amount to be financed.
   c) Find the total interest paid.
   d) Find the APR.

17. Joyce can buy a cross-country skiing outfit for $135. The store is offering the following terms: $35 down and 12 monthly payments of $9.
   a) Find the interest paid.
   b) Find the APR.

18. The Millers used their MasterCard to pay for Christmas presents costing $788.50. When the bill arrived, they could not pay the total amount that was due. They paid $288.50 the first month and decided to pay the balance in installments of $125 per month.
   a) Set up a table showing the repayment of the loan.
   b) Determine the amount of interest that was paid.

19. Determine the interest and the amount when $500 is invested for three years at 6%
   a) compounded annually,
   b) compounded semiannually,
   *c) compounded quarterly.

20. The Clars have decided to build a new house. The contractor has quoted them a price of $72,500. The taxes on the house will be $2350 per year, and fire insurance will be $350 per year. They have applied for a conventional loan from a local bank. The bank is requiring a 35% down payment, and the interest rate on the loan is 11%. The Clar's annual income is $52,000. They have more than 6 monthly payments remaining on each of the following: $218 on a car, $120 on new furniture, and $190 on a camper.
   a) Determine the required down payment.
   b) Determine their adjusted monthly income.
   c) Determine the maximum monthly payment the bank feels that they can afford.
   d) Determine the monthly payment of principal and interest for a 40-year loan.
   e) Determine their total monthly payment, including insurance and taxes.
   f) Do the Clars qualify for the mortgage?

21. The Barkers purchased a home selling for $89,900 with a 25% down payment. The period of the mortgage is 30 years and the interest rate is 13.5%. Determine the following.
   a) The amount of the down payment
   b) The monthly mortgage payment
   c) The total cost of the house
   d) The amount of the first payment that is applied to the principal
22. The Greens purchased a house for $105,000 with a down payment of $26,250. They obtained a 30-year variable rate mortgage. The terms of the mortgage are: the interest rate is based on the 6 month Treasury bill, the effective interest rate is 3.50% above the rate of the Treasury bill on the date of adjustment, interest rate is adjusted every 6 months, the interest rate will not change more than 1% (up or down) when the interest rate is adjusted, the maximum interest rate that can be charged for the duration of the loan is 16%, there is no lower limit on the interest rate, the initial mortgage interest rate is 9.5%, monthly payment of interest and principal is adjusted annually.
   a) Determine the initial monthly payments for interest and principal.
   b) Determine the interest rate in 6 months if the interest rate on the Treasury bill at the time is 5.65%.
   c) Determine the interest rate in 6 months if the interest rate on the Treasury bill at the time is 5.85%.
23. The Chus purchased a house for $95,000 with a down payment of $19,000. They obtained a 30-year variable rate mortgage. The terms of the mortgage are: the interest rate is based on the 3 month Treasury bill, the effective interest rate is 3.50% above the rate of the Treasury bill on the date of adjustment, the interest rate is adjusted every 3 months, the interest rate will not change more than 1% (up or down) when the interest rate is adjusted, the maximum interest rate that can be charged for the duration of the loan is 16%, there is no lower limit on the interest rate, the initial mortgage interest rate is 12.5%, the monthly payment of interest and principal is adjusted semiannually.
   a) Determine the initial monthly payment for interest and principal.
   b) Determine an amortization schedule for months 1–3.
   c) Determine the interest rate for months 4–6 if the interest rate on the Treasury bill at the time is 8.90%.
   d) Determine an amortization schedule for months 4–6.
   e) Determine the interest rate for months 7–9 if the interest rate on the Treasury bill at the time is 9.00%.

CHAPTER TEST

1. Property tax is $73.552 per $1000 of assessed value. If Ruby's house is assessed at $9,500, what is her property tax?
2. The average salary in Hometown USA was $28,900 in 1985 and increased to $32,400 in 1988. What was the percent change during this period?

Find the interest for the following problems using the simple interest formula.

3. $p = \$800, r = 7\%, t = 3$ years
4. $p = \$800, r = 7\%, t = 3$ months

Cory borrowed $6000 from the bank for 9 months. The rate of interest charged is 12.5%.

5. How much interest did he pay for the use of the money?
6. What is the amount he repaid to the bank on the due date of the loan?

A new car sells for $9500. To finance the car through a bank, the bank will require a down payment of 15% and monthly payments of $224.40 for 48 months.

7. How much money will the purchaser borrow from the bank?
8. How much interest will the individual pay to the bank?
9. What is the APR?

Elaine took a trip during the first week in July and charged $1375 on her MasterCard account. Assume that she made no other charges in July and had paid off all other charges.

10. What was the minimum amount due in August?
11. Assuming no other charges and that she paid $375 in August, what was the balance due in September?
12. What was the amount of interest due in September?
13. If she paid $450 in September, what part of the payment was credited to the balance due?

Brian signed a $4400 note with interest at 12.5% for 90 days on August 1. Brian made a payment of $2000 on September 15.

14. How much did he owe the bank on the date of maturity?
15. What was the total amount of interest paid on the loan?

Compute the amount and the compound interest for each of the following:

	Principal	Time	Rate	Compounded	
16.	$3600	4 years	12%	Quarterly	(by Table 8.2)
17.	$4800	2 years	16%	Semiannually	(by the Compound Interest Formula)

The Klines decided to build a new house. The contractor quoted them a price of $94,500 including the lot. The taxes on the house would be $2875 per year, and fire insurance would cost $450 per year. They have applied for a conventional loan from a bank. The bank is requiring a 25% down payment, and the interest rate is $10\frac{1}{2}\%$. The Klines' annual income is $76,500. They have more than 6 monthly payments remaining on each of the following: $220 for a car, $175 for new furniture, and $210 on a college education loan.

18. What is the required down payment?
19. Determine their adjusted monthly income.
20. What is the maximum monthly payment the bank feels the Klines can afford?
21. Determine the monthly payments of principal and interest for a 20-year loan.
22. Determine their total monthly payments including insurance and taxes.
23. Find the total cost of the house after 20 years of payments (do not include taxes and insurance).
24. How much of the total cost of the house is interest?

25. The Spinas obtain a variable-rate mortgage that includes the following terms: The interest rate is based on the 3-month Treasury bill, the interest is adjusted every three months, the interest has an add on rate of 3.25%. The initial mortgage interest rate is 8.75%. What is the interest rate for payments 4–6 if the rate on the 3 month treasury bill is 5.13% on the adjustment date?

CHAPTER 9

Mathematical Systems

9.1 MATHEMATICAL SYSTEMS

When you learned how to add integers, you were introduced to a mathematical system. When you learned how to multiply integers, you became familiar with a second mathematical system. The set of integers with the operation of subtraction and the set of integers with the operation of division are two other examples of mathematical systems.

The operations of addition, subtraction, multiplication, and division are called binary operations. A **binary operation** is an operation, or rule, that can be performed on two and only two elements of a set. The result is a single element. When we add *two* integers, the sum is *one* integer. When we multiply *two* integers, the product is *one* integer. Is finding the reciprocal of a number a binary operation? No, it is an operation on a single element of a set.

> A **mathematical system** consists of a set of elements and at least one binary operation.

Once a mathematical system is defined, there may be certain properties that are displayed in its structure. Consider the set of integers:

$$I = \{\ldots, -3, -2, -1, 0, 1, 2, 3, \ldots\}$$

We studied the integers in Section 5.2. The three dots at each end of the set indicate that the set continues in the same manner.

When we study the set of integers under the operations of addition or multiplication, we see that the commutative and associative properties hold. These properties were discussed in detail in Chapter 5. Since they are important to our discussion, let's review them. The general forms of the properties are given in Table 9.1.

TABLE 9.1

	Addition	Multiplication
Commutative property	$a + b = b + a$	$a \cdot b = b \cdot a$
Associative property	$(a + b) + c = a + (b + c)$	$(a \cdot b) \cdot c = a \cdot (b \cdot c)$

The integers are commutative under the operations of addition

$$2 + 4 = 4 + 2$$
$$6 = 6$$

and multiplication

$$2 \cdot 4 = 4 \cdot 2$$
$$8 = 8$$

but they are not commutative under subtraction

$$4 - 2 \neq 2 - 4$$
$$2 \neq -2$$

or division

$$4 \div 2 \neq 2 \div 4$$
$$2 \neq \frac{1}{2}$$

Similarly, the integers are associative under the operations of addition

$$(1 + 2) + 3 = 1 + (2 + 3)$$
$$3 + 3 = 1 + 5$$
$$6 = 6$$

and multiplication

$$(1 \cdot 2) \cdot 3 = 1 \cdot (2 \cdot 3)$$
$$2 \cdot 3 = 1 \cdot 6$$
$$6 = 6$$

Does the associative property hold for the integers under the operations of subtraction and division? See Exercises 3 and 4 at the end of this section.

When we say that a set of elements is commutative under a given operation, we mean that the commutative property holds for *any* elements a and b in the set. Similarly, when we say that a set of elements is associative under a given operation, we mean that the associative property holds for *any* elements a, b, and c in the set.

There are many different mathematical systems that can be developed. Some mathematical systems, such as addition of the integers, are commonly used, whereas others are more abstract and are used primarily in research or higher-level courses.

Let us study the mathematical system consisting of the set of integers under the operation of addition. Since there are an infinite number of integers, this mathematical system is an example of an **infinite mathematical system.** We will study certain properties of this mathematical system. The first property that we will examine is *closure*.

Closure

If we add any two integers, will the sum be an integer? The answer is yes. Some examples follow.

Addition of Integers		Sum
$3 + 9$	$=$	12
$-6 + 3$	$=$	-3
$-4 + (-3)$	$=$	-7

Whenever we add any two integers, the sum will be an integer. Since the sum of any two integers is an integer, we say that the set of integers is **closed,** or satisfies the **closure property,** under the operation of addition.

> If a binary operation is performed on any two elements of a set and the result is an element of the set, then we say that the set is **closed** (or has **closure**) under the given binary operation.

Is the set of integers closed under the operation of multiplication? The answer is yes. When any two integers are multiplied, the product will be an integer.

Is the set of integers closed under the operation of subtraction? Again the answer is yes. The difference of any two integers is an integer.

Is the set of integers closed under the operation of division? The answer is no. It is possible to find two integers whose quotient is not an integer. For example, if we select the integers 2 and 3, the quotient of 2 divided by 3 is 2/3, which is not an integer. Thus the integers are not closed under the operation of division.

We showed that the set of integers was not closed under the operation of division by finding two integers whose quotient was not an integer. A specific example illustrating that a specific property is not true in every case is called a **counterexample.** We often try to find a counterexample to confirm that a specific property is not always true.

Identity Element

Now we will discuss the identity element for the set of integers under the operation of addition. Is there an element in the set that, when added to any given integer, results in a sum that is the given integer? The answer is yes. The sum of 0 and any integer will be the given integer. For example, $1 + 0 = 0 + 1 = 1, 2 + 0 = 0 + 2 = 2$, and so on. For this reason we call 0 the

additive identity element for the set of integers. Note that for any integer a, $a + 0 = 0 + a = a$.

> An **identity element** is an element in the set such that when a binary operation is performed on it and any given element in the set, the result is the given element.

Is there an identity element for the set of integers under the operation of multiplication? The answer is yes; it is the number 1. Note that $2 \cdot 1 = 1 \cdot 2 = 2$, $3 \cdot 1 = 1 \cdot 3 = 3$, and so on. For any integer a, $a \cdot 1 = 1 \cdot a = a$. For this reason, 1 is called the **multiplicative identity element** for the set of integers.

Inverses

What integer when added to 4 gives a sum of 0; that is, $4 + \underline{\hphantom{xx}} = 0$? The blank is to be filled in with the integer -4: $4 + (-4) = 0$. We say that -4 is the additive inverse of 4, and 4 is the additive inverse of -4. Note that the sum of the element and its additive inverse gives the additive identity element 0. What is the additive inverse of 12? Since $12 + (-12) = 0$, -12 is the additive inverse of 12.

Here are some other examples of integers and their additive inverses:

Element	+	Additive Inverse	=	Identity Element
0	+	(0)	=	0
1	+	(−1)	=	0
2	+	(−2)	=	0
3	+	(−3)	=	0
−5	+	(5)	=	0

Note that for the operation of addition, every integer a has a unique inverse, $-a$, such that $a + (-a) = -a + a = 0$.

> When a binary operation is performed on two elements in a set and the result is the identity element for the binary operation, then the two elements are said to be **inverses** of each other.

Does every integer have an inverse under the operation of multiplication? For multiplication the product of an integer and its inverse must yield the multiplicative identity element, 1. What is the multiplicative inverse of 2?

That is, 2 times what number gives 1?

$$2 \cdot ? = 1$$

Since $2 \cdot \frac{1}{2} = 1$, $\frac{1}{2}$ is the multiplicative inverse of 2. However, since $\frac{1}{2}$ is not an integer, 2 does not have a multiplicative inverse in the set of integers.

Associative Property

We discussed the associative property earlier in this section. Does the associative property hold for the set of integers under the operation of addition? That is, does $(a + b) + c = a + (b + c)$ for any integers a, b, and c? Let us select some integers to see whether the associative property holds:

$$(2 + 3) + 4 = 2 + (3 + 4)$$
$$5 + 4 = 2 + 7$$
$$9 = 9 \qquad \text{True}$$

$$(-6 + 3) + (-5) = -6 + [3 + (-5)]$$
$$-3 + (-5) = -6 + (-2)$$
$$-8 = -8 \qquad \text{True}$$

By examining our specific examples we see that the associative property *appears* to hold. However, we have not proved that the associative property holds for every possible case. It has been proven that the associative property holds for the integers under the operation of addition. But this proof is beyond the scope of this book. Since the integers are associative under the operation of addition, $(a + b) + c = a + (b + c)$ for all integers a, b, and c.

Group

Let us review what we have learned about the mathematical system consisting of the set of integers under the operation of addition.

a) The set of integers is *closed* under the operation of addition.

b) The set of integers has an *identity element* under the operation of addition.

c) Each element in the set of integers has an *inverse* under the operation of addition.

d) The *associative property* holds for the set of integers under the operation of addition.

A mathematical system that is closed, has an identity element for the given operation, has an inverse element for each element in the set, and is associative under the given operation is called a **group.**

Any mathematical system that meets the following four requirements is called a **group.**

1. The set of elements is *closed* under the given operation.
2. There exists an *identity element* for the set.
3. Every element in the set has an *inverse.*
4. The set of elements is *associative* under the given operation.

Commutative Group

Does the commutative property hold for the set of integers under the operation of addition? That is, does $a + b = b + a$ for any integers a and b? If we study some examples, we see the commutative property *appears* to hold:

$$4 + 3 = 3 + 4$$
$$7 = 7 \qquad \text{True}$$
$$5 + (-2) = -2 + 5$$
$$3 = 3 \qquad \text{True}$$
$$-6 + (-7) = -7 + (-6)$$
$$-13 = -13 \qquad \text{True}$$

The commutative property does hold for the set of integers under the operation of addition. However, the proof of this is beyond the scope of this book.

A group that also satisfies the commutative property is called a **commutative** or **abelian group.**

Thus the set of integers under the operation of addition is not only a group, but is also a commutative group.

Table 9.2 summarizes the properties a mathematical system must have to be a commutative group.

To determine whether a given mathematical system is a group under a given operation, we will check, in the following order, to see whether (a) the system is closed under the given operation, (b) there is an identity element in the set for the given operation, (c) every element in the set of elements has an inverse under the given operation, and (d) the associative property holds under the given operation. If we find that any of these four requirements is *not* met, we can stop and state that the mathematical system is not a group. If we are asked to determine whether the mathematical system is a commutative group, then we also need to check to see that the commutative property holds for the given operation.

TABLE 9.2

Summary of the Properties of a Commutative Group

A mathematical system with a given set of elements and a binary operation * is a commutative group if the following conditions hold.

1. The system is closed. For any elements a and b in the set, $a * b$ is a member of the set.
2. There exists an identity element in the set. For any element a in the set, if $a * i = i * a = a$, then i is called the identity element.
3. Every element in the set has an inverse. For any element a in the set, there exists an element b such that $a * b = b * a = i$. Element b is the inverse of a, and a is the inverse of b.
4. The set is commutative under the operation *. For any elements a and b in the set, $a * b = b * a$.
5. The set is associative under the operation *. For any elements a, b, and c, in the set, $(a * b) * c = a * (b * c)$.

DID YOU KNOW?

Many famous mathematicians (including Joseph Louis Lagrange, Augustin-Louis Cauchy, and Johann Carl Friedrich Gauss) contributed to the development of the theory of groups. However, the two mathematicians who brought group theory into focus were Niels Abel (1802–1829), a Norwegian mathematician, and Evariste Galois (1811–1832), a French mathematician. Note that the abelian group was named for Niels Abel.

Both men had very short lives, and neither received much recognition until after death. Abel died at age 26 of malnutrition and chest complications. Galois died at age 21 in a duel at dawn over a woman. Galois seems to have suspected that he might be killed. On the eve of his death he wrote in a note to a friend, "Ask Jacobi or Gauss publicly to give their opinion, not as to the truth, but as to the importance of the theorems. . . . Subsequently there will be, I hope, some people who will find it to their profit to decipher all this mess."

The note was attached to what Galois thought were some new theorems in the theory of equations. They turned out to contain the essence of the theory of groups.

■ Example 1

Determine whether the set of rational numbers under the operation of multiplication form a group.

Solution: Recall from Chapter 5 that the rational numbers are the set of numbers of the form p/q where p and q are integers, $q \neq 0$. All fractions and integers are rational numbers.

Closure If we multiply any two rational numbers, the product will be a rational number. Therefore the rational numbers are closed under the operation of multiplication.

Identity Element The multiplicative identity element for the set of rational numbers is 1. Note, for example, $3 \cdot 1 = 1 \cdot 3 = 3$, $-6 \cdot 1 = 1 \cdot (-6) = -6$, and $\frac{3}{8} \cdot 1 = 1 \cdot \frac{3}{8} = \frac{3}{8}$. For any rational number a, $a \cdot 1 = 1 \cdot a = a$.

Inverse Elements For the mathematical system to be a group under the operation of multiplication, *each and every* rational number must have a multiplicative inverse in the set of rational numbers. Remember that for the operation of multiplication, the product of a number and its inverse must give the multiplicative identity element, 1. Let's check a few rational numbers:

Rational Number	·	Inverse	=	Identity Element
3	·	$\frac{1}{3}$	=	1
$\frac{2}{3}$	·	$\frac{3}{2}$	=	1
$-\frac{1}{5}$	·	-5	=	1

Does every rational number have an inverse under the operation of multiplication? By looking at the examples it may appear that each rational number does have an inverse. But there is one rational number that does not have an inverse. Do you know what that number is? It is the number 0:

$$0 \cdot ? = 1$$

Since there is no rational number that, when multiplied by 0, gives 1, 0 does not have a multiplicative inverse. Since every rational number does not have an inverse, this mathematical system is not a group.

There is no need at this point to check the associative property, since we have already shown that the mathematical system is not a group. ❖

EXERCISES 9.1

1. What is a binary operation?
2. What are the parts of a mathematical system?
3. Give an example to show that the commutative property does not hold for the set of integers under the operation of subtraction.
4. Give an example to show that the commutative property does not hold for the set of integers under the operation of division.
5. Give an example to show that the associative property does not hold for the set of integers under the operation of subtraction.
6. Give an example to show that the associative property does not hold for the set of integers under the operation of division.
7. What properties are required for a mathematical system to be a group?
8. What properties are required for a mathematical system to be a commutative group?

9. Is the set of positive integers a commutative group under the operation of addition? Explain.
10. Is the set of integers a group under the operation of multiplication? Explain.
11. Is the set of rational numbers a commutative group under the operation of addition? Explain.
12. Is the set of positive integers a group under the operation of subtraction? Explain.
13. Is the set of integers a group under the operation of subtraction? Explain.
*14. Is the set of irrational numbers a group under the operation of addition? Explain.
*15. Is the set of irrational numbers a group under the operation of multiplication? Explain.
*16. Is the set of real numbers a group under the operation of addition? Explain.
*17. Is the set of real numbers a group under the operation of multiplication? Explain.

RESEARCH ACTIVITY

18. There are other classifications of mathematical systems besides groups. For example, there are *rings* and *fields*. Do research to determine the requirements that must be met for a mathematical system to be (a) a ring or (b) a field. (c) Is the set of real numbers, under the operations of addition and multiplication, a field?

9.2 FINITE MATHEMATICAL SYSTEMS

In the previous section we studied infinite mathematical systems. In this section we will study some finite mathematical systems. A **finite mathematical system** is one whose set contains a finite number of elements.

Clock Arithmetic

Let us develop a finite mathematical system called *clock arithmetic*. The set of elements in this system will be the hours on a clock {1, 2, 3, 4, 5, 6, 7, 8, 9, 10, 11, 12}. The binary operation that we will use is addition. We will discuss the properties of this system shortly. We will first form an addition table of the elements in clock arithmetic. What is 4 + 9 equal to in clock arithmetic? To find the sum, assume that it is 4 o'clock and determine what time it will be in 9 hours (see Fig. 9.1). If we add 9 hours to 4 o'clock, the clock will read 1 o'clock. Thus 9 + 4 = 1 in clock arithmetic. Would 9 + 4 be the same as 4 + 9? Yes, 4 + 9 = 9 + 4 = 1.

FIGURE 9.1

Addition in clock arithmetic is given in Table 9.3.

TABLE 9.3

+	1	2	3	4	5	6	7	8	9	10	11	12
1	2	3	4	5	6	7	8	9	10	11	12	1
2	3	4	5	6	7	8	9	10	11	12	1	2
3	4	5	6	7	8	9	10	11	12	1	2	3
4	5	6	7	8	9	10	11	12	1	2	3	4
5	6	7	8	9	10	11	12	1	2	3	4	5
6	7	8	9	10	11	12	1	2	3	4	5	6
7	8	9	10	11	12	1	2	3	4	5	6	7
8	9	10	11	12	1	2	3	4	5	6	7	8
9	10	11	12	1	2	3	4	5	6	7	8	9
10	11	12	1	②	3	4	5	6	7	8	9	10
11	12	1	2	3	4	5	6	7	8	9	10	11
12	1	2	3	4	5	6	7	8	9	10	11	12

The binary operation of this system is defined by the table. It is denoted by the symbol +. To determine the value of $a + b$, where a and b are any two numbers in the set, find a in the left-hand column and find b along the top row. Assume that there is a horizontal line through a and a vertical line through b; the point of intersection of these two lines is where you find the value of $a + b$. For example, $10 + 4 = 2$ has been circled in Table 9.3. Note that $4 + 10$ also equals 2, but this will not necessarily be so for all examples in this chapter. Be careful when looking up answers in the table.

■ **Example 1** _____

Determine whether the clock arithmetic system under the operation of addition is a commutative group.

Solution: We check the five requirements that must be satisfied for a commutative group.

Closure   Is the set of elements in clock arithmetic closed under the operation of addition? Yes, since Table 9.3 contains only the elements in the set {1, 2, 3, 4, 5, 6, 7, 8, 9, 10, 11, 12}. If Table 9.3 had contained an element other than the numbers 1 through 12, the set would not have been closed under addition.

Identity Element   Is there an identity element for clock arithmetic? If it is currently 4 o'clock, how many hours have to pass before it is 4 o'clock again? Twelve hours: $4 + 12 = 12 + 4 = 4$. In fact, given any hour, in 12 hours the clock will return to the starting point. Therefore 12 is the additive identity element in clock arithmetic.

In examining Table 9.3 we see that the row of numbers next to the 12 in the left-hand column is identical to the row of numbers along the top. We also see that the column of numbers under the 12 in the top row is identical to the column of numbers on the left. The search for such a column and row is one technique for determining whether there is an identity element for a system defined by a table.

Inverses Is there an inverse for the number 4 in clock arithmetic for the operation of addition? Recall that the identity element in clock arithmetic is 12. What number when added to 4 gives 12; that is, $4 + \underline{\hspace{1cm}} = 12$? By examining Table 9.3 we see that $4 + 8 = 12$ and also $8 + 4 = 12$. Thus 8 is the additive inverse of 4, and 4 is the additive inverse of 8.

To find the additive inverse of 7, find 7 in the left-hand column. Look to the right of the 7 until you come to the identity element 12. Determine the number at the top of this column. The number is 5. Since $7 + 5 = 5 + 7 = 12$, 5 is the inverse of 7, and 7 is the inverse of 5. The other inverses can be found in the same way:

Element	+	Inverse	=	Identity Element
1	+	11	=	12
2	+	10	=	12
3	+	9	=	12
4	+	8	=	12
5	+	7	=	12
6	+	6	=	12
7	+	5	=	12
8	+	4	=	12
9	+	3	=	12
10	+	2	=	12
11	+	1	=	12
12	+	12	=	12

Note that each element in the set has an **inverse.**

Commutative Property Does the commutative property hold under the given operation? Does $a + b = b + a$ for all elements a and b of the set? Let us randomly select some values for a and b and see whether the commutative property appears to hold. Let $a = 5$ and $b = 8$; then from Table 9.3 we see that

$$5 + 8 = 8 + 5$$
$$1 = 1, \quad \text{True}$$

Let $a = 9$ and $b = 6$; then

$$9 + 6 = 6 + 9$$
$$3 = 3, \quad \text{True}$$

The commutative property holds for these two specific cases. In fact, if we were to randomly select *any* values for *a* and *b*, we would find that $a + b = b + a$. Thus the commutative property of addition is true in clock arithmetic. Note that if there is just one set of values *a* and *b* such that $a + b \neq b + a$, then the system is not commutative.

Associative Property Now let us look at the associative property. Does $(a + b) + c = a + (b + c)$ for all values *a*, *b*, and *c* of the set? Remember that we always evaluate the values within the parentheses first. Let us randomly select some values for *a*, *b*, and *c*. Let $a = 2, b = 6$ and $c = 8$; then

$$(2 + 6) + 8 = 2 + (6 + 8)$$
$$8 + 8 = 2 + 2$$
$$4 = 4, \quad \text{True}$$

Let $a = 5, b = 12$, and $c = 9$; then

$$(5 + 12) + 9 = 5 + (12 + 9)$$
$$5 + 9 = 5 + 9$$
$$2 = 2, \quad \text{True}$$

If we randomly select *any* elements *a*, *b*, and *c* of the set, we will see that $(a + b) + c = a + (b + c)$. Thus the system of clock arithmetic is associative under the operation of addition. Note that if there is just one set of values *a*, *b*, and *c* such that $(a + b) + c \neq a + (b + c)$, then the system is not associative.

Since this system satisfies the five properties required for a mathematical system to be a commutative group, clock arithmetic under the operation of addition is a commutative group. ❖

One method that can be used to determine whether a system defined by a table is commutative under the given operation is to determine whether the elements in the table are symmetric about the main diagonal. The main diagonal is the diagonal from the upper left-hand corner to the lower right-hand corner of the table. See Table 9.4(a).

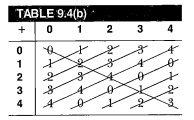

TABLE 9.4(a)					
+	**0**	**1**	**2**	**3**	**4**
0	0	1	2	3	4
1	1	2	3	4	0
2	2	3	4	0	1
3	3	4	0	1	2
4	4	0	1	2	3

TABLE 9.4(b)					
+	**0**	**1**	**2**	**3**	**4**
0	0	1	2	3	4
1	1	2	3	4	0
2	2	3	4	0	1
3	3	4	0	1	2
4	4	0	1	2	3

If the elements are symmetric about the main diagonal, then the system is commutative. If the elements are not symmetric about the main diagonal, then the system is not commutative. If you examine the system in

Tables 9.4(a) and 9.4(b), you will see that its elements are symmetric about the main diagonal. Therefore this mathematical system is commutative.

We can tell if the commutative property holds by observing the symmetry about the main diagonal. There is also, in many cases, a way of determining if the associative property holds without checking every single case.

A mathematical system defined by a table that is *less than a six-element by six-element table, where every element in the set appears exactly once in each row and each column of the table,* will be **associative** under the given operation if:

a)   the system is closed

b)   an identity element exists

c)   each element has an inverse, and

d)   the commutative property holds.

If we observe Table 9.4(a) we see that the table is a 5-element by 5-element table. By observation we can see that the system is closed and that every element in the set (0, 1, 2, 3, and 4) appears in every row and every column. We have shown that the commutative property holds. Therefore if this system has an identity element, and each element has an inverse, then the associative property must hold under the operation of $+$. By observing the table we can see that the identity element is 0. We also see that each element has an inverse, 0–0, 1–4, 2–3, 3–2, 4–1. Thus the associative property must hold under the operation of $+$.

We can conclude that if a mathematical system is defined by a less than six-element by six-element table, and each element in the set appears in each row and each column of the table, and the system is closed, and the commutative property holds, and there is an identity element, and each element has an inverse, then the associative property must hold, and the mathematical system *is a commutative group.*

It is possible to have groups that are not commutative. Such groups are called **noncommutative** or **nonabelian groups.** However, a *noncommutative group defined by a table must be at least a six-element by six-element table.* Nonabelian groups are illustrated in Exercises 51 and 52 at the end of this section. We suggest that you do at least Exercise 51.

Now we will look at another finite mathematical system.

■ **Example 2**

Consider the mathematical system determined by the table

$\odot$	1	3	5	7
1	5	7	1	3
3	7	1	3	5
5	1	3	5	7
7	3	5	7	1

a) Determine the elements in the set of this mathematical system.

b) Determine the binary operation.

c) Determine whether this mathematical system is a commutative group.

Solution

a) The set of elements for this mathematical system consists of the elements found on the top (or left-hand side) of the table. The set of elements is {1, 3, 5, 7}.

b) The binary operation is ⊙.

c) We must determine whether the five requirements for a commutative group are satisfied.

Closure   Since all the elements in the table are in the original set of elements, {1, 3, 5, 7}, the system is closed.

Identity Element   The identity element is 5. Note that the column of elements under the 5 is identical to the left-hand column *and* the row of elements to the right of the 5 is identical to the top row.

Inverse Elements   When an element operates on its inverse element, the result is the identity element. For this example the identity element is 5. To determine the inverse of 1, find the element to replace the question mark:

$$1 \odot ? = 5$$

Since $1 \odot 1 = 5$, 1 is the inverse of 1. We say that 1 is its own inverse.

To find the inverse of 3, find the element to replace the question mark:

$$3 \odot ? = 5$$

Since $3 \odot 7 = 7 \odot 3 = 5$, 7 is the inverse of 3 (and 3 is the inverse of 7). The elements and their inverses are as follows:

Element	⊙	Inverse	=	Identity Element
1	⊙	1	=	5
3	⊙	7	=	5
5	⊙	5	=	5
7	⊙	3	=	5

Every element has a unique inverse.

Commutative Property   Since the elements in the table are symmetric about the main diagonal, the commutative property holds for the operation of $\odot$. One example of the commutative property is

$$3 \odot 5 = 5 \odot 3$$
$$3 = 3 \quad \text{True}$$

Associative Property   The table is a four-element by four-element table, which is less than six by six, and each element appears in every row and column. The system is closed, has an identity element, has inverses, and is commutative. Therefore, the associative property must hold under the operation of $\odot$. One example of the associative property is

$$(7 \odot 3) \odot 1 = 7 \odot (3 \odot 1)$$
$$5 \odot 1 = 7 \odot 7$$
$$1 = 1 \quad \text{True}$$

Since the five necessary properties hold, this mathematical system is a commutative group. ❖

Mathematical Systems without Numbers

Thus far all the systems we have discussed have been based on sets of numbers. This need not be so. The example below illustrates a mathematical system using symbols other than numbers as its elements.

■ Example 3 _____

Given the mathematical system defined by the following table:

$\otimes$	#	$\triangle$	L
#	$\triangle$	L	#
$\triangle$	L	#	$\triangle$
L	#	$\triangle$	L

determine:

a)   The set of elements
b)   The binary operation
c)   Closure or nonclosure of the system
d)   The identity element
e)   The inverse of #
f) $\# \otimes L$ and $L \otimes \#$
g) $(\# \otimes \triangle) \otimes L$ and $\# \otimes (\triangle \otimes L)$

Solution:

a) The set of elements of this mathematical system is $\{\#, \triangle, L\}$.

b) The binary operation is $\otimes$.

c) Since the table does not contain any symbols except $\#$, $\triangle$, and L, the system is closed under $\otimes$.

d) The identity element is L, since the row next to L in the left-hand column is the same as the top row, and the column under L is identical to the left-hand column:

$$\text{Note that } \# \otimes L = L \otimes \# = \#,$$
$$\triangle \otimes L = L \otimes \triangle = \triangle,$$
$$\text{and } L \otimes L = L.$$

e) To find the inverse of $\#$, we must determine the element to replace the question mark:

$$\text{Element} \otimes \text{inverse element} = \text{identity element}$$
$$\# \otimes \,? = L$$

Since $\# \otimes \triangle = L$ and $\triangle \otimes \# = L$, $\triangle$ is the inverse of $\#$.

f) $\# \otimes L = \#,$ $L \otimes \# = \#.$

g) $(\# \otimes \triangle) \otimes L,$ $\# \otimes (\triangle \otimes L)$
  $= L \otimes L$ $= \# \otimes \triangle$
  $= L$ $= L$ ❖

■ **Example 4** _____

Determine whether the following mathematical system is a commutative group.

$*$	A	B	C	D
A	D	A	B	C
B	A	B	C	D
C	B	C	D	A
D	C	D	A	B

Solution

  1. The system is closed by definition of closure.

  2. The identity element is B.

3.   Each element has an inverse as illustrated below:

Element		Inverse		Identity
A	$*$	C	$=$	B
B	$*$	B	$=$	B
C	$*$	A	$=$	B
D	$*$	D	$=$	B

4.   By examining the table we can see that it is symmetric about the main diagonal. Thus the system is commutative under the given operation. One example of the commutative property is

$$D * C = C * D$$
$$A = A \qquad \text{True}$$

5.   Since each element in the set appears in every row and column, the system is closed, has an identity element, has inverses, and is commutative, the associative property holds under the operation of $*$.

The following example shows that the associative property holds for one specific case:

$$(D * A) * C = D * (A * C)$$
$$C * C = D * B$$
$$D = D \qquad \text{True}$$

Since all five properties are satisfied, the system is a commutative group. ❖

■ Example 5

Determine whether the following mathematical system is a commutative group:

$\sim$	x	y	z
x	x	z	y
y	z	y	x
z	y	x	z

Solution

1.   The system is closed.

2.   Since there is no row identical to the top row, there is no identity element. Therefore this mathematical system is *not a group. There is no need to go any further*, but for practice, let's look at a few more items.

3. Since there is no identity element, there can be no inverses.

4. The table is symmetrical about the main diagonal. Therefore the commutative property does hold for the operation of $\sim$.

5. Although the commutative property does hold, the associative property does not. The following example illustrates that the associative property is not true for each and every case:

$$(x \sim y) \sim z \neq x \sim (y \sim z)$$
$$z \sim z \neq x \sim x$$
$$z \neq x$$

Note that the associative property does not hold even though the commutative property does hold. This can happen when there is no identity element and every element does not have an inverse, as in this example. ❖

■ **Example 6** _____

Determine if the following mathematical system is a commutative group under the operation of *.

*	a	b	c
a	a	b	c
b	b	b	a
c	c	a	c

Solution: The system is closed. There is an identity element, *a*. Each element has an inverse, namely $a - a$, $b - c$, and $c - b$. The commutative property holds since there is symmetry about the main diagonal. The only item left to show is that the associative property holds. Since every element in the set does not appear in every row and every column of the table, we need to check the associative property carefully. There are many specific cases where the associative property does hold. However the following counterexample illustrates that the associative property does not hold for every case.

$$(b * b) * c \neq b * (b * c)$$
$$b * c \neq b * a$$
$$a \neq b$$

Since we have shown that the associative property does not hold under the operation of *, this system is not a commutative group. ❖

DID YOU KNOW?

The theory of groups, often considered to be one of the most complex and abstract areas in all of mathematics, has been shown to be a structural link between arithmetic, algebra, geometry, coding theory, crystallography, quantum mechanics, and elementary particle physics.

Some physicists believe that the most intimate look they are likely to get at the basic structure of the universe will be provided by group theory.

Professor Daniel Gorenstein, a mathematician at Rutgers University, stated, "When you have two elementary particles and they hit and bang off, that sounds like physics. When you want to analyze what's behind this physical event that's occurring, the mathematics very often gets expressed in group theory."

EXERCISES 9.2

Use Table 9.3 to determine the following sums in clock arithmetic.

1.	4 + 7	**2.**	9 + 7
3.	11 + 8	**4.**	12 + 4
5.	8 + 10	**6.**	11 + 11
7.	3 + (2 + 7)	**8.**	(9 + 4) + 8
9.	(2 + 9) + 11	**10.**	(3 + 11) + 12
11.	(4 + 3) + (2 + 8)	**12.**	(5 + 7) + (8 + 9)

Determine the value of the following in clock arithmetic by starting at the first number and counting counter-clockwise on the clock the number of units given by the second number.

13. 8 − 5 **14.** 12 − 8 **15.** 4 − 7
16. 5 − 8 **17.** 11 − 12 **18.** 2 − 9

19. Can you find another way to obtain the answers for Exercises 13 through 18? Explain.

20. Develop an addition table for the five-hour clock in Fig. 9.2.

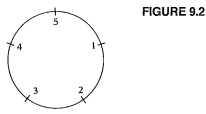

FIGURE 9.2

Find the following sums and differences in clock 5 arithmetic.

21.	3 + 4	**22.**	4 + 5	**23.**	4 + 2
24.	4 + 4	**25.**	1 + 3	**26.**	1 + 5
27.	4 − 2	**28.**	5 − 3	**29.**	2 − 4
30.	1 − 5	**31.**	(4 − 5) + 2	**32.**	3 + (4 + 1)

33. Determine whether 5-hour clock arithmetic under the operation of addition is a commutative group.

34. A mathematical system is defined by a less than six-element by six-element table where every element in the set appears in each row and column. Assuming that closure exists, there is an identity element, each element has an inverse, and the commutative property holds, what conclusions can you draw?

35. Consider the mathematical system indicated by the following table:

*	0	1	2	3
0	0	1	2	3
1	1	2	3	0
2	2	3	0	1
3	3	0	1	2

a) What are the elements of the set in this mathematical system?
b) What is the binary operation?

c) Is the system closed? Explain.

d) Is there an identity element for the system under the given operation? If so, what is it?

e) Does every element in the system have an inverse? If so, give each element and its corresponding inverse.

f) Is the system commutative? Give an example to verify your answer.

g) Is the system associative? Give an example to verify your answer.

h) Is this mathematical system a commutative group? Explain.

Repeat questions (a)–(h) in Exercise 35 for the mathematical system given by each of the following tables.

36.

∅	3	5	7	9
3	7	9	3	5
5	9	3	5	7
7	3	5	7	9
9	5	7	9	3

37.

▼	13	14	17
13	13	14	17
14	14	17	13
17	17	13	14

38.

~	2	6	8	4
2	6	8	4	2
6	8	4	2	6
8	4	2	6	8
4	2	6	8	4

39.

⊙	1	2	3	4
1	1	3	4	2
2	3	4	2	1
3	4	2	1	3
4	2	1	3	4

40. Given the mathematical system

!	z	p	o	n
z	z	p	o	n
p	p	o	n	z
o	o	n	z	p
n	n	z	p	o

determine the following.

a) The elements in the set

b) The binary operation

c) Closure or nonclosure of the system

d) $(z \mathbin{!} o) \mathbin{!} p$

e) $p \mathbin{!} (n \mathbin{!} z)$

f) The identity element

g) The inverse of z

h) The inverse of p

Determine whether the following mathematical systems are commutative groups. Explain your answer.

41.

3	○	△	□
○	○	△	□
△	△	○	□
□	□	△	○

42.

*	a	b	c
a	a	c	b
b	c	b	a
c	b	a	c

43.

ϕ	~	*	?	L	P
~	L	P	~	*	*
*	P	~	*	~	L
?	~	*	?	L	P
L	*	~	L	P	?
P	*	L	P	?	~

44.

☺	a	b	□	○	△
a	a	□	○	b	△
b	□	b	△	a	○
□	b	△	□	○	a
○	○	a	b	△	□
△	△	○	a	□	b

45.

⊙	a	b	c	d	e
a	c	d	e	a	b
b	d	e	a	b	c
c	e	a	b	c	d
d	a	b	c	d	e
e	b	c	d	e	a

46.

#	0	1	2	3	4	5
0	0	0	0	0	0	0
1	0	1	2	3	4	5
2	0	2	4	0	2	4
3	0	3	0	3	0	3
4	0	4	2	0	4	2
5	0	5	4	3	2	1

47. a) Consider the set consisting of two elements $\{E, O\}$, where E stands for an even number and O stands for an odd number. For the operation of addition, complete the following table.

+	E	O
E		
O		

b) Determine whether this mathematical system forms a commutative group under addition. Explain your answer.

48. a) Let E and O represent even numbers and odd numbers, respectively, as in Exercise 47. Complete the following table for the operation of multiplication.

$\times$	E	O
E		
O		

b) Determine whether this mathematical system forms a commutative group under the operation multiplication. Explain your answer.

Make up your own mathematical systems that are groups. List the identity element and the inverses of each element. Do this with sets containing:

49. Three elements   **50.** Four elements

PROBLEM SOLVING

52. Suppose that three books numbered 1, 2, and 3 are placed next to one another on a shelf. If we remove volume 3 and place it before volume 1, the new order of books is 3, 1, 2. Let us call this replacement R. We can write

$$R = \begin{pmatrix} 1 & 2 & 3 \\ 3 & 1 & 2 \end{pmatrix},$$

which indicates the books were switched in order from 1, 2, 3 to 3, 1, 2. Other possible replacements are S, T, U, V, and I, as indicated.

$$S = \begin{pmatrix} 1 & 2 & 3 \\ 2 & 1 & 3 \end{pmatrix}, \quad T = \begin{pmatrix} 1 & 2 & 3 \\ 3 & 2 & 1 \end{pmatrix}, \quad U = \begin{pmatrix} 1 & 2 & 3 \\ 1 & 3 & 2 \end{pmatrix},$$

$$V = \begin{pmatrix} 1 & 2 & 3 \\ 2 & 3 & 1 \end{pmatrix}, \quad I = \begin{pmatrix} 1 & 2 & 3 \\ 1 & 2 & 3 \end{pmatrix}$$

Replacement set I indicates that the books were removed from the shelves and placed back in their original order. Consider the mathematical system with the set of elements R, S, T, U, V, I, with the operation $*$.

To evaluate $R * S$, write

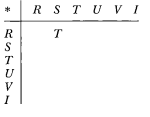

R replaces 1 with 3, and S replaces 3 with 3 (no change), so $R * S$ replaces 1 with 3 (see Fig. 9.3). R replaces 2 with 1, and S replaces 1 with 2, so

51. The following table is an example of a noncommutative or nonabelian group:

∞	1	2	3	4	5	6
1	5	3	4	2	6	1
2	4	6	5	1	3	2
3	2	1	6	5	4	3
4	3	5	1	6	2	4
5	6	4	2	3	1	5
6	1	2	3	4	5	6

a) Show that this is a group. (You cannot prove that the associative property holds, but you can give some examples to show that it appears to hold.)

b) Find a counterexample to show that the commutative property does not hold.

$R * S$ replaces 2 with 2 (no change). R replaces 3 with 2, and S replaces 2 with 1, so $R * S$ replaces 3 with 1. $R * S$ replaces 1 with 3, 2 with 2, and 3 with 1:

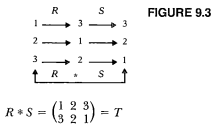

FIGURE 9.3

$$R * S = \begin{pmatrix} 1 & 2 & 3 \\ 3 & 2 & 1 \end{pmatrix} = T$$

Since this is the same as replacement set T, we write $R * S = T$.

a) Complete the following table for the operation using the procedure outlined.

$*$	R	S	T	U	V	I
R		T				
S						
T						
U						
V						
I						

b) Is this mathematical system a group? Explain.
c) Is this mathematical system a commutative group? Explain.

RESEARCH ACTIVITY _____

53. In Section 7.3 we introduced matrices. Show that
2 × 2 matrices under the operation of addition
form a commutative group.

 9.3 # MODULAR ARITHMETIC

The clock arithmetic we discussed in the previous section is an example of modular arithmetic. In this section we will discuss some other modular arithmetic systems and their properties.

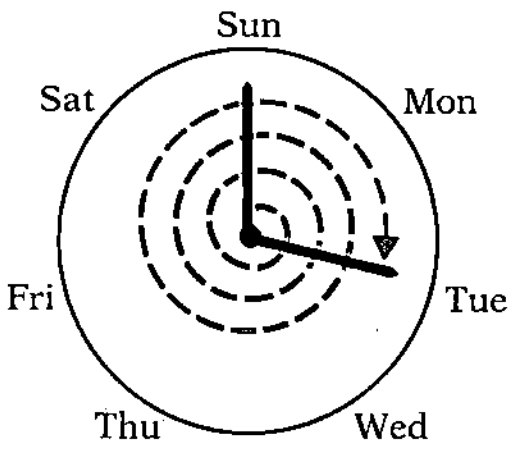

If today is Sunday, what day of the week will it be in 23 days? Many of you can correctly calculate that in 23 days it will be Tuesday. This answer was arrived at by dividing 23 by 7 and observing the remainder of 2. Twenty-three days represent three weeks plus two days. Since we are interested only in the day of the week that the twenty-third day will fall on, the three-week segment is unimportant to the answer. The remainder of 2 indicates that the answer will be two days later than Sunday. Two days after Sunday is Tuesday. This is illustrated in Fig. 9.4.

FIGURE 9.4

In 23 days the hand in Fig. 9.4 would make three complete revolutions and end on Tuesday. Figure 9.5 has the days of the week replaced by numbers: Sunday = 0, Monday = 1, Tuesday = 2, and so on. If we start at 0 and move the hand 23 places, we will end at 2. This system represents a modulo 7 arithmetic system. Table 9.5 shows a modulo 7 addition table.

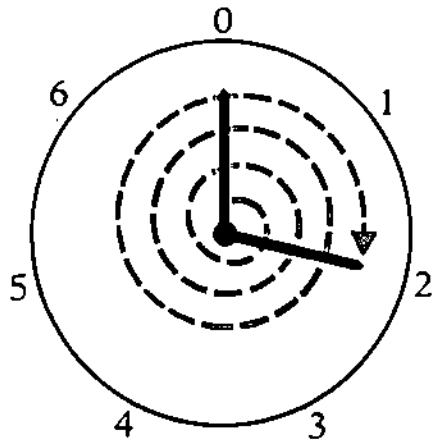

If we start at 4 and add 6, we will end at 3 on the clock in Fig. 9.6. This number is circled in Table 9.5. The other numbers can be obtained in the same way.

FIGURE 9.5

TABLE 9.5							
+	0	1	2	3	4	5	6
0	0	1	2	3	4	5	6
1	1	2	3	4	5	6	0
2	2	3	4	5	6	0	1
3	3	4	5	6	0	1	2
4	4	5	6	0	1	2	③
5	5	6	0	1	2	3	4
6	6	0	1	2	3	4	5

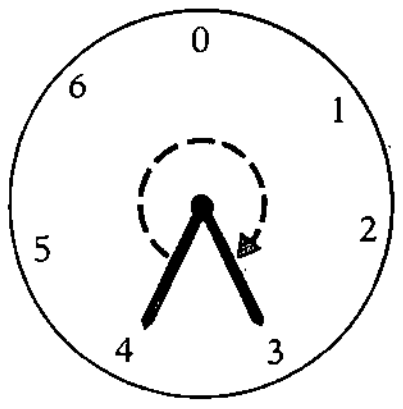

A second method of determining the sum of 4 + 6 in modulo 7 arithmetic is to divide the sum, 10, by 7 and observe the remainder:

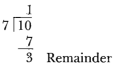

$$\begin{array}{r} 1 \\ 7\overline{)10} \\ \underline{7} \\ 3 \end{array}$$ Remainder

FIGURE 9.6

Since 10 divided by 7 has a remainder of 3, a 3 is placed at the appropriate location in the table.

Now we will introduce a new concept, the concept of congruence.

> *a* **is congruent to** *b* modulo *m*, written, $a \equiv b \pmod{m}$, if *a* and *b* have the same remainder when divided by *m*.

We can show, for example, that $10 \equiv 3 \pmod 7$ by dividing both 10 and 3 by 7 and observing that we obtain the same remainder in each case.

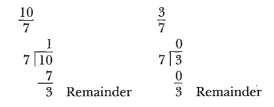

Since the remainders are the same, 3 in each case, 10 is congruent to 3 in modulo 7, and we may write $10 \equiv 3 \pmod 7$.

Now consider $37 \equiv 5 \pmod 8$. If we divide both 37 and 5 by 8, each will have the same remainder, 5.

In any modulo system we can develop a set of **modulo classes** by placing all numbers with the same remainder in the appropriate modulo class. In a modulo 7 system, every number must have a remainder of either 0, 1, 2, 3, 4, 5, or 6. Thus there are seven modulo classes in a modulo 7 system. The seven classes are given in Table 9.6.

Every number is congruent to a number from 0 to 6 in mod 7. For example, $24 \equiv 3$ mod 7 because 24 is in the same modulo class as 3.

TABLE 9.6
Modulo 7 Classes

0	1	2	3	4	5	6
0	1	2	3	4	5	6
7	8	9	10	11	12	13
14	15	16	17	18	19	20
21	22	23	24	25	26	27
28	29	30	31	32	33	34
.	.	.	.	.	.	.
.	.	.	.	.	.	.
.	.	.	.	.	.	.

When you are asked to find a solution to a problem in modular arithmetic, your answer, if it exists, should always be a number from 0 through $m - 1$, where m is the **modulus** of the system. For example, in a modulo 7 system, since 7 is the modulus, the solution will be a number from 0 through 6.

Example 1

Determine which number, from 0 through 6, the following numbers are congruent to in modulo 7.

a) 84   b) 65   c) 47

Solution: We could determine the answer by increasing the entries in Table 9.6. Another method of finding the answer is to divide the given number by 7 and observe the remainder.

a) $84 \equiv ? \pmod 7$

To determine the value that 84 is congruent to in mod 7, divide 84 by 7. The remainder will be the answer:

$$\begin{array}{r} 12 \\ 7\,\overline{)84} \\ 7 \\ \hline 14 \\ 14 \\ \hline 0 \end{array}$$

Thus $84 \equiv 0 \pmod 7$.

b) $65 \equiv ? \pmod 7$

$$\begin{array}{r} 9 \\ 7\,\overline{)65} \\ 63 \\ \hline 2 \end{array}$$

Thus $65 \equiv 2 \pmod 7$.

c) $47 \equiv ? \pmod 7$

$$\begin{array}{r} 6 \\ 7\,\overline{)47} \\ 42 \\ \hline 5 \end{array}$$

Thus $47 \equiv 5 \pmod 7$. ❖

■ **Example 2**

Evaluate each of the following in mod 5.

a) $4 + 3$ **b)** $4 - 3$ **c)** $2 \cdot 4$

Solution

a) $4 + 3 \equiv ? \pmod 5$
$7 \equiv ? \pmod 5$

$$5 \overline{\smash{\big)}\ 7} \atop {\underline{5} } \atop 2$$

with quotient 1

Since the remainder is 2, $4 + 3 \equiv 2 \pmod 5$.

b) $4 - 3 \equiv ? \pmod 5$
$1 \equiv ? \pmod 5$
$1 \equiv 1 \pmod 5$

Remember that we wish to replace the question mark by a number between 0 and 4, inclusive. Thus $4 - 3 \equiv 1 \pmod 5$.

c) $2 \cdot 4 \equiv ? \pmod 5$
$8 \equiv ? \pmod 5$

$$5 \overline{\smash{\big)}\ 8} \atop {\underline{5} } \atop 3$$

with quotient 1

$8 \equiv 3 \pmod 5$
Thus $2 \cdot 4 \equiv 3 \pmod 5$.   ❖

Notice that in Table 9.6, every number in the same modulo class differs by a multiple of the modulo, in this case a multiple of 7. By adding (or subtracting) a multiple of the modulo number to a given number, we do not change the modulo class or congruence of the given number. For example, $3, 3 + 1(7), 3 + 2(7), 3 + 3(7), \ldots, 3 + n(7)$ are all in the same modulo class, namely 3. We use this fact in the solution to Example 3.

■ **Example 3**

Find the replacement for the question mark that makes each of the following true.

a) $3 - 15 \equiv ? \pmod{7}$ **b)** $2 - 4 \equiv ? \pmod{5}$ **c)** $4 - ? \equiv 6 \pmod{8}$

Solution

a) In mod 7, if we add 7 or a multiple of 7 to a number, the resulting sum will be in the same modulo class. Thus if we add 7, 14, 21, . . . to 3, the result will be a number in the same modulo class. We wish to replace 3 with an equivalent mod 7 number that is greater than 15. If we add 14 to 3, we get a sum of 17, which is greater than 15:

$$3 - 15 \equiv ? \pmod{7}$$
$$(3 + 14) - 15 \equiv ? \pmod{7}$$
$$17 - 15 \equiv ? \pmod{7}$$
$$2 \equiv 2 \pmod{7}$$

Therefore ? = 2.

b) $2 - 4 \equiv ? \pmod{5}$

In mod 5 we can add 5 (or a multiple of 5) to a number, and the resulting sum will be in the same modulo class. Thus we can add 5 to 2 so that $2 - 4 \equiv ? \pmod{5}$ becomes $7 - 4 \equiv ? \pmod{5}$:

$$7 - 4 \equiv ? \pmod{5}$$
$$3 \equiv ? \pmod{5}$$
or
$$3 \equiv 3 \pmod{5}$$

Thus $2 - 4 \equiv 3 \pmod{5}$. This could be demonstrated with a modulo 5 clock. The minus sign indicates a counterclockwise direction.

c) $4 - ? \equiv 6 \pmod{8}$

In mod 8, if we add 8 (or a multiple of 8) to a number, the resulting sum will be in the same modulo class. Thus we can add 8 to 4 so that the statement becomes

$$(8 + 4) - ? \equiv 6 \pmod{8}$$
$$12 - ? \equiv 6 \pmod{8}$$

We can see that $12 - 6 = 6$. Therefore ? = 6.   ❖

■ **Example 4**

Find all replacements for the question mark that make the statements true.

a) $4 \cdot ? \equiv 3 \pmod{5}$ **b)** $3 \cdot ? \equiv 0 \pmod{6}$ **c)** $3 \cdot ? \equiv 2 \pmod{6}$

Solution: One method of determining the solution is to replace the question mark with the numbers 0–4 and then find the equivalent modulo class of the product. We use the numbers 0–4 because we are working in modulo 5.

a) $4 \cdot ? \equiv 3 \pmod 5$
     $4 \cdot 0 \equiv 0 \pmod 5$
     $4 \cdot 1 \equiv 4 \pmod 5$
     $4 \cdot 2 \equiv 3 \pmod 5$
     $4 \cdot 3 \equiv 2 \pmod 5$
     $4 \cdot 4 \equiv 1 \pmod 5$

Therefore $? = 2$ for $4 \cdot 2 \equiv 3 \pmod 5$.

b)   Replace the question mark with the numbers 0–5 and follow the procedure used in part (a).
     $3 \cdot ? \equiv 0 \pmod 6$
     $3 \cdot 0 \equiv 0 \pmod 6$
     $3 \cdot 1 \equiv 3 \pmod 6$
     $3 \cdot 2 \equiv 0 \pmod 6$
     $3 \cdot 3 \equiv 3 \pmod 6$
     $3 \cdot 4 \equiv 0 \pmod 6$
     $3 \cdot 5 \equiv 3 \pmod 6$

Therefore we can replace the question mark with 0, 2, or 4 and have a true statement. The answers are 0, 2, and 4.

c) $3 \cdot ? \equiv 2 \pmod 6$

By examining the products in part (b) we see that there are no values that can satisfy the statement. The answer is "no solution."   ❖

     Modular arithmetic systems under the operation of arithmetic are commutative groups as illustrated in Example 5.

■ Example 5

Construct a mod 5 addition table and show that the mathematical system is a commutative group.

Solution: The set of elements in modulo 5 arithmetic is {0, 1, 2, 3, 4}; the binary operation is $+$:

+	0	1	2	3	4
0	0	1	2	3	4
1	1	2	3	4	0
2	2	3	4	0	1
3	3	4	0	1	2
4	4	0	1	2	3

To determine whether this is a commutative group, we must see whether the five properties mentioned earlier are satisfied.

1. *Closure:* Since every entry in the table is a member of the set {0, 1, 2, 3, 4}, the system is closed under addition.

2. *Identity Element:* An easy way to determine whether there is an identity element is to see whether there is a row in the table that is identical to the elements at the top of the table. Note that the row next to 0 is identical to the top of the table. This indicates that 0 *might be* the identity element. Now look at the column under the 0 at the top of the table. If this column is identical to the left-hand column, then 0 is the identity element. Since the column under 0 is the same as the left-hand column, 0 is the additive identity element in modulo 5 arithmetic.

Element	+	Identity	=	Element
0	+	0	=	0
1	+	0	=	1
2	+	0	=	2
3	+	0	=	3
4	+	0	=	4

3. *Inverses:* Does every element have an inverse? Recall that the element plus its inverse must equal the identity element. In this example the identity element is 0.

Element	+	Inverse	=	Identity	
0	+	_?_	=	0	Since 0 + 0 = 0, 0 is its own inverse.
1	+	_?_	=	0	Since 1 + 4 = 0, 4 is the inverse of 1.
2	+	_?_	=	0	Since 2 + 3 = 0, 3 is the inverse of 2.
3	+	_?_	=	0	Since 3 + 2 = 0, 2 is the inverse of 3.
4	+	_?_	=	0	Since 4 + 1 = 0, 1 is the inverse of 4.

Note that each element has an inverse.

4. *Commutative Property:* Is $a + b = b + a$ for *all* elements a and b of the given set? From the table we can see that the system is commutative, since the elements in the table are symmetric about the main diagonal. We will give one example to illustrate the commutative property: $(4 + 2 = 2 + 4 = 1)$.

5. *Associative Property:* Is $(a + b) + c = a + (b + c)$ for *all* elements *a*, *b*, and *c* of the given set? Since the table is less than six elements by six elements and since the four properties *a–d* given on page 458 are met, the associative property holds. We will give one example to illustrate the associative property:

$$(2 + 3) + 4 = 2 + (3 + 4)$$
$$0 + 4 = 2 + 2$$
$$4 = 4$$

Since each of the five properties is satisfied, modulo 5 arithmetic under the operation of addition forms a commutative group.   ❖

EXERCISES 9.3

If today is Tuesday (day 2), determine the day of the week it will be at the end of each of the following periods. (Assume no leap years.)

1.	20 days	**2.**	123 days	**3.**	365 days
4.	2 years	**5.**	1 year, 50 days	**6.**	3 years

Consider the 12 months as a modulo 12 system. Determine the month it will be in the specified number of months. Use the current month as your reference point.

7.	15 months	**8.**	43 months
9.	3 years, 5 months	**10.**	2 years, 4 months
11.	5 years	**12.**	75 months

Determine what number the following are congruent to in mod 5.

13.	$7 + 10$	**14.**	$8 + 12$
15.	$7 + 5 + 2$	**16.**	$7 - 4$
17.	$4 - 7$	**18.**	$4 \cdot 5$
19.	$3 \cdot 4$	**20.**	$16 \cdot 10$
21.	$3 - 11$	**22.**	$7 - 9$
23.	$(16 \cdot 2) - 5$	**24.**	$5 - 12$

Find the number (less than the modulus) to which each of the following is congruent in the indicated modulo system.

25.	14, mod 6	**26.**	23, mod 8
27.	96, mod 15	**28.**	12, mod 6
29.	38, mod 9	**30.**	41, mod 2
31.	36, mod 7	**32.**	55, mod 4
33.	-6, mod 7	**34.**	-5, mod 4
35.	-13, mod 11	**36.**	-11, mod 13

Find all replacements (less than the modulus) for the question mark that make each of the following true.

37.	$4 + 3 \equiv ? \pmod 5$	**38.**	$? + 9 \equiv 6 \pmod 8$
39.	$3 - ? \equiv 4 \pmod 5$	**40.**	$4 \cdot ? \equiv 2 \pmod 6$
41.	$? - 4 \equiv 6 \pmod 8$	**42.**	$4 \cdot 5 \equiv ? \pmod 5$
43.	$2 \cdot ? \equiv 3 \pmod 4$	**44.**	$4 \cdot ? \equiv 3 \pmod 4$
45.	$2 \cdot ? \equiv 3 \pmod 9$	**46.**	$3 \cdot ? \equiv 3 \pmod{12}$
47.	$3 \cdot ? \equiv 2 \pmod 8$	**48.**	$4 - 6 \equiv ? \pmod 8$
49.	$5 - 7 \equiv ? \pmod 9$	**50.**	$6 - ? \equiv 8 \pmod 9$

51. Leap years are always presidential election years. Some leap years are . . . 1980, 1984, 1988. . . .
  a) List the next five presidential election years.
  b) What will be the first election year after the year 3000?
  c) List the election years between the years 2500 and 2525.

52. A pilot is scheduled to fly for five consecutive days and rest for three consecutive days. If today is the second day of her rest shift, determine whether she will be flying
  a) 60 days from today.
  b) 90 days from today.
  c) 240 days from today.
  d) Was she flying 6 days ago?
  e) Was she flying 20 days ago?

53. A nurse's work pattern at XYZ Hospital consists of working the 7 A.M.–3 P.M. shift for three weeks and then the 3 P.M.– 11 P.M. shift for two weeks.
  a) If this is the third week of the pattern, what shift will the nurse be working 6 weeks from now?

b) If this is the fourth week of the pattern, what shift will the nurse be working 7 weeks from now?

c) If this is the first week of the pattern, what shift will the nurse be working 11 weeks from now?

54. A truck driver's routine is as follows: drive 3 days from New York to Chicago; rest 1 day in Chicago; drive 3 days from Chicago to Los Angeles; rest 2 days in Los Angeles; drive 5 days to return to New York; rest 3 days in New York. Then the cycle begins again.

If the truck driver is starting his trip to Chicago today, what will he be doing

a) 20 days from today?

b) 60 days from today?

c) 1 year from today?

55. a) Construct a modulo 4 addition table.

b) Is the system closed? Explain.

c) Does the commutative property hold for the system? Give an example.

d) Does the associative property hold for the system? Give an example.

e) Is there an identity element for the system? If so, what is it?

f) Does every element in the system have an inverse? If so, list the elements and their inverses.

g) Is the system a commutative group?

h) Will every modulo system under the operation of addition be a commutative group? Explain.

56. Construct a modulo 8 addition table. Answer the questions in Exercise 55.

57. a) Construct a modulo 4 multiplication table.

b) Is the system closed under the operation of multiplication?

c) Does the commutative property hold for all elements a and b of the set? Give an example to support your answer.

d) Does the associative property hold for all elements a, b, and c of the set? Give an example to support your answer.

e) Is there an identity element in the system? If so, what is it?

f) Does every element in the system have an inverse element? Make a list showing the elements that have a multiplicative inverse, and list the inverses.

g) Is this mathematical system a commutative group? Explain.

58. Construct a modulo 7 multiplication table. Answer the questions in Exercise 57.

PROBLEM SOLVING

59. A perpetual calendar is to be made by separately listing all of the individual calendars needed to cover all possible situations that may occur. How many individual calendars are needed to develop the perpetual calendar?

We have not discussed division in modular arithmetic. Can you replace the question marks with the number or numbers that make the statement true?

60. $5 \div 7 \equiv ? \pmod 9$

61. $? \div 5 \equiv 5 \pmod 9$

62. $? \div ? \equiv 1 \pmod 4$

63. $1 \div 2 \equiv ? \pmod 5$

64. $2 \div ? \equiv 4 \pmod 6$

Solve for x where k is any counting number.

65. $5k \equiv x \pmod 5$

66. $5k + 3 \equiv x \pmod 5$

67. $4k - 3 \equiv x \pmod 4$

68. One important use of modular arithmetic is coding. One type of coding circle is given in Fig. 9.7. To use this code, the person you are sending the message to must know the code key to decipher the code. The code key to this message is j. Can you decipher this code? (*Hint:* Subtract the code key from the code numbers.)

19 10 22 19 21 15 10 26 19 9 9 11

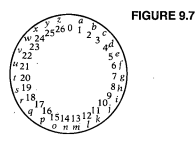

FIGURE 9.7

69. The procedure of *casting out nines* can be used to check arithmetic problems. The procedure is based on the fact that any whole number is congruent to the sum of its digits modulo 9. For example, the number 5783 and the sum of its digits, 5 + 7 + 8 + 3, or 23, are both congruent to 5 (mod 9): 5783 ≡ 5 (mod 9) and 23 ≡ 5 (mod 9).

a) Explain why the procedure works.

b) Perform the indicated operations and check your solutions by casting out nines.

1)
$$\begin{array}{r} 4236 \\ +3784 \end{array}$$

2)
$$\begin{array}{r} 8493 \\ -237 \end{array}$$

3)
$$\begin{array}{r} 187 \\ \times 85 \end{array}$$

4) 485 ÷ 23

SUMMARY

The set of integers under the operation of addition forms a mathematical system. A mathematical system consists of a set of elements and at least one binary operation. Some mathematical systems display certain mathematical properties in their structure.

One type of mathematical system is clock arithmetic under the operation of addition. Clock arithmetic is one example of a group. A mathematical system is a group if (1) it is closed, (2) it has an identity element, (3) every element has an inverse, and (4) it is associative. If a group is also commutative, it is known as a commutative or abelian group.

When a mathematical system is defined by a table, if the elements in the table are symmetric about the main diagonal, then the commutative property holds for the given operation. A mathematical system defined by a less than six-element by six-element table where every element appears in each row and column will be associative if it is closed, it has an identity element, every element has an inverse, and the commutative property holds.

Modular mathematical systems are also groups under the operation of addition. *a* is congruent to *b* modulo *m*, written $a \equiv b \pmod{m}$, if *a* and *b* have the same remainder when divided by *m*.

REVIEW EXERCISES

1. List the parts of a mathematical system.

Determine the value of the following in clock arithmetic.

2. 8 + 9   **3.** 4 + 9
4. 6 − 10   **5.** 4 + 3 + 9
6. 8 − 11   **7.** 6 − 4 + 8

8. List the properties of a group, and explain what each property means.
9. What is an abelian group?
10. Determine whether the set of integers under the operation of addition forms a group.
11. Determine whether the set of integers under the operation of multiplication forms a group.

12. Determine whether the set of rational numbers under the operation of addition forms a group.

13. Determine whether the set of rational numbers under the operation of multiplication forms a group.

Determine whether the following are commutative groups. Explain your answer.

14.

:	*	?	△	p
*	?	△	*	p
?	△	p	?	*
△	*	?	△	p
p	p	*	p	*

15.

*	a	c	b
a	a	b	c
c	b	c	a
b	c	a	b

16.

⊗	1	○	?	△
1	1	○	?	△
○	○	?	△	1
?	?	△	1	○
△	△	1	○	?

17.

?	4	#	L	P
4	#	4	P	L
#	4	P	L	#
L	P	L	#	4
P	L	#	4	P

Find the number (less than the modulus) to which each of the following is congruent in the indicated modulo system.

18. 20, modulo 3   **19.** 24, modulo 5   **20.** 19, modulo 8
21. 59, modulo 6   **22.** 74, modulo 12   **23.** 85, modulo 4

Find all replacements (less than the modulus) for the question mark that make each of the following true.

24. $5 + 8 \equiv ? \pmod 9$ **25.** $? - 2 \equiv 4 \pmod 4$ **26.** $4 \cdot ? \equiv 3 \pmod 6$
27. $4 - ? \equiv 5 \pmod 7$ **28.** $? \cdot 4 \equiv 0 \pmod 8$ **29.** $41 \equiv ? \pmod{12}$

30. Construct a modulo 6 addition table. Then determine whether the modulo 6 system forms a commutative group under the operation of addition.

31. Construct a modulo 4 multiplication table. Then determine whether the modulo 4 system forms a commutative group under the operation of multiplication.

CHAPTER TEST

1. A mathematical system consists of two items. Name them.
2. List the requirements needed for a mathematical system to be a commutative group.
3. Is the set of integers a commutative group under the operation of subtraction? Explain your answer completely.
4. Develop a 4-hour clock arithmetic addition table.
5. Is 4-hour clock arithmetic under the operation of addition a commutative group? Explain your answer completely.

Determine whether the following mathematical systems are commutative groups. Explain your answers completely.

6.

*	a	b	c
a	a	b	c
b	b	b	a
c	c	a	d

7.

?	1	2	3
1	1	3	2
2	3	2	1
3	2	1	3

8.

W	□	△	L	Z
□	Z	□	△	L
△	□	△	L	Z
L	△	L	Z	□
Z	L	Z	□	△

Determine the number to which each of the following is congruent in the indicated modular system.

9. 64, modulo 9   10. 80, modulo 6

Find the replacements less than the modulus for the question mark that make each of the following true.

11. $6 + 9 \equiv ?$ (mod 7)   12. $? - 5 \equiv 4$ (mod 8)   13. $2 \cdot ? \equiv 4$ (mod 6)
14. $5 - ? \equiv 7$ (mod 9)   15. $62 \equiv ?$ (mod 5)   16. $4 \cdot ? \equiv 3$ (mod 6)
17. **a)** Construct a modulo 5 addition table.
    b) Explain why this mathematical system is a commutative group.

Geometry

10.1

POINTS, LINES, AND PLANES

Human beings recognized shapes, sizes, and physical forms long before geometry was considered a science. Geometry as a science is said to have begun in the Nile Valley of ancient Egypt. The Egyptians used geometry to measure land and to build pyramids and other structures.

The word "geometry" is derived from two Greek words, *ge,* meaning earth, and *metron,* meaning measure. Thus geometry means earth measure or measurement of the earth.

Unlike the Egyptians, the Greeks were interested in more than just the applied aspects of geometry. The Greeks attempted to apply their knowledge of logic to geometry. Thales of Miletus around 600 B.C. was the first to be credited with using deductive methods to develop geometric concepts. Another outstanding Greek geometer was Pythagoras, who continued the systematic development of geometry that was begun by Thales.

Around 300 B.C., Euclid collected and summarized most of the Greek mathematics of his time. In a set of 13 books called *Elements,* Euclid laid the foundation for plane Euclidean geometry, which is now taught in most high schools.

The material in the first part of this chapter will be an introduction to some concepts of nonmetric geometry, that is, geometry without reference to measurements. The remainder of the chapter will be a discussion of some concepts of plane geometry, topology, and non-Euclidean geometry.

"It is the glory of geometry that from so few principles, fetched from without,. . . it is able to accomplish so much."
I. Newton

Point and Line

Three basic terms in geometry are **point, line,** and **plane.** It is not possible to define any of these terms by using simpler terms. Therefore these three terms are not given a formal definition. Although we do not define point, line, and plane, we have a common idea of what each is.

Let's discuss some properties of a line. We will assume that a line means a straight line unless otherwise stated.

1. A line is a set of points. We say that each point is on the line and that the line passes through each point.

2. Any two distinct points determine a unique line (see Fig. 10.1). The arrows at both ends of the line indicate that the line continues in each direction. The line in Fig. 10.1 is symbolized by $\overleftrightarrow{AB}$ or $\overleftrightarrow{BA}$.

3. Any point on a line separates the line into three parts—the point itself and two **half lines.** For example, in Fig. 10.2, point B separates the line into the point B, half line BC, and half line BA. Half line BC is symbolized as $\overset{\circ}{\overrightarrow{BC}}$. The open circle above the B indicates that point B is not included in the half line.

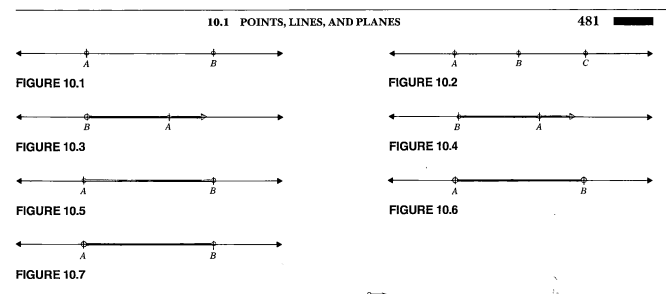

FIGURE 10.1

FIGURE 10.2

FIGURE 10.3

FIGURE 10.4

FIGURE 10.5

FIGURE 10.6

FIGURE 10.7

Consider the half line $\overset{\circ\longrightarrow}{BA}$ in Fig. 10.3. If the **end point,** B, is included with the set of points on the half line, the result is called a **ray.** Ray BA, symbolized by $\overrightarrow{BA}$, is illustrated in Fig. 10.4.

A **line segment** is that part of a line between two points, including the end points. Line segment AB, symbolized $\overline{AB}$, is illustrated in Fig. 10.5.

An open line segment is the set of points on a line between two points, excluding the end points. Open line segment AB, symbolized $\overset{\circ-\circ}{AB}$, is illustrated in Fig. 10.6.

Figure 10.7 illustrates a half open line segment, AB, symbolized $\overset{\circ-}{AB}$.

Table 10.1 summarizes the various notations we have discussed.

TABLE 10.1		
Description	**Diagram**	**Symbol**
Line AB	A ——— B	$\overleftrightarrow{AB}$
Half line AB	A ——— B	$\overset{\circ\rightarrow}{AB}$
Ray AB	A ——— B	$\overrightarrow{AB}$
Ray BA	A ——— B	$\overrightarrow{BA}$
Line segment AB	A ——— B	$\overline{AB}$
Open line segment AB	A ——— B	$\overset{\circ-\circ}{AB}$
Half open line segment AB	A ——— B	$\overset{-\circ}{AB}$
	A ——— B	$\overset{\circ-}{AB}$

In Chapter 2 we discussed the meaning of intersection of sets. Recall that the intersection (∩) of two sets is the set of elements (points in this case) common to both sets.

Consider the lines in Fig. 10.8. If we take the intersection of $\overrightarrow{AB}$ and $\overrightarrow{BA}$, we get line segment $\overline{AB}$. Thus $\overrightarrow{AB} \cap \overrightarrow{BA} = \overline{AB}$.

The union of two sets was also discussed in Chapter 2. The union (∪) of two sets is the set of elements (points in this case) that belong to either of the sets or both sets. If we take the union of $\overrightarrow{AB}$ and $\overrightarrow{BA}$, we get $\overleftrightarrow{AB}$ (Fig. 10.9). Thus $\overrightarrow{AB} \cup \overrightarrow{BA} = \overleftrightarrow{AB}$.

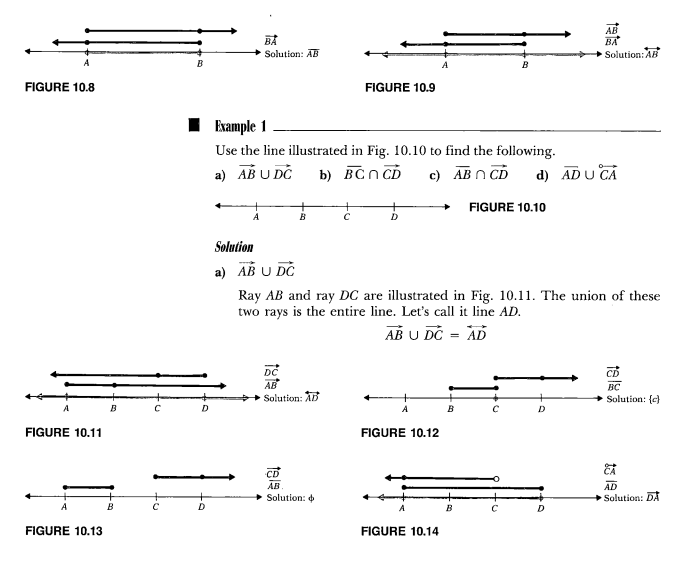

FIGURE 10.8

FIGURE 10.9

■ **Example 1** _____

Use the line illustrated in Fig. 10.10 to find the following.

a) $\overrightarrow{AB} \cup \overrightarrow{DC}$ b) $\overline{BC} \cap \overrightarrow{CD}$ c) $\overline{AB} \cap \overrightarrow{CD}$ d) $\overline{AD} \cup \overset{\circ}{\overrightarrow{CA}}$

FIGURE 10.10

Solution

a) $\overrightarrow{AB} \cup \overrightarrow{DC}$

Ray AB and ray DC are illustrated in Fig. 10.11. The union of these two rays is the entire line. Let's call it line AD.

$$\overrightarrow{AB} \cup \overrightarrow{DC} = \overleftrightarrow{AD}$$

FIGURE 10.11

FIGURE 10.12

FIGURE 10.13

FIGURE 10.14

b) $\overline{BC} \cap \overrightarrow{CD}$

The intersection of line segment BC and ray CD is point C (Fig. 10.12).

$$\overline{BC} \cap \overrightarrow{CD} = \{C\}$$

c) $\overline{AB} \cap \overrightarrow{CD}$

Since line segment AB and ray CD have no points in common, their intersection is empty (Fig. 10.13).

$$\overline{AB} \cap \overrightarrow{CD} = \emptyset$$

d) $\overline{AD} \cup \overset{\circ}{\overrightarrow{CA}}$

The union of line segment AD and half line CA is ray DA (or $\overrightarrow{DB}$ or $\overrightarrow{DC}$) (Fig. 10.14).

$$\overline{AD} \cup \overset{\circ}{\overrightarrow{CA}} = \overrightarrow{DA} \qquad \qquad ❖$$

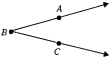

FIGURE 10.15

An **angle,** denoted by ⋇, is the union of two rays with a common end point (Fig. 10.15).

$$\overrightarrow{BA} \cup \overrightarrow{BC} = ⋇ABC \text{ (or } ⋇ CBA)$$

The point common to both rays is called the **vertex** of the angle. When you write the letters naming the angles, the vertex is always placed in the middle. The rays that make up the angle are called its **sides.** Note that an angle is not the space between the two sides.

▪ **Example 2** _____

Refer to Fig. 10.16. Find the following.

a) $\overrightarrow{BF} \cup \overrightarrow{BD}$

b) $⋇ABF \cap ⋇FBE$

c) $\overleftrightarrow{AC} \cap \overleftrightarrow{DE}$

d) $\overrightarrow{BA} \cup \overrightarrow{BC}$

Solution

a) $\overrightarrow{BF} \cup \overrightarrow{BD} = ⋇FBD$

b) $⋇ABF \cap ⋇FBE = \overrightarrow{BF}$

c) $\overleftrightarrow{AC} \cap \overleftrightarrow{DE} = \{B\}$

d) $\overrightarrow{BA} \cup \overrightarrow{BC} = \overleftrightarrow{AC} \qquad \qquad ❖$

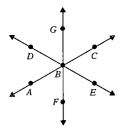

FIGURE 10.16

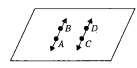

FIGURE 10.17

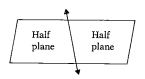

FIGURE 10.18

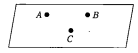

FIGURE 10.19

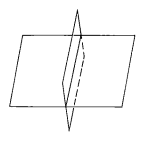

FIGURE 10.20

Plane

As we stated earlier, the term "plane" is not formally defined. We think of a plane as a two-dimensional surface that extends infinitely in both directions. The surface of an infinitely large blackboard would be an example of a plane. Euclidean geometry is called **plane geometry** because it is the study of two-dimensional figures in a plane.

Two lines in the same plane that do not intersect are called **parallel lines.** Figure 10.17 illustrates two parallel lines in a plane ($\overleftrightarrow{AB}$ is parallel to $\overleftrightarrow{CD}$).

Some of the properties of planes are as follows:

1. Any three points that are not on the same line (noncollinear points) determine a unique plane (Fig. 10.18).
2. A line in a plane divides the plane into three parts—the line and two half planes (Fig. 10.19).
3. Any line and a point not on the line determines a unique plane.
4. The intersection of two planes is a line (Fig. 10.20).

Two planes that do not intersect are said to be **parallel planes.** For example, in Fig. 10.21, plane *ABE* is parallel to plane *GHF*.

Two lines that do not lie in the same plane and do not intersect are called **skewed lines.** Figure 10.21 illustrates skewed lines ($\overleftrightarrow{AB}$ and $\overleftrightarrow{CD}$).

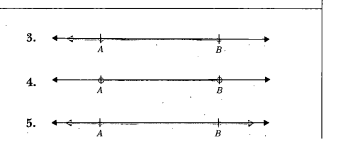

FIGURE 10.21

EXERCISES 10.1

Identify each of the following as a line, half line, ray, line segment, open line segment, or half open line segment. Denote each by its appropriate symbol.

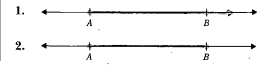

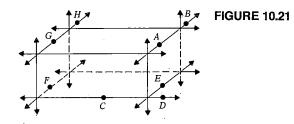

6.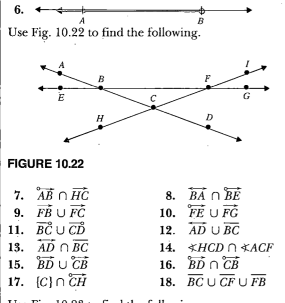
A   B

Use Fig. 10.22 to find the following.

FIGURE 10.22

7. $\overset{\leftrightarrow}{AB} \cap \overset{\rightarrow}{HC}$ **8.** $\overset{\rightarrow}{BA} \cap \overset{\circ\rightarrow}{BE}$

9. $\overset{\rightarrow}{FB} \cup \overset{\rightarrow}{FC}$ **10.** $\overset{\rightarrow}{FE} \cup \overset{\rightarrow}{FG}$

11. $\overset{\circ\rightarrow}{BC} \cup \overset{\circ\rightarrow}{CD}$ **12.** $\overset{\leftrightarrow}{AD} \cup \overline{BC}$

13. $\overset{\leftrightarrow}{AD} \cap \overline{BC}$ **14.** $\sphericalangle HCD \cap \sphericalangle ACF$

15. $\overset{\circ\rightarrow}{BD} \cup \overset{\circ\rightarrow}{CB}$ **16.** $\overset{\circ\rightarrow}{BD} \cap \overset{\circ\rightarrow}{CB}$

17. $\{C\} \cap \overset{\circ\rightarrow}{CH}$ **18.** $\overline{BC} \cup \overline{CF} \cup \overline{FB}$

Use Fig. 10.23 to find the following.

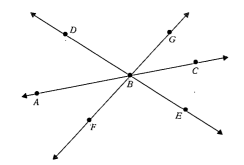

FIGURE 10.23

19. $\sphericalangle ABE \cup \overset{\circ\rightarrow}{AB}$ **20.** $\overset{\circ\rightarrow}{BF} \cup \overset{\rightarrow}{BE}$

21. $\overset{\rightarrow}{BF} \cap \overset{\rightarrow}{BE}$ **22.** $\overset{\circ\rightarrow}{BD} \cup \overset{\rightarrow}{BE}$

23. $\sphericalangle GBC \cap \sphericalangle CBE$ **24.** $\overset{\leftrightarrow}{DE} \cup \overset{\circ\rightarrow}{BE}$

25. $\sphericalangle CBE \cap \sphericalangle EBC$ **26.** $\{B\} \cap \overset{\rightarrow}{BA}$

27. $\overset{\circ\rightarrow}{AC} \cap \overline{AC}$ **28.** $\overset{\circ\rightarrow}{AC} \cap \overset{\circ\rightarrow}{BE}$

29. $\overset{\rightarrow}{EB} \cap \overset{\rightarrow}{BE}$ **30.** $\overline{GF} \cap \overline{AB}$

31. What are parallel lines?

32. What are skewed lines?

33. How many planes can be drawn through a given line?

34. How many lines can be drawn through a given point? How many planes can be drawn through a given point?

35. What is the intersection of two distinct planes?

36. What is the intersection of a plane and a line not on the plane?

37. Draw a diagram similar to Fig. 10.24 to represent a plane separated into two half planes and a line $\overset{\leftrightarrow}{AB}$. Mark two points C and D in the same half plane and draw $\overline{CD}$. Is $\overline{CD}$ entirely contained in the same half plane as C and D? Mark two points Q and R so that Q is in one half plane and R in the other. Draw $\overline{QR}$. Does $\overline{QR}$ intersect $\overset{\leftrightarrow}{AB}$? At how many points will $\overline{QR}$ intersect $\overset{\leftrightarrow}{AB}$?

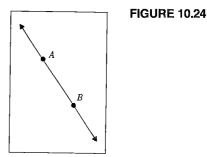

FIGURE 10.24

38. a) Will three noncollinear points A, B, and C always determine a plane?

b) Is it possible to determine more than one plane with three noncollinear points?

c) How many planes can be constructed through three collinear points?

39. Mark three noncollinear points A, B, and C. Draw the ray $\overset{\rightarrow}{AB}$ and the ray $\overset{\rightarrow}{AC}$. What geometric figure is produced?

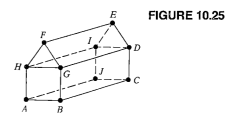

FIGURE 10.25

Figure 10.25 suggests a number of lines and planes. The lines may be described by pairs of points, and the planes may be described by naming three points. Name:

40. A pair of parallel planes.
41. A pair of planes whose intersection is a line.
42. Three planes whose intersection is a single point.
43. Three planes whose intersection is a line.
44. A line and a plane whose intersection is a point.
45. A line and plane whose intersection is a line.

46. A pair of parallel lines.
47. A pair of skew lines.
48. Three lines that intersect at a single point.
49. Four planes whose intersection is a single point.
50. What geometrical advantage is there in building chairs with three legs rather than four legs?

PROBLEM SOLVING

51. Suppose that you have three distinct lines, all lying in the same plane. Find all the possible ways in which the three lines can be related. Sketch each case (4 cases).
52. Suppose that you have three distinct planes in space. Find all the possible ways these planes can be related. Sketch each case (5 cases).

53. Suppose that you have three distinct lines in space (not necessarily in the same plane). Find all the possible ways in which these three lines can be related. Sketch each case (9 cases).

10.2 ANGLES AND POLYGONS

An angle can be formed by the rotation of a ray about a point (Fig. 10.26). An angle has an initial side and a terminal side. The initial side indicates the position of the ray prior to rotation. The terminal side indicates the position of the ray after rotation. The **measure of an angle** is the amount of rotation from its initial side to its terminal side. In Fig. 10.26 the letter x represents the measure of the angle. The point common to the initial and terminal sides of the angle is called the **vertex** of the angle.

Angles can be measured in **degrees,** radians, or gradients. In this text we will discuss only the degree unit of measurement. The symbol for degrees is the same as the symbol for temperature degrees. An angle of 45 degrees is written 45°. A protractor is used to measure angles.

Consider a circle whose circumference is divided into 360 equal parts. If a line is drawn from each mark on the circumference to the center of the circle, we get 360 wedge-shaped pieces. The measure of an angle formed by the straight sides of each wedge-shaped piece is defined to be 1 degree.

There are a number of ways to name an angle. The angle in Fig. 10.27 can be denoted as follows:

$$\angle ABC, \quad \angle CBA, \quad \angle B, \quad \angle x$$

An angle divides a plane into three distinct parts: the angle itself, its interior, and its exterior (Fig. 10.27).

There are many different types of angles. Some are summarized in Table 10.2.

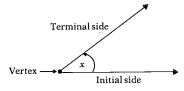

FIGURE 10.26

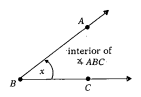

FIGURE 10.27

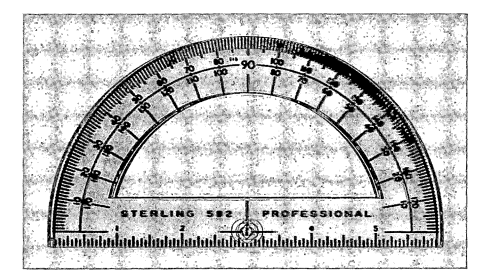

Protractors are used to measure degrees.

Photo by Marshall Henrichs

TABLE 10.2		
Name	**Figure**	**Characteristics**
Right angle		Measure of right angle is 90°. The symbol ⌐ is used to indicate right angles.
Acute angle		Measure of angle between 0° and 90°, $0° < x < 90°$
Obtuse angle		Measure of angle between 90° and 180°, $90° < x < 180°$
Straight angle		Measure of angle is 180°.

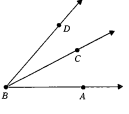

FIGURE 10.28

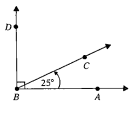

FIGURE 10.29

Two angles in the same plane are **adjacent angles** when they have a common vertex and a common side but no common interior points. $\angle DBC$ and $\angle CBA$ in Fig. 10.28 are adjacent angles. However, $\angle DBA$ and $\angle CBA$ are not adjacent angles.

Two angles the sum of whose measures is 90° are called **complementary angles.** Each angle is called the complement of the other.

■ **Example 1** _____

If $\angle ABC = 25°$, and $\angle ABC$ and $\angle CBD$ are complementary angles, determine $\angle CBD$ (Fig. 10.29).

Solution: The sum of the two angles must be 90°:

$$\angle CBD = 90° - \angle ABC$$
$$= 90° - 25°$$
$$= 65° \qquad ❖$$

■ **Example 2** _____

If $\angle ABC$ and $\angle DEF$ are complementary angles and $\angle DEF$ is twice as large as $\angle ABC$, determine the measure of each angle (Fig. 10.30).

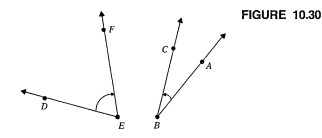

FIGURE 10.30

Solution: If $\angle ABC = x$, then $\angle DEF = 2 \cdot x$ or $2x$, since it is twice as large. The sum of the measures of these angles must be 90°:

$$\angle ABC + \angle DEF = 90°$$
$$x + 2x = 90°$$
$$3x = 90°$$
$$x = 30°$$

Therefore $\angle ABC = 30°$ and $\angle DEF = 2 \cdot 30°$, or 60°. ❖

Two angles the sum of whose measures is 180° are said to be **supplementary angles.** Each angle is called the supplement of the other.

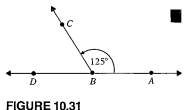

FIGURE 10.31

Example 3

If $\angle ABC$ and $\angle CBD$ are supplementary angles and $\angle ABC = 125°$, find $\angle CBD$ (Fig. 10.31).

Solution: The sum of the two angles must be 180°:

$$\angle CBD = 180° - \angle ABC$$
$$= 180° - 125°$$
$$= 55°$$ ❖

Example 4

If $\angle ABC$ and $\angle DEF$ are supplementary angles and $\angle ABC$ is three times as large as $\angle DEF$, determine $\angle ABC$ and $\angle DEF$ (Fig. 10.32).

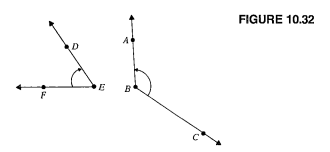

FIGURE 10.32

Solution: If $\angle DEF = x$, then $\angle ABC = 3x$:

$$\angle ABC + \angle DEF = 180°$$
$$3x + x = 180°$$
$$4x = 180°$$
$$x = 45°$$

Thus $\angle DEF = 45°$ and $\angle ABC = (3)(45°) = 135°$. ❖

When two straight lines intersect, the nonadjacent angles formed are called **vertical angles** (Fig. 10.33). In Fig. 10.33, $\angle 1$ and $\angle 3$ are vertical angles, and $\angle 2$ and $\angle 4$ are vertical angles. We can show that vertical angles have the same measure, that is, they are equal. For example, in Fig. 10.33 we see that

$$\angle 1 + \angle 2 = 180° \qquad \text{Why?}$$
$$\angle 2 + \angle 3 = 180° \qquad \text{Why?}$$

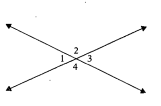

FIGURE 10.33

Since $\angle 2$ has the same measure in both cases, $\angle 1$ must equal $\angle 3$.

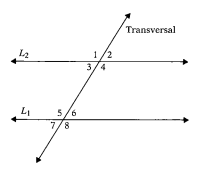

FIGURE 10.34

A line that intersects two different lines at two different points is called a **transversal.** Figure 10.34 illustrates two parallel lines cut by a transversal.

Figure 10.34 illustrates that when two parallel lines are cut by a transversal, eight angles are formed. How many pairs of supplementary angles are formed? How many pairs of vertical angles are formed?

∢3 and ∢6 are **alternate interior angles,** as are ∢4 and ∢5. ∢1 and ∢8, and ∢2 and ∢7, are **alternate exterior angles.** Angles 1 and 5, 3 and 7, 2 and 6, and 4 and 8 are called **corresponding angles.**

It can be shown that when two parallel lines are cut by a transversal, alternate interior angles have the same measure, alternate exterior angles have the same measure, and corresponding angles have the same measure.

■ Example 5

Figure 10.35 shows two parallel lines cut by a transversal. Determine the measure of ∢1 through ∢7.

Solution

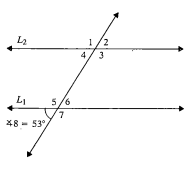

FIGURE 10.35

Measure of Angle	Reason
∢6 = 53°	∢8 and ∢6 are vertical angles.
∢5 = 127°	∢8 and ∢5 are supplementary angles.
∢7 = 127°	∢5 and ∢7 are vertical angles.
∢1 = 127°	∢1 and ∢7 are alternate exterior angles.
∢4 = 53°	∢4 and ∢6 are alternate interior angles.
∢2 = 53°	∢6 and ∢2 are corresponding angles.
∢3 = 127°	∢3 and ∢1 are vertical angles.

❖

A **polygon** is a closed figure in a plane determined by three or more straight line segments. Examples of polygons are given in Fig. 10.36. The straight line segments that form the polygon are called its **sides.** The union of the sides of a polygon and its interior is called a **polygonal region.** A **regular polygon** is one whose sides are all the same length and whose interior angles all have the same measure. Figures 10.36(b) and 10.36(d) are regular polygons.

The names of some polygons are given in Table 10.3.

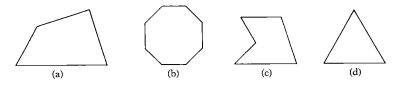

(a)   (b)   (c)   (d)

FIGURE 10.36

TABLE 10.3	
Number of Sides	Name
3	triangle
4	quadrilateral
5	pentagon
6	hexagon
7	heptagon
8	octagon
9	nonagon
10	decagon
12	dodecagon
20	icosagon

TABLE 10.4		
Sides	Triangles	Sum of the Measures of the Angles
3	1	$1(180°) = 180°$
4	2	$2(180°) = 360°$
5	3	$3(180°) = 540°$
6	4	$4(180°) = 720°$

FIGURE 10.37

The sum of the angles in a triangle is 180°. Using this fact, we can develop a formula for finding the sum of the interior angles of any polygon.

Consider the quadrilateral $ABCD$ (Fig. 10.37). If we draw a straight line segment between any two vertices, two triangles are formed (Fig. 10.38).

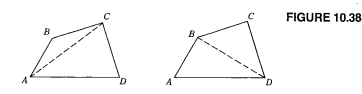

FIGURE 10.38

Since the sum of the measures of the angles of a triangle is 180°, the sum of the angles of a four-sided figure is $2 \cdot 180°$, or 360°.

Now let's examine a pentagon (Fig. 10.39). As can be seen from Fig. 10.39, we can draw two straight line segments to form three triangles. Thus the sum of the measures of the interior angles of a five-sided figure is $3 \cdot 180°$, or 540°. Figure 10.40 illustrates that four triangles can be drawn in a six-sided figure. Table 10.4 summarizes this information.

If we continue this procedure, we can see that for an n-sided polygon the sum of the measures of the interior angles is $(n - 2)180°$.

FIGURE 10.39

FIGURE 10.40

The sum of the measures of the interior angles of an n-sided polygon is $(n - 2)180°$.

FIGURE 10.41

■ **Example 6**

A gazebo is in the shape of a regular octagon (Fig. 10.41). Determine (a) the measure of an interior angle and (b) the measure of exterior angle 1.

Solution

a) Using the formula $(n - 2)180°$, we can determine the sum of the measures of the interior angles of an octagon.

The sum of the measures of the interior angles
$$= (8 - 2)180°$$
$$= 6(180°)$$
$$= 1080°$$

The measure of an interior angle of a regular polygon can be determined by dividing the sum of the interior angles by the number of angles.

The interior angle of a regular octagon $= \dfrac{1080°}{8} = 135°$

b) Since ∢1 is the supplement of an interior angle,

$$∢1 = 180° - 135° = 45° \qquad \qquad ❖$$

In order to discuss area in the next section we must be able to identify various types of triangles and quadrilaterals. Table 10.5 illustrates certain types of triangles and their characteristics.

TABLE 10.5		
Name	Figure	Characteristics
Acute triangle		All angles are acute angles.
Obtuse triangle		One angle is an obtuse angle.
Right triangle		One angle is a right angle. The side opposite the right angle is called the hypotenuse.
Scalene triangle		No two sides are equal in length.

Isosceles triangle		Two sides are equal in length. The angles opposite the sides that are equal in length have the same measure.
Equilateral triangle		All three sides are equal in length. The three angles have the same measure (60° each).

The different classifications of quadrilaterals are illustrated in Table 10.6.

TABLE 10.6

Name	Figure	Characteristics
Trapezoid		Two sides are parallel.
Parallelogram		Both pairs of opposite sides are parallel.
Rhombus		Both pairs of opposite sides are parallel. The four sides are equal in length.
Rectangle		Both pairs of opposite sides are parallel. The angles are right angles.
Square		Both pairs of opposite sides are parallel. The four sides are equal in length. The angles are right angles.

In everyday living it is often necessary to deal with geometric figures that have the "same shape" but are of different sizes. For example, an architect will make a small-scale drawing of a floor plan, or a photographer will make an enlargement of a photograph. Figures that have the same shape but may be of different sizes are called **similar figures.** Two sets of similar figures are illustrated in Fig. 10.42.

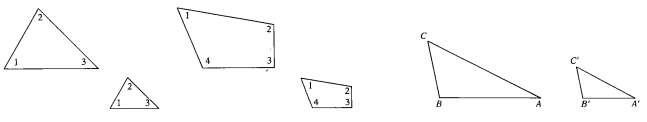

FIGURE 10.42   **FIGURE 10.43**

Each set of similar figures has **corresponding angles** and **corresponding sides.** The corresponding sides and angles of two similar triangles are illustrated in Fig. 10.43.

In Fig. 10.43, angles A, B, and C and their respective corresponding angles A', B', and C' have the same measure. The sides AB, BC, and CA and $A'B'$, $B'C'$, and $C'A'$ are the corresponding sides of $\triangle ABC$ and $\triangle A'B'C'$, respectively.

Note that similar figures have angles with the same measure but may have different-sized sides. The corresponding sides of similar polygons are in proportion.

> Two polygons are **similar** if their corresponding angles have the same measure and their corresponding sides are in proportion.

■ **Example 7** _____

Consider the similar figures in Fig. 10.44. Determine the following:

a)   The length of side BC **b)**   The length of side $A'C'$

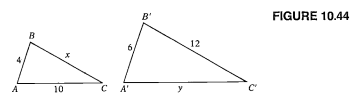

FIGURE 10.44

Solution

a) We will represent side BC with the letter x. Since the corresponding sides must be in proportion, we can write a proportion (as explained in Section 8.1) to find side BC. Since corresponding sides AB and $A'B'$ are known, we will use them as one ratio in the proportion:

$$\frac{AB}{A'B'} = \frac{BC}{B'C'}$$

$$\frac{4}{6} = \frac{x}{12}$$

Now cross-multiply as was explained earlier:

$$4 \cdot 12 = 6 \cdot x$$
$$48 = 6x$$
$$8 = x$$

Thus $BC = 8$.

b) Represent $A'C'$ with the letter y:

$$\frac{AB}{A'B'} = \frac{AC}{A'C'}$$

$$\frac{4}{6} = \frac{10}{y}$$

$$4 \cdot y = 6 \cdot 10$$
$$4y = 60$$
$$y = 15 \qquad \qquad ❖$$

■ **Example 8** ─────────────────────────────────

Consider the similar triangles ABC and $AB'C'$ (Fig. 10.45). Determine

a) the length of side BC,   **b)** the length of side AB',

c) $\angle AB'C'$,   **d)** $\angle ACB$,

e) $\angle ABC$.

FIGURE 10.45

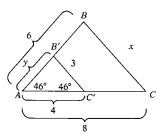

Solution

a) We will set up a proportion to find the length of side BC. Since corresponding sides AC and AC' are known, we will use them as one ratio in the proportion:

$$\frac{AC}{AC'} = \frac{BC}{B'C'}$$
$$\frac{8}{4} = \frac{x}{3}$$
$$24 = 4x$$
$$6 = x$$

Therefore the length of $BC = 6$.

b)
$$\frac{AC}{AC'} = \frac{AB}{AB'}$$
$$\frac{8}{4} = \frac{6}{y}$$
$$8y = 24$$
$$y = 3$$

Therefore $AB' = 3$.

c) Since the sum of the angles in a triangle must measure 180°, $\angle AB'C'$ must be $180° - (25° + 57°)$, or 98°.

d) Since triangles ABC and $AB'C'$ are similar, their corresponding angles must have the same measure. Therefore $\angle ACB$ must have the same measure as its corresponding angle $\angle AC'B'$. Thus $\angle ACB = 57°$.

e) Since $\angle ABC$ and $\angle AB'C'$ are corresponding angles of similar triangles, $\angle ABC = \angle AB'C'$. Using the results from part (c), the measure of $\angle ABC$ is 98°. ❖

     If the corresponding sides of two similar triangles are the same length, the figures are called **congruent figures.** Corresponding angles of congruent figures have the same measure, and the corresponding sides are equal in length. Two congruent figures when placed one upon the other would coincide.

■ **Example 9** _____

Triangles ABC and $A'B'C'$ are congruent. Find

a)   the length of side $A'C'$ b)   the length of side AB c) $\angle C'A'B'$

d) $\angle ACB$ e) $\angle ABC$

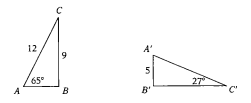

Solution: We are given that $\triangle ABC$ is congruent to $\triangle A'B'C'$. Using this fact we know that the corresponding sides and angles are equal.

a) $A'C' = AC = 12$

b) $AB = A'B' = 5$

c) $\angle C'A'B' = \angle CAB = 65°$

d) $\angle ACB = \angle A'C'B' = 27°$

e) The sum of the angles of a triangle is 180°. Since $\angle A = 65°$ and $\angle C = 27°$, $\angle B = 180 - 65 - 27 = 88°$. ❖

EXERCISES 10.2

Classify the following angles as acute, right, straight, or obtuse.

Find the complementary angle of each of the following.

7. 30°   8. 45°   9. $67\frac{1}{2}°$ 10. 34.4°
11. 87°   12. 1°   13. 5°   14. $3\frac{1}{2}°$

Find the supplementary angle of each of the following.

15. 91°   16. 4°   17. 105°   18. 179°
19. $13\frac{3}{4}°$ 20. $89\frac{1}{4}°$ 21. 114.7°   22. 137.9°

Name each of the polygons.

Identify each of the triangles as acute, obtuse, or right.

Identify each of the triangles as scalene, isosceles, or equilateral.

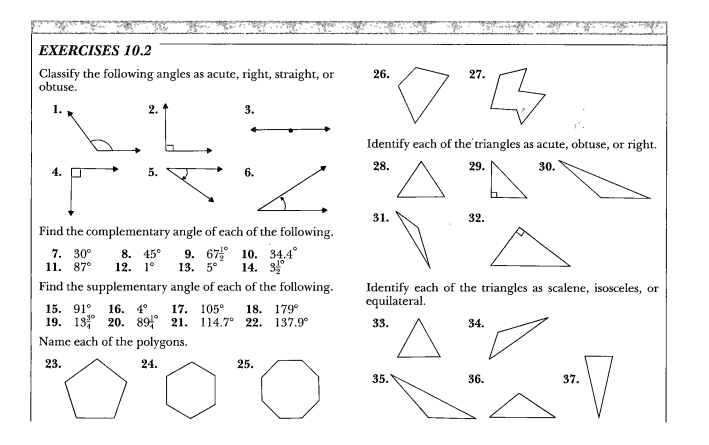

Name each of the quadrilaterals.

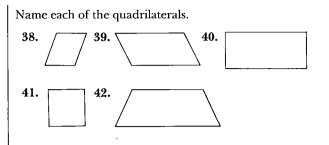

38. **39.** **40.**

41. **42.**

Match the name of the angles with the parts of Fig. 10.46.

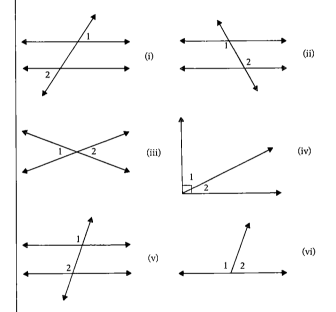

FIGURE 10.46

43. Vertical angles
44. Supplementary angles
45. Complementary angles
46. Alternate interior angles
47. Alternate exterior angles
48. Corresponding angles

49. If ∢1 and ∢2 are complementary angles and ∢1 is four times as large as ∢2, find the measures of ∢1 and ∢2.
50. If ∢1 and ∢2 are complementary angles and ∢1 is five times as large as ∢2, find the measures of ∢1 and ∢2.

51. If ∢1 and ∢2 are supplementary angles and ∢1 is four times as large as ∢2, find the measures of ∢1 and ∢2.
52. Given the set of parallel lines cut by a transversal in Fig. 10.47, determine the measures of ∢1 through ∢7.
53. Given the set of nonparallel lines cut by a transversal in Fig. 10.48, determine the measure of ∢x.

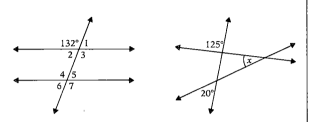

FIGURE 10.47   **FIGURE 10.48**

54. Given the set of nonparallel lines cut by a transversal in Fig. 10.49, find ∢x.
55. Given the set of nonparallel lines cut by a transversal in Fig. 10.50, find ∢x.

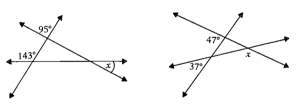

FIGURE 10.49   **FIGURE 10.50**

56. Given that l_1 and l_2 are parallel in Fig. 10.51, determine the measures of ∢1 through ∢12.

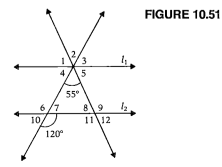

FIGURE 10.51

57. Sketch an octagon.
    a) Show that the figure can be divided into six triangles.
    b) Use the figure to determine the sum of the interior angles.
    c) Use the formula to verify the results obtained in part (b).
58. Sketch a decagon. Answer similar questions for this figure as asked in Exercise 57.

Find the sum of the interior angles of each polygon.

59. Nonagon   60. Octagon
61. Hexagon   62. Pentagon
63. Decagon   64. Icosagon

Find the measure of an interior angle of each regular polygon. If a side of the polygon is extended, find the supplement of the interior angle.

65. Triangle   66. Quadrilateral
67. Heptagon   68. Decagon
69. Pentagon   70. Octagon

In Fig. 10.52, $\angle ABC$ makes an angle of 125° with the floor. If l_1 and l_2 are parallel lines, determine the measures of the following.

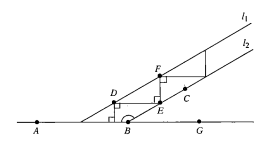

FIGURE 10.52

71. $\angle GBC$ 72. $\angle EDF$ 73. $\angle DFE$

74. The legs of a picnic table form an isosceles triangle as indicated in Fig. 10.53. If $\angle ABC = 80°$, find $\angle x$ and $\angle y$ so that the top of the table will be parallel to the ground.
75. What are similar figures?
76. What are congruent figures?
77. Triangles ABC and $A'B'C'$ are similar figures (Fig. 10.54). Find the length of the following.
    a) Side $B'C'$ b) Side $A'C'$

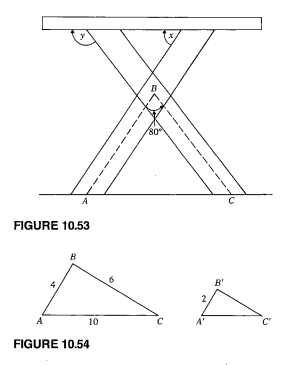

FIGURE 10.53

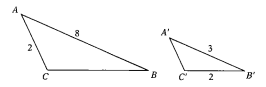

FIGURE 10.54

78. Triangles ABC and $A'B'C'$ are similar figures (Fig. 10.55). Find the length of the following.
    a) Side CB b) Side $A'C'$

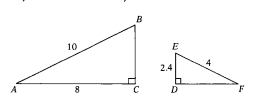

FIGURE 10.55

79. Triangles ABC and DEF are similar figures (Fig. 10.56). Find the length of the following.
    a) Side BC b) Side DF

FIGURE 10.56

80. Triangles *ABC* and *A'B'C'* are similar figures (Fig. 10.57). Find the length of the following.
    a) Side *A'C*   **b)** Side *BC*   **c)** Side *BB'*

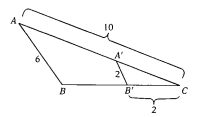

FIGURE 10.57

81. Quadrilaterals *ABCD* and *A'B'C'D'* are similar figures (Fig. 10.58). Find the length of the following.
    a) Side *AB*   **b)** Side *BC*   **c)** Side *D'C'*

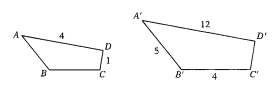

FIGURE 10.58

Pentagons *ABCDE* and *A'B'C'D'E'* are similar figures (Fig. 10.59). Find the measure of the following.

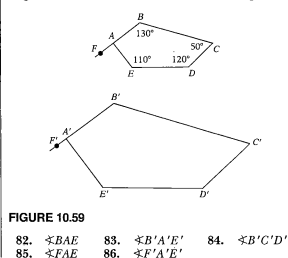

FIGURE 10.59

82. ∡*BAE*   **83.** ∡*B'A'E'*   **84.** ∡*B'C'D'*
85. ∡*FAE*   **86.** ∡*F'A'E'*

Triangles *ABC* and *A'B'C'* are congruent figures. Find each of the following.

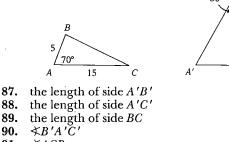

87. the length of side *A'B'*
88. the length of side *A'C'*
89. the length of side *BC*
90. ∡*B'A'C'*
91. ∡*ACB*
92. ∡*ABC*

Quadrilaterals *ABCD* and *A'B'C'D'* are congruent figures. Find each of the following.

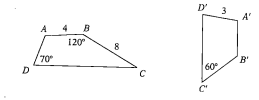

93. the length of side *A'B'*
94. the length of side *AD*
95. the length of side *B'C'*
96. ∡*BCD*
97. ∡*A'D'C'*
98. ∡*DAB*

For the supplementary angles, find ∡1 and ∡2.

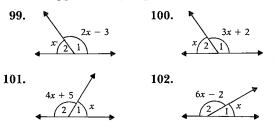

99.   **100.**

101.   **102.**

For the complementary angles, find ∡ 1 and ∡ 2.

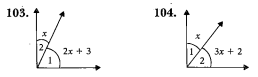

103.   **104.**

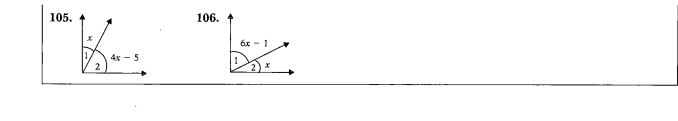

105.

x

1

2

$4x - 5$

106.

$6x - 1$

1

2 x

10.3 THE METRIC SYSTEM

The metric system was developed in France in the 18th century. However, because of advancements in the scientific community in the twentieth century, the 18th century standards had to be revised. In 1960 at the 11th General Conference of Weights and Measures a new International System of Units (SI System) was formulated. In the original metric system the base units of measurement were based upon physical objects. In the SI system they are based primarily on scientific data. For example, the meter in the 18th century system was the Delambre-Méchain survey-derived "one ten-millionth part of a meridional quadrant of the earth." In the SI system the meter is defined as 1,650,763.73 wavelengths in vacuum of the orange-red line of the spectrum of krypton-86. The only base unit still defined by an artifact in the SI system is the kilogram. The kilogram is a cylinder of platinum-iridium alloy kept by the International Bureau of Weights and Measures, located in Sevres, near Paris.

Two systems of weights and measures exist side by side in the United States today, the **U.S. customary system** and the **SI system** (System International). The SI, or metric, system is used predominantly in the automotive, construction, farm equipment, computer, and bottling industries. Many other industries are now in the process of converting to the SI system. As a result of the increased use of the metric system in the United States, we are and will be coming in contact with the system more each day. This section will introduce you to the metric terminology and symbolism. In Sections 10.4 and 10.5 we will show how the metric units are used in applied problems. In Appendix C we show how to convert from the customary system to the metric system and vice versa.

As its name implies, the customary system (inherited but now different from the British imperial system) has been customarily used throughout U.S. history. However, over the past 100 years, use of the metric system in manufacturing in the United States has slowly but steadily increased. Today it is nearly as important as the customary system.

There are many advantages to using the metric system. Some of these are summarized here.

1. The metric system is the worldwide accepted standard measurement system. All industrial nations except the United States that trade internationally use the metric system as the official unit of measurement.

2. There is only one basic unit of measurement for each physical quantity. In the customary system there are often many units used to represent the same physical quantity. For example, when discussing length, we use inches, feet, yards, rods, miles, and so on. Converting from one of these units to the other is often a tedious task (consider changing 12 miles to inches). In the metric system it is possible to make many conversions by simply moving the decimal point.

3. The SI system is based on the number 10, and there is little need for fractions, since most quantities can be expressed as decimals.

Basic Terms and Rules

Because the official definitions of many metric terms are quite technical, we will present them in lay language.

The **meter** (m) is the base unit of length in the metric system. The meter is a little more than a yard. Basketball players are often more than two meters tall.

The **kilogram** (kg) is the base unit of mass. The kilogram is a little more than two pounds. A newborn baby might have a mass of about 3 kilograms. The gram (g), a unit of mass derived from the kilogram, is used to measure small amounts. A nickel has a mass of about five grams.

The **liter** (ℓ) is not a base unit of the metric system, but it is commonly used to measure volume. A liter is a little more than a quart. The gas tank of a compact car may hold 50 liters of gasoline. The **cubic decimeter** is the official unit of volume. It is equal to a liter, but liter is more commonly used.

The term **degree Celsius** (°C) is used to measure temperature in place of the official or base unit, kelvin. The freezing point of water is 0°C, and the boiling point of water is 100°C. The temperature on a warm day may be 30°C.

The metric system is based on the number 10, and it is therefore a decimal system. Prefixes are used to denote a multiple or part of a base unit. For example, a **dekameter** represents 10 meters, and a **centimeter** represents $\frac{1}{100}$ of a meter. In large numbers, groups of three digits are separated by a space, not a comma. The number for one thousand is 1 000 and the number for nine million is 9 000 000. Commas are not used in the SI system because many countries use the comma as we use a decimal point. Table 10.7 summarizes the more commonly used prefixes and their meanings. These prefixes pertain to all the units of measurement in the metric system. Thus 1 kiloliter = 1 000 liters, 1 kilogram = 1 000 grams, 1 centimeter = $\frac{1}{100}$ meter, and 1 milliliter = $\frac{1}{1000}$ liter.

TABLE 10.7

Prefix	Symbol	Meaning
kilo	k	1 000 × base unit
hecto	h	100 × base unit
deka	da	10 × base unit
		base unit
deci	d	$\frac{1}{10}$ of base unit
centi	c	$\frac{1}{100}$ of base unit
milli	m	$\frac{1}{1000}$ of base unit

When writing metric measures the unit names, such as meter and liter, are written in lowercase letters. Write the unit names in full or use the proper symbols. Never write c meter or m. liter. The symbols are never pluralized (but full names are). For example, five milliliters is symbolized as 5 mℓ and not 5 mℓs.

DID YOU KNOW?

There are prefixes for expressing very large and very small quantities. These prefixes are often used by scientists and engineers in areas that require a high degree of accuracy.

Prefix	Symbol	Meaning
tera	T	1 000 000 000 000 × base unit
giga	G	1 000 000 000 × base unit
mega	M	1 000 000 × base unit
micro	m	$\frac{1}{1\,000\,000}$ of base unit
nano	n	$\frac{1}{1\,000\,000\,000}$ of base unit
pico	p	$\frac{1}{1\,000\,000\,000\,000}$ of base unit

Conversions within the Metric System

We will use Table 10.8 to help demonstrate how to change from one metric unit to another metric unit (meters to kilometers, and so on).

TABLE 10.8

kilometer km 1 000 m	hectometer hm 100 m	dekameter dam 10 m	meter m 1 m	decimeter dm 0.1 m	centimeter cm 0.01 m	millimeter mm 0.001 m

The meters in Table 10.8 can be replaced by grams, liters, or any other base unit of the metric system. Regardless of which of these units we choose, the procedure will be the same. For purposes of explanation we have used the meter.

The table tells us that 1 hectometer equals 100 meters and 1 millimeter is 0.001 (or $\frac{1}{1000}$) meter. The millimeter is the smallest unit in the table. A centimeter is 10 times as large as a millimeter, a decimeter is 10 times as large as a centimeter, a meter is 10 times as large as a decimeter, and so on. Since each unit is 10 times as large as the unit on its right, converting from one unit to another is simply a matter of multiplying or dividing by powers of 10.

■ Example 1

a) Change 30 meters to dekameters.

b) Change 3.7 kilometers to meters.

Solution

a) When we are changing from a smaller unit of measurement to a larger unit of measurement, the number in the answer must be smaller than the number in the original measurement. This indicates that division by a power of 10 must occur. The dekameter is one place to the left of the meter. We must therefore divide by 10 to change from meters to dekameters. We divide by 10 by moving the decimal point in the number one place to the left. Why?

$$30.\ m = 3.0\ dam$$

Therefore 30 m = 3 dam.

b) When we are changing from a larger unit of measurement to a smaller unit of measurement, the number in the answer must be larger than the number in the original measurement. This indicates that multiplication by a power of 10 must occur. The meter is three places to the right of the kilometer. We must therefore multiply by 1 000 (10 × 10 × 10) to change from kilometers to meters. We do so by moving the decimal point three places to the right. Why?

$$3.7\ km = 3\ 700.\ m$$

Therefore 3.7 km = 3 700 m. ❖

The procedure above may be summarized by the following rules:

1. To change from a smaller unit to a larger unit (for example from meters to kilometers), move the decimal point in the original quantity one place to the left for each larger unit of measurement until you obtain the desired unit of measurement.

2. To change from a larger unit to a smaller unit (for example from kilometers to meters), move the decimal point in the original quantity one place to the right for each smaller unit of measurement until you obtain the desired unit of measurement.

Now let us try another example, using grams and liters.

■ **Example 2** ─────────────────────────────────

a) Change 403.2 grams to kilograms.
b) Change 14 grams to centigrams.
c) Change 0.18 liters to milliliters.
d) Change 240 dekaliters to kiloliters.

Solution

a) To change from grams to kilograms, we note that there are three greater units: dag, hg, and kg. Thus the decimal point must be moved three places to the left:

$$403.2 \text{ g} = 0.403 \ 2 \text{ kg}$$

b) To change grams to centigrams, move the decimal point two places to the right:

$$14 \text{ g} = 1 \ 400 \text{ cg}$$

c) To change liters to milliliters, move the decimal point three places to the right:

$$0.18 \ \ell = 180 \ m\ell$$

d) To change dekaliters to kiloliters, move the decimal point two places to the left:

$$240 \text{ da}\ell = 2.40 \text{ k}\ell \qquad ❖$$

■ **Example 3** ─────────────────────────────────

Arrange in order from the smallest to largest length: 3.4 m, 3 421 mm, 104 cm.

Solution

To be compared, these lengths should all be in the same units of measure. Convert all the measures to millimeters:

$$3.4 \text{ m} = 3\ 400 \text{ mm}, \quad 3\ 421 \text{ mm}, \quad 104 \text{ cm} = 1\ 040 \text{ mm}$$

The lengths arranged in order from smallest to largest are 104 cm, 3.4 m, 3 421 mm. ❖

Length

The basiç unit of length is the meter. Other units of length that are commonly used are the kilometer, centimeter, and millimeter. The meter, which is a little longer than one yard, is used to measure items that we measure in yards and feet. A man who stands about two meters tall is a very tall man. A tractor trailer unit (18-wheeler) is about 18 meters long.

The kilometer is used to measure what we measure in miles. For example, it is about 5 120 km from New York to Seattle. One kilometer is about 0.6 miles, and one mile is about 1.6 kilometers.

Centimeters and millimeters are used to measure what we measure in inches. The centimeter is a little less than 1/2 of an inch (see Fig. 10.60), and the millimeter is a little less than 1/20 of an inch. A millimeter is about the thickness of a dime. A book may measure 20 cm by 25 cm with a thickness of about 3 cm. Millimeters are often used in scientific work and other areas in which small quantities must be measured. The length of a small insect may be measured in millimeters.

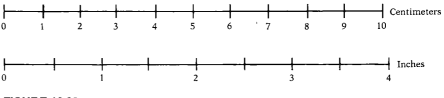

FIGURE 10.60

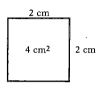

FIGURE 10.61

A football field with surrounding
track is approximately half a
hectare (5940 square yards).
*Photo by Russell A. Thompson/
Taurus Photos*

Area

The area enclosed in a square with 1 cm sides (Fig. 10.61) is 1 cm × 1 cm = 1 cm². A square whose sides are 2 cm (Fig. 10.62) has an area of 2 cm × 2 cm = 4 cm².

Areas are always expressed in square units, such as square centimeters, square kilometers, or square meters. When finding areas, we must be careful that all the numbers being multiplied are expressed in the same units.

In the metric system the square centimeter will replace the square inch. The square foot and square yard will be replaced by the square meter. In the future you might purchase carpet or other floor covering by the square meter instead of by the square yard.

For measuring large land areas the metric system uses a square unit 100 meters on each side (a square hectometer). This unit is called a **hectare** (pronounced "hectair," symbolized ha). A hectare is about 2.5 acres. One square mile of land contains about 260 hectares.

Very large units of area are measured in square kilometers. One square kilometer is about 4/10 of a square mile.

FIGURE 10.62

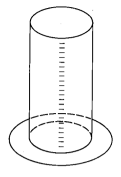

FIGURE 10.63

Volume

When a figure has only two dimensions—length and width—we can find its area. When a figure has three dimensions—length, width, and height—we can find its volume. The volume of an item can be considered the space occupied by the item.

In the metric system, volume may be expressed in terms of liters or cubic meters, depending on what is being measured.

The volume of liquids is expressed in liters. A liter is a little larger than a quart. Liters are used in place of pints, quarts, or gallons. A liter can be divided into 1000 equal parts, each of which is called a milliliter. Figure 10.63 illustrates a type of liter-container (a 1 000 mℓ graduated cylinder) that is often used in chemistry. Milliliters are used to express the volume of very small amounts of liquid. Drug dosages are often expressed in milliliters. An eight-ounce cup will hold about 240 mℓ of liquid.

The kiloliter, 1 000 liters, is used to represent the volume of large amounts of liquid. Tank trucks carrying gasoline to service stations hold about 10.5 kiloliters of gasoline.

Cubic meters are used to express the volume of large amounts of solid material. The volume of a dump truck's load of topsoil is measured in cubic meters. The volume of natural gas for heating a house may soon be measured in cubic meters instead of cubic feet.

The liquid in a liter container will fit exactly in a cubic decimeter (Fig. 10.64). Note that 1 liter = 1 000 mℓ and 1 cubic decimeter = 1 000 cm^3. Since 1 liter = 1 cubic decimeter, 1 mℓ must equal 1 cm^3. Other useful facts are illustrated in Table 10.9.

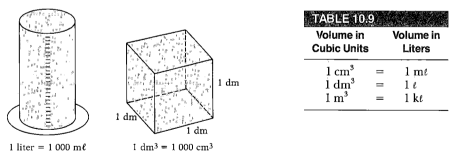

1 liter = 1 000 mℓ 1 dm3 = 1 000 cm3

TABLE 10.9	
Volume in Cubic Units	Volume in Liters
1 cm^3 =	1 mℓ
1 dm^3 =	1 ℓ
1 m^3 =	1 kℓ

FIGURE 10.64

Mass

Weight and mass are not the same. Weight is a measure of the gravitational pull on an object. The gravitational pull on the earth is about six times as great as the gravitational pull on the moon. Thus a person on the moon weighs about 1/6 as much as on earth, even though the person's mass

FIGURE 10.65

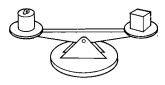

FIGURE 10.66

remains the same. In space, where there is no gravity, a person has no weight.

The gravitational pull even on the earth varies from point to point. The closer you are to the center of the earth, the greater the gravitational pull. Thus a person will weigh very slightly less on a mountain than in a nearby valley.

Mass is a measure of the amount of matter in an object. Mass is determined by the molecular structure of the object, and it will not change from place to place. For this reason, scientists use mass rather than weight.

Although weight and mass are not the same, on the earth they are proportional to each other (the greater the weight, the greater the mass). Therefore for our purposes we can treat weight and mass as the same.

Weight is measured on a spring scale (Fig. 10.65). To measure mass, we use a balance and compare the object under consideration with a known mass (Fig. 10.66).

The kilogram is the basic unit of mass in the metric system. It is a little more than two pounds. Items that we measure in pounds will usually be measured in kilograms. For example, an average-sized man has a mass of about 75 kilograms.

The gram (a unit that is 0.001 kilogram) is relatively small and is used in place of the ounce. A nickel has a mass of 5 grams, a cube of sugar has a mass of about 2 grams, and a large paper clip has a mass of about 1 gram.

The milligram is used extensively in the medical and scientific fields, as well as in the pharmaceutical industry. Practically all bottles of tablets are now labeled in either milligrams or grams.

The metric tonne (t) is used to express the mass of heavy items. A metric tonne is equal to 1 000 kilograms. It is a little larger than our customary ton of 2000 pounds. The mass of a large truck may be expressed in metric tonnes.

A kilogram of water has a volume of exactly one liter. In fact, a liter is defined to be the volume of one kilogram of water at a specified temperature and pressure. Thus mass and volume are easily interchangeable in the metric system. (How would you change pounds of water to cubic feet or gallons of water in our customary system?) Figure 10.67 illustrates the relationship between volume of water in cubic decimeters, capacity in liters, and mass in kilograms. Table 10.10 expands on this relationship between volume and mass of water.

| TABLE 10.10 | | |
Volume in Cubic Units	Volume in Liters	Mass
$1\ cm^3$	$=$ $1\ m\ell$	$=$ $1\ g$
$1\ dm^3$	$=$ $1\ \ell$	$=$ $1\ kg$
$1\ m^3$	$=$ $1\ k\ell$	$=$ $1\ t\ (1\ 000\ kg)$

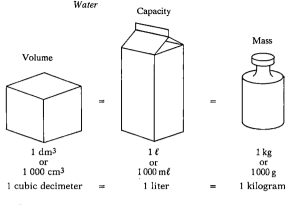

Water

Capacity

Mass

Volume

1 dm³
or
1 000 cm³

1 ℓ
or
1 000 mℓ

1 kg
or
1 000 g

1 cubic decimeter = 1 liter = 1 kilogram

FIGURE 10.67

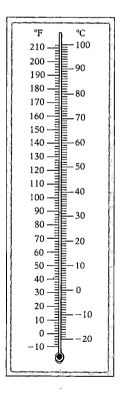

FIGURE 10.68

Temperature

The Celsius scale is used to measure daily temperatures in the metric system. Figure 10.68 shows a thermometer with the Fahrenheit scale on the left and the Celsius scale on the right.

The Celsius scale was named for the Swedish astronomer Anders Celsius (1701–1744), who first devised it in 1742. On the Celsius scale, water freezes at 0°C and boils at 100°C. The Celsius thermometer in the past was called a "centigrade thermometer." Recall that *centi* means 1/100, and there are 100 degrees between the freezing point of water and the boiling point of water. Thus 1°C is 1/100 of this interval.

Table 10.11 gives some common temperatures in both Celsius (°C) and Fahrenheit (°F).

TABLE 10.11		
−18°C	A very cold day	0°F
0°C	Freezing point of water	32°F
10°C	A warm winter day	50°F
20°C	A mild spring day	68°F
30°C	A warm summer day	86°F
37°C	Body temperature	98.6°F
100°C	Boiling point of water	212°F
177°C	Oven temperature for baking	350°F

The question may arise, "How can we change units of measure in the metric system to equivalent units of measure in the customary system or vice versa?" Appendix C gives tables and examples.

EXERCISES 10.3

1. Without referring to any table, name as many of the metric system prefixes as you can and give their meanings. If you don't already know all the prefixes in Table 10.7, memorize them now.

Fill in the blanks.

2. A meter is a little longer than a _____ .
3. A kilogram is a little more than _____ pounds.
4. A nickel has a mass of about _____ grams.
5. The temperature on a warm day may be _____ °C.

In exercises 6–11 fill in the blank with the appropriate letter, a–f.

6. deci _____   a) 1000 times base unit
7. milli _____   b) $\frac{1}{100}$ of base unit
8. hecto _____   c) $\frac{1}{1000}$ of base unit
9. kilo _____   d) 100 times base unit
10. centi _____   e) $\frac{1}{10}$ of base unit
11. deka _____   f) 10 times base unit
12. Cindy ran 100 meters, and Chuck ran 100 yards in the same length of time. Who ran faster?

Without referring to any of the tables or your notes, give the symbol and the equivalent in grams for each of the following units.

13. milligram   14. centigram
15. decigram   16. gram
17. dekagram   18. hectogram
19. kilogram

Fill in the missing values.

20. 2m = _____ cm
21. 3 dam = _____ m
22. 15.7 hg = _____ g
23. 0.024 hℓ = _____ ℓ
24. 242.6 cm = _____ hm
25. 1.34 mℓ = _____ ℓ
26. 417 g = _____ kg

27. 14.27 hℓ = _____ ℓ
28. 1.34 hm = _____ cm
29. 0.076 mm = _____ m

Convert the given units to the units indicated.

30. 130 cm to hm   31. 8.3 m to cm
32. 1 049 mm to m   33. 94.5 kg to g
34. 3 472 mℓ to dℓ   35. 895 ℓ to mℓ
36. 14 000 mℓ to ℓ   37. 24 dm to km

Arrange each of the following in order from smallest to largest.

38. 64 dm, 67 cm, 680 mm
39. 5.6 dam, 0.47 km, 620 cm
40. 4.3 ℓ, 420 cℓ, 0.045 kℓ
41. 2.2 kg, 2 400 g, 24 300 dg
42. 0.045 kℓ, 460 dℓ, 48 000 cℓ
43. Charlie cut a piece of lumber 7.8 meters in length into six equal pieces. What was the length of each section (a) in meters? (b) in centimeters?
44. A baseball diamond is 27 meters along each side. (a) How many meters does a batter run if he hits a home run? (b) How many kilometers? (c) How many millimeters?
45. How many centimeters of picture molding should be purchased to frame two pictures? One picture is 33 centimeters by 440 millimeters, and the other is 3.4 decimeters by 44 centimeters. Molding is sold in centimeters.
46. At 70 cents a meter, what will it cost Karen to sew a lace border around a bedspread that measures 190 cm by 114 cm?
47. The filter pump on an aquarium circulates 360 milliliters of water every minute. If the aquarium holds 30 liters of water, how long will it take to circulate all the water?
48. Dale drove 1 200 kilometers and used 187 liters of gasoline. What was her average rate of gas use for the trip (a) in kilometers/liter? (b) in meters/liter?
49. A mixture of 15 grams of salt and 16 grams of baking soda is poured into 250 milliliters of water. What is the total mass of the mixture in grams?

PROBLEM SOLVING

50. The following question was selected from a nursing exam. Can you answer it?

    In caring for a patient after delivery, you are to give 0.2 mg Ergotrute. The ampule is labeled $\frac{1}{300}$ grain/mℓ. How much would you draw and give? (60 mg = 1 grain)

    a) 15 cc   b) 1.0 cc   c) 0.5 cc   d) 0.01 cc

RESEARCH ACTIVITIES

51. Use an encyclopedia and other reference materials to determine how and why the different units of measurement were defined in the customary system.

52. Investigate the development of the metric system in Europe. Which groups of people had the most influence in developing the metric system?

10.4 PERIMETER AND AREA

Many household chores are simplified by the ability to calculate perimeters and areas of common geometric shapes found around the home. The **perimeter**, P, is the sum of the lengths of the sides of the figures. The **area, A,** is the total amount of surface within the figure's boundaries. Area is measured in square units. Table 10.12 gives the formulas for finding the areas and perimeters of some of the more common shapes.

TABLE 10.12

Figure	Sketch	Area (A)	Perimeter (P)
Triangle		$A = \frac{1}{2}bh$	$P = s_1 + s_2 + b$
Trapezoid		$A = \frac{1}{2}h(b_1 + b_2)$	$P = s_1 + s_2 + b_1 + b_2$
Parallelogram		$A = l \cdot h$	$P = 2l + 2w$
Rectangle		$A = l \cdot w$	$P = 2l + 2w$
Square		$A = s^2$	$P = 4s$

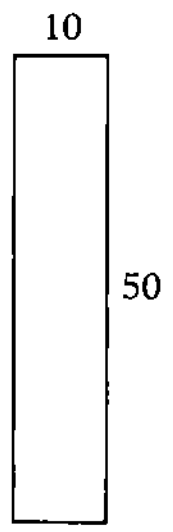

10

50

FIGURE 10.69

■ **Example 1** _____

You are coating your driveway with blacktop sealer. One can of sealer costs $9 and covers 200 square feet. If your driveway is 50 feet long and 10 feet wide (Fig. 10.69), (a) find the area of your driveway, (b) determine how many cans of blacktop sealer are needed, and (c) determine the cost of sealing your driveway.

Solution

a) The area of your driveway in square feet is

$$A = l \cdot w = 50 \cdot 100 = 500 \text{ ft}^2$$

The area of the driveway is in square feet because both the length and width are measured in feet.

b) To determine the number of cans needed, divide the area of the driveway by the area covered by one can of sealer:

$$\frac{500}{200} = 2.5$$

Since you cannot purchase a half can of sealer three cans of sealer are needed.

c) The cost of the sealer is 3 × $9 or $27.   ❖

"He is unworthy of the name man who is ignorant of the fact that the diagonal of a square is incommeasurable with its side."
Plato

The Pythagorean theorem was introduced in Chapter 5. The Pythagorean theorem is an important tool used in finding the perimeter and area of triangles. For this reason we restate it at this time.

Pythagorean theorem. The sum of the squares of the lengths of the legs of a right triangle is equal to the square of the length of the hypotenuse.
Symbolically, if a and b represent the lengths of the legs and c is the length of the hypotenuse (the side opposite the right angle), then $a^2 + b^2 = c^2$.

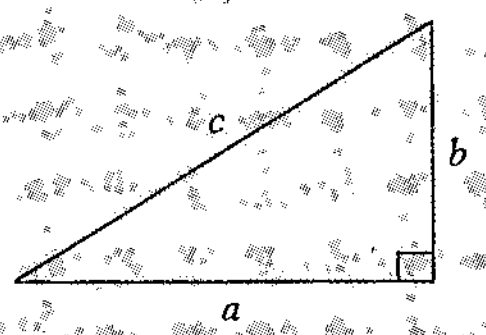

c

b

a

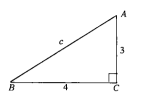

FIGURE 10.70

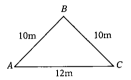

FIGURE 10.71

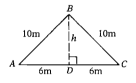

FIGURE 10.72

■ **Example 2** ───────────────────────────────

Determine c in triangle ABC (Fig. 10.70).

Solution: Since the triangle is a right triangle, we can use the Pythagorean theorem to find c:

$$a^2 + b^2 = c^2$$
$$4^2 + 3^2 = c^2$$
$$16 + 9 = c^2$$
$$25 = c^2$$
$$5 = c$$

■ **Example 3** ───────────────────────────────

a) Determine the height of the isosceles triangle ABC (Fig. 10.71).
b) Determine the area of the triangle.

Solution

a) Since the triangle is isosceles, the height (or altitude) bisects the base, or cuts the base into two equal parts. The right triangle ADB has sides of lengths 6 and h, and a hypotenuse of length 10 (Fig. 10.72).

   To determine the height, h, we can use the Pythagorean Theorem for triangle ADB (or triangle CDB):

$$a^2 + b^2 = c^2$$
$$6^2 + h^2 = 10^2$$
$$36 + h^2 = 100$$
$$h^2 = 64$$
$$h = 8$$

b) We can now find the area of triangle ABC:

$$A = \tfrac{1}{2}bh$$
$$= \tfrac{1}{2}(12)8$$
$$= 48 \text{ square meters}$$

$$3^2 + 4^2 = 5^2$$
$$9 + 16 = 25$$
$$25 = 25$$

It has been proven that there are an infinite number of solutions to the equation $a^2 + b^2 = c^2$.

In 1637 the French mathematician Pierre de Fermat while reading a book by Diophantus, scribbled a note in the margin proposing that there are no positive integer solutions to the equation.

$$a^n + b^n = c^n$$

where n is greater than 2. In other words when $n = 3$, no set of integers satisfies the equation

$$a^3 + b^3 = c^3$$

Then a tantalizing sentence was added that was to haunt mathematicians for centuries to come. Fermat added that although he had a wonderful proof for the theorem, he did not have enough room to write it out. This theorem, called **Fermat's last theorem** became one of the oldest and most famous conjectures in mathematics.

Since the theorem was proposed, many mathematicians have tried to prove the conjecture. Some mathematicians have found proofs for special cases, and a computer has been used to show that Fermat's conjecture is true for all exponents less than 150,000.

In March 1988 the mathematics community was buzzing with excitement over the prospect that Fermat's last theorem had finally been proven. Yoichi Miyaoka, a Japanese mathematician working at the Max Planck Institute in Bonn West Germany, presented what he and other experts believed was a proof of Fermat's last theorem. However, after careful scrutiny of the proposed proof certain serious flaws were discovered. That Miyaoka's initial attempt failed is hardly surprising or unusual in mathematics research. Normally mathematicians privately circulate proposed proofs and discuss possible errors and oversights for months before gaining enough confidence to announce a proof publicly. In Miyaoka's case, the fact that Fermat's last theorem was such a famous unsolved problem put him in the spotlight he had not as yet sought.

Mathematicians are quite confident that someone, if not Miyaoka, will eventually come up with the proof of Fermat's last theorem.

Circles

A commonly used plane figure that is not a polygon is a **circle.** A circle is a set of points equidistant from a fixed point called the center. A **radius** of a circle is a line segment from the center of the circle to any point on the circle (see Fig. 10.73a). A **diameter** of a circle is a line segment through the center of a circle with both end points on the circle (Fig. 10.73b).

TABLE 10.13

Area	Circumference
$A = \pi r^2$	$C = 2\pi r$

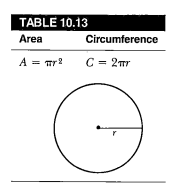

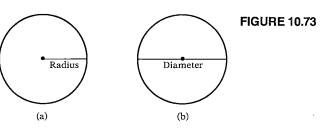

FIGURE 10.73

(a)   (b)

Note that the diameter of the circle is twice its radius. The **circumference** is the length of a simple closed curve that forms the circle. The formulas for the area and circumference of a circle are given in Table 10.13. We introduced the symbol pi, π, in Chapter 5. Recall that π is approximately 3.14.

■ **Example 4** _____

At the local pizza shop the price for a medium plain pizza is $7.50, and that for a large plain pizza is $14.00. The diameters of the large and medium pizzas are 18 inches and 12 inches, respectively. As we were enjoying a large pizza, we wondered how we could compare the medium and the large pizza. Should we have bought two medium pizzas instead of the one large pizza? The price of the large pizza is almost twice the price of the medium pizza, but the diameter is not twice as large.

"A man in an ivory tower dreamed of Pi to the 200th power. (This is apt to occur when you mix liquor with clams at a very late hour.)"
John McClellen

Solution: If we assume that both pizzas have the same thickness, then there are two measures that we can consider to compare the pizzas—the circumference and the area:

	Medium Pizza	Large Pizza
Circumference	$c = 2\pi r'$ $= 2(3.14)(6)$ $= 37.68$ in.	$c = 2\pi r$ $= 2(3.14)(9)$ $= 56.52$ in.
Area	$A = \pi r^2$ $= (3.14)(6^2)$ $= 113.04$ in.^{2}	$A = \pi r^2$ $= (3.14)(9^2)$ $= 254.34$ in.^{2}

The circumference in this case is not the best measure for comparison. It tells us only the distance around the pizzas. Since we assume that the thicknesses of the pizzas are the same, then the surface area will tell us whether two medium pizzas or one large pizza contains more to eat. The larger pizza (254.34 in.^{2}) is more than twice the size of a medium pizza (2 ×

113.04 in.^{2} = 226.08 in.^{2}) Two medium pizzas would cost $15. Thus we saved $1 and had more to eat by buying one large pizza instead of two medium pizzas. ❖

■ **Example 5** _____

You plan to fertilize your lawn. The shapes and dimensions of your lot, house, pool, and shed are shown in Fig. 10.74. Those areas will not be fertilized. One bag of fertilizer costs $9.95 and covers 5000 square feet. Determine how many bags of fertilizer are needed and the total cost of the fertilizer.

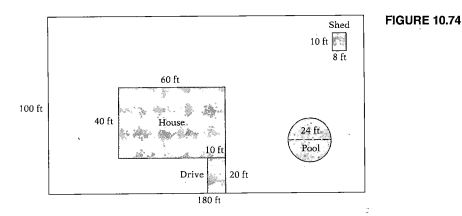

FIGURE 10.74

Solution: The total area of the lot is 100 · 180, or 18,000 square feet. From this we must subtract the area of the house, driveway, pool, and shed.

$$\text{Area of house} = 60 \cdot 40 = 2400 \text{ ft}^2$$
$$\text{Area of driveway} = 20 \cdot 10 = 200 \text{ ft}^2$$
$$\text{Area of shed} = 8 \cdot 10 = 80 \text{ ft}^2$$

Since the diameter of the pool is 24 feet, its radius is 12 feet.

$$\text{Area of pool} = \pi r^2 = 3.14(12)^2 = 3.14(144) = 452.16 \text{ square feet}$$

The total area of the house, driveway, shed, and pool is approximately 2400 + 200 + 80 + 452, or 3132 square feet. The area to be fertilized is 18,000 − 3132, or 14,868 square feet. The number of bags of fertilizer is found by dividing the total area to be fertilized by the number of square feet covered per bag.

The number of bags of fertilizer is 14,868/5000, or 2.97 bags. Therefore three bags are needed. At $9.95 per bag, the total cost is $29.85. ❖

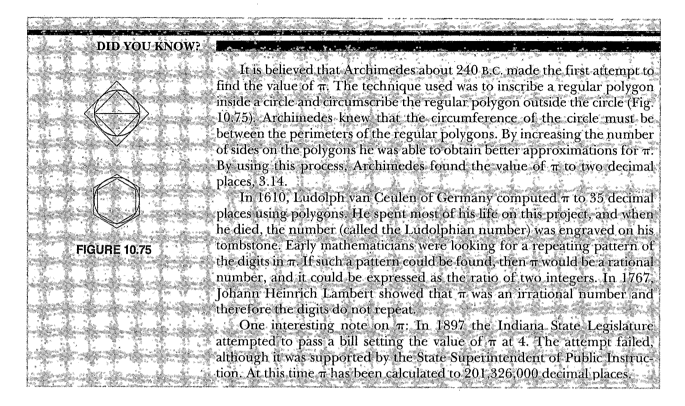

FIGURE 10.75

It is believed that Archimedes about 240 B.C. made the first attempt to find the value of π. The technique used was to inscribe a regular polygon inside a circle and circumscribe the regular polygon outside the circle (Fig. 10.75). Archimedes knew that the circumference of the circle must be between the perimeters of the regular polygons. By increasing the number of sides on the polygons he was able to obtain better approximations for π. By using this process, Archimedes found the value of π to two decimal places, 3.14.

In 1610, Ludolph van Ceulen of Germany computed π to 35 decimal places using polygons. He spent most of his life on this project, and when he died, the number (called the Ludolphian number) was engraved on his tombstone. Early mathematicians were looking for a repeating pattern of the digits in π. If such a pattern could be found, then π would be a rational number, and it could be expressed as the ratio of two integers. In 1767, Johann Heinrich Lambert showed that π was an irrational number and therefore the digits do not repeat.

One interesting note on π: In 1897 the Indiana State Legislature attempted to pass a bill setting the value of π at 4. The attempt failed, although it was supported by the State Superintendent of Public Instruction. At this time π has been calculated to 201,326,000 decimal places.

Whenever you multiply units of length, be very careful to make sure that the units are the same. You can multiply feet by feet to get square feet or yards by yards to get square yards. You cannot get a valid answer if you multiply numbers expressed in feet by numbers expressed in yards.

Sometimes it may be necessary to convert an area from one square unit of measure to a different square unit of measure. For example, you may have to change square feet of carpet to square yards of carpet before placing an order with the dealer.

■ **Example 6**

Convert one square foot to square inches.

Solution: It may be helpful to draw a diagram (Fig. 10.76).

$$1 \text{ ft} = 12 \text{ in.}$$

Therefore $1 \text{ ft}^2 = 12 \text{ in.} \times 12 \text{ in.} = 144 \text{ in.}^2$. ❖

A similar procedure can be used to make other conversions from one square unit to another.

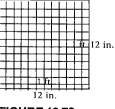

FIGURE 10.76

■ **Example 7** _____

Convert 7.5 square feet to square inches.

Solution: Since $1 \text{ ft}^2 = 144 \text{ in.}^2$,
$$7.5 \text{ ft}^2 = 7.5 \times 144 = 1080 \text{ in.}^2 \qquad ❖$$

■ **Example 8** _____

Convert 8424 square inches to square feet.

Solution: $1 \text{ ft}^2 = 144 \text{ in.}^2$

We can therefore divide 8424 in.^{2} by 144 in.^{2} to obtain the number of square feet:
$$8424 \text{ in.}^2 = 8424/144 = 58.5 \text{ ft}^2 \qquad ❖$$

■ **Example 9** _____

a) Convert 1 square meter to square centimeters.
b) Convert 4.2 square meters to square centimeters.

Solution

a) Since 1 meter equals 100 cm, we replace 1 meter with 100 cm (see Fig. 10.77). The area $1 \text{ m}^2 = 1 \text{ m} \cdot 1 \text{ m} = 100 \text{ cm} \times 100 \text{ cm} = 10\,000 \text{ cm}^2$. Thus the area of one square meter is 10 000 times the area of one square centimeter.

b) Since $1 \text{ m}^2 = 10\,000 \text{ cm}^2$,
$$4.2 \text{ m}^2 = 4.2 \times 10\,000 \text{ cm}^2$$
$$= 42\,000 \text{ cm}^2 \qquad ❖$$

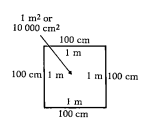

1 m² or
10 000 cm²

FIGURE 10.77

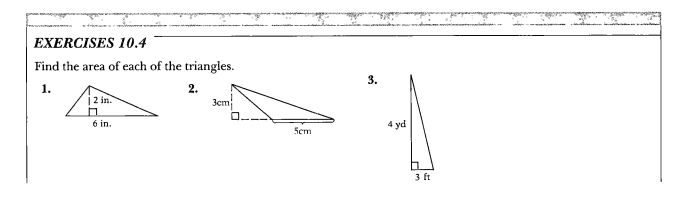

EXERCISES 10.4

Find the area of each of the triangles.

1. 2 in. 6 in.

2. 3cm 5cm

3. 4 yd 3 ft

Find the area and perimeter of each of the quadri-laterals.

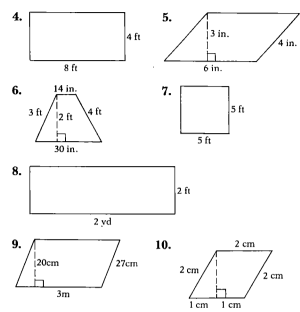

4.

8 ft, 4 ft

5.

3 in., 4 in., 6 in.

6.

14 in., 3 ft, 2 ft, 4 ft, 30 in.

7.

5 ft, 5 ft

8.

2 ft, 2 yd

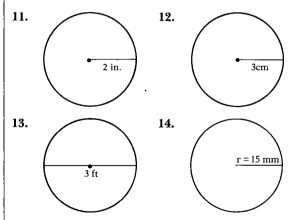

9.

20cm, 27cm, 3m

10.

2 cm, 2 cm, 2 cm, 1 cm, 1 cm

Find the area and circumference of each of the circles.

11.

2 in.

12.

3cm

13.

3 ft

14.

r = 15 mm

One square yard equals 9 square feet. Use this information to convert the following.

15. 86 ft^2 to square yards
16. 6.9 ft^2 to square yards
17. 302 ft^2 to square yards
18. 18.3 yd^2 to square feet
19. 23.6 yd^2 to square feet

One square meter equals 10,000 square centimeters. Use this information to convert the following.

20. 12.6 m^2 to cm^2
21. 21.6 m^3 to cm^2
22. 608 cm^2 to m^2
23. 1075 cm^2 to m^2

24. A building 40 m by 64 m is surrounded by a walk 1.5 m wide. Find the dimensions of the region covered by the building and the walk. Find the area of the walk.

25. A gallon of wood sealer covers 300 ft^2. Ms. Zenna plans to seal the floor of a gymnasium that is 100 ft by 160 ft. How many gallons of sealer are needed?

26. Ms. Dellen has selected carpet that costs $12.99 per yard to use in the living room and family room. The living room measures 15 ft by 21 ft, and the family room measures 18 ft by 16 ft. How many square yards of carpeting will she need? What will it cost?

27. One method of estimating the cost of building a new house is to charge by square foot of living space (using exterior dimensions). If the current rate is $55 per square foot, determine the cost of building the house in Fig. 10.78.

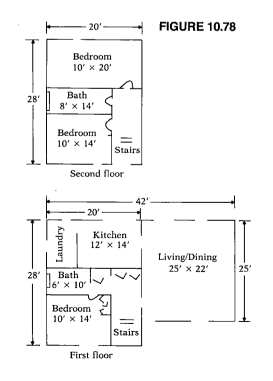

FIGURE 10.78

Second floor

First floor

28. Determine the cost of putting a hardwood floor in the living/dining area of the house in Fig. 10.78. The cost of the hardwood flooring selected is $1.99 per square foot. To find the actual cost, it is necessary to add 5% for waste.

29. Determine the cost of covering the three bedroom floors in Fig. 10.78 with carpet. The cost of the carpet selected is $14.95 per square yard. (Add 5% for waste.)

30. Asphalt floor tiles are 22 cm by 30 cm. How many tiles will be needed to cover a basement room that is 11.6 meters by 13.9 meters?

31. What is the effect on the area of a square if a side is doubled in length?

32. What is the effect on the area of a square if a side is tripled?

33. What is the effect on the area of a square if a side is cut in half?

34. What is the effect on the area of a rectangle if its length is doubled and its width is left unchanged?

35. What is the effect on the area of a rectangle if its length and width are both doubled?

36. What is the effect on the area of a rectangle if its length is doubled and its width is cut in half?

37. What is the effect on the area of a triangle if the base is doubled and its altitude is left unchanged?

38. What is the effect on the area of a triangle if the base is doubled and its altitude is cut in half?

39. An isosceles right triangle has an area of 50 cm². What are the lengths of the sides?

40. Herman's garden is 13.4 meters by 16.5 meters. How large is his garden in hectares?

41. Mr. Jones has purchased a farm that is in the shape of a rectangle. The dimensions of the piece of land are 1.4 km by 3.75 km. How many hectares of land did he purchase?

42. A free-standing fireplace is to be placed in the corner of a living room. The shape and measurements of the base for the fireplace are given in Fig. 10.79. The base will be covered with a special stone that is sold by the square yard. Determine the number of square yards of stone the mason must order.

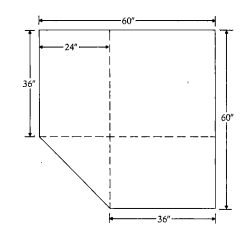

FIGURE 10.79

PROBLEM SOLVING

43. The Smiths are painting their house this summer. The dimensions of the house are given in Fig. 10.80.
   a) Find the total number of square feet to be painted. (Assume that there are 12 windows, each 2 ft 3 in. × 3 ft 4 in., and two doors, each 80 in. × 36 in. The roof will not be painted.)
   b) Determine the number of gallons needed if two coats of paint are required. One gallon of paint covers 500 square feet.
   c) Determine the cost of the paint if paint sells for $18.95 per gallon.

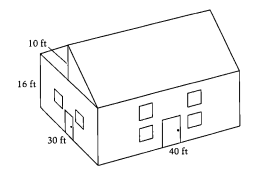

FIGURE 10.80

10.5   VOLUME

Solid geometry is the study of three-dimensional figures. When discussing a one-dimensional figure, such as a line, we can find its length. When discussing a two-dimensional figure, such as a rectangle, we can find its area. When discussing three-dimensional figures, such as those given in Table 10.14, we can find their surface area and volume. The volume of a three-dimensional figure is measured in cubic units, such as cubic feet, cubic yards, or cubic meters. The **surface area** of a figure is the sum of the areas of the surfaces of the figure and is measured in square units. The formulas for surface area are provided for reference purposes and for the Problem Solving problems.

TABLE 10.14

Figure	Sketch	Volume (V)	Surface Area (SA)
Cube		$V = s^3$	$SA = 6s^2$
Rectangular solid		$V = lwh$	$SA = 2lw + 2wh + 2lh$
Cylinder		$V = \pi r^2 h$	$SA = 2\pi rh + 2\pi r^2$
Cone		$V = \dfrac{1}{3}\pi r^2 h$	$SA = \pi r^2 + \pi rs$
Sphere		$V = \dfrac{4}{3}\pi r^3$	$SA = 4\pi r^2$

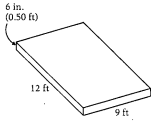

6 in.
(0.50 ft)

12 ft

9 ft

FIGURE 10.81

Example 1

A patio is to be built of concrete. The base of the patio is to be a slab of concrete 12 feet long by 9 feet wide by 6 inches thick (Fig. 10.81)

a) How many cubic yards of concrete are needed?

b) If one cubic yard of concrete costs $36.50, how much will the concrete for the patio cost?

Solution

a) When the calculations are performed, all the measurements must be in the same units. We will therefore convert all units to yards. There are 36 inches in a yard. Therefore 6 inches is 6/36 or 1/6 yard. The length of 12 feet is equal to 4 yards, and the width of 9 feet is equal to 3 yards.

$$V = l \cdot w \cdot h$$
$$= 4 \cdot 3 \cdot \tfrac{1}{6}$$
$$= 2 \text{ yd}^3$$

Note that since all three measurements were in terms of yards, the answer is in cubic yards.

b) Since one cubic yard of concrete costs $36.50, two cubic yards will cost 2 × $36.50, or $73.00.   ❖

Example 2

a) How many cubic feet of water are needed to fill a 24-foot-diameter swimming pool that is 4 feet high?

b) If one cubic foot of water weighs 62.5 pounds, what is the weight of the water in the pool?

c) If there are 7.5 gallons of water per cubic foot of water, how many gallons of water does the pool hold?

Solution

a) This figure can be considered a cylinder (see Fig. 10.82). The volume of a cylinder is $\pi r^2 h$. The radius is half the diameter, or 12 ft. The height, h, is 4 ft.

$$V = \pi(12)^2(4)$$
$$= 3.14(144)(4)$$
$$= 1809 \text{ ft}^3 \quad \text{(to the nearest foot)}$$

b) Since one cubic foot of water weighs 62.5 pounds, 1809 cubic feet of water weigh 1809 × 62.5, or 113,062.5 pounds.

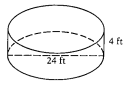

4 ft

24 ft

FIGURE 10.82

c) Since one cubic foot of water contains 7.5 gallons, 1809 cubic feet of water contain 1809 × 7.5, or 13,567.5 gallons of water. The pool holds 13,567.5 gallons of water.   ❖

A **polyhedron** is a closed surface formed by the union of polygonal regions. Each polygonal region is called a **face** of the polyhedron. Figure 10.83 illustrates some polyhedrons.

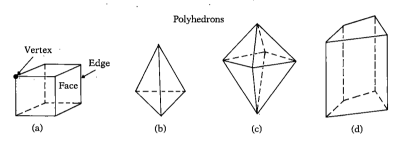

Polyhedrons

(a)   (b)   (c)   (d)

FIGURE 10.83

The line segment formed by the intersection of two faces is called an **edge.** The point at which two or more line segments intersect is called a **vertex.** In Fig. 10.83(a) there are 6 faces, 12 edges, and 8 vertices. For the polyhedron in Fig. 10.83(a) we see that,

number of vertices − number of edges + number of faces   = 2
       8   −   12   +   6   = 2

This formula is true for any polyhedron.

> Number of vertices − number of edges + number of faces = 2

We suggest that you verify that this formula holds for Fig. 10.83(b), 10.83(c), and 10.83(d).

A **regular polyhedron** is one whose faces are all regular polygons of the same size and shape. Figure 10.84(b) is a regular polyhedron.

A **prism** is a special type of polyhedron whose bases are congruent polygons and whose sides are parallelograms. These parallelogram regions are called the **lateral faces** of the prism. If all the lateral faces are rectangles, the prism is said to be a **right prism.**

Some right prisms are illustrated in Fig. 10.84.

The volume of any prism can be found by multiplying the area of the base *B* by the height *h* of the prism.

The volume of a prism $= B \cdot h$.

Prisms

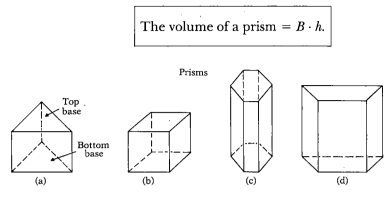

FIGURE 10.84

■ **Example 3**

Find the volume of the triangular prism shown in Fig. 10.85.

Solution: First find the area of the triangular base:

$$\text{Area of base} = \tfrac{1}{2}bh$$
$$= \tfrac{1}{2}(10)(4)$$
$$= 20 \text{ square units}$$

Now find the volume by multiplying the area of the triangular base by the height of the prism:

$$V = B \cdot h$$
$$= 20 \cdot 8$$
$$= 160 \text{ cubic units} \qquad ❖$$

■ **Example 4**

Find the volume of the remaining solid after the cylinder, triangular prism, and square prism have been cut from the solid (Fig. 10.86).

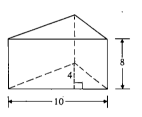

FIGURE 10.85

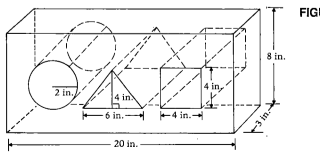

FIGURE 10.86

Solution: To find the volume of the remaining solid, find the volume of the rectangular solid. Then subtract the volume of the two prisms and the cylinder that were cut out.

$$\text{Volume of rectangular solid} = l \cdot w \cdot h$$
$$= 20 \cdot 3 \cdot 8$$
$$= 480 \text{ in.}^3$$

$$\text{Volume of circular cylinder} = \pi r^2 h$$
$$= (3.14)(2^2)(3)$$
$$= (3.14)(4)(3)$$
$$= 37.68 \text{ in.}^3$$

$$\text{Volume of triangular prism} = \text{area of the base} \cdot \text{height}$$
$$= \tfrac{1}{2}(6)(4)(3)$$
$$= 36 \text{ in.}^3$$

$$\text{Volume of square prism} = s^2 \cdot h$$
$$= 4^2 \cdot 3$$
$$= 48 \text{ in.}^3$$

$$\text{Volume of solid} = 480 - 37.68 - 36 - 48$$
$$= 358.32 \text{ in.}^3 \qquad ❖$$

Another special category of polyhedrons is **pyramids.** Unlike prisms, pyramids have only one base. Some pyramids are illustrated in Fig. 10.87.

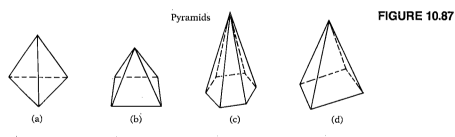

Pyramids   **FIGURE 10.87**

(a)   (b)   (c)   (d)

Note that all but one face of a pyramid intersect a common vertex. The volume of any pyramid can be found by multiplying one third of the area of the base B of the pyramid by the height h of the pyramid.

$$\text{The volume of a pyramid} = \frac{1}{3}Bh.$$

■ **Example 5** _____

Find the volume of the pyramid shown in Fig. 10.88.

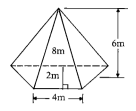

FIGURE 10.88

Solution: First find the area of the base of the pyramid. Since the base of the pyramid is a trapezoid,

$$\text{Area of base} = \tfrac{1}{2}h\,(b_1 + b_2)$$
$$= \tfrac{1}{2}(2)(4 + 8)$$
$$= 12 \text{ m}^2$$

Now find the volume:
$$V = \tfrac{1}{3}B \cdot h$$
$$= \tfrac{1}{3}(12)(6)$$
$$= 24 \text{ m}^3 \qquad ❖$$

It might be necessary in certain situations to convert volume from one cubic unit to a different cubic unit. For example, when purchasing topsoil you might have to change the amount of topsoil from cubic feet to cubic yards prior to placing your order.

■ Example 6

a) Convert one cubic yard to cubic feet.

b) Convert 8.3 cubic yards to cubic feet.

Solution

a) It may be helpful to draw a diagram (Fig. 10.89).

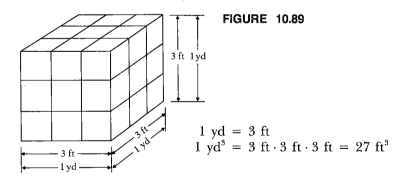

FIGURE 10.89

$$1 \text{ yd} = 3 \text{ ft}$$
$$1 \text{ yd}^3 = 3 \text{ ft} \cdot 3 \text{ ft} \cdot 3 \text{ ft} = 27 \text{ ft}^3$$

b) $1 \text{ yd}^3 = 27 \text{ ft}^3$

Thus $8.3 \text{ yd}^3 = 8.3 \times 27 = 224.1 \text{ ft}^3$.   ❖

■ **Example 7**

Convert 116.1 cubic feet to cubic yards.

Solution: Since $1 \text{ yd}^3 = 27 \text{ ft}^3$, we can divide 116.1 ft^3 by 27 ft^3 to obtain the number of cubic yards:

$$116.1 \text{ ft}^3 = \frac{116.1}{27} = 4.3 \text{ yd}^3 \qquad \diamondsuit$$

EXERCISES 10.5

Find the volume of each of the solids.

Find the volume of each of the pyramids.

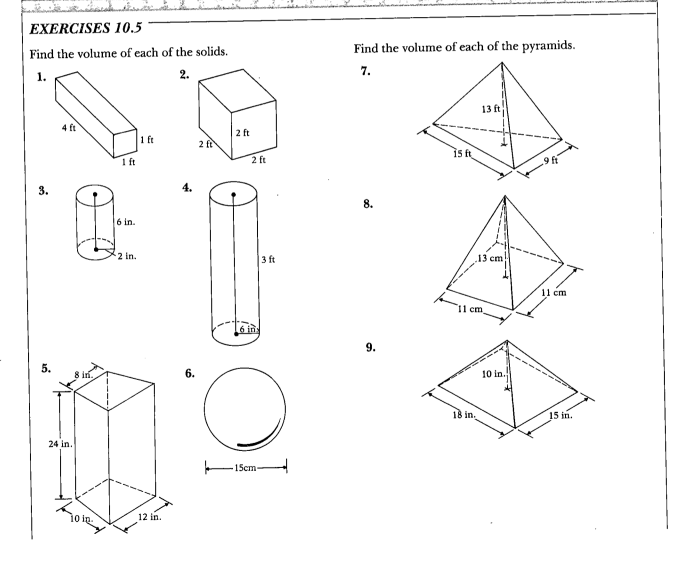

1.
4 ft
1 ft
1 ft

2.
2 ft
2 ft
2 ft
2 ft

3.
6 in.
2 in.

4.
3 ft
6 in.

5.
8 in.
24 in.
10 in.
12 in.

6.
15cm

7.
13 ft
15 ft
9 ft

8.
13 cm
11 cm
11 cm

9.
10 in.
18 in.
15 in.

10.

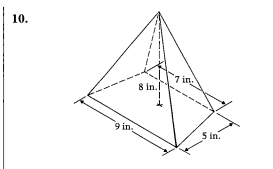

One cubic yard equals 27 cubic feet. Use this information to make the following conversions.

11. 4 yd^3 to cubic feet **12.** 3.2 yd^3 to cubic feet
13. 212 ft^3 to cubic yards **14.** 84 ft^3 to cubic yards

15. a) How many cubic centimeters of water will a rectangular fish tank hold if the tank is 60 cm long, 45 cm wide, and 20 cm high?
b) If 1 cc = 1 mℓ, how many milliliters will the tank hold?
c) How many liters will the tank hold?

16. a) What is the volume of water in a swimming pool that is 18 meters long and 10 meters wide and has an average depth of 2.5 meters? Answer in cubic meters.
b) If 1 m^3 = 1 kℓ, how many kiloliters of water will it take to fill the pool?

17. a) A milk container has a square base with 9.5-centimeter sides. Its height is 19 centimeters. Determine the volume of milk the container holds in cubic centimeters.
b) If 1 cm^3 = 1 mℓ, determine the volume in liters.

18. a) A circular cylinder containing glass cleaner has a base with a radius of 6 centimeters. Its height is 24 centimeters. Determine the number of cubic centimeters of glass cleaner in the container.
b) If 1 cm^3 = 1 mℓ, determine the number of liters of glass cleaner in the container.

19. a) How many cubic feet of dirt are needed to fill a ditch that is 6 ft wide by 8 ft long by 27 in. deep?
b) How many cubic yards of dirt are needed?

20. Mr. Lockhart's driveway is 60 ft long by 12.5 ft wide. He plans to lay 4 in. of blacktop on the driveway.
a) How many cubic feet of blacktop are needed?
b) How many cubic yards?

21. The Pyramid of Cheops in Egypt has a square base measuring 720 ft on a side. Its height is 480 ft. What is its volume?

22. A gallon is a measure of liquid volume. It is equivalent to 231 cubic inches. Measure the base and the height *of the liquid* in a half-gallon paper carton of milk and compute its volume in cubic inches. Do your results check with the value given?

23. The ball at the end of a ballpoint pen has a diameter of 1.5 mm. Find the volume of the ball.

24. What is the total piston displacement of a 2-cylinder engine if each piston has a diameter of $2\frac{1}{4}$-in. and a $3\frac{1}{2}$-in. stroke? (Fig. 10.90)

FIGURE 10.90

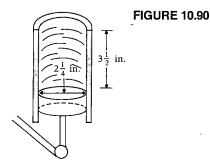

25. In baking a cake you have to choose between a round pan with a 9-in. diameter and a 7-in. × 9-in. rectangular pan.
a) Determine the area of the base of each pan.
b) If both pans are 2 in. deep, determine the volume of each pan.

26. A flower box is 4 ft long, and its ends are in the shape of a trapezoid (Fig. 10.91). The upper and lower bases of the trapezoid measure 12 in. and 8 in., respectively, and the height is 9 in. Find the volume of the flower box
a) in cubic inches; **b)** in cubic feet.

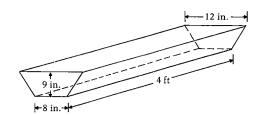

FIGURE 10.91

27. What is the smallest number of cuts that would divide a cube of wood whose edge measures 3 in. into cubes, each of whose edges measures 1 in.?

28. a) If we double the length of each edge of a cube, what effect does this have on its volume?
    b) What effect does this have on its surface area?
29. a) If we double the radius of the base of a cylinder, what effect does this have on the volume?
    *b) What effect does this have on the lateral area (the area of the side)?
30. a) If we double the radius of a sphere, what effect does this have on its volume?
    b) What effect does this have on the surface area?
31. A half-inch cube is a cube measuring $\frac{1}{2}$ in. on a side. A half cubic inch is half the volume of a cube measuring 1 in. on a side. Is the volume of a half-inch cube the same as the volume of a half cubic inch? Explain.
32. A king-size waterbed mattress is 7 ft long, 7 ft wide, and 9 in. high.
    a) How many cubic feet of water are needed to fill the waterbed?
    b) If 1 ft^3 of water weighs 62.5 lb, what is the weight of the water in the mattress?
    c) If 1 gal of water weighs 8.3 lb, how many gallons of water does the mattress hold?
33. A hot-water heater is in the shape of a circular cylinder. It has a height of 5 ft 6 in. and a diameter of 15 in.

a) How many cubic feet of water will the tank hold?
b) How many gallons of water does the tank hold? (1 cubic foot ≈ 7.5 gallons.)
34. The general procedure for determining the number of rolls of wallpaper needed to wallpaper a given rectangular room is as follows:
    a) Determine the room's overall perimeter in feet. Ignore windows and doors at this time.
    b) Multiply the perimeter by the height in feet of the room. This gives the number of square feet of wall space.
    c) Divide this number by 30, the approximate number of usable square feet in a roll of wallpaper.
    d) From this number, deduct one roll for every two average-sized openings (windows, doors, fireplace, and so on).
    Mrs. Jones plans to wallpaper her kitchen. It is 12 ft long, 10 ft wide, and 8 ft high, and it contains 2 doorways and 2 windows. How many rolls of wallpaper will she need for her kitchen?
35. One U.S. gallon of liquid has a volume of 231 cubic inches. If a cylindrical jar has a base whose radius is 2 in., how tall must the jar be in order to hold 1 gal?

PROBLEM SOLVING

36. The dimensions of a storage tank are 16 meters by 12 meters by 12 meters. If the tank is filled with water, determine the following.
    a) The volume of the tank in cubic meters
    b) The number of kiloliters of water the tank will hold
    c) The mass of the water in metric tonnes
37. A hot-water heater in the shape of a cylinder has a radius of 30 cm and a height of 150 cm. If the tank is filled with water, determine the following.
    a) The volume of the tank in cubic meters
    b) The number of liters of water the tank will hold
    c) The mass of the water in kilograms
38. The first coat of paint for the outside of a building requires 1 liter of paint for each 10 square meters. The second coat requires 1 liter for every 15 square meters. If the paint costs $1.75 per liter, what will be the cost of two coats of paint for the four outside walls of a building 20 meters long, 12 meters wide, and 6 meters high?
39. Use Fig. 10.92 to give a geometric interpretation of the equation

$$(a + b)^3 = a^3 + 3a^2b + 3ab^2 + b^3.$$

What is the volume in terms of a and b of each numbered piece in Fig. 10.92? An eighth piece is not illustrated. What is its volume?

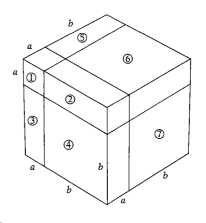

FIGURE 10.92

10.6 NETWORK THEORY

Network theory, also called **graph theory,** is a branch of a nonmetric geometry called **topology** that Leonhard Euler (1707–1783) originated approximately 200 years ago. The Swiss mathematician wondered whether there existed some route that one could take that would cross each of the seven bridges over the River Pregel (which flowed through Königsberg, then a part of Prussia) exactly once (Fig. 10.93). This problem is historically known as the "Königsberg bridge problem." At this point we suggest that you spend a few minutes attempting to find a path that will satisfy Euler's requirements.

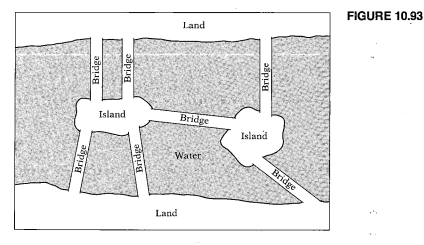

FIGURE 10.93

Before we determine whether there is a path that will cross each bridge exactly once, let us consider the following examples.

■ **Example 1** _____

Copy the shapes in Fig. 10.94 on a sheet of paper. Then, starting at any point, try to trace each figure on the paper without retracing a line and without removing your pencil from the paper. If you succeed, indicate your starting point and ending point.

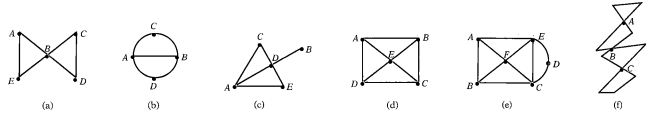

(a)   (b)   (c)   (d)   (e)   (f)

FIGURE 10.94

Solution: The figure in (a) can be traced if you start at any point. You will end at the point at which you started.

The figure in (b) can be traced, but only if you start at point A or point B. If you start at point A, you will end at point B, and vice versa.

The figure in (c) can be traced, but only if you start at point A or point B. If you start at point A, you will end at point B, and vice versa.

It is not possible to trace the figure in (d) without retracing a line.

The figure in (e) can be traced, but only if you start at point A or point B. If you start at point A, you will end at point B, and vice versa.

The figure in (f) can be traced if you start at any point. You will end at the point at which you started.   ❖

Before we can answer the Königsberg bridge problem and understand why we were able to trace only some of the figures in Example 1 without retracing a line, we must introduce a number of new terms and concepts.

A **vertex** is any designated point. An **arc** is any line, either straight or curved, that begins and ends at a vertex. Consider Figs. 10.95 and 10.96. Figure 10.95 has three designated vertices—A, B, and C—and two arcs—1 and 2. Figure 10.96 has five designated vertices—A, B, C, D, and E—and six arcs. A vertex with an odd number of attached arcs is called an **odd vertex**. Figure 10.95 has two odd vertices, A and C. Figure 10.96 also has two odd vertices, A and D. A vertex with an even number of attached arcs is called an **even vertex**. Vertex B in Fig. 10.95 and vertices B, C, and E in Fig. 10.96 are even. A **network** is any continuous (not broken) system of arcs and vertices.

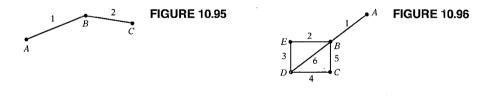

FIGURE 10.95

FIGURE 10.96

A network is said to be **traversable** if it can be traced without removing the pencil from the paper and without tracing an arc more than once.

After completing Example 1, do you have an intuitive feeling as to when a figure is traversable? Think about odd and even vertices. Leonhard Euler discovered an important scientific principle concealed in the Königsberg bridge problem. He presented his simple and ingenious solution to the Russian Academy at St. Petersburg in 1735. His method was to replace the land areas by points and the bridges by lines connecting the points. He used the terminology of odd and even vertices in developing a general principle for determining whether a figure was traversable.

The rules of traversability as developed by Euler are given below.

Rules of Traversability

1. A network with no odd (all even) vertices is traversable; you may start from any vertex, and you will end where you began.

2. A network with exactly two odd vertices is traversable; you must start at either of the odd vertices and finish at the other.

3. A network with more than two odd vertices is not traversable.

Note: It is impossible for a network to contain an odd number of odd vertices. (If you don't believe this, try to construct such a network.) Now go back to Example 1 and determine which figures are traversable, using these rules.

If we refer to Example 1 and Fig. 10.94, we can see that (a) and (f) are traversable from any point, since they contain only even vertices. The figures marked (b), (c), and (e) have exactly two odd vertices and can be traversed but only by starting at one of the odd vertices, either point *A* or point *B*. Figure (d) contains more than two odd vertices and therefore cannot be traversed.

Now let us return to the Königsberg bridge problem. If we let the islands and the mainland be represented as vertices and the bridges as arcs, Fig. 10.97 simplifies to the network in Fig. 10.98. Since the network in Fig. 10.98 has four odd vertices, it cannot be traversed. It is therefore impossible to cross each bridge only once.

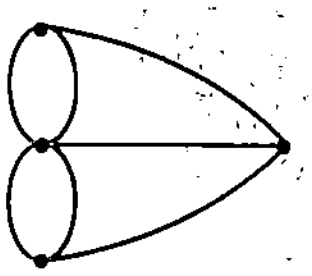

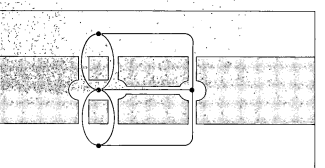

FIGURE 10.97

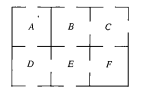

FIGURE 10.98

FIGURE 10.99

Example 2

The floor plan of a six-room house is shown in Fig. 10.99. The openings represent doors, and the letters represent rooms.

a) Determine the rooms that contain an odd number of doors; an even number of doors.

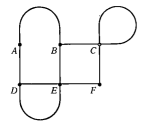

(a)

(b)

FIGURE 10.100

b) Each room in part (a) can be represented as an odd or even vertex. Use this information to determine whether it is possible to walk through the house using each door only once.

c) Determine a path to walk through each room of the house using each door only once.

d) Use the figure sketched in part (c) to develop a network that represents the path.

Solution

a) Rooms B and D contain three doors each. Rooms A and F contain two doors each. Rooms C and E have four doors.

b) Since there are only two odd vertices, B and D, the figure is traversable, and you can walk through the house using each door only once.

c) You must start in either room B or room D. If you start in B, you will end in D, and vice versa. Follow the arrows in Fig. 10.100(a). When you leave room E, you can leave by either door.

d) The network representing the path is shown in Fig. 10.100(b).   ❖

EXERCISES 10.6

1. Use the rules of traversability stated on page 533 to determine whether or not the networks are traversable. If a network is traversable, state the points from which you may start and end.

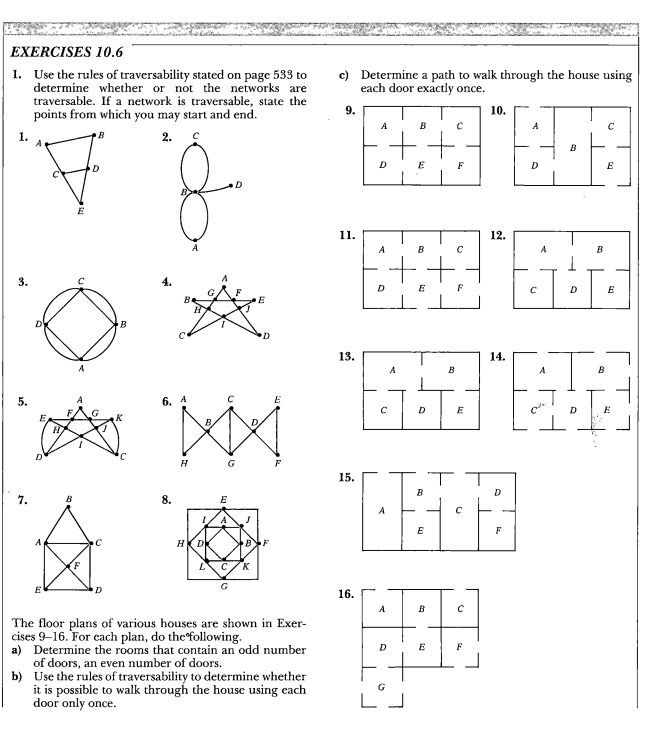

1.

2.

3.

4.

5.

6.

7.

8.

The floor plans of various houses are shown in Exercises 9–16. For each plan, do the following.

a) Determine the rooms that contain an odd number of doors, an even number of doors.

b) Use the rules of traversability to determine whether it is possible to walk through the house using each door only once.

c) Determine a path to walk through the house using each door exactly once.

9.

10.

11.

12.

13.

14.

15.

16.

17. Refer to Fig. 10.101. Is it possible to cross each of the eight bridges exactly once? If so, explain why and show the path.

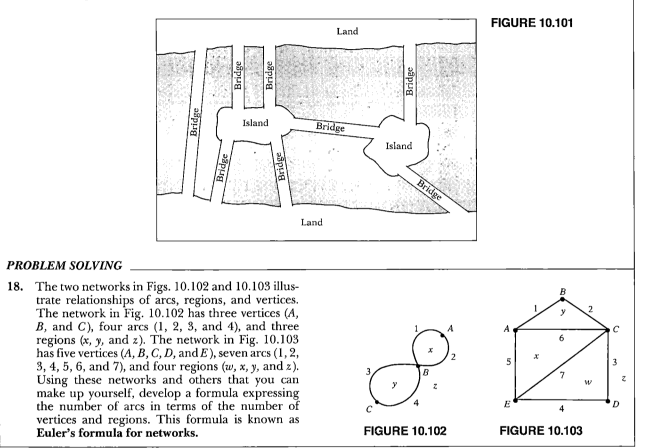

FIGURE 10.101

PROBLEM SOLVING

18. The two networks in Figs. 10.102 and 10.103 illustrate relationships of arcs, regions, and vertices. The network in Fig. 10.102 has three vertices (A, B, and C), four arcs (1, 2, 3, and 4), and three regions (x, y, and z). The network in Fig. 10.103 has five vertices (A, B, C, D, and E), seven arcs (1, 2, 3, 4, 5, 6, and 7), and four regions (w, x, y, and z). Using these networks and others that you can make up yourself, develop a formula expressing the number of arcs in terms of the number of vertices and regions. This formula is known as **Euler's formula for networks.**

FIGURE 10.102   **FIGURE 10.103**

10.7   THE MÖBIUS STRIP, KLEIN BOTTLE, AND MAPS

One of the first pioneers of topology was the German astronomer and mathematician August Ferdinand Möbius (1790–1866). Möbius, the director of the University of Leipzig's observatory, spent a great deal of time studying geometry. He discussed the configuration known as the Möbius net, which later played an essential part in the systematic development of projective geometry. In a memoir presented to the Académie des Sciences, he discussed the properties of one-sided surfaces, including the Möbius strip.

Möbius strip
Photo by Marshall Henrichs

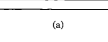

(a)

(b)

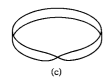

(c)

FIGURE 10.104

If you place a pencil on one surface of a sheet of paper and do not remove it from the sheet, you must cross the edge to get to the other surface. Thus a sheet of paper has one edge and two surfaces. The sheet retains these properties even when crumpled into a ball. The Möbius strip is a one-sided, one-edged surface. You can construct one, as shown in Fig. 10.104, by (a) taking a strip of paper, (b) giving one end a half twist, and (c) taping the ends together.

The Möbius strip has some very interesting properties. In order to better understand these properties, perform the following experiments.

FIGURE 10.105

Experiment 1

Make a Möbius strip, using a strip of paper and scotch tape. Place a magic marker on the edge of the strip (Fig. 10.105). Pull the strip slowly so that the magic marker marks the edge—do not remove the marker from the edge. Continue pulling the strip and observe what happens.

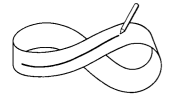

FIGURE 10.106

FIGURE 10.107

FIGURE 10.108

Experiment 2

Make a Möbius strip. Place a magic marker on the surface of the strip (Fig. 10.106). Pull the strip slowly so that the magic marker marks the surface. Continue and observe what happens.

Experiment 3

Make a Möbius strip. With a scissors make a small hole in the middle of the strip. Cut along the strip, keeping the scissors in the middle of the strip (Fig. 10.107). Continue cutting and observe what happens.

Experiment 4

Make a Möbius strip. Make a small hole at a point about one third of the width of the strip. Cut along the strip, keeping the scissors the same distance from the edge (Fig. 10.108). Continue cutting and observe what happens.

Another topological object is the Klein bottle. The Klein bottle, named after Felix Klein (1849–1925), is a bottle with only one side. Nobody will ever see an actual Klein bottle, since it exists only in the topologist's imagination. A true Klein bottle passes through itself without the existence of a hole, which is of course a physical impossibility.

A model of the Klein bottle can be created as follows.

a) ·A stretchable piece of glass tubing is flared (Fig. 10.109).

b) The neck, or thin end, is passed inside (Fig. 10.110)

c) and is joined to the base (Fig. 10.111).

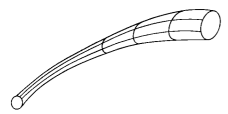

FIGURE 10.109

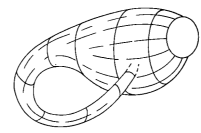

FIGURE 10.110

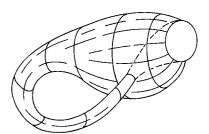

FIGURE 10.111

In examining the model of the Klein bottle in Fig. 10.111, one can see that the bottle would not hold any water. Therefore one might conclude that the bottle has only one side.

FIGURE 10.112

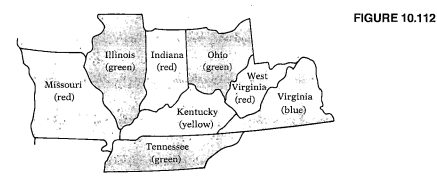

Limerick from an unknown writer:

"A mathematician confided
That a mobius band is one-sided,
And you'll get quite a laugh
If you cut one in half,
For it stays in one piece
when divided.

Another mathematician writer
added this:

A mathematician named
Klein
Thought the mobius band
was divine.
He said, 'If you glue
The edges of two
You'll get a weird bottle like
mine.'"

Maps have fascinated topologists for years because of the many challenging problems they present. Mapmakers have known for a long time that whether a map is drawn on a flat surface or a sphere, four colors are sufficient to differentiate each country from its immediate neighbors.

In Fig. 10.112 we can see that there are no two states with a common border that are marked with the same color. Regardless of the complexity of the map, only four colors are needed to show any two subdivisions with a common boundary in different colors. Regions that meet at only one point (such as the states of Arizona, Colorado, Utah, and New Mexico) are not considered to have a common boundary.

There are certain cases in which more than four colors may be needed to draw a map. For example, it has been proved that a map drawn on a Möbius strip requires a maximum of six colors. When a topologist rolls and bends a flat map into the shape of a doughnut (called a torus), the map that once required only four colors may now require seven.

Topologists proved some time ago that a maximum of six colors is all that is needed to draw a map on a Möbius strip, and a maximum of seven colors is needed to draw a map on a torus. However, mathematicians were becoming very frustrated at their inability to prove that only four colors are needed to draw a map on a flat surface. Then, in 1976, Kenneth Appel and Wolfgang Haken of the University of Illinois succeeded in proving the theorem, using their ingenuity, logic, and a computer. Appel and Haken used 1200 hours of computer time to prove the theorem. One step in solving the four-color problem was to reduce any map to a graph. This was done by replacing each subdivision with a point. Then if two subdivisions had a common boundary, they were connected with a straight line. By doing this, the four-color problem could then be stated in the following manner.

The points of any graph in the plane can be colored with four colors in such a way that no two points connected by a line are the same color.

Figure 10.113 illustrates how the map in Fig. 10.112 can be reduced to a graph.

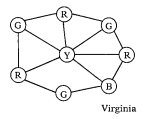

Virginia

FIGURE 10.113

DID YOU KNOW?

A **Jordan Curve** can be considered a circle that has been twisted out of shape. Like a circle, it has an inside and an outside. To get from one side to the other at least one line must be crossed. Consider the Jordan Curve in Fig. 10.114. Are points A and B inside or outside the curve?

FIGURE 10.114

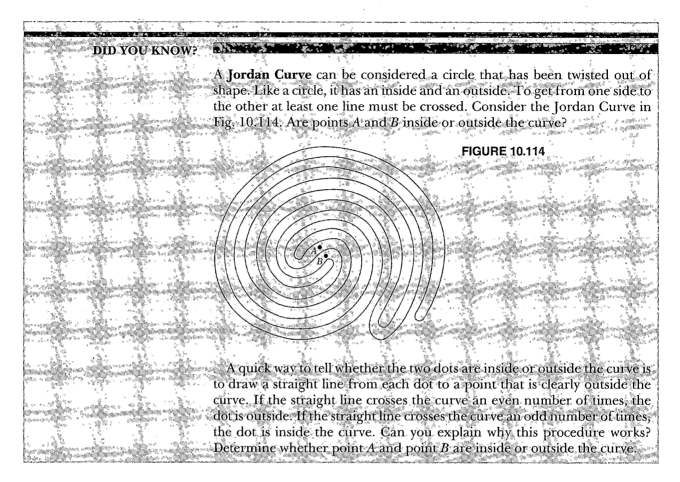

A quick way to tell whether the two dots are inside or outside the curve is to draw a straight line from each dot to a point that is clearly outside the curve. If the straight line crosses the curve an even number of times, the dot is outside. If the straight line crosses the curve an odd number of times, the dot is inside the curve. Can you explain why this procedure works? Determine whether point A and point B are inside or outside the curve.

EXERCISES 10.7

1. Use the result of Experiment 1 to find the number of edges on a Möbius strip.
2. Use the result of Experiment 2 to find the number of surfaces on a Möbius strip.
3. How many separate strips are obtained in Experiment 3?
4. How many separate strips are obtained in Experiment 4?
5. **a)** Take a strip of paper, give it one full twist, and connect the ends. Is this a Möbius strip with only one side? Explain.

   b) Determine the number of edges, as in Experiment 1.

   c) Determine the number of surfaces, as in Experiment 2.

   d) Cut the strip down the middle. What is the result?

6. Make a Möbius strip. Cut it one third of the way from the edge, as in Experiment 4. You should get two loops, one going through the other. Determine whether either (or both) of these loops is itself a Möbius strip.

7. Take a strip of paper, make one whole twist and another half twist, and then tape the ends together. Test by a method of your choice to see whether this has the same properties as a Möbius strip.

For each map in Figure 10.115
  (a) Color the maps in Fig. 10.115, using a maximum of four colors, so that no two regions with a common border have the same color.
  (b) Show how each map in Fig. 10.115 can be reduced to a graph.

| **8.** | **9.** | **10.** | **11.** | **12.** |

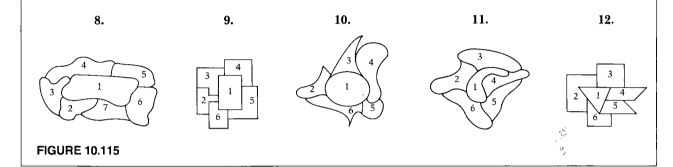

FIGURE 10.115

10.8 NON-EUCLIDEAN GEOMETRY (OPTIONAL)

The geometry that is traditionally taught in high schools is plane or Euclidean geometry. The foundations for this course were presented by Euclid around 330 B.C. in a set of 13 books called *Elements*.

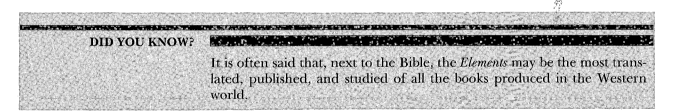

DID YOU KNOW?

It is often said that, next to the Bible, the *Elements* may be the most translated, published, and studied of all the books produced in the Western world.

Euclid was the first mathematician to use the **axiomatic method** in developing a branch of mathematics. First, Euclid introduced **undefined terms,** such as point, line, plane, and angle. He related these to physical space by such statements as "A line is length without breadth," so that we may intuitively understand them. However, since such statements play no further role in his system, they constitute primitive or undefined terms.

Second, Euclid introduced certain **definitions.** The definitions are introduced when needed and are often based on the undefined terms. Some terms that Euclid introduced and defined include triangle, right angle, and hypotenuse.

Third, Euclid asserted certain primitive propositions called **postulates** (now called **axioms**[*]) about the undefined terms and definitions. The reader is asked to accept these statements as true on the basis of their "obviousness" and their relationship with the physical world. For example, the Greeks accepted all right angles as being equal. This is Euclid's fourth postulate.

Fourth, Euclid proved, using deductive reasoning,[†] other propositions called **theorems.** One theorem that Euclid proved is known as the Pythagorean theorem: "The sum of the areas of the squares constructed on the arms of a right triangle is equal to the area of the square constructed on the hypotenuse." He also proved that the sum of the angles of a triangle is 180°.

Using only the ten axioms given in Table 10.15, Euclid deduced 465 propositions (or theorems) in plane and solid geometry, number theory, and Greek geometrical algebra.

Euclid's fifth axiom may be better understood by observing Fig. 10.116.The sum of angles A and B is less than the sum of two right angles (180°). Therefore the two lines will meet if extended.

John Playfair (1748–1819), a Scottish physicist and mathematician, wrote a geometry book that was published in 1795. In this book, Playfair gave a logically equivalent interpretation of Euclid's fifth postulate. This version is often referred to as Playfair's postulate or the Euclidean parallel postulate.

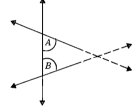

FIGURE 10.116

> **The Euclidean parallel postulate.** "Given a line and a point not on the line, one and only one line can be drawn through the given point parallel to the given line."

The Euclidean parallel postulate may be better understood by observing Fig. 10.117.

←——————————→ Given line   **FIGURE 10.117**

←——————•——————→ Line parallel to
        Given   given line through
        point   given point

---

[*]The concept of the axiom has changed significantly since Euclid's time. Nowadays any statement may be designated as an axiom, whether it is self-evident or not. All axioms are *accepted* as true. A set of axioms forms the foundation for a mathematical system.

[†]Deductive reasoning is discussed in Section 3.6.

TABLE 10.15	
Euclid's axioms	**Present-day Interpretation**
1. Given two points, there is an interval that joins them.	Two points determine a straight line.
2. An interval can be prolonged indefinitely.	A straight line extends indefinitely (in both directions).
3. A circle can be constructed when its center and a point on it are given.	Same
4. All right angles are equal.	Same
5. If a straight line falling on two straight lines makes the interior angles on the same side less than two right angles, the two straight lines, if produced indefinitely, meet on that side on which the angles are less than the two right angles.	Given a line and a point not on the line, one and only one line can be drawn through the given point parallel to the given line.
6. Things equal to the same thing are equal.	Same
7. If equals are added to equals, the sums are equal.	Same
8. If equals are subtracted from equals, the remainders are equals.	Same
9. Things that coincide with one another are equal.	Two figures that can be moved to coincide with each other are congruent.
10. The whole is greater than a part.	Same

Euclid

Many mathematicians after Euclid believed that this postulate was not as self-evident as the other nine. Others believed that this postulate could be proved from the other nine postulates and therefore was not needed at all. Of the many attempts to prove that the fifth postulate was not needed, the most noteworthy one was presented by Girolamo Saccheri (1667–1733), a Jesuit in Italy. In the course of his elaborate chain of deductions, Saccheri proved many of the theorems of what is now called hyperbolic geometry. However, Saccheri did not realize what he had done. He believed that Euclid's geometry was the only "true" geometry and concluded that his own work was in error. Thus Saccheri narrowly missed receiving credit for a great achievement—the founding of non-Euclidean geometry.

In 1763 the German mathematician Georg Simon Klügel listed nearly 30 attempts to prove axiom 5 from the remaining axioms, and he rightly concluded that all the alleged proofs were unsound. Fifty years later a new generation of geometers were becoming more and more frustrated. One of them, a Hungarian named Farkos Bolyai, wrote a letter to his son, Janos Bolyai. "I entreat you leave the science of parallels alone. . . . I have traveled past all reefs of this infernal dead sea and have always come back with a broken mast and torn sail." The son, refusing to heed his father's advice, continued to think about parallels until, in 1823, he saw the whole truth and

Ivanovich Lobachevsky
Courtesy of The Granger Collection

G. F. Bernhard Riemann
Courtesy of The Granger Collection

enthusiastically declared, "I have created a new universe from nothing." He saw that geometry branches out in two directions, depending on whether Euclid's fifth postulate is applied or not. He recognized two different geometries and published his discovery as a 24-page appendix to a textbook written by his father. The famous mathematician George Bruce Halsted called it "the most extraordinary two dozen pages in the whole history of thought." Farkos Bolyai proudly presented a copy of his son's work to his friend Carl Friedrich Gauss, then Germany's greatest mathematician, whose reply to the father had a devastating effect on Janos. Gauss wrote, "I am unable to praise this work. . . . To praise it would be to praise myself. Indeed, the whole content of the work, the path taken by your son, the results to which he is led, coincides almost entirely with my meditations which occupied my mind partly for the last thirty or thirty-five years." We now know from his earlier correspondence that Gauss had indeed been familiar with **hyperbolic geometry** even before Janos was born. In his letter, Gauss also indicated that it was his intention not to let his theory be published during his lifetime, but to record it so that the theory would not perish with him. It is believed that the reason Gauss did not publish his work was that he feared being ridiculed by other prominent mathematicians of his time.

At about the same time as Bolyai's publication, the Russian Nikolay Ivanovich Lobachevsky published a paper that was remarkably like Bolyai's, although it was quite independent of it. Lobachevsky made a deeper investigation and wrote several books. In marked contrast to Bolyai, who received no recognition during his lifetime, Lobachevsky received great praise and became a professor at the University of Kazan.

After the initial discovery, little attention was paid to the subject until 1854, when G. F. Bernhard Riemann (1826–1866), a student of Gauss, suggested a second type of non-Euclidean geometry, which is now called **spherical, elliptical,** or **Riemannian geometry.** The hyperbolic geometry of his predecessors was synthetic; that is, it was not based on or related to any concrete model when it was developed. Riemann's geometry was closely related to the theory of surfaces. A **model** may be considered a physical interpretation of the undefined terms that satisfies the axioms. A model may be a picture or an actual physical object.

The two types of non-Euclidean geometries that we have mentioned are elliptical geometry and hyperbolic geometry. The major difference among the three geometries lies in the fifth axiom. The fifth axiom of the three geometries is given in Table 10.16.

Since you probably conceive of a line in the same way Euclid did, you might not understand the fifth axiom of the two non-Euclidean geometries. It is important to remember that the term "line" is undefined. Thus a line can be interpreted differently in different geometries. A model for Euclidean geometry is a plane, such as a blackboard (Fig. 10.118a). A model

TABLE 10.16		
Euclidean	**Elliptical**	**Hyperbolic**
Given a line and a point not on the line, one and only one line can be drawn parallel to the given line through the given point.	Given a line and a point not on the line, no line can be drawn through the given point parallel to the given line.	Given a line and a point not on the line, two or more lines can be drawn through the given point parallel to the given line.

for elliptical geometry is a sphere (Fig. 10.118b). A model for hyperbolic geometry is a pseudosphere (Fig. 10.118c). A pseudosphere is similar to two trumpets placed back to back. It should be obvious that a line on a plane cannot be the same as a line on any of the other two figures.

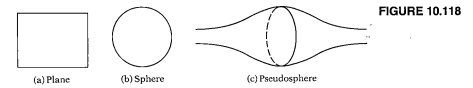

FIGURE 10.118

(a) Plane   (b) Sphere   (c) Pseudosphere

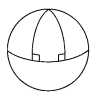

FIGURE 10.119

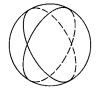

FIGURE 10.120

A circle on the surface of a sphere is called a great circle (or a geodesic) if it divides the sphere into two equal parts. If we were to cut through a sphere along a great circle, we would have two identical pieces. If we interpret a line to be a great circle, we can see that the fifth axiom of elliptical geometry is true. Two great circles on a sphere must intersect. Hence there can be no parallel lines (see Fig. 10.119).

If we were to construct a triangle on a sphere, the sum of its angles would be greater than 180° (see Fig. 10.120). The theorem "The sum of the angles of a triangle is greater than 180°" has been proved by means of the axioms of elliptical geometry. The sum of the angles varies with the area of the triangle and gets closer to 180° as the area decreases.

The lines in hyperbolic geometry are represented by geodesics on the surface of the pseudosphere. A geodesic is the shortest and least-curved arc between two points on a surface. Figure 10.121 illustrates two lines on the surface of a pseudosphere. Figure 10.122 illustrates a line from another perspective.

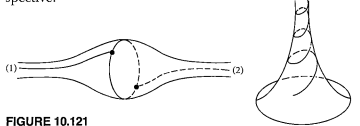

(1) ——— (2)

FIGURE 10.121

FIGURE 10.122

Figure 10.123 illustrates the fifth axiom of hyperbolic geometry.* Note that through the given point, two lines are drawn parallel to the given line.

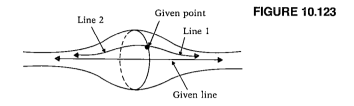

FIGURE 10.123

If we were to construct a triangle on a pseudosphere, the sum of the angles would be less than 180° (Fig. 10.124). The theorem "The sum of the angles of a triangle is less than 180°" has been proved by means of the axioms of hyperbolic geometry.

We have stated that the sum of the angles of a triangle is 180°, is greater than 180°, and is less than 180°. Which statement is correct? Each statement is correct *in its own system.*

There are many theorems that hold true for all three geometries—vertical angles still have the same measure, one can uniquely bisect a line segment with a straightedge and compass alone, and so on. In fact, all the theorems deduced from the first four axioms alone are true in all the geometries.

FIGURE 10.124

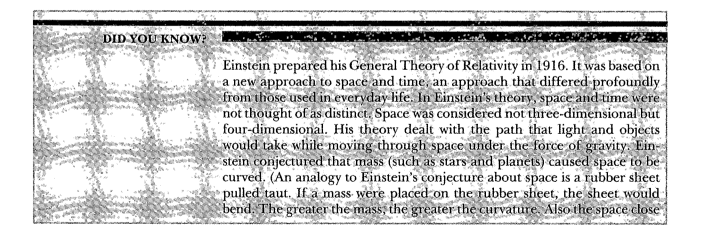

*A formal discussion of hyperbolic geometry is beyond the scope of this text.

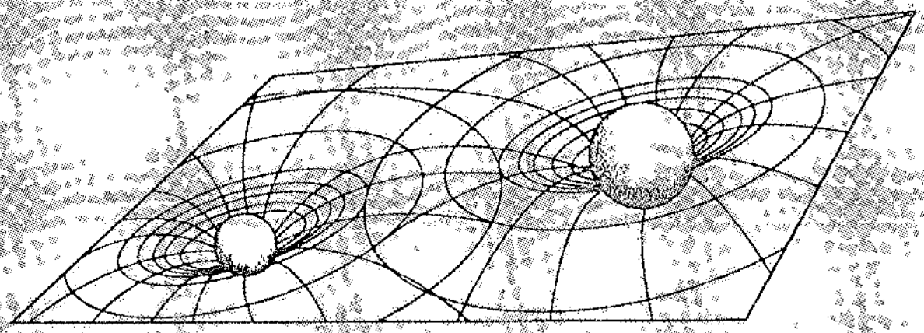

FIGURE 10.125

to the mass would have a greater curvature, as suggested in Fig. 10.125.)

To prove his conjecture, Einstein exposed himself to the non-Euclidean geometry created by Riemann some 60 years earlier. Einstein felt that the trajectory of a particle in space would then not represent a straight line but the straightest curve possible, a geodesic. He also conjectured that space-time has geometric properties that vary from point to point.

Einstein's theory was confirmed by the solar eclipses of 1919 and 1922, and it is now widely accepted.

Mathematicians and astronomers have discovered that there are only three essentially different ways in which space can curve. Space can curve as a sphere. This type of curvature, called positive, is described by Riemannian geometry. Space can also be flat or without curvature. This type, referred to as zero curvature, is described by Euclidean geometry. The third possibility is that space can be curved like a saddle. This type of curvature, called negative, is described by hyperbolic geometry. Space-time is now thought to be a combination of all three curvatures and to be constantly changing. (When a moon revolves about a planet and when a planet revolves about a star, the space configuration in that area is changing.)

"The Great Architect of the universe now appears to be a great mathematician."
British physicist Sir James Jeans

The many theorems based on the fifth postulate may differ in each geometry. It is important for the student to realize that each theorem proved is true *in its own system* because each is logically deduced from the given set of axioms of the system. No one system is the "best" system. Euclidean geometry may appear to be the one to use in the classroom, where the blackboard is flat. However, in discussions involving the earth as a whole, elliptical geometry may be the most useful, since the earth is a sphere. If the object under consideration has the shape of a saddle or pseudosphere, hyperbolic geometry may be the most useful. Albert Einstein, in the exposition of his General Theory of Relativity, described space as being non-Euclidean in nature and used non-Euclidean geometry to explain his theory.

EXERCISES 10.8

1. When were the *Elements* written?
2. Why are the *Elements* important?
3. **a)** List the four parts of an axiomatic system as used by Euclid in developing Euclidean geometry.
   b) Discuss each of the four parts.
4. Theorems are proved by using deductive reasoning. Read Section 3.6 and explain the deductive reasoning process.
5. List the accomplishments of each of the following.
   a) Janos Bolyai
   b) Carl Friedrich Gauss
   c) Nikolay Ivanovich Lobachevsky
   d) Girolamo Saccheri
   e) G. F. Bernhard Riemann
6. State the fifth axiom of each of the following.
   a) Euclidean geometry
   b) Hyperbolic geometry
   c) Elliptical geometry

7. State the theorem concerning the sum of the angles of a triangle in each of the following.
   a) Euclidean geometry
   b) Hyperbolic geometry
   c) Elliptical geometry
8. What model is often used in describing and explaining Euclidean geometry?
9. What model is often used in describing and explaining elliptical geometry?
10. What model is often used in describing and explaining hyperbolic geometry?
11. What do we mean when we say that there is no one axiomatic system of geometry that is "best"?
12. Explain why Einstein used non-Euclidean geometry to explain his General Theory of Relativity? (See the Did You Know? section.)
13. List the three types of curvature of space and the types of geometry that correspond to them.

RESEARCH ACTIVITIES

14. Write a paper on the life and achievements of Euclid. (Use encyclopedias and books on the history of mathematics as sources.)

15. There are other types of geometries; one example is projective geometry. Determine what projective geometry is, and investigate some of its common uses. (An art teacher or encyclopedia may offer information and other references.)

SUMMARY

The terms "point," "line," and "plane" are not given formal mathematical definitions. A point divides a line into three parts—two half lines and the point. An angle is formed by the rotation of a ray about a point. Angles may be measured in degrees.

The metric system has many advantages over the customary system. Some advantages are: (1) there is only one basic unit for each physical quantity; (2) conversions within the metric system are simpler than those within the customary system; (3) there is little need for common fractions in the metric system, since most numbers can be expressed as decimals.

Conversions within the metric system are performed by multiplication or division by powers of 10. To multiply or divide, you simply move the decimal point to the right or left, respectively.

A polygon is a closed figure on a plane determined by three or more straight lines. The sum of the interior angles of an n-sided polygon equals $(n - 2) \cdot 180°$.

Two polygons are similar if their corresponding angles have the same measure and their corresponding sides are proportional. Two similar polygons are congruent when their corresponding sides have the same length. Two congruent figures when placed one upon the other would coincide.

The ability to find perimeters, areas, and volumes is often helpful in doing household projects. A theorem that is sometimes helpful in finding areas is the Pythagorean theorem ($a^2 + b^2 = c^2$).

Network theory is a branch of topology. The study of topology began with the consideration of the Königsberg bridge problem in the eighteenth century. A network is traversable if it contains zero or two odd vertices. The Möbius strip and Klein bottle are interesting examples of topological figures.

The non-Euclidean geometries differ from Euclidean geometry in their treatment of the parallel postulate. Einstein used non-Euclidean geometry when he developed his General Theory of Relativity.

REVIEW EXERCISES

With reference to Fig. 10.126, find the following.

1. $\overleftrightarrow{HB} \cap \overline{BF}$
2. $\angle ABH \cap \angle HBC$
3. $\overleftrightarrow{EG} \cap \overleftrightarrow{BD}$
4. $\overline{BF} \cup \overline{FC} \cup \overline{BC}$
5. $\overset{\circ}{\overrightarrow{AB}} \cup \overset{\circ}{\overrightarrow{DC}}$
6. $\overset{\circ}{\overrightarrow{FI}} \cap \overset{\circ}{\overrightarrow{FB}}$

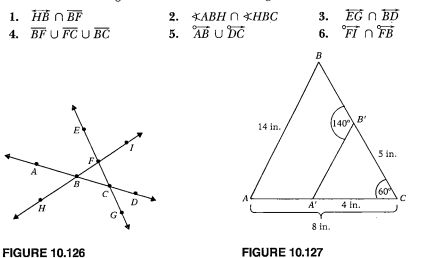

FIGURE 10.126   **FIGURE 10.127**

Given similar triangles ABC and $A'B'C$ in Fig. 10.127, find the following.

7. The length of $A'B'$
8. The length of BC
9. The measure of $\angle ABC$
10. The measure of $\angle BAC$

11. In Fig. 10.128, l_1 and l_2 are parallel lines. Find ∢1 through ∢6 as indicated.

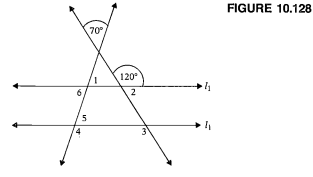

FIGURE 10.128

Complete the following.

12. 230 mℓ = _____ cℓ   **13.** 15 ℓ = _____ mℓ
14. 6.7 mℓ = _____ cℓ   **15.** 3 kg = _____ g
16. 310 mg = _____ g   **17.** 0.7 g = _____ mg
18. 37 g = _____ kg   **19.** 9 807 kg = _____ g

20. **a)** What is the volume in cubic meters of water in a rectangular shaped swimming pool that is 8 meters long, 4 meters wide and 2 meters deep?
   b) If 1 m³ = 1 kℓ, what is the mass of the water in kilograms?

Find the area of each of the figures.

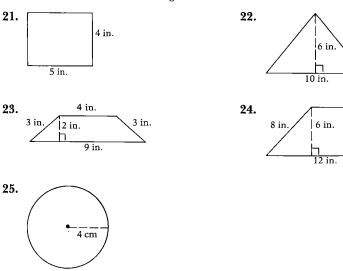

Find the volume of each of the figures.

26.

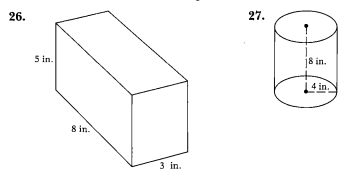

5 in.

8 in.

3 in.

27.

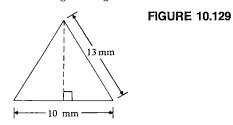

8 in.

4 in.

28. Find the area of the isosceles triangle in Fig. 10.129.

FIGURE 10.129

13 mm

10 mm

Find the volume of each solid.

29.

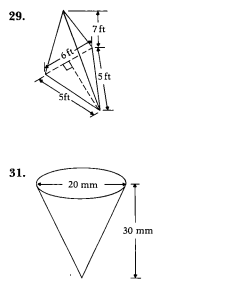

7 ft

6 ft

5 ft

5ft

30.

10 m   13 m

9 m

13 m

31.

20 mm

30 mm

32.

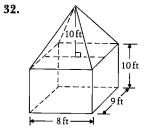

10 ft

10 ft

9 ft

8 ft

33. Determine the total cost of covering the kitchen and both bathroom floors in Fig. 10.78 (p. 520) with tile. The cost of the tile selected is $18.50 per square yard for the bathrooms and $16.95 per square yard for the kitchen. (Add 5% for waste.)

34. Farmer Jones has a water trough whose ends are trapezoids and whose sides are rectangles (Fig. 10.130). He is afraid that the base it is sitting on will not support the weight of the trough when it is filled with water.

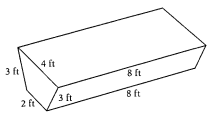

FIGURE 10.130

a) Find the number of cubic feet of water contained in the trough.
b) Find the total weight, assuming that the trough weighs 375 lb and the water weighs 62.5 lb per cubic foot.
c) If 1 gal of water weighs 8.3 lb, how many gallons of water will the trough hold?

Determine whether the networks are traversable. If they are, indicate from which points you may start.

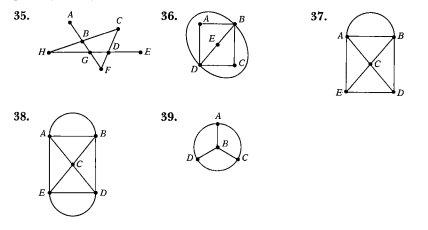

35.   36.   37.

38.   39.

40. Summarize the history of non-Euclidean geometry.
41. State the fifth axiom of Euclidean, elliptical, and hyperbolic geometry.

10 *CHAPTER TEST*

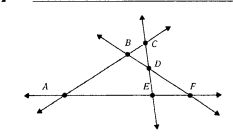

Use the figure above to describe the following sets of points:

1. $\overset{\circ\!-\!\rightarrow}{AF} \cap \overline{AE}$
2. $\overset{\circ\!-\!\circ}{BC} \cup \overline{BC}$
3. $\sphericalangle ABF \cap \sphericalangle CBD$
4. $\overrightarrow{AE} \cup \overrightarrow{EA}$
5. $\sphericalangle A = 37.4°$. Find the measure of the complement of $\sphericalangle A$.
6. $\sphericalangle B = 93.7°$. Find the measure of the supplement of $\sphericalangle B$.

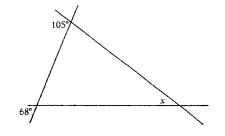

7. Find the measure of $\sphericalangle x$.
8. Find the sum of the interior angles of a pentagon.
9. Triangles ABC and $A'B'C'$ are similar figures. Find the length of side $B'C'$.

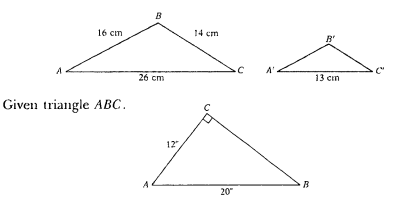

Given triangle ABC.

10. Find the length of side *CB*.
11. Find the perimeter of △ *ABC*.
12. Find the area of △ *ABC*.

The bathroom floor is in the shape of a trapezoid. A contractor will cover the floor with tile and place a molding around the room where the floor meets the walls. There is only one door, which is 76.2 cm wide. See the figure for dimensions.

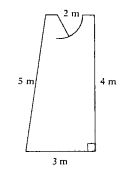

13. Find the number of square meters of floor that will be covered.
14. Find the number of meters of molding needed to go all the way around the room.

Convert each metric measurement to the indicated equivalent metric measurement.

15. 2 dm = _____ m
16. 5 kg = _____ g
17. 8 ℓ = _____ mℓ
18. 10 dam = _____ km
19. 3ℓ = _____ kℓ
20. 1000 mm = _____ m

21. A sketch showing the dimensions of an ice skating rink with semi-circular ends is illustrated below. How many cubic feet of water are needed to fill the rink to a depth of 4 inches?

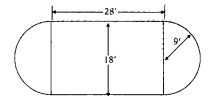

22. Find the volume of the pyramid.

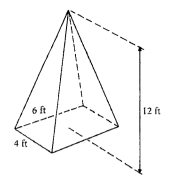

Determine whether the network is traversable. If the network is traversable, state at which points you may start.

23.

24.

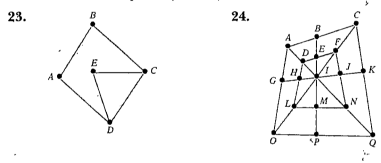

CHAPTER 11

Probability

11.1 THE NATURE OF PROBABILITY

History

Probability originated from the study of games of chance. Archaeologists have found artifacts used in games of chance in Egypt that are estimated to have been used around 3000 B.C. Paintings in tombs and drawings on pottery, among other things, provide illustrations of people or gods tossing astragals (knuckle bones of dogs or sheep) and using counting boards to record totals. It has been suggested that people were tossing these bones in games of chance as much as 40,000 years ago.

The six-sided die evolved long before the birth of Christ. The oldest dice that have been found date back to about 3000 B.C. Around 1400 B.C. the arrangements of the points were developed (top and bottom surfaces sum to 7) that are still used today.

Mathematical problems relating to games of chance were studied by a number of mathematicians of the Renaissance. Italy's Girolamo Cardano (1501–1576) in his *Liber de Ludo Aleae* [Book on the games of chance] presents the first systematic computations of probabilities. Although it is basically a gambler's manual, many consider it the first book ever written on probability. A short time later, two French mathematicians, Blaise Pascal (1623–1662) and Pierre de Fermat (1601–1665), worked together studying "the geometry of the die." In 1657 the Dutch mathematician Christian Huygens (1629–1695) published *De Ratiociniis in Luno Aleae* [On ratiocination in dice games], which contained the first documented reference to the concept of mathematical expectation (see Section 11.7). The Swiss mathematician Jacob Bernoulli (1654–1705), whom many consider the founder of probability theory, is said to have fused pure mathematics with the empirical methods used in statistical experiments. The works of Pierre-Simon de Laplace (1749–1827) dominated probability throughout the nineteenth century.

The concepts that developed provided the tools for handling statistical materials in connection with public finance, health administration, insurance, and elections. Since the middle of the nineteenth century, probability has gradually gained ground as part of physical theory. James Clerk Maxwell, a Scottish physicist, used probability to deduce gas laws. Probability has also helped in the discoveries of atomic theory and genetics. It is used today in many areas, including opinion polls, mortality tables, quality control, and gambling.

The Nature of Probability

Before we discuss what the word "probability" means and learn how to calculate probabilities, we must introduce a few definitions.

Jacob Bernoulli
The Bettmann Archive

Each of us has an intuitive idea of what the word "experiment" means. An **experiment** is a controlled operation that yields a set of results. The process in which medical researchers administer experimental drugs to patients to determine their reaction is one type of experiment. Rolling dice, tossing coins, and so on, are other types. The possible results of an experiment are called its **outcomes.**

When you roll a die and observe the number of points that face up, the possible outcomes are 1, 2, 3, 4, 5, and 6. It is equally likely that you will roll any one of the possible numbers.

> If each outcome of an experiment has the same chance of occurring as any other outcome, they are said to be **equally likely outcomes.**

Can you think of a second set of equally likely outcomes when a die is rolled? An odd number is as likely to be rolled as an even number. Therefore odd and even are another set of equally likely outcomes.

An **event** is a subcollection of the outcomes of an experiment. For example, when a die is rolled, the event of rolling a number greater than 2 can be satisfied by any one of four outcomes—3, 4, 5, or 6. The event of rolling a 5 can be satisfied by only one outcome, the 5 itself. The event of rolling an even number can be satisfied by any of three outcomes—2, 4, or 6.

Most of us accept the fact that if a "fair coin" is tossed many, many times, it will land heads up approximately half of the time. Intuitively, one can guess that the probability that a fair coin will land heads up is $\frac{1}{2}$. Does this mean that if a coin is tossed twice, it will land heads up exactly once? If a fair coin is tossed ten times, will there necessarily be five heads? The answer is clearly no. What then does it mean when we state that the probability that a fair coin will land heads up is $\frac{1}{2}$? To answer this question, let us examine Table 11.1, which shows what may occur when a fair coin is tossed a given number of times.

"The laws of probability, so true in general, so fallacious in particular."
Edward Gibbon, 1796

Ancient gambling devices (left) and modern dice (right).
Courtesy of U.S. Department of Justice

TABLE 11.1

Number of Tosses	Expected Number of Heads	Actual Number of Heads Observed	Relative Frequency of Heads
10	5	4	$\dfrac{4}{10} = 0.4$
100	50	43	$\dfrac{43}{100} = 0.43$
1,000	500	540	$\dfrac{540}{1,000} = 0.54$
10,000	5,000	4,852	$\dfrac{4,852}{10,000} = 0.4852$
100,000	50,000	49,770	$\dfrac{49,770}{100,000} = 0.49770$

The last column of Table 11.1, the relative frequency of heads, is a ratio of the number of heads observed to the total number of tosses of the coin. Note that as the number of tosses increases, the relative frequency of heads gets closer and closer to $\frac{1}{2}$, or 0.5. In general, the **relative frequency of event** E is a ratio of the number of times event E has occurred to the total number of times the experiment was performed.

The nature of probability is summarized by the law of large numbers.

> The **law of large numbers** states that probability statements apply in practice to a large number of events—not to a single event. It is the relative frequency over the long run that is accurately predictable, not the individual events or precise totals.

Probability is classified as either **empirical** (experimental) or **theoretical** (mathematical). *Empirical probability* is the relative frequency of occurrence of the event and is determined by actual observations of an experiment. *Theoretical probability* is determined through a study of the possible *outcomes* that can occur for the given experiment. The empirical probability of an event E is symbolized by $P'(E)$. The theoretical probability of event E is symbolized by $P(E)$.

Both empirical probability and theoretical probability will be discussed in detail in later sections. To give you an intuitive idea of the difference between them, let's examine how the probability of tossing a head with a fair coin might be determined both empirically and theoretically.

Empirical	Theoretical
Toss a coin 100 times. Observe how many times heads occur. If heads occurred 43 times, the empirical probability is 0.43:	A coin has two possible outcomes: heads or tails. Thus one of every two tosses should be a head:

$$P'(E) = \frac{43}{100} = 0.43 \qquad\qquad P(E) = \frac{1}{2} = 0.5$$

Notice that the probabilities differ slightly. If we increased the number of tosses, the empirical probability would get closer to the theoretical probability of 0.5, as explained by the law of large numbers. There are many instances when only empirical probability can be used. For example, in testing the effectiveness of a new drug the only way to determine its effectiveness is to give it to patients and observe the results. Can you find other areas in which only empirical probability can be used?

EXERCISES 11.1

1. A probability statement is often interpreted differently by different people. Consider the statement "The National Weather Service predicts that the probability of rain today in the Denver, Colorado, area is 80 percent." Below are five possible interpretations of that statement. Which, if any, do you think is its actual meaning?
   a) In 80 percent of the Denver area there will be rain.
   b) There is an 80 percent chance that at least any one point in the Denver area will receive rain.
   c) It has rained on 80 percent of the days with similar weather conditions in the Denver area.
   d) There is an 80 percent chance that a specific location in the Denver area will receive rain.
   e) There is an 80 percent chance that the entire Denver area will receive rain.

2. What mathematician is considered by many to be the founder of probability theory?

3. In order to determine premiums, life insurance companies must compute the probable date of death. On the basis of a great deal of research it is determined that Mr. Bennett, age 35, will live another 42.34 years. Does this mean that Mr. Bennett will live until he is 77.34 years old? If not, what does it mean?

4. The probability of rolling a 2 on a die is $\frac{1}{6}$. Does this mean that if a die is rolled six times, one 2 will appear? If not, what does it mean?

5. If you roll a die many times, what would you expect to be the relative frequency of rolling the number 5?

6. If you roll a die many times, what would you expect to be the relative frequency of rolling an even number?

7. Briefly explain the difference between empirical probability and theoretical probability.

8. Explain how you would find the empirical probability of rolling a 4 on a die.

9. Find the empirical probability of rolling a 4 by rolling a die 50 times.

10. Explain how you would determine the theoretical probability of rolling a 4 on a die.

11. Find the theoretical probability of rolling a 4 on a die.

EMPIRICAL OR EXPERIMENTAL PROBABILITY

Empirical probability is based on past experiences and is used to determine what "probably" will or will not occur in the future. Empirical probability is used in manufacturing, educational tests and measurements, genetics, weather forecasting, insurance, investments, opinion polls, and other areas in which present-day data are used to predict future trends.

Empirical probability can only suggest what should happen on the basis of present-day knowledge. It cannot guarantee that an event will or will not occur. Old Faithful, a geyser at Yellowstone National Park, has erupted about once every hour since records of its activities have been kept. It will "probably" continue to erupt once every hour today and for many years to come. However, the fact that it has always erupted in the past is no guarantee that it will continue to erupt in the future.

The empirical probability of an event is the *relative frequency of the event,* as discussed in Section 11.1. The empirical probability of event E, symbolized by $P'(E)$, is determined by the formula.

Empirical Probability

$$P'(E) = \frac{\text{number of times event } E \text{ has occurred}}{\text{total number of times the experiment has been performed}}$$

A probability will always be a number between 0 and 1, inclusive, and may be expressed as a decimal, a fraction, or a percent. An empirical probability of 0 indicates that the event has never occurred. An empirical probability of 1 indicates that the event has never failed to occur.

■ **Example 1** _____

In 100 tosses of a fair coin, 58 landed heads up. Find the empirical probability of the coin landing heads up.

Solution $P'(E) = \dfrac{58}{100} = 0.58.$ ❖

■ **Example 2** _____

A survey of 500 randomly selected people is taken to determine whether they favor candidate A, B, C, or none of the three. The results of the survey are indicated in Table 11.2. Find the empirical probability that an individual selected at random from the 500 people surveyed favors (a) candidate A, (b) candidate B, and (c) none of the candidates.

TABLE 11.2				
A	B	C	None	Total Surveyed
140	180	130	50	500

Solution

a) $P'\left(\begin{array}{c}\text{favors}\\\text{candidate } A\end{array}\right) = \dfrac{\text{number favoring } A}{\text{total number surveyed}} = \dfrac{140}{500} = \dfrac{7}{25}$ (or 0.28)

b) $P'\left(\begin{array}{c}\text{favors}\\\text{candidate } B\end{array}\right) = \dfrac{180}{500} = \dfrac{9}{25}$ (or 0.36)

c) $P'\left(\begin{array}{c}\text{favors none}\\\text{of the}\\\text{candidates}\end{array}\right) = \dfrac{50}{500} = \dfrac{1}{10}$ (or 0.10) ❖

Opinion pollsters often use empirical methods with small samples to estimate the feelings of a large group (see Chapter 12, Statistics).

DID YOU KNOW?

In the United States there are usually more males born than females. However, there are usually more females alive in the United States because of their longer life span.

Year	Males	Females	Males per 1000 Females	P'(males)
1981	1,860,272	1,768,966	1052	0.513
1982	1,885,676	1,794,861	1051	0.512
1983	1,865,553	1,773,380	1052	0.513
1984	1,879,490	1,789,651	1050	0.512

Empirical Probability in Genetics

Using empirical probability, Gregor Mendel (1822–1884) developed the laws of heredity. These laws became the foundation for the study of genetics.

Mendel crossbred different types of "pure" pea plants and observed the resulting offspring. For example, when he crossbred a pure yellow pea plant and a pure green pea plant, the resulting offspring (the first generation) were always yellow (see Fig. 11.1a). When he crossbred a pure round-seeded

Photo by Marshall Henrichs

pea plant and a pure wrinkled-seeded pea plant, the resulting offspring (the first generation) were always round (Fig. 11.1b).

Traits such as yellow color and round seeds Mendel called **dominant** because they overcame or "dominated" the other trait. The green and wrinkled traits he labeled **recessive.**

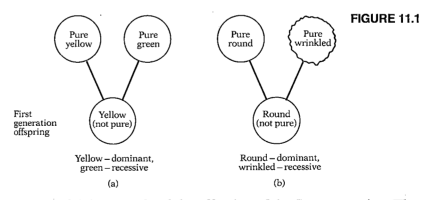

FIGURE 11.1

First generation offspring

Yellow – dominant, green – recessive

(a)

Round – dominant, wrinkled – recessive

(b)

Mendel then crossbred the offspring of the first generation. The resulting second generation had both the dominant and the recessive traits of their grandparents (see Fig. 11.2a and b). What's more, these traits always appeared in approximately a 3 to 1 ratio of dominant to recessive.

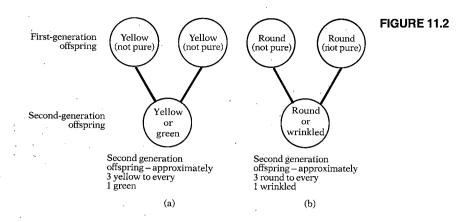

FIGURE 11.2

Table 11.3 shows the actual results of Mendel's experiments in genetics. Note that the ratio of dominant trait to recessive trait in the second generation offspring is about 3 to 1 for each of the four experiments. The empirical probability of the dominant trait has also been calculated. How would you find the empirical probability of the recessive trait?

TABLE 11.3

Second Generation Offspring

Dominant Trait	Number with Dominant Trait	Recessive Trait	Number with Recessive Trait	Ratio of Dominant to Recessive	P'(dominant trait)
Yellow seeds	6022	Green seeds	2001	3.01 to 1	$\frac{6022}{8023} = 0.75$
Round seeds	5474	Wrinkled seeds	1850	2.96 to 1	$\frac{5474}{7324} = 0.75$
Tall plants	787	Short plants	277	2.84 to 1	$\frac{787}{1064} = 0.74$
Purple flowers	705	White flowers	224	3.15 to 1	$\frac{705}{929} = 0.76$

Mendel used the information from charts like Table 11.3 to theorize about the outcomes of his experiments. His theories of genetics were developed by using empirical probability. Mendel theorized that a pure plant or animal has two identical genes.* A hybrid (not pure) plant or animal has two different genes. The hybrid has the same characteristic as the dominant gene.

We will represent dominant genes with capital letters and recessive genes with small letters. For example, a pure dominant yellow pea plant will be labeled *YY*, and a pure recessive green pea plant will be labeled *yy*. A

---

* The word "gene" was not coined until some years after Mendel's death. The German word *Mendel* actually used is translated as "factor."

hybrid yellow pea plant will be labeled *Yy*. There can be no hybrid green pea plant. Why?

When a pure yellow, *YY*, is crossbred with a pure green, *yy*, the offspring received one gene (either *Y* or *y*) from each parent. The offspring of the resulting first generation will be yellow, with genes *Yy* (see Fig. 11.3a). Similarly, crossbreeding a pure round, *RR*, with a pure wrinkled, *rr*, results in round offspring with genes *Rr* (Fig. 11.3b).

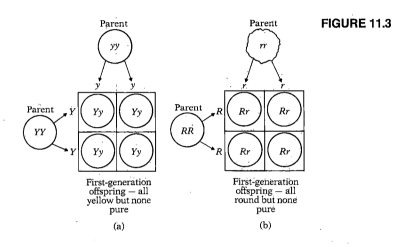

FIGURE 11.3

First-generation offspring — all yellow but none pure

(a)

First-generation offspring — all round but none pure

(b)

Mendel theorized that when he crossbred the offspring of the first generation, the resulting second generation would receive one gene from each parent. In Fig. 11.4(a), each parent has *Yy* genes. The possible offspring are *YY, yY, Yy*, and *yy*. Since each of these possible outcomes is equally likely, and since *Y* is dominant over *y*, the ratio of the dominant yellow (*YY, yY*, and *Yy*) to the recessive green (*yy*) is 3 to 1. The ratio of dominant to recessive in Table 11.3 supports Mendel's theories.

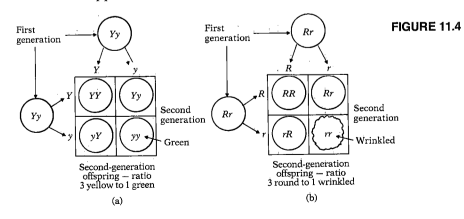

FIGURE 11.4

Second-generation offspring — ratio 3 yellow to 1 green

(a)

Second-generation offspring — ratio 3 round to 1 wrinkled

(b)

Note how Mendel used empirical probability (observed values) and theoretical probability (possible outcomes) to develop his laws of genetics.

Even though we have not formally discussed theoretical probability, we can use the ideas developed in this section to do some problems involving genetics.

■ **Example 3** _____

Brown hair is dominant over blonde hair in humans. If Stuart, who has brown hair, with genes *Bb,* marries Joyce, who has blonde hair, find the probability that their first child will have blonde hair.

Solution: Since Joyce has blonde hair, she must have *bb* genes. Why? We can make a chart (Fig.11.5) to determine the possible outcomes and probabilities. Note that two of the four possible outcomes (*bb, bb*) will result in blonde hair. Thus the probability that the first child born has blonde hair is 2/4 = 1/2 or 0.5.

FIGURE 11.5

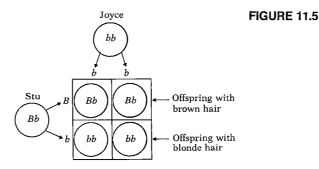

EXERCISES 11.2

1. Roll a die 60 times and record the results. Find the empirical probability of rolling (a) a 1 and (b) a 6. (c) Does it appear that the probability of rolling a 1 is the same as the probability of rolling a 6?

2. Roll a pair of dice 60 times and record the sums. Compute the empirical probability of rolling a sum of (a) 2 and (b) 7. (c) Does it appear that the probability of rolling a sum of 2 is the same as the probability of rolling a sum of 7?

3. Find the empirical probability of tossing exactly one head when two coins are tossed 50 times.

4. Refer to the Did You Know? on p. 563 and explain how the empirical probability of males was found.

5. No baseball player has ever hit more than 61 home runs in a single season in the major leagues. What is the empirical probability that a baseball player in either of the major leagues will hit more than 61 home runs next season? Is it possible that a baseball player will hit more than 61 home runs next season?

6. Old Faithful, a geyser in Yellowstone National Park, has erupted hot water and steam about

once every hour since records of its activities have been kept. What is the empirical probability of Old Faithful's erupting within the next hour? What is the empirical probability that Old Faithful will fail to erupt within the next hour?

7. The last 20 birds that fed at the Zwick's bird feeder were 12 sparrows, 5 cardinals, and 3 chickadees. Use this information to determine the empirical probability that the next bird to feed from the feeder is a
   a) sparrow, b) cardinal, c) chickadee.

8. In a sample of 10,000 births, 75 were found to have syndrome A. Find the empirical probability that Mrs. Gardina's first child will be born with this syndrome.

9. Of 100 brand A carrot seeds planted, 75 germinated and produced carrots.
   a) Find the empirical probability of germination of brand A carrot seeds.
   b) How many of these seeds should you plant if you want to grow 300 carrots?

10. A certain community recorded 3000 births last year, of which 1460 were females. If Mrs. Chang lives in that community, what is the empirical probability that her next child will be (a) female or (b) male.

11. There is a faulty candy machine that requires exact change next to the classroom. A student obtains the following information by observing fellow students' experience with the machine.

Outcome	Frequency of Occurrence
Gives the candy only	25
Gives the candy and returns the money	6
Does not give the candy and keeps the money	9
	40

   a) If the machine is tried a forty-first time, what is the empirical probability that it will work properly?
   b) What is the empirical probability of obtaining candy from the machine?

12. In a recent survey, a sample of 1400 people were asked whether the president was performing above average, average, or below average. The results were as follows.

Number of People	Rating
562	above average
304	average
400	below average
134	no opinion

Find the empirical probability that the next person surveyed
   a) rates the president above average,
   b) rates the president below average,
   c) has no opinion.

13. An experimental serum was injected into 500 guinea pigs. Initially, 150 of the guinea pigs had circular cells, 250 had elliptical cells, and 100 had irregularly shaped cells. After the serum was injected, it was observed that none of the guinea pigs with circular cells were affected, 50 with elliptical cells were affected, and all of those with irregular cells were affected. Find the empirical probability that a guinea pig with (a) circular cells, (b) elliptical cells, and (c) irregular cells will be affected by injection of the serum.

14. Jim finds an irregularly shaped five-sided rock. He labels each side and tosses the rock 100 times. The result of his tosses is illustrated below. Find the empirical probability that the rock will land on side 4 if tossed again.

Side   1   2   3   4   5
Frequency   32   18   15   13   22

15. If a round pea plant, Rr, is crossed with a second round pea plant, Rr, what is the probability that the resulting offspring will be wrinkled rr? (Refer to Figs. 11.3 and 11.4.)

16. If a pure dominant yellow pea plant, YY, is crossed with a pure recessive green pea plant, yy, what is the probability that the offspring will be green?

17. If a dominant yellow pea plant, Yy, is crossed with a pure recessive green pea plant, yy, what is the probability that the offspring will be green?

18. A pure dominant purple flower plant, PP, is allowed to self-pollinate. Find the probability that the offspring will have white flowers.

19. In human eyes the color brown is dominant and the color blue is recessive. Let B represent a gene for brown eyes, and let b represent a gene for blue eyes. A person with Bb genes (B from one parent and b from the other parent) will there-

fore have brown eyes, but a person with *bb* genes (*b* from both parents) will have blue eyes. If Carlos, who is known to have *Bb* genes, marries Maria, who has *bb* genes, what is the probability that their first child will have (a) blue eyes; (b) brown eyes?

20. Donald, who has brown eyes, plans to marry Jennifer, who also has brown eyes. Is it possible that their children will have blue eyes? Explain.

21. The Rh blood factor is a blood antigen that is passed to offspring by the parents. A person with the Rh factor is said to be Rh positive (about 85 percent of American Caucasians and 93 percent of American blacks are born with this substance), and one lacking the Rh factor is said to be Rh negative. Rh positive is dominant, and Rh negative is recessive. Thus if one parent or both pass this Rh antigen to the child, the child will be Rh positive.

   a) If the mother is pure Rh positive and the father is pure Rh negative, what is the probability that the offspring will be Rh negative?

   b) If both parents are Rh positive, but neither is purely so, what is the probability that the offspring will be Rh negative?

   c) If both parents are Rh negative, what is the probability that the child will be Rh negative?

22. Many types of disorders, such as albinism, hemophilia, Tay-Sachs disease, and sickle-cell anemia are genetic in nature. Albinism is a known recessive trait that causes children to be born without skin color, hair color, or eye color. Assuming that both parents are normal and their firstborn is albino, determine the probability that their next-born will also be albino.

23. "Investors root for the Redskins." This quote appeared in many newspapers and magazines before the 1988 Super Bowl. In 13 of the 14 years that a former National Football League (NFL) team has won the Super Bowl the stock market has increased. In 6 of the 7 years that a former American Football League (AFL) team has won the Super Bowl the stock market has decreased. This phenomenon, known as "the Super Bowl predictor," has been one of the most

reliable indicators of whether the stock market will increase or decrease in the year that follows. In the 1988 Super Bowl the Washington Redskins (NFL) defeated the Denver Broncos (AFL).

The table that follows indicates the winners of all the Super Bowls up to 1988, their former affiliation, and the New York Stock Exchange Composite Index (average gain or loss) for the year following the Super Bowl.

Year	Super Bowl Winner	Team Roots	NYSE Stocks Average Gain/Loss
1967	Green Bay Packers	NFL	+23.1%
1968	Green Bay Packers	NFL	+9.4%
1969	New York Jets	AFL	−12.5%
1970	Kansas City Chiefs	AFL	−2.5%
1971	Baltimore Colts	NFL	+12.3%
1972	Dallas Cowboys	NFL	+14.3%
1973	Miami Dolphins	AFL	−19.6%
1974	Miami Dolphins	AFL	−30.3%
1975	Pittsburgh Steelers	NFL	+31.9%
1976	Pittsburgh Steelers	NFL	+21.5%
1977	Oakland Raiders	AFL	−9.3%
1978	Dallas Cowboys	NFL	+2.1%
1979	Pittsburgh Steelers	NFL	+15.5%
1980	Pittsburgh Steelers	NFL	+25.7%
1981	Oakland Raiders	AFL	−8.7%
1982	San Francisco 49ers	NFL	+14.0%
1983	Washington Redskins	NFL	+17.5%
1984	Los Angeles Raiders	AFL	+1.3%
1985	San Francisco 49ers	NFL	+26.2%
1986	Chicago Bears	NFL	+14.0%
1987	New York Giants	NFL	−0.3%
1988	Washington Redskins	NFL	?

   a) Determine the empirical probability that the stock market will increase in 1988.

   b) Determine the empirical probability that the stock market will decrease in 1988.

   c) Do you believe that there is any logical connection between the winner of the Super Bowl and the performance of the stock market in the following year? Explain your answer.

RESEARCH ACTIVITY

24. Determine how insurance companies use empirical probabilities in determining insurance premi-

ums. An insurance agent may be able to direct you to a source of information to answer this question.

THEORETICAL PROBABILITY

When we determine the probability of an event through observations of experiments, we are finding an empirical probability. When we determine the probability of an event by determining and analyzing the possible outcomes of an experiment, we are finding a theoretical or mathematical probability. Theoretical probability is often used in gambling because it indicates what will happen *over the long run*. You will soon be able to determine that the theoretical probability of tossing a head with a fair coin is 1/2. This means that over the long run the ratio of heads to total tosses should approach 1/2.

In the remaining sections the word "probability" will refer to theoretical probability unless otherwise noted. If an event has *equally likely* outcomes, the probability of event E, symbolized by $P(E)$, may be calculated with the following formula.

> **Probability**
> $$P(E) = \frac{\text{total number of outcomes favorable to } E}{\text{total number of possible outcomes}}$$

Let us do a problem illustrating how this formula is used.

■ **Example 1** _____

A die is rolled. Find the probability of rolling

a) an odd number,   **b)** a number greater than 4,
c) a 7,   **d)** a number less than 7.

Solution

a) There are six possible outcomes: 1, 2, 3, 4, 5, and 6. The event of rolling an odd number can occur in three ways (1, 3, 5):

$$P(\text{odd number}) = \frac{\text{total no. of outcomes that result in an odd number}}{\text{total no. of possible outcomes}}$$
$$= \frac{3}{6} = \frac{1}{2}$$

b) There are two numbers greater than 4, namely, 5 and 6:

$$P(\text{number greater than 4}) = \frac{2}{6} = \frac{1}{3}$$

c) There are no outcomes that will result in a 7:

$$P(7) = \frac{0}{6} = 0$$

d) All of the outcomes 1 through 6 are less than 7:

$$P(\text{number less than } 7) = \frac{6}{6} = 1$$

❖

This example illustrates three important facts.

---

1. The probability of an event that cannot occur is 0.
2. The probability of an event that must occur is 1.
3. Every probability will be a number between 0 and 1 inclusive, that is, $0 \leq P(E) \leq 1$.

Let us consider a second example.

■ **Example 2** _____

Twenty-five famous astronauts are listed in Table 11.4. Each name is written on a slip of paper, and the 25 slips are deposited in a bag. One name is to be selected at random from the bag. Find the probability of the following. That

a) the last name starts with C;

b) the first and last names start with the same letter;

c) the name of the first person to step on the moon is selected;

d) the name of one of the three astronauts on the first moon-walk mission is selected;

e) an astronaut with a last name starting with P is selected.

TABLE 11.4

Twenty-five Famous Astronauts

Edwin Aldrin, Jr.	John Glenn
Neil Armstrong	Richard Gordon, Jr.
Alan Bean	Vergil Grissom
Guion Bluford, Jr.	Fred Haise, Jr.
Frank Borman	Jim Lovell
Scott Carpenter	James McDivitt
Eugene Cernan	Sally Ride
Michael Collins	Walter Schirra
Charles Conrad, Jr.	Alan Shepard
L. Gordon Cooper	Thomas Stafford
Robert Crippen	Edward White
John Engle	John Young
James Erwin	

Solution

a) Six of the 25 astronauts have last names that start with C:

$$P\text{(last name starts with C)} = \frac{6}{25}$$

b) Charles Conrad is the only astronaut whose first and last names start with the same letter:

$$P\text{ (astronaut's first and last names start with the same letter)} = \frac{1}{25}$$

c) Neil Armstrong was the first person to step on the moon on July 20, 1969:

$$P\text{(the name of the first person to step on the moon)} = \frac{1}{25}$$

d) Neil Armstrong; Edwin Aldrin, Jr.; and Michael Collins were on the first moon-walk mission:

$$P\text{(one of the three astronauts on the first moon-walk mission)} = \frac{3}{25}$$

e) There are no astronauts on the list wth the last name starting with P:

$$P\text{(astronaut has a last name starting with P)} = \frac{0}{25} = 0 \qquad \clubsuit$$

In any given experiment an event must either occur or not occur. *The sum of the probability that an event will occur and the probability that it will not occur is 1.* Thus for any event *A* we conclude that

$$P(A) + P(\text{not } A) = 1.$$

Using algebra, we can show that

$$P(\text{not } A) = 1 - P(A).$$

For example, if the probability that event *A* will occur is 5/12, the probability that event *A* will not occur is $1 - 5/12$, or 7/12. Similarly, if the probability that event *A* will not occur is 0.3, the probability that event *A* will occur is $1 - 0.3 = 0.7$ or 7/10. We make use of this concept in the following example.

Example 3

A deck of cards is illustrated in Fig. 11.6. There are four suits: spades, diamonds, hearts, and clubs. Each suit has 13 cards. Hearts and diamonds are red; clubs and spades are black. The jacks, queens, and kings are picture cards. There are 12 picture cards in the deck of 52 cards. One card is to be selected at random from the deck of cards. Find the probability that the card selected is

a) a 3,

b) not a 3,

c) a spade,

d) a picture card,

e) a card greater than 6 and less than 9.

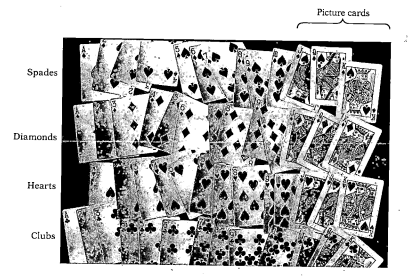

FIGURE 11.6
Deck of cards.
Photo by Marshall Henrichs

Solution

a) There are four 3s in a deck of 52 cards:

$$P(3) = \frac{4}{52} = \frac{1}{13}$$

b) $P(\text{not a 3}) = 1 - P(3)$
$$= 1 - \frac{1}{13} = \frac{12}{13}$$

This could also have been found by noting that there are 48 cards that are not 3s in a deck of 52 cards:

$$P(\text{not a 3}) = \frac{48}{52} = \frac{12}{13}$$

c) There are 13 spades in the deck:

$$P(\text{spade}) = \frac{13}{52} = \frac{1}{4}$$

d) There are 12 picture cards:

$$P(\text{picture card}) = \frac{12}{52} = \frac{3}{13}$$

e) The cards that are greater than 6 and less than 9 are 7s and 8s. There are four 7s and four 8s, or a total of eight cards:

$$P(\text{greater than 6 and less than 9}) = \frac{8}{52} = \frac{2}{13} \qquad ❖$$

In Section 11.5 we will discuss "and" and "or" probability problems in depth. In Example 4 we will try to give you a basic understanding of the meaning of the words "and" and "or" when used in probability problems.

■ Example 4 ───────────────────────────────

A bouquet of roses contains five red roses, three yellow roses, and two white roses. If one rose is selected at random, find the probability the rose is a) red or white, b) red and white.

Solution

a) We shall be successful if we select one of the five red roses *or* one of the two white roses. Thus there are seven outcomes that would be favorable out of ten possible outcomes:

$$P(\text{red or white rose}) = \frac{7}{10}$$

b) To be successful, we must select one rose that is *both* red *and* white. Since none of the roses are both red and white, the probability of selecting a red and white rose is 0:

$$P(\text{red and white rose}) = \frac{0}{10} = 0 \qquad ❖$$

Note that for an "or" statement to be satisfied, only one of the criteria in the statement needs to be satisfied (in some cases, both may be satisfied, but this is not necessary). For an "and" statement to be satisfied, both criteria in the statement must be satisfied.

DID YOU KNOW?

There is no Nobel Prize in mathematics. The Field medals, the equivalent in mathematics of Nobel Prizes, are awarded every four years at meetings of the International Congress of Mathematicians. A small number (two to four) have been awarded every four years since 1936—by tradition to mathematicians under the age of 40. In 1986, three Field medals were awarded. The recipients were Gerd Faltings of West Germany, Michael Freedman of the United States, and Simon Donaldson of Great Britain.

EXERCISES 11.3

One card is selected at random from a deck of cards. Find the probability that the card selected is:

1. a 7
2. a picture card
3. not a picture card
4. the ace of spades
5. a red card
6. a red card or a black card
7. a red card and a black card
8. a card greater than 4 and less than 9
9. a king and a heart
10. not the 5 of clubs

Present U.S. currency consists of $1, $2, $5, $10, $20, $50, $100, $500, $1000, $5000, $10,000, and $100,000 bills. (The Treasury Department discontinued printing all bills over $100 in July, 1969, but the larger bills printed earlier are still being used.) One of each of these bills is placed in a hat, and one bill is selected at random. Find the probability that the bill selected is:

11. a $10 bill
12. a $10 or $20 bill
13. a bill greater than $100
14. a bill between $1 and $100

Fifteen record albums, including six Bruce Springsteen, five Whitney Houston, three Pink Floyd, and one New York Philharmonic are wrapped in plain brown paper. Ms. Gilligan selects one album at random. Find the probability that the album selected is:

15. a Bruce Springsteen album
16. a New York Philharmonic album

17. either a Bruce Springsteen or a Whitney Houston album
18. not a Pink Floyd album

On Monday, October 19, 1987 (now known as Black Monday), the stock market dropped a record 508 points. On the New York Stock Exchange, 1973 stocks dropped, 56 stocks remained the same, and 52 stocks increased in value. If one stock on the New York Stock Exchange was selected at random, find the probability that on Black Monday the stock:

19. increased in value
20. remained the same
21. decreased in value
22. did not decrease in value

A package of flower seeds contains 100 seeds for white flowers, 50 for red flowers, 25 for pink flowers, and 25 for blue flowers. One seed is selected at random. Find the probability that it will produce:

23. a red flower
24. a red or white flower
25. a green flower
26. a red and white flower

A pond is stocked with 300 fish: 50 bass, 100 trout, 80 pike, and 70 perch. Janet Jones plans to fish until she catches one fish. Find the probability that it will be:

27. a bass
28. a trout
29. a pike or perch
30. a shark

Use Table 11.4 to determine the probability that an astronaut selected at random:

31. has a last name beginning with the letter G
32. has a first name beginning with the letter G
33. is listed alphabetically before Jim Lovell
34. is listed alphabetically between Guion Bluford, Jr., and Sally Ride
35. has a first name of John
36. has a first and last name that do not start with the same letter

A traffic light is red for 30 seconds, yellow for 5 seconds, and green for 40 seconds. What is the probability that when you reach the light,

37. the light is red
38. the light is yellow
39. the light is not red
40. the light is not red or yellow

Each individual letter of the word "pfeffernuesse" is placed on a piece of paper, and all 13 pieces of paper are placed in a hat. If one letter is selected at random from the hat find the probability that

41. the letter f is selected
42. a letter f is not selected
43. a vowel is selected
44. the letter f or e is selected
45. the letter e is not selected
46. the letter g is selected

There are 12 pencils in a cup. Five are sharpened and have erasers, four are sharpened but do not have erasers, and three are not sharpened but have erasers. If one pencil is selected at random from the cup, find the probability that the pencil

47. is sharpened
48. is not sharpened
49. has an eraser
50. does not have an eraser

The minimum age required to obtain a regular driver's license (without restrictions) in the 50 states and the District of Columbia is summarized as follows:

Minimum Age for License	Number of States
15	1
16	20
17	4
18	23
19	1
21	2

If one of these states is selected at random, find the probability that the minimum age needed for a regular driver's license is

51. 15
52. less than 17
53. between 15 and 17
54. greater than 16
55. greater than or equal to 16

 TREE DIAGRAMS AND THE COUNTING PRINCIPLE

We stated earlier that the possible results of an experiment are called its outcomes. In order to solve more difficult probability problems we must first be able to determine all the possible outcomes of the experiment. A list of all the possible outcomes of an experiment is called a **sample space.**

Tree diagrams are often helpful in determining sample spaces. To construct a tree diagram, as in Fig. 11.7, indicate a starting point. From this starting point, make a branch to each possible outcome of the first stage of the experiment. From each possible outcome of the first stage, branch all the possible outcomes of the second stage. Continue the process for each stage of the experiment.

■ **Example 1** ————————————————————————

a) Two coins are tossed. Construct a tree diagram to determine the sample space of all the possible outcomes.

Use the sample space to find the probability of tossing

b) no heads when two coins are tossed,

c) exactly one head when two coins are tossed,

d) two heads when two coins are tossed.

Solution

a) For clarity, we will call one coin a penny and the other coin a nickel. Suppose we toss the penny first (it makes no difference which coin is tossed first or whether the two coins are tossed at the same time, since each coin is *independent* of the other). The penny has two possible outcomes: heads or tails (see Fig. 11.7).

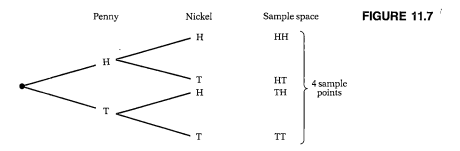

FIGURE 11.7

If the penny lands heads up, the nickel can land either heads up or tails up. If the penny lands tails up, the nickel can land heads up or tails up. Note that the sample space consists of four **sample points:** HH, HT, TH, TT.

b) One outcome has no heads (TT). Since there are four possible outcomes,

$$P(\text{no heads}) = \frac{1}{4}$$

c) Two outcomes have exactly one head (HT, TH). Thus

$$P(\text{exactly 1 head}) = \frac{2}{4} = \frac{1}{2}$$

d) One outcome has two heads (HH). Thus

$$P(\text{two heads}) = \frac{1}{4}$$

❖

Let us look at the results of Example 1 again:

$$P(0 \text{ heads}) = \tfrac{1}{4}$$
$$P(\text{exactly 1 head}) = \tfrac{1}{2}$$
$$P(2 \text{ heads}) = \tfrac{1}{4}$$
$$\text{Sum of probabilities} = \overline{1}$$

This example illustrates that when all the possible outcomes of an experiment are considered, the sum of their probabilities is 1.

Some important rules of probability are summarized below.

> **1.** Every probability will be a number from 0 to 1. A probability of 0 indicates that the event cannot occur. A probability of 1 indicates that the event must occur.
>
> **2.** The sum of the probabilities of all possible outcomes of an experiment is equal to 1.

When it is necessary only to know the number of possible outcomes for an experiment, the **counting principle** may be used. The counting principle is also an aid in constructing a tree diagram because it lets you know how many branches you will need in your diagram. We introduce the counting principle here and study it again in Section 11.8.

> **Counting principle.** If a first task can be performed in M distinct ways and a second task can be performed in N distinct ways, then the two tasks in that specific order can be performed in $M \cdot N$ distinct ways.

If we wish to find the number of possible outcomes when two coins are tossed, we can reason like this. The first coin has two possible outcomes: heads or tails. The second coin also has two possible outcomes: heads or tails. Thus the two coins together have $2 \cdot 2$ or four possible outcomes. If you examine the sample space in Example 1, you will see there four sample points. Also note that the first set of branches in the tree diagram consists of two branches (one branch for each of the two possible outcomes of the first coin). Then each of these branches leads to two other branches (one for each of the two possible outcomes of the second coin).

■ **Example 2** _____

A hat contains three chips: one red, one blue, and one green. A bowl contains each of the four numbers 1, 2, 3, and 4 listed on a separate sheet of folded paper. You select one chip at random from the hat and select one number at random from the bowl.

a) Use the counting principle to determine the number of possible outcomes.

b) Construct a tree diagram and determine the sample space.

c) Find the probability that you select a red chip *and* the number 4.

d) Find the probability that you select a red chip *or* the number 4.

e) Find the probability of selecting a red chip.

Solution

a) There are three possible outcomes when you select the chip. There are
four possible outcomes when you select the number. Thus when considered together, there are $3 \cdot 4$ or 12 possible outcomes.

b) From the counting principle we know that the first set of branches will
contain three branches (one for each of the three possible outcomes,
R, B, G) and *each* of these branches will lead to four other branches
(one for each of the four possible outcomes 1, 2, 3, 4). (See Fig. 11.8.)

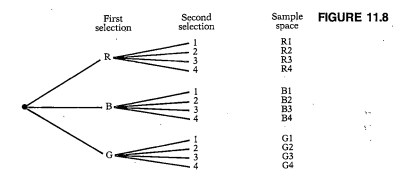

FIGURE 11.8

c) There is only one sample point that contains both a red chip *and* the
number 4 (R4). Thus

$$P(\text{select red chip and number 4}) = \frac{1}{12}$$

d) There are six sample points that contain either a red chip *or* the number 4 (R1, R2, R3, R4, B4, G4). Thus

$$P(\text{select red chip or 4}) = \frac{6}{12} = \frac{1}{2}$$

e) There are four sample points that contain a red chip (R1, R2, R3, R4).
Thus

$$P(\text{select red chip}) = \frac{4}{12} = \frac{1}{3} \qquad \qquad ❖$$

■ **Example 3** ————————————————————————————

A pair of dice is rolled.

a) Use the counting principle to determine the number of sample points in the sample space.

b) Determine the sample space.

Use the sample space to find the probability of obtaining

c) a sum of 7,   d)   a sum of 7 or 11,   e)   a double.

Solution

a) The first die has six possible outcomes, and the second die has six possible outcomes. Thus when considered together, there are $6 \cdot 6$ or 36 possible outcomes.

b) The sample space is illustrated in Fig. 11.9. We suggest that you construct a tree diagram to make sure you understand how this sample space was determined. Note that the sample space contains 36 sample points.

FIGURE 11.9

1, 1	2, 1	3, 1	4, 1	5, 1	6, 1
1, 2	2, 2	3, 2	4, 2	5, 2	6, 2
1, 3	2, 3	3, 3	4, 3	5, 3	6, 3
1, 4	2, 4	3, 4	4, 4	5, 4	6, 4
1, 5	2, 5	3, 5	4, 5	5, 5	6, 5
1, 6	2, 6	3, 6	4, 6	5, 6	6, 6

c) There are six ways to roll a sum of 7—(1, 6), (2, 5), (3, 4), (4, 3), (5, 2), and (6, 1)—out of 36 possible outcomes.

$$P(\text{sum of } 7) = \frac{6}{36} = \frac{1}{6}.$$

d) There are eight outcomes that will give a sum of 7 or 11: The six listed in part (c) and (6, 5) and (5, 6).

$$P(7 \text{ or } 11) = \frac{8}{36} = \frac{2}{9}.$$

e) There are six possible ways to roll a double: (1, 1), (2, 2), (3, 3), (4, 4), (5, 5), (6, 6).

$$P(\text{double}) = \frac{6}{36} = \frac{1}{6}.$$

❖

■ **Example 4** _____

Two balls are to be selected without replacement from a bag that contains one red, one blue, one green, and one orange ball. (*Note:* The term "without replacement" means that a ball is not returned to the bag after it is selected.)

a) Use the counting principle to determine the number of points in the sample space.

b) Construct a tree diagram to determine the sample space.

Solution

a) The first selection may be any one of the four balls. Once the first ball is selected, only three balls remain for the second selection. Thus there are 4 · 3, or 12 sample points in the sample space.

b) The first ball selected can be red, blue, green, or orange. Since this experiment is done without replacement, the same colored ball cannot be selected twice. For example, if the first ball selected is red, the second ball selected must be either blue, green, or orange. The tree diagram and sample space are given in Fig. 11.10. The sample space contains 12 points. That result checks with the answer obtained in part (a).

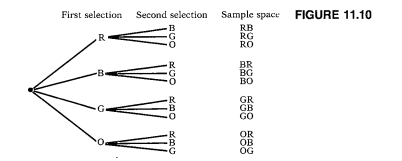

First selection   Second selection   Sample space   **FIGURE 11.10**

The counting principle can be extended to any number of experiments, as illustrated in Example 5.

■ Example 5

There are two highways from New York to Cleveland, three highways from Cleveland to Chicago, and two highways from Chicago to San Francisco as illustrated in Fig. 11.11.

New York   Cleveland   Chicago   San Francisco   **FIGURE 11.11**

a) Use the counting principle to determine the number of different routes from New York to San Francisco.

b) Use a tree diagram to determine the routes.

Solution

a) There are two routes from New York to Cleveland, three routes from Cleveland to Chicago, and two routes from Chicago to San Francisco. Thus there are $2 \cdot 3 \cdot 2$ or 12 routes from New York to San Francisco.

b) The tree diagram illustrating the 12 possibilities is given in Fig. 11.12.

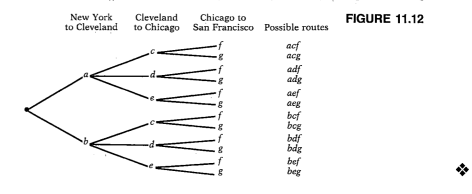

FIGURE 11.12

■ Example 6

A license plate is to consist of two letters followed by three digits. Determine how many different license plates are possible if

a) repetition of letters and digits is permitted;

b) repetition of letters is not permitted, but repetition of digits is permitted;

c) repetition of letters and digits is not permitted;

d) the first letter must be a vowel and the first digit cannot be a 0, and repetition of letters and digits is not permitted.

Solution: There are 26 letters and 10 digits (0–9). We have five positions to fill as indicated below:

$$\overline{\text{L}}\ \overline{\text{L}}\ \overline{\text{D}}\ \overline{\text{D}}\ \overline{\text{D}}$$

a) Since repetition is permitted, there are 26 possible choices for both the first and second positions. There are 10 possible choices for the third, fourth, and fifth positions:

$$\frac{26}{\text{L}}\ \frac{26}{\text{L}}\ \frac{10}{\text{D}}\ \frac{10}{\text{D}}\ \frac{10}{\text{D}}$$

Since $26 \cdot 26 \cdot 10 \cdot 10 \cdot 10 = 676{,}000$, there are 676,000 different possible arrangements.

b) There are 26 possibilities for the first position. Since repetition of letters is not permitted, we cannot use the same letter selected for the first position for the second position. Therefore there are only 25 possibilities for the second position:

$$\frac{26}{\text{L}}\ \frac{25}{\text{L}}\ \frac{10}{\text{D}}\ \frac{10}{\text{D}}\ \frac{10}{\text{D}}$$

Since $26 \cdot 25 \cdot 10 \cdot 10 \cdot 10 = 650{,}000$, there are 650,000 different possible arrangements.

c) Since repetition of neither letters nor digits is permitted we have

$$\frac{26}{\text{L}}\ \frac{25}{\text{L}}\ \frac{10}{\text{D}}\ \frac{9}{\text{D}}\ \frac{8}{\text{D}}$$

Since $26 \cdot 25 \cdot 10 \cdot 9 \cdot 8 = 468{,}000$, there are 468,000 different possible arrangements.

d) Since the first letter must be a vowel, it must be either an *a, e, i, o,* or *u.* Thus there are five possible choices for the first position. The second position can be filled by any of the letters except for the vowel selected for the first position. Therefore there are 25 possibilities for the second position.

   Since the first digit cannot be a 0, there are nine possibilities for position 3. The fourth position can be filled by any digit except the one selected for position three. Thus there are nine possibilities for position four. Since the last position cannot be filled by any of the two digits previously used, there are eight possibilities for the last position:

$$\frac{5}{\text{L}}\ \frac{25}{\text{L}}\ \frac{9}{\text{D}}\ \frac{9}{\text{D}}\ \frac{8}{\text{D}}$$

Since $5 \cdot 25 \cdot 9 \cdot 9 \cdot 8 = 81{,}000$, there are 81,000 different arrangements that meet the conditions specified. ❖

EXERCISES 11.4

1. The daily double at most race tracks consists of selecting the winning horse in both the first and second races. If the first race has seven entries and the second race has eight entries, how many daily double tickets must you purchase to guarantee a win?

2. The trifecta at most race tracks consists of selecting the first-, second-, and third-place finishers in a particular race in their proper order. If there are six entries in the trifecta race, how many tickets must you purchase to guarantee a win?

3. The operators of a stereo business are planning a grand opening. They wish to advertise that they have many different sound systems available. They stock 8 different turntables, 6 different cartridges, 10 different receivers, 12 different speakers, and 4 different tape decks. Assuming that a sound system will consist of one of each and that all pieces are compatible, how many different sound systems can they make?

4. In how many different ways can five children line up in a straight line?

5. In how many different ways can six books be arranged on a shelf?

6. If a club consists of eight members, how many different arrangements of president and vice-president are possible?

7. A $100, a $50, and a $20 prize are to be awarded to three different people. If seven people are being considered for the prizes, how many different arrangements are possible?

An entry code is to consist of four digits (0–9).

8. How many different entry codes are possible if repetition is permitted?

9. How many different entry codes are possible if repetition is not permitted?

10. How many different entry codes are possible if the first digit cannot be a 0 or 1?

11. How many different entry codes are possible if the first two entries must both be even and repetition is not permitted? (*Note:* 0 is considered to be even.)

12. A telephone number consists of seven digits with the restriction that the first digit cannot be 0 or 1.
    a) How many distinct telephone numbers are possible?
    b) Since there are 262 million telephones in the United States, it became necessary to increase the capacity of the numbering system. To accomplish this goal, area codes were developed. How many distinct telephone numbers are now possible with three-digit area codes, where the first digit of the area code is not 0 or 1?

A license plate is to consist of four digits followed by two letters. Determine the number of different license plates possible under the following conditions.

13. Repetition of numbers and letters is permitted.
14. Repetition of numbers and letters is not permitted.
15. The first and second digits must be odd, and repetition is not permitted.
16. The first digit cannot be zero, and repetition is not permitted.

17. Find the number of heart flushes possible when there are 13 hearts in a deck of cards. (A heart flush is any set of five hearts.)

In Exercises 18–27, use the counting principle to determine the number of sample points in the sample space. Use these results as an aid in constructing the tree diagram.

18. a) A couple plan to have exactly two children. Construct a tree diagram and list the sample space of the possible arrangements of boys and girls.
    b) Find the probability that the family has two girls.
    c) Find the probability that the family has at least one girl.

19. a) Construct a tree diagram and list the sample space when three coins are tossed.
    b) Find the probability that exactly three heads are tossed.
    c) Find the probability that exactly one head is tossed.

20. a) A couple plan to have exactly three children. Construct a tree diagram and list the sample space.
    b) Find the probability that the family has no boys.
    c) Find the probability that the family has at least one girl.

21. a) A hat contains 4 marbles, one yellow, one red, one blue, and one green. Two marbles are to be selected at random without replacement from the hat. Construct a tree diagram and list the sample space of all possible outcomes.
b) Find the probability of selecting at least one marble that is not red.
c) Find the probability of selecting no green marbles.

22. a) Two coins are tossed and a die is rolled. Construct a tree diagram and list the sample space.
b) Find the probability of tossing two heads and rolling the number 4.
c) Find the probability of rolling exactly one head.

23. a) A bag contains three chips: one red, one blue, and one green. Two chips are to be selected at random with replacement. Construct a tree diagram and list the sample space. (The term "with replacement" means that the first object is replaced before the second object is selected.)
b) Find the probability of selecting two red chips.
c) Find the probability of selecting a red chip followed by a blue chip.

24. Repeat Exercise 23 without replacement.

25. An individual can be classified as male or female with red, brown, black, or blonde hair and with brown, blue, or green eyes.
a) How many different classifications are possible?
b) Construct a tree diagram to determine the sample space (for example, red-headed, blue-eyed, and male).
c) If each outcome is equally likely, find the probability that the individual will be a male with black hair and blue eyes.
d) Find the probability that the individual will be a female with blonde hair.

26. A pea plant must have exactly one of each of the following pairs of traits: short (s) or tall (t); round (r) or wrinkled (w) seeds; yellow (y) or green (g) peas; and white (wh) or purple (p) flowers (for example, short, wrinkled, green pea with white flowers).
a) How many different classifications of pea plants are possible?
b) Use a tree diagram to determine all the classifications possible.
c) If each characteristic is equally likely, find the probability that the pea plant will have round peas.
d) Find the probability that the pea plant will be short, be wrinkled, have yellow seeds, and have purple flowers.

27. A coin is flipped, and then a die is rolled. Construct a tree diagram and list the sample space.

Use the sample space in Exercise 27 to find the probability of obtaining:

28. a head;
29. a 3;
30. a head and a 3;
31. a tail and an odd number;
32. a tail and a number greater than 4;
33. a tail or a 3.
34. If the die is rolled before the coin was flipped, would any of these probabilities change? Explain.

35. A pair of dice is rolled. Construct a tree diagram and list the sample space.

Use the sample space in Exercise 35 to find the probability that when a pair of dice is rolled:

36. Exactly one 2 is rolled.
37. Their sum is 2.
38. Their sum is over 10.
39. At least one 3 is rolled.
40. Exactly one 3 is rolled.
41. A double is rolled.
42. A sum of 7 or a sum of 11 is rolled.
43. An odd sum is rolled.
44. A sum of 2, 3, or 12 is rolled.
45. A double or a sum of 10 is rolled.

11.5 **"OR" AND "AND" PROBLEMS**

To be able to do the problems in this section and the remainder of the chapter, you must have a thorough understanding of addition, subtraction, multiplication, and division of fractions. If you have forgotten how to use fractions, we strongly suggest that you review Section 5.3 before beginning this section.

In Section 11.4 we showed how to work probability problems by constructing sample spaces. Often it is inconvenient or too time-consuming to solve a problem by first constructing a sample space. For example, if an experiment consists of selecting two cards without replacement from a deck of 52 cards, there would be 52 · 52 or 2704 points in the sample space. Trying to list all these sample points could take hours. In this section we will learn how to solve probability problems that contain the words "and" or "or" without the need for constructing a sample space. To solve such problems, it will be necessary for you to learn and understand two basic formulas, as will be explained shortly. Probability problems that contain the word "and" or "or" are often referred to as **compound probability problems.**

"Or" Problems

The "or" type of problem requires obtaining a "successful" outcome for *at least one* of the given events. We will develop a formula for finding the probability of event A or event B, symbolized P(A or B), by studying and analyzing Example 1.

■ Example 1 _____

Each of the numbers 1, 2, 3, 4, 5, 6, 7, 8, 9, and 10 is written on a separate piece of paper. The ten pieces of paper are then placed in a hat, and one piece is randomly selected. Find the probability that the piece of paper selected contains an even number or a number greater than 6.

Solution: There are ten possible outcomes. The numbers 2, 4, 6, 8, and 10 are even, so P(even number) = 5/10. The numbers 7, 8, 9, and 10 are greater than 6, so P(greater than 6) = 4/10. You may have a feeling from the examples given in the previous sections that to find P(A or B) you simply add P(A) + P(B) to obtain your answer. If we do this here, P(even number) + P(number greater than 6) = 5/10 + 4/10 = 9/10. However, by checking the sample space, we observe that only seven of the ten numbers, 1, ②, 3, ④, 5, ⑥, ⑦, ⑧, ⑨, ⑩ (those circled) satisfy the criteria. Thus the probability must be 7/10 rather than 9/10. Can you find the discrepancy? The discrepancy arises in that two numbers, 8 and 10, satisfy both events (Fig. 11.13).

When adding the probability of an even number to the probability of a number greater than 6, we counted those two numbers twice. To rectify the situation, we must now subtract the probability of selecting a number that satisfies both events from the sum of the probabilities. Since two of the ten numbers satisfy both events, P(selecting a number that is both even and greater than 6) is 2/10.

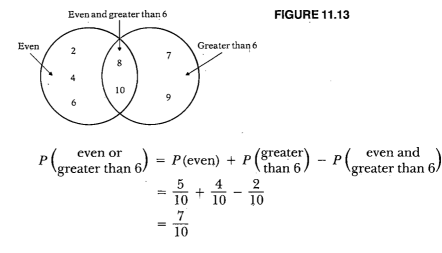

FIGURE 11.13

$$P\left(\begin{array}{c}\text{even or}\\\text{greater than 6}\end{array}\right) = P(\text{even}) + P\left(\begin{array}{c}\text{greater}\\\text{than 6}\end{array}\right) - P\left(\begin{array}{c}\text{even and}\\\text{greater than 6}\end{array}\right)$$

$$= \frac{5}{10} + \frac{4}{10} - \frac{2}{10}$$

$$= \frac{7}{10}$$

This answer, 7/10, checks with the answer we obtained previously.   ❖

From this example we can see that the probability of event A or event B can be calculated by the addition formula:

$$\boxed{P(A \text{ or } B) = P(A) + P(B) - P(A \text{ and } B).}$$

■ **Example 2** ────────────────────────────

Consider the same sample space as in the preceding problem. If one piece of paper is selected, find the probability that it contains a number less than 4 or greater than 6.

Solution

$$P(\text{selecting a number less than 4}) = \frac{3}{10}$$

$$P(\text{selecting a number greater than 6}) = \frac{4}{10}$$

Since there are no numbers that are both less than 4 and greater than 6, $P(\text{selecting a number less than 4 and greater than 6}) = 0$. Therefore

$$P\left(\begin{array}{c}\text{less than 4 or}\\\text{greater than 6}\end{array}\right) = P(\text{less than 4}) + P\left(\begin{array}{c}\text{greater}\\\text{than 6}\end{array}\right) - P\left(\begin{array}{c}\text{less than 4 and}\\\text{greater than 6}\end{array}\right)$$

$$= \frac{3}{10} + \frac{4}{10} - 0$$

$$= \frac{7}{10}$$ ❖

In Example 2 it is impossible to select a number that is both less than 4 and greater than 6 when only one number is to be selected. Events such as these are said to be mutually exclusive.

> Two events A and B are **mutually exclusive** if it is impossible for both events to occur simultaneously.

If events A and B are mutually exclusive, then $P(A \text{ and } B) = 0$, and the addition formula simplifies to $P(A \text{ or } B) = P(A) + P(B)$.

Example 3

One card is selected from a deck of cards. Determine whether the following pairs of events are mutually exclusive, and find $P(A \text{ or } B)$.

a) $A =$ an ace, $B =$ a king
b) $A =$ an ace, $B =$ a spade
c) $A =$ a red card, $B =$ a black card
d) $A =$ a picture card, $B =$ a red card

Solution

a) It is impossible to select both an ace and a king when only one card is selected. Therefore these events are mutually exclusive.

$$P(\text{ace or king}) = P(\text{ace}) + P(\text{king}) = \frac{4}{52} + \frac{4}{52} = \frac{8}{52} = \frac{2}{13}$$

b) The ace of spades is both an ace and a spade; therefore these events are not mutually exclusive.

$$P(\text{ace}) = \frac{4}{52}, \qquad P(\text{spade}) = \frac{13}{52}, \qquad P(\text{ace and spade}) = \frac{1}{52}$$

$$P(\text{ace or spade}) = P(\text{ace}) + P(\text{spade}) - P(\text{ace and spade})$$

$$= \frac{4}{52} + \frac{13}{52} - \frac{1}{52} = \frac{16}{52} = \frac{4}{13}$$

c) It is impossible to select both a red card and a black card when only one card is selected. Therefore the events are mutually exclusive.

$$P(\text{red or black}) = P(\text{red}) + P(\text{black}) = \frac{26}{52} + \frac{26}{52}$$

$$= \frac{52}{52}$$

$$= 1$$

(A red card or a black card must be selected.)·

d) There are six picture cards that are red: jack, queen, and king of hearts and jack, queen, and king of diamonds. Thus these events are not mutually exclusive.

$$P\left(\begin{array}{c}\text{picture card}\\\text{or red card}\end{array}\right) = P\left(\begin{array}{c}\text{picture}\\\text{card}\end{array}\right) + P\left(\begin{array}{c}\text{red}\\\text{card}\end{array}\right) - P\left(\begin{array}{c}\text{picture card}\\\text{and red card}\end{array}\right)$$

$$= \frac{12}{52} + \frac{26}{52} - \frac{6}{52}$$

$$= \frac{32}{52}$$

$$= \frac{8}{13} \qquad \qquad ❖$$

"And" Problems

A second type of problem is the "and" type of problem, which requires obtaining a favorable outcome in *each* of the given events.

Recall the counting principle given in Section 11.4: If a first task can be performed in M distinct ways and a second task can be performed in N distinct ways, then the two tasks in the specific order can be performed in $M \cdot N$ distinct ways. We can sometimes use this basic principle in determining the probability of event A and event B, symbolized by $P(A$ and $B)$.

For example, the probability of selecting two aces from a deck of cards is found by multiplying the probability of selecting the first ace by the probability of selecting the second ace. $P(\text{two aces}) = P(\text{first ace}) \cdot P(\text{second ace})$. This experiment will be successful only if *both* aces are selected.

The probability of event A and event B can be calculated by the formula

$$\boxed{P(A \text{ and } B) = P(A) \cdot P(B).}$$

Since this type of problem requires obtaining a favorable outcome in *both* of the given events, **we must assume that event A has occurred before calculating the probability of event B.**

■ **Example 4** _____

Two cards are to be selected with replacement from a deck of cards. Find the probability that two aces will be selected on only two draws.

Solution: Since the deck of 52 cards contains four aces, the probability of selecting an ace on the first draw is 4/52. The card selected is then returned to the deck. Therefore the probability of selecting an ace on the second draw remains 4/52.

$$
\begin{aligned}
P(2 \text{ aces}) &= P(\text{ace } 1) \cdot P(\text{ace } 2) \\
&= \frac{4}{52} \cdot \frac{4}{52} \\
&= \frac{1}{13} \cdot \frac{1}{13} \\
&= \frac{1}{169}
\end{aligned}
$$
❖

■ **Example 5** _____

Repeat Example 4 without replacement.

Solution: The probability of selecting an ace on the first draw is 4/52. When calculating the probability of selecting the second ace, we must assume that the first ace has been selected. Once this first ace has been selected, only 51 cards remain in the deck. Three of these cards are aces. The probability of selecting an ace on the second draw becomes 3/51. The probability of selecting two aces without replacement is

$$
\begin{aligned}
P(2 \text{ aces}) &= P(\text{ace } 1) \cdot P(\text{ace } 2) \\
&= \frac{4}{52} \cdot \frac{3}{51} \\
&= \frac{1}{13} \cdot \frac{1}{17} \\
&= \frac{1}{221}.
\end{aligned}
$$
❖

Event A and event B are **independent events** if the occurrence of either in no way effects the probability of occurrence of the other. Rolling dice and tossing coins are examples of independent events.

In Example 4 the probability of selecting an ace on the first draw was 4/52. The probability of selecting an ace on the second draw was also 4/52. Since the first card was returned to the deck, the probability of selecting an ace on the second draw was not affected by the first selection. Thus these events are independent. Are the events in Example 5 independent?

■ **Example 6** _____

A package of 25 zinnia seeds contains 8 seeds for red flowers, 12 seeds for white flowers, and 5 seeds for yellow flowers. Three seeds are randomly selected and planted. Find the probability of each of the following.

a) All three seeds will produce red flowers.

b) The first seed selected will produce a red flower, the second seed will produce a white flower, and the third seed will produce a red flower.

c) None of the seeds will produce red flowers.

d) If we consider events *A, B,* and *C* as the process of selecting and planting the three seeds, respectively, are the three events independent?

Solution: Each time a seed is selected and planted, the number of seeds remaining decreases by one.

a) The probability that the first seed selected produces a red flower is 8/25. If the first seed selected is red, only 7 red seeds in 24 are left. The probability of selecting a second red seed is 7/24. If the second seed selected is red, only 6 red seeds in 23 are left. The probability of selecting a third red seed is 6/23.

$$P(3 \text{ red seeds}) = P(\text{red seed } 1) \cdot P(\text{red seed } 2) \cdot P(\text{red seed } 3)$$
$$= \frac{8}{25} \cdot \frac{7}{24} \cdot \frac{6}{23}$$
$$= \frac{14}{575}$$

b) The probability that the first seed selected produces a red flower is 8/25. Once a seed for a red flower is selected, only 24 seeds are left. Twelve of the remaining 24 will produce white flowers. Thus the probability that the second seed selected will produce a white flower is 12/24. After the second seed had been selected, there are 23 seeds left, 7 of which will produce red flowers. The probability that the third seed produces a red flower is therefore 7/23.

$$P(\text{first red and second white and third red})$$
$$= P(\text{first red}) \cdot P(\text{second white}) \cdot P(\text{third red})$$
$$= \frac{8}{25} \cdot \frac{12}{24} \cdot \frac{7}{23} = \frac{28}{575}$$

c) If no flowers are to be red, they must either be white or yellow. There are 17 seeds that will not produce red flowers (12 for white and 5 for yellow). The probability that the first seed does not produce a red flower is 17/25. After the first seed has been selected, 16 of the remaining 24 seeds will not produce red flowers. After the second seed has been selected, 15 of the remaining 23 seeds will not produce red flowers.

$$P(\text{none red}) = P(\text{first not red}) \cdot P(\text{second not red}) \cdot P(\text{third not red})$$
$$= \frac{17}{25} \cdot \frac{16}{24} \cdot \frac{15}{23}$$
$$= \frac{34}{115}$$

d) The events are not independent, since the selection of the first seed has an effect on the probability of selecting the remaining seeds. The events are dependent. ❖

DID YOU KNOW?

There is an interesting and instructive probability question that has become known as the "birthday problem." Among 24 people chosen at random, what would you guess is the probability that at least 2 of them will have the same birthday? Contrary to intuition, the probability is slightly larger than 1/2 and rises rapidly if the number of people chosen is increased. The probability that at least 2 out of 40 randomly selected people will have the same birthday is 4/5. The proof illustrates a principle on which the solution of many probability problems is based: If P is the probability that an event will occur, then $1 - P$ is the probability that the event will not occur. Thus if the probability that an event will occur is 0.3, the probability that the event will not occur is $1 - 0.3 = 0.7$. The ways in which it is possible that no 2 of the 24 people share a birthday can be examined (for simplicity, anyone born on February 29 may call the birthday February 28). There are 365 days on which the first person selected can have a birthday, and the probability that that person's birthday is on any of those days is 365/365. The probability that the second person has a birthday on any other day, given specification of the first birthday, is 364/365. The probability that the third person has a birthday on a day other than the first two is 363/365, and so on. Thus the probability that of 24 people no 2 have the same birthday is $P = 365/365 \cdot 364/365 \cdot 363/365 \cdot \ldots \cdot 342/365 = 0.462$. This number is slightly less than 1/2. Therefore the probability that at least 2 people of the 24 chosen have the same birthday, $1 - P = 1 - 0.462 = 0.538$, is slightly larger than 1/2.

EXERCISES 11.5

1. An individual is selected at random. Let event A be that the individual is happy. Let event B be that the individual is healthy.

   a) Are events A and B mutually exclusive events? Explain.
   b) Are they independent events? Explain.

2. A family is selected at random. Let event A be that the father smokes. Let event B be that the mother smokes.
  a) Are events A and B mutually exclusive events? Explain.
  b) Are they independent events? Explain.
3. If events A and B are mutually exclusive events, why can the formula $P(A \text{ or } B) = P(A) + P(B) - P(A \text{ and } B)$ be simplified to $P(A \text{ or } B) = P(A) + P(B)$?

A single die is rolled. Find the probability of rolling each of the following.

4. A 2
5. A 2 or 3
6. An odd number
7. An odd number or a number greater than 2
8. A number greater than 5 or less than 3
9. A number greater than 3 or less than 5
10. An odd number or a number that is divisible by 3

A bag contains eight bills as follows: three $1 bills, two $5 bills, two $10 bills, and one $20 bill. One bill is randomly selected from the bag. Find the probability of selecting each of the following.

11. A bill of an even denomination
12. A $1 or a $5 bill
13. A bill greater than $5
14. A bill other than a $1 bill

One card is selected from a deck of cards. Find the probability of selecting each of the following.

15. An ace or a king
16. A king or a heart
17. A picture card or a black card
18. A club or a red card
19. A card less than 8 or a diamond (*Note:* The ace is considered a low card.)
20. A card greater than 9 or a black card.

Two cards are selected from a deck of cards. Find the probability of the following (a) with replacement, (b) without replacement.

21. They are both jacks.
22. They are both red.
23. They are both spades.
24. The first is a king and the second is a queen.
25. The first is a heart and the second is a black card.
26. They are both picture cards.
27. Neither is a picture card.

28. The first is a picture card and the second is not a picture card.

29. The Haefners plan to have five children. Find the probability that all their children will be boys. (Assume that $P(\text{boy}) = \frac{1}{2}$, and assume independence.)
30. The McDonalds presently have five boys. Mrs. McDonald is expecting another child. Find the probability that it will be a boy.

A family has three children. Assuming independence, find the probability of the following.

31. All three are girls.
32. All three are boys.
33. The youngest child is a boy, and the older children are girls.

A box contains four marbles: two red, one blue, and one green. Two marbles will be selected at random. Find the probability of selecting each of the following (a) with replacement, (b) without replacement.

34. A red marble and then a blue marble
35. Two blue marbles
36. No red marbles
37. No green marbles

Each individual letter of the word "pfeffernuesse" is placed on a piece of paper, and all 13 pieces of paper are placed in a hat. Three letters are selected at random from the hat. Find the probability of selecting each of the following (a) with replacement, (b) without replacement.

38. three fs
39. no fs
40. three vowels
41. The first letter selected is an f, the second letter is an e, and the third letter is an e.
42. The first letter selected is not a vowel and the second and third letters selected are vowels.

A sample of 150 people indicates that 50 favor Candidate A, 65 favor Candidate B, 20 favor Candidate C, and 15 have no opinion. Three *different* people from the sample are randomly selected. Find the probability of each of the following.

43. The first favors Candidate A, the second favors Candidate B, and the third favors Candidate C.
44. The first two have no opinion, and the third favors Candidate C.
45. They all favor Candidate A.
46. None favor Candidate C.

Refer to the list of astronauts in Table 11.4 on page 571. If the names of three astronauts are selected at random without replacement, what is the probability of each of the following?

47. The three selected were on the first moon-walk mission.
48. None of those selected was on the first moon-walk mission.
49. They all have last names beginning with the letter C.
50. They all have last names beginning with the letter B.

An experimental drug was given to a sample of 100 hospital patients with an unknown sickness. Of the total, 70 patients reacted favorably, 10 reacted unfavorably, and 20 were unaffected by the drug. Assume that this sample is representative of the entire population. If this drug is given to Mr. and Mrs. Jones and their son Mickey, what is the probability of each of the following? (Assume independence.)

51. Mrs. Jones reacts favorably.
52. Mr. and Mrs. Jones react favorably, and Mickey is unaffected.
53. They all react favorably.
54. None reacts favorably.

A pair of dice is made so that each die contains one point on one side, two points on two sides, and three points on three sides. When these dice are rolled, what is the probability of rolling each of the following?

55. A pair of 1s   **56.** A pair of 2s
57. A pair of 3s   **58.** A pair
59. A sum of 3

Each question of a five-question multiple-choice exam has four possible answers. Peter picks an answer at random for each question. Find the probability that he selects the correct answer on

60. any one given question;
61. only the first question;
62. only the third and fourth questions;
63. all five questions;
64. none of the questions.

Eduardo and Eva purchase tickets in advance to go on a bus tour up Pikes Peak. When they arrive to board the bus, the driver informs the ticket holders that the tour was oversold. There were 22 tickets sold, and only 20 seats are available. The bus driver informs the ticket holders that two ticket stubs of the 22 that were collected will be selected at random, and those two ticket holders will have their money refunded. Find the probability that

65. Eduardo is the first selected and Eva is the second selected;
66. neither Eduardo nor Eva is selected.

67. Certain birth defects and syndromes are polygenetic in nature. In the typical polygenetic affliction, the chance that an offspring will be born with one of these afflictions is small. Once an offspring is born with the affliction, the probability that future offspring of the same parents will be born with the same affliction increases. Let us assume that the probability of a child's being born with affliction A is 0.001. If a child is born with this affliction, the probability of a future child's being born with the same affliction becomes 0.04.
   a) Are the events of the births of two children in the same family with affliction A independent?
   b) A couple plans to have one child. Find the probability that the child will be born with this affliction.
   c) A couple plans to have two children. Find the probability that (1) both children will be born with the affliction; (2) the first has the affliction and the second does not; (3) neither has the affliction.

The probability that a heat-seeking torpedo will hit its target is 0.4. If the first torpedo hits its target, the probability that the second torpedo will hit the target increases to 0.9 because of the extra heat generated by the first explosion. If two heat-seeking torpedos are fired at a target, find the probability that

68. neither hits the target;
69. the first hits the target and the second misses the target;
70. both hit the target;
71. the first misses the target and the second hits the target.

The Internal Revenue Service claims that 24 in every 1000 people in the $10,000–$50,000 income bracket are audited yearly. Assuming that the returns to be audited are selected at random and that each year's selections are independent of the previous year's selections, find the probability that a person in this income bracket will be audited

72. this year;
73. next year;
74. the next two years in succession;
75. this year but not next year;
76. neither this year nor next year.

77. Ms. Runningdeer has a lottery ticket with a three-digit number in the range 000 to 999. Three balls are to be selected at random with replacement from a bin. There are an equal number of balls marked with the digits 0, 1, 2, . . ., 9. Find the probability that Ms. Runningdeer's number is selected.

PROBLEM SOLVING

78. A bag contains five red chips, three blue chips, and two yellow chips. Two chips are selected from the bag without replacement. Find the probability that two chips of the same color are selected.

79. Bob has ten coins from Japan: three 1-yen coins, one 10-yen coin, two 20-yen coins, one 50-yen coin, and three 100-yen coins. Two coins are selected at random without replacement. Assuming that each coin is equally likely to be selected, find the probability that Bob selects at least one 1-yen coin.

80. An investment advisor is considering the purchase of stock from five drug companies and ten computer companies. Knowing that each company had the same record of gain over the last year, the investment advisor decides to select three stocks at random from the fifteen stocks. What is the probability that two stocks are computer companies and one is a drug company?

81. Two playing cards are dealt to you from a well-shuffled deck of 52 cards. If either card is a diamond, or both are, you win; otherwise, you lose. Determine whether this game favors you, is fair, or favors the dealer. Explain your answer.

82. You have three cards: an ace, a king, and a queen. A friend shuffles the cards, selects two of them at random, and discards the third. You ask your friend to show you a picture card, and she turns over the king. What is the probability that she also has the queen?

 ## ODDS

The odds against catching the flu are 30 to 1; the odds in favor of winning the lottery are 1 to 78,000; the odds against being audited by the IRS this year are 50 to 1. We see the word "odds" daily in newspapers and magazines and often use it ourselves. Yet there is a widespread misunderstanding of its meaning. This section will explain the differences and relationships of "odds against," "odds in favor," and "probability."

The odds at horse races, the odds given at craps, and the odds at all gambling games in Las Vegas and other casinos throughout the world are always odds against unless they are otherwise specified. The odds *against* an event is a ratio of the probability that the event will fail to occur (failure) to the probability that the event will occur (success).

$$\text{Odds against event} = \frac{P\,(\text{event fails to occur})}{P\,(\text{event occurs})} = \frac{P\,(\text{failure})}{P\,(\text{success})}$$

In order to find odds against or odds in favor, you must first determine the probability of success and the probability of failure.

Photo by Marshall Henrichs

■ **Example 1** _____

Find the odds against rolling a 3 on one roll of a die.

Solution: When a die is rolled, the possible outcomes are 1, 2, 3, 4, 5, 6. The probability of rolling a 3 is 1/6. The probability of failing to roll a 3 is $1 - 1/6$, or 5/6.

$$\text{Odds against rolling a 3} = \frac{P(\text{fails to roll a 3})}{P(\text{rolls a 3})}$$

$$= \frac{5/6}{1/6} = \frac{5}{6} \cdot \frac{6}{1} = \frac{5}{1}$$

The ratio 5/1 is commonly written as 5:1 and reads "5 to 1." Thus the odds against rolling a 3 are 5 to 1.   ❖

In Example 1, for each dollar bet in favor of the rolling of a 3, five dollars should be bet against the rolling of a 3 if it is to be a "fair game." Consider the possible outcomes of the die—1, 2, 3, 4, 5, 6. Over the long run, one out of every six rolls will result in a 3, and five out of every six rolls will result in a number other than 3. The person betting in favor of the rolling of a 3 will either lose one dollar (if a number other than a 3 is rolled) or win five dollars (if a 3 is rolled). The person betting against the rolling of a 3 will either win one dollar (if a number other than a 3 is rolled) or lose five dollars (if a 3 is rolled). If this game is played for a long enough period, each player will theoretically break even.

■ **Example 2** _____

It is estimated that three out of every ten new businesses in a specific location go bankrupt within their first year of operation. A new business just opened next door. What are the odds against its going bankrupt?

Solution: The probability that the business will go bankrupt is 3/10. Therefore the probability that the business will not go bankrupt is $1 - 3/10$, or 7/10.

$$\text{Odds against the business's going bankrupt} = \frac{P\left(\begin{array}{c}\text{the business fails}\\\text{to go bankrupt}\end{array}\right)}{P\left(\begin{array}{c}\text{the business goes}\\\text{bankrupt}\end{array}\right)}$$

$$= \frac{7/10}{3/10}$$

$$= \frac{7}{10} \cdot \frac{10}{3}$$

$$= \frac{7}{3} \quad \text{or} \quad 7:3 \qquad ❖$$

Note: The denominators of the probabilities in an odds problem will always divide out.

WHAT WERE ODDS ON THIS ONE?

Never underestimate the ability of the American motorist as a newsmaker. He is a singular operator and totally unpredictable.

One only has to go back into the records for proof of this, and this is from the records:

In the year 1905, only two automobiles were registered in the entire state of Missouri.

In the year 1905, Missouri's only two registered automobiles were involved in a head-on collision!

Source: The *Traveler*, Automobile Club of Rochester, N.Y., Inc.

The odds *in favor of* an event are expressed as a ratio of the probability that the event will occur to the probability that the event will fail to occur.

$$\boxed{\text{Odds in favor of event} = \frac{P(\text{event occurs})}{P(\text{event fails to occur})} = \frac{P(\text{success})}{P(\text{failure})}}$$

If the odds *against* an event are $a:b$, then the odds *in favor of* the event will be $b:a$.

■ **Example 3** _____

Two percent of the U.S. population is born gifted. Find

a) the odds against an individual's being born gifted;

b) the odds in favor of an individual's being born gifted.

Solution

a) The probability of being born gifted is 0.02 or 2/100. The probability of not being born gifted is therefore $1 - 2/100 = 98/100$.

$$\text{Odds against being born gifted} = \frac{P(\text{not born gifted})}{P(\text{born gifted})}$$

$$= \frac{\dfrac{98}{100}}{\dfrac{2}{100}}$$

$$= \frac{\overset{49}{\cancel{98}}}{\underset{1}{\cancel{100}}} \cdot \frac{\overset{1}{\cancel{100}}}{\underset{1}{\cancel{2}}}$$

$$= \frac{49}{1} \quad \text{or} \quad 49:1$$

b) The odds in favor of being born gifted are 1:49.   ❖

DID YOU KNOW?

You chances of getting hit by lightning are better than winning the lottery, but if you want to take a chance here are the odds:

Picking six of 40	3.8 million to 1
Picking six of 44	7.06 million to 1
Picking six of 49	14 million to 1
Picking seven of 40	18.6 million to 1
Being struck by lightning this year	0.7 million to 1
	(1 in 701,537 people)

■ **Example 4** _____

Odds given at racetracks are always odds against. At a local racetrack the odds against the horse Sloppy Joe's winning are 7:2. George purchases a $2 win ticket on Sloppy Joe, and Sloppy Joe wins.

a) What will George's net (actual) winnings be?

b) How much will be returned to him at the ticket window?

Solution

a) As explained after Example 1, the 7:2 odds against mean that George will win $7 for each $2 he bet in favor of Sloppy Joe's winning. Thus for a $2 bet George's net winnings is $7.

b) At the ticket window he will receive his original bet of $2 plus his net winnings of $7, a total of $9. (In actuality he may receive more but never less than the amount specified on the tote board, where the track gives only rounded odds.) ❖

In Example 4 if George had bet $10 on Sloppy Joe to win, what would have been his net winnings?

When odds are given either in favor or against a particular event, it is possible to determine the probabilities of that event.

■ **Example 5** _____

The odds against an individual's receiving a promotion in any given year are 17:2. Find the probability that

a) Mr. Jones, an individual selected at random, receives a promotion this year.

b) Mr. Jones does not receive a promotion this year.

Solution

a) We have been given odds against and been asked to find probabilities. The denominators of the probabilities are found by adding the numbers in the odds statement. In this example the denominators must be 17 + 2, or 19. The numerators of the probabilities are the numbers given in the odds statements.

$$\text{Odds against being promoted} = \frac{P(\text{fails to be promoted})}{P(\text{promoted})}$$

Thus the ratio of failure to success must be

$$\frac{17/19}{2/19}$$

The probability of being promoted is 2/19.

b) The probability of not being promoted is 17/19. ❖

EXERCISES 11.6

A die is tossed. Find the odds against rolling

1. a 4
2. an even number
3. a number less than 3
4. a number greater than 4

A card is picked from a deck of cards. Find the odds against and the odds in favor of selecting

5. a 3
6. a picture card
7. a club
8. a black card
9. the ace of spades
10. a 3 or a 4

11. A million tickets are sold for a lottery. If you purchase one ticket, what are your odds (a) against winning, (b) in favor of winning?
12. One person is selected at random from a class of 18 males and 11 females. Find the odds against selecting (a) a female, (b) a male.
13. The odds against Man of Peace winning the fifth race are 5:2.
    a) Find the probability that Man of Peace wins.
    b) Find the probability that Man of Peace loses.
14. The odds against Mashed Potatoes winning the sixth race are 9:2. John places a $2 bet on Mashed Potatoes to win, and Mashed Potatoes wins. Find the following.
    a) John's net winnings
    b) The amount John receives at the winning ticket window
15. A recent newspaper article claims that the probability of an individual's catching type A flu is 1/6.
    a) Find the odds against an individual's catching type A flu.
    b) Find the odds in favor of an individual's catching type A flu.

The results of test scores for a class of 28 students are five As, six Bs, thirteen Cs, three Ds, and one F. If one student is selected at random, find

16. the odds in favor of the student receiving a grade of C
17. the odds in favor of the student passing the test
18. the odds against a student receiving a grade of C or higher
19. the odds against a student receiving a grade of A
20. the odds against a student receiving a grade of F.

21. The odds in favor of an event are 8:3. Find the probability that the event does not occur.
22. The odds in favor of Tasha winning a racquetball tournament are 1:8. Find the probability that Tasha will win the tournament.
23. The odds against Kathy getting a teaching position next fall are 2:9. Find the probability that Kathy gets a teaching position next fall.
24. Gout constitutes about 5 percent of all systemic arthritis, and it is uncommon in women. The male to female ratio of gout is estimated as 20 to 1.
    a) If J. Douglas has gout, what are the odds against J. Douglas's being female?
    b) If J. Douglas has gout, what is the probability that J. Douglas is a male?
25. One in 40 individuals in the $10,000–$40,000 tax range will be randomly selected to have their income tax returns audited for this year. Mr. Frank is in this income tax range. Find
    a) the probability that Mr. Frank is audited;
    b) the odds against Mr. Frank's being audited.

 EXPECTED VALUE (EXPECTATION)

Expected value is often used to determine the expected results of an experiment or business venture *over the long run*. Expectation is used to make important decisions in many different areas. In business, for example, expectation is used to predict future profits of a new product. In the insurance industry, expectation is used to determine how much each insurance policy should cost for the company to make an overall yearly profit. When a toll bridge is constructed, expectation is used to determine what amounts should be charged to make a given dollar profit. Expectation is also used to determine the expected results in games of chance such as the lottery, roulette, craps, and slot machines. Following is an illustration of how expectation is used.

Tim and Barbara are trying an experiment. Tim tells Barbara he will give her $1 if she can roll an even number on a single die. If she fails to roll an even number, she must give Tim $1. Who would win money in the long run if this game were played many times? The probability that an odd number will occur is 1/2; the probability that an even number will occur is also 1/2. Let us see how Tim would do. We would expect in the long run that half the time he would win $1 and half the time he would lose $1, therefore breaking even. Mathematically, we could find Tim's expected gain or loss by the following procedure:

$$\text{Tim's expected gain or loss} = P\binom{\text{Tim}}{\text{wins}}\binom{\text{amount}}{\text{Tim wins}} + P\binom{\text{Tim}}{\text{loses}}\binom{\text{amount}}{\text{Tim loses}}$$
$$= \tfrac{1}{2}(\$1) + \tfrac{1}{2}(-\$1) = \$0$$

Note that the loss is written as a negative number. This procedure indicates that Tim has an expected gain or loss (or expected value) of $0. The expected value of zero indicates that he would indeed break even, as we had anticipated. If his expected value was positive, it would indicate a gain; if negative, a loss.

The **expected value** can be used to determine the expected gain or loss of an experiment or business venture *over the long run*. The expected value, E, is calculated by multiplying the probability of an event occurring by the **net** amount of money that will be gained or lost if the event occurs. If there are a number of different events and amounts to be considered, we use the following formula.

Expected Value

$$E = P_1A_1 + P_2A_2 + P_3A_3 + \cdots + P_nA_n$$

In the formula the numbers on the bottom of the letters are called *subscripts*. The symbol P_1 represents the probability that the first event will occur; A_1 represents the net amount won or lost if the first event occurs; P_2 is the probability of the second event; A_2 is the net amount won or lost if the second event occurs; and so on. The sum of these products of the probabilities and their respective amounts is the expected value. The expected value is the average (mean) result that would be obtained if the experiment was performed a great many times.

■ **Example 1** ————————————————————————

Smitty's Construction Company, after considerable research, is planning to bid on a building contract. From past experience, Smitty estimates that if the company's bid is accepted, there is a 50 percent chance of making a $500,000 profit, a 10 percent chance of breaking even, and a 40 percent chance of losing $200,000, depending on weather conditions, inflation, and a possible strike. How much can Smitty "expect" to make or lose on this contract if his bid is accepted?

Solution: There are three amounts to be considered: a gain of $500,000, breaking even at $0, and a loss of $200,000. The probability of gaining $500,000 is 0.5, the probability of breaking even is 0.1, and the probability of losing $200,000 is 0.4.

$$
\text{Smitty's expectation} = \overbrace{P_1A_1}^{\text{gain}} + \overbrace{P_2A_2}^{\substack{\text{break}\\\text{even}}} + \overbrace{P_3A_3}^{\text{loss}}
$$
$$
= (0.5)(\$500,000) + (0.1)(\$0) + (0.4)(-\$200,000)
$$
$$
= \$250,000 + \$0 - \$80,000
$$
$$
= \$170,000
$$

In the long run, Smitty would have an average gain of $170,000 on each bid of this type. However, there is still a 40 percent chance that he would lose $200,000 on this *particular* bid.   ❖

■ **Example 2** ————————————————————————

Malcolm rolls a die. If he rolls a 3, Charles will give him $4. If Malcolm does not roll a 3, he must give Charles $1. Find Malcolm's expectations.

Solution: There are two possible outcomes. Malcolm can either win $4 or lose $1. The probability of rolling a 3 is 1/6. The probability of not rolling a three is 5/6.

$$
E_{\text{Malcolm}} = P(\text{Malcolm wins})(\text{amount wins}) + P(\text{Malcolm loses})(\text{amount loses})
$$
$$
= \frac{1}{6}(\$4) + \frac{5}{6}(-\$1)
$$
$$
= \frac{4}{6} - \frac{5}{6}
$$
$$
= -\$\frac{1}{6} \quad \text{or} \quad -16.7\cancel{c}
$$

Malcolm can expect to lose and Charles can expect to win about 17¢ each time Malcolm rolls the die.   ❖

The amounts in the expectation formula are **net** amounts. The net amounts are the actual amounts won or lost. Consider the following example, which illustrates the use of the net amounts.

■ Example 3

One hundred lottery tickets are sold for $2 each. One prize of $50 will be awarded. John purchases one ticket. Find his expectation.

Solution: There are two amounts to be considered: the net amount John may win and the cost of the ticket. If John wins, his net or actual winnings are $48 (the $50 awarded him minus his $2 cost of the ticket). If he does not win, he loses the $2 paid for the ticket. Since only one prize will be awarded, John's probability of winning is 1/100. His probability of losing is therefore 99/100.

$$E = P(\text{John wins})(\text{amount wins}) + P(\text{John loses})(\text{amount loses})$$
$$E = \frac{1}{100}(48) + \frac{99}{100}(-2)$$
$$= \frac{48}{100} - \frac{198}{100}$$
$$= \frac{-150}{100}$$
$$= -\$1.50$$

John's expectation is $-\$1.50$ per ticket. ❖

In the last example, John's expectation is $-\$1.50$ when purchasing a $2 ticket. The **fair price** of the ticket is the amount that should be charged for the ticket if the expectation is to be 0. Since John can expect to lose $1.50 per ticket, the price of the ticket should be reduced by $1.50 if he is to break even over the long run. The fair price of the ticket is $2.00 − 1.50, or $0.50.

■ Example 4

Find John's expectation in Example 3 if the ticket costs 50¢.

Solution

$$E = P(\text{John wins})(\text{amount wins}) + P(\text{John loses})(\text{amount loses})$$
$$E = \frac{1}{100}(49.50) + \frac{99}{100}(-0.50)$$
$$= \frac{49.50}{100} - \frac{49.50}{100} = 0$$

This example illustrates that 50¢ is the fair price for a ticket. ❖

■ **Example 5** _____

One thousand lottery tickets are sold for \$1 each. One grand prize of \$500 and two consolation prizes of \$100 will be awarded. Find

a) Carol's expectation if she purchases one ticket;

b) Carol's expectation if she purchases five tickets.

c) What is the fair price of a ticket?

Solution

a) There are three amounts to be considered in this problem: the net gain in winning the grand prize, the net gain in winning the consolation prize, and the loss of the cost of the ticket. If Carol wins the grand prize, her net gain is \$499 (\$500 minus \$1 spent for the ticket). If Carol wins the consolation prize, her net gain is \$99 (\$100 minus \$1 spent for the ticket). We will assume that the winners' names are replaced in the pool after being selected. The probability that Carol wins the grand prize is 1/1000. Since two consolation prizes will be awarded, the probability that she wins a consolation prize is 2/1000. The probability that she does not win either prize is $1 - 3/1000 = 997/1000$.

$$E = P_1A_1 + P_2A_2 + P_3A_3$$
$$E = \frac{1}{1000}(\$499) + \frac{2}{1000}(\$99) + \frac{997}{1000}(-\$1)$$
$$= \frac{499}{1000} + \frac{198}{1000} - \frac{997}{1000}$$
$$= \frac{-300}{1000}$$
$$= -\$0.30$$
$$= -30¢$$

b) Carol loses 30¢ on each ticket purchased. On five tickets her expectation is $(-\$0.30)(5)$, or $-\$1.50$.

c) We learned in part (a) that Carol loses 30¢ per ticket and the vendor makes 30¢ per ticket. To break even, the vendor has to charge 30¢ less per ticket. The fair price of the ticket is $\$1 - 30¢$, or 70¢.   ❖

Expectation can also be used in problems that do not involve an exchange of money. If an experiment yielding numerical data is performed many times, the numerical average of those trials can also be considered as an expectation. In problems of this type, the *A*s in the expectation formula will represent the numerical amounts observed.

■ **Example 6** _____

A highway crew repairs 30 potholes a day in dry weather and 12 potholes a day in wet weather. If the weather in this region is wet 40 percent of the time, find the expected (average) number of potholes that can be repaired per day.

Solution: Since the weather is wet 40 percent of the time, it will be dry 60 percent (100% − 40%) of the time.

$$E = P(\text{dry})(\text{amount repaired}) + P(\text{wet})(\text{amount repaired})$$
$$= 0.6(30) + 0.4(12) = 18.0 + 4.8 = 22.8$$

Thus the average or expected number of potholes repaired per day is 22.8. ❖

DID YOU KNOW?

Carrying two brown satchels, one filled with $777,000 in $100 bills and the other empty, an unidentified man, dressed in jeans and cowboy boots, walked into Binion's Horseshoe Casino in Las Vegas last week. He exchanged his money for $500 chips, strode to the craps table and put all of the chips on the back line, which meant that he was betting against the woman who happened to be rolling the dice. She first threw a six, then a nine, and finally a seven. Said the dealer: "Pay the back line." The man scooped up his chips, traded them at the casino cage for $1,554,000 in cash and shook hands with Jack Binion, the stunned president of the casino. Said Binion: "It was the biggest bet in a gambling house that I have ever heard of." As the man walked out of the casino with his two brown satchels, both now stuffed with $100 bills, and climbed into his car, he told Binion: "You know, this damned inflation was just eroding this money. I figured I might as well double it or lose it." With that, he drove off into the night, still unidentified.

Source: ©1980 Time Inc. All rights reserved. Reprinted by permission from TIME.

EXERCISES 11.7

1. An investment club is considering purchasing a given stock option. After considerable research the club members determine that there is a 40 percent chance of making $5000, a 10 percent chance

of breaking even, and a 50 percent chance of losing $3500. Find the expectation of this purchase.

2. An investment counselor is advising her client on a particular investment. She estimates that if the tax

law does not change the client will make $60,000, but if the tax law changes the client will lose $15,000. Find the client's expected value if there is a 60% chance that the tax law will change.

3. In July in Rochester, NY the grass grows 1/2 inch a day on a sunny day and 1/4 inch a day on a cloudy day. In Rochester, NY in July 75% of the days are sunny and 25% of the days are cloudy.
   a) Find the expected amount of grass growth on a typical day in July in Rochester, NY.
   b) Find the expected total grass growth in the month of July in Rochester, NY.

4. Bob and Larry play the following game: Larry picks a card from a deck of cards. If he selects a club, Bob gives him $10. If not, he gives Bob $4.
   a) Find Larry's expectation.
   b) Find Bob's expectation.

5. A cookie jar contains six $1 bills, three $5 bills, and one $10 bill. Aaron will select and keep one bill from the jar.
   a) Find Aaron's expected gain.
   b) What is a fair price for Aaron to pay to play this game?

6. A multiple-choice exam has five possible answers for each question. For each correct answer you are awarded 2 points. For each incorrect answer, 1/2 point is subtracted from your score. For answers left blank, no points are added or subtracted.
   a) If you do not know the correct answer to a particular question, is it to your advantage to guess?
   b) If you do not know the correct answer but can eliminate one possible choice, is it to your advantage to guess?

7. One thousand raffle tickets are sold for $1 each. One prize of $500 is to be awarded.
   a) Pete purchases one ticket. Find his expected value.
   b) Find the fair price of a ticket.
   c) If the vendor sells all 1000 tickets, how much profit will he make?

8. Ten thousand raffle tickets are sold for $5 each. Four prizes will be awarded—one for $10,000, one for $5,000, and two for $1,000. Sidhardt purchases one of these tickets.
   a) Find his expected value.
   b) Find the fair price of a ticket.

9. One million tickets are sold for 50¢ each for a weekly New York State lottery. One winner will be selected and awarded $50,000.
   a) Find the expected value of a person who purchases one ticket.

   b) How much will New York State make weekly on this lottery?
   c) Find the fair price of a ticket.

10. One million tickets are sold for a New York State lottery. Each ticket sells for $1. One grand prize of $100,000, three second prizes of $10,000, and ten third prizes of $1000 will be awarded.
    a) Find the expectation of a person who purchases one ticket.
    b) Find the fair price of a ticket.

11. Carnivals and bazaars often have a game known as "under and over." The player selects a sum under 7, a sum of 7, or a sum over 7. The dice are rolled. If the player selected the proper sum, he receives 1:1 odds on under 7, 1:1 odds on over 7, and 4:1 odds on the 7. (*Hint:* Refer to sample space on p. 580.)
    a) What is the expectation of a player who bets $1 on over 7?
    b) What is the expectation of a player who bets $1 on under 7?
    c) What is the expectation of a player who bets $1 on the 7?
    d) Which of these games has the greatest expectation?

12. According to a mortality table, the probability that a 20-year-old woman will survive one year is 0.994, and the probability that she will die within one year is 0.006. If she buys a $10,000 one-year term policy for $100, what is the company's expected gain or loss?

13. A championship tennis match is to be played in one of two locations: Flushing, New York, or Dallas, Texas. Weather permitting, the match will be played outdoors. If the weather is inclement, the match will be played indoors. The weather in Flushing is inclement 30 percent of the time in the month when the match is to be played. In the same month it is inclement 10 percent of the time in Dallas. Flushing's outdoor courts have a seating capacity of 2000; its indoor courts have a seating capacity of 600. Dallas's outdoor courts can seat 1500; its indoor courts can seat 800.
    a) What is the expected number of seats available in Flushing? In Dallas?
    b) If all seats sell for $6 and the match is sold out, what is the expected revenue if the match is played in Flushing? In Dallas?

14. A die is rolled many times, and the points facing up are recorded. Find the expected (average) number of points facing up over the long run.

15. American Airlines is planning its staffing needs

for next year. On January 1 the Civil Aeronautics Board will inform American Airlines whether they will be granted the new routes they have requested. If the new routes are approved, they will hire 850 new employees. If the new routes are not granted, they will hire only 150 new employees. If the probability that the Board will grant American Airlines' request is 0.25, what is the expected number of new employees to be hired by American Airlines?

16. Manufacturer *XYZ* is negotiating a contract with its employees. The probability that the union will go on strike is 5 percent. If the union goes on strike, the company estimates that it will lose $345,000 for the year. If the union does not go on strike, the company estimates that it will make $1,200,000. Find the company's expected gain or loss for the year.

17. On a clear day in Reading the AAA makes an average of 125 service calls for motorist assistance; on a rainy day it makes an average of 190 service calls; and on a snowy day it makes an average of 280 service calls. If the weather in Reading is clear 200

days of the year, rainy 100 days of the year, and snowy 65 days of the year, find the expected number of service calls made by the AAA in a given day.

18. During those years when financial responsibilities are greatest, such as when children are in school or there is a mortgage on the house, a person may choose to buy a term life insurance policy. The insurance company will pay the face value of the policy if the insured person dies during the term of the policy. For how much should an insurance company sell a 10-year term policy with a face value of $40,000 to a 30-year-old male in order for the company to make a profit? The probability of a 30-year-old man living to age 40 is 0.97.

*19. A football team has scored the number of points given in the table during the last 12 games. What is the expected number of points per game?

Number of points	10	16	21	28	44
Number of games	3	2	4	2	1

PROBLEM SOLVING

20. One of the more popular sweepstakes is the *Reader's Digest* sweepstakes. The prizes and the approximate probability of winning the prize listed in the sweepstakes package are:

$10 — one in 824
$100 — one in 21,446
$500 — one in 160,852
$5,000 — one in 8,042,600
$10,000 — one in 8,042,600
$25,000 — one in 16,085,000
$250,000 — one in 23,423,000

Find your expectation if you must use a 25¢ stamp to enter the sweepstakes.

21. The dealer shuffles five black cards and five red cards and spreads them out on the table. You choose two at random. If both cards are red or both cards are black, you win a dollar. Otherwise, you lose a dollar. Determine whether the game favors you, is fair, or favors the dealer. Explain your answer.

22. A dealer has three cards. One is red on both sides, another is black on both sides, and the third is red on one side and black on the other side. The cards are shuffled in a hat, and one is drawn at random and placed flat on the table. The side showing is red. If the other side is red, you win a dollar. If the other side is black, you lose a dollar. Determine whether the game favors the player, is fair, or favors the dealer.

The Way to Play Roulette

One of the most entertaining and exciting of all casino games is roulette, with its fascinating little white ball spinning hypnotically until it finds a resting place in a winning slot (see chapter opening photo). Roulette offers a variety of bets—straight-up bets, in which you bet on a single number . . . Red or Black . . . Odd or Even . . . and combination bets, in which you split your bets on a combination of adjoining numbers.

A number of typical bets are indicated by means of white-lettered spots on the adjoining Roulette Layout. The winning odds for each letter are listed below. These bets can be made on any corresponding combinations of numbers. For example: Bet "F"—the same placement on number "28" would pay off on 28, 29, and 30. The same principle applies to all other combination bets. On the roulette wheel the odd numbers are colored red, the even numbers are colored black, and the 0 and 00 are colored green. The 0 and 00 are considered neither odd nor even numbers.

Straight Bets

A—35 to 1	STRAIGHT-UP—All numbers, 0 and 00.
B—2 to 1	COLUMN BET—Pays off on any number in that horizontal column.
C—2 to 1	1st DOZEN—Pays off on any number 1 through 12. Same for second and third dozen.
D—Even money	ODD (or EVEN) 1–18 (or 19–36) RED (or BLACK)

Combination Bets

E—17 to 1	SPLIT—Pays off on 11 and 12.
F—11 to 1	Pays off on 28, 29, and 30.
G—8 to 1	CORNER—Pays off on 17, 18, 20, and 21.
H—6 to 1	Pays off on 0, 00, 1, 2, and 3.
I—5 to 1	Pays off on 22, 23, 24, 25, 26, and 27.

23. Round and round it goes, and where it stops nobody knows! When the roulette wheel is spun, the ball can drop into any one of 38 slots. The 38 slots are marked with the numbers 1 through 36 and the numbers 0 and 00. Eighteen slots are colored red, eighteen are black, and the 0 and 00 are green.

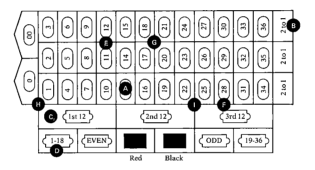

a) From the information given above, compute the probability of winning the following bets.
    1) Straight bet (selecting the correct number, including 0 and 00)
    2) Column bet (0 and 00 not considered to be in any of the three columns)
    3) First dozen
    4) Even-money bets (0 and 00 do not satisfy any of these events)
       a. 1–18 (or 19–36)
       b. even number (or odd number)
       c. red (or black)

    5) Split (see above)
    6) Bet F
    7) Corner bet
    8) Bet H
    9) Bet I
b) Compute your expectation at the odds quoted above for each of the games listed in part (a). Assume that you bet $1 to win.
c) In which games do you have (1) the highest expectation, (2) the lowest expectation?
d) The odds quoted for each game are not the "true" odds. Find the true odds for each game.

11.8 COUNTING PROBLEMS

The **counting principle** discussed in Section 11.4 is used to determine the number of possible ways two tasks can be performed. If a first task can be performed in M distinct ways and a second task can be performed in N distinct ways, then the two tasks in that specific order can be performed in

$M \cdot N$ distinct ways. For example, if there are five routes from New York to Chicago and six routes from Chicago to Los Angeles, then there are $5 \cdot 6$, or 30, distinct routes from New York to Los Angeles via Chicago. The counting principle can be expanded for three or more tasks, as illustrated below.

Example 1

A large company is given the sole use of the 345 telephone exchange by the Southern Bell Telephone Company. How many different telephone numbers (or extensions) can the company use if the phone number is to begin with 345 followed by four digits?

Solution: The phone numbers will look like this:

$$345-\ \underline{\quad}\ \underline{\quad}\ \underline{\quad}\ \underline{\quad}$$

We need to determine in how many different ways the last four digits can be arranged. Consider the last four positions:

$$\underline{\quad}\ \underline{\quad}\ \underline{\quad}\ \underline{\quad}$$

Since there are ten digits and we are given no restrictions, each of these positions can be filled with any of the ten digits:

$$\underline{10}\ \ \underline{10}\ \ \underline{10}\ \ \underline{10}$$

Thus there are $10 \cdot 10 \cdot 10 \cdot 10 = 10,000$ different possible phone numbers. ❖

Example 2

At a baseball card show, Kristen DiMarco wishes to display five different Mickey Mantle cards.

a) In how many different ways can she place the five cards in a straight row?

b) If she wishes to place Mickey Mantle's rookie card in the middle, in how many ways can she arrange the cards?

Solution

a) There are five positions to fill using the five cards. In the first position, on the left, she can use any one of the five cards. In the second position she can use any of the four remaining cards. In the third position she can use any of the three remaining cards, and so on. The number of distinct possible arrangements is

$$\underline{5} \cdot \underline{4} \cdot \underline{3} \cdot \underline{2} \cdot \underline{1} = 120$$

b) We begin by satisfying the specific requirements stated. In this case the rookie card must be placed in the middle. Therefore there is only one possibility for the middle position:

$$\underline{}\ \ \underline{}\ \ \underline{1}\ \ \underline{}\ \ \underline{}$$

For the first position there are now four possibilities. For the second position, there will be three possibilities. For the fourth position there will be two possibilities. Finally, in the last position, there is only one possibility:

$$\underline{4}\cdot\underline{3}\cdot\underline{1}\cdot\underline{2}\cdot\underline{1}=24$$

Thus under the condition stated there are 24 different possible arrangements. ❖

Permutations of Distinct Objects

Permutations and combinations are two other techniques that may be used to determine the number of possible outcomes that exist for a given situation. We will discuss permutations now and combinations shortly.

> A **permutation** of n objects is any *ordered arrangement* of the n objects.

■ **Example 3** ————————————————————

How many distinct ways can the letters a, b, c, d be arranged if repetition of the letters is not permitted?

Solution: Using the counting principle, we can use any of the four letters as the first letter. Once the first letter is selected, the second selection can be any one of the remaining three letters. After the first two letters are selected, there are only two letters remaining for the third selection. There will be only one left for the final selection. Thus by the counting principle there are $4\cdot3\cdot2\cdot1$, or 24, different possible arrangements. They are

a, b, c, d	b, a, c, d	c, a, b, d	d, a, b, c
a, b, d, c	b, a, d, c	c, a, d, b	d, a, c, b
a, c, b, d	b, c, a, d	c, b, a, d	d, b, a, c
a, c, d, b	b, c, d, a	c, b, d, a	d, b, c, a
a, d, b, c	b, d, a, c	c, d, a, b	d, c, a, b
a, d, c, b	b, d, c, a	c, d, b, a	$d, c, b, a.$

❖

Each of the 24 distinct ordered arrangements in Example 3 is called a permutation. Note that, for example, a, b, c, d is a different permutation than a, b, d, c, since the *order* of the letters is different.

When using the counting principle in Example 3, we evaluated $4 \cdot 3 \cdot 2 \cdot 1$ to get 24. The expression $4 \cdot 3 \cdot 2 \cdot 1$ can be represented as 4 factorial, symbolized by 4!.

$$4! = 4 \cdot 3 \cdot 2 \cdot 1 = 24$$

A second example of factorial notation is

$$6! = 6 \cdot 5 \cdot 4 \cdot 3 \cdot 2 \cdot 1 = 720$$

> For any number n, **n factorial, symbolized $n!$**, is
>
> $$n! = n(n-1)(n-2) \cdots (3)(2)(1).$$

Note that 0! is defined to be 1.

In general, the number of distinct ordered arrangements, or permutations, of n objects is **n factorial.**

■ **Example 4** _____

In how many ways can eight different books be arranged on a shelf?

Solution: Since there are eight different books, 8 factorial permutations are possible:

$$8! = 8 \cdot 7 \cdot 6 \cdot 5 \cdot 4 \cdot 3 \cdot 2 \cdot 1 = 40{,}320. \qquad \diamondsuit$$

■ **Example 5** _____

Consider the same four letters used in Example 3, namely, a, b, c, d. How many different arrangements, or permutations, are possible if only two letters of the four are to be used and repetition is not allowed?

Solution: We can again use the counting principle to find the solution. The first selection can be any one of four possible letters. The second selection can be any one of the remaining three letters. Thus there will be $4 \cdot 3$, or 12, different arrangements or permutations:

a, b	b, a	c, a	d, a
a, c	b, c	c, b	d, b
a, d	b, d	c, d	d, c

$\diamondsuit$

Note that in Example 5 a, a b, b c, c and d, d are not considered permutations, since repetition of an object is not permitted.

■ Example 6 ───────────────────────────────

Consider the five letters p, q, r, s, t. In how many distinct ways can three letters be selected and arranged if repetition is not allowed?

Solution: Using the counting principle, we find that there are five possible letters for the first choice, four possible letters for the second choice, and three possible letters for the third choice:

$$5 \cdot 4 \cdot 3 = 60$$

Thus there are 60 different possible ordered arrangements, or permutations.   ❖

In Example 6 we determined the number of different ways in which we could select and arrange three of the five items. We can indicate this using the notation $_5P_3$. The notation $_5P_3$ is read "the number of permutations of five items taken three at a time." In general, the notation $_nP_r$, read "the number of permutations of n items taken r at a time," is used to indicate the number of distinct ways in which r of the n items can be selected and arranged.

Using the counting principle, we can see that the number of permutations when r objects are selected from a total of n objects, symbolized $_nP_r$, is

$$_nP_r = \underbrace{n(n - 1)(n - 2) \cdots (n - r + 1)}_{r \text{ factors}}$$

For example, if there are seven distinct objects $(n = 7)$ and four are to be selected $(r = 4)$, the number of possible permutations is

$$_nP_r = n(n - 1) \cdots (n - r + 1),$$
$$_7P_4 = 7 \cdot 6 \cdot \ldots \cdot (7 - 4 + 1),$$
$$= 7 \cdot 6 \cdot 5 \cdot 4 = 840$$

Thus $_7P_4 = 840$.

Now let us develop an alternative formula that can be used to find the number of permutations possible when r objects are selected from n objects:

$$_nP_r = n(n - 1)(n - 2) \cdots (n - r + 1)$$

or

$$_nP_r = n(n - 1)(n - 2) \cdots (n - r + 1) \times \frac{(n - r)!}{(n - r)!}$$

or

$$_nP_r = \frac{n(n-1)(n-2)\cdots(n-r+1)\overbrace{(n-r)(n-r-1)\cdots(3)(2)(1)}^{(n-r)!}}{(n-r)!}$$

Since the numerator of the expression above is $n!$, we can see that

$$_nP_r = \frac{n!}{(n-r)!}.$$

> The number of permutations possible when r objects are selected from n objects is found by the **permutation formula**
>
> $$_nP_r = \frac{n!}{(n-r)!}$$

In Example 6 we found that when selecting 3 out of 5 letters there were 60 permutations. We can obtain the same result using the permutation formula.

$$_5P_3 = \frac{5!}{(5-3)} = \frac{5!}{2!} = \frac{5\cdot4\cdot3\cdot\cancel{2!}}{\cancel{2!}} = 60$$

Example 7

A club consisting of eight people must choose a president, vice-president, and a secretary. How many different arrangements or permutations are possible?

Solution: Since there are eight people, $n = 8$, of which three are to be selected; thus $r = 3$.

$$_8P_3 = \frac{8!}{(8-3)!} = \frac{8!}{5!} = \frac{8\cdot7\cdot6\cdot\cancel{5\cdot4\cdot3\cdot2\cdot1}}{\cancel{5\cdot4\cdot3\cdot2\cdot1}} = 336$$

Thus with eight people there can be 336 different arrangements for president, vice-president, and secretary.   ❖

Example 8

A go-cart track has ten cars available. If six people must select a car in which to ride, how many different permutations of car and driver are possible?

Solution: There are ten cars from which six are to be selected. Thus $n = 10$ and $r = 6$.

$$_{10}P_6 = \frac{10!}{(10 - 6)!} = \frac{10!}{4!} = \frac{10 \cdot 9 \cdot 8 \cdot 7 \cdot 6 \cdot 5 \cdot \cancel{4!}}{\cancel{4!}}$$
$$= 151,200$$

There are 151,200 different possible arrangements of car and driver. ❖

We have worked permutation problems (selecting and arranging, without replacement, r items out of n *distinct* items) using the counting principle and using the permutation formula. When given a permutation problem, unless specified by your instructor, you may use either technique to work the problem.

Permutations of Nondistinct Objects

Now we will consider permutation problems in which every item is not distinct. Consider the word "BOB." How many permutations of the word "BOB" are possible? Although the Bs are identical, we will call the first one B_1 and the second one B_2. If the Bs were distinguishable, there would be six permutations:

$$\begin{array}{ccc} B_1, O, B_2 & O, B_1, B_2 & B_2, B_1, O \\ B_1, B_2, O & O, B_2, B_1 & B_2, O, B_1 \end{array}$$

Now let's remove the subscripts:

$$\begin{array}{ccc} B, O, B & O, B, B & B, B, O \\ B, B, O & O, B, B & B, O, B \end{array}$$

Note that there are only three distinct permutations:

$$\begin{array}{ccc} B, O, B & O, B, B & B, B, O \end{array}$$

If we divide the number of permutations possible, assuming that all letters are distinct, 6 or 3!, by the number of ways the two Bs can be arranged ($B_1 B_2$ or $B_2 B_1$), 2 or 2!, we get the answer 3:

$$\frac{3 \cdot 2 \cdot 1}{2 \cdot 1} = 3$$

The same reasoning can be used to determine the formula for the number of distinct permutations of n objects, where any number of the objects are identical.

> The number of distinct permutations of n objects where n_1 of the objects are identical, n_2 of the objects are identical, . . . , n_r of the objects are identical is found by the formula
>
> $$\frac{n!}{n_1!\, n_2! \cdots n_r!}$$

■ Example 9

In how many different ways can the numbers 1, 1, 1, 2, 2 be arranged?

Solution: There are five numbers including three 1s and two 2s. The number of different permutations is

$$\frac{5!}{3!2!} = \frac{5 \cdot 4 \cdot 3 \cdot 2 \cdot 1}{3 \cdot 2 \cdot 1 \cdot 2 \cdot 1} = \frac{5 \cdot \overset{2}{\cancel{4}} \cdot \cancel{3 \cdot 2 \cdot 1}}{\cancel{3 \cdot 2 \cdot 1} \cdot \cancel{2} \cdot 1} = 10$$

The ten different permutations are

1, 1, 1, 2, 2	1, 2, 2, 1, 1
1, 1, 2, 1, 2	2, 1, 1, 1, 2
1, 1, 2, 2, 1	2, 1, 1, 2, 1
1, 2, 1, 1, 2	2, 1, 2, 1, 1
1, 2, 1, 2, 1	2, 2, 1, 1, 1

❖

■ Example 10

In how many different ways can the letters of the word "statistics" be arranged?

Solution: There are ten letters including three t's, three s's, and two i's. The number of possible arrangements is

$$\frac{10!}{3!3!2!} = \frac{10 \cdot 9 \cdot 8 \cdot 7 \cdot 6 \cdot 5 \cdot 4 \cdot 3 \cdot 2 \cdot 1}{3 \cdot 2 \cdot 1 \cdot 3 \cdot 2 \cdot 1 \cdot 2 \cdot 1} = 50{,}400.$$

There are 50,400 different possible arrangements of the letters in the word "statistics." ❖

Combination Problems

Counting problems in which the arrangement of the objects is unimportant are called combination problems.

> A combination of n objects is any *unordered arrangement* of the n objects.

Let us first consider two questions. In how many ways can a committee of three people be selected from a total of eight people? In how many ways can exactly two questions in a six-question multiple-choice exam be answered correctly? A committee consisting of Jan, Suheyla, and Alice is the same as a committee consisting of Suheyla, Alice, and Jan. The order of selection is not important. The same reasoning can be used to show that the order in which the questions on the exam are answered is not important to the final grade.

In order to answer the two questions, let us develop a formula for finding the number of possible combinations, using our knowledge of permutations.

The notation $_nC_r$ is used to indicate the number of combinations when r objects are selected from n objects. For example, the number of combinations when two objects are selected from five objects is written $_5C_2$.

Given the set $\{a, b, c, d, e\}$, how many permutations of two letters are possible? How many combinations of two letters are possible?

Permutations		Combinations	
ab, ba, ac, ca, ad, da, ae, ea,		ab, ac, ad, ae, bc, bd,	
bc, cb, bd, db, be, eb, cd, dc,	20	be, cd, ce, de	10
ce, ec, de, ed			

There are twice as many permutations as combinations. This result is not surprising because from one combination of two letters two permutations can be formed. For example, the combination ab gives the permutations ab and ba. Thus $_5P_2 = 2(_5C_2)$. Since $2 = 2!$, we may write $_5P_2 = 2!(_5C_2)$. If we repeat the same procedure selecting three letters out of the given five, we find that there are six times as many permutations as combinations, because one combination of three letters gives six distinct permutations.

Combination	Permutations	
$a\,b\,c$	$a\,b\,c$	$a\,c\,b$
	$b\,a\,c$	$b\,c\,a$
	$c\,a\,b$	$c\,b\,a$

Thus there are six times as many permutations as there are combinations: $_5P_3 = 6(_5C_3)$. Since 6 can be represented as $3!$, $_5P_3 = 3!(_5C_3)$. The same procedure will show that $_5P_4 = 4!(_5C_4)$, and in general, $_nP_r = r!(_nC_r)$. Dividing both sides of the equations by $r!$ gives

$$_nC_r = \frac{_nP_r}{r!}$$

Since $_nP_r = n!/(n - r)!$, the combination formula may be expressed as

$$_nC_r = \frac{n!/(n - r)!}{r!} = \frac{n!}{(n - r)!r!}$$

> The number of combinations possible when r objects are selected from n objects is found by the **combination formula**
>
> $$_nC_r = \frac{n!}{(n - r)!r!}$$

■ Example 11

In how many ways can a committee of three people be selected from a total of eight people?

Solution: This problem is a combination problem because the order in which the committee members are selected is unimportant. There are a total of eight people; thus $n = 8$. Three are to be selected; thus $r = 3$.

$$_8C_3 = \frac{8!}{(8 - 3)!3!} = \frac{8!}{5!3!} = \frac{8 \cdot 7 \cdot \cancel{6} \cdot \cancel{5 \cdot 4 \cdot 3 \cdot 2 \cdot 1}}{\cancel{5 \cdot 4 \cdot 3 \cdot 2 \cdot 1} \cdot \cancel{3} \cdot \cancel{2} \cdot 1} = 56$$

Thus 56 different combinations are possible.   ❖

■ Example 12

A multiple-choice exam consists of six questions. In how many ways can exactly two questions be answered correctly?

Solution: This is a combination problem because the order in which the questions are answered has no effect on the number of questions answered correctly.

$$_6C_2 = \frac{6!}{(6 - 2)!2!} = \frac{6!}{4!2!} = \frac{\overset{3}{\cancel{6}} \cdot 5 \cdot \cancel{4 \cdot 3 \cdot 2 \cdot 1}}{\cancel{4 \cdot 3 \cdot 2 \cdot 1} \cdot \cancel{2} \cdot 1} = 15$$

There are 15 distinct combinations:

1, 2	2, 3	3, 4	4, 5	5, 6
1, 3	2, 4	3, 5	4, 6	
1, 4	2, 5	3, 6		
1, 5	2, 6			
1, 6				

Since order is not important in combination problems, 3, 2, for example, is the same as 2, 3, and therefore it is counted only once. ❖

Example 13

Claudia must select three flavors of ice cream for her banana split. If the restaurant has seven different flavors, how many different banana splits can she order?

Solution: Since a banana split containing vanilla, chocolate, and strawberry ice cream, for example, is the same as a banana split with chocolate, strawberry, and vanilla ice cream, this is a combination problem.

$$_7C_3 = \frac{7!}{(7-3)!3!} = \frac{7!}{4!3!}$$
$$= \frac{7 \cdot \cancel{6} \cdot 5 \cdot \cancel{4 \cdot 3 \cdot 2 \cdot 1}}{4 \cdot 3 \cdot 2 \cdot 1 \cdot \cancel{3} \cdot \cancel{2} \cdot 1} = 35$$

Claudia can order 35 different combinations of ice cream flavors.   ❖

Example 14

At Su Wong's Chinese Restaurant, dinner for eight consists of three items from column A, four items from column B, and three items from column C. If columns A, B, and C have five, seven, and six items, respectively, how many different dinner combinations are possible?

Solution: For column A, three items out of five items must be selected. This can be represented as $_5C_3$. For column B, four items out of seven must be selected. This can be represented as $_7C_4$. For column C, three items out of six must be selected, or $_6C_3$.

$$_5C_3 = 10, \qquad _7C_4 = 35, \qquad _6C_3 = 20$$

Using the counting principle, we can determine the total number of dinner combinations by multiplying the number of choices from columns A, B, and C together:

$$\text{Total number of dinner choices} = {_5C_3} \cdot {_7C_4} \cdot {_6C_3}$$
$$= 10 \cdot 35 \cdot 20 = 7000$$

Therefore 7000 different combinations are possible under these conditions.   ❖

EXERCISES 11.8

1. Explain the difference between a permutation and a combination.
2. Explain how to find $n!$ for any whole number n.

Classify each of the following as a permutation or a combination problem.

3. The number of arrangements of the letters in the word "HELP" (for example, PLEH)
4. An ice cream parlor has 15 different flavors. George orders a sundae and has to select 3 flavors. How many different selections are possible?
5. A teacher decides to give six identical prizes to 6 of the 30 students in his class. In how many ways can he do this?
6. A teacher decides to give six different prizes to 6 of the 30 students in her class. In how many ways can she do this?
7. A bookcase contains 20 different books. Three books will be selected at random and given away as prizes to three individuals. In how many ways can this be done?
8. A bookcase contains 20 different books. Three books will be selected at random and given away as a prize to one individual. How many different sets of 3 books can the individual receive?
9. A night guard visits ten different offices every hour. The pattern is varied each night so that the guard will not follow a specific routine. In how many different ways can this be done?
10. A student must select and answer four out of five questions on an exam. In how many ways can this be done?

Evaluate the following.

11. $5!$
12. $_8P_3$
13. $_8C_3$
14. $0!$
15. $_7P_7$
16. $_7C_7$
17. $\dfrac{7!}{5! \cdot 2!}$
18. $_5P_2$
19. $_4C_0$
20. $_6C_4$
21. $_6P_2$
22. $_6P_4$
23. $\dfrac{n!}{(n-1)!}$
24. $\dfrac{_5P_2}{_5C_2}$

25. Use the counting principle to determine how many different identification numbers are possible if an identification number is to have a letter followed by four digits (a) if repetition is allowed; (b) if repetition is not allowed.
26. Use the counting principle to determine how many different Social Security numbers are possible if each one consists of nine digits. The first digit may be zero, and repetition is allowed.
27. Six different books are placed on a shelf. How many different arrangements are possible?
28. How many different ways are there to line up eight dogs in a dog show?
29. A man has 8 pairs of pants, 12 shirts, 15 ties, and 6 sport coats. How many different outfits can he wear?
30. Three of a sample of ten radios will be selected at random and tested for defects. In how many ways can this be done?
31. Three of five finalists will be awarded a scholarship. In how many ways can the scholarships be awarded if (a) they are all for the same amount? (b) they are all for different amounts?
32. a) To open a combination lock, you must know the lock's three-number sequence in its proper order. Why should this be called a permutation lock rather than a combination lock?
    b) Assuming that a combination lock has 40 numbers, determine how many different three-number arrangements are possible if repetition of numbers is allowed?
    c) Answer the question in part (b) if repetition is not allowed.
33. In how many ways can the letters in the word "HORSE" be arranged?
34. In how many ways can the letters in the word "WINNING" be arranged?
35. In how many ways can the letters in the word "MISSISSIPPI" be arranged?
36. A basketball coach must assign five positions to nine players. In how many ways can this be done if you assume that each player can play any position and each of the five positions is considered unique?
37. One prize of $100, two prizes of $50, and three prizes of $10 are to be given away to six different people. In how many ways can this be done?
38. Three of 15 people in a class must receive a grade of A. In how many ways can the A's be distributed?
39. In how many ways can an individual select exactly five out of six questions to be answered on an exam?
40. In how many ways can the winner and first and second runners-up be selected from seven remaining finalists in a science project contest?

41. In how many ways can five drivers be assigned to seven different cars?

42. A night watchman at an office building must punch his time clock at each of eight stations as he makes his rounds. For security purposes he does not want to check the stations in the same order every time. How many different routes are available to the watchman?

43. In one question of a history test the student is asked to match ten dates with ten events; each date can only be matched with one event. In how many different ways can this question be answered?

44. A five-card poker hand is to be dealt from a deck of 52 cards. How many different poker hands are possible?

45. Five different colored flags will be placed on a pole, one beneath another. The arrangement of the colors indicates the message. How many messages are possible if five flags are to be selected from eight different colored flags?

46. A quinella bet consists of selecting the first- and second-place winners, in any order, in a particular event. For example, suppose you select a 2–5 quinella. If 2 wins and 5 finishes second, or if 5 wins and 2 finishes second, you win. In the game of jai alai, 8 two-man teams play against one another. How many quinella tickets are necessary to guarantee a win?

47. In how many ways can the manager of a baseball team arrange his batting order of nine players if the pitcher must bat last?

48. Five Rembrandts are to be displayed in a museum.
  a) In how many different ways can they be arranged if they must be next to one another?
  b) In how many different ways can they be displayed if a specific one is to be in the middle?

49. The singing group Bon Jovi is planning to make a record album consisting of six fast songs and four slow songs. If they have ten fast songs and seven slow songs to choose from, how many different possible combinations do they have?

50. On an English test, Mary must write an essay for three of the five questions in Part 1 and four out of six questions in Part 2. How many different combinations of questions can she answer?

51. An editor has eight manuscripts for mathematics books and five manuscripts for computer science books. If he is to select five mathematics and three computer science books for publication, how many different choices does he have?

52. How many ways are there of selecting three kings and two queens when five cards are selected at random from a deck of cards without replacement?

53. How many ways are there of selecting three diamonds and two clubs when five cards are selected at random from a deck of cards without replacement?

54. Bob is sent to the store to get five different bottles of regular soda and three different bottles of diet soda. If there are ten different regular sodas and seven different diet sodas to choose from, how many different choices does Bob have?

55. How many different committees can be formed from 6 teachers and 50 students if the committee is to consist of two teachers and three students?

56. A teacher is constructing a mathematics test consisting of 10 questions. She has a pool of 28 questions, which are classified as to level of difficulty as follows: 6 difficult questions, 10 average questions, and 12 easy questions. How many different 10-question tests can she construct from the pool of 28 questions if her test is to have 3 difficult, 4 average, and 3 easy questions?

57. General Mills is testing six oat cereals, five wheat cereals, and four rice cereals. If it plans to market three of the oat cereals, two of the wheat cereals, and two of the rice cereals, how many different combinations are possible?

58. A catering service is making up trays of hors d'oeuvres. The hors d'oeuvres are categorized as inexpensive, average, and expensive. If the client must select three of the seven inexpensive, five of the eight average, and two of the four expensive hors d'oeuvres, how many different choices are possible?

59. Dr. Berry is rearranging the medicine bottles in her medicine chest. She has four bottles of cough syrup, three bottles of pain reliever, and five bottles of aspirin. In how many ways can she rearrange the bottles for each of the following conditions.
  a) The bottles are arranged in any order (assume that each bottle is distinct).
  b) The medications of the same type are to be grouped together, and each bottle is considered to be distinct.
  c) All the bottles of the same medication are considered identical, and the bottles can be arranged in any order.
  d) All the bottles of the same medicine are considered identical, and the medications of the same type are to be grouped together.

PROBLEM SOLVING

60. On a preferential ballot, a person is asked to rank three of seven candidates for the office of chairperson, giving first, second, and third choices. What is the minimum number of ballots that must be cast in order to guarantee that at least two ballots are the same?

61. Consider a ten-question test in which each question can be answered either right or wrong.

    a) How many different ways are there to answer the questions so that exactly eight are right and two are wrong?

    b) How many ways are there to answer the questions so that at least eight are right?

62. The notation $_nC_r$ may be written $\binom{n}{r}$.

    a) Using this notation, evaluate each of the combinations in the following array. Form a triangle of the results, similar to the one given, by

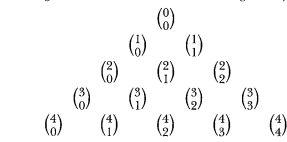

placing the answer to each combination in the same relative position in the triangle.

    b) Using the number pattern in (a), can you find the next row of numbers of the triangle (known as **Pascal's triangle**)?

63. a) Four people at dinner make a toast. If each person is to tap glasses with each other person, how many taps will take place?

    b) Repeat part (a) with five people.

    c) How many taps will there be if there are n people at the dinner table?

64. In how many ways can two people be assigned to two chairs at a circular table? Assume that the various places at the table are indistinguishable and that only the relative positions of the people are considered. In how many ways can three people be assigned to three chairs at a circular table? Repeat the process for four chairs. Can you find a formula that will give you the number of permutations when n people are to be assigned to n chairs around a circular table?

 ## PROBABILITY PROBLEMS USING COMBINATIONS

In Section 11.4 we discussed counting techniques. Now we will use the counting techniques to work some probability problems.

Suppose we want to find the probability of selecting two picture cards (jacks, queens, or kings) when two cards are selected, without replacement, from a deck of 52 cards. Using the "and" probability formula discussed in Section 11.5, we could reason as follows:

$$P(2 \text{ picture cards}) = P(1\text{st picture card}) \cdot P(2\text{nd picture card})$$
$$= \frac{12}{52} \cdot \frac{11}{51} = \frac{132}{2652} \quad \text{or} \quad \frac{11}{221}$$

Since the order of the two picture cards that are selected is not important to the final answer, this can be considered a combination probability problem. We can find the probability of selecting two picture cards, using combinations, by finding the number of possible successful outcomes (selecting two picture cards) and dividing that answer by the total number of possible outcomes (selecting any two cards).

The number of ways in which two picture cards can be selected from the 12 picture cards in a deck is $_{12}C_2$:

$$_{12}C_2 = \frac{12!}{(12-2)!2!} = \frac{\overset{6}{\cancel{12}} \cdot 11 \cdot \cancel{10!}}{\cancel{10!} \cdot \cancel{2} \cdot 1} = \frac{66}{1} = 66$$

The number of ways in which two cards can be selected from a deck of 52 cards is $_{52}C_2$:

$$_{52}C_2 = \frac{52!}{(52-2)!2!} = \frac{\overset{26}{\cancel{52}} \cdot 51 \cdot \cancel{50!}}{\cancel{50!} \cdot \cancel{2} \cdot 1} = 1326$$

Thus

$$P(\text{selecting 2 picture cards}) = \frac{_{12}C_2}{_{52}C_2} = \frac{66}{1326} = \frac{11}{221}$$

Notice that the same answer is obtained by using either method. To give you more practice with counting techniques, we will work the problems in this section using counting techniques.

 Example 1 ————————————————————————————

A club consists of four men and five women. Three members are to be selected at random to form a committee. What is the probability that the committee will consist of three women?

Solution

$$P(\text{committee consisting of 3 women}) = \frac{\text{number of possible committees with 3 women}}{\text{total number of possible 3-member committees}}$$

To solve this problem, we must determine the number of possible subcommittees with three women and the total number of possible three-member committees.

Since it does not matter in which order the individuals are selected to serve on the committee, the number of possible committees can be found by using the combination formula.

Since there are five women and we wish to choose three, the number of possible committees with three women is $_5C_3$:

$$_5C_3 = \frac{5!}{(5-3)!3!} = \frac{5!}{2!3!}$$

$$= \frac{5 \cdot \cancel{4}^{2} \cdot \cancel{3 \cdot 2 \cdot 1}}{\cancel{2} \cdot 1 \cdot \cancel{3 \cdot 2 \cdot 1}} = 10$$

Thus there are ten possible committees consisting of three women.

Since there are a total of nine members, the total number of three-member committees is $_9C_3$:

$$_9C_3 = \frac{9!}{(9-3)!3!} = \frac{9!}{6!3!}$$

$$= \frac{\cancel{9}^{3} \cdot \cancel{8}^{4} \cdot 7 \cdot \cancel{6 \cdot 5 \cdot 4 \cdot 3 \cdot 2 \cdot 1}}{6 \cdot 5 \cdot 4 \cdot 3 \cdot 2 \cdot 1 \cdot \cancel{3} \cdot \cancel{2} \cdot 1} = 84$$

$$P(\text{committee consisting of 3 women}) = \frac{_5C_3}{_9C_3}$$

$$= \frac{10}{84} = \frac{5}{42}$$

Thus the probability of randomly selecting a committee of three women is $\frac{5}{42}$. ❖

■ Example 2

There are ten seats left aboard an airplane, six of which are aisle seats and four of which are window seats. If three people about to board the plane are given seat numbers at random, what is the probability that they are all given aisle seats?

Solution

$$P(\text{all 3 given aisle seats}) = \frac{\begin{array}{c}\text{number of possible}\\\text{combinations of aisle seats}\\\text{for 3 people}\end{array}}{\begin{array}{c}\text{total number of possible}\\\text{combinations of seats}\\\text{for 3 people}\end{array}}$$

Since the order in which the passengers are seated is immaterial, the number of possible arrangements can be found by the combination formula.

There are six aisle seats, and three people are to be seated. Thus the number of possible combinations of aisle seats is $_6C_3$:

$$_6C_3 = \frac{6!}{(6-3)!3!} = \frac{6!}{3!3!}$$
$$= \frac{\cancel{6} \cdot 5 \cdot 4 \cdot \cancel{3 \cdot 2 \cdot 1}}{\cancel{3} \cdot \cancel{2} \cdot 1 \cdot \cancel{3 \cdot 2 \cdot 1}}$$
$$= 20$$

Since a total of ten seats are available, the total number of different arrangements for the three people is $_{10}C_3$:

$$_{10}C_3 = \frac{10!}{(10-3)!3!} = \frac{10!}{7!3!}$$
$$= \frac{10 \cdot \overset{3}{\cancel{9}} \cdot \overset{4}{\cancel{8}} \cdot \cancel{7 \cdot 6 \cdot 5 \cdot 4 \cdot 3 \cdot 2 \cdot 1}}{\cancel{7 \cdot 6 \cdot 5 \cdot 4 \cdot 3 \cdot 2 \cdot 1} \cdot \cancel{3} \cdot \cancel{2} \cdot 1}$$
$$= 120$$

$$P(\text{all 3 given aisle seats}) = \frac{_6C_3}{_{10}C_3} = \frac{20}{120}$$
$$= \frac{1}{6}$$

Thus the probability of all three passengers getting aisle seats is $\frac{1}{6}$.   ❖

■ **Example 3** _____

You are dealt a flush in the game of poker when you are dealt five cards of the same suit. If you are dealt a five-card hand, find the probability that you will be dealt a flush in hearts.

Solution

$$P(\text{flush in hearts}) = \frac{\text{number of possible 5-card heart flushes}}{\text{total number of 5-card hands possible}}$$

The order in which the five cards are received is immaterial. Thus the number of possible hands can be found by the combination formula.

Since there are 13 hearts in a deck of cards, the number of possible five-card flush hands is $_{13}C_5$:

$$_{13}C_5 = \frac{13!}{(13-5)!5!} = \frac{13!}{8!5!}$$

$$= \frac{13 \cdot 12 \cdot 11 \cdot 10 \cdot 9 \cdot 8!}{8! \cdot 5 \cdot 4 \cdot 3 \cdot 2 \cdot 1} = 1287$$

The total number of possible five-card hands in a deck of 52 cards is $_{52}C_5$:

$$_{52}C_5 = \frac{52!}{(52-5)!5!} = \frac{52!}{47!5!}$$
$$= \frac{52 \cdot 51 \cdot 50 \cdot 49 \cdot 48 \cdot 47!}{47! \cdot 5 \cdot 4 \cdot 3 \cdot 2 \cdot 1} = 2{,}598{,}960$$

$$P(\text{flush in hearts}) = \frac{_{13}C_5}{_{52}C_5} = \frac{1287}{2{,}598{,}960} = \frac{33}{66{,}640}$$

The probability of being dealt a heart flush is $\frac{33}{66{,}640}$.   ❖

■ Example 4

The Kellogg Company is testing 12 new cereals for possible production. They are testing three oat cereals, four wheat cereals, and five rice cereals. If we assume that each of the 12 cereals has the same chance of being selected and four new cereals will be produced, find the probability that

a) no wheat cereals are selected,

b) at least one wheat cereal is selected,

c) two wheat cereals and two rice cereals are selected.

Solution

a) If no wheat cereals are to be selected, then only oat and rice cereals must be selected. There are a total of eight cereals that are oat or rice. Thus the number of ways that four oat or rice cereals may be selected out of the eight possible oat or rice cereals is $_8C_4$. The total number of possible selections is $_{12}C_4$.

$$P(\text{no wheat cereals}) = \frac{_8C_4}{_{12}C_4} = \frac{70}{495} = \frac{14}{99}$$

b) When four cereals are selected, the choice must contain either no wheat cereal or at least one wheat cereal. Since one of these outcomes must occur, the sum of the probabilities must be 1:

$$P(\text{no wheat cereal}) + P(\text{at least 1 wheat cereal}) = 1$$

Therefore

$$P(\text{at least one wheat cereal}) = 1 - P(\text{no wheat cereal})$$
$$= 1 - \frac{14}{99} = \frac{99}{99} - \frac{14}{99} = \frac{85}{99}$$

Note that the probability of selecting no wheat cereals was found in part (a).

c) The number of ways of selecting two wheat cereals out of four wheat cereals is $_4C_2$. The number of ways of selecting two rice cereals out of five rice cereals is $_5C_2$. The total number of possible selections when four cereals are selected from the 12 choices is $_{12}C_4$. Since both the two wheat *and* the two rice cereals must be selected, the probability is calculated as follows:

$$P(2 \text{ wheat and 2 rice}) = \frac{_4C_2 \cdot {_5C_2}}{_{12}C_4} = \frac{6 \cdot 10}{495} = \frac{60}{495} = \frac{12}{99} \qquad \diamondsuit$$

EXERCISES 11.9

In Exercises 1 through 41 the problems are to be done without replacement. Use combinations to determine probabilities.

1. Each of the numbers 1 through 6 is written on a piece of paper, and the six pieces of paper are placed in a hat. If two numbers are selected at random, find the probability that both numbers selected are even.

2. An urn contains five red balls and four blue balls. You plan to draw three balls at random. Find the probability of selecting three red balls.

3. A hat contains 4 one-dollar bills, 3 five-dollar bills, and 1 ten-dollar bill. If you select 2 bills at random from the hat, find the probability of selecting 2 one-dollar bills.

4. There are four good and four defective batteries in a box. If you select three at random, find the probability that you select three good batteries.

5. Each of the digits 0, 1, 2, 3, 4, 5, 6, 7, 8, and 9 is placed on a slip of paper, and the slips are placed in a hat. If three slips of paper are selected at random, find the probability that the three numbers selected are greater than 4.

6. The ten finalists in a scholarship contest consist of six women and four men. If two equal prizes are to be awarded, find the probability that both prizes are awarded to women. Assume that each applicant has an equal chance of being selected.

7. A three-person committee is to be selected at random from five Democrats and three Republicans. Find the probability that all three selected are Democrats.

8. A committee of four is to be randomly selected from a group of seven teachers and eight students.

Find the probability that the committee will consist of four students.

9. You are dealt five cards from a deck of 52 cards. Find the probability that you are dealt five red cards.

10. You are dealt five cards from a deck of 52 cards. Find the probability that you are dealt no aces.

A television game show has five doors of which the contestant must pick two. Behind two of the doors are expensive cars, and behind the other three doors are consolation prizes. The contestant gets to keep the items behind the two doors she selects. Find the probability that the contestant wins

11. both cars
12. no cars
13. at least one car

Two chambers of a "six-shooter" are loaded at random. If three shots are to be fired in succession, find the probability that

14. two bullets are fired
15. no bullets are fired
16. at least one bullet is fired

A hat contains three red, four white, and five blue chips. If three chips are selected at random, find the probability of selecting

17. two red chips and one blue chip
18. two blue chips and one white chip
19. at least one red chip
20. one red, one white, and one blue chip

Drug A is given to five patients. Drug B is given to four patients, and drug C is given to six patients. If four of these 15 patients are selected at random, find the probability that

21. two were given drug A and two were given drug C;
22. three were given drug C and one was given drug A;
23. at least one patient was given drug C;
24. one was given drug A, two were given drug B, and one was given drug C.

Five men and six women are going to be assigned to a specific row of seats in the theatre. If the 11 tickets for the numbered seats are given out at random, find the probability that

25. five women are given the first five seats next to the center aisle;
26. at least one woman is in one of the first five seats;
27. exactly one woman is in one of the first five seats;
28. three women are seated in the first three seats and two men are seated in the next two seats.

PROBLEM SOLVING

The first wheel of a three-wheel slot machine consists of three oranges, four cherries, two lemons, six plums, two bells, and one bar. The second wheel consists of two oranges, five cherries, three lemons, five plums, two bells, and one bar. The third wheel consists of four oranges, four cherries, three lemons, three plums, two bells, and two bars. When you pull the handle of the slot machine, find the probability of obtaining

34. three cherries
35. cherry, cherry, bell
36. three bars
37. at least one cherry

38. If three men and three women are to be assigned at random to six seats in a row at a theatre, find the probability that they will alternate by sex.
39. A club consists of 15 people including Ali, Kendra, Ted, Alice, Marie, Dan, Linda, and Frank. From the 15 members a president, vice-president, and treasurer will be selected at random, and an advisory committee of five other individuals will also be selected at random.
   a) Find the probability that Ali is selected president, Kendra is selected vice-president, Ted is selected treasurer, and the other five individuals named above form the advisory committee.

Determine the probability of winning the state lottery if you must select

29. exactly 6 specific numbers out of a total of 40 numbers
30. exactly 6 specific numbers out of a total of 44 numbers
31. any 6 of 8 specific numbers out of a total of 40 numbers

32. A full house in poker consists of getting three of one card and two of another card in a five-card hand. For example, if a hand contains three kings and two 5s, it is a full house. If five cards are dealt at random from a deck of 52 cards, without replacement, find the probability of getting three kings and two 5s.
33. A royal flush consists of the ace, king, queen, jack, and 10 all in the same suit. If seven cards are dealt at random from a deck of 52 cards, find the probability of getting a
   a) royal flush in spades
   b) royal flush in any suit

   b) Find the probability that three of the eight individuals named above are selected for the three officers' positions and the other five are selected for the advisory board.
40. A pair of aces and a pair of 8s is often referred to as the "dead man's hand." This is the poker hand held by James "Wild Bill" Hickok when he was shot in the back and killed in Deadwood, South Dakota.
   a) Find the probability of being dealt the dead man's hand when dealt five cards, without replacement, from a deck of 52 cards.
   b) The actual cards Hickok was holding when shot were the aces of spades and clubs, the 8s of spades and clubs, and the 9 of diamonds. If you are dealt five cards without replacement, find the probability of being dealt this exact hand.
41. A number is written with a magic marker on each card of a deck of 52 cards. The numeral 1 is put on the first card, 2 on the second, and so on. The cards are then shuffled and cut. What is the probability that the top four cards will be in ascending order? (For example, the top card is 12, the second 22, the third 41, and the fourth 51.)

SUMMARY

Probability is either empirical or theoretical. Empirical probability is based on past observations and is calculated by

$$P'(E) = \frac{\text{number of times event } E \text{ has occurred}}{\text{total number of times the experiment has been performed}}$$

Theoretical probability is based on the possible outcomes of an experiment and is calculated by

$$P(E) = \frac{\text{total number of outcomes favorable to } E}{\text{total number of possible outcomes}}$$

The total number of possible outcomes is called the sample space and is often found by means of a tree diagram. There are two general types of compound probability problems, the "or" type and the "and" type. The formulas for calculating these probabilities are

$$P(A \text{ or } B) = P(A) + P(B) - P(A \text{ and } B)$$

and

$$P(A \text{ and } B) = P(A) \cdot P(B)$$

where it is assumed that event A has occurred before the probability of event B is calculated.

Two events are mutually exclusive when both events cannot happen in the same trial. Two events are independent if the outcome of either event has no effect on the outcome of the other event.

Odds are quoted either against or in favor of the occurrence of an event. Odds against an event are

$$\text{Odds against event} = \frac{P(\text{event fails to occur})}{P(\text{event occurs})}$$

If the odds against an event are $a:b$, the odds in favor of the event will be $b:a$. The expectation of an experiment is an indication of the average gain or loss that will occur if the experiment is performed many times. The expectation, E, is calculated by the formula

$$E = P_1 A_1 + P_2 A_2 + \cdots + P_n A_n$$

where P represents probability and A represents net amount gained or lost.

The number of possible permutations or ordered arrangements when r objects are selected from n objects is

$$_nP_r = \frac{n!}{(n-r)!}$$

where $n! = n(n-1)(n-2)\cdots(3)(2)(1)$.

The number of distinct permutations of n objects where $n_1, n_2, \cdots, n_r$ objects are identical is found by the formula

$$\frac{n!}{n_1!n_2! \cdots n_r!}$$

The number of possible combinations or unordered arrangements when r objects are selected from n objects is

$$_nC_r = \frac{n!}{(n-r)!r!}$$

Combinations and permutations are often used in working probability problems.

REVIEW EXERCISES

1. Of 100 tosses of a trick coin, 80 resulted in heads. Find the empirical probability of the coin's landing heads up.
2. Select a card from a deck of cards 100 times with replacement, and compute the empirical probability of selecting a spade.
3. Tina, Pierre, Gina, and Carla form a club. They plan to select a president and a vice-president.
    a) Construct a tree diagram showing all the possible outcomes.
    b) List the sample space.

Each of the digits 0, 1, 2, 3, 4, 5, 6, 7, 8, 9 is written on a piece of paper, and all the pieces of paper are placed in a hat. One number is selected at random. Find the probability that the number selected is

4. even   5. odd
6. even or greater than 4   7. prime
8. prime or even   9. greater than 4 or less than 6

Each of the digits 0, 1, 2, 3, 4, 5, 6, 7, 8, 9 is written on a piece of paper, and all the pieces of paper are placed in a hat. Two pieces of paper are to be selected. Find the following probabilities (a) with replacement, (b) without replacement.

10. Both numbers are even.
11. Both are the number 2.
12. Both numbers are prime.
13. Both numbers are less than 3.

A dog pound has six dogs to give away—two collies, one German shepherd, and three poodles. A family decides to select at random two of these dogs. Find the probability of each of the following.

14. They are both collies.
15. They are both German shepherds.
16. The first selected is a poodle and the second a collie.

17. A Radio Shack science fair kit has 150 projects, of which 12 are defective. If one project is selected at random, find the odds against selecting a defective project.

18. Assume that the odds against winning the game of Monopoly are 1:2. Find the probability of winning the game.

19. A thousand raffle tickets are sold at $2 each. Three prizes of $200 and two prizes of $100 will be awarded.
    a) Find the expectation of a person who purchases a ticket.
    b) Find the expectation of a person who purchases three tickets.
    c) Find the fair price to pay for a ticket.

20. If Pete selects a picture card from a deck of cards, George will give him $6. If Pete does not select a picture card, he must give George $2.
    a) Find Pete's expectation.
    b) Find George's expectation.
    c) If Pete plays this game 100 times, how much can he expect to lose or gain?

21. Five finalists remain in a lottery drawing. The prizes to be awarded among the finalists are $100, $1000, $10,000, $100,000, $1,000,000.
    a) In how many different ways can the winners be selected?
    b) What is the expectation of a finalist?

22. Six finalists remain in a lottery drawing. Three will receive $1000 each, two will receive $10,000 each, and one will receive $100,000. How many different arrangements of prizes are possible?

23. Each of Mr. Gonzole's three children is planning to select his or her own pet rabbit from a litter of five rabbits. In how many ways can this be done?

24. Three astronauts out of nine must be selected for a mission. One will be the captain, one will be the navigator, and one will explore the surface of the moon. In how many ways can these selections be made?

25. Dr. Fox has three doses of serum for influenza type A. Six patients in the office require the serum. In how many ways can Dr. Fox dispense the serum?

26. **a)** Ten of 15 huskies are to be selected to pull a dogsled. In how many ways can this selection be made?
    b) How many different arrangements of the ten huskies on a dogsled are possible?

27. In the New York State lottery game called Lotto, you must select 6 from 45 numbers to win the grand prize. Find the probability of winning the grand prize at Lotto.

28. A club consists of six males and eight females. Three males and three females are to be selected to represent their club at a dinner. How many different combinations are possible?

29. In a medical research laboratory, one cage contains eight rats, numbered 1 through 8, and another cage contains five guinea pigs, numbered 1 through 5. Three rats and two guinea pigs are to be selected at random to be given an experimental drug. How many different combinations of animals are possible?

30. Two cards are selected at random, without replacement, from a deck of 52 cards. Find the probability that two aces are selected (use combinations).

A bag contains five red chips, three white chips, and two blue chips. Three chips are to be selected at random, without replacement. Find the probability that:

31. They are all red.

32. Two are red and one is blue.

33. One is red, one is white, and one is blue.

34. At least one is red.

CHAPTER TEST

1. In a certain community, of the last 180 births, 85 were male. Find the empirical probability that the next child born in this community is a male.

Each of the numbers 1 through 8 is written on a sheet of paper, and the eight sheets of paper are placed in a hat. If one sheet of paper is selected at random from the hat, find the probability that the number selected is

2. greater than 4
3. even
4. even or greater than 4
5. even and greater than 4
6. greater than 3 or less than 5
7. greater than 3 and less than 5

If two sheets of paper are selected, without replacement, from the hat discussed above, find the probability that

8. both numbers are greater than 4
9. both numbers are even
10. the first number is odd and the second number is even
11. neither of the numbers is greater than 6

12. One card is selected at random from a deck of cards. Find the probability that the card selected is a red card of a picture card.

One die is rolled and one letter—either a, b, or c—is selected at random.

13. Use the counting principle to determine the number of sample points in the sample space.
14. Construct a tree diagram illustrating all the possible outcomes, and list the sample space.

By observing the sample space, determine the probability of obtaining

15. the number 6 and the letter a
16. the number 6 or the letter a
17. an even number or the letter c

18. A code is to consist of three digits followed by two letters. Find the number of possible codes if the first digit cannot be a 0 or 1 and replacement is not permitted.
19. A small class consist of 16 females and 11 males. If one person is selected at random from the class, find the odds against the person's being female.

20. The probability that Jane wins the raquetball tournament is $\frac{2}{7}$. Find the odds against Jane's winning the tournament.

21. You get to select one card at random from a deck of cards. If you pick a club, you win $10. If you pick a heart, you win $5. If you pick any other suit, you lose

21. You get to select one card at random from a deck of cards. If you pick a club, you win $10. If you pick a heart, you win $5. If you pick any other suit, you lose $7. Find your expectation for this game.

22. There are six people from whom three are to be selected and given small prizes. One will be given a book, one will be given a calculator, and the third will be given a $10 bill. In how many different ways can these prizes be awarded?

A bin has a total of 15 batteries of which six are defective. If you select two at random without replacement, find the probability that

23. none are good
24. at least one is good

25. There are five green apples and seven red apples in a bucket. Five apples are to be selected at random without replacement. Find the probability that three red apples and two green apples are selected.

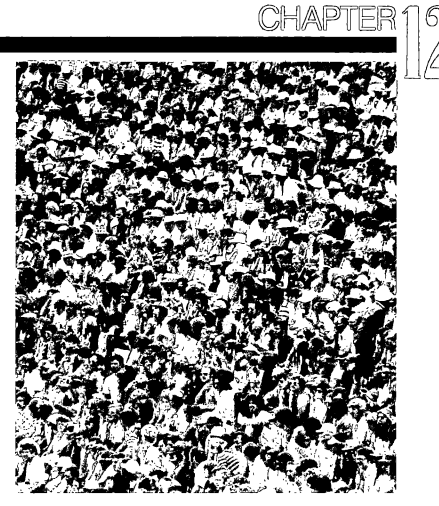

CHAPTER 12

Statistics

12.1

SAMPLING TECHNIQUES

Statistics is the art and science of gathering, analyzing, and making inferences (predictions) from numerical information obtained in an experiment. This numerical information is referred to as **data.** Statistics, originally associated with numbers gathered for governments, has grown significantly and is now used in all walks of life.

Governments use statistics to estimate the amount of unemployment and the cost of living. Thus statistics has become an indispensable tool in attempts to regulate the economy. In psychology and education the statistical theory of tests and measurements has been developed to compare achievements of individuals from diverse places and backgrounds. Another use of statistics that we are all familiar with is the public opinion poll. Newspapers and magazines carry the results of different Harris and Gallup polls on topics ranging from the president's popularity to the number of cans of beer consumed. In recent years these polls have had a high degree of accuracy. The A. C. Nielson rating is a public opinion poll that determines the country's most and least watched TV shows. Statistics is used in scores of other professions; in fact, it is difficult to find one that does not depend on some aspect of statistics.

Statistics is divided into two main branches: descriptive and inferential. **Descriptive statistics** is concerned with the collection, organization, and analysis of data. **Inferential statistics** is concerned with making generalizations or predictions from the data collected.

Probability and statistics are closely related. Someone in the field of probability is interested in computing the chance of occurrence of a particular event when all the possible outcomes are known. A statistician's interest lies in drawing conclusions about what the possible outcomes may be through observations of only a few particular events. If a probability expert and a statistician found identical boxes, the probability expert might open the box, observe the contents, replace the cover, and proceed to compute the probability of randomly selecting a specific object from the box. The statistician, on the other hand, might select a few items from the box without looking at the contents and make a prediction as to the total contents of the box. The few items the statistician randomly selected from the box constitute what is referred to as a **sample.** The entire contents of the box constitute what is referred to as the **population.** The statistician often uses a subset of the population, called a sample, to make predictions concerning the population.

When a statistician draws a conclusion from a sample, there is always the possibility that the conclusion is incorrect. For example, suppose a box contains 90 blue marbles and 10 red marbles. If the statistician selects a random sample of 5 marbles from the box and they are all blue, he or she may wrongly conclude that the box contains all blue marbles. If the statistician takes a larger sample, say 15 marbles, he or she is likely to select some red marbles. At that point the statistician may make a prediction about the contents of the box based on the sample selected. Of course, the most accurate result would occur if every object in the box, the entire population, were sampled. However, in most statistical experiments this is not practical.

Consider the task of a polling organization that has been commissioned to determine the political strength of a certain candidate running in a national election. It is not possible for pollsters to ask each of the millions of registered voters his or her preference for candidates. Thus pollsters must select and use a sample of the population to obtain their information. How large a sample do you think they use to make predictions about an upcoming national election? (Keep in mind that there are approximately 113 million registered voters.) You might be surprised to learn that pollsters use only about 1600 registered voters in their national sample. How can a pollster using such a small percentage of the population make an accurate prediction?

The answer lies in the fact that when pollsters select a sample, they use sophisticated statistical techniques to obtain a sample that is unbiased. An **unbiased sample** is one that is representative of the entire population. The sample they use must be like a small replica of the entire population with regard to income, education, sex, race, religion, political affiliation, age, and so on. The procedures that statisticians use to obtain reliable samples are quite complex. The following sampling techniques will give you a brief idea of how statisticians obtain unbiased samples.

Random Sampling

If a sample is drawn in such a way that each time an item is selected, each item in the population has an equal chance of being drawn, the sample is said to be a **random sample.** Under these conditions, one combination of a specified number of items will have the same probability of being selected as any other combination. When all of the items in the population are similar with regard to the specific characteristic we are interested in, a random sample can be expected to produce satisfactory results. For example, consider a large container holding 300 tennis balls that are identical except for color. One-third of the balls are red, one-third are white, and one-third are blue. If the balls can be thoroughly mixed between each draw of a tennis ball so that each ball has an equally likely chance of being selected, then randomness is not difficult to achieve. However, if the objects or items are not all the same size, shape, or texture, it might be impossible to obtain a random sample by reaching into a container and selecting an object.

The best procedure for selecting a random sample is to use a random number generator or a table of random numbers. A random number generator is a device, usually a calculator or computer program, that produces a list of random numbers. A random number table is a collection of random digits in which each digit has an equal chance of appearing. To select a random sample, first assign a number to each element in the population. Numbers are usually assigned in order. Then select the number of random numbers needed, which is determined by the sample size. Each numbered element from the population that corresponds to a selected random number becomes part of the sample.

Systematic Sampling

When a sample is obtained by drawing every nth item on a list or production line, the sample is a **systematic sample.** The first item should be determined by using a random number.

It is important that the list from which a systematic sample is chosen include the entire population one wants to study. The failure of the *Literary Digest* to correctly forecast the 1936 presidential election results was due to the fact that its systematic sample of more than 2,300,000 voters was not selected from an appropriate list. The voters were selected from lists of automobile owners and telephone subscribers, which in 1936 failed to include enough people in lower-income groups.

Another problem that one must avoid when using this method of sampling is the constantly recurring characteristic. For example, on an assembly line, every tenth item could be the work of robot X. If only every tenth item is checked, the work of other robots doing the same job may not be checked and may be defective.

Cluster Sampling

The **cluster sample** is sometimes referred to as an *area sample* because it is frequently applied on a geographical basis. Essentially, the sampling consists of a random selection of groups of units. For example, on a geographical basis we might select blocks of a city or some other geographical subdivision to use as a sample unit. Another example would be to select x boxes of screws from a whole order, count the number of defective screws in the x boxes, and use this number to determine the expected number of defective screws in the whole order.

Stratified Sampling

When a population is divided into parts, called strata, for the purpose of drawing a sample, the procedure is known as **stratified sampling.** Stratified sampling involves dividing up the population by characteristics such as sex, race, religion, or income. When a population has varied characteristics, it is desirable to separate the population into classes with similar characteristics and then take a random sample from each stratum (or class). Frequently, a proportionately stratified sample is used. A **proportionately stratified sample** is one in which each sample stratum has the same proportion as the stratum in the population. The criteria for deciding the appropriate stratum for a sampling unit are termed the stratifying factors. To be effective, the stratifying factors should divide the population into homogeneous groups. The strata selected must be relevant to the problem under study; otherwise, the introduction of stratification will not increase the reliability of the sample compared with other sampling techniques.

The use of stratified sampling requires some knowledge of the population. For example, to obtain a cross section of voters in a city, we must know where various groups are located and the approximate numbers in each location.

For more complete information on sampling techniques, consult a statistics book.

DID YOU KNOW?

Public opinion polls, such as those taken by the Harris and Gallup agencies, use only a small percentage of the population as a sample when making generalizations about the entire population. For example, in national surveys, where the population consists of millions, a sample of 1500–1700 is usually sufficient—if selected properly. Very sophisticated statistical techniques are used to attain a sample that is representative of the population from which it was drawn. When selected properly, the sample should resemble a small replica of the population.

The A. C. Nielson Company, which determines the popularity of TV shows, uses a sample of only 1170 families to draw conclusions about 83 million viewers. The A. C. Nielson families are paid $1 a month to have their TV sets metered. The meter, called an audiometer, is connected to a control computer, which determines the percentage of families watching each show. Each year, 20 percent of these families are changed so that no family is monitored for more than five years. (In contrast, most statistical surveys, such as Harris and Gallup, never use the same person more than twice.)

In any statistical survey, no matter how carefully planned, there is always a small possibility that the sample selected will be biased (not representative) and will result in inaccurate predictions.

EXERCISES 12.1

1. Explain the difference between descriptive and inferential statistics.
2. Define statistics in your own terms.
3. When you hear the word "statistics," what specific words or ideas come to mind?
4. Name five areas other than those mentioned in which statistics is used.
5. Attempt to list at least two professions in which no aspect of statistics is used.
6. Explain the difference between probability and statistics.
7. a) What is a sample?
   b) What is a population?
8. a) What is a random sample?
   b) How might a random sample be selected?
9. a) What is a systematic sample?
   b) How might a systematic sample be selected?
10. a) What is a cluster sample?
    b) How might a cluster sample be selected?
11. a) What is a stratified sample?
    b) How might a stratified sample be selected?
12. Discuss why a proportionally stratified sampling technique may be the best one to use to determine election results.
13. a) Select a topic of interest to which a random sampling technique can be applied to obtain data.
    b) Explain how you would obtain a random sample for your topic of interest.
    c) Actually obtain the sample by the procedure stated in part (b).
14. The principal of an elementary school wishes to determine the "average" family size of the children who attend the school. To obtain a sample, the principal visits each room and selects the four students closest to each corner of the room. The principal asks each of these students how many people are in his or her family.
    a) Will this technique result in an unbiased sample? Explain your answer.
    b) If the sample is biased, will the mean be greater than or less than the true family size? Explain.

 12.2 ## THE MISUSES OF STATISTICS

Statistics, when used properly, is a valuable tool to society. However, many individuals, businesses, and advertising firms misuse statistics to their own advantage. Each individual should examine statistical statements very carefully before accepting them as fact. Two questions you may ask yourself are: Was the sample used to gather the statistical data unbiased and of sufficient size? Is the statistical statement ambiguous; that is, can it be interpreted in more than one way?

Let us examine two advertisements. "Three out of five dentists recommend sugarless gum for their patients who chew gum." In this advertisement, it is unknown how large a sample was used or how many times this experiment was performed before the desired results were found. The advertisement does not mention that possibly only one of one hundred dentists recommended gum at all.

In a golf ball commercial, a "type A" ball is hit, and a second ball is hit in the same manner. The type A ball travels farther. From this the public is to conclude that the type A is the better ball. The advertisement makes no mention of the number of times the experiment was previously performed or the results of the earlier experiments. Possible sources of bias include (1) wind speed and direction, (2) the fact that no two swings are identical, and (3) the fact that the ball may land on a rough or smooth surface.

Vague or ambiguous words also lead to statistical misuses or misinterpretations. The word "average" is one such culprit. There are at least four different "averages," some of which are discussed in Section 12.5. Each is calculated differently, and each may have a different value for the same sample. During contract negotiations it is not uncommon for the employer to state publicly that the average salary of its employees is $25,000, while the employees' union states that the average is $20,000. Who is lying? It is possible that both are telling the truth. Each will use the average that best suits its needs to arrive at the stated figures. Advertisers also use the average that most enhances their product. Consumers often misinterpret this average as the one with which they are most familiar.

Another vague word is the word "largest." For example, ABC claims that it is the largest department store in the United States. What does this mean—largest profit, largest sales, largest building, largest staff, largest acreage, or largest number of outlets?

Still another deceptive technique used in advertising is to state a claim from which the public may draw irrelevant conclusions. For example, a disinfectant company claims that its disinfectant killed 40,760 germs in a laboratory in five seconds. "To prevent colds, use disinfectant A." It may well be that the germs killed in the laboratory were not related to any type of cold germ. In another example, company C claims that its paper towels are heavier than its competition's towels. Therefore they will hold more water. Is weight a measure of absorbency? A rock is heavier than a sponge, yet a sponge is more absorbent.

An insurance advertisement claims that in Rochester, New York, 212 people switched to insurance company A. One may conclude that this company is offering something special to attract these people. What may have been omitted from the advertisement is that 415 people in Rochester, New York, dropped insurance company A during the same period.

A foreign car manufacturer claims that nine out of ten of a particular model that it sold in the United States during the previous ten years were still on the road. From this statement the public is to conclude that this foreign car is well manufactured and would last for many years. The commercial neglects to state that this particular car has been selling in the United States for only a few years. The manufacturer could just as well have stated that nine out of ten of these cars sold in the United States in the previous 100 years were still on the road.

Charts and graphs can also be misleading or deceptive. Figure 12.1 (see page 640) gives two graphs showing performance of stocks. Which stock would you purchase? Actually, both graphs present identical information; the only difference is that the vertical scale of stock B has been exaggerated.

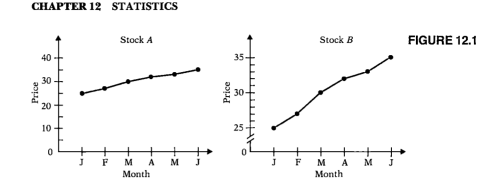

FIGURE 12.1

Figure 12.2 contains two graphs that show the same change. However, the graph on the left appears to show a greater increase than the graph on the right, once again because of a different scale.

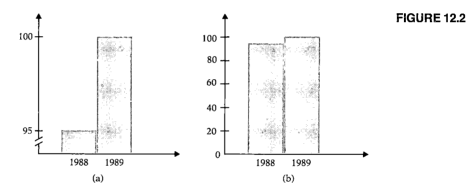

FIGURE 12.2

If you invest $1, by next year you will have $2. This is sometimes illustrated as in Fig. 12.3. Actually your investment has only doubled, but the area of the square on the right is four times that of the square on the left. By expressing the amounts as cubes (Fig. 12.4), you increase the volume eightfold.

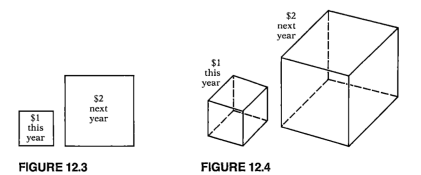

FIGURE 12.3   FIGURE 12.4

We do not want to leave you with the impression that statistics is used solely for the purpose of misleading or cheating the consumer. As stated earlier, there are many important and necessary uses of statistics. Most statistical reports are accurate and useful.

We hope this section has made you a more aware consumer.

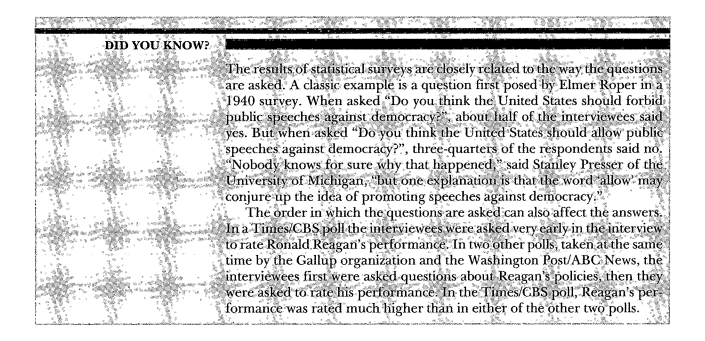

DID YOU KNOW?

The results of statistical surveys are closely related to the way the questions are asked. A classic example is a question first posed by Elmer Roper in a 1940 survey. When asked "Do you think the United States should forbid public speeches against democracy?", about half of the interviewees said yes. But when asked "Do you think the United States should allow public speeches against democracy?", three-quarters of the respondents said no. "Nobody knows for sure why that happened," said Stanley Presser of the University of Michigan, "but one explanation is that the word 'allow' may conjure up the idea of promoting speeches against democracy."

The order in which the questions are asked can also affect the answers. In a Times/CBS poll the interviewees were asked very early in the interview to rate Ronald Reagan's performance. In two other polls, taken at the same time by the Gallup organization and the Washington Post/ABC News, the interviewees first were asked questions about Reagan's policies, then they were asked to rate his performance. In the Times/CBS poll, Reagan's performance was rated much higher than in either of the other two polls.

EXERCISES 12.2

1. Find five advertisements or commercials that may be statistically misleading. Explain why each may be misleading.

Discuss each of the following statements, and tell what possible misuses or misinterpretations may exist.

2. Most accidents occur on Saturday evenings. Therefore people do not drive carefully on Saturday evenings.
3. Morgan's is the largest department store in New York. So shop at Morgan's and save money.

4. Most doctors who smoke use brand X cigarettes. Therefore it is safe to smoke brand X.
5. Eighty percent of all automobile accidents occur within 10 miles of the driver's home. Therefore it is safer to take long trips.
6. In Olean High School, half the students are below average in mathematics. Therefore the school should receive more federal aid to raise standards.
7. Florida has the highest death rate in the country. Therefore Florida should be condemned as unsafe.

8. Arizona has the highest death rate for asthma in the country. Therefore it is unsafe to go to Arizona if you have asthma.

9. Four out of six dentists preferred Calgonite toothpaste. Therefore it is the best toothpaste.

10. More men than women are involved in automobile accidents. Therefore women are better drivers.

11. Brand X has 80 percent fewer calories than the best-selling brand. So purchase brand X and stay healthy.

12. The average depth of the pond is only three feet, so it is safe to go wading.

13. A recent survey showed that 60 percent of those surveyed preferred Coors beer to Budweiser beer. Therefore more people buy Coors than Budweiser.

14. Males have a higher mean average than females on the mathematics part of the Scholastic Aptitude Test. Therefore on this test a particular male selected at random will outperform a particular female selected at random.

15. Below is a chart of the profits of the ABC Company from January to December of last year.

Month	Profit (percent)
January	6.0
February	5.0
March	4.5
April	4.3
May	4.8
June	5.0
July	5.5
August	6.0
September	7.0
October	6.6
November	7.2
December	7.5

a) Draw a line graph that makes the profit appear stable.

b) Draw a line graph that makes the profit appear to have a high gain.

16. The population of the United States in 1970 was 203.3 million. In 1980 the population was 226.5 million.

a) Draw a bar graph that appears to show a small population increase.

b) Draw a bar graph that appears to show a large population increase.

17. In 1988 the leading magazine in circulation was *TV Guide,* with a circulation of 17.0 million. The magazine with the second largest circulation was *Modern Maturity,* with a circulation of 16.7 million.

a) Draw a bar graph that appears to show a small difference in circulation.

b) Draw a bar graph that appears to show a large difference in circulation.

18. The following chart shows the average price paid for a man's suit from 1983 to 1987.

Year	Price
1983	$213
1984	$218
1985	$225
1986	$242
1987	$254

a) Draw a line graph that makes the increase in price appear to be small.

b) Draw a line graph that makes the increase in price appear to be large.

19. The following advertisement appeared in many newspapers.

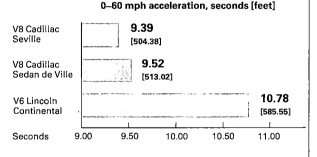

0–60 mph acceleration, seconds [feet]

a) Draw a graph that shows the entire scale from 0 to 11 seconds.

b) Does the new graph give a different impression?

12.3 FREQUENCY DISTRIBUTIONS

It is not uncommon for statisticians and others to have to analyze thousands of pieces of data. A **piece of data** is a single response to an experiment. When the amount of data is large, it is usually advantageous to construct a frequency distribution. A **frequency distribution** is a listing of the observed values and the corresponding frequency of occurrence of each value.

Example 1

The number of children per family is recorded for 64 families surveyed. Construct a frequency distribution of the following data:

$$
\begin{array}{cccccccc}
0 & 1 & 1 & 2 & 2 & 3 & 4 & 5 \\
0 & 1 & 1 & 2 & 2 & 3 & 4 & 5 \\
0 & 1 & 1 & 2 & 2 & 3 & 4 & 6 \\
0 & 1 & 2 & 2 & 2 & 3 & 4 & 6 \\
0 & 1 & 2 & 2 & 2 & 3 & 4 & 7 \\
0 & 1 & 2 & 2 & 3 & 3 & 4 & 8 \\
0 & 1 & 2 & 2 & 3 & 3 & 5 & 8 \\
0 & 1 & 2 & 2 & 3 & 3 & 5 & 9
\end{array}
$$

Solution: Listing the number of children (observed values) and the number of families (frequency) gives the following frequency distribution:

Number of Children (observed values)	Number of Families (frequency)
0	8
1	11
2	18
3	11
4	6
5	4
6	2
7	1
8	2
9	1
	64

There were 8 families with no children, 11 families with one child, 18 families with two children, and so on. Note that the sum of the frequencies is equal to the original number of pieces of data, 64. ❖

Often data are grouped in "classes" to provide information about the distribution that would be difficult to observe if the data were ungrouped.

Graphs called histograms and frequency polygons can be made of grouped data. These graphs also provide a great deal of useful information.

When data are grouped in classes, certain rules should be followed.

1. The classes should be of the same "width."
2. The classes should not overlap.
3. Each piece of data should belong to one and only one class.

In addition it is often suggested that a frequency distribution should be constructed with 5 to 12 classes. If there are too few or too many classes, the distribution may become difficult to interpret.

To understand these rules, consider a set of observed values that go from a low of 0 to a high of 26. Let's assume that the first class is arbitrarily selected to go from 0 through 4. Thus any of the data with values of 0, 1, 2, 3, 4 would belong in this class. We say that this class has a "width" of 5, since there are 5 integral values that belong to it. Since this first class ended with 4, the second class must start with 5. If this class is to have a width of 5, at what value must it end? The answer is 9 (5, 6, 7, 8, 9). The second class is 5–9. Continuing in the same manner we obtain the following set of classes:

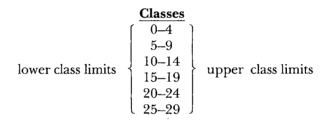

We need not go beyond the 25–29 class because the largest value we are considering is 26. There is no overlap among the classes. Each of the values from a low of 0 to a high of 26 belongs to one and only one class. The choice of the first class, 0–4, was arbitrary. If we wanted to have more classes or fewer, we would make the class widths smaller or larger, respectively.

The numbers 0, 5, 10, 15, 20, 25 are called the **lower class limits,** and the numbers 4, 9, 14, 19, 24, 29 are called the **upper class limits.** Each class has a width of 5. Notice that the class width, 5, can be obtained by subtracting the first lower class limit from the second lower class limit: $5 - 0 = 5$. The difference between any two consecutive lower class limits and the difference between any two consecutive upper class limits is also 5.

■ **Example 2** _____

The number of delegates at the 1988 Republican National Convention, held in New Orleans, is indicated in Table 12.1. Use Table 12.1 to construct a frequency distribution. Let the first class be 4–26.

TABLE 12.1							
Delegates to the 1988 Republican National Convention							
Alabama	38	Illinois	92	Nebraska	25	Rhode Island	21
Alaska	19	Indiana	51	Nevada	20	South Carolina	37
Arizona	33	Iowa	37	New Hampshire	23	South Dakota	18
Arkansas	27	Kansas	34	New Jersey	64	Tennessee	45
California	175	Kentucky	38	New Mexico	26	Texas	111
Colorado	36	Louisiana	41	New York	136	Utah	26
Connecticut	35	Maine	22	North Carolina	54	Vermont	17
Delaware	17	Maryland	41	North Dakota	16	Virgin Islands	4
D.C.	14	Massachusetts	52	Ohio	88	Virginia	50
Florida	82	Michigan	77	Oklahoma	36	Washington	41
Georgia	48	Minnesota	31	Oregon	32	West Virginia	28
Guam	4	Mississippi	31	Pennsylvania	96	Wisconsin	47
Hawaii	20	Missouri	47	Puerto Rico	14	Wyoming	18
Idaho	22	Montana	20				

Solution: There are 54 pieces of data. Rewriting the data in ascending order, that is, low to high, gives the following arrangement of the data:

$$
\begin{array}{cccccc}
4 & 19 & 26 & 35 & 41 & 64 \\
4 & 20 & 26 & 36 & 45 & 77 \\
14 & 20 & 27 & 36 & 47 & 82 \\
14 & 20 & 28 & 37 & 47 & 88 \\
16 & 21 & 31 & 37 & 48 & 92 \\
17 & 22 & 31 & 38 & 50 & 96 \\
17 & 22 & 32 & 38 & 51 & 111 \\
18 & 23 & 33 & 41 & 52 & 136 \\
18 & 25 & 34 & 41 & 54 & 175 \\
\end{array}
$$

The first class is 4–26. The second class must therefore start at 27. If we subtract 4 from 27 we obtain a class width of 23 units. The upper class limit of the second class must therefore be 26 + 23 or 49.

$$
\begin{array}{ll}
4\text{--}26 & \text{first class} \\
27\text{--}49 & \text{second class}
\end{array}
$$

The remaining classes will be 50–72, 73–95, 96–118, 119–141, 142–164, and 165–187. Since the highest observed value is 175, there is no need to go any further. Note that each two consecutive lower class limits differ by 23, as do each two consecutive upper limits. There are 20 pieces of data in the 4–26 class (4, 4, 14, 14, 16, 17, 17, 18, 18, 19, 20, 20, 20, 21, 22, 22, 23, 25, 26, and 26). There are 21 pieces of data in the 27–49 class, 5 pieces in the 50–72 class, and so on. The complete frequency distribution is as shown in the following chart.

Delegates	Number of States or Protectorates
4–26	20
27–49	21
50–72	5
73–95	4
96–118	2
119–141	1
142–164	0
165–187	1

The number of states or protectorates is 54, so we have included each piece of data. There are eight classes in the frequency distribution. ❖

The **modal class** of a frequency distribution is the class with the greatest frequency. In Example 2 the modal class is 27–49. The **midpoint** of a class, also called the **class mark,** is found by adding the lower and upper class limits and dividing the sum by 2. The midpoint of the first class in Example 2 is

$$\frac{4 + 26}{2} = \frac{30}{2} = 15$$

DID YOU KNOW?

Statistical errors often result from careless observations. Carefully determine the number of Fs that appear in the statement below. The answer is in the answer section at the back of the book.

FINISHED FILES ARE THE RE-
SULT OF YEARS OF SCIENTIF-
IC STUDY COMBINED WITH
THE EXPERIENCE OF YEARS.

EXERCISES 12.3

1. Given the frequency distribution on the right determine the following.
   a) The total number of observations
   b) The width of each class
   c) The midpoint of the second class
   d) The modal class
   e) The class limits of the next class if an additional class has to be added

Class	Frequency
7–13	4
14–20	2
21–27	3
28–34	0
35–41	4
42–48	6

2. Given the frequency distribution

Class	Frequency
20–29	8
30–39	6
40–49	4
50–59	3
60–69	8
70–79	2

determine the following.
a) The total number of observations
b) The width of each class
c) The midpoint of the second class
d) The modal class(es)
e) The class limits of the next class if an additional class has to be added

3. The results of a quiz given to a statistics class are given below. Construct a frequency distribution, letting each class have a width of 1 (as in Example 1).

0	1	2	3	5	6	7	8	8	9
0	1	2	3	5	6	7	8	8	9
0	1	2	4	5	6	7	8	8	9
0	1	3	4	6	6	8	8	8	10
1	1	3	4	6	6	8	8	8	10

4. The heights of a class of first-grade children, rounded to the nearest inch, are given below. Construct a frequency distribution, letting each class have a width of 1.

30	33	34	36	37
32	34	35	36	37
32	34	35	37	38
33	34	35	37	38
33	34	35	37	38

Note: There is no height of 31 inches. However, it is customary to include a missing value as an observed value and assign to it a frequency of 0.

A sample of 50 sixth-grade students were given IQ tests. The results of the test are given below.

80	89	92	95	97	100	102	106	110	120
81	89	93	95	98	100	103	108	113	120
87	90	94	97	99	100	103	108	114	122
88	91	94	97	100	100	103	108	114	128
89	92	94	97	100	101	104	109	119	135

Construct a frequency distribution with a first class of

5. 80–84 **6.** 80–88 **7.** 80–90 **8.** 80–92

Below are the ages of the U.S. presidents at their first inauguration.

57	57	49	52	50	51	51	56
61	61	64	56	47	56	60	61
57	54	50	46	55	55	62	52
57	68	48	54	54	51	43	69
58	51	65	49	42	54	55	

Construct a frequency distribution with a first class of

9. 40–45 **10.** 42–47 **11.** 42–46 **12.** 40–44

Use the data in Example 2 on page 644 to construct a frequency distribution with a first class of

13. 4–25 **14.** 4–30 **15.** 0–25 **16.** 0–23

The number of delegates for the 1988 Democratic National Convention, held in Atlanta, are given in Table 12.2.

TABLE 12.2

Delegates to the 1988 Democratic National Convention			
Alabama	56	Montana	19
Alaska	12	Nebraska	25
Arizona	36	Nevada	16
Arkansas	38	New Hampshire	18
California	314	New Jersey	109
Colorado	45	New Mexico	24
Connecticut	52	New York	255
Delaware	15	North Carolina	82
D.C.	16	North Dakota	15
Florida	136	Ohio	159
Georgia	77	Oklahoma	46
Guam	3	Oregon	45
Hawaii	20	Pennsylvania	178
Idaho	18	Puerto Rico	51
Illinois	173	Rhode Island	22
Indiana	79	South Carolina	44
Iowa	52	South Dakota	15
Kansas	39	Tennessee	70
Kentucky	55	Texas	183
Louisiana	63	Utah	23
Maine	23	Vermont	14
Maryland	76	Virgin Islands	3
Massachusetts	98	Virginia	75
Michigan	138	Washington	65
Minnesota	78	West Virginia	37
Mississippi	40	Wisconsin	81
Missouri	77	Wyoming	13

Construct a frequency distribution with a first class of

17. 3–43   **18.** 3–40   **19.** 0–40   **20.** 0–45

The population of the world's 35 largest urban areas is given below in millions (rounded to the nearest 100,000).

16.2	7.8	7.0	5.4	4.4	3.8	3.1
11.3	7.7	7.0	5.0	4.3	3.7	3.1
10.8	7.6	7.0	4.8	4.3	3.2	3.1
9.3	7.4	6.0	4.7	4.1	3.2	3.0
8.6	7.3	5.5	4.6	4.1	3.1	2.9

Construct a frequency distribution with a first class of

21. 2.0–3.9   **22.** 2.5–4.4
23. 2.9–5.8   **24.** 2.8–5.9

Table 12.3 indicates the number of visitors for each of the United States 49 national parks in 1987. Use Table 12.3 to construct a frequency distribution of *visits* with a first class of:

25. 0–999,999
26. 0–1,999,999
27. 0–1,500,000

Use Table 12.3 to construct a frequency distribution of *acres* with a first class of:

28. 0–999,999
29. 0–2,000,000
30. 0–1,500,000

TABLE 12.3

Annual number of visitors to national parks—

	Visits	Acres		Visits	Acres
Great Smoky Mountains, Tenn-N.C.	10,209,800	520,269	Bryce Canyon, Utah	718,342	35,835
Acadia, Maine	4,288,154	41,357	North Cascades, Wash.	651,606	504,781
Grand Canyon, Ariz.	3,513,030	1,218,375	Redwood, Calif.	610,897	110,178
Yosemite, Calif.	3,152,275	761,170	Biscayne, Fla.	607,968	173,039
Olympic, Wash.	2,822,850	921,935	Denali, Alaska	575,013	4,716,726
Yellowstone, Wyo.-Mont.-Idaho	2,573,194	2,219,785	Wind Cave, S.D.	563,720	28,292
			Lassen Volcanic, Calif.	472,431	106,372
Rocky Mountain, Colo.	2,531,864	265,200	Arches, Utah	468,916	73,379
Zion, Utah	1,777,619	146,598	Crater Lake, Ore.	460,550	183,224
Shenandoah, Va.	1,767,727	195,382	Capitol Reef, Utah	428,808	241,904
Glacier, Mont.	1,660,737	1,013,572	Theodore Roosevelt, N.D.	424,846	70,416
Mammoth Cave, Ky.	1,636,300	52,428	Big Bend, Tex.	227,921	735,416
Grand Teton, Wyo.	1,450,800	310,521	Voyageurs, Minn.	201,727	218,056
Haleakala, Hawaii	1,333,900	28,655	Channel Islands, Calif.	174,607	249,354
Mount Rainier, Wash.	1,292,027	235,404	Canyonlands, Utah	172,384	357,570
Badlands, S. Dak.	1,174,398	243,302	Guadalupe Mountains, Tex.	156,344	76,293
Sequoia, Calif.	1,139,389	402,482	Glacier Bay, Alaska	130,926	3,225,284
Hot Springs, Ark.	1,101,242	5,839	Great Basin, Nev.	63,532	77,109
Kings Canyon, Calif.	1,081,172	461,901	Kenai Fjords, Alaska	60,428	669,541
Hawaii Volcanoes, Hawaii	1,006,058	229,117	Katmai, Alaska	38,212	3,716,000
			Isle Royale, Mich.	31,760	571,790
Everglades, Fla.	787,493	1,398,938	Wrangell-Saint Elias, Alaska	29,191	8,331,604
Virgin Islands, V.I.	785,354	14,689	Lake Clark, Alaska	16,418	2,636,839
Carlsbad Caverns, N.M.	781,300	46,755	Gates of the Arctic, Alaska	1,060	7,523,888
Petrified Forest, Ariz.	758,082	93,532	Kobuk Valley, Alaska	230	1,750,421
Mesa Verde, Colo.	728,566	52,085			

Note: Figures are for 1987.
USN&WR—Basic data: National Park Service

12.4 STATISTICAL GRAPHS

In this section we will study three types of graphs: the circle graph, the histogram, and the frequency polygon. Each can be used to represent statistical data.

Circle graphs are often used to compare parts of one or more components of the whole to the whole. An example is shown in Fig. 12.5.

The circle graph in Fig. 12.5 gives the percentages of the various shades of tinted contact lenses sold by all manufacturers in 1987. The sector indicated by Baby blue should have an area that is 25% of the entire circle. The area of the Aqua sector should be 15% of the total area, and so on. The total area of the circle will be 100%.

In order to construct circle graphs we must be able to find central angles. A **central angle** of a circle is an angle whose vertex is at the center of the circle. Since each circle contains 360°, we find the central angles of each class (sector) by multiplying the percent of the total of each class by 360°.

Percentage of sales of colored contact lenses in 1987

FIGURE 12.5

■ Example 1

A survey of 2000 business executives yielded the following information:

Vacation Time (weeks)	Number of Executives
2	322
3	180
4	940
5	320
Other	238

Using this information as a sample construct a circle graph illustrating the percent in each category.

Solution: First determine the percent of the total for each of the categories. Then determine the measure of the corresponding central angle, as illustrated in Table 12.4.

TABLE 12.4			
Weeks of Vacation	Number of Executives	Percent of Total (to nearest percent)	Measure of Central Angle
2	322	$\frac{322}{2000} \times 100 = 16.1\%$	$0.161 \times 360° = 58.0°$
3	180	$\frac{180}{2000} \times 100 = 9.0\%$	$0.09 \times 360° = 32.4°$
4	940	$\frac{940}{2000} \times 100 = 47.0\%$	$0.47 \times 360° = 169.2°$
5	320	$\frac{320}{2000} \times 100 = 16.0\%$	$0.16 \times 360° = 57.6°$
Other	238	$\frac{238}{2000} \times 100 = 11.9\%$	$0.119 \times 360° = 42.8°$
Total	2000	100.0%	360.0°

Vacation time of executives

FIGURE 12.6

Now use a protractor to construct the circle graph and label it properly, as illustrated in Fig. 12.6. The measure of the central angle for two-week vacations is 58.0°, the three-week sector is 32.4°, and so on.   ❖

Note that the total percent in Example 1 is 100%. Occasionally, the total percent may be slightly less than or greater than 100% due to an error caused by rounding off decimals. The total degree measure may sometimes differ slightly from 360° because of errors due to rounding off.

Histograms and frequency polygons are statistical graphs used to illustrate frequency distributions. A **histogram** is a graph with observed values on its horizontal scale and frequencies on its vertical scale. A bar is constructed above each observed value (or class when classes are used) indicating the frequency of that value. The horizontal and vertical scales are not axes. The horizontal scale need not start at zero, and the calibrations on the horizontal scale do not have to be the same as the calibrations on the vertical scale. The vertical scale must start at zero. To accommodate large frequencies on the vertical scale, it may be necessary to break the scale. Because histograms and other bar graphs are easy to interpret visually, they are used a great deal in newspapers and magazines.

■ **Example 2** _____

The frequency distribution developed in Example 1 in Section 12.3 is given below. Construct a histogram of this frequency distribution.

Number of Children (observed values)	Number of Families (frequency)
0	8
1	11
2	18
3	11
4	6
5	4
6	2
7	1
8	2
9	1

Solution: The vertical scale must extend at least to the number 18, since that is the greatest recorded frequency (see Fig. 12.7). The horizontal scale must include the numbers 0 through 9, the number of children observed. There are 8 families with no children. We indicate this by constructing a bar above the number 0 on the horizontal scale extended up to 8 on the vertical scale. Since 11 families have one child, a bar extending up to 11 is constructed above the number 1 on the horizontal scale. We continue this procedure for each observed value. The horizontal and vertical scales should both be labeled, the bars should be the same width, and the histogram should have a title.

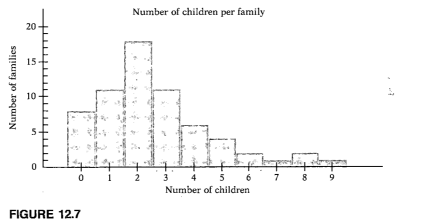

FIGURE 12.7

Frequency polygons are line graphs with scales the same as those of the histogram; that is, the horizontal scale indicates observed values and the vertical scale indicates frequency. To construct a frequency polygon, place a dot at the corresponding frequency above each of the observed values. Then connect the dots with straight line segments. In constructing frequency polygons, always put in two additional class marks, one at the lower end and one at the upper end on the horizontal scale (values for these added class marks are not needed on the frequency polygon). Since the frequency at these added class marks is 0, the endpoints of the frequency polygon will always be on the horizontal scale.

■ **Example 3** ————————————————————————

Construct a frequency polygon of the frequency distribution in Example 2.

Solution: Since there are 8 families with no children, place a mark above the 0 at 8 on the vertical scale, as shown in Fig. 12.8. There are 11 families with one child. Place a mark above the 1 at the 11 on the vertical scale, and so on. Connect the dots by straight line segments, and bring the endpoints of the graph down to the horizontal scale, as shown.

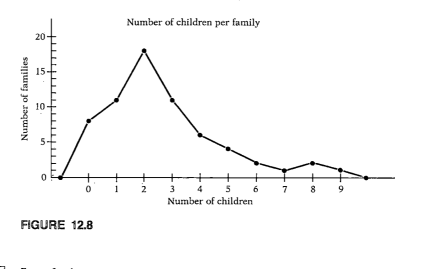

FIGURE 12.8

Example 4

Given the following frequency distribution of average gas mileage for 1988 automobiles, construct a histogram and then construct a frequency polygon on the histogram.

Mileage (mpg)	Number of Cars
10–14	4
15–19	32
20–24	36
25–29	8
30–34	5
35–39	4
40–44	1

Solution: The histogram can be constructed with either class limits or class marks on the horizontal scale. Frequency polygons are constructed with class marks on the horizontal scale. Since we will construct a frequency polygon on the histogram, we will use class marks. Recall that class marks, or class midpoints, are found by adding the lower class limit and upper class limit and dividing the sum by 2. For the first class, the class mark is $(10 + 14)/2$, or 12. Since the class widths are 5 units, the class marks will also differ by 5 units (see Fig. 12.9).

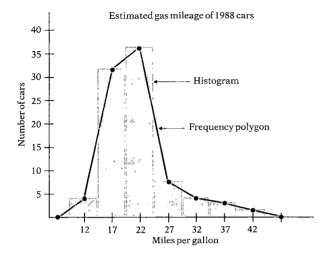

Estimated gas mileage of 1988 cars

Histogram

Frequency polygon

FIGURE 12.9

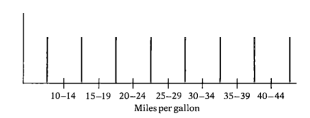

FIGURE 12.10

When class limits are used on the horizontal scale of a histogram, they are placed under the appropriate bars. The class limits should be placed so that they would fall between the bars if the bars were to be extended downward. Fig. 12.10 indicates the horizontal scale of the histogram in Fig. 12.9 using class limits.

Example 5 _____

Given the histogram in Fig. 12.11, construct the frequency distribution.

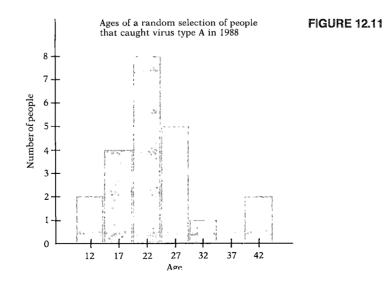

Ages of a random selection of people that caught virus type A in 1988

FIGURE 12.11

Solution: Since there are 5 units between class midpoints, each class width must also be 5 units. Since 12 is the midpoint of the first class, there must be 2 units below and 2 above it. The first class must be 10–14. The second class must therefore be 15–19, and so on.

Age	Number of People
10–14	2
15–19	4
20–24	8
25–29	5
30–34	1
35–39	0
40–44	2

❖

EXERCISES 12.4

1. The circle graph in Fig. 12.12 illustrates the percentage (to the nearest tenth of one percent) of students receiving degrees in various majors at a given university over the past 10 years. If the total number of degrees awarded over the past 10 years is 3600, determine the approximate number of degrees awarded for each sector of the circle graph.

2. The Penfield School District's budget of $20,962,415 for the 1988–89 school year will be distributed as indicated in Fig. 12.13. Determine to the nearest dollar the amount allocated for each sector of the circle.

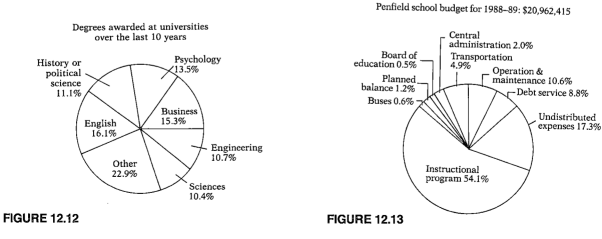

FIGURE 12.12

Degrees awarded at universities over the last 10 years

Penfield school budget for 1988–89: $20,962,415

FIGURE 12.13

3. Table 12.5 contains information on the market share of the top 10 best selling soft drinks in 1987. Construct a circle graph indicating the market share of soft drinks. Include a sector of the circle for other soft drinks.

TABLE 12.5

Soft Drink	Percent of Market
Coca-Cola Classic	19.8%
Pepsi	18.7%
Diet Coke	7.6%
Diet Pepsi	4.8%
Dr. Pepper	4.0%
Sprite	3.6%
Mountain Dew	3.3%
Caffeine Free Diet Coke	1.7%
Coca-Cola	1.7%
RC Cola	1.7%

4. The following chart indicates the number of franchised fast food restaurants as of January 1988, according to the primary type of food they sell.

Type of Food	Number of Restaurants
Chicken	8,683
Hamburgers	29,600
Pizza	11,593
Mexican food	3,620
Seafood	2,630
Pancakes	1,630
Steak	10,240
All other	2,266

Construct a circle graph of the preceding information.

5. The frequency distribution below indicates the number of years the 38 employees of the ABC Company have worked for that company.

Years	Number of Employees
0	6
1	4
2	2
3	1
4	6
5	8
6	3
7	3
8	5

a) Construct a histogram of the frequency distribution.
b) Construct a frequency polygon of the frequency distribution.

6. The frequency distribution below indicates the ages of a group of 40 people attending a party.

Age	Number of People
20	6
21	3
22	0
23	4
24	6
25	3
26	8
27	10

a) Construct a histogram of the frequency distribution.
b) Construct a frequency polygon of the frequency distribution.

7. The frequency distribution below represents the annual salaries in thousands of dollars of the people in management positions at the ABC Corporation.

Salary (in $1000)	Number of People
20–25	4
26–31	6
32–37	8
38–43	9
44–49	8
50–55	5
56–61	3

a) Construct a histogram of the frequency distribution.
b) Construct a frequency polygon of the frequency distribution.

8. The frequency distribution below indicates the gross weekly salaries of families whose children receive free lunch at Portersville Elementary School.

Salary	Number of Families
133–141	8
142–150	6
151–159	8
160–168	4
169–177	3
178–186	1

a) Construct a histogram of the frequency distribution.

b) Construct a frequency polygon of the frequency distribution.

9. Using the histogram in Fig. 12.14, answer the following questions.

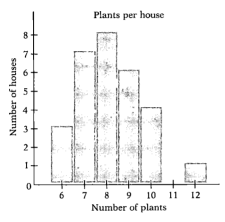

FIGURE 12.14

a) How many houses were surveyed?

b) In how many houses were nine plants observed?

c) What is the modal class?

d) How many plants were observed?

e) Construct a frequency distribution from this histogram.

10. Using the histogram in Fig. 12.15, answer the following questions.

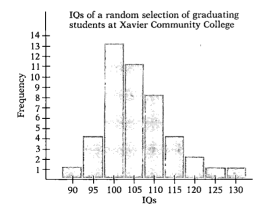

FIGURE 12.15

a) What are the upper and lower class limits of the first and second classes?

b) How many students had IQs in the class with a class mark of 100?

c) How many students were surveyed?

d) Construct a frequency distribution.

11. Use the frequency polygon in Fig. 12.16 to answer the following questions.

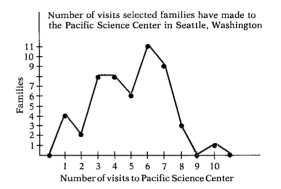

FIGURE 12.16

a) How many families visited the Pacific Science Center four times?

b) How many families visited the Pacific Science Center at least six times?

c) How many families were surveyed?

d) Construct a frequency distribution from the frequency polygon.

e) Construct a histogram from the frequency distribution in part (d).

12. Construct a histogram and a frequency polygon from the frequency distribution given in Exercise 1 for Section 12.3. See page 646.

13. Construct a histogram and a frequency polygon from the frequency distribution given in Exercise 2 for Section 12.3. See page 647.

14. Construct a histogram and a frequency polygon from the frequency distribution given in the solution of Example 2 in Section 12.3.

15. Starting salaries (rounded to the nearest thousand) for chemical engineers with B.S. degrees and no experience are given below for 25 different companies.

20	21	22	24	26
21	21	22	24	26
21	21	23	25	26
21	22	23	25	27
21	22	23	25	27

a) Construct a frequency distribution.
b) Construct a histogram.
c) Construct a frequency polygon.
16. The ages of a random sample of U.S. ambassadors are

40	43	45	50	52	55	59	64
41	43	46	50	53	55	60	65
41	44	47	50	54	55	60	65
42	44	48	51	54	57	60	66
43	45	48	51	54	58	62	67

a) Construct a frequency distribution with the first class 40–44.
b) Construct a histogram.
c) Construct a frequency polygon.
17. The world's 50 busiest airports in 1987 are given in Table. 12.6.
a) Construct a frequency distribution with the first class 8.0–13.0 million passengers.
b) Construct a histogram.
c) Construct a frequency polygon.

TABLE 12.6

Airport	Passengers[1]	Airport	Passengers[1]
1. O'Hare International: Chicago	53,338,056	26. Intercontinental: Houston	13,996,015
2. Hartsfield Atlanta International: Atlanta	45,191,480	27. Tacoma International: Seattle	13,642,666
3. International: Los Angeles	41,417,867	28. Sky Harbor Airport: Phoenix	13,274,015
4. International: Dallas/Ft. Worth	39,945,326	29. International: Philadelphia	12,780,306
5. Stapleton International: Denver	34,685,944	30. International: Orlando	12,495,346
6. Heathrow Airport: London	31,315,300	31. McCarran International: Las Vegas	12,303,400
7. International: Newark	29,433,046	32. Fiumicino Airport: Rome	12,241,145
8. International: San Francisco	28,607,363	33. Charlotte/Douglas International	11,987,339
9. Kennedy International: New York	27,223,733	34. Benito Juarez Airport: Mexico City	11,310,871
10. Tokyo International (Haneda): Japan	27,217,761	35. Stockholm-Arlanda Airport	10,599,000
11. La Guardia: New York	22,188,871	36. Kingsford Smith Airport: Australia	10,114,958
12. International: Miami	21,947,368	37. Salt Lake City International	9,990,986
13. International: Boston-Logan	21,862,718	38. Copenhagen Airport	9,971,012
14. International: Lambert-St. Louis	20,352,383	39. Athens Airport	9,599,651
15. Frankfurt-Main: West Germany	19,802,229	40. Zurich Airport	9,250,967
16. Orly Airport: Paris	18,543,670	41. International: Tampa	9,198,139
17. International: Honolulu	18,235,154	42. International: San Diego	9,084,438
18. Osaka International: Japan	17,694,649	43. Dulles International: Washington, D.C.	8,962,346
19. Detriot: (Wayne County)	17,604,583	44. Changi Airport: Singapore	8,912,233
20. International: Toronto	17,136,147	45. Memphis International Airport	8,725,359
21. International: Minneapolis-St. Paul	17,073,605	46. International: Baltimore	8,670,506
22. Gatwick Airport: London	16,309,300	47. Dusseldorf Airport	8,493,402
23. International: Pittsburgh	15,989,507	48. Vancouver International Airport	8,385,000
24. Charles De Gaulle: Paris	14,427,026	49. International: Kansas City	8,309,567
25. National Airport: Washington, D.C.	14,307,980	50. Amsterdam Airport Schiphol	8,207,969

[1]Enplaned, deplaned, and transfer, in millions. *Source:* Airport Operators Council International, June 1987.

 12.5 MEASURES OF CENTRAL TENDENCY

Most people have an intuitive idea of what is meant by an average. The term "average" is used daily in many familiar ways: This car averages 13 miles per gallon. The average test grade was 76. Her batting average is .313. On the average, one out of four will be selected.

An average is a number that is representative of a group of data. There are at least four different averages. We will discuss the mean, median, mode, and midrange. Each is calculated differently and may yield different results for the same set of data. Each will result in a number near the center of the data; and for this reason, averages are commonly referred to as the measures of central tendency.

The first average to be considered is the arithmetic mean, or simply the mean. It is symbolized either by $\bar{x}$ (read "x bar") or by the Greek letter mu, μ. The symbol $\bar{x}$ is used when the mean of a *sample* of the population is calculated. The symbol μ is used when the mean of the *entire population* is calculated. We will assume that the data studied in this book represents samples, and therefore we will always use $\bar{x}$ for the mean.

A symbol often used in statistics that we will use in explaining the mean is the Greek letter sigma, symbolized Σ. The symbol Σ is used to indicate "summation." The notation Σx, read "the sum of x," is used to indicate the sum of all the data. For example, if there are six pieces of data, 4, 6, 1, 0, 5, 3, then $\Sigma x = 4 + 6 + 1 + 0 + 5 + 3 = 19$.

Now we can discuss the procedure for finding the mean of a set of data.

The **mean** is found by summing all the pieces of data and dividing this sum by the number of pieces of data.

$$\text{Mean} = \frac{\text{sum of the data}}{\text{number of pieces of data}}$$

If we let $\bar{x}$ represent the mean, Σx represent the sum of all the data, and n represent the number of pieces of data, the formula for the mean becomes

$$\textbf{Mean}$$
$$\bar{x} = \frac{\Sigma x}{n}$$

■ **Example 1** _____

Find Tom's mean grade if he received grades of 49, 75, 40, 78, and 78 on his statistics exams.

Solution

$$\text{Mean} = \frac{\text{sum of the data}}{\text{number of pieces of data}}$$

$$= \frac{49 + 75 + 40 + 78 + 78}{5} = \frac{320}{5} = 64$$

Therefore the mean, $\bar{x}$, is 64.   ❖

The mean represents "the balancing point" of a set of scores. Consider a seesaw. If a seesaw is pivoted at the mean and uniform weights are placed at points corresponding to the test scores, as illustrated in Fig. 12.17, the seesaw will balance.

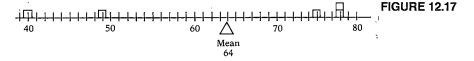

FIGURE 12.17

Mean
64

A second average is the median.

The **median** is the value in the middle of a set of *ranked data*.

Ranked data are data that are listed from smallest values to largest values, or vice versa.

To find the median of a set of data, rank the data and determine the value in the middle; this value will be the median.

■ Example 2 _____

Find the median of Tom's grades in Example 1.

Solution: Ranking the data from smallest to largest gives 40, 49, 75, 78, 78. Since 75 is the value in the middle of this set of ranked data (two pieces above and two pieces below), 75 is the median.   ❖

If there are an even number of pieces of data, the median will be halfway between the two middle pieces. The median will be found by adding the two middle pieces and dividing this sum by 2.

■ **Example 3** _____

Find the median of the set of data 1, 4, 6, 8, 9, 8, 3, 3.

Solution: Ranking the data gives 1, 3, 3, 4, 6, 8, 8, 9. There are eight pieces of data. Therefore the median will lie halfway between the two middle pieces, the 4 and the 6. The median is (4 + 6)/2, or 10/2, or 5.   ❖

■ **Example 4** _____

Find the median of the set of data 2, 3, 3, 3, 4, 5.

Solution: There are six pieces of data. Therefore the median will lie halfway between the two middle pieces. Both middle pieces are 3s. The median is (3 + 3)/2, or 6/2, or 3.   ❖

A third average is the mode.

| The **mode** is the piece of data that occurs most frequently.

■ **Example 5** _____

Find the mode of Tom's grades in Example 1.

Solution: Tom's grades were 49, 75, 40, 78, 78. The grade 78 is the mode, since it occurs twice and the other values occur only once.   ❖

If no one piece of data occurs more frequently than every other piece of data, the set of data has no mode. For example, neither of the following two sets of data has a mode:

$$1, 2, 3, 4, 5 \qquad \text{no mode}$$
$$1, 1, 2, 3, 3, 4, 5 \qquad \text{no mode*}$$

The last average that we will discuss is the midrange.

| The **midrange** is the value halfway between the lowest and the highest values.

_____

*Some textbooks refer to sets of data of this type as bimodal.

The midrange is found by adding the lowest and highest values and dividing the sum by 2.

$$\text{Midrange} = \frac{\text{lowest value} + \text{highest value}}{2}$$

■ **Example 6** _____

Find the midrange of Tom's grades given in Example 1: 49, 75, 40, 78, 78.

Solution: Tom's lowest grade is 40, and his highest grade is 78.

$$\text{Midrange} = \frac{\text{lowest} + \text{highest}}{2}$$

$$= \frac{40 + 78}{2} = \frac{118}{2} = 59 \qquad \text{❖}$$

   Tom's "average" grade for the exam scores of 49, 75, 40, 78, 78 can be considered any one of the following values: 64 (mean), 75 (median), 78 (mode), or 59 (midrange). Which average do you feel is most representative of his grades? We will discuss this question later in this section.

■ **Example 7** _____

Find the mean, median, mode, and midrange of the following set of data: 40, 25, 28, 35, 42, 60, 60, 73.

Solution

a)   The mean $= \dfrac{40 + 25 + 28 + 35 + 42 + 60 + 60 + 73}{8} = \dfrac{363}{8} =$
     45.375.   (The mean will be rounded to the nearest tenth, or 45.4).

b)   Listing the data from the smallest to largest gives 25, 28, 35, 40, 42, 60, 60, 73. Since there are an even number of pieces of data, the median is halfway between 40 and 42. The median $= (40 + 42)/2 = 82/2 = 41$.

c)   The mode is the piece of data that occurs most frequently; the mode is 60.

d)   The midrange $= (L + H)/2 = (25 + 73)/2 = 98/2 = 49$.   ❖

At this point you should be able to calculate the four measures of central tendency. Now let's examine the circumstances in which each is used.

The mean is used when each piece of data is to be considered and "weighed" equally. It is the most commonly used average. It is the only average that can be affected by *any* change in the set of data; for this reason it is the most sensitive of all the measures of central tendency (see Exercise 14).

Occasionally, one or more pieces of data may be much greater or much smaller than the rest of the data. When this occurs, these "extreme" values have the effect of increasing or decreasing the mean significantly so that the mean will not be representative of the total set of data. Under these circumstances the median is often used in place of the mean. The median is often used in describing average family incomes, because a relatively small number of families have extremely large incomes. These few families would inflate the mean income, making it nonrepresentative of the millions of families in the population.

Consider a set of exam scores from a mathematics class: 0, 16, 19, 65, 65, 65, 68, 69, 70, 72, 73, 73, 75, 78, 80, 85, 88, 92. Which average would best represent these grades? The mean is 64.06. The median is 71. Since only three scores out of eighteen fall below the mean, the mean would not be considered a good representative score, and the median of 71 would probably be the better average to use.

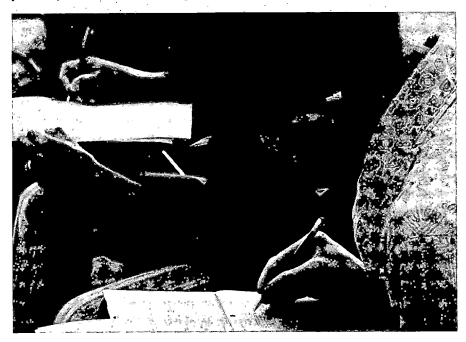

Standardized tests are often marked in percentiles.

Photo by Monkmeyer

The mode is the piece of data, if any, that occurs most frequently. Builders planning houses are interested in the most common family size. Retailers ordering shirts are interested in the most common shirt size. An individual purchasing a thermometer might select one with the most common reading. These examples illustrate how the mode may be used.

The midrange is sometimes used as the average when the item being studied is constantly fluctuating. Average daily temperatures are used to compare temperatures in different areas. They are calculated by adding the lowest and highest temperatures for the day and dividing the sum by 2. The midrange is actually the mean of the high value and the low value of a set of data. Occasionally, the midrange is used to estimate the mean, since it is much easier to calculate.

Sometimes an average itself is of little value. One must be very careful in interpreting its meaning. Jim is told that the average depth of Willow Pond is only three feet. He is not a good swimmer but decides that it is safe to go out a short distance in this shallow pond. After he is rescued, he exclaims, "I thought this pond was only three feet deep." Jim didn't realize that an average does not indicate extreme values or spread of the values. The spread of data will be discussed in Section 12.6.

DID YOU KNOW?

In addition to the measures of central tendencies there are also **measures of location.** The measures of location are often used to make comparisons. For example, measures of location may be used to compare two or more individuals from different populations. The measures of location are generally used only when the amount of data is large.

Two measures of location are **percentiles** and **quartiles.** Figure 12.18 indicates Scholastic Aptitude Test scores given as percentiles.

FIGURE 12.18

From ATP Guide for High Schools and Colleges, College Entrance Examination Board, 1988. Reprinted by permission of Educational Testing Service.

SCORES		SAT						ACHIEVEMENT TESTS			
CURRENT SCORES & PERCENTILES		VERBAL SUBSCORES						ACH 1	ACH 2	ACH 3	
TEST DATE	GRADE LEVEL	VERBAL	READING	VOCAB		MATH	TSWE				ACH AVG
NOV 83	12	480	45	50		490	48				500
PERCENTILES											
NATIONAL H S SAMPLE		82				78					
COLLEGE-BOUND SENIORS		68	58	74		56	64	25	31	39	

Percentiles divide the set of data into 100 equal parts, as in Fig. 12.19. An individual whose score falls in the seventy-sixth percentile has outperformed about 76 percent of all those taking the exam. James scored in the forty-fifth percentile on an exam. What does this mean? It *does not* mean that 45 percent of the questions were answered correctly. It *does not* mean that the grade on the exam was 45 percent. It *does* mean that James's score, regardless of what it was, was higher than about 45 percent of all the other scores, or that James performed better than 45 percent of all those taking the exam.

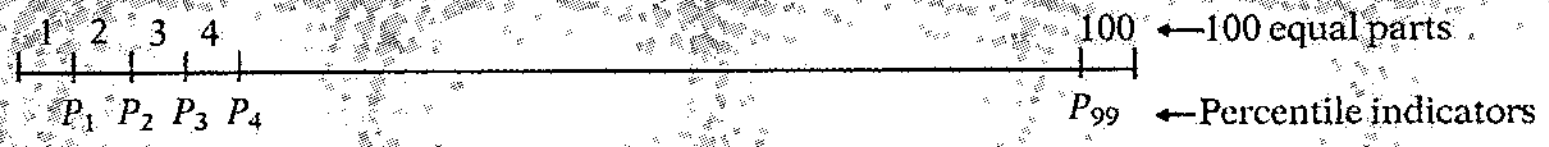

FIGURE 12.19

In general, if someone scores in the nth percentile, that person has outperformed about n percent of the population who took the exam. Therefore about $(100 - n)$ % of the population has outperformed the person.

A second measure of location is the quartile. **Quartiles** divide data into four equal parts (see Fig. 12.20).

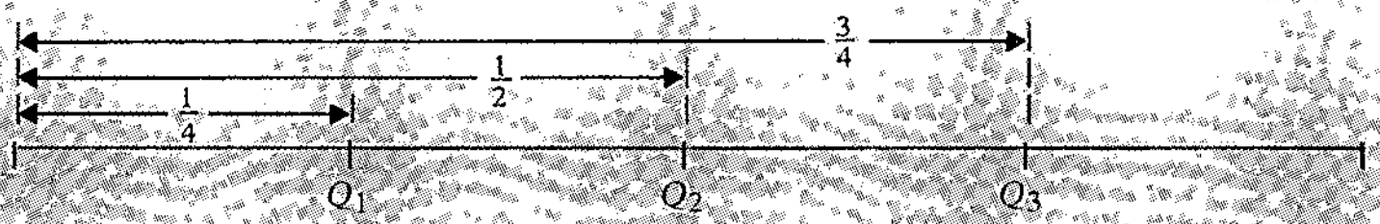

FIGURE 12.20

The first quartile, Q_1, is the value higher than 1/4 or 25 percent of the total data; the second quartile, Q_2, is higher than 1/2 the data; and the third quartile, Q_3, is higher than 3/4 of the data. Thus Q_1 is the twenty-fifth percentile, Q_2 is the median (or fiftieth percentile), and Q_3 is the seventy-fifth percentile.

EXERCISES 12.5

Find the mean, median, mode, and midrange of each of the following sets of data. Round answers off to the nearest tenth.

1. 1, 3, 3, 4, 5, 7, 8, 8, 8
2. 4, 5, 10, 12, 10, 9, 8, 372, 40, 37
3. 60, 72, 80, 84, 86, 45, 96
4. 5, 3, 6, 6, 6, 9, 11
5. 1, 3, 5, 7, 9, 11, 13, 15

6. 1, 7, 11, 27, 36, 14, 12, 9, 1
7. 40, 50, 30, 60, 90, 100, 140
8. 1, 1, 1, 1, 4, 4, 4, 4, 6, 8, 10, 12, 15, 21
9. 5, 7, 11, 12, 10, 9, 12, 14, 16
10. 1, 1, 1, 1, 9, 9, 9, 9
11. 237, 463, 812, 487, 387
12. 003, 2432, 16203, 962, 72
13. 78, 60, 72, 83, 92, 96

14. The mean is the "most sensitive" average, since it is affected by any change in the data.
    a) Determine the mean, median, mode, and midrange for 1, 2, 3, 5, 5, 7, 11.
    b) Change the 7 in the data above to a 10. Find the mean, median, mode, and midrange.
    c) Which averages were affected by changing the 7 to a 10 in part (b)?
    d) Which averages will be affected by changing the 11 to a 10?

15. In 1981 the National Center for Health Statistics indicated a new record "average life expectancy" of 74.7 years for the total U.S. population. The average life expectancy for men was 71.2 years and for women, 78.2 years. Which "average" do you think the National Center for Health is using? Explain your answer.

16. To get a grade of B, a student must have a mean average of 80. Jim has a mean average of 79 for ten quizzes. He approaches his teacher and asks for a B, explaining that he missed a B by only one point. What is wrong with Jim's reasoning?

17. The Webers' monthly gas and electric bills for the year are indicated below. Find the mean, median, mode, and midrange.

    $83.64, $84.72, $72.18, $82.74, $80.50, $76.96
    $75.96, $57.95, $49.41, $81.08, $80.60, $81.72

18. Below are the Ranieris' monthly telephone charges for a one-year period. Find the mean, median, mode, and midrange.

    $35.11, $33.48, $34.33, $32.13, $37.15, $45.53
    $34.68, $44.19, $45.29, $27.78, $28.94, $47.15

The chart below indicates the total number of medals won in all Winter Olympic games from 1924 up to and including 1988, by the top 10 winning teams.

		Gold	Silver	Bronze	Total
1.	Soviet Union	85	67	68	220
2.	Norway	57	62	58	177
3.	East Germany	48	49	41	134
4.	United States	46	51	34	131
5.	Finland	37	46	40	123
6.	Sweden	40	27	33	100
7.	Austria	28	38	33	99
8.	West Germany	28	27	24	79
9.	Switzerland	25	27	26	78
10.	Canada	16	13	19	48

Find the mean, median, mode, and midrange for the following.

19. Gold medals
20. Silver medals
21. Bronze medals
22. Total medals

23. Table 12.7 gives the annual cinema attendance and the per person cinema attendance for the 22 leading countries in each category.

TABLE 12.7

Annual cinema attendance (in millions)		Number of Visits per Person*	
USSR	4,200.0	Singapore	16.8
India	2,920.0	USSR	15.9
United States	1,022.0	Brunei	14.7
Philippines	318.0	Taiwan	13.0
Italy	276.3	Guyana	12.9
Mexico	269.8	Hong Kong	12.6
Vietnam	253.0	Grenada	12.5
Taiwan	229.0	Iceland	11.4
Burma	222.5	Bulgaria	10.7
Spain	200.5	United Arab	
Romania	193.6	Emirates	9.8
Pakistan	187.4	Macau	9.3
France	176.4	Mongolia	9.3
Brazil	164.8	Mauritius	8.7
Japan	164.0	Romania	8.7
West Germany	135.5	Malta	8.5
Indonesia	123.6	Burma	8.1
Poland	98.9	Philippines	7.5
Great Britain	98.8	Western Samoa	7.4
Bulgaria	95.9	Bolivia	6.8
Canada	93.2	Israel	6.5
Czechoslovakia	82.3	Chad	6.0
		Ireland	5.8

* There are 4.5 visits per person in the U.S., 4.1 in Canada, 1.8 in Great Britain, and 1.4 in Japan.

a) Find the mean, median, mode, and midrange of annual cinema attendance for the 22 countries listed.
b) Find the mean, median, mode, and midrange of the number of visits per capita for the 22 countries listed.
c) Which country would you say has the greatest moviegoers? Explain your answer.

24. Table 12.8 is a listing of Babe Ruth's home runs for the years he played baseball.

TABLE 12.8

Year	Club	Number of Home Runs
1914	Boston	0
1915	Boston	4
1916	Boston	3
1917	Boston	2
1918	Boston	11
1919	Boston	29
1920	New York	54
1921	New York	59
1922	New York	35
1923	New York	41
1924	New York	46
1925	New York	25
1926	New York	47
1927	New York	60
1928	New York	54
1929	New York	46
1930	New York	49
1931	New York	46
1932	New York	41
1933	New York	34
1934	New York	22
1935	Boston	6

a) Determine the mean, median, mode, and midrange of home runs hit over his 22 years of play.

b) Which average do you feel is most representative? Explain why. (Ruth was a pitcher until 1919. The first year he batted over 150 times was 1919, when he became a full-time outfielder.)

25. Paula's mean average on six exams is 83. Find the sum of her scores.

26. Mary's mean average on five exams is 72. Find the sum of her scores.

27. *Sports Illustrated* reported that the lowest salary of any professional baseball player of the 624 players on the major league roster on opening day of 1987 season was $62,500 (the rookie salary). The greatest salary was $2.46 million (Eddie Murray, Baltimore Orioles). The total salary for all 624 ball players was $256,296,770. Determine whether it is possible to find the measure of central tendency indicated. If so, find the value.

a) Mean
b) Median
c) Mode
d) Midrange

28. Construct a set of five pieces of data in which the mode has a lower value than the median and the median has a lower value than the mean.

29. Construct a set of five pieces of data with a mean of 70 with no two pieces of data the same.

30. Construct a set of six pieces of data with a mean, median, and midrange of 70 with no two pieces of data the same.

31. A mean average of 80 for five exams is needed for a final grade of B. George's first four exam grades are 70, 76, 83, 80. What grade does George need on the fifth exam to get a B in the course?

32. A mean average of 60 on seven exams is needed to pass a course. On her first six exams, Sheryl received grades of 49, 72, 80, 60, 57, and 69.

a) What grade must she receive on her last exam to pass the course?

b) An average of 70 is needed to get a C in the course. Is it possible for Sheryl to get a C? If so, what grade must she receive on the seventh exam?

c) If her lowest grade is to be dropped and only her six best exams are counted, what grade must she receive on her last exam to pass the course?

d) If her lowest grade is to be dropped and only her six best exams are counted, what grade must she receive on her last exam to get a C in the course?

33. Which of the measures of central tendency *must* be an actual piece of data in the distribution?

34. Construct a set of six pieces of data such that if only one piece of data is changed, the mean, median, and mode will all change.

35. Is it possible to construct a set of six pieces of data such that by changing only one piece of data you cause the mean, median, mode, and midrange to change? Explain your answer.

Refer to the Did You Know? on p. 663, and then answer the following questions.

36. When given a set of data, what must be done to the data before percentiles can be determined?

37. Mary scored in the 85th percentile on the verbal part of her College Board Test. Explain what this means.

38. When a national sample of the heights of kindergarten children were taken, Kevin was told that he was in the 43rd percentile. Explain what this means.

39. A union leader is told that when all workers' salaries are considered, the first quartile is $13,750. Explain what this means.

40. The third quartile for the time it takes students to travel one way to Central Valley College is 47.6 minutes. Explain what this means.

41. Give the names of two other statistics that have the same value as the fiftieth percentile.

42. Jim took an admission test for the University of California and scored in the 85th percentile. The following year Jim's sister Kendra took a similar admission test for the University of California and scored in the 90th percentile.

　　a) Is it possible to determine which of the two answered the higher percentage of questions correctly on their respective exams? Explain your answer.

　　b) Is it possible to determine which of the two was in a better relative position with respect to their respective populations? Explain your answer.

43. The following statistics represent monthly salaries at the Midtown Construction Company:

mean	$460	first quartile	$420
median	$450	third quartile	$485
mode	$440	83rd percentile	$525

　　a) What is the most common salary?
　　b) What salary did half the employees surpass?
　　c) About what percentage of employees surpassed $485?
　　d) About what percentage of employees' salaries was below $420?
　　e) About what percentage of employees surpassed $525?
　　f) If the company has 100 employees, what is the total monthly salary of all employees?

RESEARCH ACTIVITY _____

44. Two other measures of location that we did not mention in the Did You Know? on p. 663 are stanines and deciles. Use statistics books and books on educational testing and measurements to determine what stanines and deciles are and when percentiles, quartiles, stanines, and deciles are used.

12.6 MEASURES OF DISPERSION

The measures of central tendency by themselves do not always give sufficient information to analyze a situation and make necessary decisions. As an example, two manufacturers of airplane engines are under consideration for a contract. Manufacturer A's engines have an average (mean) life of 1000 hours of flying time before they must be rebuilt. Manufacturer B's engines have an average life of 950 hours of flying time before they must be rebuilt. If you assume that both cost the same, which engines should be purchased? Obviously, there are too many items to be considered here to answer this question. However, the average engine life may not be the most important factor. Manufacturer A's engines have an average life of 1000 hours. This could mean that half will last about 500 hours and the other half will last about 1500 hours. If in fact all of manufacturer B's engines have a life span of between 900 and 1000 hours, then B's engines are more consistent and reliable. If A's engines were purchased, they would all have to be rebuilt every 300 hours or so, because it would be impossible to determine which ones would fail first. If B's engines were purchased, they could go much longer before having to be rebuilt. This example is of course an exaggeration, used to illustrate the importance of knowing something about the *spread*, or *variability*, of the data.

The measures of dispersion are used to give indications of the spread of the data. The range and standard deviation are the measures of dispersion that will be discussed in this text.

The **range** is the difference between the highest and lowest values; it indicates the total spread of the data.

$$\boxed{\text{Range} = \text{highest value} - \text{lowest value}}$$

■ **Example 1** _____

Find the range of the following weights: 140, 195, 203, 175, 85, 197, 220, 212, 415.

Solution: Range = highest value − lowest value = 220 − 85 = 135. The range of these data is 135 pounds.   ❖

The second measure of dispersion is the **standard deviation.** It is symbolized either by the letter s or by the Greek letter sigma, σ*. The s is used when the standard deviation of a sample is calculated. The σ is used when the standard deviation of the entire population itself is calculated. Since we are assuming that all data presented here are for samples, we will always use s to represent the standard deviation in this book (note, however, that on the doctors' charts in Fig. 12.22, σ is used). The standard deviation is a measure of how much the data *differ from the mean*. The larger the spread of the data about the mean, the larger the standard deviation. Consider the following two sets of data:

$$5, 8, 9, 10, 12, 13 \qquad 8, 9, 9, 10, 10, 11$$

Both have a mean of 9.5. Which set of values on the whole do you feel differs less from the mean of 9.5? Figure 12.21 may make the answer more apparent.

The scores in the second set of data are closer to the mean and have a smaller standard deviation. You will soon be able to verify such relationships yourself.

---

*Our alphabet uses both uppercase and lowercase letters—for example, A and a. The Greek alphabet also uses both uppercase and lowercase letters. The symbol Σ is the capital Greek letter sigma, and σ is the lowercase Greek letter sigma.

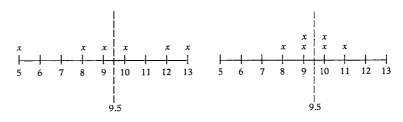

FIGURE 12.21

Sometimes only a very small standard deviation is desirable or acceptable. Consider a cereal manufacturer that is to put eight ounces of cereal into a box. If the amount of cereal put into the box varies too much—sometimes underfilling, sometimes overfilling—the manufacturer will soon be in trouble with consumer groups and government agencies.

At other times a larger spread of data is desirable and expected. In considering intelligence quotients (IQs) it is expected that there will be considerable spread about the mean, since we are all different.

Example 2 illustrates the procedure to follow to find the standard deviation of a set of data.

■ Example 2

Find the standard deviation of the following set of data:

$$7, 9, 11, 15, 18$$

Solution: First, determine the mean:

$$\bar{x} = \frac{7 + 9 + 11 + 15 + 18}{5} = \frac{60}{5} = 12$$

Next, construct a table with three columns, as illustrated in Table 12.9, and list the data in the first column (it is often helpful to list the data in ascending or descending order).

TABLE 12.9

Data	Data − mean	(Data − mean)²
7		
9		
11		
15		
18		

Complete the second column by subtracting the mean, 12 in this case, from each piece of data in the first column (Table 12.10).

TABLE 12.10

Data	Data − mean	(Data − mean)²
7	7 − 12 = −5	
9	9 − 12 = −3	
11	11 − 12 = −1	
15	15 − 12 = 3	
18	18 − 12 = 6	
	0	

The sum of the values in column 2 should always be zero; if not, you have made an error. (If the mean is a decimal, there may be a slight round-off error.)

Next, square the values in column 2 and place the squares in column 3 (Table 12.11).

TABLE 12.11

Data	Data − mean	(Data − mean)²
7	−5	$(-5)^2 = (-5)(-5) = 25$
9	−3	$(-3)^2 = (-3)(-3) = 9$
11	−1	$(-1)^2 = (-1)(-1) = 1$
15	3	$(3)^2 = (3)(3) = 9$
18	6	$(6)^2 = (6)(6) = 36$
	0	80

Add the numbers in column 3. In this case the sum is 80. Divide this sum by one less than the number of pieces of data $(n - 1)$.* In this case the number of pieces of data is 5. Therefore we divide by 4:

$$\frac{80}{4} = 20$$

Finally, take the square root† of this number. The square root of 20 can be found by using a calculator or the square root table in the appendix. Since $\sqrt{20} = 4.5$, thus the standard deviation, symbolized s, is 4.5 (to the nearest tenth). ❖

---

*When finding the standard deviation of a sample, we divide the sum of (data − mean)² column by $n - 1$. When finding the standard deviation of a population, we divide the sum by n. In this text we will always assume that the set of data represents a sample and divide by $n - 1$.

†Square root is discussed in Chapter 5.

Now we will develop a formula for finding the standard deviation of a set of data. If we call the individual data x and the mean $\bar{x}$, then the three column heads in Table 12.11 could be written

$$x \qquad x - \bar{x} \qquad (x - \bar{x})^2$$

Let us follow the procedure we used to obtain the standard deviation in Example 2. We found the sum of the (Data − mean)^2 column, which is the same as the sum of the $(x - \bar{x})^2$ column. We can represent the sum of the $(x - \bar{x})^2$ column by using the summation notation, $\Sigma(x - \bar{x})^2$. Thus in Table 12.11, $\Sigma(x - \bar{x})^2 = 80$. We then divided this number by one less than the number of pieces of data, $n - 1$. Thus we have

$$\frac{\Sigma(x - \bar{x})^2}{n - 1}$$

Finally, we took the square root of this value to obtain the standard deviation:

Standard Deviation

$$s = \sqrt{\frac{\Sigma(x - \bar{x})^2}{n - 1}}$$

■ **Example 3** ——————————————————————————————

Find the standard deviation of the following set of data:

$$20, 30, 32, 35, 40, 40, 41, 42, 43, 47$$

Solution: The mean, $\bar{x}$, is

$$\bar{x} = \frac{\Sigma x}{n} = \frac{20 + 30 + 32 + 35 + 40 + 40 + 41 + 42 + 43 + 47}{10} = \frac{370}{10}$$
$$= 37$$

Now find $\Sigma(x - \bar{x})^2$ (see Table 12.12 on page 672).

$$s = \sqrt{\frac{\Sigma(x - \bar{x})^2}{n - 1}} = \sqrt{\frac{562}{9}} = \sqrt{62.4} \approx 7.9$$

Note that there are ten pieces of data. Thus $n - 1$ is $10 - 1$ or 9. The standard deviation is 7.9.

TABLE 12.12

x	x − x̄	(x − x̄)²
20	20 − 37 = −17	(−17)² = 289
30	− 7	49
32	− 5	25
35	− 2	4
40	3	9
40	3	9
41	4	16
42	5	25
43	6	36
47	10	100
	0	$\Sigma(x - \bar{x})^2 = 562$

❖

Standard deviation will be used in Section 12.7 to find the percentage of data between any two values in a normal curve. Standard deviations are also often used in determining norms for a population (see Exercise 22).

DID YOU KNOW?

By convention, scientists and statisticians usually do not report something publicly until they are at least 95 percent certain of it. For example, when pollsters give the results of a poll, they are generally at least 95 percent certain of its accuracy.

EXERCISES 12.6

1. Without actually doing the calculations, decide which of the following two sets of data will have the greater standard deviation: 5, 8, 9, 10, 12, 16 or 8, 9, 9, 10, 10, 11. Explain why.
2. Of the following two sets of data, which would you expect to have the larger standard deviation? Explain.

   2, 4, 6, 8, 10   or   102, 104, 106, 108, 110

3. What does it mean when the standard deviation of a set of data is zero?
4. Can you think of any situations in which a large standard deviation may be desirable?
5. Can you think of any situations in which a small standard deviation may be desirable?

Find the range and standard deviations of the following sets of data.

6. 4, 3, 0, 8, 10
7. 6, 6, 10, 12, 3, 5
8. 120, 121, 122, 123, 124, 125, 126
9. 4, 0, 3, 6, 9, 12, 2, 3, 4, 7
10. 4, 8, 9
11. 6, 6, 6, 6, 6, 6
12. 10, 9, 14, 15, 17
13. 5, 6, 5, 4, 4, 5, 6
14. 42, 40, 44, 49, 30, 33, 54, 52
15. 7, 4, 7, 4, 7, 4, 7, 8
16. 5, 7, 12, 20, 17, 8, 13, 10, 10, 8
17. 60, 72, 84, 88, 76

18. 3, 4, 5, 9, 3, 7, 4, 4, 9, 2
19. 103, 106, 109, 112, 115, 118, 121

20.
  a) Pick any five numbers. Compute the mean and the standard deviation of this distribution.
  b) Add 20 to each of the numbers in your original distribution and compute the mean and the standard deviation of this new distribution.
  c) Subtract 5 from each number in your original distribution and compute the mean and standard deviation of this new distribution.
  d) Can you draw any conclusions about changes in the mean and the standard deviation when the same number is added to or subtracted from each piece of data in a distribution?
  e) How will the mean and standard deviation of the numbers 6, 7, 8, 9, 10, 11, 12 differ from the mean and standard deviation of the numbers 596, 597, 598, 599, 600, 601, 602? Find the mean and standard deviation of the latter set of numbers.

21.
  a) Pick any five numbers. Compute the mean and standard deviation of this distribution.
  b) Multiply each number in your distribution by 4 and compute the mean and the standard deviation of this new distribution.
  c) Multiply each number in your original distribution by 9 and compute the mean and the standard deviation of this new distribution.
  d) Can you draw any conclusions about changes in the mean and the standard deviation when each score in a distribution is multiplied by the same number?
  e) The mean and standard deviation of the distribution 1, 3, 4, 4, 5, 7 are 4 and 2, respectively. Use the conclusion drawn in part (d) to determine the mean and standard deviation of the distribution 5, 15, 20, 20, 25, 35.

*22. The chart in Fig. 12.22 uses the symbol σ to represent the standard deviation. Note that 2σ represents the value that is two standard deviations above the mean; −2σ represents the value that is two standard deviations below the mean. The unshaded areas, from two standard deviations below the mean to two standard deviations above the mean, are considered the normal range. For example, the average (mean) eight-year-old boy has a height of about 50 inches. But any heights between approximately 46 inches and 55 inches are considered normal for eight-year-old boys.

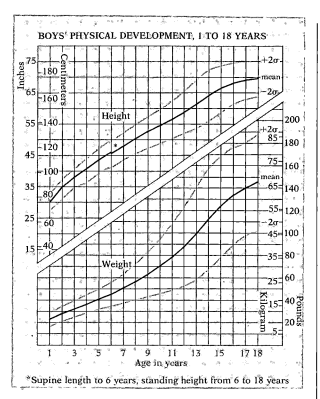

FIGURE 12.22

Refer to Fig. 12.22 to answer the following questions.

  a) What happens to the standard deviation for weights of boys as the age of boys increases? What is the significance of this fact?
  b) At age 16, what is the mean weight, in pounds, of boys?
  c) What is the approximate standard deviation of boys' weights at age 16?
  d) Find the mean weight and normal range for boys at age 13.
  e) Find the mean height and normal range for boys at age 13.
  f) Assuming that these charts were constructed so that approximately 95 percent of all children are always in the normal range, determine what percentage of children will not be in the normal range.

12.7   THE NORMAL CURVE

Certain shapes of distributions of data are more common than others. A few of the more common ones are illustrated below. In each distribution that we will discuss, the vertical scale is once again the frequency, and the horizontal scale is the observed values.

In a **rectangular distribution** (Fig. 12.23), all the observed values occur with the same frequency. If a die is rolled many times, one would expect the numbers 1–6 to occur with about the same frequency. The distribution representing the outcomes of the die would be rectangular.

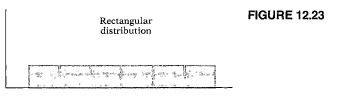

FIGURE 12.23

Rectangular
distribution

In **J-shaped distributions** the frequency is either constantly increasing (Fig. 12.24a) or constantly decreasing (Fig. 12.24b). The number of packages of cigarettes smoked per day by individuals shows a distribution like Fig. 12.24(b). The bars from left to right might represent 0 packs, 1 pack, 2 packs, and so on.

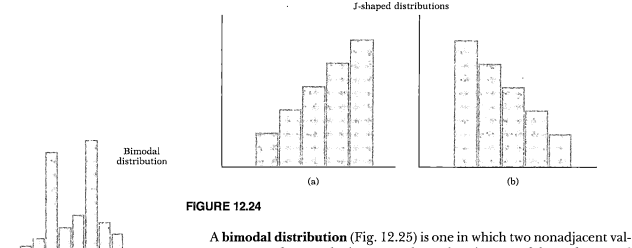

J-shaped distributions

(a)   (b)

FIGURE 12.24

Bimodal
distribution

FIGURE 12.25

A **bimodal distribution** (Fig. 12.25) is one in which two nonadjacent values occur more frequently than any other values in a set of data. If an equal number of men and women were weighed, the distribution of their weights would probably be bimodal, with one mode for the women's weights and the second for the men's weights.

Another distribution, called a **skewed distribution,** has more of a "tail" on one side than the other. A skewed distribution with a tail on the right (Fig. 12.26a) is said to be skewed to the right. If the tail is on the left (Fig. 12.26b), the distribution is referred to as skewed to the left.

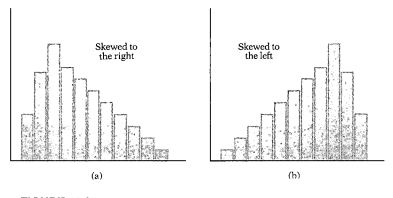

(a)   (b)

FIGURE 12.26

The number of children per family might be a distribution that is skewed to the right. A number of families have no children, more families may have one child, the greatest percentage may have two children, fewer three children, still fewer four children, and so on.

Since few families have relatively high incomes, distributions of family incomes might be skewed to the right.

If the histograms of the skewed distributions in Fig. 12.26 are smoothed out to form curves, we get the curves illustrated in Fig. 12.27.

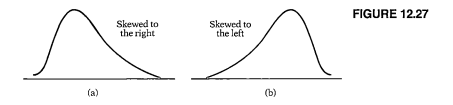

FIGURE 12.27

(a)   (b)

In Fig. 12.27(a) the greatest frequency appears on the left side of the curve, and the frequency decreases from left to right. Since the mode is the value with the greatest frequency, the mode would appear on the left side of the curve.

Every value in the set of data is considered in determining the mean. The values on the far right of the curve in Fig. 12.27(a) would tend to increase the value of the mean. Thus the value of the mean would be farther to the right than the mode. The median would be between the mode and the median. The relationship between the mean, median, and mode for curves that are skewed to the right and left is given in Fig. 12.28.

Each of these distributions is useful in describing sets of data. However, the most important distribution is the **normal** or **Gaussian distribution.** The histogram of a normal distribution is illustrated in Fig. 12.29.

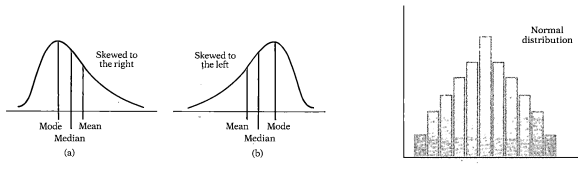

FIGURE 12.28   **FIGURE 12.29**

The normal distribution is important because many sets of data are normally distributed, or they closely resemble a normal distribution. Such distributions include intelligence quotients, heights and weights of males, heights and weights of females, lengths of full-grown boa constrictors, weights of watermelons, wear-out mileage of automobile brakes, and life spans of General Electric refrigerators—to name a few.

The normal distribution is symmetric about the mean. If you were to fold the histogram down the middle, the left side would fit exactly on the right side. *In a normal distribution the mean, median, and mode will all have the same value.*

When the histogram of a normal distribution is smoothed out to form a curve, the curve is bell-shaped. The bell may be high and narrow or short and wide. Each of the three curves in Fig. 12.30 represents a normal curve. The curve on the left has the smallest standard deviation (spread from the mean); the curve on the right has the largest.

Since the curve is symmetric, there will always be 50 percent of the data above (to the right of) the mean and 50 percent of the data below (to the left of) the mean. In addition, every normal distribution has approximately 68 percent of the data between the value that is one standard deviation below the mean and the value that is one standard deviation above the mean (see Fig. 12.31). Approximately 95 percent of the data will be between the value that is two standard deviations below the mean and the value that is two standard deviations above the mean.

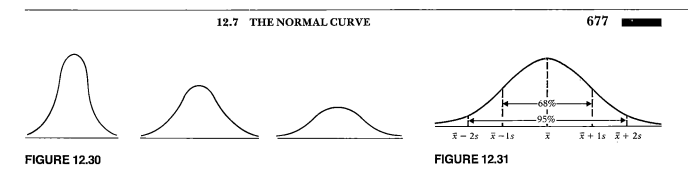

FIGURE 12.30   **FIGURE 12.31**

Thus if a normal distribution has a mean of 100 and a standard deviation of 10, approximately 68 percent of all the data will be between $100 - 10$ and $100 + 10$, or between 90 and 110. Approximately 95 percent of the data will be between $100 - 20$ and $100 + 20$, or between 80 and 120. In fact, given any normal distribution with a known standard deviation and mean, it is possible through the use of Table 12.13 to determine the percentage of data between any two given values.

The table gives z-**scores (or standard scores)** and the areas between the z-scores and the mean. We use z-scores to determine how far, in terms of standard deviations, a given score is from the mean of the distribution. For example, a score that has a z-value of 1.5 means that the score is 1.5 standard deviations above the mean. The standard or z-score is calculated by the formula:

$$z = \frac{\text{value of the piece of data } - \text{ mean}}{\text{standard deviation}}$$

Letting x represent the value of the given piece of data, $\bar{x}$ the mean, and s the standard deviation, the formula above can be symbolized as

$$z = \frac{x - \bar{x}}{s}$$

In this text we shall use the notation z_x to represent the z-score, or standard score, of the value x. For example, if a normal distribution has a mean of 86 with a standard deviation of 12, a score of 110 has a standard or z-score of

$$z_{110} = \frac{110 - 86}{12} = \frac{24}{12} = 2$$

Therefore a value of 110 in this distribution has a z-score of 2. The score of 110 is thus two standard deviations above the mean.

TABLE 12.13

Areas under the Standard Normal Curve (the z table)

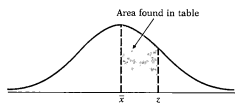

Area found in table

The column under A gives the proportion of the area under the entire curve that is between $z = 0$ and a positive value of z.

$\bar{x}$ z

z	A	z	A	z	A	z	A	z	A	z	A
.00	.000	.28	.110	.56	.212	.84	.300	1.12	.369	1.40	.419
.01	.004	.29	.114	.57	.216	.85	.302	1.13	.371	1.41	.421
.02	.008	.30	.118	.58	.219	.86	.305	1.14	.373	1.42	.422
.03	.012	.31	.122	.59	.222	.87	.308	1.15	.375	1.43	.424
.04	.016	.32	.126	.60	.226	.88	.311	1.16	.377	1.44	.425
.05	.020	.33	.129	.61	.229	.89	.313	1.17	.379	1.45	.427
.06	.024	.34	.133	.62	.232	.90	.316	1.18	.381	1.46	.428
.07	.028	.35	.137	.63	.236	.91	.319	1.19	.383	1.47	.429
.08	.032	.36	.141	.64	.239	.92	.321	1.20	.385	1.48	.431
.09	.036	.37	.144	.65	.242	.93	.324	1.21	.387	1.49	.432
.10	.040	.38	.148	.66	.245	.94	.326	1.22	.389	1.50	.433
.11	.044	.39	.152	.67	.249	.95	.329	1.23	.391	1.51	.435
.12	.048	.40	.155	.68	.252	.96	.332	1.24	.393	1.52	.436
.13	.052	.41	.159	.69	.255	.97	.334	1.25	.394	1.53	.437
.14	.056	.42	.163	.70	.258	.98	.337	1.26	.396	1.54	.438
.15	.060	.43	.166	.71	.261	.99	.339	1.27	.398	1.55	.439
.16	.064	.44	.170	.72	.264	1.00	.341	1.28	.400	1.56	.441
.17	.068	.45	.174	.73	.267	1.01	.344	1.29	.402	1.57	.442
.18	.071	.46	.177	.74	.270	1.02	.346	1.30	.403	1.58	.443
.19	.075	.47	.181	.75	.273	1.03	.349	1.31	.405	1.59	.444
.20	.079	.48	.184	.76	.276	1.04	.351	1.32	.407	1.60	.445
.21	.083	.49	.188	.77	.279	1.05	.353	1.33	.408	1.61	.446
.22	.087	.50	.192	.78	.282	1.06	.355	1.34	.410	1.62	.447
.23	.091	.51	.195	.79	.285	1.07	.358	1.35	.412	1.63	.449
.24	.095	.52	.199	.80	.288	1.08	.360	1.36	.413	1.64	.450
.25	.099	.53	.202	.81	.291	1.09	.362	1.37	.415	1.65	.451
.26	.103	.54	.205	.82	.294	1.10	.364	1.38	.416	1.66	.452
.27	.106	.55	.209	.83	.297	1.11	.367	1.39	.418	1.67	.453

Data below the mean will always have a negative z-scores; data above the mean will always have positive z-scores. The mean will always have a z-score of 0.

 Example 1

A normal distribution has a mean of 100 and a standard deviation of 10. Find z-scores for the following values.

TABLE 12.13 (continued)

z	A	z	A	z	A	z	A	z	A	z	A
1.68	.454	1.96	.475	2.24	.488	2.52	.494	2.80	.497	3.08	.499
1.69	.455	1.97	.476	2.25	.488	2.53	.494	2.81	.498	3.09	.499
1.70	.455	1.98	.476	2.26	.488.	2.54	.495	2.82	.498	3.10	.499
1.71	.456	1.99	.477	2.27	.488	2.55	.495	2.83	.498	3.11	.499
1.72	.457	2.00	.477	2.28	.489	2.56	.495	2.84	.498	3.12	.499
1.73	.458	2.01	.478	2.29	.489	2.57	.495	2.85	.498	3.13	.499
1.74	.459	2.02	.478	2.30	.489	2.58	.495	2.86	.498	3.14	.499
1.75	.460	2.03	.479	2.31	.490	2.59	.495	2.87	.498	3.15	.499
1.76	.461	2.04	.479	2.32	.490	2.60	.495	2.88	.498	3.16	.499
1.77	.462	2.05	.480	2.33	.490	2.61	.496	2.89	.498	3.17	.499
1.78	.463	2.06	.480	2.34	.490	2.62	.496	2.90	.498	3.18	.499
1.79	.463	2.07	.481	2.35	.491	2.63	.496	2.91	.498	3.19	.499
1.80	.464	2.08	.481	2.36	.491	2.64	.496	2.92	.498	3.20	.499
1.81	.465	2.09	.482	2.37	.491	2.65	.496	2.93	.498	3.21	.499
1.82	.466	2.10	.482	2.38	.491	2.66	.496	2.94	.498	3.22	.499
1.83	.466	2.11	.483	2.39	.492	2.67	.496	2.95	.498	3.23	.499
1.84	.467	2.12	.483	2.40	.492	2.68	.496	2.96	.499	3.24	.499
1.85	.468	2.13	.483	2.41	.492	2.69	.496	2.97	.499	3.25	.499
1.86	.469	2.14	.484	2.42	.492	2.70	.497	2.98	.499	3.26	.499
1.87	.469	2.15	.484	2.43	.493	2.71	.497	2.99	.499	3.27	.500
1.88	.470	2.16	.485	2.44	.493	2.72	.497	3.00	.499	3.28	.500
1.89	.471	2.17	.485	2.45	.493	2.73	.497	3.01	.499	3.29	.500
1.90	.471	2.18	.485	2.46	.493	2.74	.497	3.02	.499	3.30	.500
1.91	.472	2.19	.486	2.47	.493	2.75	.497	3.03	.499	3.31	.500
1.92	.473	2.20	.486	2.48	.493	2.76	.497	3.04	.499	3.32	.500
1.93	.473	2.21	.487	2.49	.494	2.77	.497	3.05	.499	3.33	.500
1.94	.474	2.22	.487	2.50	.494	2.78	.497	3.06	.499		
1.95	.474	2.23	.487	2.51	.494	2.79	.497	3.07	.499		

a) 110   **b)** 115   **c)** 100   **d)** 88

Solution

a)
$$z = \frac{\text{value} - \text{mean}}{\text{standard deviation}}$$

$$z_{110} = \frac{110 - 100}{10} = \frac{10}{10} = 1$$

(A score of 110 is one standard deviation above the mean.)

b)
$$z_{115} = \frac{115 - 100}{10} = \frac{15}{10} = 1.5$$

c)
$$z_{100} = \frac{100 - 100}{10} = \frac{0}{10} = 0$$

(The mean always has a z-score of 0.)

d)
$$z_{88} = \frac{88 - 100}{10} = \frac{-12}{10} = -1.2$$

(A score of 88 is 1.2 standard deviations below the mean.)   ❖

The total area under the normal curve is 1.00. Table 12.13 will be used to determine the area under the normal curve between any two given points. The values in the table have been rounded off.

Table 12.13 gives the area under the normal curve from the mean (a z-value of 0) *to a z-value to the right of the mean.* For example, between the mean and z = 2.00 there is 0.477 of the total area under the curve (Fig. 12.32). To change this area of 0.477 to a percentage, simply multiply by 100: 0.477 × 100 is 47.7 percent. Thus 47.7 percent of all scores will be between the mean and the score that is two standard deviations above the mean.

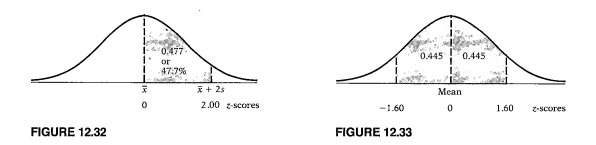

FIGURE 12.32   **FIGURE 12.33**

When you are finding the area under the normal curve, it is often helpful to draw a picture such as the one in Fig. 12.32, indicating the area or percentage to be found.

The normal curve is symmetric about the mean. Thus the same percentage of data is between the mean and a positive z-score as between the mean and the corresponding negative z-score. For example, there is the same area under the normal curve between a z of 1.60 and the mean as between a z of −1.60 and the mean. Both have an area of 0.445 (Fig. 12.33).

At this time we have the necessary knowledge to find the percentage of data between any two values in a normal distribution. To find the percentage of data between any two values, convert the values to z-scores, using the formula $z = (x - \bar{x})/s$, and look up the percentage in Table 12.13. When finding the percentage of data between two values on opposite sides of the mean, you must add the individual percentages. When finding the percentage of data between two values on the same side of the mean, you must subtract the smaller percentage from the larger percentage.

■ **Example 2** ——————————————————————————

Intelligence quotients are normally distributed with a mean of 100 and a standard deviation of 10. Find the percentage of individuals with IQs in the following ranges.

a) Between 100 and 110   b) Between 80 and 100
c) Between 80 and 110   d) Between 110 and 120
e) Below 120   f) Above 115

Solution

a) The area under the normal curve between the values of 100 and 110 is illustrated in Fig. 12.34.

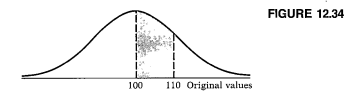

FIGURE 12.34

Converting 100 to a z-score yields a z-score of 0:

$$z_{100} = \frac{100 - 100}{10} = \frac{0}{10} = 0$$

Converting 110 to a z-score yields a z-score of 1:

$$z_{110} = \frac{110 - 100}{10} = \frac{10}{10} = 1.00$$

The percentage of individuals with IQs between 100 and 110 is the same as the percentage of data between z-scores of 0 and 1 (see Fig. 12.35).

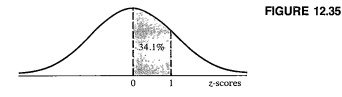

FIGURE 12.35

From Table 12.13 we determine that 0.341 of the area, or 34.1 percent of all the data, is between z-scores of 0 and 1.00. Therefore 34.1 percent of individuals have IQs between 100 and 110.

b)
$$z_{80} = \frac{80 - 100}{10} = \frac{-20}{10} = -2.00$$
$$z_{100} = 0 \text{ (from part a)}$$

The percentage of data between scores of 80 and 100 is the same as the percentage between $z = -2$ and $z = 0$ (Fig. 12.36). The percentage of data between the mean and two standard deviations below the mean is the same as between the mean and two standard deviations above the mean. The percentage of data between $z = 0$ and $z = 2$ is 47.7 percent. Thus 47.7 percent of the data is also between $z = -2$ and $z = 0$. Therefore 47.7 percent of all individuals have IQs between 80 and 100.

FIGURE 12.36

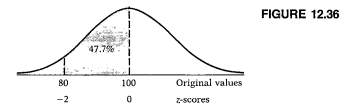

c) In parts (a) and (b) we determined that $z_{110} = 1$ and $z_{80} = -2$. Since the values are on opposite sides of the mean, the percentage of data between the two values is found by adding the individual percentages: $34.1 + 47.7 = 81.8$ (Fig. 12.37). Thus 81.8 percent of the IQs are between 80 and 110.

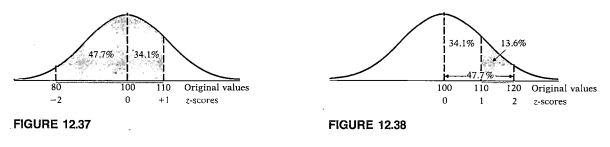

FIGURE 12.37   FIGURE 12.38

d)
$$z_{120} = \frac{120 - 100}{10} = \frac{20}{10} = 2.00$$
$$z_{110} = 1.00 \text{ (from above)}$$

Since both values are on the same side of the mean (Fig. 12.38), the smaller percentage must be subtracted from the larger percentage: 47.7% − 34.1% is 13.6%. Thus 13.6 percent of all the individuals have IQs between 110 and 120.

e) The percentage of IQs below 120 is the same as the percentage of data below a z of 2. Between $z = 2$ and the mean is 47.7 percent of the data (Fig. 12.39). To this percentage we add the 50 percent of the data below the mean to give 97.7 percent. Thus 50% + 47.7%, or 97.7%, of all IQs are below 120.

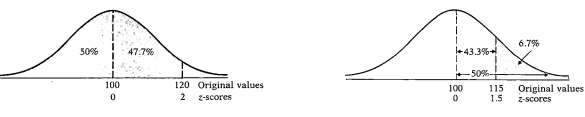

FIGURE 12.39   **FIGURE 12.40**

f)

$$z = \frac{115 - 100}{10} = \frac{15}{10} = 1.50$$

The percentage of IQs above 115 is the same as the percentage of data above $z = 1.5$ (Fig. 12.40). Fifty percent of the data is to the right of the mean. Since 43.3 percent of the data is between the mean and $z = 1.5$, 50% − 43.3%, or 6.7%, of the data is greater than $z = 1.5$. Thus 6.7 percent of all IQs are greater than 115.   ❖

■ **Example 3** ───

The wear-out mileage of a certain tire is normally distributed with a mean of 35,000 miles and standard deviation of 2500 miles.

a) Find the percentage of tires that will last at least 35,000 miles.

b) Find the percentage of tires that will last between 30,000 and 37,500 miles.

c) Find the percentage of tires that will last at least 39,000 miles.

d) If the manufacturer guarantees the tires to last at least 30,000 miles, what percentage of tires will fail to live up to the guarantee?

e) If 200,000 tires are produced, how many will last at least 39,000 miles?

Solution

a) In a normal distribution, half the data are always above the mean. Since 35,000 miles is the mean, half or 50 percent of the tires will last at least 35,000 miles.

b) Convert 30,000 miles and 37,500 miles to z-scores:

$$z_{30,000} = \frac{30,000 - 35,000}{2500} = -2.00$$

$$z_{37,500} = \frac{37,500 - 35,000}{2500} = 1.00$$

Now look up the areas in Table 12.13. The percentage of tires that will last between 30,000 and 37,500 miles is 47.7% + 34.1%, or 81.8 percent (Fig. 12.41).

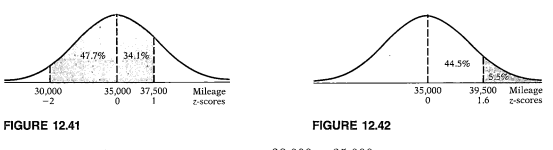

FIGURE 12.41	**FIGURE 12.42**

c)
$$z_{39,000} = \frac{39,000 - 35,000}{2500} = 1.60$$

The percentage of data between the mean and z = 1.60 is 44.5% (Fig. 12.42). Therefore the percentage of data above z = 1.60 is 50% − 44.5% = 5.5%. Thus 5.5 percent of the tires will last at least 39,000 miles.

d) To solve this problem, we must find the percentage of tires that last less than 30,000 miles: $z_{30,000} = -2.00$. From Table 12.13 we determine that 47.7 percent of the data is between z = −2.00 and z = 0, (Fig. 12.43). The percentage to the left of z = −2 is found by subtracting 47.7% from 50% to obtain 2.3%. Thus 2.3 percent of all the tires will last less than 30,000 miles and fail to live up to the guarantee.

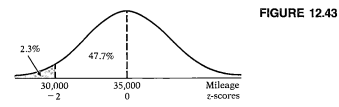

FIGURE 12.43

e) In part (c) we determined that 5.5 percent of all tires will last at least 39,000 miles. We now multiply 0.055 times 200,000 to determine the number of tires that will last at least 39,000 miles: $0.055 \times 200,000 = 11,000$ tires. ❖

Sometimes the results of a statistical distribution are clear and can be used to make specific predictions about individuals in the distribution. Other times, the results of a distribution are clear, yet the results cannot be used to make specific predictions regarding the individuals in the population.

Consider Fig. 12.44, which indicates the survival time of individuals who contract anthrax. Anthrax is an infectious disease of animals such as cattle or sheep, which can be transmitted to humans who consume or handle infectious products, including hair. By observing Fig. 12.44 we can clearly make the prediction that an individual who contracts anthrax and is treated will live longer than an individual who contracts anthrax and is not treated.

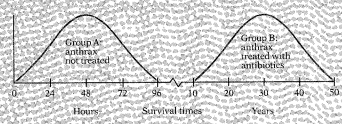
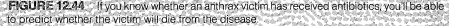

FIGURE 12.44 If you know whether an anthrax victim has received antibiotics, you'll be able to predict whether the victim will die from the disease.

Now consider Fig. 12.45, which shows the Scholastic Aptitude Test scores in mathematics for males and females. Notice that the mean score is slightly higher for males than females. Yet the distributions are so close and overlap so much that it is impossible to use these distributions to show that a particular male will outperform a particular female. By looking at the overlapping distributions it should be clear that there are many females who outperformed males on this test, just as there are many males who outperformed many females.

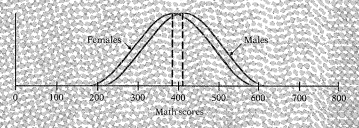
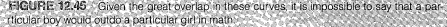

FIGURE 12.45 Given the great overlap in these curves, it is impossible to say that a particular boy would outdo a particular girl in math.

EXERCISES 12.7

Try to list any distributions other than those noted that are:

1. rectangular
2. J-shaped
3. skewed
4. bimodal

5. In a distribution that is skewed to the right, which has the greatest value—the mean, median, or mode? Which has the smallest value?
6. In a distribution that is skewed to the left, which has the greatest value—the mean, median, or mode? Which has the smallest value?
7. List three populations other than those noted that may be normally distributed.
8. List three populations other than those noted that may not be normally distributed.

Using Table 12.13, find the area specified in each of the following.

9. Above the mean
10. Below the mean
11. Between two standard deviations below the mean and one standard deviation above the mean
12. Between 1.2 and 1.9 standard deviations above the mean
13. To the right of $z = 1.42$
14. To the left of $z = 1.42$
15. To the left of $z = -1.68$
16. To the right of $z = -1.68$
17. To the right of $z = 2.03$
18. To the left of $z = 1.96$
19. To the left of $z = -1.21$
20. To the left of $z = -0.20$

Using Table 12.13, find the percentage of data specified in each of the following.

21. Between $z = 0$ and $z = 1.4$
22. Between $z = -0.2$ and $z = -0.84$
23. Between $z = -1.3$ and $z = 2.2$
24. Less than $z = -1.8$
25. Greater than $z = -1.8$
26. Greater than $z = 2.66$
27. Less than $z = 1.96$
28. Between $z = 0.74$ and $z = 1.65$
29. Between $z = -2.65$ and $z = -0.92$
30. Between $z = -2.15$ and $z = 3.31$

Heights of kindergarten children are normally distributed with a mean of 40 inches and standard deviation of 6 inches.

31. What percentage of kindergarten children are between 40 and 46 inches tall?
32. What percentage of kindergarten children are less than 46 inches tall?
33. What percentage of kindergarten children are between 34 inches and 46 inches tall?
34. If 1000 kindergarten children are selected at random, how many will be less than 43 inches tall?

The results of a statistics examination are normally distributed with a mean of 74 and a standard deviation of 8. Find the percentage of individuals that scored as indicated below.

35. Between 58 and 82
36. Less than 78
37. Greater than 78
38. Between 78 and 86
39. Between 60 and 72
40. Greater than 70

The diameters of red blood cells are normally distributed with a mean diameter of 0.008 millimeters and a standard deviation of 0.002 millimeters.

41. Find the percentage of red blood cells with diameters greater than 0.009 millimeters.
42. Find the percentage of red blood cells with diameters between 0.004 and 0.013 millimeters.
43. Find the percentage of red blood cells with diameters greater than 0.003 millimeters.
44. If 500 cells are isolated, how many will have diameters greater than 0.009 millimeters?

The weights of full-grown Brontosaurus, the most massive dinosaur, were normally distributed with a mean of 33 tons and a standard deviation of 4 tons. Find the percentage of Brontosaurus in each of the following categories.

45. Greater than 33 tons
46. Between 31 and 35 tons
47. Between 34 and 36 tons
48. Less than 30 tons
49. Greater than 30 tons
50. Less than 39 tons

A vending machine is designed to dispense a mean of 6.7 ounces of coffee into a 7-ounce cup. If the standard deviation of the amount of coffee dispensed is 0.4

ounces and the amount of coffee dispensed is normally distributed, find each of the following.

51. The percentage of times the machine will dispense from 6.5 to 6.8 ounces
52. The percentage of times it will dispense less than 6.0 ounces
53. The percentage of times it will dispense less than 6.8 ounces
54. The percentage of times the 7-ounce cup will overflow

The life expectancy of brand *A* light bulbs is normally distributed with a mean life of 1500 hours and a standard deviation of 100 hours.

55. Find the percentage of bulbs that will last more than 1450 hours.
56. Find the percentage of bulbs that will last between 1400 and 1550 hours.
57. Find the percentage of bulbs that will last fewer than 1480 hours.
58. If 80,000 of these bulbs are produced, how many will last 1500 hours or more?
59. If 80,000 of these bulbs are produced, how many will last between 1400 and 1600 hours?

60. A weight watchers' clinic guarantees that its new customers will lose at least 5 pounds by the end of their first month of participation or their money will be refunded. If the loss of weight of customers at the end of their first month is normally distributed with a mean of 6.7 pounds and a standard deviation of 0.81 pounds, find the percentage of customers that will be able to claim a refund.

61. The warranty on the motor in a certain model of Kitchen Aid dishwasher is good for 8 years. If the breakdown times of this motor are normally distributed with a mean of 10.2 years and a standard deviation of 1.8 years, find the percentage of motors that can be expected to require repair or replacement under warranty.
62. A vending machine that dispenses coffee does not appear to be working correctly. The machine rarely gives the proper amount of coffee. Some of the time the cup is underfilled, and some of the time the cup overflows. Does this indicate that the mean number of ounces dispensed has to be adjusted or that the standard deviation of the amount of coffee dispensed by the machine is too large? Explain your answer.
63. Mr. Brittain marks his class on a normal curve. Those with *z*-scores above 1.8 will receive A, those between 1.8 and 1.1 will receive B, those between 1.1 and −1.2 will receive C, those between −1.2 and −1.9 will receive D, and those under −1.9 will receive F. Find the percentage of grades that will be A, B, C, D, and F.
*64. Professor Hart marks his classes on the normal curve. His statistics class has a mean of 72 with a standard deviation of 8. He has decided that 10 percent of the class will get A, 20 percent will get B, 40 percent will get C, 20 percent will get D, and 10 percent will get F. Find the following
 a) The minimum grade needed to get an A
 b) The minimum grade needed to pass the course (D or better)
 c) The range of grades that will result in a C grade

PROBLEM SOLVING

65. How can one tell if a distribution is approximately normal? A statistical theorem called **Chebyshev's theorem** states that the minimum percent of data between plus and minus *K* standard deviations from the mean ($K > 1$) in *any distribution* can be found by the formula

$$\text{Minimum percent} = 1 - \frac{1}{K^2}$$

Thus, for example, between ±2 standard deviations from the mean, there will always be a minimum of 75 percent of the data. This minimum percent is true for any distribution:

For $K = 2$, $1 - \dfrac{1}{2^2} = 1 - \dfrac{1}{4} = \dfrac{3}{4}$ or 75%

Likewise, between ±3 standard deviations from the mean, there will always be a minimum of 89 percent of the data:

For $K = 3$, $1 - \dfrac{1}{3^2} = 1 - \dfrac{1}{9} = \dfrac{8}{9}$ or 89%

Table 12.14 gives the minimum percentages of data in *any distribution* and the actual percentages of data in *the normal distribution* between ±1.1, ±1.5, ±2.0, and ±2.5 standard deviations from

TABLE 12.14				
	K = 1.1	**K = 1.5**	**K = 2**	**K = 2.5**
Minimum % (for any distribution)	17.4%	55.6%	75%	84%
Normal distribution %	72.8%	86.6%	95.4%	99.8%
Given distribution %				

the mean. The minimum percentages of data in any distribution were calculated by using Chebyshev's theorem. The actual percentages of data for the normal distribution were calculated by using the area given in the standard normal, or z, table.

The third row of Table 12.14 has been left blank for you to fill in percentages of data, as will be explained shortly.

Consider the following 30 pieces of data obtained from a quiz:

1, 1, 1, 1, 2, 2, 2, 2, 3, 3, 4, 4, 4, 5, 6, 6, 6, 7, 7, 7, 7, 8, 8, 8, 8, 9, 9, 9, 10, 10

a) Find the mean of the set of scores.
b) Find the standard deviation of the set of scores.
c) Determine the values that correspond to 1.1, 1.5, 2, and 2.5 standard deviations above the mean. (For example, the value that corresponds to 1.5 standard deviations above the mean is $\bar{x} + 1.5s$.)

Then determine the values that correspond to 1.1, 1.5, 2, and 2.5 standard deviations below the mean. (For example, the value that corresponds to 1.5 standard deviations below the mean is $\bar{x} - 1.5s$.)

d)   By observing the 30 pieces of data, determine the actual percentage of quiz scores between

± 1.1 standard deviations from the mean
± 1.5 standard deviations from the mean
± 2 standard deviations from the mean
± 2.5 standard deviations from the mean

e) Place the percentages found in part (d) in the third row of Table 12.14.
f) Compare the percentage of data in the third row of Table 12.14 with the minimum percentage in the first row and the normal percentages in the second row, then make a judgment as to whether this set of 30 scores is approximately normally distributed. Explain your answer.

RESEARCH ACTIVITIES

66. Obtain a set of test scores from your teacher.
  a) Find the mean, median, mode, and midrange of the test scores.
  b) Find the range and standard deviation of the set of scores. (You may round the mean off to the nearest tenth when finding the standard deviation.)
  c) Construct a frequency distribution of the set of scores. Select your first class so that there will be between 5 and 12 classes.
  d) Construct a histogram and frequency polygon of the frequency distribution in part (c).
  e) Does the histogram in part (d) appear to represent a normal distribution?
  f) Use the procedure explained in Exercise 65 to determine whether the set of scores represented is a normal distribution.
67. In this project you actually become the statistician.
  a) Select a project of interest to you in which data must be collected.
  b) Write a proposal to be submitted to your teacher for approval. In your paper, discuss

the aims of your project and how you plan to gather the data to make your sample unbiased.
  c) After your proposal is approved, gather 50 pieces of data by the method stated in your proposal.
  d) Rank the data from smallest to largest.
  e) Compute the mean, median, mode, and midrange.
  f) Determine the range and standard deviation of the data. You may round the mean off to the nearest tenth when computing the standard deviation.
  g) Construct a frequency distribution, histogram, and frequency polygon of your data. Select your first class so that there will be between 5 and 12 classes. Make sure to label your histogram and frequency polygon.
  h) Does your distribution appear to be normal? Explain your answer. Does it appear to be another type of distribution discussed? Explain.
  i) Determine whether your distribution is approximately normal by using the technique discussed in Exercise 65.

SUMMARY

Statistics is a tool that is essential to modern society. However, great care must be taken to ensure that it is used properly. Statistical statements can often be misleading and should be carefully evaluated before being accepted.

A sample is a subset of the total population. Information gathered from the sample is used to make generalizations about the population from which the sample was drawn. If the generalizations are to be reliable, the sample selected must be representative of the population.

A frequency distribution is a listing of observed values along with their corresponding frequencies. Frequency distributions simplify the statistician's work when the amount of data is large. Two graphs that are used to illustrate frequency distributions are the histogram and the frequency polygon. The histogram is a bar graph; the frequency polygon is a line graph. Both have observed values on the horizontal scale and frequencies on the vertical scale.

Four measures of central tendency that we have discussed are the mean, median, mode, and midrange. The mean, $\bar{x}$, is the sum of the data divided by the number of pieces of data. The mean can be found by the formula

$$\bar{x} = \frac{\Sigma x}{n}$$

The median is the piece of data in the middle of a set of ranked data. The mode is the most common piece of data. The midrange is halfway between the lowest piece of data and the highest piece of data. The midrange is found by adding the lowest and the highest pieces of data and dividing this sum by 2.

Two measures of location are percentiles and quartiles. The nth percentile is the value that is higher than n percent of the data. The twenty-third percentile is the value higher than 23 percent of all the data in the distribution. Quartiles divide the data into four equal parts. The first quartile, Q_1, is the value higher than 25 percent of the data, the second quartile is the same as the median, and the third quartile is the value higher than 75 percent of all the data.

Two measures of dispersion are the range and standard deviation. The range is found by subtracting the lowest value from the highest value. The standard deviation is a measure of the spread of the data about the mean. The standard deviation is found by subtracting the mean from each individual value, squaring the differences, summing the squares of the differences, dividing that sum by one less than the number of pieces of data, and finally taking the square root of that number. The standard deviation can be found by using the formula

$$s = \sqrt{\frac{\Sigma(x - \bar{x})^2}{n - 1}}$$

Of the many types of frequency distributions, the normal distribution is the most important. The normal curve is a bell-shaped symmetrical curve. The percentage of data between any two values in a normal distribution can be found when the standard deviation and the mean of the distribution are known. To find the percentage of data between any two values, convert the values to z-scores, using the formula

$$z = \frac{x - \bar{x}}{s}$$

then look up the corresponding areas in Table 12.13, and convert those areas to percentages.

REVIEW EXERCISES

Given the set of data 60, 62, 70, 78, 81, 87, find each of the following:

1. Mean
2. Median
3. Mode
4. Midrange
5. Range
6. Standard deviation

Given the set of data 1, 2, 9, 11, 16, 4, 9, 20, 4, 14, 12, 18, find each of the following.

7. Mean
8. Median
9. Mode
10. Midrange
11. Range
12. Standard deviation

13. Consider the following set of data:

$$
\begin{array}{ccccccc}
20 & 21 & 22 & 23 & 26 & 28 & 30 \\
20 & 21 & 23 & 24 & 26 & 29 & 30 \\
20 & 21 & 23 & 24 & 26 & 30 & 30 \\
20 & 21 & 23 & 25 & 27 & 30 & 30 \\
21 & 21 & 23 & 26 & 28 & 30 & 30
\end{array}
$$

   a) Construct a frequency distribution.
   b) Construct a histogram.
   c) Construct a frequency polygon.

14. Consider the following set of data:

$$
\begin{array}{cccccccc}
30 & 40 & 52 & 58 & 63 & 72 & 78 & 84 \\
32 & 44 & 53 & 60 & 67 & 72 & 78 & 86 \\
36 & 46 & 54 & 60 & 68 & 74 & 78 & 88 \\
38 & 48 & 55 & 60 & 70 & 76 & 80 & 92 \\
39 & 50 & 57 & 62 & 72 & 78 & 82 & 98
\end{array}
$$

   a) Construct a frequency distribution with the first class 30–38.
   b) Construct a histogram.
   c) Construct a frequency polygon.

Anthropologists have determined that a certain type of primitive human had a mean head circumference of 40 centimeters with a standard deviation of 5 centimeters. Given that head sizes were normally distributed, determine the percentage of heads that are

15. Between 35 and 45 centimeters   **16.** Between 30 and 50 centimeters
17. Less than 48 centimeters   **18.** Greater than 48 centimeters
19. Greater than 37 centimeters

The heights of fully grown redwood trees are normally distributed with a mean height of 270 feet and a standard deviation of 30 feet. Find the percentage of redwood trees that are

20. Greater than 270 feet   **21.** Between 270 feet and 290 feet
22. Less than 275 feet   **23.** Between 210 and 300 feet

A study was made on the weights of adult men. The following statistics were obtained from that study:

mean	187 lb	first quartile	173 lb
median	180 lb	third quartile	227 lb
mode	175 lb	86th percentile	234 lb
standard deviation	23 lb		

24. What is the most common weight?
25. What weight did half of those surveyed surpass?
26. About what percentage of those surveyed surpassed 227 lb?
27. About what percentage of those surveyed weighed below 173 lb?
28. About what percentage of those surveyed weighed more than 234 lb?
29. If 100 men were surveyed, what is the total weight of all the men?
30. What weight represents 2 standard deviations above the mean?
31. What weight represents 1.8 standard deviations below the mean?

Below is a listing of U.S. presidents and the number of children in their families:

Washington	0	Pierce	3	Wilson	3
J. Adams	5	Buchanan	0	Harding	0
Jefferson	6	Lincoln	4	Coolidge	2
Madison	0	A. Johnson	5	Hoover	2
Monroe	2	Grant	4	F. D. Roosevelt	6
J. Q. Adams	4	Hayes	8	Truman	1
Jackson	0	Garfield	7	Eisenhower	2
Van Buren	4	Arthur	3	Kennedy	3
W. H. Harrison	10	Cleveland	5	L. B. Johnson	2
Tyler	14	B. Harrison	3	Nixon	2
Polk	0	McKinley	2	Ford	4
Taylor	6	T. Roosevelt	6	Carter	4
Fillmore	2	Taft	3	Reagan	4

Determine the following.

32. Mean number of children   **33.** Mode
34. Median   **35.** Midrange
36. Range
37. Standard deviation (round mean off to nearest tenth)
38. Construct a frequency distribution; let the first class be 0–1.
39. Construct a histogram.   **40.** Construct a frequency polygon.
41. Does this distribution appear to be normal? Explain.
42. On the basis of this sample, do you think that the number of children per family in the United States would have a normal distribution? Explain.
43. Do you feel that this sample is representative of the population? Explain.

CHAPTER TEST

Given the set of data 3, 10, 12, 15, 15, find the following.

1. Mean
2. Median
3. Mode
4. Midrange
5. Range
6. Standard deviation

Consider the following set of data:

6	8	15	26	29	36
6	10	16	26	29	38
6	12	20	27	30	38
6	12	24	27	32	42
7	15	26	27	34	46

7. Construct a frequency distribution. Let the first class be 5–10.
8. Construct a histogram of the frequency distribution.
9. Construct a frequency polygon of the frequency distribution.

The following represent statistics of monthly salaries at the Hoperchang Publishing Company:

Mean	$480	First quartile	$450
Median	$450	Third quartile	$485
Mode	$475	79th percentage	$500
Standard Deviation	$40		

10. What is the most common salary?
11. What salary did half the employees surpass?
12. About what percentage of employees surpassed $450?
13. About what percentage of employees' salaries was below $500?
14. If the company has 100 employees, what is the total monthly salary of all employees?
15. What salary represents 1 standard deviation above the mean?
16. What salary represents 1.5 standard deviations below the mean?

The useful life of an electric motor is normally distributed with a mean of 7.3 years and a standard deviation of 0.5 years. Find the percentage of motors with a useful life:

17. Between 7.3 and 7.8 years
18. Less than 7.1 years
19. Between 7.5 and 8.5 years
20. If the manufacturer warranties its motors for 7 years, what percentage will need to be replaced under warranty?

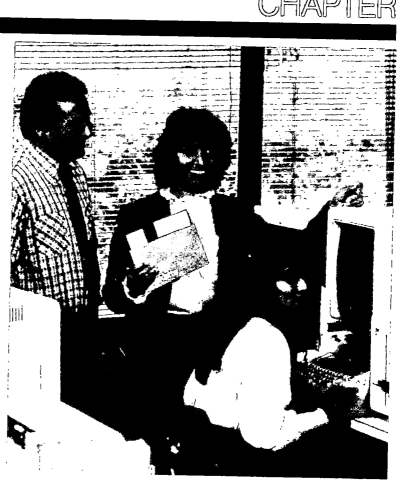

CHAPTER 13

Computers

13.1 GENERATIONS OF THE COMPUTER

We are now in the midst of a computer revolution. The computer has made and will continue to make profound changes in our lives. Charles F. Kettering, inventor and scientist, once wrote: "We should all be concerned about the future because we will have to spend the rest of our lives there." Computers, says astronomer Carl Sagan, rival the invention of writing as one of the most profound innovations in human history, for they are "remaking the world at a phenomenal rate."

DID YOU KNOW?

Students are learning to communicate with computers at a very early age. Many adults may have trouble grasping the fact that in the future they may need more than the ability to read and write to be considered literate. Literacy may soon include being able to access, manipulate, and store information in a computer.

As times change, so does the definition of literacy. For example, in twelfth-century England, literacy meant the ability to compose and recite in Latin. Eventually, English replaced Latin as the language one had to know to be literate in England. But even then, because of the difficulty of using a quill on parchment, writing in the Middle Ages was considered a special skill that was not automatically coupled with the ability to read. Thus reading and dictating (to a specialist who could write) were typically paired. With the invention of the printing press the basic skills of modern literacy—reading and writing—became widespread.

Today, computers are becoming so simple and inexpensive that millions of people can be computer literate without understanding how computers actually work. The time is fast approaching when our definition of literacy may be the ability to read, write, and use a computer.

The computer age is said to have begun on June 14, 1951, when the Sperry-Rand Corporation delivered the first commercially available computer, the UNIVAC (*Univ*ersal *A*utomated *C*omputer), to a client, the U. S. Bureau of the Census. This was the first time a computer had been built for data processing applications rather than military, scientific, or engineering use. In 1953, IBM delivered the first computer capable of storing a program (or series of instructions), the IBM 650. The name most commonly associated with the development of the stored program is John von Neumann (Fig. 13.1).

FIGURE 13.1
John von Neumann
Institute for Advanced Study,
Princeton, N.J.

"It is unworthy of excellent men to
lose hours like slaves in the labor
of calculations."

Gottfried Wilhelm Leibniz

The computers built before 1958 used vacuum tubes, which were subject to frequent burnout. The computers in this period are called **first-generation computers.**

In 1947, Robert Shockley, J. Bardeen, and H. W. Brattain of Bell Laboratories invented a tiny, deceptively simple device called the transistor. Transistors were much smaller than vacuum tubes, worked much faster, and had fewer failures. They gave off very little heat and therefore could be spaced closer together than vacuum tubes. They were also quite cheap to make.

Within a few years, scientists at Bell Labs built the first fully transistorized (solid state) computer, a machine they called the Leprechaun. The computers built from 1958 to 1964 used transistors. These computers are called **second-generation computers.**

The next breakthrough came in the late 1950s. Working independently, Jack Kilby of Texas Instruments and Robert Noyce of Fairchild Semiconductors almost simultaneously realized that any number of transistors could be etched directly on a single piece of silicon along with the conductors that connect them. Such an **integrated circuit (IC)** is a complete electronic circuit on a small chip of silicon. The **chip** (Fig. 13.2) is less than 1/8-inch square and contains hundreds of electronic components.

The era of **third-generation computers** is said to have begun on April 7, 1964, when IBM first announced the IBM® 360 series. The 360 series was launched with an all-out massive marketing effort to make computers a business tool. The result of this effort went beyond IBM's wildest dreams.

Today, hundreds of thousands of transistors can be etched on a tiny silicon chip. Integrated circuits, first used in 1965, evolved into **large-scale**

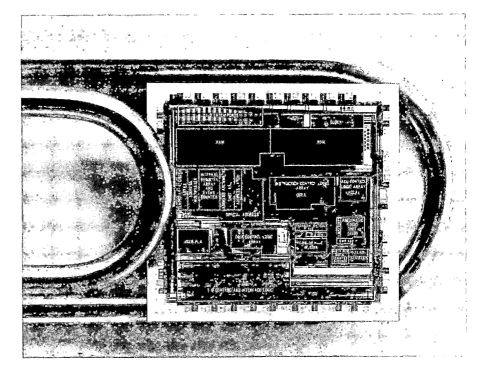

FIGURE 13.2
The chip. This single computer chip contains everything needed for processing and storing data, yet it is smaller than a standard paper clip.

Courtesy of AT&T Bell Laboratories.

integration of circuits (LSI) in 1970. Thousands of ICs were crammed onto a tiny 1/4-inch square of silicon. In 1975, **very large scale integration (VLSI)** was achieved.

In 1970 a young engineer named Ted Hoff designed the microprocessor. The **microprocessor** is a computer on a silicon chip—it contains the entire processing unit of a computer, consisting of the *control* and *arithmetic and logic* units. Such a processing unit, called a **central processing unit (CPU),** is assembled on the chip with thousands of microscopic electronic components. The chip illustrated in Fig. 13.2 is a microprocessor.

The era of **fourth-generation computers** began around 1970 with the development of the microprocessor and the use of very large scale integrated circuits. The microprocessor on a single chip can be programmed to do any number of tasks, from running a watch to steering a spacecraft. The microprocessor is also the heart of the personal computer. The first personal computer, the Altair 880C, cost $395 unassembled and $621 assembled. The Altair soon vanished but was replaced by others, including one bearing an odd symbol, an apple with a bite taken out of it. Suddenly, the future was now.

Table 13.1 summarizes the important characteristics of the various computer generations, and Fig. 13.3 on page 699 illustrates components used in the different generations.

TABLE 13.1

Characteristics of the Various Computer Generations

	First Generation (1951–1958)	Second Generation (1958–1964)	Third Generation (1964–1970)	Fourth Generation (1970–)
Technology	Vacuum tubes; card-oriented	Transistors; tape-oriented	Integrated circuits; time-sharing; disk-oriented	Very large scale integrated circuits; microprocessors; bubble memory
Operation time	Milliseconds (thousandths of a second)	Microseconds (millionths of a second)	Nanoseconds (billionths of a second)	Nanoseconds or picoseconds (trillionths of a second)
Cost	$5/function	$0.50/function	$0.05/function	$0.01 to $0.0001/function
Processing speed	2000 instructions/second	1 million instructions/second	10 million instructions/second	100 million to 1 billion instructions/second
Memory size (bytes)	1000–4000	4000–32,000	32,000–3,000,000	3,000,000+
Mean time between failures	Minutes–hours	Days	Days–weeks	Weeks
Auxiliary units	Punched card–oriented	Tape-oriented	Disk-oriented	Disk and mass storage

The **fifth-generation** computer project was started in Japan in 1982. The fifth-generation project aims specifically to produce artificial intelligence (AI) computers that will use logical rules rather than arithmetic operations to act on information. This system will be based on the technology of parallel processing (the manipulation of data by many different processors at the same time) and will require new computer languages. The resulting machine is expected to be capable of making one billion inferences per second, working 2000 times faster than today's most advanced computers.

Japanese scientists define artificial intelligence as the ability of machines to produce "humanlike responses," including understanding natural language, recognizing objects visually, and answering questions on the basis of a body of knowledge.

The project at this time has been a disappointment to some, since there have been few breakthroughs, and some work has been abandoned. There is hope that the computer's main subsystems, where logical inferences are made, will be completed by 1989.

Thomas Watson, often referred to as "the Old Man."
Courtesy of International Business Machines Corporation.

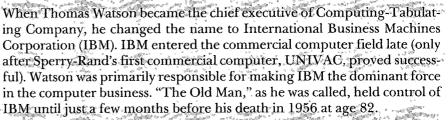

When Thomas Watson became the chief executive of Computing-Tabulating Company, he changed the name to International Business Machines Corporation (IBM). IBM entered the commercial computer field late (only after Sperry-Rand's first commercial computer, UNIVAC, proved successful). Watson was primarily responsible for making IBM the dominant force in the computer business. "The Old Man," as he was called, held control of IBM until just a few months before his death in 1956 at age 82.

On May 28 and 29, 1959, a meeting was held at the Pentagon to consider establishing a common programming language for business applications of electronic computers that would not be directly associated with any manufacturer. As a result of this meeting, COBOL (*Co*mmon *B*usiness *O*riented *L*anguage) was invented. Grace Hopper, a graduate of Vassar and Yale, worked continuously to develop and test COBOL compilers. In 1969 she received the Data Processing Management "Man of the Year" award.

In February 1984, President Reagan promoted Grace Hopper from captain to commodore in the U. S. Navy. Commodore Hopper is credited with coining the word "bug" for an error in a program. In August 1945, while she and some associates were working on an experimental machine called the Mark I, a circuit malfunctioned. A researcher using tweezers located and removed the problem: a 2-inch-long moth. Hopper taped the offending insect into her logbook. She said, "From then on, when anything went wrong with a computer, we said it had a bug in it."

Commodore Grace Hopper.
Courtesy of Sperry Corporation

Japanese scientists have announced that they hope to create a sixth-generation computer after the fifth-generation computer is completed. The sixth-generation computer would include biological elements. This project would be pursuing a technology that is comparable to the human brain. The project has been named The Human Frontiers Program.

Japan is not the only country working on fifth-generation computers. The U. S. government has allocated billions of dollars toward the develop-

FIGURE 13.3
Components that have
been used in computers.
*Courtesy of International
Business Machines Corporation.*

ment of fifth-generation computers. Some U.S. companies that are currently working on developing the true supercomputer are Cray Computers,
IBM, and Control Data Corporation.

EXERCISES 13.1

In Exercises 1 through 6, indicate the significant contribution of the individuals named to the development of
computers.

1. Robert Shockley, J. Bardeen, and H. W. Brattain
2. Jack Kilby and Robert Noyce
3. Ted Hoff
4. John von Neumann
5. Thomas Watson
6. Grace Hopper

7. What event is said to have begun the computer
   age?
8. What company developed the first transistorized
   computer?What was its name?
9. Indicate the components used in each of the four
   generations of computers.
10. What is an integrated circuit (IC)?
11. What are large-scale integrated circuits?
12. What is a microprocessor?

RESEARCH ACTIVITY

Research and explain briefly what influence the following individuals had on the development of the computer.

13. John Napier   14. Blaise Pascal
15. Gottfried Leibniz 16. Charles Babbage
17. Ada, Countess of Lovelace

18. Joseph Jacquard
19. George Boole
20. Herman Hollerith
21. Howard Aiken
22. John Atanasoff and Clifford Berry
23. John Mauchly and J. Presper Eckert, Jr.

AN INTRODUCTION TO THE COMPUTER

The basic functions of any digital computer system, regardless of its size or sophistication, are (1) input, (2) memory or storage, (3) control, (4) processing (arithmetic and logic unit), and (5) output. Figure 13.4 shows the relationship among the five parts of the computer system.

Arithmetic/Logic unit

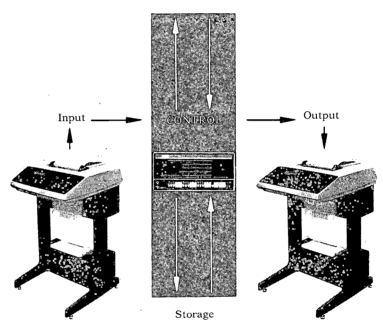

Input

Output

Storage

FIGURE 13.4
Basic components of the computer.
*Courtesy of International
Business Machines Corporation.*

Input

In order for a computer to follow instructions and perform calculations, the instructions and data must be entered into the computer. Various methods can be used to input such instructions and data.

The keyboard is one popular way to input data. When information typed on a keyboard can be displayed on a monitor, the combination of keyboard and monitor is referred to as a **video display unit** (VDU). Optical scanning devices, including those now used in many grocery and department stores, are becoming more and more popular for inputting data (Fig. 13.5). Optical scanners can read more than 1500 characters per second. Input can also be generated photoelectrically by using a cathode-ray tube (CRT) and electronic pointers or light pens (styli) that can erase, alter, or add to memory any information displayed on the CRT. Data and instructions can be fed into the computer by many other methods, including tape recorders, punched cards, special telephones called modems (Fig. 13.6), and teletypewriters. Typed print and voice input with a limited vocabulary are now being perfected.

FIGURE 13.5
Wand reader.   The wand is a hand-held photoelectric scanning device that can read special optical-character recognition (OCR) characters.
NCR Corporation

FIGURE 13.6
Modems (or acoustic couplers) make it possible for traveling profession-als to get up-to-date information from their home office computer.
Photo courtesy of Hewlett-Packard Company

Storage

Consider a situation in which you want to add many numbers. Writing the numbers down on a sheet of paper is a form of memory, or storage. The computer also needs to store the instructions and data it receives before it can act on them. This information is stored on integrated circuits. All of the inputted information is converted to binary (base 2) notation before it is stored. The storage area, or memory, of a computer is referred to as **primary storage** or **main storage.**

The size of a computer's memory is generally given in **kilobytes. A bit,** short for *bi*nary dig*it,* is a single piece of information in the form of a 1 (an electronic pulse) or a 0 (no electronic pulse). A **byte** is usually a group of eight bits taken as a unit. For example, the byte 00010010 consists of eight bits. Each byte has a specific meaning. A memory of 1 kilobyte (1K) can store approximately 1000 bytes of information. (One kilobyte is actually 1024 bytes.) A 64K computer can store about 64,000 bytes of information in its main storage.

The seven-bit ASCII (American Standard Code of Information Interchange) code gives a seven-bit designation of the digits, letters, and symbols used by most computer manufacturers. For example, the byte 1000111 is interpreted as the letter G in many digital computers. The eighth bit (the leftmost digit) of the byte, called the *parity bit,* may be used by the computer for checking purposes.

Each computer has two types of memory, random access memory (RAM) and read-only memory (ROM). Using **random access memory,** you can recall and read what is in memory and write new information into memory. RAM can have its contents altered; that is, new data can be added and old data can be read from RAM. RAM is volatile—its contents are lost when the power to the computer is turned off. If there is a momentary power shortage, the entire contents of RAM may be lost. When you see an ad for a computer with a 256K or 512K memory, the advertisement is generally referring to RAM.

Read-only memory (ROM) cannot be altered. The contents of ROM are placed there by the manufacturer when the computer is built. ROM is nonvolatile—when the power is turned off, ROM is not affected. Generally, a computer has more RAM than ROM.

The storage unit stores the input information, programming steps, and calculations that have been performed in the arithmetic unit and returned to storage. The computer may have internal storage or may use external storage units or both. The information remains in storage until called on by the control unit.

Modern-day computers use semiconductors for primary storage. Since semiconductor memory can be mass-produced economically, the cost of primary storage chips that once cost $80 to $90 each to build can now be made in volume for less than $1 apiece.

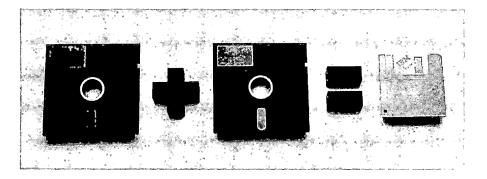

FIGURE 13.7
Floppy and hard disks.
Courtesy of International Business Machines Corporation

Primary storage should not be confused with secondary or auxiliary storage. **Secondary storage** is not a part of the computer. Secondary storage is a medium, such as magnetic disks or magnetic tape, that is used to store large amounts of data. Magnetic disks can be of two types, floppy (Fig. 13.7) and hard. Hard disks generally hold more information, and the information can be retrieved at a much faster rate than from floppy disks. The data on a disk are read and written by a disk drive, and the data on tape are read and written by a tape drive.

A promising storage device is the magnetic bubble, which was introduced by Bell Laboratories in 1966. A magnetic bubble memory chip now exists that has more than ten times the capacity of the most powerful integrated circuit chip.

Control

The heart of the computer is the **central processing unit (CPU)**. The CPU is composed of the control and processing unit. Data or programs are taken from storage or input units by the central processing unit for manipulation. Then they are returned to storage to await further orders about output. The CPU's control system can be compared to a railroad switchyard control tower. There must be a path or a wire between every two computer elements that will ever be connected (there are millions), and there must be a switching element that is capable of opening and closing them. Each switching element in a computer is called a "gate." A gate can open to allow data signals to use that path or close to prevent them from using that path. The control unit must open the proper combination of gates and keep all other gates closed. The control unit fetches instructions from the program, one after the other. As each new instruction comes up, the control unit must set up the computer to perform that instruction by opening the appropriate set of gates in the required sequence. The program itself is usually stored in the same memory as the data. The program is written in a code and special format that the control interprets as instructions. The program itself is referred to as **software** and the equipment is referred to as **hardware.**

The various parts of a computer often work at different rates. For example, the *input and output (I/O)* devices are often slower than the processing unit. The control unit also serves to synchronize the various speeds of the input and output units to those of the arithmetic and logic unit.

Processing

In a digital computer, calculations and arithmetical decisions take place in the *arithmetic/logic unit (ALU)*. Four kinds of arithmetical operations, or mathematical calculations, can be performed on data: addition, subtraction, multiplication, and division. Some computers perform all calculations using only the operation of addition. (See the Did You Know? on page 165.)

Operations in the logic unit of the ALU are usually operations of comparison. The ALU is able to compare numbers, letters, or special characters and act according to the outcome of the comparison. It is by comparison that a computer is able to tell, for example, whether there are unfilled seats on an airplane or whether your checking account is overdrawn.

The three basic operations of comparison are

1. Equal to ($=$). If the number of tickets sold for an airplane flight equals the number of seats on the airplane, the flight is sold out.

2. Less than ($<$). If the number of tickets sold for an airplane flight is less than the number of seats available, the flight is not sold out.

3. Greater than ($>$). If the number of tickets sold for an airplane flight is greater than the number of seats, the flight is overbooked.

There are six combinations of operations: "is equal to" ($=$), "is less than" ($<$), "is greater than" ($>$), "is less than or equal to" ($\leq$), "is greater than or equal to" ($\geq$), and "is less than or greater than" ($<>$). Note that the last operation is the same as "is not equal to" ($\neq$).

Output

The output is the means by which the operator or user obtains the results of the computations performed by the computer. The various types and devices of output include typewriters, printers, paper tape, cathode-ray tubes (monitors), and modems.

The input/output (I/O) devices are relatively slow (speeds are measured in milliseconds or thousandths of a second) in comparison with the speed of the CPU (speeds are measured in nanoseconds, or billionths of a second). The CPUs were inefficient because they could calculate quickly (less than a second to a few seconds), but the mechanism for displaying the results was comparatively slow (often many minutes). Thus the CPU had to "wait" with the solution until the results could be displayed.

To solve this problem, computer scientists and engineers designed several techniques. In one of them, **time-sharing,** many jobs enter the computer simultaneously, each isolated and each solved individually. The computer solves the problems according to a programmed schedule of priorities and makes the solutions available to the appropriate sources, which may be many miles away, as soon as the answers are available. Thus the CPU is used continuously to do the calculations required by each of the jobs. A source may be connected to the master computer by telephone lines.

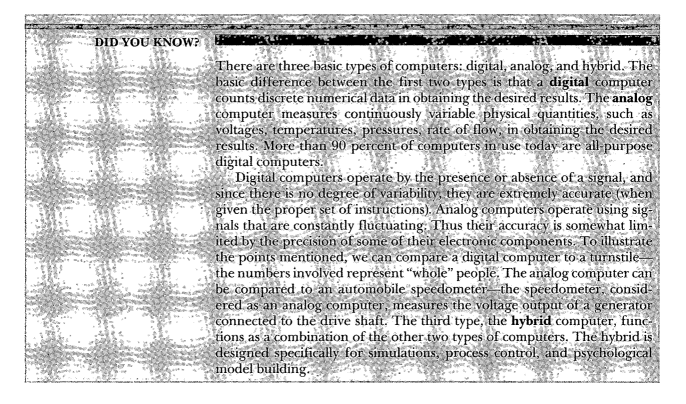

DID YOU KNOW?

There are three basic types of computers: digital, analog, and hybrid. The basic difference between the first two types is that a **digital** computer counts discrete numerical data in obtaining the desired results. The **analog** computer measures continuously variable physical quantities, such as voltages, temperatures, pressures, rate of flow, in obtaining the desired results. More than 90 percent of computers in use today are all-purpose digital computers.

Digital computers operate by the presence or absence of a signal, and since there is no degree of variability, they are extremely accurate (when given the proper set of instructions). Analog computers operate using signals that are constantly fluctuating. Thus their accuracy is somewhat limited by the precision of some of their electronic components. To illustrate the points mentioned, we can compare a digital computer to a turnstile—the numbers involved represent "whole" people. The analog computer can be compared to an automobile speedometer—the speedometer, considered as an analog computer, measures the voltage output of a generator connected to the drive shaft. The third type, the **hybrid** computer, functions as a combination of the other two types of computers. The hybrid is designed specifically for simulations, process control, and psychological model building.

EXERCISES 13.2

1. List the five basic functions of a digital computer and describe the purpose of each function.
2. List some input devices.
3. List some output devices.
4. What is the central processing unit? What takes place in the CPU?
5. What happens to information entered in the computer before it is stored in memory?
6. What is a bit? What is a byte?
7. What is the name given to approximately 1000 bytes of information?
8. What are some differences between RAM and ROM? What does each term stand for?
9. What is the difference between primary and secondary storage?

10. What do most modern-day computers use for primary storage?
11. What are some secondary storage devices?
12. What are the four arithmetic operations that take place in the arithmetic unit of the ALU?
13. What are the three basic operations of comparison that take place in the logic unit of the ALU?
14. Give a situation illustrating how the three basic comparisons listed in Exercise 13 can be used.
15. Using the three basic comparisons listed in Exercise 13, list all the possible combinations of comparisons that the computer can make.
16. What is time-sharing?

13.3 ORDER OF OPERATIONS AND FLOWCHARTS

More and more schools, from elementary through college, are requiring that students learn to communicate with computers by using a programming language. In Sections 13.4–13.6 we will introduce the BASIC—*B*eginners *A*ll-purpose *S*ymbolic *I*nstruction *C*ode—programming language. BASIC is the most commonly used programming language for personal computers. In Section 13.7 we will introduce a second programming language called LOGO.

Figure 13.8 illustrates the keyboard of a personal computer. The symbols used for mathematical operations in BASIC are illustrated in Table 13.2. A symbol used for exponentiation on many computers, including IBM PCs and compatibles, is $\uparrow$. On computers that use this symbol, $3 \uparrow 2$ means 3^2. Before we learn to do some elementary programming, we must know the order in which the computer will perform the various mathematical operations. This information is given in Table 13.3. Note that the order of operations used by a computer is identical to the order of operations used by a scientific calculator.

TABLE 13.2

Operation	BASIC Symbol	Example	Written in BASIC	Result
Addition	+	$5 + 3$	$5 + 3$	8
Subtraction	−	$6 - 2$	$6 - 2$	4
Multiplication	*	$5 \cdot 4$	$5 * 4$	20
Division	/	$30 \div 5$	$30 / 5$	6
Exponentiation	$\wedge$(or $\uparrow$)	3^2	$3 \wedge 2$ (or $3 \uparrow 2$)	9
Square root	SQR ( )	$\sqrt{25}$	SQR(25)	5

FIGURE 13.8
The Apple IIe keyboard.
Courtesy of Apple Computer, Inc.

TABLE 13.3
Priority of Operations
1. The computer first performs the operations within parentheses or brackets. If there are nested parentheses or brackets (one pair within another), the information in the inner ones is evaluated first.
2. The computer then perform all exponential operations (raising to powers or finding roots).
3. Next it performs all multiplications and divisions, from left to right.
4. Then it performs all additions and subtractions, from left to right.

■ **Example 1** ─────────────────────────────────

What results will the computer give for the following arithmetic expressions?

a) $2 + 3 * 5$ **b)** $4 + 3 \wedge 2 * 5$ **c)** $6 + 4 / 2$ **d)** $4 / 2 * 3$

Solution

a) Since the computer will perform multiplication before addition, the computer will multiply 3 times 5 and then add 2, for an answer of 17:

$$2 + 3 * 5 = 2 + (3 \cdot 5) = 2 + 15 = 17$$

b) The computer will first square 3 ($3 \wedge 2 = 3^2 = 9$). It will then multiply this number by the 5 and add 4:

$$4 + 3 \wedge 2 * 5 = 4 + [(3)^2 \cdot 5] = 49$$

c) The computer will first divide 4 by 2, and to this quotient it will add 6:

$$6 + 4 / 2 = 6 + (4 \div 2) = 6 + 2 = 8$$

d) Since multiplication and division are equal in priority, the computer works from left to right:

$$4 / 2 * 3 = (4 \div 2) \cdot 3 = 6 \qquad \qquad ❖$$

In Section 6.1 we stated that $-6^2 = -(6)^2 = -36$. The computer, however, interprets $-6 \wedge 2$ at the beginning of a calculation to mean $(-6)^2 = 36$. If you key in either $-(6 \wedge 2)$ or $0 - 6 \wedge 2$, the computer will display the answer -36.

■ **Example 2** ─────────────────────────────────

What results will the computer give for the following arithmetic expressions?

a) $-6 \wedge 2 + (4 - 3) \wedge 3 * 4$

b) $(2 * (4 + 2)) \wedge 2 + SQR(36)$

Solution

a) The computer will first evaluate $(4 - 3)$, since it is within parentheses. It then performs the exponentiation:

$$\begin{aligned}
-6 \wedge 2 + (4 - 3) \wedge 3 * 4 &= (-6)^2 + [(4 - 3)^3 \cdot 4] \\
&= (-6)^2 + (1^3 \cdot 4) \\
&= 36 + (1 \cdot 4) \\
&= 36 + 4 \\
&= 40
\end{aligned}$$

b) This problem contains nested parentheses. The inner parentheses are evaluated first:

$$\begin{aligned}
(2 * (4 + 2)) \wedge 2 + SQR(36) &= (2 \cdot (4 + 2))^2 + \sqrt{36} \\
&= (2 \cdot 6)^2 + \sqrt{36} \\
&= (12)^2 + \sqrt{36} \\
&= 144 + 6 \\
&= 150 \qquad \qquad ❖
\end{aligned}$$

■ **Example 3** _____

Write each of the following in BASIC symbols.

a) $3^4 + (5 \cdot 3)$

b $(\sqrt{49} + 3) \cdot 5$

c) $[(3 \cdot 6) + (5 \cdot 7)] \div 3^2$

d) $2x + 5\sqrt{x}$

e) $3x^2 - 4x - 3$

Solution

a) $3^4 + (5 \cdot 3) = 3 \wedge 4 + 5 * 3$. Parentheses are not needed.

b) $(\sqrt{49} + 3) \cdot 5 = (SQR(49) + 3) * 5$.

c) $[(3 \cdot 6) + (5 \cdot 7)] \div 3^2 = (3 * 6 + 5 * 7) / 3 \wedge 2$.

d) $2x + 5\sqrt{x} = 2 * x + 5 * SQR(x)$. Remember that the multiplication symbol must be used every time we want to multiply.

e) $3x^2 - 4x - 3 = 3 * x \wedge 2 - 4 * x - 3$.   ❖

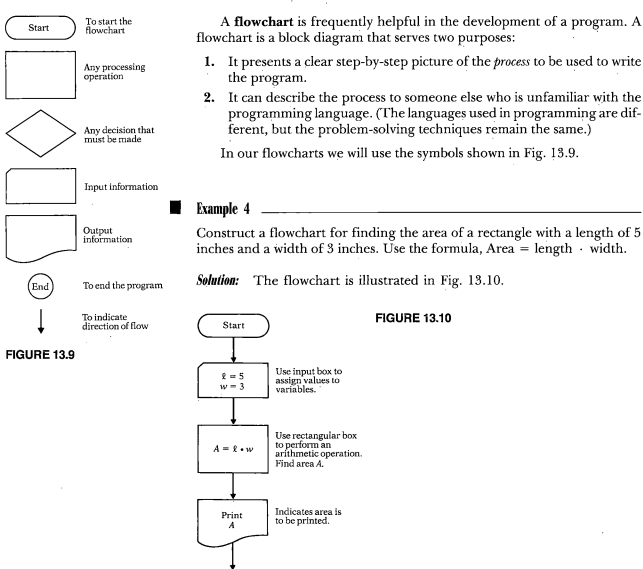

FIGURE 13.9

A **flowchart** is frequently helpful in the development of a program. A flowchart is a block diagram that serves two purposes:

1. It presents a clear step-by-step picture of the *process* to be used to write the program.
2. It can describe the process to someone else who is unfamiliar with the programming language. (The languages used in programming are different, but the problem-solving techniques remain the same.)

In our flowcharts we will use the symbols shown in Fig. 13.9.

■ **Example 4** _____

Construct a flowchart for finding the area of a rectangle with a length of 5 inches and a width of 3 inches. Use the formula, Area = length · width.

Solution: The flowchart is illustrated in Fig. 13.10.

FIGURE 13.10

In certain situations the values of the variables are not known and will have to be inserted later. When constructing flowcharts for such situations, place the required variables in the input box. This procedure is illustrated in Example 5.

Start

b_1, b_2, h — Input variables needed to find area.

$A = \frac{1}{2}h\,(b_1 + b_2)$ — Find area A.

Print A — Print area A.

End

FIGURE 13.11

■ **Example 5** ——————————————————

Construct a flowchart for finding the area of a trapezoid. Use the formula Area $= \frac{1}{2} \cdot$ height $\cdot$ (base 1 + base 2).

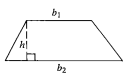

Solution: The flowchart is illustrated in Fig. 13.11.   ❖

Now we will examine some flowcharts in which a decision must be made.

■ **Example 6** ——————————————————

Construct a flowchart that will print the sum of two numbers, x and y, if the sum is less than 30. If the sum is greater than or equal to 30, the program is to end with nothing being printed.

Solution: We start the flowchart with a *Start* symbol (Fig. 13.12). Next we use the input symbol to input the values of x and y. Then we compute the sum of x and y, as indicated by the rectangle. In the next step a decision must be made. If the sum is less than 30, we want the computer to print the sum, s. If the sum is greater than or equal to 30, we want the flowchart to end without printing the sum. Note that in a decision step the "yes" is generally represented by a horizontal line from the side of the diamond and the "no" is a vertical line extended from the bottom of the diamond. If the decision is yes (the sum is less than 30), a **loop** is created.   ❖

■ **Example 7** ——————————————————

Construct a flowchart to illustrate the following. Multiply two numbers, x and y. If their product is greater than or equal to 50, add 6 to the product. If their product is less than 50, subtract 3 from the product. Print the answer.

Solution: The flowchart is illustrated in Fig. 13.13. After the decision on whether the product is greater than or equal to 50 has been made, a second arithmetic operation must be performed. We arbitrarily selected the letter A to represent the sum or difference. The final answer, A, is printed. If, for example, x is given a value of 12 and y is given a value of 5, the product is greater than or equal to 50 and the answer 66, $[(12 \cdot 5) + 6]$, will be printed. If x is given a value of 12 and y a value of 4, the product is less than 50 and the answer 45, $[(12 \cdot 4) - 3]$, will be printed.   ❖

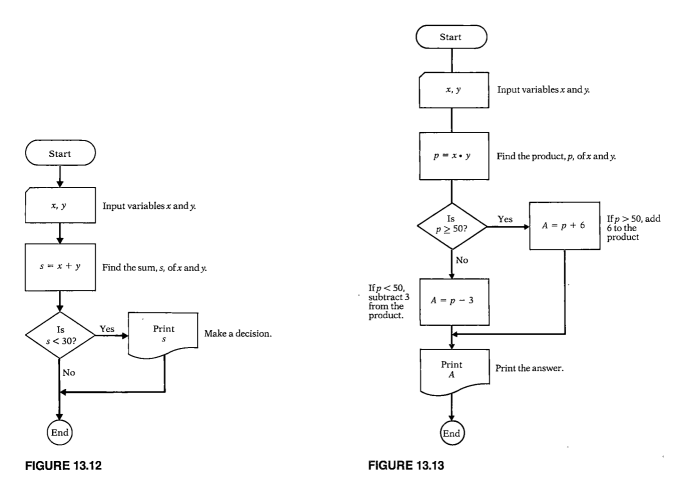

FIGURE 13.12

FIGURE 13.13

EXERCISES 13.3

1. What does BASIC stand for?

Write the answer the computer will give for each of the following.

2. $2 * 6 + 3 - 4$
3. $2 + 3 * 5$
4. $2 / 3 + 4$
5. $2 + 4 / 2$
6. $2 * 3 \wedge 2 + 4$
7. $16 * 4 / 8$
8. $9 * SQR(16) + 3$
9. $6 - 2 - 4 * 3$
10. $(-4 * -3) + 12$
11. $(2 + 4) \wedge 2 - 3 * 4$
12. $((3 + 4 / 2)) \wedge 3 - SQR(36)$
13. $-4 \wedge 2 + 3 - 5 * 7$
14. $(5 - 3 * 2) \wedge 2 / 2 * 4$
15. $4 + 10 / 5 * 2 + 7$
16. $8 - 6 - 4 * 5 + 7 - 3 * 5$
17. $4 + 10 / (5 * 2) + 7$
18. $(3 + (5 \wedge 2 - 4)) / 4 * 3$
19. $2 \wedge 3 - 4 * SQR(25)$
20. $((3 + 5) \wedge 2 - 4) / 4 * 3$

Convert each of the following to BASIC symbols. Then find the answers.

21. $2 + (3 \cdot 4)$
22. $(2 + 3) \cdot 4$
23. $(2 + 3^2) \cdot 5$
24. $\sqrt{25} + (6 \div 2)$

25. $(\sqrt{25} + 6) \div 2$ **26.** $(3 - 2) \cdot 6$

27. $3^2 \div 2^2$ **28.** $4 - 2 + (3 \cdot 5)$

29. $(4^2 - 6) \div 5\sqrt{36}$

30. $[3 \cdot (2 + 5)]^2$

31. $(4 \cdot 3) - (5 \cdot 6)^2$

32. $\sqrt{25} \cdot \sqrt{4} - 2 \cdot 6^2$

33. $-3^2 + (4 \cdot 7) - (30 \div 5)$

34. $[(4 \div 8) \cdot 16] \div (8 \cdot 3)$

35. $[(-2)^2 + 3]^2 \div (4 + 5)$

36. $(4 \div 8) \cdot (16 \div 8) \cdot 3$

Write the following statements in algebraic notation.

37. $4 * x + 4$ **38.** $2 * x - 7$

39. $x \wedge 2$ **40.** $x \wedge 3 - 4 * x$

41. $3 * SQR(x) + 4$ **42.** $(x + 3) / x \wedge 3$

43. $4 * x \wedge 2 - 5 * x + 3$

44. $-2 * x \wedge 4 - 3 * x / x \wedge 3$

45. $5 / x + 2 * x - 3$

46. $(5 + x) \wedge 2 * (4 - x) 3$

47. $-3 * SQR(x) - 5 * x$

48. $4 * x \wedge 3 - 5 * x \wedge 2 + 4 * x - 7$

49. $3 * SQR(x \wedge 3) - 6 * x \wedge 3$

50. $(x - 7) * (x + 6) \wedge 4 + 2$

51. $3 * (7 * x \wedge 2 + 4 * x)$

52. $4 / (x + 3) - (6 * x) / (2 * x - 3)$

Write the following statements in BASIC symbols.

53. $3x + 4$

54. $-5x - 7$

55. $3x^2$

56. $4x^2 - 3x$

57. $(x + 4) \div (2x + 3)$

58. $\sqrt{x} + 6/x$

59. $3\sqrt{x} - 7x + 1$

60. $(2x + 3)^2$

61. $6/x^2 + 5x + 4$

62. $2x^3 + 7x^2 + 5x - 4$

63. $(2x^2 + 6x - 4) \div (x - 3)$

64. $3(2x^2 + 3x + 4)$

65. $[(-2x)^3 + 4]^2$

66. $\frac{1}{2}(3^2 + 2x)(x - 1)$

67. $(2x - 3)^2(x^2 + 5)$

68. $2[(x + 4)(2x - 3)] \div (x^2 + 5)$

In Exercises 69 through 82, construct a flowchart to evaluate each of the equations.

69. Area, $A = l \cdot w$ **70.** Interest, $i = prt$

71. The value of y in the equation $y = 3x + 4$, when $x = 4$

72. The area of a square, $A = s^2$

73. The value of y in the equation $y = 2x^2 - 3x - 5$

74. The average speed, $S = d/t$

75. The number of seconds in x minutes, $s = 60x$

76. The number of seconds, s, in y hours and x minutes. Use the equation $s = 3600y + 60x$, and evaluate when $y = 3$ hours and $x = 5$ minutes.

77. The perimeter of a rectangle, $P = 2l + 2w$

78. The area of a circle, $A = \pi r^2$ (let $\pi = 3.14$)

79. The amount to be repaid, $A = P(1 + r)^n$

80. The discriminate d in the quadratic formula, $d = b^2 - 4ac$

81. The value of C in $C = \dfrac{5}{9}(F - 32)$

82. The value of F in $F = \dfrac{9}{5}C + 32$ when $C = 37°$

In Exercises 83 through 90, construct a flowchart to solve the problem.

83. Select a number x. If the number is greater than 12, add 3 to the number. If the number is less than or equal to 12, subtract 4 from the number. Print the answer.

84. Select a number x. If the number is even, divide it by 2. If the number is odd, multiply it by 3. Print the answer.

85. Select two numbers, x and y. If their product is positive, add 1 to the answer. If their product is zero or a negative number, add 2 to the answer. Print the answer.

86. Multiply two numbers, x and y. If the product is greater than or equal to 50, add 10 to the product. If the product is less than 50, subtract 5 from the product. Print the answer.

87. Multiply a number N by 3. If the product is even, subtract 1. If the product is odd, add 2. Print the answer.

88. Select a number M. If M is less than or equal to 20, divide M by 20 and print the result. If M is greater than 20, multiply M by 4 and then end the flowchart without printing the answer.

89. Select a number. If its square is even, multiply the number by 3 and print the product. If the square is not even, end the flowchart without printing any value.

90. Evaluate $b^2 - 4ac$ for values of a, b, and c. If the answer is greater than or equal to 0, print the answer. If the answer is less than 0, have the flowchart end without printing the answer.

13.4 INTRODUCTORY BASIC PROGRAMS

It is not essential to have a computer available to learn to program in BASIC. However, hands-on experience is extremely helpful. If a computer is available, it may be advantageous to use it along with your textbook. Do not be afraid to experiment on the computer—you will not damage it. Before you begin to program in BASIC, you should have an understanding of the following: how to get on the computer, how to use the keyboard, how to make typing corrections, how to format a disk, how to save a program on a disk, and how to recall a program from the disk. These procedures vary from computer to computer. Consult a manual or your instructor for specific instructions for your computer.

Some BASIC programming statements (like different dialects) vary from computer to computer. In this text we will introduce the BASIC language used on the Apple II e microcomputers. If you are using a computer other than the Apple II e, check with your instructor to determine how the BASIC statement and computer operations differ for the specific computer available to you.

All computers inform you when you have made an error in programming, the most common of which is a syntax error. A **syntax error** means that an "illegal" statement has been entered. Usually, you have typed something wrong or left something out. For example, to print a statement, we use a PRINT statement. If we type in

    10 PRIMT "GOOD"

the computer will display a syntax error message because PRINT is spelled incorrectly. When typing in numbers, do not use commas. If you do, you may be given a syntax error message. All errors must be corrected in order to obtain results from the computer.

Instructions given to a computer as part of a program are called *statements*. In writing a program in BASIC, each statement must be preceded by a line number. A computer reacts differently to operating system commands and program statements. *Operating system commands,* such as LIST, LOAD, RUN, do not require line numbers and are executed immediately after the return key is pressed. Statements in programs are not executed until the program is complete and the RUN command is typed. Then the computer executes the statements in the order given by the line numbers, unless instructed to do otherwise.

In this text we will often label our first line with the number 10 and number successive lines in multiples of 10. The computer executes statements in the order of the line numbers unless instructed to do otherwise. By using 10, 20, 30, 40, . . . , we leave space for additional instructions that we may want to insert at a later time.

Many programs contain variables. **Variables** are letters that are used to represent numbers. Often we will have to assign a numerical value to a vari-

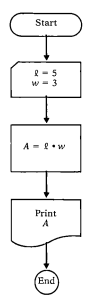

FIGURE 13.14

able. In this section we will discuss three methods for doing this: the LET statement, READ and DATA statements, and the INPUT statement.

Consider the flowchart in Fig. 13.14, which can be used to find the area of a rectangle with a length of 5 inches and a width of 3 inches.

We can assign values to variables with **LET** statements. For example, to assign the length a value of 5 and the width a value of 3, we can type the following:

```
10 LET L = 5
20 LET W = 3
```

On most computers it is not necessary to type the word LET. Thus 10 L = 5 also assigns the variable L a value of 5. For the sake of clarity we will use LET in our programs. Make sure you use 10 LET L = 5 and not 10 LET 5 = L. To name a variable, you may use either a letter or a letter followed by a number, such as in 10 LET A1 = 5.

Another method for assigning values to variables is by means of **READ** and **DATA** statements. These statements are often used when a large amount of data is to be fed into the computer. They are often part of a "loop," as will be illustrated in Section 13.6. Whenever there is a READ statement, there must be a DATA statement. The relative location of the DATA statement in the program has no effect on the program's execution. Whenever the computer encounters its first READ statement, it looks for the lowest-numbered DATA statement. The computer then takes the values in the DATA statement and assigns these numbers to the variables in the READ statement. The DATA is often placed at the beginning of the program so that if you decide to change the data it is easy to locate.

In this illustration we can assign the length a value of 5 and the width a value of 3 by writing the following:

```
10 DATA 5, 3
20 READ L, W
```

Two other methods can be used to assign the length a value of 5 and the width a value of 3:

```
10 READ L, W       or      10 READ L, W
20 DATA 5, 3              20 DATA 5
                          30 DATA 3
```

When the computer comes to the READ statement, it automatically looks for the DATA statement. The first piece of data, 5, is assigned to the first variable, L, and the second value, 3, is assigned to the second variable, W. Note that commas are used to separate the values in the DATA statement and the variables in the READ statement.

Here are two programs that can be used to find the area of a rectangle with a length of 5 inches and a width of 3 inches:

```
10 LET L = 5          10 DATA 5, 3
20 LET W = 3          20 READ L, W
30 LET A = L * W      30 LET A = L * W
40 PRINT A            40 PRINT A
50 END                50 END
```

After you complete the program, you must press the return key, type RUN, and press the return key again. The monitor will then display the answer, 15.

Note how similar the programs are to the flowchart in Fig. 13.14.

One error that is commonly made by students is to assign the variables values after using them in an assigned statement. Consider the following programs:

```
10 LET A = L * W      10 DATA 5, 3
20 LET L = 5          20 LET A = L * W
30 LET W = 3          30 READ L, W
40 PRINT A            40 PRINT A
50 END                50 END
```

In the program on the left the first step asks the computer to multiply L times W. But how can the computer do this if you have not yet told it the values of L and W? Some computers will assign a value of zero to any variables not given specific values. On such computers the answer printed would be 0. Other computers may print an error message, letting you know that you have not assigned values to L and W.

In the program on the right, the computer has again been instructed to multiply L and W without knowing their values. This program does not assign values to L and W until line 30, which comes after the use of the variables in line 20. If the READ statement had preceded the LET statement, the program would run properly.

■ **Example 1** _____

a) Construct a flowchart to find the interest I for a principal (p) of $2000, rate ($r$) of 12%, and time ($t$) of 2 years. Use $I = prt$, where r is the rate in decimal form.

b) Write a program using LET statements for finding the interest for the given values of p, r, and t.

c) Write a program using READ and DATA statements for finding the interest for the given values of p, r, and t.

Solution: The flowchart is illustrated in Fig. 13.15.

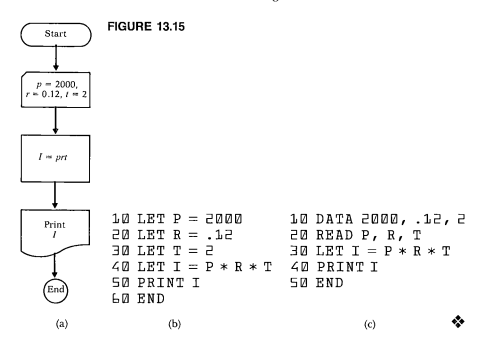

FIGURE 13.15

```
10 LET P = 2000      10 DATA 2000, .12, 2
20 LET R = .12       20 READ P, R, T
30 LET T = 2         30 LET I = P * R * T
40 LET I = P * R * T 40 PRINT I
50 PRINT I           50 END
60 END
```

(a)   (b)   (c)   ❖

A third method of assigning values to variables is by means of the **INPUT** statement. INPUT statements can be used when the variables are to be assigned values after the program is written and while it is running.

The flowchart in Fig. 13.16 can be used to find the area of a triangle.

Note that specific values have not been assigned to the base b or the height h. If we wish to write a program in which b and h will be assigned values at a later time, we can write the following:

```
10 INPUT B, H
```

This statement tells the computer that we will assign B and H values at a later time. The program that corresponds to this flowchart is

```
10 INPUT B, H
20 LET A = 1/2 * B * H
30 PRINT A
40 END
```

When we type RUN and press the return key, the computer will show a question mark, indicating that we must now assign specific values to the variables. If we then type in specific values for B and H separated by a comma

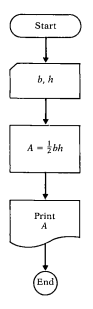

FIGURE 13.16

and press the return key, the computer will display the answer. The process might look like this:

RUN ← You type (Then press the return key.)

computer displays → ? 6,9 ← You type (Then press the return key.)

computer displays → 27

The first value that you input is assigned to the first variable in the INPUT statement, the second value is assigned to the second variable, and so on.

If you type RUN a second time and then press the return key, the computer will again display a question mark, which indicates that it is waiting for more data to be entered.

Example 2

a) Construct a flowchart that can be used to find the value of y for given values of x in the equation $y = 3x^2 - 6x + 5$.

b) Write a program, using an INPUT statement, to find the value of y for any given value of x in the equation $y = 3x^2 - 6x + 5$.

Solution: (a) See Fig. 13.17.

FIGURE 13.17

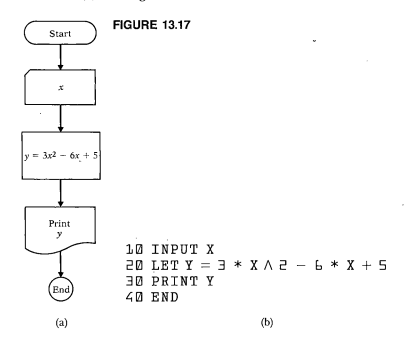

```
10 INPUT X
20 LET Y = 3 * X ∧ 2 - 6 * X + 5
30 PRINT Y
40 END
```

(a)   (b)

❖

Note that when you use an INPUT statement, you are asked by the computer to supply information to the computer. This mode of operation is often referred to as the **interactive mode.**

Another useful statement is the **REM** (REMARK) statement. The REM statement is used to insert explanatory remarks within a program (internal documentation) to make the program more understandable to those who may look at it. During the execution of a program, the computer ignores REM statements and moves on to the next line. The following example illustrates how REM statements may be used:

```
10 REM THIS PROGRAM CHANGES FEET TO METERS
20 REM F = FEET, M = METERS
30 INPUT F
40 LET M = .304 * F
50 PRINT M
60 END
```

If you look at the program later, the REM statements may help you remember its purpose.

EXERCISES 13.4

Write a program for finding each of the following. Unless otherwise specified, use INPUT statements.

1. The area of a square, $A = s^2$

2. The area of a rectangle, $A = l \cdot w$, where $l = 9$ and $w = 5$. Use LET statements.

3. The area of a rectangle, where $l = 9$ and $w = 5$. Use READ and DATA statements.

4. The perimeter of a rectangle, $P = 2l + 2w$

5. The circumference of a circle, $c = 2\pi r$, where π has a value of 3.14.

6. The value of y in the equation $y = 3x^2 - 4$, where x has a value of 5. Use a LET statement.

7. The value of z in $z = 4x^2y + 5y$, where x has a value of 6 and y has a value of 4. Use READ and DATA statements.

8. The Celsius temperature, $C = \frac{5}{9}(F - 32)$

9. The Fahrenheit temperature, $F = \frac{9}{5}C + 32$

10. The amount to be repaid, $A = P(1 + r)^n$ where $P = 2500$, $r = 0.12$, and $n = 4$. Use READ and DATA statements.

11. The annual percentage rate,

$$A = \frac{2i}{(f + l)t}$$

12. The value of x in $x = \sqrt{b^2 - 4ac}$, where $b = 5$, $a = 12$, and $c = -2$. Use READ and DATA statements.

13. The area of a trapezoid, $A = \frac{1}{2}h(b_1 + b_2)$, where $h = 10$, $b_1 = 8$, and $b_2 = 20$. Use LET statements.

14. The volume of a sphere, $v = \frac{4}{3}\pi r^3$, where $r = 4$. Use LET statements.

15. The average of five exams,

$$A = (B + C + D + E + F)/5$$

 13.5 # THE PRINT STATEMENT

The **PRINT** statement can be used (1) to print the value of a variable, as in Examples 1 and 2 in the previous section, (2) to type a message, (3) to provide a blank line on the printout, and (4) to cause the computer to perform a calculation and output the result of that calculation.

Here are some general rules regarding PRINT statements.

1. When the word PRINT appears by itself on a line, the computer prints a blank line.

2. When an item is typed in quotation marks after the word PRINT the computer prints exactly what appears between the quotation marks.

3. When a *comma* is typed between the responses in a PRINT statement, the computer types the responses on the same line, with a fixed space between the responses.

4. When a *semicolon* is typed between the responses in a PRINT statement, the computer types the responses on the same line, with no space between the responses.

The uses of the PRINT statement are illustrated in the following examples.

 ## Example 1

Determine the output of the following program:

```
10 PRINT "THE"
30 PRINT "ARE"
40 PRINT "BITING"
50 END
20 PRINT "FISH"
```

Solution: After RUN is typed the computer will display the following:

```
THE
FISH
ARE
BITING                                              ❖
```

Note that the computer performs the statements in the order of the line numbers, not the order in which the statements are entered.

◼ Example 2 _____

Determine the output of the following program:

```
10 DATA 3, 5, 7
20 READ X, A, B
30 PRINT A
40 PRINT A + B
50 PRINT
60 PRINT B + 1
70 PRINT "B + 1"
80 END
```

Solution

	Corresponds to Line Number
5	30
12	40
	50
8	60
B + 1	70

Note that the PRINT statement in line 30 printed the value of a variable. The PRINT statement in line 40 is used to perform a calculation and print its answer. The PRINT statement in line 50 leaves a blank line in the program. The PRINT statement in line 60 performs a calculation (adding 1 to the value of B) and prints the result. The PRINT statement in line 70 prints the information within quotation marks.   ❖

■ **Example 3** ───

Determine the printout of the following program:

```
 10 LET A = 6
 20 LET B = 4
 30 PRINT A
 40 PRINT A, B
 50 LET C = 12
 60 PRINT 25 + 58
 70 PRINT "HELLO"
 80 PRINT "A + C = "; A + C
 90 PRINT "A + C = ", A + C
100 PRINT
110 PRINT A + B, A * B, A/B
120 PRINT " ", "HELLO"
130 END
```

Solution

			Corresponds to Line Number
6			30
6	4		40
83			60
HELLO			70
A + C = 18			80
A + C =	18		90
			100
10	24	1.5	110
	HELLO		120

Let us analyze the computer's execution of the program just given. Line 30 has the computer print the value of A, 6. Line 40 has the computer print the values of A and B with a fixed space between the values. Line 60 has the computer calculate and print the sum of 25 plus 58, or 83. Line 70 has the computer type HELLO (note the quotation marks). Line 80 has the computer print $A + C =$ (note the quotation marks), then calculate the sum of $A + C$, and record that sum directly after the $A + C =$. Line 90 does basically the same. However, since a comma is used in place of a semicolon, the sum is placed a fixed distance to the right. Line 100 causes the computer to leave a blank line in the printout. Line 110 will have the computer print all three calculations with a fixed space between each response. Line 120 will have the computer print the word HELLO near the center of the page. It will do this because there is nothing within the first set of quotation marks. If we wished to move the HELLO farther to the right, we could use

```
120 PRINT " ", " ", "HELLO".
```
❖

■ **Example 4** ───────────────────────────────

John's test scores on five exams were 78, 96, 83, 59, 75. Write a program for finding his average test score using

a) the INPUT statement,

b) the READ and DATA statements,

c) the LET statement (without INPUT or READ and DATA statements).

Have the answer read THE AVERAGE IS _____.

Solution

a)
```
10 INPUT B, C, D, E, F
20 LET A = (B + C + D + E + F) / 5
30 PRINT "THE AVERAGE IS"; A
40 END
RUN
? 78, 96, 83, 59, 75
THE AVERAGE IS 78.2
```

b)
```
10 READ B, C, D, E, F
20 LET A = (B + C + D + E + F) / 5
30 PRINT "THE AVERAGE IS"; A
40 DATA 78, 96, 83, 59, 75
50 END
RUN
THE AVERAGE IS 78.2
```

c)
```
10 LET B = 78
20 LET C = 96
30 LET D = 83
40 LET E = 59
50 LET F = 75
60 LET A = (B + C + D + E + F) / 5
70 PRINT "THE AVERAGE IS"; A
80 END
RUN
THE AVERAGE IS 78.2
```
❖

■ **Example 5** ───────────────────────────────────

Write a BASIC program using the READ and DATA statements (assigning
x a value of 5 and y a value of 7) that will result in the following outputs.

a) THE SUM IS 12
    THE PRODUCT IS 35.

b) SUM   PRODUCT
    12   35

c) SUM   PRODUCT

    12   35

Solution

a) 10 READ X, Y
    20 DATA 5, 7
    30 LET S = X + Y
    40 LET P = X * Y
    50 PRINT "THE SUM IS"; S
    60 PRINT "THE PRODUCT IS"; P
    70 END

b) 10 READ X, Y
    20 DATA 5, 7
    30 LET S = X + Y
    40 LET P = X * Y
    50 PRINT "SUM", "PRODUCT"
    60 PRINT S, P
    70 END

c) 10 READ X, Y
    20 DATA 5, 7
    30 LET S = X + Y
    40 LET P = X * Y
    50 PRINT "SUM", "PRODUCT"
    60 PRINT
    70 PRINT S, P
    80 END   ❖

    When a LET statement contains a variable on both sides of the equals
sign, only the value of the variable on the left side is changed. Statements
such as 10 LET A = A + 1 are referred to as *assignment statements* and are

not equations. When the computer executes the statement 1Ø LET A = A + 1, it first evaluates the expression on the right-hand side, A + 1. It then replaces the old value of A with the value of the expression it just found (A + 1). If the same variable appears on both sides of the equals sign in a LET statement (such as 1Ø LET A = A + 1 or 2Ø LET A = A + B), the value of the variable (or variables) on the right side of the equals sign are used in determining the new value of the variable on the left side of the equals sign. Note that other programming languages use different symbols, such as ← or := , for assignments to avoid confusing them with equations.

■ **Example 6** ——————————————————————————

Determine the output of the following program:

```
1Ø DATA 3, 7
2Ø READ A, B
3Ø LET A = A + 1
4Ø PRINT A
5Ø LET A = A + 1
6Ø PRINT A
7Ø LET A = A + B
8Ø PRINT A
9Ø END
```

Solution

	Corresponds to Line Number
4	4Ø
5	6Ø
12	8Ø

In line 2Ø, A is assigned a value of 3. In line 3Ø, A = 3 + 1 = 4. Line 4Ø has the computer print the value of A, 4. Line 5Ø assigns a new value of 5 (A = A + 1 = 4 + 1) to A. Line 6Ø has the computer print the new value of A, 5. In line 7Ø the variable A is assigned a new value of 12 (A = A + B = 5 + 7). Line 8Ø has the computer print the new value of A, 12.   ❖

EXERCISES 13.5

Each of the following statements or group of statements contains an error. Find the error and indicate how it may be corrected.

1. 1Ø LET 5 = X
2. 1Ø PRINT 2X

3. `10 INPUT XY`
4. `10 INPUT 5`
5. `10 LET Y = 4X`
6. `10 READ 5, 10`
   `20 DATA X, Y`
7. `10 PRINT Y = 4 + 5`
8. `10 PRINT I LIKE MATH`
9. `10 PRINT "I LIKE MATH`
10. `10 DATA 5, 10`
    `20 READ X, Y, Z`
11. `10 DATA 23, 400, 75`
    `20 READ X,Y`
12. `10 DATA 5 20`
    `20 READ X Y`
13. `10 PRINT Y = "2 + 3"`
14. `READ A, B`
    `DATA 5, 10`

15. Indicate what will be printed when each of the lines (a)–(i) of the following program are executed.

    ```
    10 DATA 3, 4
    20 READ X, Y
    ```

    a) `30 PRINT Y`
    b) `40 PRINT 4 + 6`
    c) `50 PRINT "X = "; X`
    d) `60 PRINT "Y = ", Y`
    e) `70 PRINT`
    f) `80 PRINT "X + Y = "; X + Y`
    g) `90 PRINT Y + 9`
    h) `100 REM PRINT "X * Y"`
    i) `110 PRINT X + Y, X * Y`
    `120 END`

What is the computer output for the following programs?

16. ```
    10 READ S
    20 PRINT S∧2
    30 DATA 5
    40 END
    ```
17. ```
    10 READ X, Y
    20 PRINT X * Y / 2
    30 DATA 15, 16
    40 END
    ```

18. ```
    10 INPUT X, Y, Z
    20 LET R = (X + Y) * Z
    30 PRINT R
    40 END
    RUN
    ? 5, 6, 7
    ```
19. ```
    10 INPUT X, Y, Z
    20 PRINT (X + Y) * Z
    30 END
    RUN
    ? 5, 6, 7
    ```
20. ```
    10 READ B, H
    20 DATA 8, 12
    30 PRINT "AREA IS" ; (B * H) / 2
    40 END
    ```
21. ```
    10 READ P, R, N
    20 DATA 100, .05, 2
    30 PRINT "AMOUNT IS "; P *
       (1 + R)∧N
    40 END
    ```
22. ```
    10 INPUT A, B, C
    20 LET D = (A + B + C) / 3
    30 PRINT   "AVERAGE IS" ; D
    40 PRINT "GOOD-BYE"
    50 END
    RUN
    ? 8, 10, 12
    ```

Write a BASIC program using READ and DATA statements to evaluate each of the following.

23. The average speed (s) when the distance (d) is 13.3 and time (t) is 4.6. Use $s = d/t$.
24. The volume (V) when the length (l) is 17.2, the width (w) is 9.43, and the height (h) = 4.21. Use $V = l \cdot w \cdot h$.
25. The volume (V) when the height (h) is 6.7 and the radius (r) is 3.4. Use $V = \frac{1}{3}\pi r^2 h$.
26. The area (A) of a trapezoid when the height (h) is 10, one base (b_1) is 12, and the other base (b_2) is 20. Use $A = \frac{1}{2}h(b_1 + b_2)$.
27. Write the computer output for the following program:

    ```
    10 PRINT "MY NAME IS GEORGE"
    20 PRINT 20 * 15
    30 PRINT "20 × 15 = ", 20 * 15
    40 LET X = 5
    50 LET Y = 8
    60 PRINT "X + Y = "; X + Y
    70 PRINT "X + Y = ", X + Y
    ```
    (continued on the next page)
    ```

```
 80 PRINT
 90 PRINT (15 * 20) / X + Y
100 END
```

28. Write a BASIC program using the READ and DATA statements (assigning A a value of 16 and B a value of 20) that will result in the following outputs.

    a) ```
       THE SUM IS 36
       THE PRODUCT IS 320
       ```

    b) ```
       SUM            PRODUCT

       36             320
       ```

29. Write a BASIC program for converting miles to kilometers (1 mile = 1.6 km). Use the INPUT statement. Write the programs that will result in the following outputs.

    a) ```
       ? 6
       6 MILES = 9.6 KILOMETERS
       ```

    b) ```
       ? 6
       MILES          KILOMETERS

       6              9.6
       ```

30. Write a program that will give the following output:

    ```
    PRODUCT OF TWO NUMBERS
    PICK TWO NUMBERS
    ? 7, 9
    THEIR PRODUCT IS 63
    ```

31. Write a program using INPUT statements that will result in the following output:

    ```
    PICK ANY TWO NUMBERS
    ? 5, 12
    THEIR PRODUCT IS 60
    PICK TWO DIFFERENT NUMBERS
    ? 9, 6
    THEIR SUM IS 15
    ```

32. Determine the output of the following program:

    ```
    10 LET A = 5
    20 LET A = A + 1
    30 LET A = 2 * A
    40 LET A = A / 4
    50 PRINT A
    60 END
    ```

33. Determine the output of the following program.

Assume that after RUN is typed, 3 is used for X and 5 is used for Y.

```
10 INPUT X
20 PRINT X∧2
30 INPUT Y
40 PRINT X * Y
50 END
```

34. Determine the output of the following program:

    ```
    10 DATA 3, 7
    20 READ A, B
    30 LET B = A + 1
    40 PRINT B
    50 LET B = A + B
    60 PRINT B
    70 PRINT A
    80 END
    ```

35. Write a program using READ and DATA statements to change from Fahrenheit temperature to Celsius temperature. Use $C = 5/9(F - 32)$. Write the program in two ways so that the output reads as follows:

    a) 41 Fahrenheit = 5 Celsius

    b) Fahrenheit   Celsius
       41   5

36. Write a program using an INPUT statement for computing the percent increase (or decrease), using the formula

    $$\text{percent increase } (P) = \frac{\text{new value } (X) - \text{previous value } (Y)}{\text{previous value } (Y)} \times 100.$$

    Write the program so that the output reads as follows:

    ```
    ? 10, 4
    NEW VALUE IS 10
    OLD VALUE IS 4
    PERCENT INCREASE IS 150%
    ```

37. Another important statement is INT(X). This statement has the computer determine the greatest whole integer less than or equal to X. For example, INT (3.14) = 3, INT (0.69) = 0, and INT (−2.48) = −3. Determine the output of the following program:

    ```
    10 LET A = 5.62
    20 PRINT A, INT(A),
       INT(SQR(A)), (INT(A))∧2
    30 END
    ```

38. Mr. and Mrs. Zimmer have decided that Mrs. Zimmer's salary for the next few years will be used to purchase government bonds to pay for their daughter's college education. Mrs. Zimmer works on commission; therefore her take-home pay varies weekly.

a) Construct a flowchart to determine the number of bonds she can purchase weekly and her remaining balance if the bonds cost $25 each (she cannot purchase fractional parts of a bond).
b) Construct a program to determine the number of bonds she can purchase and the remaining balance.

PROBLEM SOLVING

39. Write a program to generate the pattern shown.

```
XXXXX XXXXXX XXXXXXXXX XXXXXXX XXXXX
XX   X X         X       X X       X
XXXXX XXXXXX XXXXXXXXX X           XXXXX
XX     X         X       X X       X
XX     X         X       X X       X
XX     XXXXXX X           X XXXXXXX XXXXX
```

 LOOPS

A great advantage of the computer over some calculators is that the computer has the ability to perform repeated calculations when programmed to do so. Thus a teacher calculating students' averages need only enter all the data; the computer will compute the average of each student. On a calculator (except on the programmable types) the teacher would have to perform each calculation separately. The computer saves people many hours of tedious calculations.

As was stated earlier, a computer executes a program line by line in numerical order unless it is instructed to do otherwise. The loop is one method of instructing the program to modify the ordinary order.

Example 1

Construct a flowchart that can be used to develop a program that will instruct the computer to print the results of multiplying each of the first ten positive integers by 5.

Solution: The flowchart in Fig. 13.18(a) will continue multiplying the consecutive integers by 5 and printing the products. This flowchart can be simplified to Fig. 13.18(b).

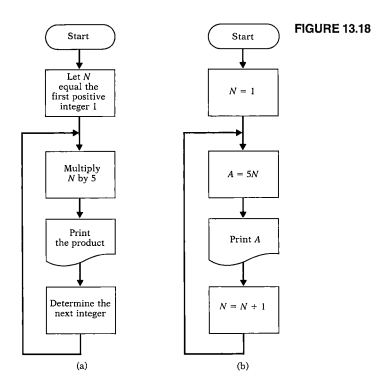

FIGURE 13.18

If we leave the flowchart as it is, we will obtain the following output:

5
10
15
.
.
.
5N
.
.
.

We must provide instructions to tell the computer when to stop. One method of achieving this is to insert a decision step, as shown in Fig. 13.19.

For $N = 1$, the value of A is 5, which is printed. Since $N = 1$, which is not equal to 10, we proceed to $N = N + 1$. This has the effect of raising the value of N to 2. The flowchart then loops back to $A = 5N$. Since N is now 2, the value of A that is printed is $5 \cdot 2$, or 10. This process continues in the same manner until N has a value of 10. When N is 10, the product of $5 \cdot 10 = 50$ is printed. At this point, since $N = 10$, the decision box "flags" the program to end. ❖

FIGURE 13.19

Before we discuss how to form a loop in BASIC programming, we must understand how the computer makes a decision when it comes to a decision box. The computer has the ability to evaluate mathematical relations involving equalities and inequalities. Table 13.4 indicates these mathematical relations and the BASIC symbols for them. For example, the expression $9 < = 11$ means 9 is less than or equal to 11, that is, $9 \leq 11$; and $5 < > 7$ means 5 is not equal to 7, that is, $5 \neq 7$.

TABLE 13.4

Mathematical Symbols	Meaning	BASIC Symbols
$=$	equals	$=$
$<$	less than	$<$
$\leq$	less than or equal to	$< =$
$>$	greater than	$>$
$\geq$	greater than or equal to	$> =$
$\neq$	not equal to	$< >$

The most common methods used to instruct a computer not to follow the numerical order in executing a BASIC program are (1) GO TO statements, (2) IF–THEN statements, and (3) FOR and NEXT statements.

Let us now examine each of these three instructions. The **GO TO** statement instructs the computer to GO TO a specific line number in the program.

■ **Example 2** _____

Construct a BASIC program of the flowchart in Fig. 13.18 using a GO TO statement.

Solution

```
10 LET N = 1
20 LET A = 5 * N
30 PRINT A
40 LET N = N + 1
50 GO TO 20
60 END
```

Just as the flowchart will continue indefinitely, this program when run will continue indefinitely. Note that the GO TO statement is placed after the LET $N = N + 1$. After the new value of N is calculated, this new value is looped back to line 20, where a new value of A is calculated. Since there is no exit criterion, this process will continue. ❖

Another method used in forming loops is the **IF–THEN** statement. When the computer executes an IF–THEN statement like IF $X = 5$ THEN 60, if the current value of X is equal to 5, the computer will jump to line 60. If the current value of X is not equal to 5, the computer will simply go to the next line in the usual numerical order. Often the line number at the end of the IF–THEN statement is inserted after you have finished writing the program and know the line numbers of each statement.

The IF–THEN (or IF–GO TO) command gives the computer a method of transferring from one point of the program to another point if the stated criterion is met. IF–THEN statements are often used to terminate a program, as illustrated in Example 3.

Example 3

Figure 13.19 illustrates a flowchart that can be used to multiply each of the first ten integers by 5 and print the result. Construct a program, using an IF–THEN statement, that will generate this output.

Solution: We will use the IF–THEN statement, as illustrated below:

```
10 LET N = 1
20 LET A = 5 * N
30 PRINT A
40 IF N = 10 THEN 70
50 LET N = N + 1
60 GO TO 20
70 END
```

When $N = 1$, A becomes 5, and line 30 has the computer print the number 5. Since $N = 1$, the conditions of line 40 are not met, and the computer goes directly to line 50. At line 50, N is assigned a value of $1 + 1$, or 2. Line 60 has the computer loop back to line 20. The value of A now becomes $5 \cdot 2$, or 10. Line 30 has the computer print the number 10. Since $N = 2$, the conditions of line 40 are not met. The computer goes directly to line 50, and N now becomes 3. This program will continue sequentially until $N = 10$. When $N = 10$, the condition of the IF–THEN statement (line 40) is met, and the program goes directly to line 70, which ends the program. The computer will print or display the numbers 5, 10, 15, . . ., 50 vertically.   ❖

We have previously discussed the READ and DATA statements. In the examples presented, we were careful to give the same number of pieces of data as there were variables to be read. This need not be so, as is illustrated in Example 4.

■ **Example 4** _____

Determine the output of the following BASIC program:

```
10 REM FIND AVERAGE GRADES
20 DATA 56, 75, 83, 94, 62, 58, 78, 88
30 READ A, B
40 LET M = (A + B) / 2
50 PRINT "A = "; A, "B = "; B, "AVERAGE = "; M
60 GO TO 30
70 END
```

Solution: At first, A is assigned a value of 56, and B is assigned a value of 75. Line 40 assigns M to be the average of A and B. Line 50 will have the computer print

```
A = 56      B = 75      AVERAGE = 65.5
```

Line 60 will return the program to line 30. Since the values of 56 and 75 were previously used, the computer will disregard them and assign the next value in the list, 83, to A and the following value, 94, to B. The computer will then print

```
A = 83      B = 94      AVERAGE = 88.5
```

The computer printout will read

```
A = 56      B = 75      AVERAGE = 65.5
A = 83      B = 94      AVERAGE = 88.5
A = 62      B = 58      AVERAGE = 60
A = 78      B = 88      AVERAGE = 83
```

After this point the computer may print out the message OUT OF DATA ERROR IN 30. ❖

The error message in Example 4 occurred because when the computer tried to assign new values to A and B in line 30, it realized that all the data in the DATA statement had been used. A good program will have a step to terminate the program after the last piece of data is used. Example 5 shows how this can be done.

■ **Example 5** _____

Modify the program in Example 4 so that it will terminate after all the given data have been used.

Solution

```
10 REM FIND AVERAGE GRADES
20 DATA 56, 75, 83, 94, 62, 58, 78, 88, -1, -1
30 READ A, B
35 IF A = -1 THEN 70
40 LET M = (A + B) / 2
50 PRINT "A = "; A, "B = "; B, "AVERAGE = "; M
60 GO TO 30
70 END
```

We added two pieces of data, -1 and -1, to the original set of data in Example 4. These pieces of data, called **flags** or **sentinels,** will tell the computer when to terminate the program. When A is assigned a value of -1, the new line (35) will cause the computer to go to line 70, which will end the program. Any values may be designated as flags. The computer output will be the same as in Example 4, but this program will end after the computer prints the last line.

```
A = 78     B = 88     AVERAGE = 83
```
❖

■ **Example 6** ——————————————————————————

A variation of the BASIC program in Example 5 is given below. Determine the computer output.

```
 10 REM FIND AVERAGE GRADES
 20 DATA 56, 75, 83, 94, 62, 58, 78, 88, -1, -1
 30 LET C = 0
 40 READ A, B
 50 IF A = -1 THEN 100
 60 LET C = C + 1
 70 LET M = (A + B) / 2
 80 PRINT "STUDENT NO"; C, "A = "; A, "B = "; B,
 85 PRINT "AVERAGE = "; M
 90 GO TO 40
100 END
```

Solution: We added the statements 30 LET C = 0 and 60 LET C = C + 1, and we changed the print message in line 80. By adding lines 30 and 60, we created a *counter.* When the program begins, C is assigned an initial value of 0. (Some computers will automatically assign a value of 0 to any unassigned variable. However, it is good programming practice to assign a value to each variable.) When line 60 is reached, C is increased by 1. Line 70 calculates the average of the first two pieces of data (belonging to student 1). When the program is run, lines 80 and 85 will have the computer print

STUDENT NO 1 ┌A = 56┐ B = 75 AVERAGE = 65.5

Note the comma at the end of the PRINT statement in line 80. This comma tells the computer that the information in the following PRINT statement, line 85, is to be printed as if it were a continuation of the PRINT statement in line 80. Thus if space permits, the information from lines 80 and 85 will be printed on the same line. Line 90 will cause the program to loop back to line 40. Then A will be assigned a value of 83 and B a value of 94. Since C had a value of 1 in the first loop, it will become 2 in the second loop, and the computer will print

STUDENT NO 2 A = 83 B = 94 AVERAGE = 88.5

The computer printout will be

```
STUDENT NO 1    A = 56    B = 75    AVERAGE = 65.5
STUDENT NO 2    A = 83    B = 94    AVERAGE = 88.5
STUDENT NO 3    A = 62    B = 58    AVERAGE = 60
STUDENT NO 4    A = 78    B = 88    AVERAGE = 83  ❖
```

Example 7

a) Write a program for the flowchart given in Fig. 13.20.

b) Determine the output of the program (and flowchart) if N, the value inputted, is less than 20, equal to 20, and greater than 20.

FIGURE 13.20

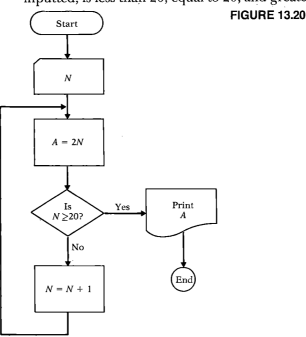

Solution

a)
```
10 INPUT N
20 LET A = 2 * N
30 IF N > = 20 THEN 60 (exit criterion)
40 LET N = N + 1
50 GO TO 20
60 PRINT A
70 END
```

b) If a value less than 20 is substituted for N, the computer will give the answer 40. For example, suppose N is assigned a value of 18 then A will have a value of 36. This value of A is not printed because before the computer reaches the PRINT statement (line 60) it is looped back to line 20.

N	A
18	36

Since N is not greater than or equal to 20, N keeps increasing by 1 until it reaches 20, as follows:

N	A
19	38
20	40

When N reaches 20 the computer prints the value of A, which is 40.

If N is initially assigned a value of 20, the computer will again print the answer 40.

If N is initially assigned a value greater than 20, the computer will print A, which is twice N.

Thus if $N = 25$, the computer will print the answer 50.　　　❖

Another method of creating a loop is to use the **FOR** and **NEXT** statements. The FOR statement cannot be used without an accompanying NEXT statement. When the computer comes across a FOR statement in a BASIC program, it executes (as a block) all the statements between the FOR and NEXT statements. It then loops back to the FOR statement and again executes the statements between the FOR and NEXT statements. This process is continued for as many times as specified in the FOR statement. After this loop has been evaluated the specified number of times, the computer continues to the next line in the program. Consider the following example.

■ **Example 8** ————————————————————————

Determine the output of the following program:

```
10 FOR I = 1 TO 100
20 PRINT "GOOD MORNING"
30 NEXT I
40 END
```

Solution: This BASIC program includes a FOR–NEXT loop. At first *I*, the loop control variable, is assigned a value of 1. Line 20 will have the computer print the message GOOD MORNING. Line 30 tells the computer to go back to the FOR statement and take the next integral value of *I*, which is 2. The computer then prints GOOD MORNING again. This procedure will continue until *I* has a value of 100. After the hundredth GOOD MORNING is printed, the program will end. The output of this program will consist of 100 lines of GOOD MORNING, one below the other. ❖

Example 9

Construct a program using the FOR and NEXT commands to print the first 100 integers (as an index), one below the other.

Solution

```
10 FOR I = 1 to 100
20 PRINT I
30 NEXT I
40 END
```

Example 10

Write a program using a FOR–NEXT loop that will result in the following output:

```
5
10
15
20
```

Solution

```
10 FOR I = 1 to 4
20 PRINT 5 * I
30 NEXT I
40 END
```

Example 11

Determine the output of the following program:

```
10 FOR J = 3 TO 5
20 PRINT J
30 PRINT J;" SQUARED = "; J∧2
40 NEXT J
50 PRINT "GOODBYE"
60 END
```

Solution

```
3
3 SQUARED = 9
4
4 SQUARED = 16
5
5 SQUARED = 25
GOODBYE
```

❖

A STEP statement is often used in conjunction with a FOR statement. A typical program follows:

```
10 FOR X = 1 TO 3 STEP .1
20 LET Y = 2 * X∧2 + 3 * X
30 PRINT "X = "; X,   "Y = "; Y
40 NEXT X
50 END
```

This program will calculate and print the values of y in the equation $y = 2x^2 + 3x$ for values of x from 1 to 3 in multiples of one tenth (0.1). The output will begin as follows:

```
X = 1         Y = 5
X = 1.1       Y = 5.72
X = 1.2       Y = 6.48
```

and so on.

Example 12

Write a BASIC program for finding the sum of the first 100 counting numbers without using the FOR and NEXT statements.

Solution: Let us develop a flowchart for this one, as shown in Fig. 13.21.

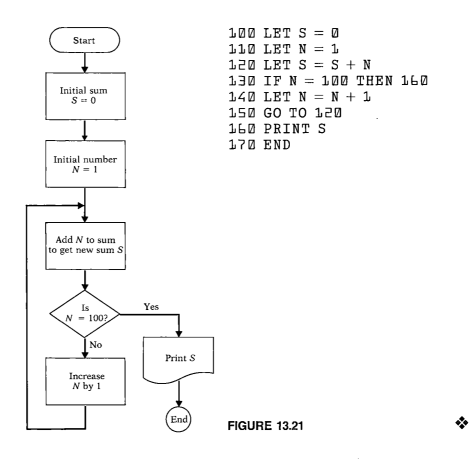

```
100 LET S = 0
110 LET N = 1
120 LET S = S + N
130 IF N = 100 THEN 160
140 LET N = N + 1
150 GO TO 120
160 PRINT S
170 END
```

FIGURE 13.21   ❖

■ **Example 13** _____

Write a BASIC program for finding the sum of the first 100 counting numbers, using the FOR and NEXT statements.

Solution

```
10 LET S = 0
20 FOR I = 1 TO 100
30 LET S = S + I
40 NEXT I
50 PRINT S
60 END
```
                                                            ❖

Note that the program in Examples 12 and 13 will produce the same outputs.

☐ **Example 14** _____

Write a BASIC program for adding the first N integers in which the user interactively enters N.

Solution: As you probably realize by now, many different programs can be written to do the same thing. The one we will write is a variation of the one in Example 13. Since the value of N is to be inserted at a later time, we will have to use an INPUT statement.

```
 5 REM SUM OF FIRST N INTEGERS
10 LET S = 0
20 INPUT N
30 FOR I = 1 TO N
40 LET S = S + I
50 NEXT I
60 PRINT "THE SUM IS"; S
70 END
```

When a value of N is inserted, the computer will go to work and determine the sum of the first N integers.

☐ **Example 15** _____

Determine the output of the following program:

```
10 DATA 2, .30, 5, 1.05, 10, .72, 2, 4.85
15 DATA 7, 1.60
20 PRINT "NO ITEMS", "UNIT PRICE", "COST",
25 PRINT "SUBTOTAL"
30 LET T = 0
40 READ N, P
50 LET T = T + N * P
60 PRINT N, P, N * P, T
70 GO TO 40
80 END
```

Solution: Lines 20 and 25 will have the computer print

```
NO ITEMS      UNIT PRICE      COST      SUBTOTAL
```

Lines 10 and 15 are DATA statements. The computer will consider the DATA command in line 15 to be an extension of the DATA command in line 10. Line 30 assigns the subtotal, T, an initial value of 0. Line 40 will result in assigning to N a value of 2 and to P a value of .30. Line 50 will calculate the subtotal, T. For $N = 2$ and $P = .30$, T has a value of .60. Line 60 will have the computer print

    ₂   .∃   .Ь   .Ь

Line 70 will refer the program back to line 40. This time N will be assigned a value of 5, and P will be assigned a value of 1.05. Line 50, T = T + N ∗ P, will have a new value for T, $T = .60 + (5 \times 1.05) = 5.85$. Line 60 will have the computer print

    5   1.05   5.25   5.85

This process will continue in the following manner to give the following results:

```
NO ITEMS      UNIT PRICE     COST      SUBTOTAL
₂             .∃             .Ь        .Ь
5             1.05           5.25      5.85
10            .72            7.2       13.05
₂             4.85           9.7       22.75
7             1.Ь            11.2      33.95
OUT OF DATA ERROR IN 40
```
❖

Example 16

Determine the output of the following program:

```
10 FOR I = 2 TO 4
20 LET S = I + 1
30 NEXT I
40 PRINT S
50 END
```

Solution: Since the two variables, I and S, are repeatedly changing, we will make a table listing I, S, and comments.

I	S	Comment
2	3	(Initial value of I is 2, $S = 2 + 1$, or 3)
3	4	(I becomes 3, $S = 3 + 1$, or 4)
4	5	(I becomes 4, $S = 4 + 1$, or 5)

The computer now exits the loop and reaches line 40. It prints the value of S, which is 5.
❖

■ **Example 17** _____

Determine the output of the following program:

```
10 LET S = 0
20 FOR I = 2 TO 4
30 LET S = S + I
40 NEXT I
50 PRINT S
60 END
```

Solution: Since the two variables, S and I, are repeatedly changing, we will make a table to help find the answer. Notice that line 10 gives S an initial value of 0 but is not part of the FOR–NEXT loop.

I	S	Comment
2	2	(Initially $S = 0, I = 2$, line 30 makes $S = 0 + 2$, or 2)
3	5	($S = S + I = 2 + 3$, or 5)
4	9	($S = S + I = 5 + 4$, or 9)

The computer now exits the loop and reaches line 50. Line 50 has the computer print the value of S, which is 9.   ❖

EXERCISES 13.6

1. Determine the output of the following program:

```
10 PRINT "THE"
20 GO TO 70
30 PRINT "IS"
35 PRINT "HERE"
40 END
50 PRINT "MAN"
60 GO TO 30
70 PRINT "YOUNG"
80 GO TO 50
```

Determine the output of the following programs.

2.
```
10 READ X
20 PRINT X, SQR (X)
30 GO TO 10
40 DATA 1, 4, 9, 16
50 END
```

3.
```
10 FOR X = 1 TO 5
20 PRINT X, X ∧ 2 - 1
```

```
30 NEXT X
40 END
```

4.
```
10 LET A = 8
20 LET C = 0
30 IF A > 5 THEN 60
40 PRINT A
50 GO TO 70
60 PRINT C
70 END
```

5.
```
10 LET A = 0
20 FOR I = 1 TO 10
30 LET A = A + I
40 NEXT I
50 PRINT A
60 END
```

6.
```
10 READ A, B, C
20 DATA 1, 5, 4
30 PRINT SQR (B ∧ 2 - 4 * A * C)
40 END
```

7. Determine the output of the following program.

```
10 READ X
20 DATA 1, 2, 3, 4
30 LET Y = X ^ 3
40 PRINT Y
50 GO TO 10
60 END
```

8. Revise the program in Exercise 7 to eliminate the OUT OF DATA message that the computer will print.

9. a) Write a program for each of the flowcharts in Fig. 13.22.
   b) Write the computer printout for each program.
   c) How do the printouts compare?

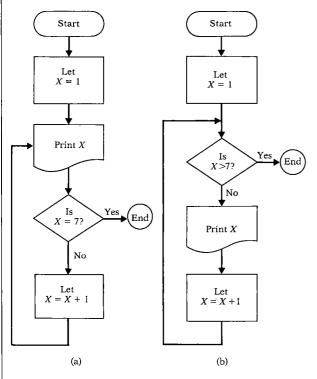

(a)   (b)

FIGURE 13.22

10. Revise the program in Example 15 to eliminate the OUT OF DATA message.

11. Construct a program from the flowchart in Fig. 13.23.

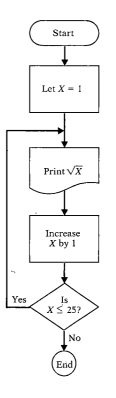

FIGURE 13.23

12. a) Construct a flowchart to use to determine what one cent will grow to if it is doubled every day for 30 days.
   b) Write a program of the flowchart constructed in part (a).

13. a) Write a program that will find the value of y in the equation $y = 3x + 4$ for any given value of x.
   b) Write a program that will find the values of y in the equation $y = 3x + 4$ when x is assigned integral values of 1 through 5. Use the READ and DATA statements.
   c) Repeat part (b) using FOR and NEXT statements in place of the READ and DATA statements.

14. Write a program that will have the computer change feet into yards if the number of feet is less than 5280. If the number of feet is greater than or

equal to 5280, have the computer convert feet to miles.

15. Determine the output of the following program if

a) A is assigned a value of 10 and B is assigned a value of 10;

b) A is assigned a value of 6 and B is assigned a value of 12;

c) A is assigned a value of 15 and B is assigned a value of 4:

```
 5 REM THIS PROGRAM
   COMPARES TWO NUMBERS
10 PRINT "INPUT ANY TWO
   NUMBERS"
20 INPUT A, B
30 IF A < B THEN 60
40 IF A > B THEN 80
50 IF A = B THEN 100
60 PRINT A; "IS LESS THAN "; B
70 GO TO 10
80 PRINT A; "IS GREATER
   THAN"; B
90 GO TO 10
100 PRINT A; "IS EQUAL TO"; B
110 GO TO 10
120 END
```

16. Use the formula $i = prt$ to write a program to find the interest i on a principal p of $1000 for a time t of 4 years at rates r from 5% to 14%, in steps of 0.1%.

17. Write a program that will generate the following output:

```
RUN
FIRST NUMBER
? 40
LAST NUMBER
? 45

N    SQUARE ROOT OF N    N-SQUARED
40   6.3245553           1600
41   6.4031242           1681
42   6.4807407           1764
43   6.5574385           1849
44   6.6332496           1936
45   6.7082039           2025
```

18. a) Write a program for the flowchart in Fig. 13.24.

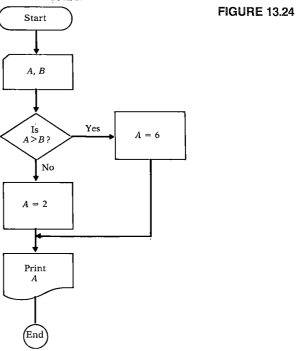

FIGURE 13.24

b) Determine the output if
  1) $A < B$,
  2) $A = B$,
  3) $A > B$.

19. Determine the output of the following program:

```
10 FOR I = 1 TO 3
20 LET P = 4 * I
30 NEXT I
40 PRINT P
50 END
```

20. Determine the output of the following program:

```
10 FOR I = 1 TO 3
20 LET P = 4 * I
30 PRINT I, P
40 NEXT I
50 END
```

21. Determine the output of the following program:

```
10 LET R = 2
20 FOR I = 4 TO 7
30 LET R = I - R
40 NEXT I
50 PRINT R
60 END
```

22. Determine the output of the following program:

```
10 LET W = 3
20 FOR J = 1 TO 4
30 LET W = W + J
40 PRINT J, W
50 NEXT J
60 END
```

PROBLEM SOLVING

23. In Chapter 8 we explained that when you repay a bank loan or mortgage, the amount of money that goes to repay the principal changes each month. The interest, i, owed each month is found by the formula $i = prt$, where p is the amount owed on the balance, r is the rate, and t is the time. The rate and time must be expressed in the same units of time. Consider the following example.

You borrow $7000 to buy a car. You finance the balance for 36 months at a rate of 1 percent per month (12% per year). Your monthly payments are $232.50.

The interest owed for the first month of the loan is $i = 7000(0.01)(1) = \$70$. Of the first payment of $232.50, $70 is interest and $232.50 - \$70 = \$162.50$ is applied to the balance. Thus after the first payment you owe a balance of $7000 - 162.50 = 6837.50$. The interest on the second payment is $i = (6837.50)(0.01)(1) = \$68.375$. A chart or table illustrating the interest owed each month and the balance remaining after each payment is called an amortization table.

Develop a program using a FOR–NEXT loop for finding the interest and balance remaining for each of the 36 months of this loan.

24. The amount P dollars will grow to if invested at $r\%$ for n periods is determined by the formula $A = P(1 + r)^n$.

   a) Construct a program to find A when P, r, and n are known. Use INPUT commands.
   b) Construct a program to determine A for the following values.

$P = 400$	$r = 0.04$	$n = 6$
$P = 1200$	$r = 0.11$	$n = 2$
$P = 20{,}000$	$r = 0.14$	$n = 4$

   Have output read

   P   R   N   A

   with the appropriate values beneath the letters.

25. a) Write a program to find the mean of a set of five pieces of data, using the formula

$$\text{Mean} = \frac{\text{sum of data}}{\text{number of pieces of data}}$$

   b) Write a program to find the standard deviation of a set of five pieces of data, using the formula

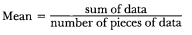

$$\text{Standard deviation} = \sqrt{\frac{\Sigma(\text{data} - \text{mean})^2}{\text{one less than the number of pieces of data}}}$$

   Refer to Sections 12.5 and 12.6 for an explanation of the mean, the standard deviation, and the Σ symbol.

26. One method that is commonly used in computing depreciation on income tax returns is the straight-line method. With the straight-line method the amount to be depreciated is divided evenly over the useful life of the item. Thus if the amount to be depreciated is $640 for 8 years, a deduction of $640/8 = \$80$ would be taken for each of the 8 years. Write a program that will result in the following computer output:

```
STRAIGHT-LINE
DEPRECIATION METHOD
```

YEAR	DEPRECIATION	BALANCE
1	80	560
2	80	480
3	80	400
4	80	320
5	80	240
6	80	160
7	80	80
8	80	0

27. Write a program to list the first 100 terms of the Fibonacci series, as described in Section 5.5.

28. Write a program to find the sum of the first 100 terms of the Fibonacci series.

13.7 LOGO

Logo was developed by the Logo Group at the Massachusetts Institute of Technology. The key individual in the project was Seymour Papert, the author of *Mindstorms: Children, Computers and Powerful Ideas.* The word Logo is derived from the Greek word for "thought." The goal of Papert and the MIT Logo Group was to place Piaget's theories about natural learning in a powerful artificial intelligence environment. Logo has no bounds. It can be used by preschool children as well as by computer scientists.

In working with Logo we can be in the draw mode or the edit mode. In the **edit mode,** one can write programs and store the programs for future use. In the **draw mode** it is possible to draw all types of geometric shapes and patterns. A little triangle, which is called a **turtle,** is moved around the screen. The path of the turtle is marked with lines that create the patterns.

To get started, we insert a Logo Language diskette into the disk drive of the computer. The message "Welcome to Logo" will appear on the screen followed by a question mark and the cursor. The question mark indicates that the computer is waiting for a Logo command. The cursor indicates where the typing will appear on the screen. Type the command "SHOW TURTLE" (which we abbreviate ST), press return, and the computer will be in the draw mode. At this point the turtle appears in the middle of the screen. This position of the turtle will be referred to as **home** (Fig. 13.25).

The turtle is directed with a set of word commands. A list of word commands that can be used is given in Table 13.5. The commands in Table 13.5 are called **primitives.** In writing a set of commands the abbreviations are

Children are learning to use computers at an early age.
Courtesy of International Business Machines Corporation

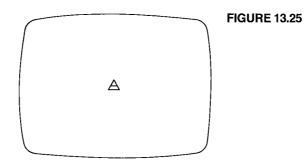

FIGURE 13.25

used and not the words. These commands are for Apple Logo. Apple Logo is a version of Logo developed for the Apple Computer. MIT Logo runs on an Apple or a Commodore 64 computer.

The draw mode is also called the splitscreen mode. In the splitscreen mode the prompt (a question mark) and the cursor are located at the bottom

TABLE 13.5		
Graphics Command	Abbreviation	Effect
FORWARD	FD	Sends the turtle forward a given number of turtle steps
BACK	BK	Sends the turtle backward a given number of turtle steps
RIGHT	RT	Turns the turtle to the right a given number of degrees
LEFT	LT	Turns the turtle to the left a given number of degrees
HIDE TURTLE	HT	The turtle does not appear on screen; the turtle draws faster
SHOW TURTLE	ST	Puts the computer in draw mode and exhibits turtle
CLEAR SCREEN	CS	Clears the screen, sends the turtle home, and sets it facing north
PRINT	PR	Prints the words or symbols to the right of double quotation marks on the screen
HEADING		In the draw mode, tells the turtle's heading in degrees
PENUP	PU	In the draw mode, allows the turtle to move without leaving a trail
PENDOWN	PD	In the draw mode, causes the turtle to leave a trail
PRINT POSITION	PRPOS	Prints coordinates of turtle
SET POSITION	SETPOS	Places turtle at indicated coordinates
HOME		Returns the turtle to the center of the screen heading north

of the screen. The bottom four lines of the screen are for typing commands. The top part of the screen is for graphics.

In Table 13.5 we indicated that FD is forward. The command FD 40 means that the turtle will move forward 40 turtle steps (40 units) in the forward direction. To move the turtle backward 15 turtle steps, we type the command BK 15. The command RT 45 changes the direction of the turtle by turning it 45 degrees to the right. The command LT 80 changes the direction of the turtle by turning it 80 degrees to the left. To demonstrate how one can draw with the turtle, we will illustrate basic commands in Example 1.

▦ Example 1 ─────────────────────────────────

Illustrate the movement of the turtle for the list of commands given.

a) CS   b) FD 50   c) LT 90   d) FD 60   e) RT 45

f) FD 40   g) RT 45   h) BK 60   i) HOME

Solution: The results of these commands are shown in Figs. 13.26(a)–13.26(i).

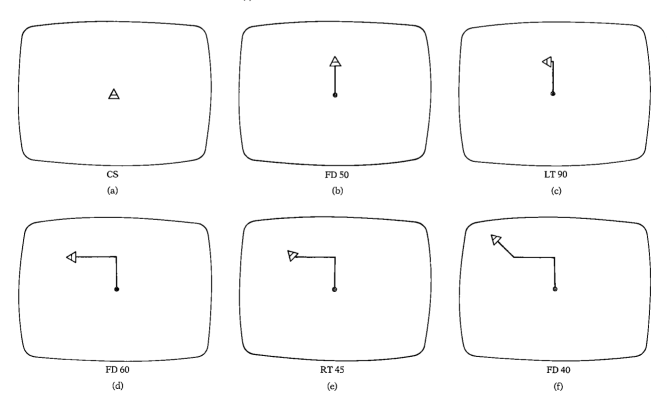

CS	FD 50	LT 90
(a)	(b)	(c)
FD 60	RT 45	FD 40
(d)	(e)	(f)

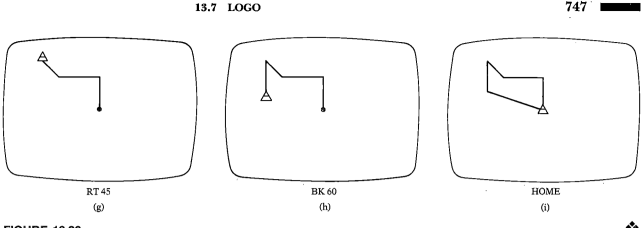

| RT 45 | BK 60 | HOME |
| (g) | (h) | (i) |

FIGURE 13.26 ❖

■ **Example 2** _____

List the sequence of LOGO commands that will have the turtle draw a square whose sides have a length of 80 units.

Solution: The list of commands that will result in drawing the square in Fig. 13.27 is

```
CS
FD 80
RT 90
FD 80
RT 90
FD 80
RT 90
FD 80
RT 90
```

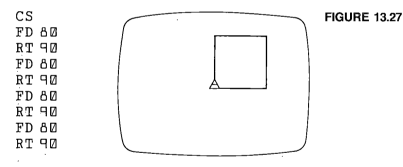 **FIGURE 13.27**

Note that it is necessary to press the return key after each command when you wish to list the command vertically. ❖

You can type a whole list of commands on a single line. After you have completed the list and pressed the return key, the turtle will carry out the list of commands in their proper order. For example, if we type the following list of commands,

```
FD 40  RT 45  FD 50  RT 90  FD 50  LT 45  BK 50
```

and then press the return key, the pattern in Fig. 13.28(a) will be displayed. Figure 13.28(b) is a detailed drawing to show exactly what happens.

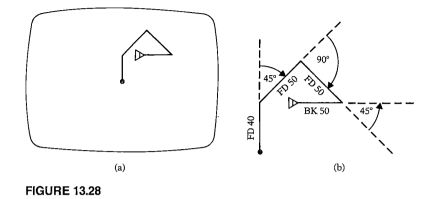

FIGURE 13.28   ❖

What happens when the turtle is in standard position (at home, heading north) and we enter FD 300? The turtle disappears off the top of the screen and reappears at the bottom of the screen. This is called **wraparound.** If you move too far to the right, the turtle appears on the left of the screen. If you move too far down, the turtle appears at the top of the screen. The turtle is limited to about 240 steps from the top to the bottom of the screen. It is limited by about 280 steps from the left to the right of the screen.

In Example 2 we typed each of the commands FD 80 and RT 90 four times. Logo has a command called **REPEAT** that reduces the work. The instruction

```
REPEAT 4[FD 80 RT 90]
```

tells the turtle to repeat the movement four times, FORWARD 80 (steps) then RIGHT 90 (degrees). Press the return key, and the square is displayed, as in Fig. 13.27. Also note that the turtle is at home and heading north (original position). The turtle made four 90° turns, or a total of 360°.

The symbols for arithmetic calculations for Logo are similar to the symbols for BASIC. The symbols for Logo are given in Table 13.6.

TABLE 13.6			
Operation	Logo Symbol	Logo Example	Result
Addition	+	2 + 3	5
Subtraction	−	5 − 4	1
Multiplication	*	9 * 3	27
Division	/	15 / 3	5
Exponentiation	∧	5∧2	25
Square root	SQRT	SQRT(16)	4

Example 3

Write a set of commands that instruct the turtle to draw a regular pentagon. A regular pentagon is a figure with five equal sides and five equal angles. The length of each side is to be 40 turtle steps.

Solution: Since the sum of the interior angles of a regular polygon is 360 degrees, the turtle must make five equal turns whose sum is 360°. Dividing 360 degrees by 5, we have 72 degrees. Therefore the turtle must make 5 turns of 72 degrees. The arithmetic for calculating the angle can be included as part of the instruction. A single instruction that will result in a pentagon is

```
REPEAT 5[FD 40  RT 360/5]
```

Press the return key, and the pentagon given in Fig. 13.29 is drawn.

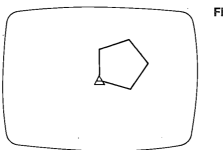

FIGURE 13.29

Note that in Example 3, part of the command is RT 360/5. The computer divides 360 by 5 to obtain 72 before the first right turn is executed.

The words (commands) in Table 13.5 are called primitives. These commands are placed in the memory of the computer by the software. We can develop a new set of instructions and assign a name to these instructions. The computer can recall these instructions from its memory when the name is typed. Each of these new sets of instructions is called a **procedure.** Creating a procedure in Logo is the same as writing a program.

To create a procedure, we must leave the draw mode and enter the edit mode. To enter the edit mode, we type **EDIT** followed by a space, a single set of double quotation marks, and the name of the procedure, and then press the return key.

To illustrate the steps for writing a procedure, we will define a procedure called PENTAGON.

1.	Type EDIT, double quotation marks, and the procedure name.	EDIT "PENTAGON
2.	Type the body of the procedure.	REPEAT 5[FD 40 RT 360/5]
3.	Type END (on a separate line) to indicate that the procedure is complete.	END
4.	Press CTRL-C to send the procedure to memory. (Hold down the control key while pressing the C key.) The computer will exit the edit mode.	

When the computer leaves the edit mode, the procedure is entered into the computer's memory. Logo then reenters the edit mode and responds with the message: PENTAGON DEFINED.

An alternative way to enter the edit mode to define a procedure is to type EDIT (ED) and press the return key. The edit mode is entered, and TO PENTAGON can be typed, followed by the lines in Steps 2–4.

To have the computer execute the instructions of a procedure in memory (when it is in draw mode), we do the following: (1) Type the name of the procedure and (2) press the return key. For example, for the procedure PENTAGON, type PENTAGON and press the return key, and the figure will be drawn as in Fig. 13.29.

The procedure called PENTAGON can now be used in the same manner as the primitives FORWARD, BACK, RIGHT, and LEFT, in drawing pictures. For example, the command

REPEAT 4[FD 40 PENTAGON BK 40 RT 360/4]

will generate the design in Fig. 13.30.

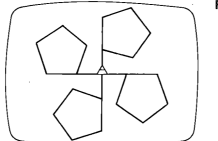

FIGURE 13.30

■ **Example 4** _____

a) Write a procedure called SEGMENT that will generate a line segment whose length is 30 turtle steps and return the turtle to the original position.

b) Write a procedure called SQUARE that will generate a square with sides of length 30 turtle steps.

c) Write a procedure called TRIANGLE that will generate an equilateral triangle with sides of length 30 turtle steps. An equilateral triangle has three equal sides and three equal angles.

Solution

a) ```
    EDIT "SEGMENT
    FD 30 BK 30
    END
    ```

b)  ```
    EDIT "SQUARE
    REPEAT  4[FD 30  RT 360/4]
    END
    ```

c) ```
    EDIT "TRIANGLE
    REPEAT 3[FD 30 RT 360/3]
    END
    ```

There is no limit to the patterns and figures you can construct with the procedures PENTAGON, TRIANGLE, SEGMENT, and SQUARE and the primitives.

## Example 5

Predict what happens when each of the following is executed:

a)  `REPEAT  12[SEGMENT  RT 360/12]`

b)  `REPEAT  4[SQUARE  RT 360/4]`

c)  `REPEAT  4[FD 20  TRIANGLE  BK 20  RT 360/4]`

*Solution:*  Figure 13.31 illustrates the solution.

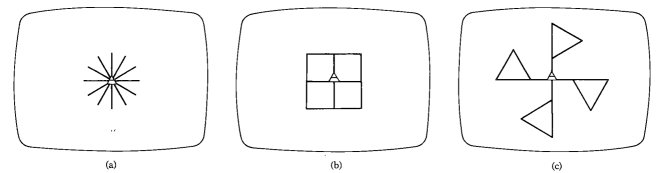

(a)                          (b)                          (c)

FIGURE 13.31
    ```

It is possible to draw a figure that looks like a circle with Logo. It will not be a true circle, since each point on the circumference will not be exactly the same distance from the center. To construct the circle, we instruct the turtle to draw a regular polygon with 360 sides. Since the computer must draw the turtle for each turn, it will take a long time to draw the circle. The time can be shortened by adding the command HIDE TURTLE (HT) to our procedure. When the command HIDE TURTLE is used, the turtle will not be visible on the screen, and the movements are made more quickly. A procedure for drawing a circle is given below, and the result is illustrated in Fig. 13.32.

```
ED  "CIRCLE
HT
REPEAT  360[FD 1  RT 1]
END
```

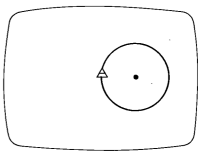

FIGURE 13.32

Plotting Points with Logo

Logo is a means of introducing the student to plotting points on a coordinate plane. Each point on the screen is associated with an ordered pair of numbers (a, b). The number a is the x-coordinate. A positive value of a indicates a point to the right of the vertical line through home, and a negative value of a indicates a point to the left of the vertical line. The number b is the y-coordinate. A positive value of b indicates a point above the horizontal line through home, and a negative value of b indicates a point below the horizontal line. The origin, the home position of the turtle, has coordinates $(0, 0)$.

To move the turtle in a horizontal direction to a particular position, we input an x-coordinate. The command that we use is SETX. For example, if the turtle is at home and we type SETX 30, the turtle moves to the right horizontally to the point where the x-coordinate is 30. Likewise, if we type SETX -40, the turtle moves horizontally to the left to the point on the screen where the x-coordinate is -40.

The command SETY works in a similar manner with the y-coordinate. For example, if the turtle is at home and we type SETY 35, the turtle moves vertically (up) to the point where the y-coordinate is 35. If we type SETY -40, the turtle will move down to the point where the y-coordinate is -40. With SETX and SETY the turtle's heading is not changed.

Example 6

Use SETX and SETY to graph a rectangle through the points $(-30, 40)$ and $(0, 0)$. Start with the turtle in home position.

Solution: Type

```
SETX − 3Ø  SETY 4Ø  SETX Ø  SETY Ø
```

The path of the turtle is given in Fig. 13.33.

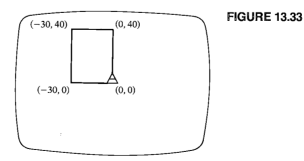

FIGURE 13.33

The final position of the turtle can be determined by typing the primitive POS. **POS** (or POSITION) is a primitive that can provide a list of two items: the first is the x-coordinate, and the second is the y-coordinate of the turtle's current position. If we type **PRPOS**, which stands for *PR*INT *POS*ITION, for Example 6, we obtain the output (0,0).

The primitive SETPOS can place the turtle at $(-30, 40)$ in one step. SETPOS requires two numbers as inputs; the first number is the x-coordinate, and the second number is the y-coordinate. For example, typing SETPOS $[-30, 40]$ moves the turtle from its present position to the point with coordinates $(-30, 40)$ without changing the turtle's heading (see Fig. 13.34).

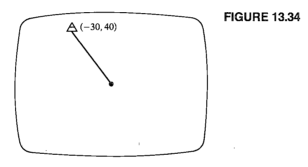

FIGURE 13.34

■ **Example 7** _____

Write a set of commands that will generate a triangle with coordinates (30, 0), (20, 40), (60, 0).

Solution: The following commands will result in the triangle in Fig. 13.35.

```
PENUP
SETPOS [30, 0]
PENDOWN
SETPOS [20, 40]
SETPOS [60, 0]
SETPOS [30, 0]
```

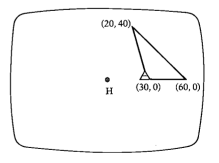

FIGURE 13.35

With the primitives and the procedures developed in this section you are now ready to try Logo. If you have access to a Logo disk, check your work with the computer.

EXERCISES 13.7

In Exercises 1–10 we will assume that each sketch starts with the turtle at its home position. Sketch the turtle's trail as a result of the commands. Use the scale given in Fig. 13.36 for turtle distances. If possible, check your answers with a computer.

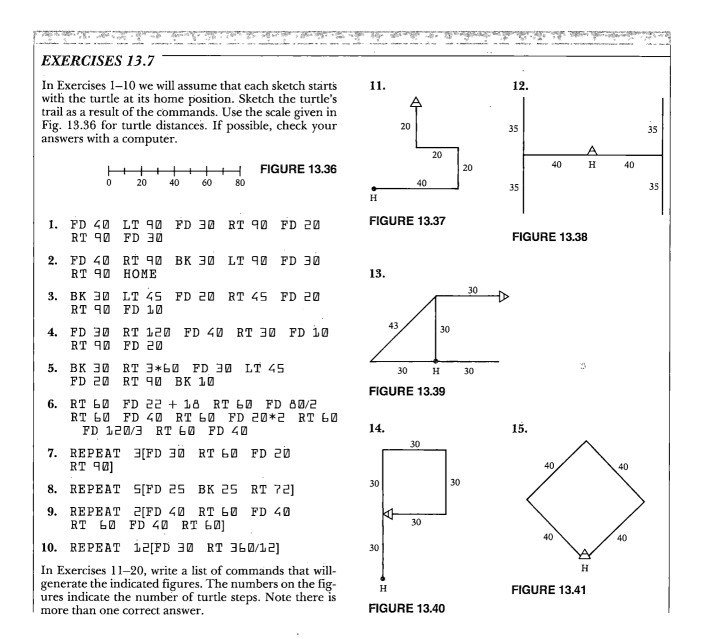

FIGURE 13.36

1. FD 40   LT 90   FD 30   RT 90   FD 20
   RT 90   FD 30

2. FD 40   RT 90   BK 30   LT 90   FD 30
   RT 90   HOME

3. BK 30   LT 45   FD 20   RT 45   FD 20
   RT 90   FD 10

4. FD 30   RT 120   FD 40   RT 30   FD 10
   RT 90   FD 20

5. BK 30   RT 3*60   FD 30   LT 45
   FD 20   RT 90   BK 10

6. RT 60   FD 22 + 18   RT 60   FD 80/2
   RT 60   FD 40   RT 60   FD 20*2   RT 60
     FD 120/3   RT 60   FD 40

7. REPEAT 3[FD 30   RT 60   FD 20
   RT 90]

8. REPEAT 5[FD 25   BK 25   RT 72]

9. REPEAT 2[FD 40   RT 60   FD 40
   RT 60   FD 40   RT 60]

10. REPEAT 12[FD 30   RT 360/12]

In Exercises 11–20, write a list of commands that will generate the indicated figures. The numbers on the figures indicate the number of turtle steps. Note there is more than one correct answer.

11.

FIGURE 13.37

12.

FIGURE 13.38

13.

FIGURE 13.39

14.

FIGURE 13.40

15.

FIGURE 13.41

16.

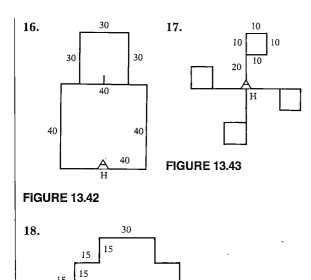

FIGURE 13.42

17.

FIGURE 13.43

18.

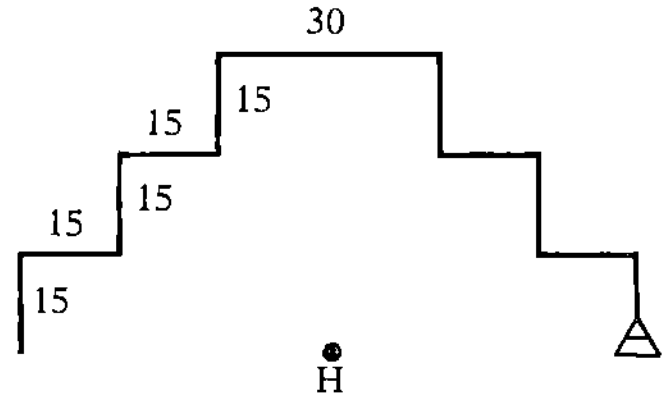

FIGURE 13.44

19. A rectangle with a width of 30 units and a length of 40 units (where home is in the center of the line segment representing the length).

20. A regular hexagon (six equal sides) with sides of length 30 units.

In Exercises 21–30, assume that the procedures PEN-TAGON, TRIANGLE, SEGMENT, SQUARE, and CIRCLE as defined in this section are in the memory of the computer. Predict the design that will result from each exercise. If possible check your results on a computer.

21. SEGMENT FD 30 TRIANGLE

22. SQUARE FD 30 RT 30 TRIANGLE

23. RT 90 TRIANGLE FD 20 RT 180 CIRCLE

24. REPEAT 6[FD 40 TRIANGLE BK 40 RT 360/6]

25. REPEAT 4[SQUARE RT 360/4]

26. REPEAT 20[SEGMENT RT 360/20]

27. REPEAT 5[FD 30 PENTAGON BK 30 RT 360/5]

28. REPEAT 4[TRIANGLE FD 30 LT 90]

29. REPEAT 6[SQUARE RT 60]

30. REPEAT 360[FD 2 RT 1]

For the given ordered pairs in Exercises 31–34, (a) write a set of commands using SETX and SETY that will instruct the turtle to move from Home to each of the points in the order given, and (b) illustrate the path of the turtle.

31. (50, 30), (−40, −10)
32. (−30, −20), (30, −20), (30, 0), (0, 0)
33. (40, 40), (0, 40), (0, 10), (20, 10)
34. (0, 50), (40, 50), (40, 30), (0, 30)

In Exercises 35–38, (a) write a set of commands using SETPOS to generate the figure (only lines that are part of the figure should show), and (b) have the turtle trace the figure that is generated by the commands.

35. Triangle with coordinates (−10, 0), (−40, 0), (−40, 50)
36. Parallelogram with coordinates (0, 0), (20, 30), (60, 30), (40, 0)
37. Trapezoid with coordinates (0, 0), (20, 40), (70, 40), (90, 0)
38. Square with coordinates (0, −40), (−40, 0), (0, 40), (40, 0)

In Exercises 39–44, use the procedures developed in this section and any primitives to write instructions to generate each figure. In the figures home is indicated with the letter H. Note there is more than one correct answer.

39.   **40.**

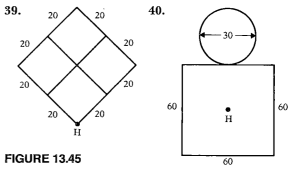

FIGURE 13.45

FIGURE 13.46

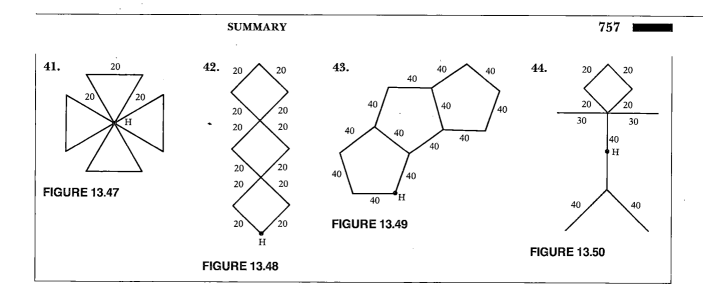

41.

FIGURE 13.47

42.

FIGURE 13.48

43.

FIGURE 13.49

44.

FIGURE 13.50

SUMMARY

John von Neumann explained a way to store instructions and data in the computer that revolutionized the computer industry. Since that time, many advances have occurred in computer technology.

The basic parts of a computer system are the input, storage, control, processing, and output. There are many types of input/output devices. The heart of the computer is the central processing unit (CPU), which contains the control unit and the arithmetic/logic unit. A computer has two kinds of memories: random access memory (RAM) and read-only memory (ROM).

An error that is often made in writing a program in BASIC is a syntax error. Programs can be saved on a disk or cassette tape.

A computer performs operations in the following order:

1. Evaluation of information within parentheses
2. Exponentiation
3. Multiplication and division from left to right
4. Addition and subtraction from left to right.

Flowcharts are often helpful in writing programs. Three methods that can be used to assign values to variables are the LET statement, READ and DATA statements, and the INPUT statement.

The PRINT statement can be used to leave a blank line in a program, print an item, or perform and print a calculation. When a semicolon is placed between items in a print statement, the items are printed next to each other. When a comma is used between the items, the items are printed a fixed distance apart.

A great advantage of the computer over some calculators is that the computer has the ability to perform repeated calculations. To accomplish this, loops are often used. Three types of statements that can be used to create loops are the GO TO statement, IF–THEN statement, and FOR and NEXT statements.

Another programming language is LOGO. LOGO is presently being used to familiarize young children with the computer and certain geometric concepts. With Logo in the edit mode we can write programs that will instruct the computer to draw figures and patterns in the draw mode.

REVIEW EXERCISES

1. Name some ways in which the computer has changed our lives.
2. Discuss the four computer generations. Describe the advances made in each generation.
3. Discuss the contribution of John von Neumann in the development of computers.
4. List the five parts of a computer system. Explain the purpose of each part.

Write each of the following in BASIC symbols.

5. $3 + 4^2$
6. $4 - (6^2 \cdot 3) + 4$
7. $3 (\sqrt{25} + 6) \div (4 - 3^2)$
8. $5 - 3x^2$
9. $2x^2 + 4x - 6$
10. $\dfrac{3xy - x^2}{x - 4}$

Evaluate each of the following.

11. $3 + 5 * 4$
12. $4 * 3 - 2$
13. $3 \wedge 3 * 2$
14. $SQR(4) * 3$
15. $6 + 4 / 2 / 2$
16. $4 / 2 * 3 + 4$

17. Construct a flowchart for finding the area of a triangle.
18. Write a program to find the area of a triangle.
19. Write a program to evaluate $x^2 + 4x - 6$ for any given x.
20. Write a program to evaluate $x^3 - 2x + 4$ for $x = 1.5$.
21. What is the computer output for the following program?

```
10 PRINT "GOOD MORNING"
20 LET X = 5
30 LET Y = 7
40 PRINT "THE SUM OF X AND Y
   IS" ; X + Y
50 PRINT "THE PRODUCT IS",
   X * Y
60 PRINT
70 PRINT "GOOD DAY"
80 END
```

22. Write a program to compute and list the squares of the first five numbers in the following form:

```
X       X SQUARED
1       1
2       4
3       9
4       16
5       25
```

23. A salesperson's salary is determined in the following manner. Salary = 15% of sales up to and including $10,000 plus 20% of all sales above $10,000.

   a) Construct a flowchart that can be used to find a salesperson's salary.

   b) Write a program that can be used to find the salesperson's salary.

24. Write a program that will give the following output:

```
INTEGER     INTEGER      MULTIPLIED BY 5
1           5
2           10
3           15
4           20
5           25
```

25. Given the equation $y = 2x^2 - 4x + 5$, write a program that will evaluate y for each value of x from 2 to 5 in intervals of 0.1.

26. Write a program that will compute the sum of the first five even numbers with an output that looks like the following:

```
NUMBER    SUM
2         2
4         6
6         12
8         20
10        30
```

Assume that the Logo turtle starts at the home position. Sketch the turtle's trail as a result of the commands. Use the following scale for turtle distances.

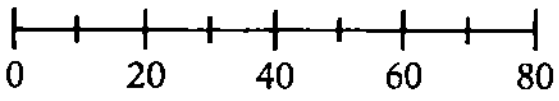

FIGURE 13.51

27. FD 40 RT 90 FD 50 RT 90 FD 40

28. RT 180 FD 80 LT 90 FD 50 RT 90 BK 40

29. PENUP FD 40 RT 90 PENDOWN FD 50 RT 90 FD 30

30. RT 270/3 FD 160/4 LT 30*3 FD (90 + 10) RT (90 − 60)

31. REPEAT 2[FD 50 LT 90 FD 40 RT 90 FD 40 RT 90
        FD 40 BK 10 RT 90]

13 CHAPTER TEST

1. Who coined the phrase "bug in a computer program"?
2. Briefly explain the differences among the four computer generations.
3. List the five basic functions of any digital computer system. Briefly explain the purpose of each part

In Exercises 4–6, indicate the result that the computer will give.

4. $4/2*8+8$
5. $5*SQR(36) - 5 \wedge 2$
6. $6 \wedge 2/2 - 2 \wedge 2$

Convert Exercises 7–9 to BASIC symbols and then give the answers.

7. $5 + (7 \cdot 9)$
8. $(5 + 7) \cdot 9^2$
9. $\sqrt{64} + (15 \div 3)$

Write a program that corresponds to the given flowchart.

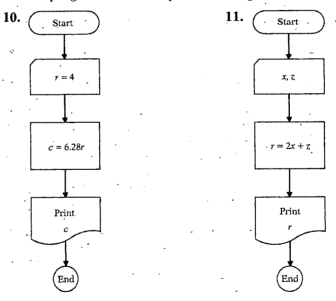

10.
Start

$r = 4$

$c = 6.28r$

Print
c

End

11.
Start

x, z

$r = 2x + z$

Print
r

End

What is the computer output for the programs in Exercises 12–14?

12.
```
1Ø READ A, B
2Ø DATA 9, 12
3Ø PRINT (B - A) * 3
4Ø END
   RUN
```

13. ```
    10 READ L, W, H
    20 DATA 6, 7, 8
    30 PRINT "THE VOLUME OF THE BOX IS"; L * W * H
    40 END
       RUN
    ```

14. ```
    10 INPUT A, B, C
    20 LET R = A∧2 + 3 * B + C
    30 PRINT R
    40 END
       RUN
       ? 2, 3, 4
    ```

15. Construct a flowchart and a BASIC program to find the value of D where $D = R \cdot T$ when $R = 55$ and $T = 3.5$.

16. Write a program for the flow chart in Exercise 15.

17. Construct a flowchart and a BASIC program with a loop to find the squares of the first ten counting numbers.

18. Write a program for the flow chart in Exercise 17.

Assume that the Logo turtle starts at the home position. Sketch the turtle's trail as a result of the commands in Exercises 19 and 20. Use the scale in Figure 13.36.

19. ```
    FD 80  RT 90  FD 60  RT 90  BK 30
    ```
20. ```
    FD 50  RT 90  FD 30  RT 90  FD 30  RT 90  FD 30
    RT 90  BK 20
    ```

APPENDIX A: MEASUREMENT CONVERSIONS

The question may arise, How can we change units of measure in the metric system to equivalent units of measure in the customary system or vice versa? Table 1 is used to convert customary units of measurement to metric units of measurement. Table 2 is used to convert metric units of measurement to customary units of measurement.

TABLE 1
Conversions to Metric

Length
1 inch (in.) = 2.54 centimeters (cm)
1 foot (ft) = 30 centimeters (cm)
1 yard (yd) = 0.9 meter (m)
1 mile (mi) = 1.6 kilometers (km)

Area
1 square inch (in^2) = 6.5 square centimeters (cm^2)
1 square foot (ft^2) = 0.09 square meter (m^2)
1 square yard (yd^2) = 0.8 square meter (m^2)
1 square mile (mi^2) = 2.6 square kilometers (km^2)
1 acre = 0.4 hectare (ha)

Volume
1 teaspoon (tsp) = 5 milliliters (mℓ)
1 tablespoon (Tbsp) = 15 milliliters (mℓ)
1 fluid ounce (fl oz) = 30 milliliters (mℓ)
1 cup (c) = 0.24 liter (ℓ)
1 pint (pt) = 0.47 liter (ℓ)
1 quart (qt) = 0.95 liters (ℓ)
1 gallon (gal) = 3.8 liters (ℓ)
1 cubic foot (ft^3) = 0.03 cubic meter (m^3)
1 cubic yard (yd^3) = 0.76 cubic meter (m^3)

Weight (mass)
1 ounce (oz) = 28 grams (g)
1 pound (lb) = 0.45 kilogram (kg)
1 ton (T) = 0.9 tonne (t)

Temperature
$C = \frac{5}{9}(F - 32)$

TABLE 2
Conversions to Customary

Length
1 millimeter (mm) = 0.04 inch (in.)
1 centimeter (cm) = 0.4 inch (in.)
1 meter (m) = 3.3 feet (ft)
1 meter (m) = 1.1 yards (yd)
1 kilometer (km) = 0.6 mile (mi)

Area
1 square centimeter (cm^2) = 0.16 square inch (in^2)
1 square meter (m^2) = 1.2 square yards (yd^2)
1 square kilometer (km^2) = 0.4 square mile (mi^2)
1 hectare (ha) = 2.5 acres

Volume
1 milliliter (mℓ) = 0.03 fluid ounce (fl oz)
1 liter (ℓ) = 2.1 pints (pt)
1 liter (ℓ) = 1.06 quarts (qt)
1 liter (ℓ) = 0.26 gallon (gal)
1 cubic meter (m^3) = 35 cubic feet (ft^3)
1 cubic meter (m^3) = 1.3 cubic yards (yd^3)

Weight (mass)
1 gram (g) = 0.035 ounce (oz)
1 kilogram (kg) = 2.2 pounds (lb)
1 tonne (t) = 1.1 tons (T)

Temperature
$F = \frac{9}{5}C + 32$

To change from an English unit of measurement to its equivalent metric unit, we multiply the English unit by the appropriate conversion factor given in Table 1. For example, the conversion factor for converting from inches to centimeters is 2.54. Thus, for example, 5 in. = 5 × 2.54 = 12.7 cm.

■ **Example 1** _____

Convert each of the following customary units to the equivalent metric measurement.

a)   11 in. = _____ cm   b)   28 mi = _____ km

c)   15 yd^2 = _____ m^2 d)   150 lb = _____ kg

Solution

a)   11 in. = 11 × 2.54 = 27.94 cm

b)   28 mi = 28 × 1.6 = 44.8 km

c)   15 yd^2 = 15 × 0.8 = 12 m^2

d)   150 lb = 150 × 0.45 = 67.5 kg   ❖

   To change from a metric unit of measurement to its equivalent customary unit, we multiply the metric unit by the appropriate conversion factor given in Table 2. For example, the conversion factor for converting from millimeters to inches is 0.04. Thus, for example, 150 mm = 150 × 0.04 = 6 inches.

■ **Example 2** _____

Change the following measurements in the metric system to the customary system.

a)   765 mm = _____ in.   b)   2 375 km = _____ mi

c)   675 ha = _____ acres   d)   346 g = _____ oz

Solution

a)   765 mm = 765 × 0.04 = 30.6 in.

b)   2 375 km = 2375 × 0.6 = 1425 mi

c)   675 ha = 675 × 2.5 = 1687.5 acres

d)   346 g = 346 × 0.035 = 12.11 oz   ❖

APPENDIX B: SQUARES AND SQUARE ROOTS

No.	Sq.	Sq. Root	No.	Sq.	Sq. Root	No.	Sq.	Sq. Root	No.	Sq.	Sq. Root
1	1	1.000	31	961	5.568	61	3,721	7.810	91	8,281	9.539
2	4	1.414	32	1,024	5.657	62	3,844	7.824	92	8,464	9.592
3	9	1.732	33	1,089	5.745	63	3,969	7.937	93	8,649	9.644
4	16	2.000	34	1,156	5.831	64	4,096	8.000	94	8,836	9.695
5	25	2.236	35	1,225	5.916	65	4,225	8.062	95	9,025	9.747
6	36	2.449	36	1,296	6.000	66	4,356	8.124	96	9,216	9.798
7	49	2.646	37	1,369	6.083	67	4,489	8.185	97	9,409	9.849
8	64	2.828	38	1,444	6.164	68	4,624	8.246	98	9,604	9.899
9	81	3.000	39	1,521	6.245	69	4,761	8.307	99	9,801	9.950
10	100	3.162	40	1,600	6.325	70	4,900	8.367	100	10,000	10.000
11	121	3.317	41	1,681	6.403	71	5,041	8.426			
12	144	3.464	42	1,764	6.481	72	5,184	8.485			
13	169	3.606	43	1,849	6.557	73	5,329	8.544			
14	196	3.742	44	1,936	6.633	74	5,476	8.602			
15	225	3.873	45	2,025	6.708	75	5,625	8.660			
16	256	4.000	46	2,116	6.782	76	5,776	8.718			
17	289	4.123	47	2,209	6.856	77	5,929	8.775			
18	324	4.243	48	2,304	6.928	78	6,084	8.832			
19	361	4.359	49	2,401	7.000	79	6,241	8.888			
20	400	4.472	50	2,500	7.071	80	6,400	8.944			
21	441	4.583	51	2,601	7.141	81	6,561	9.000			
22	484	4.690	52	2,704	7.211	82	6,724	9.055			
23	529	4.796	53	2,809	7.280	83	6,889	9.110			
24	576	4.899	54	2,916	7.348	84	7,056	9.165			
25	625	5.000	55	3,025	7.416	85	7,225	9.220			
26	676	5.099	56	3,136	7.483	86	7,396	9.274			
27	729	5.196	57	3,249	7.550	87	7,569	9.327			
28	784	5.292	58	3,364	7.616	88	7,744	9.381			
29	841	5.385	59	3,481	7.681	89	7,921	9.434			
30	900	5.477	60	3,600	7.746	90	8,100	9.487			

Answers

CHAPTER 1

EXERCISES 1.1

1. $(1234 \times 9) + 5 = 11111$

3. $1 + 3 + 5 + 7 + 9 = 25$ **5.** 1 5 10 10 5 1

7. 4, 2, 0   **9.** $-1, 1, -1$ **11.** $\dfrac{1}{16}, \dfrac{1}{32}, \dfrac{1}{64}$

13. 34, 55, 89   **15.** 89991   **17.** 25 or 5^2, 36 or 6^2

19. a) The products will consist of the digits 1, 2, 4, 5, 7, and 8.

  b) $5 \times 142,857 = 714,285$; the conjecture appears to be correct.

  c) We may conjecture that the product will contain the digits 1, 2, 4, 5, 7, and 8.

  d) $6 \times 142,857 = 857,142$
  $7 \times 142,857 = 999,999$
  $8 \times 142,857 = 1,142,856$
  The conjecture does not hold for 7 and 8.

21. a) 36, 49, 64

  b) Square the next five natural numbers.

  c) No, no number squared is 72.

23. a) Same number.

  b) End with the same number with which you started.

  c) This procedure will always end with the number with which you started.

25. A counterexample is $3 \times 5 = 15$; 15 is not divisible by 2.

27. A counterexample is $5 \times 5 = 25$; 25 is not an even number.

29. a) 360°; **b)** Yes

  c) The sum of the interior angles of a quadrilateral is 360°.

31. b) 18

  c) The sum of the digits in the product of any one- or two-digit number and 99 is 18.

EXERCISES 1.2

1. 12   **3.** No difference

5. Give five children an ice cream bar, and give the sixth child the ice cream bar in the box (or put the sixth child in the box with the ice cream bar).

7. 6   **9.** 5 cents   **11.** 50 feet

13. $7 + 7 - (7 \div 7) = 13$

15. Set both timers; when the sand for the 5-minute timer is all gone, there will be exactly three minutes left on the 8-minute timer.

17. 36 squares

19. a) Place 5 g, 1 g, and 3 g on one side and 9 g on the other side.

  b) Place 16 g, 9 g, and 3 g on one side and 27 g and 1 g on the other side.

21. 8 days

23.

11	1	15
13	9	5
3	17	7

25. Four times the center number will equal the sum of the four corners of the magic square.

27. Nine times the center number will equal the sum of all the numbers of the magic square.

29. $505.62   **31.** $120.75   **33.** 24

35. a) 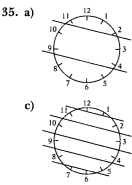   **b)** Cannot be done.

c)

37.

	Over	Return	Other Side River
Trip 1	woman & goat	woman	goat
Trip 2	woman & cabbage	woman & goat	cabbage
Trip 3	woman & wolf	woman	cabbage & wolf
Trip 4	woman & goat	—	woman, goat, wolf, cabbage

39. $(7 \times 7) + 7 + (7 + 7 + 7) = 77$ or $(7 + 7) \times 7 - (7 + 7 + 7) = 77$

41.
$$
\begin{array}{r}
67 \\
23 \\
5 \\
4 \\
1 \\
\hline
100
\end{array}
$$
Other answers are possible.

43. a) 7 moves; **b)** 15 moves **45.** $33\frac{1}{3}$ mph

47. a) 1; **b)** 1.1670817 **49. a)** 0; **b)** 0.1

REVIEW EXERCISES

1. 999 **2.** 987 **3.** 999 **4.** 975 **5.** 11 **6.** 25
7. 64 **8.** 25 **9.** 10 **10.** 21
11. a) The quotient is decreased.
 b) The quotient is increased.

12. a) 6, 10, 15, 55; **b)** Sum $= \dfrac{n(n + 1)}{2}$; **c)** 210
 d) 1275

13.

8	1	6
3	5	7
4	9	2

14.

21	7	8	18
10	16	15	13
14	12	11	17
9	19	20	6

15.

23	25	15
13	21	29
27	17	19

16.

120	140	40
20	100	180
160	60	80

17. a) 10890;
 c) The final result will be 10,890.

18.
$$
\begin{array}{l}
\$25 \text{ cost of room} \\
3 \text{ refund to the men} \\
\underline{2 \text{ bellhop}} \\
\$30
\end{array}
$$

19. 3. The answer will always be 3.
20. a) 111; **b)** The answer will always be 111.
 c)
$$
\begin{array}{l}
ABC \\
BCA \\
+\,\underline{CAB} \\
(A + B + C)100 + (B + C + A)10 + (C + A + B) \\
= (A + B + C)(100 + 10 + 1) = (A + B + C)(111) \\
\quad \dfrac{(A + B + C)(111)}{A + B + C} = 111
\end{array}
$$

21. 364 **22.** 18 **23.** $2\frac{4}{7}$ qt water or 2.57 qt water
24. Higher if there is no change
25. \$315 maximum monthly payment
26. a) **b)** **c)**

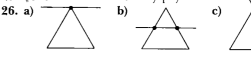

27.

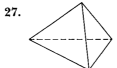

28. 140 lb **29.** 59 min, 59 sec
30. Same amount of wine in red cup as water in blue cup.

31. 125,250
32. All proper fractions (any number between -1 and 1).
33. One example: $11 + 13 = 24$
34. 40 posts **35.** 3 ducks
36.

A	B	C	D	E
E	A	B	C	D
D	E	A	B	C
C	D	E	A	B
B	C	D	E	A

37. Place six coins in each pan with one coin off to the side. If it balances, then the heavier coin is the one on the side. If the pan does not balance, then take the six coins on the heavier side and split them into two groups of three. Select the three heavier coins and weigh two coins. If the pan balances, then it is the third coin. If the pan does not balance, you can identify the heavier coin.

38. 0

39. If each number is written out in words, then the numbers of letters in the words form a magic square, each row column and diagonal having a sum of 17.

40.
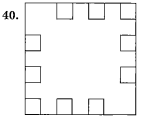

41. a) 2; **b)** 6; **c)** 24; **d)** 120
   e) $n(n - 1)(n - 2)(n - 3)\cdots 1$(or $n!$); n = number of people

42.

CHAPTER 1 TEST

1. 16, 20, 24 **2.** $\frac{1}{81}, \frac{1}{243}, \frac{1}{729}$ **3.** 3 **4.** $7\frac{1}{2}$ min
5. $579.38

6.

40	15	20
5	25	45
30	35	10

7. Four times as large
8. Less time if she had driven at 45 mph
9. 2, 6, 8, 9, 13; 11 does not divide 11,232.
10. $\frac{161}{13} \approx 12.4$ sec

CHAPTER 2

EXERCISES 2.1

1. Not well defined **3.** Well defined
5. Well defined **7.** Infinite **9.** Infinite
11. Infinite **13.** Finite
15. {Ontario, Erie, Superior, Michigan, Huron}
17. {5,6,7, ..., 350} **19.** {4}
21. {red, orange, yellow, blue, green, violet, indigo}
23. {2,3,4,5,6, ...} **25.** $B = \{x | x < 8 \text{ and } x \in N\}$
27. $C = \{x | x > 6 \text{ and } x \in N\}$
29. $E = \{x | 1 < x < 4 \text{ and } x \in N\}$
31. $G = \{x | x \geq 7 \text{ and } x \in N\}$
33. N is the set of natural numbers
35. V is the set of permanent members of the United Nations Security Council
37. C is the set of the last four letters of the English alphabet
39. J is the set of natural numbers between 1 and 7
41. False **43.** True **45.** True **47.** False **49.** 3
51. 0 **53.** Both **55.** Neither **57.** Equivalent
59. a) A is the set of natural numbers greater than 2. B is the set of all the numbers greater than 2.
   b) Set A contains only natural numbers. Set B contains other types of numbers, including fractions and decimals.
   c) $A = \{3,4,5,6,7, ...\}$
   d) Cannot write set B in roster form.
61. a) finite, $n(A) = 0$;
   b) infinite

EXERCISES 2.2

1. False **3.** True **5.** False **7.** True
9. True **11.** True **13.** None
15. $B \subseteq A, B \subset A$ **17.** None
19. $A = B, A \subseteq B, B \subseteq A$ **21.** { ⊠ }, { }
23. {red, green, yellow}, {red, green}, {red, yellow}, {green, yellow}, {red}, {yellow}, {green}, { }
25. a) {*}, {?}, {#}, {1}, {*, ?}, {*, #}, {*, 1}, {?, #}, {?, 1}, {#, 1}, {*, ?, #}, {*, ?, 1}, {?, #, 1}, {*, #, 1}, {*, ?, #, 1}, { }
   b) {*, ?, #, 1}
27. False **29.** False **31.** True
33. False **35.** False
37. True **39.** $2^5 = 32$

EXERCISES 2.3

1. *A'* is the set of all the people in the state of Pennsylvania in 1988 who have been married 25 years or more.

3. The set of college graduates with an AA degree and a BS degree.

5. The set of college graduates without an AA degree or with a BA degree.

7. The set of college graduates with an AA degree, a BS degree, and a BA degree.

9. The set of cities in the United States with a population over 100,000 or with a philharmonic orchestra.

11. The set of cities in the United States without a subway system and with a philharmonic orchestra.

13. The set of cities in the United States with a population not over 100,000 or without a subway system.

15. {1,2,3,4,5,6,8}   17. {1,5,7,8}   19. {7}   21. { }

23. {7}   25. {O,□,△,#,α,β}   27. {α,β,#}

29. {O,□,△}   31. U   33. {O,□,△,#}   35. {α,β}

37. { }   39. {*b,d,e,h,j,k*}   41. {*a,f,i*}   43. {*a,b,c,d,f,g,i*}

45. {*a,c,d,e,f,g,h,i,j,k*}   47. {*b*}   49. *A*   51. *B*

53. *C*   55. {2,6,10,14,18,...}

57. {2,6,10,14,18,...}   59. *U*   61. *A*   63. { }

65. *A*   67. {*e, f, h*}   69. {*d, j, k*}   71. {13}

73. {1,2,3,4,5,6,7,8,9,10,11,12,14,15}

75. {2,3,4,5,7,9,10,11,12,13,14,15}

77. {(*a*, 1), (*a*, 2), (*b*, 1), (*b*, 2), (*c*, 1), (*c*, 2)}

79. No   81. 6

83. {(1, 1), (1, 2), (1, 3), (1, 4), (2, 1), (2, 2), (2, 3), (2, 4), (3, 1), (3, 2), (3, 3), (3, 4), (4, 1), (4, 2), (4, 3), (4, 4)}

85. *A* = *B* or *B* ⊂ *A*   87. *A* and *B* are disjoint sets.

89. *A* = *B* or *A* ⊂ *B*

91. *B* = ∅ or *A* and *B* are disjoint sets.

93. Always true   95. Always true   97. Always true

EXERCISES 2.4

1.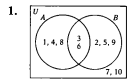

3. III   5. VI   7. III   9. II   11. IV   13. II

15. VIII   17. VIII   19. II   21. I   23. V

25. {1,2,3,4,6}   27. {1,2,3,4,5,6,7,8,9,10}   29. {2,3}

31. {7,8}   33. {1,2,3,4,5,6}   35. {4,5,6,7,8,10}

37. {3,4,5}   39. {1,2,3,6,9,10,11,12}

41. {1,2,3,4,5,6,7,8,9,12}   43. {9,11,12}

45. {7,8,9,10,11,12}

47.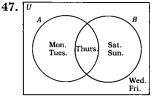

49. 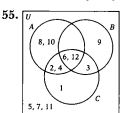   51.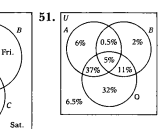

53. a) $n(A \cup B) = n(A) + n(B) - n(A \cap B)$
$$8 = 4 + 6 - 2$$
$$8 = 8$$

55.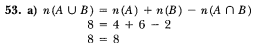

57.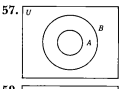

59.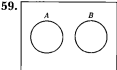

61. a)

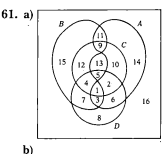

b)

1. $A \cap B \cap C \cap D$	**9.** $A \cap B \cap C \cap D'$
2. $A \cap B' \cap C \cap D$	**10.** $A \cap B' \cap C \cap D'$
3. $A \cap B \cap C' \cap D$	**11.** $A \cap B \cap C' \cap D'$
4. $A' \cap B \cap C \cap D$	**12.** $A' \cap B \cap C \cap D'$
5. $A' \cap B' \cap C \cap D$	**13.** $A' \cap B' \cap C \cap D'$
6. $A \cap B' \cap C' \cap D$	**14.** $A \cap B' \cap C' \cap D'$
7. $A' \cap B \cap C' \cap D$	**15.** $A' \cap B \cap C' \cap D'$
8. $A' \cap B' \cap C' \cap D$	**16.** $A' \cap B' \cap C' \cap D'$

EXERCISES 2.5

1. Equal   **3.** Not equal   **5.** Not equal   **7.** Equal
9. Not equal   **11.** Equal   **13.** Equal   **15.** Equal
17. Not equal   **19. a)** $\{6, 7\} = \{6, 7\}$
c)

$(A \cup B) \cap C$ $(A \cap C) \cup (B \cap C)$

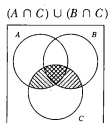

EXERCISES 2.6

1.

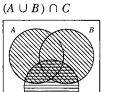

a) 57; **b)** 17; **c)** 53

3.

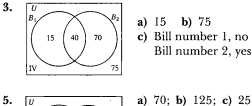

a) 15   **b)** 75
c) Bill number 1, no majority;
Bill number 2, yes majority

5.

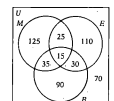

a) 70; **b)** 125; **c)** 25
d) 260; **e)** 430

7.

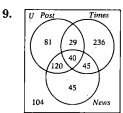

a) 185; **b)** 10
c) 25; **d)** 401

9.

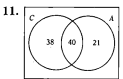

a) 236; **b)** 596
c) 466; **d)** 29
e) 346; **f)** 362

11.

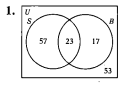

Only 99 people accounted
for out of 100 interviewed

13.

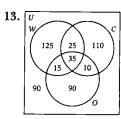

a) 410; **b)** 35;
c) 90; **d)** 50

15.

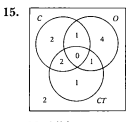

13 children

EXERCISES 2.7

1. $\{3, 4, 5, 6, \ldots, \quad n \quad, \ldots\}$
$\downarrow \downarrow \downarrow \downarrow \qquad \downarrow$
$\{4, 5, 6, 7, \ldots, n + 1, \ldots\}$

3. $\{2, 3, 4, 5, \ldots, \quad n \quad, \ldots\}$
$\downarrow \downarrow \downarrow \downarrow \qquad \downarrow$
$\{3, 4, 5, 6, \ldots, n + 1, \ldots\}$

5. $\{4, 7, 10, 13, \ldots, \quad n \quad, \ldots\}$
$\downarrow \downarrow \downarrow \downarrow \qquad \downarrow$
$\{7, 10, 13, 16, \ldots, n + 3, \ldots\}$

7. $\{8, 10, 12, 14, \ldots, \quad n \quad, \ldots\}$
$\downarrow \quad \downarrow \quad \downarrow \quad \downarrow \qquad \downarrow$
$\{10, 12, 14, 16, \ldots, n + 2, \ldots\}$

9. $\{1, \frac{1}{3}, \frac{1}{5}, \frac{1}{7}, \ldots, \quad \frac{1}{n} \quad, \ldots\}$
$\downarrow \downarrow \downarrow \downarrow \qquad \downarrow$
$\{\frac{1}{3}, \frac{1}{5}, \frac{1}{7}, \frac{1}{9}, \ldots, \frac{1}{n+2}, \ldots\}$

11. $\{1, 2, 3, 4, \ldots, n, \ldots\}$
$\downarrow \downarrow \downarrow \downarrow \qquad \downarrow$
$\{3, 6, 9, 12, \ldots, 3n, \ldots\}$

13. $\{1, 2, 3, 4, \ldots, \quad n \quad, \ldots\}$
$\downarrow \downarrow \downarrow \downarrow \qquad \downarrow$
$\{4, 6, 8, 10, \ldots, 2n + 2, \ldots\}$

15. $\{1, 2, 3, 4, \ldots, \quad n \quad, \ldots\}$
$\downarrow \downarrow \downarrow \downarrow \qquad \downarrow$
$\{2, 5, 8, 11, \ldots, 3n - 1, \ldots\}$

17. $\{1, 2, 3, 4, \ldots, \quad n \quad, \ldots\}$
$\downarrow \downarrow \downarrow \downarrow \qquad \downarrow$
$\{5, 8, 11, 14, \ldots, 3n + 2, \ldots\}$

19. $\{1, 2, 3, 4, \ldots, \quad n \quad, \ldots\}$
$\downarrow \downarrow \downarrow \downarrow \qquad \downarrow$
$\{\frac{1}{3}, \frac{1}{4}, \frac{1}{5}, \frac{1}{6}, \ldots, \frac{1}{n+2}, \ldots\}$

21. $\{1, 2, 3, 4, \ldots, n, \ldots\}$
$\downarrow \downarrow \downarrow \downarrow \qquad \downarrow$
$\{1, 4, 9, 16, \ldots, n^2, \ldots\}$

23. $\{1, 2, 3, 4, \ldots, n, \ldots\}$
$\downarrow \downarrow \downarrow \downarrow \qquad \downarrow$
$\{3, 9, 27, 81, \ldots, 3^n, \ldots\}$

REVIEW EXERCISES

1. False **2.** True **3.** False **4.** False **5.** True
6. False **7.** False **8.** True **9.** True **10.** False
11. False **12.** False **13.** $A = \{2,4,6,8,10,12\}$
14. $\{5,6,7,8,9,\ldots\}$ **15.** $\{x \mid x \geq 3, x \in N\}$
16. $\{x \mid 4 < x < 20,$ and x is an odd natural number$\}$
17. The set of natural numbers between 4 and 11
18. The set of names of the 50 states in the United States
19. $\{\ \}$ **20.** $\{1,2,3,4,9\}$ **21.** $\{3,4\}$ **22.** $\{3,4,5,6,9\}$
23. $\{1,2\}$ **24.** 16
25.

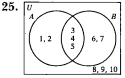

26. $\{a,c,d,e,f,g,i,k\}$ **27.** $\{i,d\}$ **28.** $\{a,b,c,d,e,f,g,i,k\}$
29. $\{e\}$ **30.** $\{a,d,e\}$ **31.** $\{a,b,d,e,f,g\}$ **32.** True
33. True **34.** \$450 **35.** 315 **36.** 10 **37.** 30
38. 110
39. $\{2, 4, 6, 8, \ldots, \quad n \quad, \ldots\}$
$\downarrow \downarrow \downarrow \downarrow \qquad \downarrow$
$\{4, 6, 8, 10, \ldots, n + 2, \ldots\}$

40. $\{3, 5, 7, 9, \ldots, \quad n \quad, \ldots\}$
$\downarrow \downarrow \downarrow \downarrow \qquad \downarrow$
$\{5, 7, 9, 11, \ldots, n + 2, \ldots\}$

41. $\{1, 2, 3, 4, \ldots, \quad n \quad, \ldots\}$
$\downarrow \downarrow \downarrow \downarrow \qquad \downarrow$
$\{5, 8, 11, 14, \ldots, 3n + 2, \ldots\}$

42. $\{1, 2, 3, 4, \ldots, \quad n \quad, \ldots\}$
$\downarrow \downarrow \downarrow \downarrow \qquad \downarrow$
$\{4, 9, 14, 19, \ldots, 5n - 1, \ldots\}$

CHAPTER 2 TEST

1. True **2.** True **3.** True **4.** False **5.** True
6. True **7.** True **8.** False **9.** True **10.** True
11. $\{x \mid 4 \leq x \leq 8, x \in N\}$ **12.** $\{x \mid x \geq 8, x \in N\}$
13. $\{1,2,3,4,5,6,7,8\}$ **14.** $\{1,2,4,5,7,9,10\}$ **15.** $\{\ \}$
16. 3
17.

18. Not equal

19.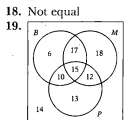

20. 14 **21.** 18 **22.** 27 **23.** 85

24. $\{4, 5, 6, 7, \ldots, \quad n \quad, \ldots\}$
  $\quad\downarrow\ \downarrow\ \downarrow\ \downarrow \qquad\qquad \downarrow$
  $\{5, 6, 7, 8, \ldots, n + 1, \ldots\}$

25. $\{1, 2, 3, 4, \ldots, \quad n \quad, \ldots\}$
  $\quad\downarrow\ \downarrow\ \downarrow\ \downarrow \qquad\qquad \downarrow$
  $\{1, 3, 5, 7, \ldots, 2n - 1, \ldots\}$

CHAPTER 3

EXERCISES 3.1

1. No **3.** No **5.** Yes
7. Compound, conjunction
9. Compound, biconditional
11. Compound, disjunction
13. Compound, conjunction
15. Compound, negation
17. Compound, conditional **19.** $p \wedge q$
21. $p \vee \sim q$ **23.** $p \to q$ **25.** $\sim p \wedge \sim q$
27. Joan is not a football official.
29. Joan is a football official and Charles is a tennis player.
31. If Joan is not a football official, then Charles is a tennis player.
33. It is false that Charles is a tennis player or Joan is a football official.
35. Joan is not a football official or Charles is not a tennis player.
37. If Charles is a tennis player, then Joan is not a football official.
39. a) $(p \wedge q) \vee r$ **41.** $p \to (q \vee \sim r)$ **43.** $\sim p \leftrightarrow \sim q$
45. $(r \leftrightarrow q) \wedge p$ **47.** $(r \vee \sim q) \leftrightarrow p$
49. The rose is red or the stem is green, and today is Saturday.
51. If the stem is green then the rose is red, or today is Saturday.
53. If today is not Saturday, then the stem is green and the rose is red.
55. If today is Saturday then the stem is green, and the rose is red.
57. The stem is green if and only if the rose is red, and today is Saturday.
59. $\sim p \to (q \wedge \sim r)$ **61.** $\sim p \leftrightarrow (q \vee r)$
63. $(q \to p) \leftrightarrow r$

65. $(p \wedge q) \wedge r$ (p: the president be at least 35 years old; q: the president be a natural-born citizen; r: the president be a resident of the country for 14 years.)
67. $p \to q$ (p: you want to kill an idea in the world today; q: get a committee working on it.)
69. $\sim p \wedge q$ (p: I know the dignity of their birth; q: I know the glory of their death.)
71. $(p \wedge c) \to (d \vee e)$

EXERCISES 3.2

1.	**3.**	**5.**	**7.**	**9.**	**11.**	**13.**
F	T	T	T	F	F	F
F	F	F	T	T	T	F
	T	T	T	T	T	T
	T	T	F	F	F	F

15.	**17.**	**19.**	**21.**	**23.**
T	T	F	T	T
T	T	T	F	T
F	T	T	F	T
T	T	T	F	T

25.	**27.**	**29.**	**31.**
T	T	T	T
F	T	T	F
F	T	F	T
F	T	T	F
F	F	T	T
F	T	T	T
F	F	T	F
F	T	T	F
F	F	T	F

33.	**35.**	**37.**
T	T	T
T	T	T
F	T	T
F	F	F
F	T	T
T	F	T
T	T	F
F	F	T

39. A tautology is a statement that is true in every case of the truth table.
41. Tautology   43. Neither   45. Neither
47. Self-contradiction   49. No   51. Yes   53. Yes
55. F   57. F   59. T   61. T   63. T   65. T
67. T   69. F   71. T   73. T   75. F   77. { }
79. $x = 4$ 81. $x = 6$ 83. Any number
85. Any number   87. $x = 2$
89. T   91. T
    F   T
    T   T
    T   T
             F
             T
             T
             F
93. 1, (2, or 3), 4, 5, 8, (11, or 12)

EXERCISES 3.3

1. They are equivalent.   3. No   5. No   7. Yes
9. No   11. Yes   15. No plants are green.
17. All birds fly.   19. Some snow is not white.
21. All instruments have strings.
23. It is false that today is not Saturday and I am not in school.
25. It is false that the Philadelphia Eagles will not play in the Super Bowl and they did win enough games.
27. It is false that the test was hard and I did study.
29. The computer does not know the answers and the operator does not strike the wrong key.
31. A man is not healthy or he is happy.
33. $x + 3 \neq 4$ or $3 + 1 = 4$
35. If the White House is not in Washington, then the U.S. Capitol is not in Chicago.
37. The grass is green or it did not rain.
39. a) Gus did it.   b) Dave did it.

EXERCISES 3.4

1. If the mouse is gray, then the cat is red.
3. If I own a motorbike, then I will buy a helmet.
5. If I plant seeds, then I will need to water the garden.
7. If I have a license, then I can go fishing.
9. If the mouse kisses the cat, then the cat loves the mouse.

11. If they win all their games, then the Eagles will win the Super Bowl.
13. If you are in favor of waste management, then you will be elected.
15. If the pumpkins turn orange, then the leaves turn red.
17. *Converse:* If I am happy, then today is Friday.
    Inverse: If today is not Friday, then I am not happy.
    Contrapositive: If I am not happy, then today is not Friday.
19. *Converse:* If I did not study, then the test was hard.
    Inverse: If the test was not hard, then I studied.
    Contrapositive: If I studied, then the test was not hard.
21. *Converse:* If the door will not open, then the key does not fit.
    Inverse: If the key fits, then the door will open.
    Contrapositive: If the door will open, then the key fits.
23. *Converse:* If we can go swimming, then there is no lightning.
    Inverse: If there is lightning, then we cannot go swimming.
    Contrapositive: If we cannot go swimming, then there is lightning.
25. *Converse:* If the opposite sides are parallel and the angles are right angles, then the figure is a rectangle.
    Inverse: If the figure is not a rectangle, then the opposite sides are not parallel or the angles are not right angles.
    Contrapositive: If the opposite sides are not parallel or the angles are not right angles, then the figure is not a rectangle.
27. If the meat is not tender, then the cattle are not fed the proper diet.
29. If you are not speeding, then you are not going faster than 55 miles per hour.
31. If the penguin can fly, then the penguin is a bird.
33. If a figure is not a rhombus, then the figure does not have equal sides.
35. If the cat is not gray and the mouse is not green, then the dog does not eat carrots.
37. If 2 divides the units digit of the counting number, then 2 divides the counting number. True.
39. If two lines are not parallel, then the two lines intersect in at least one point. True.

EXERCISES 3.5

1. Valid   **3.** Valid   **5.** Fallacy   **7.** Valid   **9.** Fallacy
11. Fallacy   **13.** Fallacy   **15.** Valid   **17.** Fallacy
19. Fallacy   **21.** Valid

EXERCISES 3.6

1. Valid   **3.** Invalid   **5.** Invalid   **7.** Valid
9. Valid   **11.** Valid   **13.** Invalid   **15.** Valid

17. $w \to p$

$$\frac{w}{\therefore p}, \text{ valid}$$

19. $t \vee p$

$$\frac{\sim p}{\therefore t}, \text{ valid}$$

21. $r \vee p$

$$\frac{p \to \sim r}{\therefore \sim p}, \text{ invalid}$$

23. $\sim w$

$$\frac{w \to m}{\therefore m}, \text{ invalid}$$

25. $r \vee s$

$$\frac{\sim r \to l}{\therefore s \vee l}, \text{ invalid}$$

27. $g \to j$

$$\frac{j \to (\sim g \vee c)}{\therefore j}, \text{ invalid}$$

29. $t \vee h$

$$\frac{\sim h}{\therefore \sim t}, \text{ invalid}$$

31. $f \to t$

$$\frac{f}{\therefore t}, \text{ valid}$$

33. $s \wedge o$

$$\frac{\sim o \vee s}{\therefore \sim s}, \text{ invalid}$$

35. $b \wedge c$

$$\frac{\sim c \vee b}{\therefore \sim b}, \text{ invalid}$$

37. $a \to b$
$b \to c$

$$\frac{a}{\therefore c}, \text{ valid}$$

39. You will be healthy.
41. If you like to eat, then you are an outdoors person.
43. If the logs are not dry, then you will have to dry the logs.
51. Tiger   Boots   Sam   Sue
Blue   Yellow   Red   Green
Nine Lives   Puss'n Boots   Friskies   Meow Mix
53. If $(P' \cup Q) \cap (P \cup Q) \subseteq P'$, then the argument is valid.

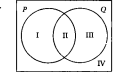

$(P' \cup Q) \cap (P \cup Q)$ is regions II, III. P' is regions III, IV. Since region II is not in P', the argument is invalid.

EXERCISES 3.7

1. $p \wedge q$ **3.** $(p \vee r) \wedge \sim r$
5. $(p \wedge q) \wedge [(p \wedge \sim q) \vee r]$ **7.** $p \vee q \vee (r \wedge \sim p)$
9.   **11.**

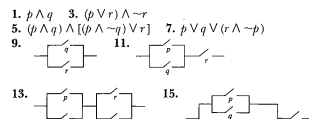

13.   **15.**

17. Not equivalent   **19.** Equivalent
21. Not equivalent
23.

25. Series
27. Parallel. Some strings of lights may be wired in series.
29. Series. Some cars may be wired in parallel.

REVIEW EXERCISES

1. The calculator is correct and I entered the correct number.
2. The calculator is not correct or I touched the wrong keys.
3. If the calculator is correct, then I entered the correct number and I did not touch the wrong keys.
4. The calculator is correct if and only if I entered the correct number.
5. The calculator is correct or I did not enter the correct number, and I did not touch the wrong keys.
6. The calculator is not correct, if and only if I touched the wrong keys and I did not enter the correct number.
7. $q \vee \sim r$ **8.** $p \wedge r$ **9.** $r \to (\sim p \vee q)$
10. $(q \leftrightarrow p) \wedge \sim r$ **11.** $(r \wedge p) \vee q$ **12.** $\sim(q \vee \sim r)$
13. T   **14.** F
     T   F
     F   F
     F   T
15. T   **16.** F
     T   T
     T   T
     F   F

17. T   **18.** F   **19.** T   **20.** F

F	T	T	T
T	F	T	T
T	F	T	T
F	T	T	T
F	T	F	T
F	T	F	T
F	T	T	T

21. T   **22.** T   **23.** T   **24.** T   **25.** T
26. Any value of x **27.** Any value of x **28.** $x \neq 7$
29. { }   **30.** 2   **31.** T   **32.** T   **33.** F   **34.** F
35. T   **36.** Equivalent   **37.** Not equivalent
38. Equivalent   **39.** Not equivalent
40. Some wood is not hard.
41. Some birds are yellow.   **42.** All apples are red.
43. No squirrels fly.   **44.** All people wear glasses.
45. It is false that the pencil is yellow and the desk is not orange.
46. It is not true that the Flyers do not score goals or they do not win games.
47. New Orleans is not in California and Chicago is a city.
48. The water is not blue or the fish will not bite.
49. *Converse:* If I get a good grade, then I have studied.
   Inverse: If I do not study, then I will not get a good grade.
   Contrapositive: If I did not get a good grade, then I did not study.
50. Invalid   **51.** Invalid   **52.** Invalid   **53.** Valid
54. Invalid
55.

56. $[p \vee (\sim p \wedge q)] \wedge r$

CHAPTER 3 TEST

1. $(p \wedge q) \vee r$ **2.** $\sim q \to r$
3. $\sim(p \leftrightarrow q)$
4. If the chair is blue, then the sofa is red or the teapot is black.
5. It is false that the chair is blue if and only if the sofa is red.
6. F   **7.** T   **8.** T

F	F	T
F	T	F
T	T	T
	F	T
	F	T
	T	T
	T	F

9. T   **10.** T   **11.** T   **12.** F
13. All numbers.   **14.** $x = 6$
15. If the dog is green, then the mouse is gray.
16. If the dog is not green, then the mouse is not gray.
17. If the mouse is not gray, then the dog is not green.
18. If the chipmunk is eating dinner, then the sun is shining.
19. If the bird has a red head, then the bird is a woodpecker.
20. Valid   **21.** Valid
22. Not equivalent
23. Some jokes are not funny.
24. No test questions are easy.

25.

CHAPTER 4 _____

EXERCISES 4.1

1. 211   **3.** 2323   **5.** 254,214   **7.** 9999∩∩ⅠⅠⅠⅠⅠ
9. ↑99∩∩∩ⅠⅠⅠⅠ   **11.** ∝ⅧⅧⅧ↑↑ 9999∩∩∩∩∩Ⅰ
13. 64   **15.** 117   **17.** 1892   **19.** 1999   **21.** 7602
23. 10,679   **25.** XXXVIII   **27.** CXLVIII
29. MCMLXXVIII   **31.** MMDCVI
33. MMMMCCCXXI or $\overline{\text{IV}}$CCCXXI
35. 57   **37.** 1976   **39.** 2605

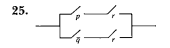

47. 364   **49.** 42,505   **51.** 9007   **53.** γ
55. ρκγ   **57.** λ'ε'ιβ
63. Needs more symbols, more difficult to perform operations

65. 1,021, MXXI, $\frac{\not{\neq}}{\not{+}}$, α′κα

67. 527, 𝟿𝟿𝟿𝟿𝟿∩∩ IIIIIII , DXXVII, φκζ

EXERCISES 4.2

1. $(4 \times 10) + (8 \times 1)$
3. $(9 \times 100) + (4 \times 10) + (2 \times 1)$
5. $(3 \times 1000) + (4 \times 100) + (5 \times 10) + (2 \times 1)$
7. $(6 \times 10{,}000) + (4 \times 1000) + (5 \times 100) +$
$(2 \times 10) + (1 \times 1)$
9. $(2 \times 100{,}000) + (4 \times 10{,}000) + (5 \times 1000) +$
$(6 \times 100) + (7 \times 10) + (2 \times 1)$
11. 32 **13.** 724 **15.** 5528 **17.** ∀ ◁
19. ∀∀ ∀ **21.** ∀ ∀∀∀ ◁◁◁◁∀∀ ∀

23. 70 **25.** 6841 **27.** 4000
29. ≡ **31.** ≡ **33.** ⋯

35. No clear way of indicating place value
39. 1,944 ◁◁◁∀∀ ◁◁∀∀∀

EXERCISES 4.3

1. Because the third position is not $(20)^2$ **3.** 7
5. 11 **7.** 79 **9.** 553 **11.** 1367 **13.** 193
15. 83 **17.** 4135 **19.** 1000_2 **21.** 10111_2
23. 626_8 **25.** $E93_{12}$ **27.** 1022_6 **29.** $1ET_{12}$
31. 1111110011_2 **33.** 4550_8 **35.** 1844
37. 27,963 **39.** 133_{16} **41.** 1566_{16}
43. 11111001000_2 **45.** 30432_5 **47.** 3710_8
49. No 6 in base 5 **51.** Correct **53.** 10213_5
55. 1373_8 **57.** 10121_3 **59.** $2^7 = 128$ orderings

EXERCISES 4.4

1. 122_4 **3.** 11412_5 **5.** $6T3_{12}$ **7.** 2100_3
9. 24001_7 **11.** 100_4 **13.** 1041_5 **15.** 325_{12}
17. 11_2 **19.** 3616_7 **21.** 231_5 **23.** 3504_7
25. 20212_4 **27.** 5503_9 **29.** 110001_2 **31.** 42_6R4
33. 63_8R4 **35.** 86_{12}R1 **37.** 13_5R4
39. $\begin{array}{r} 101_2 \\ +100_2 \\ \hline ①001 \\ \quad\downarrow 1 \\ \hline 10_2 \end{array}$ **41.** $\begin{array}{r} 10110_2 \\ +11000_2 \\ \hline ①01110 \\ \quad\downarrow 1 \\ \hline 1111_2 \end{array}$ **43.** $\begin{array}{r} 423_5 \\ +122_5 \\ \hline ①100 \\ \quad\downarrow 1 \\ \hline 101_5 \end{array}$ **45.** $\begin{array}{r} 3401_5 \\ +4221_5 \\ \hline ②3122 \\ \quad\downarrow 1 \\ \hline 3123_5 \end{array}$

EXERCISES 4.5

1. <s>18 − 30</s> **3.** <s>8 − 145</s> **5.** 43 − 221 **7.** 75 − 82
 9 − 60 <s>4 − 290</s> 21 − 442 37 − 164
 <s>4 − 120</s> <s>2 − 580</s> <s>10 − 884</s> <s>18 − 328</s>
 <s>2 − 240</s> 1 − 1160 5 − 1768 9 − 656
 $\underline{1 − 480}$ $\underline{}$ <s>2 − 3536</s> <s>4 − 1312</s>
 540 1160 $\underline{1 − 7072}$ <s>2 − 2624</s>
 9503 $\underline{1 − 5248}$
 6150

9. 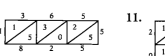 **11.**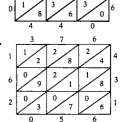

13. 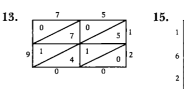 **15.**

17. 112 **19.** 1316 **21.** 625 **23.** 18,760

REVIEW EXERCISES

1. 2101 **2.** 2211 **3.** 2022 **4.** 2114 **5.** 3214
6. 2312 **7.** *bbbbaaa* **8.** *cbbbbbbaaaaaaa*
9. *cccbbbbbbbbaaaaaaaaa* **10.** *dccccccccccbbbbbbbaaaaaaaaa*
11. *dddddaaaa* **12.** *dddccccbba* **13.** 93 **14.** 27
15. 648 **16.** 273 **17.** 7482 **18.** 9574 **19.** *dxc*
20. *ayfxg* **21.** *dyhxi* **22.** *aziygxh* **23.** *fzd*
24. *bza* **25.** 93 **26.** 407 **27.** 8298 **28.** 46,883
29. 70,082 **30.** 60,529 **31.** *mc* **32.** *sog*
33. *vqi* **34.** *BAph* **35.** *LDxqe* **36.** *QFvrf*
37. ∮𝟿𝟿𝟿𝟿∩∩∩∩∩ II **38.** MCDLXII
39. $\begin{array}{l} \frac{\ }{\mp} \\ \not{σ} \\ \mathrm{p} \\ \overline{\pi} \\ \pm \end{array}$ **40.** α′υξβ **41.** ◁◁∀∀∀∀ ◁◁∀∀ **42.** ⋯ ⋮ ⋯

43. 122,025 **44.** 8254 **45.** 585 **46.** 1991
47. 1277 **48.** 1971 **49.** 34 **50.** 5 **51.** 79
52. 1459 **53.** 1451 **54.** 186 **55.** 3323_5
56. 2051_6 **57.** 111001111_2 **58.** 13033_4
59. 327_{12} **60.** 717_8 **61.** 137_8 **62.** 100011_2
63. 135_{12} **64.** 1023_7 **65.** 13101_5 **66.** 10423_8
67. 3611_7 **68.** 100_2 **69.** $3E4_{12}$ **70.** 3324_5

71. 1704_8 **72.** 1022_3 **73.** 202_5 **74.** 1232_4
75. 5050_{12} **76.** 12221_3 **77.** 110111_2 **78.** 13632_8
79. 1011_2 **80.** 112_4 **81.** 30_5 **82.** 433_6
83. 211_4R10 **84.** 12_7R12
85. ~~142~~ − ~~24~~   **86.** 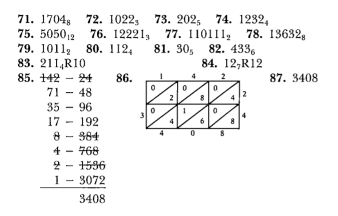   **87.** 3408
    $71 - 48$
    $35 - 96$
    $17 - 192$
    ~~8~~ − ~~384~~
    ~~4~~ − ~~768~~
    ~~2~~ − ~~1536~~
    $\underline{1 - 3072}$
        3408

CHAPTER 4 TEST

1. A number is a quantity. A numeral is the symbol used to represent the quantity.
2. 1231   **3.** 1999   **4.** 1991   **5.** 9835
6. 999∩∩ⅠⅠⅠⅠ   **7.**
8. ∀ ◁∀ ∀∀∀   **9.** ψπϵ
10. *Additive system:* Symbols are added to form numbers. *Multiplicative system:* Numbers are multiplied by the powers of the base, where the powers of the base are illustrated next to the number. *Ciphered system:* A unique numeral is used to represent each number. The numerals are placed next to one another to obtain the number. *Place value system:* Each number is multiplied by a power of the base. The position of the number indicates the power of the base by which it is multiplied.
11. 19   **12.** 59   **13.** 686   **14.** 21   **15.** 100011_2
16. 333_5 **17.** 943_{12} **18.** 7072_8 **19.** 1110_3
20. 227_8 **21.** 2412_7
22. $15 - 17$
    $7 - 34$
    $3 - 68$
    $\underline{1 - 136}$
        255

23.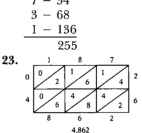
    4,862

CHAPTER 5

EXERCISES 5.1

1. A prime number is a number that is divisible by only two factors, itself and 1.
3. Divisible by 2, 3, 4, 6, 8, 9, 12
5. Divisible by 3, 5
7. Divisible by 2, 3, 4, 6, 12
9. 60, other answers are possible
11. 2, 3, 5, 7, 11, 13, 17, 19, 23, 29, 31, 37, 41, 43, 47, 53, 59, 61, 67, 71, 73, 79, 83, 89, 97
13. $2 \cdot 13$ **15.** $2^2 \cdot 3^2 \cdot 5$ **17.** $2^4 \cdot 5^2$
19. $2^3 \cdot 3^2 \cdot 5 \cdot 7$ **21.** $3^3 \cdot 37$ **23.** $2^4 \cdot 73$ **25.** 15
27. 26   **29.** 30   **31.** 4   **33.** 4   **35.** 150
37. 156   **39.** 1800   **41.** 6500   **43.** 72
45. 11, 13; 17, 19   **47.** 16   **49.** 10   **51.** 16
53. 70   **55.** Yes   **57.** Yes   **59.** No   **61.** No
65. a) All of them do.
   b) Conjecture: Every prime greater than 3 differs by 1 from a multiple of the number 6.

EXERCISES 5.2

1.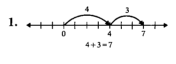
    $4 + 3 = 7$

3.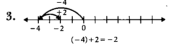
    $(-4) + 2 = -2$

5.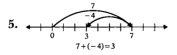
    $7 + (-4) = 3$

7.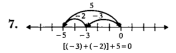
    $[(-3) + (-2)] + 5 = 0$

9.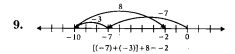

$[(-7)+(-3)]+8=-2$

11.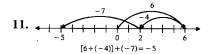

$[6+(-4)]+(-7)=-5$

13.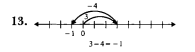

$3-4=-1$

15.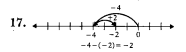

$6-3=3$

17.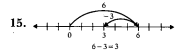

$-4-(-2)=-2$

19.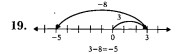

$3-8=-5$

21.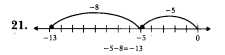

$-5-8=-13$

23.

$[4+(-2)]-3=-1$

25. 24   **27.** 9   **29.** 18   **31.** −18   **33.** −120
35. −216   **37.** 3   **39.** −5   **41.** −9   **43.** −8
45. 18   **47.** −51   **49.** 4   **51.** −14   **53.** −4
55. −3   **57.** −25

59. $\dfrac{a}{b} = \dfrac{(-1)a}{(-1)b} = \dfrac{-a}{-b}$

61. 65,226 ft   **63.** 5 yd loss (or −5 yd)
65. 881 ft below sea level (or −881 ft)
67. Perfect number   **69.** Not perfect number
71. a) 35, 51, 70
   b) 1, 5, 12, 22, 35, 51, 70
      4 7 10 13 16 19 ; each difference is 3 greater than the preceding difference.
   c) No

1. Terminating   **3.** Repeating   **5.** Repeating
7. Terminating   **9.** Repeating   **11.** Repeating
13. $0.\overline{6}$ **15.** $0.\overline{7}$ **17.** $0.41\overline{6}$ **19.** 0.4375
21. $3.\overline{5428571}$ **23.** $5.\overline{857142}$ **25.** 4/10 = 2/5
27. 432/1000 = 54/125   **29.** 2001/1000
31. 375/1000 = 3/8   **33.** 8759/10000
35. 2396/1000 = 599/250   **37.** 5/9   **39.** 47/99
41. 146/999   **43.** 411/999 = 137/333   **45.** 13/90
47. 79/30   **49.** 10/21   **51.** 8/45   **53.** 25/64
55. 16/15   **57.** 1/18   **59.** 1/5   **61.** 9/56
63. 1/10   **65.** 2/3   **67.** 3/20   **69.** 1/15
71. 24/53   **73.** 3/11   **75.** 7/12   **77.** 9/50
79. 7/156   **81.** 47/240   **83.** 797/1800   **85.** 17/12
87. 41/28   **89.** 19/24   **91.** 23/60   **93.** 4/11
95. 23/42   **97.** 0.455   **99.** −1.045
101. 0.904005   **103.** 3.4715   **105.** 11/200
107. 3/40   **109.** 9/40   **111.** 5/14
113. Yes, between any two distinct rational numbers there is another distinct rational number.
115. 1 cup rice and water, $\frac{3}{8}$ tsp salt, $1\frac{1}{2}$ tsp butter
117. $7\frac{11}{16}$ ft.   **119.** $7\frac{11}{16}$ ft.

1. Rational   **3.** Rational   **5.** Irrational
7. Rational   **9.** Rational   **11.** Irrational
13. Irrational   **15.** 6   **17.** 3   **19.** −13
21. −9   **23.** 14   **25.** Rational, integer, natural
27. Rational   **29.** Rational   **31.** Rational
33. Irrational   **35.** Irrational   **37.** Rational
39. 4   **41.** $2\sqrt{6}$ **43.** 9   **45.** $4\sqrt{2}$ **47.** $5\sqrt{3}$
49. $6\sqrt{2}$ **51.** $5\sqrt{3}$ **53.** $15\sqrt{7}$ **55.** $11\sqrt{2}$
57. $-12\sqrt{3} - 3$ **59.** $-25\sqrt{3} + 8$ **61.** $\sqrt{6}$
63. $4\sqrt{2}$ **65.** $6\sqrt{2}$ **67.** $\sqrt{2}$ **69.** 3   **71.** $2\sqrt{3}$
73. $3\sqrt{2}/2$ **75.** $\sqrt{14}/2$ **77.** 2   **79.** $3\sqrt{2}/2$
81. $\sqrt{15}/3$
83. a) The sum of the areas of the two smaller squares equals the area of the larger square.
   b) The sum of the areas of the two smaller semicircles equals the area of the larger semicircle.
   c) When similar figures are constructed on the two legs and hypotenuse of a right triangle, the sum of the areas of the figures on the legs will equal the area of the figure on the hypotenuse.

EXERCISES 5.5

1. Yes **3.** Yes **5.** Yes **7.** Yes **9.** Yes **11.** Yes
13. No; for example, $\sqrt{2} + (-\sqrt{2}) = 0$ **15.** No
17. Yes **19.** Yes
21. Commutative property; $4 + 5 = 5 + 4$ is the commutative property of addition.
23. $(-1)(-2) = (-2)(-1)$ **25.** No; $1 - 2 \neq 2 - 1$
$\qquad 2 = 2 \qquad\qquad\qquad -1 \neq 1$
27. No; $4 \div 2 \neq 2 \div 4$
$\qquad 2 \neq 1/2$
29. $[(-1)(-2)](-3) = (-1)[(-2)(-3)]$
$\qquad\qquad -6 = -6$
31. No; $(4 \div 2) \div 2 \neq 4 \div (2 \div 2)$
$\qquad\qquad 1 \neq 4$
33. No; $(2 \div 1) \div 2 \neq 2 \div (1 \div 2)$
$\qquad\qquad 1 \neq 4$
35. Associative property of addition
37. Associative property of addition
39. Associative property of multiplication
41. Commutative property of addition
43. Commutative property of addition
45. Distributive property of multiplication over addition
47. Commutative property of addition
49. Commutative property of multiplication
51. $4x + 20$ **53.** $2\sqrt{3} + 3\sqrt{2}$ **55.** $x\sqrt{3} + 3$
57. Distributive property **59.** Distributive property
61. Commutative property of addition
63. Distributive property
65. Associative property of addition
67. No **69.** Yes **71.** No **73.** No **75.** No

EXERCISES 5.6

1. 16 **3.** 4 **5.** 1 **7.** 25/36 **9.** 576 **11.** 27
13. 1/8 **15.** 1 **17.** 16 **19.** 1/4 **21.** $3^6 = 729$
23. 25 **25.** $1/6^3 = 1/216$ **27.** $3^4 = 81$ **29.** 3
31. 5.5×10^4 **33.** 9.0×10^2 **35.** 5.3×10^{-2}
37. 1.9×10^4 **39.** 1.86×10^{-6} **41.** 4.23×10^{-6}
43. 1.07×10^2 **45.** 1.53×10^{-1} **47.** 3100
49. 60,000,000 **51.** 0.0000213 **53.** 0.312
55. 9,000,000 **57.** 231 **59.** 35,000 **61.** 10,000
63. 120,000,000 **65.** 0.0153 **67.** 320
69. 0.0021 **71.** 20 **73.** 4.2×10^{12}
75. 4.5×10^{-7} **77.** 2.0×10^3 **79.** 2.0×10^{-7}
81. 3.0×10^8

83. $9.2 \times 10^{-5}, 1.3 \times 10^{-1}, 8.4 \times 10^3, 6.2 \times 10^4$
85. 3.2×10^7 sec **87.** 8.64×10^9 cu ft

EXERCISES 5.7

1. 4, 7, 10, 13, 16 **3.** $-4, 1, 6, 11, 16$
5. 6, 4, 2, 0, -2 **7.** 24, 16, 8, 0, -8 **9.** $\frac{1}{2}, 1, \frac{3}{2}, 2, \frac{5}{2}$
11. 36 **13.** 52 **15.** -138 **17.** $69\frac{1}{2}$ or $\frac{139}{2}$
19. $a_n = 4n$ **21.** $a_n = 10n - 2$ **23.** $a_n = n - \frac{7}{2}$
25. $a_n = \frac{5}{2}n - \frac{15}{2}$ **27.** $a_n = -\frac{1}{2}n - \frac{5}{2}$ **29.** 78
31. 110 **33.** -36 **35.** $-\frac{171}{2}$ **37.** -35
39. 3, 6, 12, 24, 48 **41.** 5, $-10, 20, -40, 80$
43. $-2, 2, -2, 2, -2$ **45.** $-2, -1, -\frac{1}{2}, -\frac{1}{4}, -\frac{1}{8}$
47. $5, 3, \frac{9}{5}, \frac{27}{25}, \frac{81}{125}$ **49.** 486 **51.** 48 **53.** 216
55. $\frac{3}{4}$ **57.** 8 **59.** $a_n = 2(2)^{n-1}$ **61.** $a_n = 1(3)^{n-1}$
63. $a_n = -8\left(\frac{1}{2}\right)^{n-1}$ **65.** $a_n = -6(-3)^{n-1}$
67. $a_n = -4\left(\frac{2}{3}\right)^{n-1}$ **69.** $S_4 = 30$ **71.** $S_5 = 2343$
73. $S_5 = 484$ **75.** $S_5 = -33$ **77.** $S_{50} = 2550$
79. $S_{20} = 630$ **81.** a) $a_{12} = 63$ in.; b) $S_{12} = 954$ in.
83. $S_{31} = 496$ **85.** $45,218.08 **87.** 12.288 ft
89. a) 21, 34; b) 1.619 **91.** 161.4375
93. $r = 3, a_1 = 8$ **95.** 376 pairs

REVIEW EXERCISES

1. 2, 3, 4, 6, 8, 12 **2.** 2, 3, 4, 6, 9, 12 **3.** $2^6 \cdot 3$
4. $2^4 \cdot 3 \cdot 5$ **5.** $2^2 \cdot 3^2 \cdot 5$ **6.** $2^2 \cdot 3^2 \cdot 5 \cdot 7$
7. $2^6 \cdot 3 \cdot 5$ **8.** 4, 160 **9.** 5, 240 **10.** 4, 7992
11. 40, 6720 **12.** 4, 480 **13.** 36, 432 **14.** 13/15
15. 1/10 **16.** 41/36 **17.** 28/45 **18.** 35/54
19. 53/28 **20.** 1/6 **21.** 13/40 **22.** 8/15 **23.** 7
24. 2 **25.** -2 **26.** -5 **27.** -7 **28.** 1 **29.** 4
30. 4 **31.** -6 **32.** 12 **33.** -15 **34.** -24
35. 2 **36.** -4 **37.** 6 **38.** 6 **39.** 3 **40.** 0.625
41. 0.8 **42.** 0.75 **43.** 3.75 **44.** $0.\overline{428571}$
45. $0.41\overline{6}$ **46.** 0.375 **47.** 0.875 **48.** $0.2\overline{85714}$
49. 624/1000 = 78/125 **50.** 2/3 **51.** 243/100
52. 61/33 **53.** 12083/1000 **54.** 21/5,000
55. 97/45 **56.** 232/99 **57.** 2506/495
58. 0.00425 **59.** 27/40 **60.** 2.4065 **61.** 19/24
62. $0.5\overline{05}$ **63.** $0.26\overline{6}$ **64.** $2\sqrt{3}$ **65.** $6\sqrt{2}$
66. $4\sqrt{2}$ **67.** $-3\sqrt{3}$ **68.** $8\sqrt{2}$ **69.** $-20\sqrt{3}$
70. $8\sqrt{3}$ **71.** $3\sqrt{2}$ **72.** $4\sqrt{3}$ **73.** 3 **74.** $2\sqrt{7}$
75. $(3\sqrt{2})/2$ **76.** $\sqrt{15}/5$ **77.** $15 + 5\sqrt{5}$
78. $4\sqrt{3} + 3\sqrt{2}$
79. Commutative property of addition

80. Commutative property of multiplication
81. Associative property of addition
82. Distributive property
83. Commutative property of addition
84. Commutative property of addition
85. Associative property of multiplication .
86. Commutative property of multiplication
87. Distributive property
88. Commutative property of multiplication
89. Closed   **90.** Closed   **91.** No   **92.** Closed
93. No   **94.** No   **95.** 8   **96.** $\frac{1}{9}$ **97.** 5   **98.** 125
99. 1   **100.** $\frac{1}{125}$ **101.** $2^6 = 64$ **102.** $3^4 = 81$
103. 3.2×10^3 **104.** 4.23×10^{-5}
105. 1.68×10^{-3} **106.** 4.95×10^6 **107.** 420
108. 0.0000387   **109.** 0.000175   **110.** 100,000
111. 6.4×10^2 **112.** 1.38×10^5 **113.** 2.1×10^1
114. 3.0×10^0 **115.** 33,600,000,000   **116.** 22.5
117. 3200   **118.** 5   **119.** 25   **120.** Arithmetic, 23, 28
121. Geometric, $-324, 972$
122. Arithmetic, $-16, -20$
123. Geometric, 1/16, 1/32   **124.** Arithmetic, 16, 19
125. Geometric, $-2, 2$ **126.** 4, 9, 14, 19, 24

127. $-6, -9, -12, -15, -18$
128. $-\frac{1}{2}, -\frac{5}{2}, -\frac{9}{2}, -\frac{13}{2}, -\frac{17}{2}$ **129.** 4, 8, 16, 32, 64
130. 16, 8, 4, 2, 1   **131.** $\frac{1}{2}, -\frac{1}{4}, \frac{1}{8}, -\frac{1}{16}, \frac{1}{32}$ **132.** 14
133. -36 **134.** 92   **135.** 216   **136.** $\frac{1}{4}$ **137.** -48
138. -84 **139.** -25 **140.** 632   **141.** 52
142. 170   **143.** 33   **144.** Arithmetic, $a_n = -3n + 10$
145. Arithmetic, $a_n = 5n - 5$
146. Arithmetic, $a_n = -\frac{3}{2}n + \frac{11}{2}$
147. Geometric, $a_n = 3(2)^{n-1}$
148. Geometric, $a_n = 4(-1)^{n-1}$
149. Geometric, $a_n = 5(\frac{1}{3})^{n-1}$

CHAPTER 5 TEST

1. 2, 3, 4, 6, 8, 9, 12   **2.** $2^3 \cdot 3^2 \cdot 5$ **3.** -6 **4.** -22
5. -50 **6.** 0.4355   **7.** $\frac{17}{144}$ **8.** 0.375   **9.** $\frac{49}{20}$
10. $\frac{121}{240}$ **11.** $\frac{19}{30}$ **12.** $8\sqrt{2}$ **13.** $\frac{\sqrt{15}}{3}$
14. Yes, the product of any two integers is an integer.
15. Associative property of addition
16. Distributive property   **17.** 64   **18.** 32   **19.** $\frac{1}{49}$
20. 8.0×10^6 **21.** $a_n = -4n$ **22.** -187
23. 243   **24.** 1023   **25.** $a_n = 3(2)^{n-1}$

CHAPTER 6

EXERCISES 6.1

1. 66   **3.** 6   **5.** 88   **7.** 26   **9.** 23   **11.** 346
13. 9   **15.** -49 **17.** 125   **19.** 100   **21.** 0
23. -48 **25.** 27   **27.** 7   **29.** 20   **31.** 10
33. $-\frac{83}{18}$ **35.** $\frac{5}{2}$ **37.** 0   **39.** 0   **41.** Yes   **43.** Yes
45. No   **47.** Yes   **49.** Yes

EXERCISES 6.2

1. $-x$ **3.** $4x - 7$ **5.** $5x + 3y + 3$ **7.** $-8x + 2$
9. $-5x + 3$ **11.** $r = 8$ **13.** $x = 2$ **15.** $x = -1$
17. $x = 2/3$ **19.** $x = -2$ **21.** $t = 3$ **23.** $x = 9$
25. $x = -15$ **27.** No solution   **29.** All real numbers
31. $x = 56/11$ **33.** $x = 3$ **35.** 20   **37.** 120
39. 10   **41.** 108   **43.** 240   **45.** $33.3\overline{3}$ **47.** 8
49. 50   **51.** $y = -\frac{3}{2}x + \frac{5}{2}$ **53.** $y = -\frac{4}{7}x + 2$
55. $y = \frac{2}{3}x + 2$ **57.** $y = -\frac{4}{3}x + \frac{2}{3}z + \frac{4}{3}$
59. $y = \frac{2}{3}x + \frac{5}{3}z$

EXERCISES 6.3

1. $11 - 7x$ **3.** $6y + 5$ **5.** $4r - 11$
7. $2(5 - x)$ **9.** $(6 + x)/7$ **11.** $(7y - 5) + 11$
13. $x + 11 = 19, x = 8$ **15.** $x/7 = 3, x = 21$
17. $11 + 2x = 9, x = -1$ **19.** $3x + 4 = 16, x = 4$
21. $x + 11 = 3x + 1, x = 5$
23. $x + 3 = 5(x + 7), x = -8$ **25.** 4 5/6 cups
27. 888 miles   **29.** $6.40 per hour
31. a) 11 shares of stock B; 66 shares of stock A;
   b) $353
33. $12,225   **35.** $w = 4\frac{2}{3}$ ft, $h = 6\frac{2}{3}$ ft
37. 6.42 months   **39.** $1000   **41.** 250 mi
43. 75,669 mi

EXERCISES 6.4

1.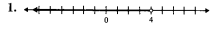

3. 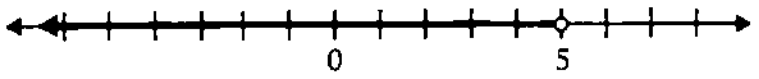

5. 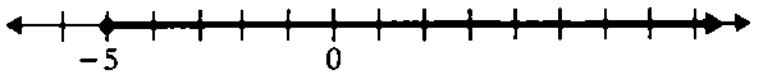

7. 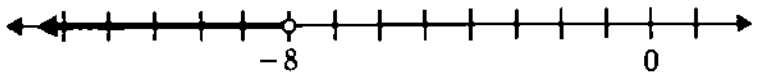

9. 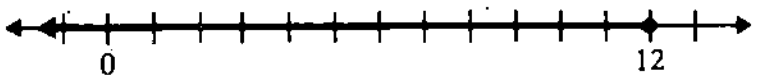

11. 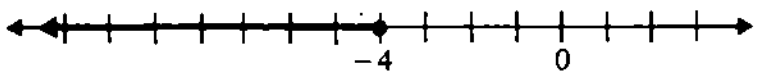

13. 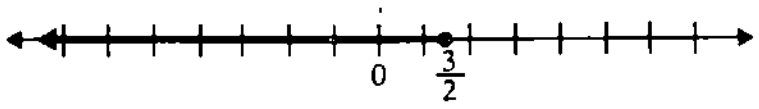

15. 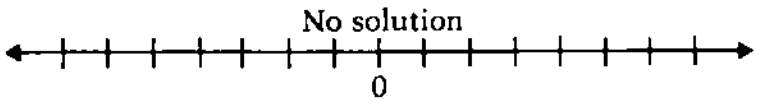

17. 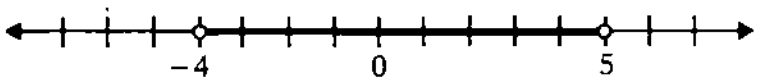

19. 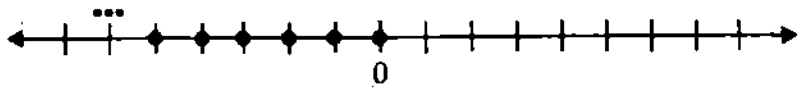

21. 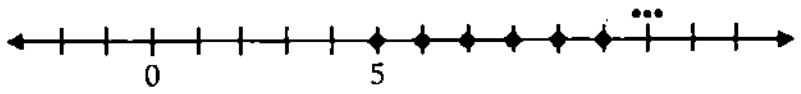

23. 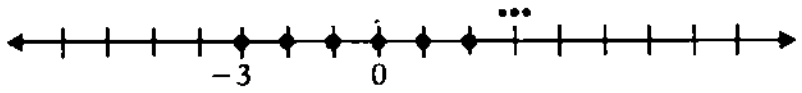

25. 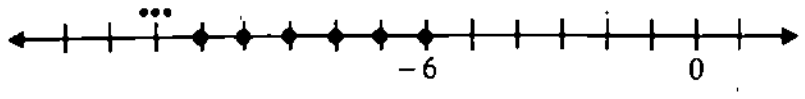

27. 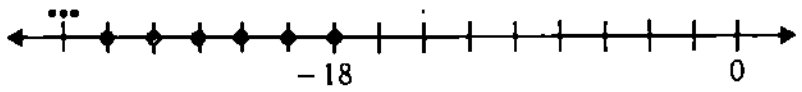

29. 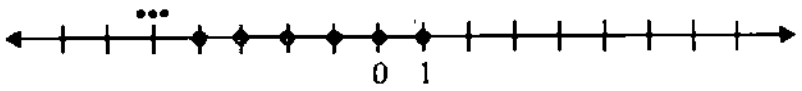

31. 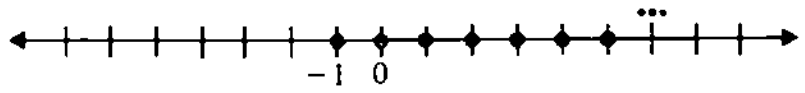

33. 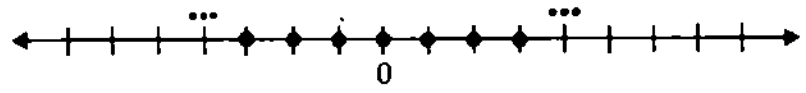

35. 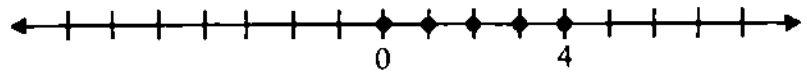

37. 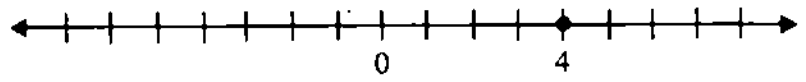

39. **a)** $94 \leq x \leq 100$, assuming 100 is the maximum score;
b) $44 \leq x < 94$
41. **a)** $400 + 670 + x \leq 1500$;
b) $x \leq 430$ pounds
43. $5800
45. $x > 600$

EXERCISES 6.5

1. 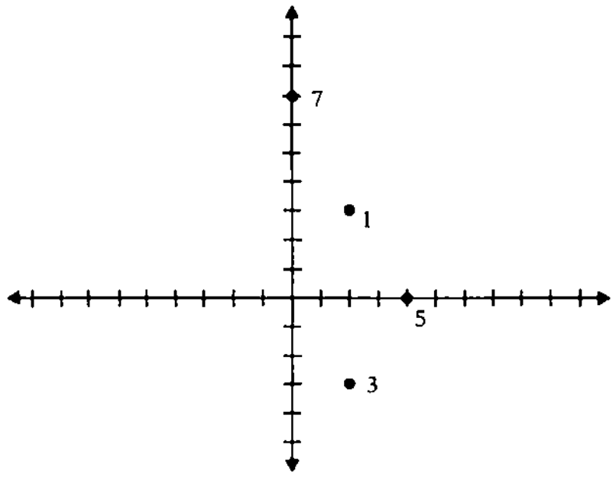

3. 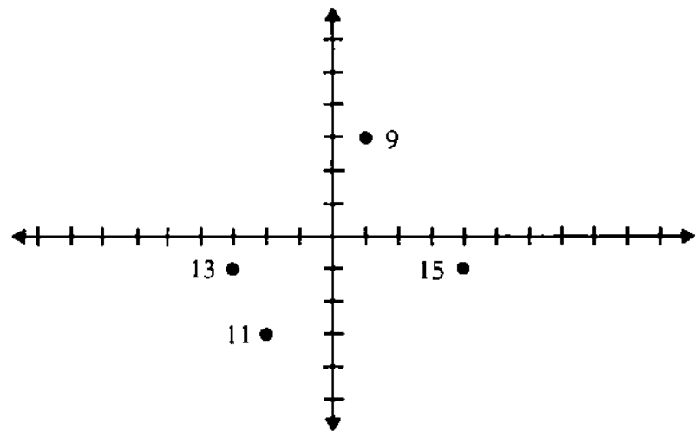

17. $A(4, 3)$, $B(2, 2)$, $C(-2, 4)$, $D(0, 2)$, $E(-3, 1)$, $F(-2, 0)$, $G(-5, -3)$, $H(0, -3)$, $I(2, -3)$, $J(3, -2)$, $K(4, 0)$
19. $D(2, 5)$, $A = 10$ square units
21. $(-1, 2)$ or $(7, 2)$
23. $b = -4$ **25.** $b = 3$

EXERCISES 6.6

1. $(3, -2)$, $(0, 0)$ **3.** $(0, 5)$, $(-10/3, 0)$
5. $(0, 4/3)$, $(8, 0)$ **7.** $(0, 1)$, $(-4, -2)$

9. 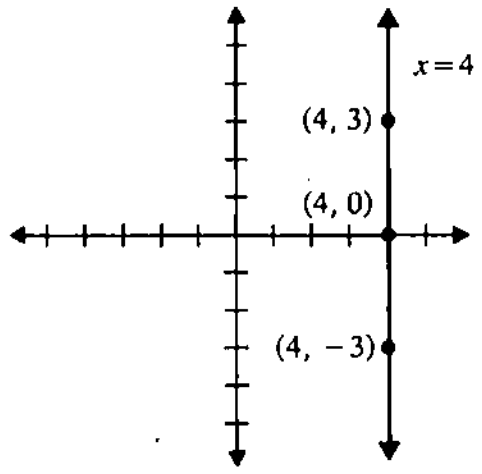

11. 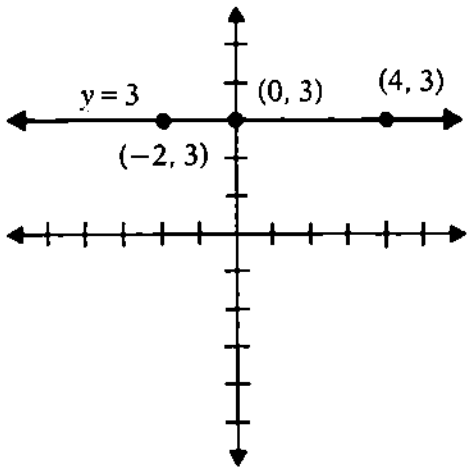

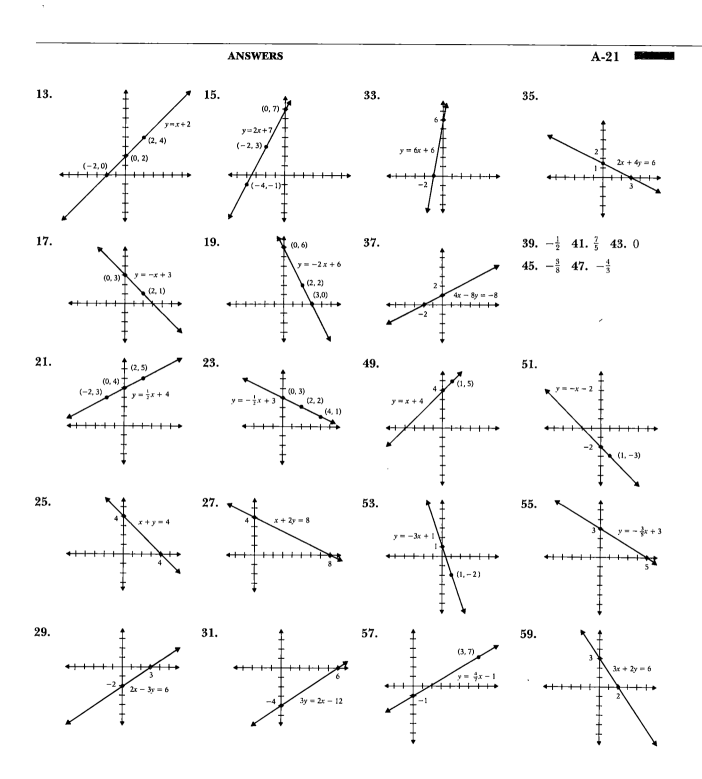

39. $-\frac{1}{2}$ **41.** $\frac{7}{5}$ **43.** 0

45. $-\frac{3}{8}$ **47.** $-\frac{4}{3}$

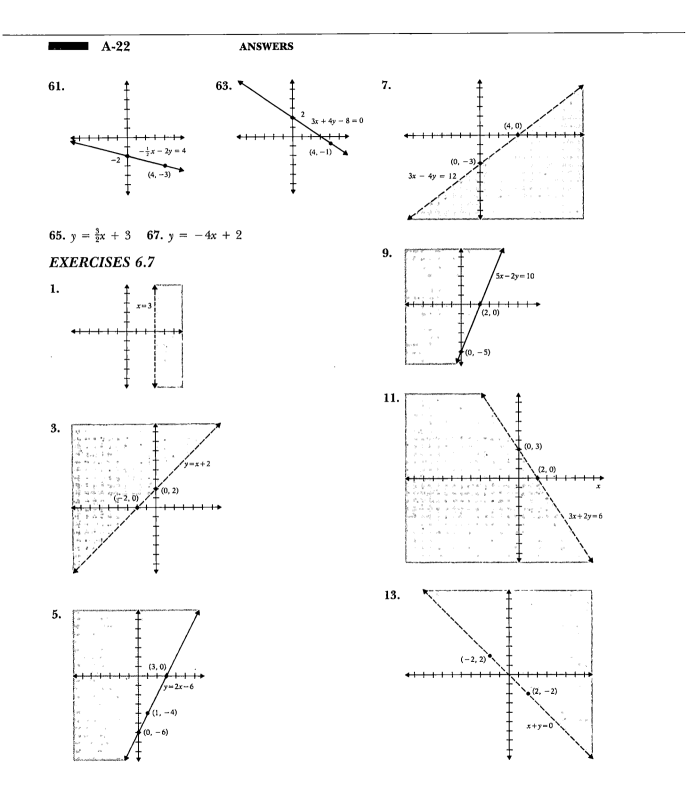

61. $-\frac{1}{2}x - 2y = 4$ $(4, -3)$ -2

63. $3x + 4y - 8 = 0$ 2 $(4, -1)$

7. $(4, 0)$ $(0, -3)$ $3x - 4y = 12$

65. $y = \frac{3}{2}x + 3$ **67.** $y = -4x + 2$

EXERCISES 6.7

1. $x = 3$

9. $5x - 2y = 10$ $(2, 0)$ $(0, -5)$

3. $y = x + 2$ $(-2, 0)$ $(0, 2)$

11. $(0, 3)$ $(2, 0)$ x $3x + 2y = 6$

5. $(3, 0)$ $y = 2x - 6$ $(1, -4)$ $(0, -6)$

13. $(-2, 2)$ $(2, -2)$ $x + y = 0$

15.

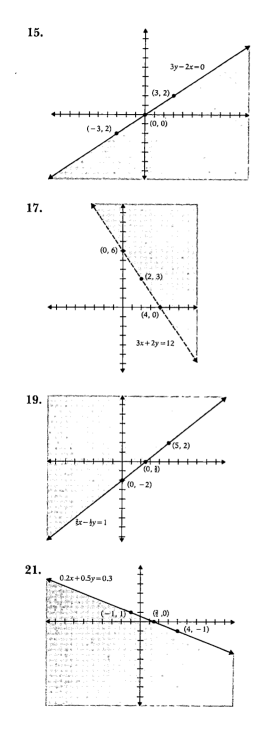

17.

19.

21.

23. a) $2l + 2w \leqslant 30$

b)

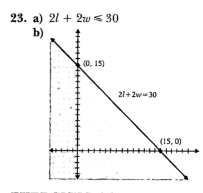

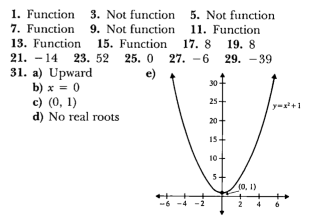

EXERCISES 6.8

1. $(x + 2)(x + 3)$ **3.** $(x + 4)(x - 3)$
5. $(x - 4)(x - 2)$ **7.** $(x - 1)(x - 1)$ or $(x - 1)^2$
9. $(x - 5)(x + 5)$ **11.** $(x - 6)(x + 5)$
13. $(x + 8)(x - 4)$ **15.** $(x - 10)(x - 4)$
17. $(2x - 1)(x + 2)$ **19.** $(2x + 1)(2x + 1)$ or $(2x + 1)^2$
21. $(5x + 2)(x + 2)$ **23.** $(5x + 3)(x - 2)$
25. $(2x - 5)(x - 6)$ **27.** $(3x + 4)(x - 6)$ **29.** $3, -4$
31. $-\frac{5}{2}, \frac{3}{4}$ **33.** $-1, -2$ **35.** $2, -4$ **37.** $1, 6$
39. $1, 3$ **41.** $4, 7$ **43.** $-2, -10$ **45.** $1, -\frac{3}{2}$
47. $-\frac{1}{5}, -2$ **49.** $1, \frac{1}{3}$ **51.** $2, \frac{1}{4}$ **53.** $7, -3$
55. $-3, -5$ **57.** $-1, 9$ **59.** No real solution
61. $-1, \frac{3}{2}$ **63.** No real solution   **65.** $\dfrac{5 \pm \sqrt{73}}{8}$
67. $\dfrac{9 \pm \sqrt{21}}{6}$ **69.** $1, \frac{7}{3}$ **71.** $\dfrac{-3 \pm \sqrt{15}}{2}$
73. a) Two; **b)** One; **c)** None

EXERCISES 6.9

1. Function   **3.** Not function   **5.** Not function
7. Function   **9.** Not function   **11.** Function
13. Function   **15.** Function   **17.** 8   **19.** 8
21. -14 **23.** 52   **25.** 0   **27.** -6 **29.** -39
31. a) Upward   **e)**
b) $x = 0$
c) $(0, 1)$
d) No real roots

33. a) Upward
   b) $x = 0$
   c) $(0, -25)$
   d) $5, -5$

e)

35. a) Downward
   b) $x = 0$
   c) $(0, -9)$
   d) No real
       roots

e)

37. a) Downward
   b) $x = 0$
   c) $(0, 2)$
   d) $\sqrt{2}, -\sqrt{2}$

e)

39. a) Upward
   b) $x = 0$
   c) $(0, -3)$
   d) $\dfrac{\sqrt{6}}{2}, \dfrac{-\sqrt{6}}{2}$

e)

41. a) Upward
   b) $x = 1$
   c) $(1, -7)$
   d) $1 + \sqrt{7},$
       $1 - \sqrt{7}$

e)

43. a) Upward
   b) $x = -\frac{5}{2}$
   c) $(-2.5, -0.25)$
   d) $-2, -3$

e)

45. a) Downward
   b) $x = 2$
   c) $(2, -2)$
   d) No real roots

e)

47. a) Upward
   b) $x = 2$
   c) $(2, -14)$
   d) $\dfrac{6 + \sqrt{42}}{3}, \dfrac{6 - \sqrt{42}}{3}$

e)

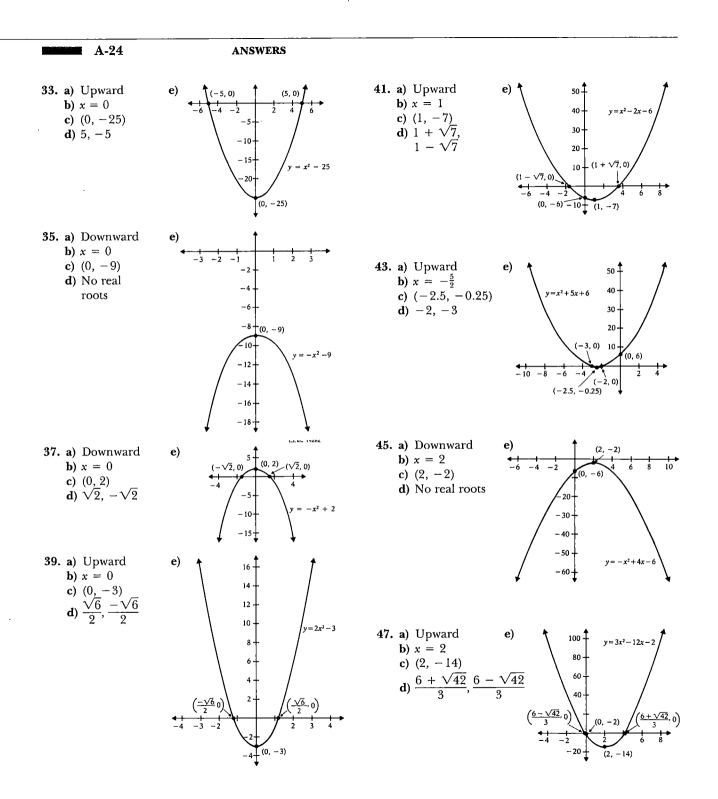

49. a) Downward

b) $x = -\frac{5}{6}$

c) $\left(-\frac{5}{6}, \frac{49}{12}\right)$

d) $-2, \frac{1}{3}$

e)

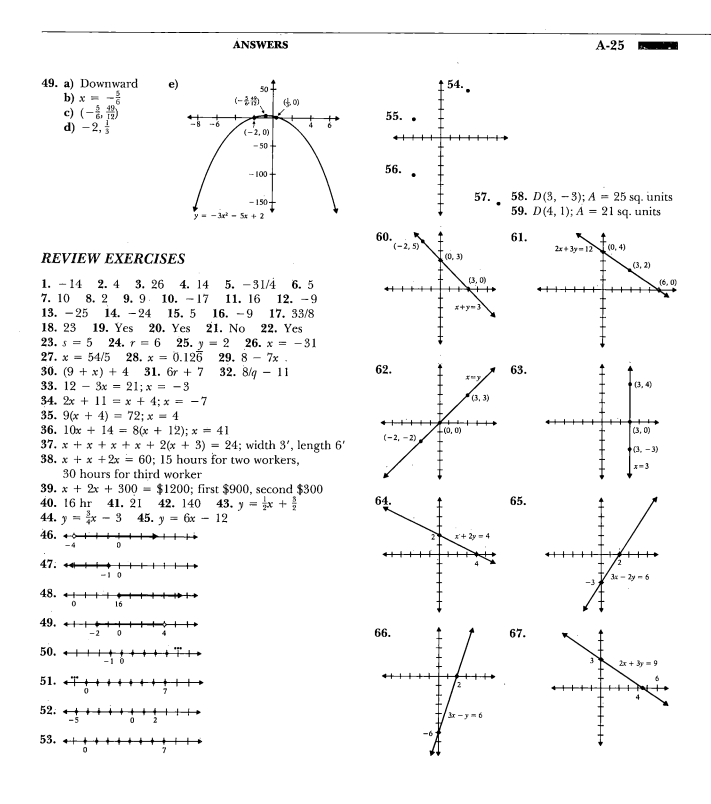

REVIEW EXERCISES

1. -14 **2.** 4 **3.** 26 **4.** 14 **5.** $-31/4$ **6.** 5

7. 10 **8.** 2 **9.** 9 **10.** -17 **11.** 16 **12.** -9

13. -25 **14.** -24 **15.** 5 **16.** -9 **17.** $33/8$

18. 23 **19.** Yes   **20.** Yes   **21.** No   **22.** Yes

23. $s = 5$ **24.** $r = 6$ **25.** $y = 2$ **26.** $x = -31$

27. $x = 54/5$ **28.** $x = 0.12\overline{6}$ **29.** $8 - 7x$

30. $(9 + x) + 4$ **31.** $6r + 7$ **32.** $8/q - 11$

33. $12 - 3x = 21; x = -3$

34. $2x + 11 = x + 4; x = -7$

35. $9(x + 4) = 72; x = 4$

36. $10x + 14 = 8(x + 12); x = 41$

37. $x + x + x + x + 2(x + 3) = 24$; width $3'$, length $6'$

38. $x + x + 2x = 60$; 15 hours for two workers, 30 hours for third worker

39. $x + 2x + 300 = \$1200$; first \$900, second \$300

40. 16 hr   **41.** 21 **42.** 140 **43.** $y = \frac{1}{2}x + \frac{3}{2}$

44. $y = \frac{3}{4}x - 3$ **45.** $y = 6x - 12$

46.

47.

48.

49.

50.

51.

52.

53.

54.

55.

56.

57.

58. $D(3, -3); A = 25$ sq. units

59. $D(4, 1); A = 21$ sq. units

60.

61.

62.

63.

64.

65.

66.

67.

68. $-\frac{1}{4}$ **69.** $-\frac{5}{2}$ **70.** $\frac{7}{6}$ **71.** $\frac{1}{3}$

72.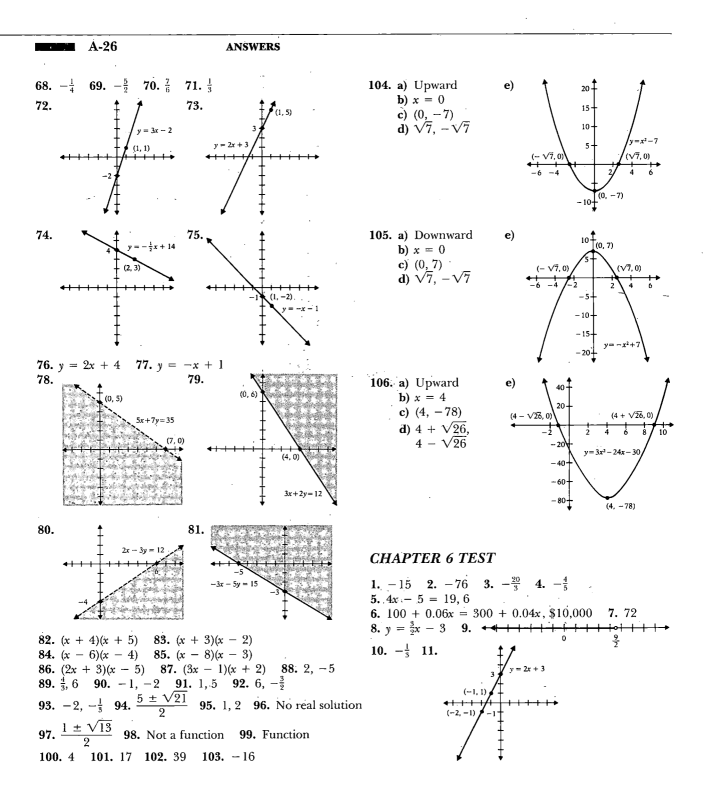

73.

74.

75.

76. $y = 2x + 4$ **77.** $y = -x + 1$

78.

79.

80.

81.

82. $(x + 4)(x + 5)$ **83.** $(x + 3)(x - 2)$

84. $(x - 6)(x - 4)$ **85.** $(x - 8)(x - 3)$

86. $(2x + 3)(x - 5)$ **87.** $(3x - 1)(x + 2)$ **88.** $2, -5$

89. $\frac{4}{3}, 6$ **90.** $-1, -2$ **91.** $1, 5$ **92.** $6, -\frac{3}{2}$

93. $-2, -\frac{1}{3}$ **94.** $\dfrac{5 \pm \sqrt{21}}{2}$ **95.** $1, 2$ **96.** No real solution

97. $\dfrac{1 \pm \sqrt{13}}{2}$ **98.** Not a function   **99.** Function

100. 4 **101.** 17 **102.** 39 **103.** -16

104. a) Upward
     b) $x = 0$
     c) $(0, -7)$
     d) $\sqrt{7}, -\sqrt{7}$
     e)

105. a) Downward
     b) $x = 0$
     c) $(0, 7)$
     d) $\sqrt{7}, -\sqrt{7}$
     e)

106. a) Upward
     b) $x = 4$
     c) $(4, -78)$
     d) $4 + \sqrt{26}$, $4 - \sqrt{26}$
     e)

CHAPTER 6 TEST

1. -15 **2.** -76 **3.** $-\frac{20}{3}$ **4.** $-\frac{4}{5}$

5. $4x - 5 = 19, 6$

6. $100 + 0.06x = 300 + 0.04x$, \$10,000   **7.** 72

8. $y = \frac{3}{2}x - 3$ **9.**

10. $-\frac{1}{3}$ **11.**

12.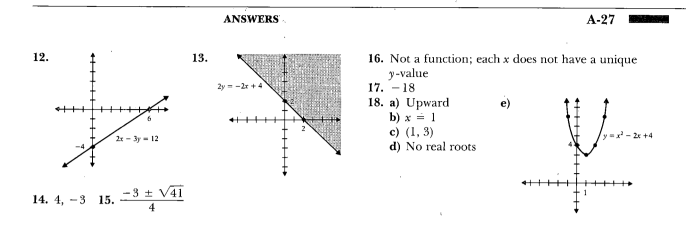

13.

$2y = -2x + 4$

16. Not a function; each x does not have a unique y-value

17. -18

18. a) Upward **e)**
  b) $x = 1$
  c) $(1, 3)$
  d) No real roots

$y = x^2 - 2x + 4$

14. $4, -3$ **15.** $\dfrac{-3 \pm \sqrt{41}}{4}$

CHAPTER 7

EXERCISES 7.1

1. $y = 5$; $(-4, 5)$, $(0, 5)$, $(4, 5)$; $x = 4$; $(4, 0)$; $(4, -2)$

3. $y = 5$; $(-4, 5)$, $(0, 5)$, $(5, 5)$; $(-4, 0)$; $x = -4$; $(-4, -4)$

13. $x - 2y = 3$; $(-5, -2)$, $(-3, -3)$, $(3, 0)$, $(1, -1)$, $(0, -4\frac{1}{2})$; $x + 2y = -9$

15. $x = 1$; $(1, 3)$; $(-3, 0)$, $(1, 0)$; $x + y + 3 = 0$; $(0, -3)$; $(1, -4)$

5. $y = x + 3$; $(0, 3)$; $(-3, 0)$; $(-5, 0)$; $(-4, -1)$; $y = -x - 5$; $(0, -5)$

7. $2x - y = 6$; $x + 2y = -2$; $(3, 0)$; $(-2, 0)$; $(2, -2)$; $(6, -4)$; $(0, -6)$

17. $(1, 3)$; $(-1, 0)$; $3x - 2y = 6$; $2y - 3x = 3$; $(2, 0)$; $(0, -3)$; Inconsistent

19. $(6, 0)$; $x - 3y = 6$; $y = \frac{1}{3}x - 2$; $(0, -2)$; Dependent

9. $y - x = 3$; $(0, 3)$; $(-3, 0)$; $(4, 0)$; $x - y = 4$; $(0, -4)$; Inconsistent

11. $(-3, 6)$; $2x + y = 0$; $y = 2x - 4$; $(0, 0)$; $(2, 0)$; $(1, -2)$; $(0, -4)$

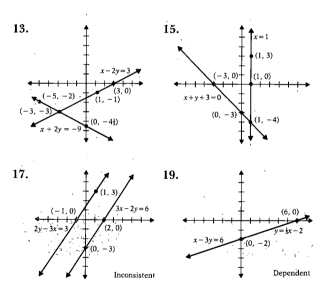

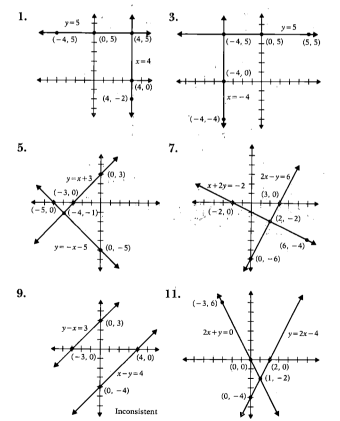

21. a) One solution; the lines will intersect at exactly one point.
  b) No solution; the two lines are parallel and will not intersect.
  c) An infinite number of solutions; both equations represent the same line, and any point on the line satisfies both equations.

23. No solution   **25.** One solution

27. Infinite number of solutions   **29.** No solution

31. Infinite number of solutions
33. Not perpendicular　　**35.** Perpendicular
37. a) $c = 10,000 + 3000n$　**39. a)** $s = 0.05d$
　　　$c = 5,000 + 4000n$　　　　　$s = 150 + 0.02d$

b)

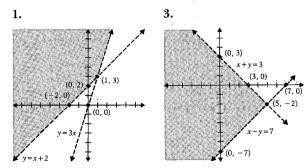

Number of terminals　　　Dollar sales ($1000s)

c) 5 terminals　　　　**c)** $5,000
　　　　　　　　　　　　d) Option 1

EXERCISES 7.2

1. Inconsistent　**3.** $(0, -3)$　**5.** Inconsistent
7. $(-2, 1)$　**9.** $(-3, -3)$　**11.** $(-4, 1)$　**13.** $(\frac{11}{2}, 19)$
15. $(\frac{2}{5}, -\frac{1}{5})$　**17.** No solution, inconsistent
19. $(3, -2)$　**21.** $(2, 2)$　**23.** $(-2, 0)$　**25.** $(-1, 8)$
27. $(-1, -2)$　**29.** $(2, 7)$
31. No solution, inconsistent　**33.** $(6, -5)$
35. $c = 30 + 0.15m$, 200 mi
　　　$c = 36 + 0.12m$
37. $s = 200 + 0.05d$, $2000
　　　$s = 0.15d$
39. $x + y = 100$, 70 g of 5%, 30 g of skim milk
　　　$0.05x + 0.00y = 100(0.035)$
41. $0.10A + 0.20B = 20$, 80 g of A, 60 g of B
　　　$0.06A + 0.02B = 6$

EXERCISES 7.3

3. $\begin{bmatrix} -2 & 1 \\ 6 & 5 \end{bmatrix}$　**5.** $\begin{bmatrix} 2 & 4 \\ 3 & 9 \\ 4 & 8 \end{bmatrix}$　**7.** $\begin{bmatrix} 3 & 11 \\ 6 & -6 \end{bmatrix}$　**9.** $\begin{bmatrix} 1 & 0 & -7 \\ 9 & 8 & -7 \\ 6 & 1 & -9 \end{bmatrix}$

11. $\begin{bmatrix} 8 & 14 \\ 0 & 4 \end{bmatrix}$　**13.** $\begin{bmatrix} 2 & 23 \\ 12 & 4 \end{bmatrix}$　**15.** $\begin{bmatrix} 16 & 15 \\ -8 & 6 \end{bmatrix}$

17. $\begin{bmatrix} 8 & 13 \\ 24 & 18 \end{bmatrix}$　**19.** $\begin{bmatrix} 15 \\ 22 \end{bmatrix}$　**21.** $\begin{bmatrix} 1 & 0 \\ 0 & 1 \end{bmatrix}$

23. Matrices A and B do not have the same dimensions
　and cannot be added. $A \times B = \begin{bmatrix} 6 & -6 & 9 \\ 10 & -4 & 18 \end{bmatrix}$

25. Cannot be added, $A \times B = \begin{bmatrix} 26 & 38 \\ 24 & 24 \end{bmatrix}$

27. Cannot be added, $A \times B = \begin{bmatrix} 17 \\ 39 \end{bmatrix}$

29. $A + B = B + A = \begin{bmatrix} 10 & 8 \\ 5 & 5 \end{bmatrix}$

31. $A + B = B + A = \begin{bmatrix} 8 & 0 \\ 6 & -8 \end{bmatrix}$

33. $(A + B) + C = A + (B + C) = \begin{bmatrix} 9 & 0 \\ 7 & 10 \end{bmatrix}$

35. $(A + B) + C = A + (B + C) = \begin{bmatrix} 5 & 5 \\ 7 & -4 \end{bmatrix}$

37. No　**39.** Yes　**41.** Yes

43. $(A \times B) \times C = A \times (B \times C) = \begin{bmatrix} 25 & 11 \\ 68 & 28 \end{bmatrix}$

45. $(A \times B) \times C = A \times (B \times C) = \begin{bmatrix} 16 & -10 \\ -24 & 2 \end{bmatrix}$

47. $(A \times B) \times C = A \times (B \times C) = \begin{bmatrix} 17 & 0 \\ -7 & 0 \end{bmatrix}$

49. Large Small　**51.** $[2546 \quad 3358]$
　　$\begin{bmatrix} 38 & 50 \\ 56 & 72 \\ 17 & 26 \\ 10 & 14 \end{bmatrix}$

53. Yes　**55.** Yes　**57.** Yes

EXERCISES 7.4

1. $(1, 6)$　**3.** $(-2, -1)$　**5.** $(3, 3)$　**7.** $(-2, -1)$
9. $(1, 1)$　**11.** $(2, -7)$　**13.** Cherries: $5 per lb;
　　　　　　　　　　　　　　　　　　mints: $1 per lb

EXERCISES 7.5

1.　　　　　　　　　　**3.**

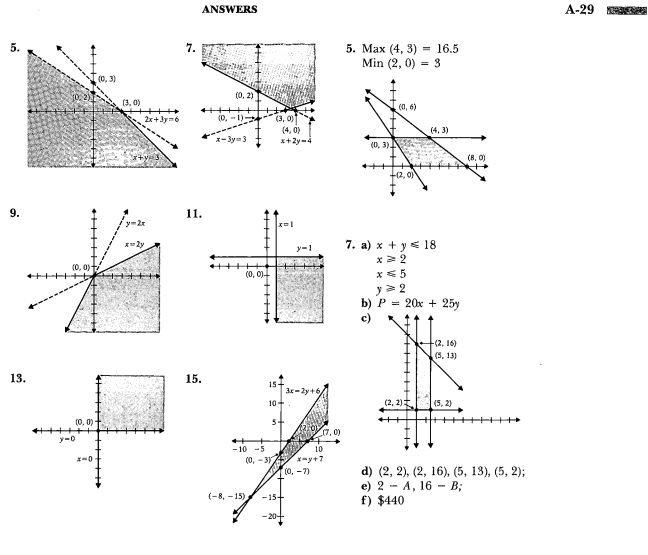

5. Max $(4, 3) = 16.5$
Min $(2, 0) = 3$

7. a) $x + y \leqslant 18$
$x \geqslant 2$
$x \leqslant 5$
$y \geqslant 2$
b) $P = 20x + 25y$
c)

d) $(2, 2), (2, 16), (5, 13), (5, 2);$
e) $2 - A, 16 - B;$
f) \$440

EXERCISES 7.6

1. Max $(3, 2) = 14$
Min $(0, 0) = 0$

3. Max $(3, 1) = 25$
Min $(0, 0) = 0$

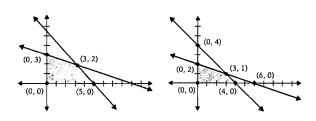

REVIEW EXERCISES

1.

2.

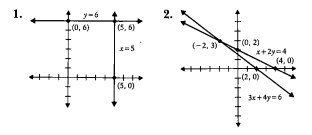

3.
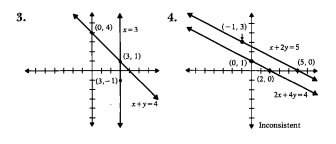

4.

Inconsistent

35. Max (9, 0) = 54

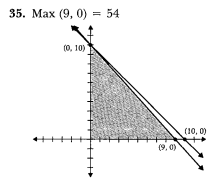

5. One solution   **6.** Infinite number of solutions
7. No solution   **8.** One solution   **9.** (12, −6)
10. (0, −2)   **11.** (−2, −8)
12. No solution, inconsistent   **13.** (9, 3)   **14.** (−5, 3)
15. (−1, −1)   **16.** (2, 0)   **17.** (30, 15)
18. Infinite number of solutions, dependent
19. (0, 2)   **20.** (−2, 2)   **21.** (1, 2)   **22.** (−4, 8)
23. Fixed: $7, mile: $1.50
24. 3 gal of 30%, 2 gal of 80% solution

25. $\begin{bmatrix} 1 & -8 \\ 7 & 7 \end{bmatrix}$ **26.** $\begin{bmatrix} 7 & 4 \\ -1 & 3 \end{bmatrix}$ **27.** $\begin{bmatrix} 12 & -6 \\ 9 & 15 \end{bmatrix}$

28. $\begin{bmatrix} 17 & 14 \\ -6 & 4 \end{bmatrix}$ **29.** $\begin{bmatrix} -20 & -28 \\ 11 & -8 \end{bmatrix}$ **30.** $\begin{bmatrix} -30 & -24 \\ 22 & 2 \end{bmatrix}$

31.   **32.**

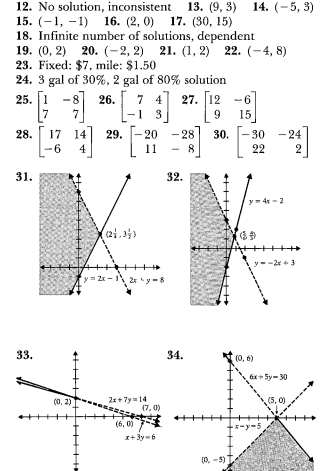

33.   **34.**

CHAPTER 7 TEST

1.

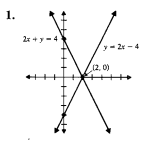

2. One solution   **3.** (3, −2)   **4.** (4, 10)
5. (−1, 5)   **6.** (−1, 3)   **7.** $(\frac{56}{23}, -\frac{1}{23})$

8. (3, −1)   **9.** $\begin{bmatrix} -7 & 7 \\ 7 & -7 \end{bmatrix}$

10. $\begin{bmatrix} -17 & 1 \\ 5 & -13 \end{bmatrix}$ **11.** $\begin{bmatrix} 14 & -34 \\ -23 & 25 \end{bmatrix}$

12.

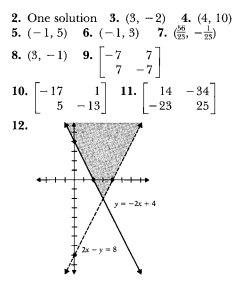

13. 5 lb of each

CHAPTER 8

EXERCISES 8.1

1. 3:4 **3.** 3:7 **5.** 53:44 **7.** 44:53 **9.** $6, $9, $12
11. 27-oz can **13.** 5-lb ham **15.** 24-oz package
17. a) 72-oz size; **b)** 17-oz size
19. a) One-liter container; **b)** one-liter container;
  c) one-liter container
21. 4 **23.** 0.21 **25.** 19.44 **27.** 4.384 **29.** $26.35
31. 19,355 gal **33.** $757.77 **35.** $126.11
37. $52.36 **39.** $199.26 **41.** $113.40 **43.** 0.3 cc
45. 0.625 cc **47.** 375 min or 6 hr 15 min
49. 336 mi **51.** 1.7 ft **53.** 24 tsp
55. $5\frac{1}{3}$ packets Bisquick baking mix, $5\frac{1}{3}$T vegetable oil,
  $2\frac{2}{3}$ eggs, $3\frac{5}{9}$c milk

EXERCISES 8.2

1. 75.0% **3.** 91.7% **5.** 34.8% **7.** 1638.0%
9. 20,034.0% **11.** 0.273 **13.** 0.00003
15. 0.031416 **17.** 0.0075 **19.** 20.2%
21. a) 0.43%; **b)** 38.33%; **c)** 61.23%
23. a) 61.2%; **b)** 105.0% **25.** 6.64% decrease
27. 118.4% markup **29.** 62.5% **31.** 36
33. 35.2% **35.** 68 **37.** 38.9 **39.** 20%
41. $3805.56 **43.** No, for a 15% discount the sales
price should be $97.75.

EXERCISES 8.5

1. $66; **b)** 16.75% **3. a)** $21.75; **b)** 17.0%
5. $35.42
7. a) $480; **b)** $74.25; **c)** 11.0%; **d)** $70.95

EXERCISES 8.3

1. $2 **3.** $15.17 **5.** $15.85 **7.** $27.60
9. $1312.50 **11.** $600 **13.** 1/4 year **15.** 6.5%
17. 13.0% **19. a)** $22.50; **b)** $1522.50
21. a) $135.42; **b)** $2364.58; **c)** 13.74%
23. a) $437.50; **b)** 7 1/4%; **c)** $362.69 **25.** 138 days
27. 136 days **29.** 134 days **31.** October 6
33. October 31 **35.** $1253.61 **37.** $2163.02
39. $712.45 **41.** $424.78 **43.** $1289.10

EXERCISES 8.4

1. $191, $1191 **3.** $412.80, $2812.80
5. $562.50, $3062.50 **7.** $5801.60, $9001.60
9. $247.20, $2247.20 **11.** $252.99, $2252.99
13. $508.80, $2508.80 **15.** $533.54, $2533.54
17. a) $3172;
  b) $3243
19. $3516 **21.** $8556
23. a) $220, $1220;
  b) $486, $1486;
  c) $1191, $2191;
  d) No predictable outcome
27. $7209.56 **29.** 8.84%

9. a)

Payment Number	Balance Due	Interest on Unpaid Balance	Payment Including Interest	Amount Paid on Balance	Unpaid Balance
1	$1046.55	—	$300.00	$300.00	$746.55
2	746.55	$9.71	200.00	190.29	556.26
3	556.26	7.23	200.00	192.77	363.49
4	363.49	4.73	200.00	195.27	168.22
5	168.22	2.19	170.41	168.22	0

  b) $23.86
11. a) $5.13; **b)** $380.13

13. a)

Payment Number	Balance Due	Interest on Unpaid Balance	Payment Including Interest	Amount Paid on Balance	Unpaid Balance
1	$1975.00	—	$350.00	$350.00	$1625.00
2	1625.00	$21.13	300.00	278.87	1346.13
3	1346.13	17.50	300.00	282.50	1063.63
4	1063.63	13.83	300.00	286.17	777.46
5	777.46	10.11	300.00	289.89	487.57
6	487.57	6.34	300.00	293.66	193.91
7	193.91	2.52	196.43	193.91	—

b) $71.43
15. a) $1215; **b)** $221.25; **c)** $8715; **d)** 11.0%

EXERCISES 8.6

1. a) $11,250; **b)** $342.90
3. a) $5800; **b)** $52,200; **c)** $1044
5. a) $601.25; **b)** Does not qualify
7. a) $205,742.40; **b)** $147,742.40; **c)** $13.17
9. a) $19,000; **b)** $924.16; **c)** $407,147.20; **d)** $5.83
11. a) $15,000; **b)** $791.25; **c)** $681.00; **d)** $784.00;
   e) yes; **f)** $6.00; **g)** $301,020; **h)** $226,020
13. a) $399.88; **b)** 8.9%; **c)** 8.1%
15. a) $412.09

b)

Payment	Interest	Principal	Balance on Loan
1	$387.92	$24.17	$48,975.83
2	387.73	24.36	48,951.47
3	387.53	24.56	48,926.91

c) 9.38%

d)

Payment	Interest	Principal	Balance on Loan
4	$382.45	$29.64	$48,897.27
5	382.21	29.88	48,867.39
6	381.98	30.11	48,837.28

e) 9.46%

REVIEW EXERCISES

1. Two 32-oz bottles   **2.** 63.4 gal   **3.** 1.8″
4. 3.64   **5.** 240   **6.** 9.7% decrease   **7.** $5.00
8. $6.75   **9.** $51.00   **10.** $114.75   **11.** $316.25
12. $3791.67   **13. a)** $60.38; **b)** $2360.38
14. a) $690; **b)** $3310; **c)** 13.9% **15. a)** 7 1/2%; **b)** $830;
c) $941.18   **16. a)** $1660; **b)** $4980; **c)** $1593.60;
d) 14.25%   **17. a)** $8; **b)** 14.5%

18. a) Payment Number	Balance Due	Interest on Unpaid Balance	Payment Including Interest	Amount Paid on Balance	Unpaid Balance
1	$788.50	—	$288.50	$288.50	$500.00
2	500.00	$6.50	125.00	118.50	381.50
3	381.50	4.96	125.00	120.04	261.46
4	261.46	3.40	125.00	121.60	139.86
5	139.86	1.82	125.00	123.18	16.68
6	16.68	0.22	16.90	16.68	—

b) 16.90

19. a) $95.50, $595.50; b) $97, $597;
c) $97.81, $597.81
20. a) $25,375; b) $3805.33; c) $951.33; d) $437.32;
e) $662.32; f) yes
21. a) $22,475; b) $772.02; c) $300,402.20; d) $13.49
22. a) $662.29; b) 9.15%; c) 9.35%
23. a) $810.92

b)

Payment	Interest	Principal	Balance on Loan
1	791.67	19.25	75,980.75
2	791.47	19.45	75,961.30
3	791.26	19.66	75,941.64

c) 12.4%

d)

Payment	Interest	Principal	Balance on Loan
4	784.73	26.19	75,915.45
5	784.46	26.46	75,888.99
6	784.19	26.73	75,862.26

e) 12.5%

CHAPTER 8 TEST

1. $698.74   2. 12.1% increase   3. $168.00
4. $14.00   5. $562.50   6. $6562.50   7. $8075.00
8. $2696.20   9. 15.00%   10. $39.00   11. $1013.00
12. $13.17   13. $436.83   14. $2507.32
15. $107.32   16. $I = $2178.00, A = $5778.00
17. $I = $1730.35, A = $6530.35   18. $23,625.00
19. $5770.00   20. $1442.50   21. $707.33
22. $984.41   23. $193,384.20   24. $98,884.20
25. 8.38%

CHAPTER 9

EXERCISES 9.1

1. An operation that can be performed on only two elements
3. $1 - 2 \neq 2 - 1$ 5. $(1 - 2) - 3 \neq 1 - (2 - 3)$
   $-1 \neq 1$ $-4 \neq 2_i$
7. Closed; associative property; identity element; inverses
9. No; no identity element   11. Yes
13. No, not associative
15. No, not closed. For example, $\sqrt{2} \cdot \sqrt{2} = 2$, and 2 is a rational number.
17. No, 0 has no multiplicative inverse.

EXERCISES 9.2

1. 11   3. 7   5. 6   7. 12   9. 10   11. 5   13. 3
15. 9   17. 11
19. Add 12 to first number; for example,
    $4 - 7 = (4 + 12) - 7 = 16 - 7 = 9 \cdot$
21. 2   23. 1   25. 4   27. 2   29. 3   31. 1
33. Yes, satisfies the five required properties
35. a) {0,1,2,3}; b) *; c) Yes; d) Yes; 0; e) Yes; 0–0, 1–3, 2–2, 3–1; f) Yes; g) Yes; h) Yes
37. a) {13,14,17}; b) ▼; c) Yes   d) Yes, 13; e) Yes; 13–13, 14–17, 17–14; f) Yes; g) Yes; h) Yes
39. a) {1,2,3,4}; b) ⊙; c) Yes; d) No; e) No; f) Yes; g) No; for example, $(1 \odot 2) \odot 3 \neq 1 \odot (2 \odot 3)$; h) No
41. No, not associative: $(\triangle 3 \,\square)3\triangle \neq \triangle 3(\square 3 \,\triangle)$
43. No, no inverse for ∼ or for *
45. Yes, satisfies the five properties. Identity element is d.
47. a)

+	E	O
E	E	O
O	O	E

   b) Yes, satisfies the five properties
51. a) Is closed; identity element is 6; inverses: 1–5, 2–2, 3–3, 4–4, 5–1, 6–6; is associative — for example, $(2 \infty 5) \infty 3 = 2 \infty (5 \infty 3)$
   $$3 \infty 3 = 2 \infty 2$$
   $$6 = 6$$
   b) $3 \infty 1 \neq 1 \infty 3$
   $2 \neq 4$

EXERCISES 9.3

1. Monday   3. Wednesday   5. Thursday   13. 2
15. 4   17. 2   19. 2   21. 2   23. 2   25. 2
27. 6   29. 2   31. 1   33. 1   35. 9   37. 2
39. 4   41. 2   43. { }   45. 6   ·47. 6   49. 7
51. a) 1992, 1996, 2000, 2004, 2008; b) 3004; c) 2504, 2508, 2512, 2516, 2520, 2524
53. a) 3 P.M.–11 P.M.; b) 7 A.M.–3 P.M.; c) 7 A.M.–3 P.M.
55. a)

+	0	1	2	3
0	0	1	2	3
1	1	2	3	0
2	2	3	0	1
3	3	0	1	2

   b) Yes; c) Yes; d) Yes; e) Yes, 0; f) Yes, 0-0, 1-3, 2-2, 3-1; g) Yes; h) Yes
57. a)

×	0	1	2	3
0	0	0	0	0
1	0	1	2	3
2	0	2	0	2
3	0	3	2	1

   b) Yes; c) Yes; d) Yes; e) Yes, 1; f) No; g) No
59. 14 calendars   61. ? = 7   63. ? = 3   65. $x = 0$
67. $x = 1$
69. b) 1)
   $$\begin{array}{r} 4236 \equiv \quad 6 \ (\text{mod } 9) \\ + \ 3784 \equiv + \ 4 \ (\text{mod } 9) \\ \hline 8020 \equiv \quad 1 \ (\text{mod } 9) \end{array}$$
   2)
   $$\begin{array}{r} 8493 \equiv \quad 6 \ (\text{mod } 9) \\ - \quad 237 \equiv -3 \ (\text{mod } 9) \\ \hline 8256 \equiv \quad 3 \ (\text{mod } 9) \end{array}$$
   3)
   $$\begin{array}{r} 187 \ \equiv \quad 7 \ (\text{mod } 9) \\ \times \quad 85 \ \equiv \times \ 4 \ (\text{mod } 9) \\ \hline 15895 \ \equiv \quad 1 \ (\text{mod } 9) \end{array}$$
   4)
   $$\begin{array}{r} 21 \\ 23 \overline{)485} \end{array} \quad \begin{array}{l} 486 \equiv 8 \ (\text{mod } 9) \\ 23 \equiv 5 \ (\text{mod } 9) \\ 21 \equiv 3 \ (\text{mod } 9) \end{array}$$

   with
   $$\begin{array}{r} 46 \\ \hline 25 \\ 23 \\ \hline 2 \end{array}$$

   $$485 \equiv (23 \cdot 21) + 2$$
   $$8 \ (\text{mod } 9) \equiv [5 \ (\text{mod } 9) \cdot 3 \ (\text{mod } 9)] + 2 \ (\text{mod } 9)$$
   $$\equiv 6 \ (\text{mod } 9) + 2 \ (\text{mod } 9)$$
   $$\equiv 8 \ (\text{mod } 9)$$

REVIEW EXERCISES

1. Set of elements, binary operation
2. 5   **3.** 1   **4.** 8   **5.** 4   **6.** 9   **7.** 10
8. Closure, associative, identity element, inverses
9. A commutative group   **10.** Yes
11. No, no inverse for 0   **12.** Yes
13. No, no inverse for 0   **14.** No, p has no inverse.
15. No, no identity element   **16.** Yes
17. No, no identity element   **18.** 2   **19.** 4   **20.** 3
21. 5   **22.** 2   **23.** 1   **24.** 4   **25.** 2   **26.** { }
27. 6   **28.** 0, 2, 4, 6   **29.** 5

30.

+	0	1	2	3	4	5
0	0	1	2	3	4	5
1	1	2	3	4	5	0
2	2	3	4	5	0	1
3	3	4	5	0	1	2
4	4	5	0	1	2	3
5	5	0	1	2	3	4

Yes, a commutative group

31.

×	0	1	2	3
0	0	0	0	0
1	0	1	2	3
2	0	2	0	2
3	0	3	2	1

No, no inverse for 0

CHAPTER 9 TEST

1. A set of elements and a binary operation
2. Closure, identity element, inverses, commutative property, associative property
3. No, not associative and not commutative

4.

+	1	2	3	4
1	2	3	4	1
2	3	4	1	2
3	4	1	2	3
4	1	2	3	4

5. Yes   **6.** No, not closed
7. No, no identity element
8. Yes, is a commutative group
9. 1   **10.** 2   **11.** 1
12. 1   **13.** 2, 5   **14.** 7   **15.** 2
16. No solution

17. a)

+	0	1	2	3	4
0	0	1	2	3	4
1	1	2	3	4	0
2	2	3	4	0	1
3	3	4	0	1	2
4	4	0	1	2	3

b) Satisfies the five properties needed to be a commutative group.

CHAPTER 10

EXERCISES 10.1

1. Ray, $\overrightarrow{AB}$ **3.** Ray, $\overrightarrow{BA}$ **5.** Line, $\overleftrightarrow{AB}$ **7.** {C}
9. ∢BFC   **11.** $\overline{BD}$ **13.** $\overline{BC}$ **15.** $\overrightarrow{BC}$ **17.** { }
19. ∢ABE   **21.** { }   **23.** $\overline{BC}$ **25.** ∢EBC
27. $\overleftrightarrow{AC}$ **29.** $\overline{BE}$ **31.** Two lines in the same plane that do not intersect   **33.** An infinite number   **35.** A line
37. Yes, yes, one   **39.** An angle (∢BAC)
(There may be more than one answer for Exercises 41–49)
41. *BCDG* and *EFGD*   **43.** HGDI, FGDE and
BGDE   **45.** $\overleftrightarrow{HI}$ ∩ *HFEI*   **47.** $\overleftrightarrow{HG}$ and $\overleftrightarrow{BC}$
49. *ABGH*, *GFH*, *GDEF*, *GBCD*
51.

53.

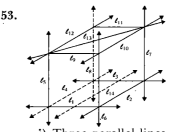

i) Three parallel lines, two lines on the same plane ($\ell_1 \parallel \ell_3 \parallel \ell_9$).
ii) Three lines on the same plane intersect in three distinct points (ℓ_9, ℓ_{10}, and ℓ_{13}).
iii) Three lines in the same plane; two are parallel, and the third line intersects the other two ($\ell_2 \parallel \ell_4$, and ℓ_1 intersects ℓ_2 and ℓ_4).

iv) Three lines intersect at a point, two lines on the same plane (ℓ_1, ℓ_2, ℓ_6).

v) Two lines intersect. A third line is parallel to the first line. The third line and the second line are skew lines (ℓ_9 and ℓ_{10} intersect, $\ell_2 \parallel \ell_{10}$, ℓ_2 and ℓ_9 are skew lines).

vi) Three lines that are skew to each other ($\ell_3, \ell_6, \ell_{13}$).

vii) Three lines are parallel to each other, and all three are in the same plane ($\ell_2 \parallel \ell_4 \parallel \ell_{14}$).

viii) Three lines in the same plane intersect at a single point ($\ell_{10}, \ell_{11}, \ell_{13}$).

ix) Two lines are parallel, and a third line has a skew relationship with the other two lines ($\ell_6 \parallel \ell_8$, and ℓ_{13} is skew to ℓ_6 and ℓ_8).

EXERCISES 10.2

1. Obtuse **3.** Straight **5.** Acute **7.** 60°
9. 22 1/2° **11.** 3° **13.** 85° **15.** 89° **17.** 75°
19. $166\frac{1}{4}°$ **21.** 65.3° **23.** Pentagon **25.** Octagon
27. Heptagon **29.** Right **31.** Obtuse
33. Equilateral **35.** Scalene **37.** Isosceles
39. Parallelogram **41.** Square **43.** iii
45. iv **47.** i **49.** ∡1 = 72°, ∡2 = 18°
51. ∡1 = 144°, ∡2 = 36° **53.** 35° **55.** 170°
57. a)

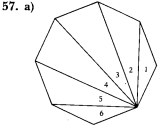

b) 180° × 6 = 1080°; **c)** (8 − 2)180° = (6)(180°) = 1080°
59. 1260° **61.** 720° **63.** 1440° **65.** 60°; 120°
67. 128.6°; 51.4° **69.** 108°; 72° **71.** 55° **73.** 35°
75. Two figures are similar if their corresponding angles have the same measure and their corresponding sides are in proportion.
77. a) 3; **b)** 5 **79. a)** 6; **b)** 3.2
81. a) $1\frac{2}{3}$; **b)** $1\frac{1}{3}$; **c)** 3 **83.** 130° **85.** 50° **87.** 5
89. 12 **91.** 30° **93.** 4 **95.** 8 **97.** 70°
99. a) ∡1 = 119°; ∡2 = 61°
101. ∡1 = 35°; ∡2 = 145°

103. ∡1 = 61°, ∡2 = 29°
105. ∡1 = 19°, ∡2 = 71°

EXERCISES 10.3

1. See Table 10.7 on page 503. **3.** 2 **5.** 30° Celsius
7. c **9.** a **11.** f **13.** mg, 0.001 g **15.** dg, 0.1 g
17. dag, 10 g **19.** kg, 1 000 g **21.** 30 m **23.** 2.4 l
25. 0.001 34 l **27.** 1 427 l **29.** 0.000 076 m
31. 830 cm **33.** 94 500 g **35.** 895 000 ml
37. 0.002 4 km **39.** 620 cm, 5.6 dam, 0.47 km
41. 2.2 kg, 2 400 g, 24 300 dg **43. a)** 1.3 m; **b)** 130 cm
45. 310 cm **47.** 1 hr 23.3 min **49.** 281 g

EXERCISES 10.4

1. 6 in.^{2} **3.** 18 ft^2 **5.** 18 in^2, 20 in **7.** 25 ft^2, 20 ft
9. 6 000 cm^2 or 0.6 m^2, 6.54 m or 654 cm
11. 12.56 in.^{2}; 12.56 in. **13.** 7.065 ft^2, 9.42 ft
15. 9.6 yd^2 **17.** 33.6 yd^2 **19.** 212.4 ft^2
21. 216 000 cm^2 **23.** 0.107 5 m^2 **25.** 53.3 gal
27. $91,850 **29.** $837.20 **31.** 4 times the area
33. 1/4 the area **35.** 4 times the area **37.** Twice the area **39.** 10 cm, 10 cm, $\sqrt{200}$ cm **41.** 525 hectares
43. a) 2410 ft^2; **b)** 10 gal **c)** $189.50 + tax

EXERCISES 10.5

1. 4 ft^3 **3.** 75.36 in.^{3} **5.** 2400 in.^{3} **7.** 292.5 ft^3
9. 900 in.^{3} **11.** 108 ft^3 **13.** 7.85 yd^3
15. a) 54 000 cm^3; **b)** 54 000 ml; **c)** 54l
17. a) 1714.75 cm^3; **b)** 1.71l **19. a)** 108 ft^3; **b)** 4 yd^3
21. 82,944,000 ft^3 **23.** 1.77 mm^3
25. a) A_{round} = 63.6 in.^{2}, A_{rect} = 63 in.^{2}
b) V_{round} = 127.2 in.^{3}, V_{rect} = 126 in.^{3}
27. 6 cuts
29. a) 4 times the volume;
b) Doubles the area of the side
31. No, a half-inch cube has a volume of 1/8 in.^{3}. A half cubic inch has a volume of 1/2 in.^{3}.
33. a) 6.75 ft^3; **b)** 50.6 gal **35.** 18.4 in.
37. a) 0.4239 m^3; **b)** 423.9l; **c)** 423.9 kg
39. ① a^3 ② $a^2 b$ ③ $a^2 b$ ④ $a b^2$ ⑤ $a^2 b$ ⑥ ab^2 ⑦ b^3 ⑧ ab^2

EXERCISES 10.6

1. Yes; start at C or D; end at D or C.
3. Yes; start at any point. **5.** No; 4 odd vertices

7. Yes; start at E or D; end at D or E.

9. a) 2 odd, 4 even; **b)** Yes; **c)** Start at B or E; end at E or B.

11. a) 2 odd, 4 even; **b)** Yes; **c)** Start at B or F; end at F or B.

13. a) 4 odd, 1 even; **b)** No

15. a) 5 odd, 1 even; **b)** No

17. Yes; must start on an island

EXERCISES 10.7

1. One edge   **3.** One strip

5. a) No; **b)** 2; **c)** 2;

  d) Two strips, one inside the other

9. b)   **11. b)**

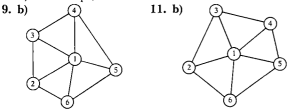

EXERCISES 10.8

1. 330 B.C.

3. a) Undefined terms, definitions, axioms (postulates), theorems

9. Sphere

11. Each type of geometry can be used in its own frame of reference.

13. Positive — elliptical geometry; zero — Euclidean; negative — hyperbolic

REVIEW EXERCISES

1. $\overline{BF}$ **2.** $\overrightarrow{BH}$ **3.** $\{C\}$ **4.** $\triangle BFC$ **5.** $\overleftrightarrow{DC}$

6. $\{\ \}$ **7.** 7 in.   **8.** 10 in.   **9.** 40°   **10.** 80°

11. $\measuredangle 1 = 50°$; $\measuredangle 2 = 60°$; $\measuredangle 3 = 120°$; $\measuredangle 4 = 130°$; $\measuredangle 5 = 50°$; $\measuredangle 6 = 50°$

12. 23 cl **13.** 15 000 ml **14.** 0.67 cl **15.** 3 000 g

16. 0.310 g   **17.** 700 mg   **18.** 0.037 kg

19. 9 807 000 g   **20. a)** 64 m³; **b)** 64 000 kg

21. 20 in.²   **22.** 30 in.²   **23.** 13 in.²   **24.** 72 in.²

25. 50.24 cm²   **26.** 120 in.³   **27.** 401.92 in.³

28. 60 mm²   **29.** 28 ft³   **30.** 540 m³

31. 3140 mm³   **32.** 960 ft³   **33.** $703.45

34. $48\sqrt{2}$ or 67.9 ft³; **b)** 4618.8 lb; **c)** 556.5 gal

35. Yes; start at A or E.   **36.** Yes; start at B or D.

37. Yes; start at D or E.   **38.** Yes; start at any point.

39. No

CHAPTER 10 TEST

1. $\overset{\circ}{AE}$ **2.** $\overline{BC}$ **3.** $\overline{BF}$ **4.** $\overrightarrow{AE}$ **5.** 52.6°

6. 86.3°   **7.** 37°   **8.** 540°   **9.** 7 cm   **10.** 16 in.

11. 48 in.   **12.** 96 in.²   **13.** 10 m²   **14.** 13.238 m

15. 0.2   **16.** 5 000   **17.** 8 000   **18.** 0.10

19. 0.003   **20.** 1   **21.** 252.8 ft³   **22.** 96 ft³

23. Yes, C or D **24.** No

CHAPTER 11

EXERCISES 11.1

1. The official definition by the Weather Service is (d). The specific location is the place where the rain gauge is. The Weather Service uses sophisticated mathematical equations to calculate these probabilities.

5. About 1 in every 6 rolls

7. Empirical probability is the relative frequency of the event. It is based upon past observations. Theoretical probability is based upon the possible outcomes of an experiment.

11. $\frac{1}{6}$

EXERCISES 11.2

5. 0; yes   **7. a)** 3/5; **b)** 1/4; **c)** 3/20

9. a) 0.75; **b)** 400   **11. a)** $\frac{5}{8}$; **b)** $\frac{31}{40}$

13. a) 0; **b)** 0.20; **c)** 1   **15.** 1/4   **17.** 1/2

19. a) 1/2; **b)** 1/2   **21. a)** 0; **b)** 1/4; **c)** 1

23. a) $\frac{13}{14}$; **b)** $\frac{1}{14}$

EXERCISES 11.3

1. 1/13   **3.** 10/13   **5.** 1/2   **7.** 0   **9.** 1/52   **11.** $\frac{1}{12}$

13. $\frac{5}{12}$ **15.** 2/5   **17.** 11/15   **19.** 52/2081

21. 1973/2081   **23.** 1/4   **25.** 0   **27.** $\frac{1}{6}$ **29.** $\frac{1}{2}$

31. $\frac{3}{25}$ **33.** $\frac{17}{25}$ **35.** 3/25   **37.** 2/5   **39.** 3/5
41. 3/13   **43.** 5/13   **45.** 9/13   **47.** 3/4   **49.** 2/3
51. 1/51   **53.** 20/51   **55.** 50/51

EXERCISES 11.4

1. 56   **3.** 23,040   **5.** 720   **7.** 210   **9.** 5040
11. 1120   **13.** 6,760,000   **15.** 728,000   **17.** 154,440
19. a) $2 \times 2 \times 2 = 8$ sample points

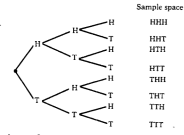

b) $\frac{1}{8}$; **c)** $\frac{3}{8}$

21. a) $4 \times 3 = 12$ sample points

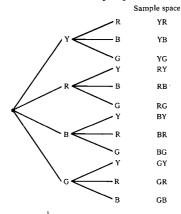

b) 1; **c)** $\frac{1}{2}$

23. a) $3 \times 3 = 9$ sample points

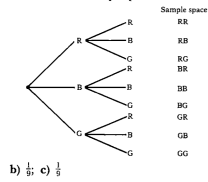

b) $\frac{1}{9}$; **c)** $\frac{1}{9}$

25. a) $2 \times 4 \times 3 = 24$ sample points

b)

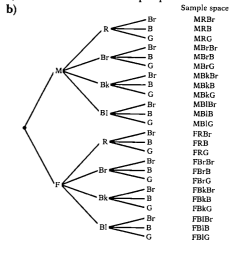

c) $\frac{1}{24}$; **d)** $\frac{1}{8}$

27. H1 T1   **29.** 1/6   **31.** 1/4   **33.** 7/12
    H2 T2
    H3 T3
    H4 T4
    H5 T5
    H6 T6

35. $6 \times 6 = 36$ sample points

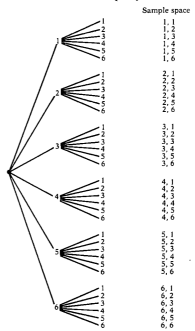

37. $\frac{1}{36}$ **39.** $\frac{11}{36}$ **41.** $\frac{1}{6}$ **43.** $\frac{1}{2}$ **45.** $\frac{2}{9}$

EXERCISES 11.5

1. a) No, both cannot occur at the same time.
 b) No, being healthy can affect your happiness.
3. Events A and B cannot occur at the same time.
 Therefore $P(A \cap B) = 0$.
5. 1/3 **7.** 5/6 **9.** 1 **11.** 3/8 **13.** 3/8 **15.** 2/13
17. 8/13 **19.** 17/26 **21. a)** 1/169; **b)** 1/221
23. a) 1/16; **b)** 1/17 **25. a)** 1/8; **b)** 13/102
27. a) 100/169; **b)** 10/17 **29.** 1/32 **31.** 1/8
33. 1/8 **35. a)** 1/16; **b)** 0 **37. a)** 9/16; **b)** 1/2
39. a) 1000/2197; **b)** 60/143
41. a) 48/2197; **b)** 3/143 **43.** 325/16539 or 0.0197
45. 196/5513 or 0.0356 **47.** 1/2300 **49.** 1/115
51. 0.7 **53.** 0.343 **55.** 1/36 **57.** 1/4 **59.** 1/9
61. 81/1024 **63.** 1/1024 **65.** 1/462
67. a) No; **b)** 0.001; **c1)** 0.00004;
 c2) 0.00096; **c3)** 0.998001
69. 0.04 **71.** 0.24 **73.** 0.024
75. 0.023424 **77.** 1/1000 **79.** 8/15
81. Favors the dealer; the probability of losing is
 (39/52)(38/51) = 0.56.

EXERCISES 11.6

1. 5:1 **3.** 2:1 **5.** 12:1, 1:12 **7.** 3:1, 1:3 **9.** 51:1,
1:51 **11. a)** 999,999:1; **b)** 1:999,999 **13. a)** 2/7;
b) 5/7 **15. a)** 5:1; **b)** 1:5 **17.** 27:1 **19.** 23:5
21. 3/11 **23.** 9/11 **25. a)** 1/40; **b)** 39:1

EXERCISES 11.7

1. $250 **3. a)** $\frac{7}{16}$ in.; **b)** $\frac{217}{16}$ or 13.56 in.
5. a) $3.10 **b)** $3.10
7. a) $-$0.50; **b)** $0.50; **c)** $500
9. a) $-$0.45; **b)** $450,000; **c)** $0.05
11. a) $-$0.17; **b)** $-$0.17; **c)** $-$0.17; **d)** All the same
13. a) 1580; 1430; **b)** $9480; $8580
15. 325 **17.** 170.41 **19.** 20.5 points
21. Favors the dealer. After the first card is chosen, the
 player must choose from the nine remaining cards.
 While four of these lead to a win, five of them lead
 to a loss. The chance of losing is $\frac{5}{9}$ or 55.6%.

23. a1) $\frac{1}{38}$; **a2)** $\frac{6}{19}$; **a3)** $\frac{6}{19}$; **a4)** all $\frac{9}{19}$; **a5)** $\frac{1}{19}$; **a6)** $\frac{3}{38}$;
 a7) $\frac{2}{19}$; **a8)** $\frac{5}{38}$; **a9)** $\frac{3}{19}$;
 b) All have expectation -5.3 cents per dollar
  except bet H, which is -7.9 cents per dollar.
 c) Highest expectation -5.3 cents at all games
  except bet H. Lowest expectation -7.9 cents
  with bet H.
 d1) 37:1; **d2)** 13:6; **d3)** 13:6; **d4)** 10:9; **d5)** 18:1;
 d6) 35:3; **d7)** 17:2; **d8)** 33:5; **d9)** 16:3

EXERCISES 11.8

3. Permutation **5.** Combination **7.** Permutation
9. Permutation **11.** 120 **13.** 56 **15.** 5040
17. 21 **19.** 1 **21.** 30 **23.** n
25. a) 260,000; **b)** 131,040 **27.** 720 **29.** 8640
31. a) 10; **b)** 60 **33.** 120 **35.** 34,650 **37.** 60
39. 6 **41.** 2520 **43.** 3,628,800 **45.** 6,720
47. 40,320 **49.** 7350 **51.** 560 **53.** 22,308
55. 294,000 **57.** 1200
59. a) 12! = 479,001,600; **b)** 103,680; **c)** 27,720; **d)** 6
61. a) 45; **b)** 56 **63. a)** 6; **b)** 10; **c)** nC_2

EXERCISES 11.9

1. 1/5 **3.** 3/14 **5.** 1/12 **7.** 5/28 **9.** 253/9996
11. 1/10 **13.** 7/10 **15.** 1/5 **17.** 3/44 **19.** 34/55
21. $\frac{10}{91}$ **23.** 59/65 **25.** 1/77 **27.** 5/77
29. $\dfrac{1}{3,838,380}$ **31.** $\dfrac{1}{137,085}$
33. a) 1/123,760; **b)** 1/30,940 **35.** 5/729
37. 821/1458 **39. a)** 1/6435; **b)** 56/6435 **41.** 1/24

REVIEW EXERCISES

1. 0.8
3. a)

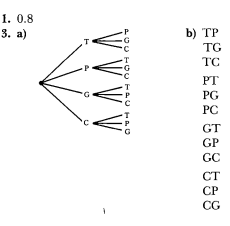

b) TP
 TG
 TC
 PT
 PG
 PC
 GT
 GP
 GC
 CT
 CP
 CG

4. 1/2   **5.** 1/2   **6.** 4/5   **7.** 2/5   **8.** 4/5   **9.** 1
10. a) $\frac{1}{4}$; **b)** $\frac{2}{9}$ **11. a)** $\frac{1}{100}$; **b)** 0   **12. a)** $\frac{4}{25}$; **b)** $\frac{2}{15}$
13. a) $\frac{9}{100}$; **b)** $\frac{1}{15}$ **14.** 1/15   **15.** 0   **16.** 1/5
17. 23:2   **18.** 2/3
19. a) $-\$1.20$; **b)** $-\$3.60$; **c)** $\$0.80$
20. a) $-\$0.15$; **b)** $\$0.15$; **c)** $-\$15.38$
21. a) 120; **b)** $\$222,220$ **22.** 60   **23.** 60
24. 504   **25.** 20   **26. a)** 3003; **b)** 3,628,800
27. 1/8,145,060   **28.** 1120   **29.** 560   **30.** 1/221
31. 1/12   **32.** 1/6   **33.** 1/4   **34.** 11/12

CHAPTER 11 TEST

1. 17/36   **2.** 1/2   **3.** 1/2   **4.** 3/4   **5.** 1/4   **6.** 1
7. 1/8   **8.** 3/14   **9.** 3/14   **10.** 2/7   **11.** 15/28
12. 8/13   **13.** 18

14.

Sample space

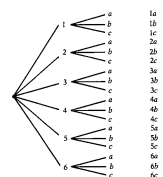

15. 1/18   **16.** 4/9   **17.** 2/3   **18.** 374,400
19. 11:16   **20.** 5:2   **21.** 25 cents   **22.** 120
23. 1/7   **24.** 6/7   **25.** 175/396

CHAPTER 12

EXERCISES 12.1

7. a) A subset of the population
   b) The entire group of items being studied
9. a) A sample obtained by selecting every nth item on a list or production line
   b) Use the random number table to select the first item, then select every nth item after that.
11. a) and **b)** Divide the population into parts (strata), then draw random samples from each stratum.

EXERCISES 12.2

3. Being the largest department store does not imply that it is inexpensive.
5. Most driving is done close to home.
7. Many elderly people retire to Florida.
9. Because four out of six dentists prefer a toothpaste, it does not mean it is the best. Also, the total number of dentists surveyed is not given, and what does "best" mean in this context?
11. Because it has fewer calories one cannot conclude that one will stay healthy by purchasing it.
13. Because most prefer it does not mean that they buy it. Other things such as cost must be considered.

15. a)

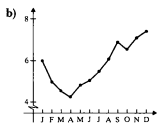

b)

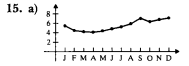

17. a)

Circulation of leading magazines

b)

Circulation of leading magazines

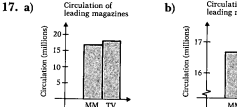

19. a)

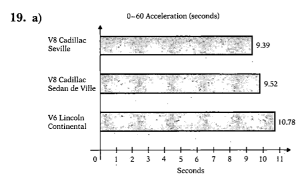

0–60 Acceleration (seconds)

V8 Cadillac Seville — 9.39
V8 Cadillac Sedan de Ville — 9.52
V6 Lincoln Continental — 10.78

Seconds

b) Yes

EXERCISES 12.3

1. a) 19; **b)** 7; **c)** 17; **d)** 42–48; **e)** 49–55

3.

Grade	Number of Students
0	4
1	6
2	3
3	4
4	3
5	3
6	7
7	3
8	12
9	3
10	2

5.

IQ	Number of Students
80– 84	2
85– 89	5
90– 94	8
95– 99	8
100–104	12
105–109	5
110–114	4
115–119	1
120–124	3
125–129	1
130–134	0
135–139	1

7.

IQ	Number of Students
80– 90	8
91–101	22
102–112	11
113–123	7
124–134	1
135–145	1

9.

Age	Number of Presidents
40–45	2
46–51	11
52–57	16
58–63	6
64–69	4

11.

Age	Number of Presidents
42–46	3
47–51	10
52–56	12
57–61	9
62–66	3
67–71	2

13.

Delegates	Number of States
4– 25	18
26– 47	22
48– 69	6
70– 91	3
92–113	3
114–135	0
136–157	1
158–179	1

15.

Delegates	Number of States
0– 25	18
26– 51	25
52– 77	4
78–103	4
104–129	1
130–155	1
156–181	1

17.

Delegates	Number of States
3– 43	24
44– 84	20
85–125	2
126–166	3
167–207	3
208–248	0
249–289	1
290–330	1

19.

Delegates	Number of States
0– 40	24
41– 81	19
82–122	3
123–163	3
164–204	3
205–245	0
246–286	1
287–327	1

21.

Population (millions)	Number of Urban Areas
2.0– 3.9	10
4.0– 5.9	11
6.0– 7.9	9
8.0– 9.9	2
10.0–11.9	2
12.0–13.9	0
14.0–15.9	0
16.0–17.9	1

23.

Population (millions)	Number of Urban Areas
2.9– 5.8	21
5.9– 8.8	10
8.9–11.8	3
11.9–14.8	0
14.9–17.8	1

25.

Number of Visits	Number of Parks
0– 999,999	30
1,000,000– 1,999,999	12
2,000,000– 2,999,999	3
3,000,000– 3,999,999	2
4,000,000– 4,999,999	1
5,000,000– 5,999,999	0
6,000,000– 6,999,999	0
7,000,000– 7,999,999	0
8,000,000– 8,999,999	0
9,000,000– 9,999,999	0
10,000,000–10,999,999	1

27.

Number of Visits	Number of Parks
0– 1,500,000	38
1,500,001– 3,000,001	7
3,000,002– 4,500,002	3
4,500,003– 6,000,003	0
6,000,004– 7,500,004	0
7,500,005– 9,000,005	0
9,000,006–10,500,006	1

29.

Number of Acres	Number of Parks
0– 2,000,000	42
2,000,001– 4,000,001	4
4,000,002– 6,000,002	1
6,000,003– 8,000,003	1
8,000,004–10,000,004	1

DID YOU KNOW? There are 6 f's

EXERCISES 12.4

1. Psychology — 486, Business — 551, Engineering — 385, Sciences — 374, English — 580, History and Political Science — 400, Other — 824.

3.

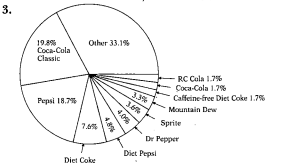

5. a) and **b)**

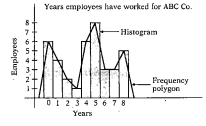

7. a) and **b)**

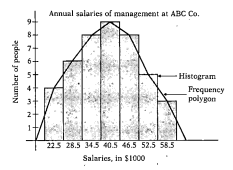

9. a) 29; **b)** 6; **c)** 8; **d)** 237;

e)	Number of Plants	Number of Houses
	6	3
	7	7
	8	8
	9	6
	10	4
	11	0
	12	1

11. a) 8; **b)** 24; **c)** 52,

d)	Visits	Number of Families
	1	4
	2	2
	3	8
	4	8
	5	6
	6	11
	7	9
	8	3
	9	0
	10	1

e)
Number of visits selected families have made to the Pacific Science Center in Seattle, Washington

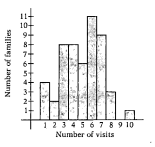

13.

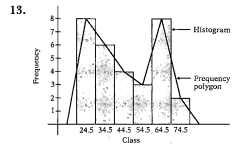

15. a)

	Number of
Salaries	Companies
20	1.
21	7
22	4
23	3
24	2
25	3
26	3
27	2

b) and c)

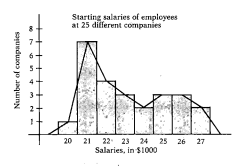

17. a)

Passengers (millions)	Number of Airports
8.0–13.0	22
13.1–18.1	11
18.2–23.2	7
23.3–27.3	2
28.4–33.4	3
33.5–38.5	1
38.6–43.6	2
43.7–48.7	1
48.8–53.8	1

b) and c)

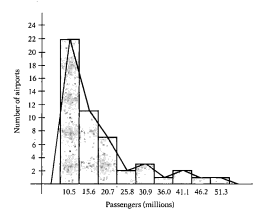

EXERCISES 12.5

1. 5.2, 5, 8, 4.5   **3.** 74.7, 80, none, 70.5
5. 8, 8, none, 8   **7.** 72.9, 60, none, 85
9. 10.7, 11, 12, 10.5   **11.** 477.2, 463, none, 524.5
13. 80.2, 80.5, none, 78
17. $75.62, $80.55, none, $67.07
19. 41, 38.5, 28, 50.5   **21.** 37.6, 33.5, 33, 43.5
23. a) 523.89, 190.5, none, 2141.5;
   b) 10.13, 9.3, none, 11.3; **c)** Singapore
25. 498
27. a) 410,732; **b)** Cannot be determined;
   c) Cannot be determined; **d)** 1,261,250
29. One example: 50, 60, 70, 80, 90
31. 91 or better   **33.** Mode   **35.** No
37. Mary performed better than approximately 85% of those who took the exam.
39. About 25% of the workers had salaries less than $13,750.

41. Median and second quartile
43. a) \$440; **b)** \$450; **c)** 25%; **d)** 25%; **e)** 17%;
  f) 46,000

EXERCISES 12.6

1. First set, because the scores have a greater spread from the mean.
3. All the data are the same.
7. 9, $\sqrt{11.2} = 3.35$
9. 12, $\sqrt{12.67} = 3.56$
11. 0, 0
13. 2, $\sqrt{0.67} = 0.82$
15. 4, $\sqrt{2.86} = 1.69$
17. 28, $\sqrt{120} = 10.95$ **19.** 18, $\sqrt{42} = 6.48$
21. d) When every piece of data in a distribution is multiplied by some number n, the mean and standard deviation of the new set of data will be n times the mean and standard deviation of the original set of data.
  e) $\bar{x} = 20, s = 10$

EXERCISES 12.7

5. The mode is the lowest value, the median is greater than the mode, and the mean is greater than the median.
9. 0.50   **11.** 0.818   **13.** 0.078   **15.** 0.046
17. 0.021   **19.** 0.113   **21.** 41.9%   **23.** 88.9%
25. 96.4%   **27.** 97.5%   **29.** 17.5%   **31.** 34.1%
33. 68.2%   **35.** 81.8%   **37.** 30.8%   **39.** 36.1%
41. 30.8%   **43.** 99.4%   **45.** 50%   **47.** 17.4%
49. 77.3%   **51.** 29.1%   **53.** 59.9%   **55.** 69.2%
57. 42.1%   **59.** 54,560   **61.** 11.1%
63. 3.6%—A, 10%—B, 74.9%—C, 8.6%—D, 2.9%—F
65. a) $\bar{x} = 5.33$; **b)** 3.00;
  c) $\bar{x} + 1.1s = 8.63$
   $\bar{x} - 1.1s = 2.03$
   $\bar{x} + 1.5s = 9.83$
   $\bar{x} - 1.5s = 0.83$
   $\bar{x} + 2s = 11.33$
   $\bar{x} - 2s = -0.67$
   $\bar{x} + 2.5s = 12.83$
   $\bar{x} - 2.5s = -2.17$
  d) $\frac{17}{30} = 56.7\%$, $\frac{28}{30} = 93.3\%$, $\frac{30}{30} = 100\%$, $\frac{30}{30} = 100\%$

	1.1	1.5	2	2.5
e) Minimum (Chebyshev's)	17.4	55.6	75	84
Normal	72.8	86.6	95.4	99.8
Given Distribution	56.7	93.3	100	100

f) Not quite normal; the percent between ± 1.1 is too low.

REVIEW EXERCISES

1. 73   **2.** 74   **3.** None
4. 73.5   **5.** 27   **6.** $\sqrt{116.8} = 10.81$
7. 10   **8.** 10   **9.** None
10. 10.5   **11.** 19
12. $\sqrt{40} = 6.32$

13. a)

Class	Frequency
20	4
21	6
22	1
23	5
24	2
25	1
26	4
27	1
28	2
29	1
30	8

b) and c)

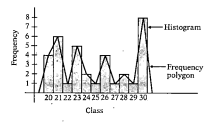

14. a)

Class	Frequency
30–38	4
39–47	4
48–56	6
57–65	7
66–74	7
75–83	7
84–92	4
93–101	1

b) and **c)**

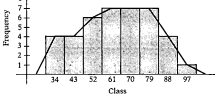

15. 68.2% **16.** 95.4% **17.** 94.5% **18.** 5.5%
19. 72.6% **20.** 50% **21.** 24.9% **22.** 56.8%
23. 81.8% **24.** 175 lb **25.** 180 lb **26.** 25%
27. 25% **28.** 14% **29.** 18,700 lb **30.** 233 lb
31. 145.6 lb **32.** 3.62 **33.** 2 **34.** 3 **35.** 7
36. 14 **37.** $\sqrt{8.35} = 2.89$

38.

Class	Frequency
0–1	7
2–3	14
4–5	10
6–7	5
8–9	1
10–11	1
12–13	0
14–15	1

39. and **40.**

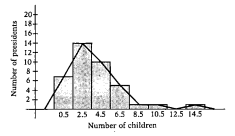

41. No **42.** No **43.** No

CHAPTER 12 TEST

1. 11 **2.** 12 **3.** 15
4. 9 **5.** 12
6. $\sqrt{24.5} = 4.95$

7.

Class	Frequency
5–10	7
11–16	5
17–22	1
23–28	7
29–34	5
35–40	3
41–46	2

8.

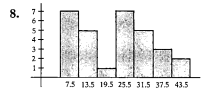

9.

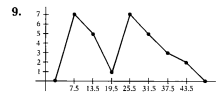

10. $475 **11.** $450 **12.** 75%
13. 79% **14.** $48,000 **15.** $520
16. $420 **17.** 34.1% **18.** 34.5% **19.** 33.7%
20. 27.4%

CHAPTER 13

EXERCISES 13.1

1. Developed the transistor
3. Developed the microprocessor
5. Watson was responsible for making IBM the dominant force in the computer business.
7. The delivery of the first UNIVAC to a client, June 14, 1951.
9. *First* — vacuum tubes; *second* — transistors; *third* — integrated circuits; *fourth* — very large-scale integrated circuits.
11. Many integrated circuits etched onto a single chip.

EXERCISES 13.2

1. *Input* — enter data and instructions; *memory or storage* — store data and instructions; *control* — open and close gates and synchronize various speeds of different parts of the computer; *processing (ALU)* — perform calculations and comparisons; *output* — obtain information from the computer.
3. Printer, monitor, paper, tape, typewriter, punched cards
5. It is converted to binary notation
7. Kilobyte
9. *Primary storage* — storage built into the computer; *secondary storage* — add-on storage devices, magnetic disks and tapes
11. Magnetic disk, magnetic tapes, magnetic bubbles
13. Equal to, less than, greater than
15. $=, <, >, <=, >=, <>$ (or $\neq$)

EXERCISES 13.3

1. Beginners All-purpose Symbolic Instruction Code
3. 17   5. 4   7. 8   9. -8 11. 24   13. -16
15. 15   17. 12   19. -12 21. $2 + 3 * 4 = 14$
23. $(2 + 3\hat{\ }2) * 5 = 55$ 25. $(\text{SQR}(25) + 6)/2 = 5.5$
27. $3\hat{\ }2/2\hat{\ }2 = 2.25$
29. $(4\hat{\ }2 - 6)/(5 * \text{SQR}(36)) = .333333333$
31. $4 * 3 - (5*6)\hat{\ }2 = -888$
33. $-(3\hat{\ }2) + 4 * 7 - 30/5 = 13$
35. $(-2\hat{\ }2 + 3)\hat{\ }2/(4 + 5) = 5.44444445$ 37. $4x + 4$
39. x^2 41. $3\sqrt{x} + 4$ 43. $4x^2 - 5x + 3$
45. $\dfrac{5}{x} + 2x - 3$ 47. $-3\sqrt{x} - 5x$ 49. $3\sqrt{x^3} - 6x^3$

51. $3(7x^2 + 4x)$ 53. $3*X + 4$ 55. $3*X\hat{\ }2$
57. $(X + 4)/(2*X + 3)$ 59. $3*\text{SQR}(X) - 7*X + 1$
61. $6/X\hat{\ }2 + 5*X + 4$ 63. $(2*X\hat{\ }2 + 6*X - 4)/(X - 3)$
65. $((-2*X)\hat{\ }3 + 4)\hat{\ }2$ 67. $(2*X - 3)\hat{\ }2*(X\hat{\ }2 + 5)$

69.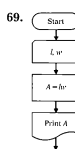

71.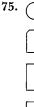

73.

75.

77.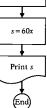

79.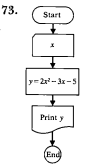

81.

83.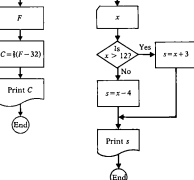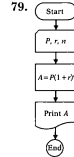

85.

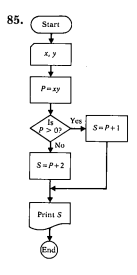

87.

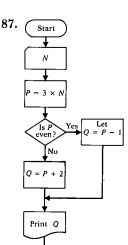

89.

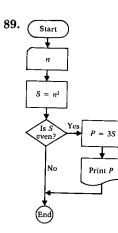

7. 10 DATA 6, 4
20 READ X, Y
30 LET Z = 4 * X^2 * Y + 5 * Y
40 PRINT Z
50 END

9. 10 INPUT C
20 LET F = 9 / 5 * C + 32
30 PRINT F
40 END

11. INPUT I, F, L, T
20 LET A = 2 * I / ((F + L) * T)
30 PRINT A
40 END

13. 10 LET H = 10
20 LET B1 = 8
30 LET B2 = 20
40 LET A = 1 / 2 * H * (B1 + B2)
50 PRINT A
60 END

15. 10 INPUT B, C, D, E, F
20 LET A = (B + C + D + E + F) / 5
30 PRINT A
40 END

EXERCISES 13.4

1. 10 INPUT S
20 LET A = S^2
30 PRINT A
40 END

3. 10 DATA 9, 5
20 READ L, W
30 LET A = L * W
40 PRINT A
50 END

5. 10 INPUT R
20 LET C = 2 * 3.14 * R
30 PRINT C
40 END

EXERCISES 13.5

1. 10 LET X = 5   **3.** 10 INPUT X, Y

5. 10 LET Y = 4*X

7. 10 PRINT "Y = 4 + 5" or
10 PRINT "Y="; 4 + 5

9. 10 PRINT "I LIKE MATH"

11. 10 DATA 23, 400, 75
20 READ X, Y, Z

13. 10 PRINT "Y = 2 + 3" or
10 PRINT "Y = "; 2 + 3

15. a) 4; **b)** 10; **c)** X=3; **d)** Y=   4;
e) (Blank line); **f)** X + Y = 7;
g) 13; **i)** 7   12
Computer will not print anything for line h.

17. 120   **19.** 77   **21.** AMOUNT IS 110.25

23. 10 DATA 13.3, 4.6
20 READ D, T
30 LET S = D/T
40 PRINT S
50 END

25.
```
10 DATA 3.4, 6.7
20 READ R,H
30 LET V = 1/3*3.14*R^2*H
40 PRINT V
50 END
```

27. MY NAME IS GEORGE
```
300
20 × 15 =    300
X + Y = 13
X + Y =      13
68
```

29. a)
```
10 INPUT M
20 LET K = 1.6 * M
30 PRINT M; "MILES ="; K; "
   KILOMETERS"
40 END
```
b)
```
10 INPUT M
20 LET K= 1.6 * M
30 PRINT "MILES", "KILOMETERS"
40 PRINT
50 PRINT M, K
60 END
```

31.
```
10 PRINT "PICK ANY TWO NUMBERS"
20 INPUT X,Y
30 PRINT "THEIR PRODUCT IS"; X*Y
40 PRINT "PICK TWO DIFFERENT
   NUMBERS"
50 INPUT A,B
60 PRINT "THEIR SUM IS "; A+B
70 END
```

33.
```
?3
9
?5
15
```

35. a)
```
10 DATA 41
20 READ F
30 LET C = 5 / 9 * (F − 32)
40 PRINT F;" FAHRENHEIT = ";C;
   "CELSIUS"
50 END
```
b)
```
10 DATA 41
20 READ F
30 LET C = 5 / 9 * (F − 32)
40 PRINT "FAHRENHEIT", "CELSIUS"
50 PRINT
60 PRINT "  ";F,"  ";C
70 END
```

37. 5.62   5   2   25

```
39. 10 PRINT "X X X X X   X X X X X   X X X X X X X   X X X X X X X   X X X X X"
    20 PRINT "X X     X X             X             X X             X        "
    30 PRINT "X X X X X   X X X X X   X X X X X X X   X             X X X X X"
    40 PRINT "X X         X           X             X X             X        "
    50 PRINT "X X         X           X             X X             X        "
    60 PRINT "X X         X X X X X   X             X   X X X X X X   X X X X X"
    70 END
```

EXERCISES 13.6

1. THE
   YOUNG
   MAN
   IS
   HERE

3. 1   0
   2   3
   3   8
   4   15
   5   24

5. 55

7. 1
   8
   27
   64
   OUT OF DATA LINE 10

9. a)
```
10 LET X = 1
20 PRINT X
30 IF X = 7 THEN 60
40 LET X = X + 1
50 GO TO 20
60 END

10 LET X = 1
20 IF X > 7 THEN 60
30 PRINT X
40 LET X = X + 1
50 GO TO 20
60 END
```

b) 1   1
   2   2
   3   3
   4   4
   5   5
   6   6
   7   7

c) Same

11.
```
10 LET X = 1
20 PRINT SQR (X)
30 LET X = X + 1
40 IF X < = 25 THEN 20
50 END
```
Alternative Solution:
```
10 FOR I = 1 TO 25
20 PRINT SQR (I)
30 NEXT I
40 END
```

13. a)
```
10 INPUT X
20 LET Y = 3 * X + 4
30 PRINT Y
40 END
```
b)
```
10 DATA 1, 2, 3, 4, 5, −1
20 READ X
30 IF X = −1 THEN 70
40 LET Y = 3 * X + 4
50 PRINT Y
60 GO TO 20
70 END
```
c)
```
10 FOR X = 1 TO 5
20 LET Y = 3 * X + 4
30 PRINT Y
40 NEXT X
50 END
```

15. a) INPUT ANY TWO NUMBERS
   ?10, 10
   10 IS EQUAL TO 10
b) INPUT ANY TWO NUMBERS
   ?6, 12
   6 IS LESS THAN 12
c) INPUT ANY TWO NUMBERS
   ?15, 4
   15 IS GREATER THAN 4

17.
```
10 REM THIS PROGRAM FINDS SQUARE
   ROOTS AND SQUARES
20 PRINT "FIRST NUMBER"
30 INPUT A
40 PRINT "LAST NUMBER"
50 INPUT B
60 PRINT
70 PRINT "N", "SQUARE ROOT OF N",
   "N − SQUARED"
80 FOR N = A TO B
90 LET R = SQR (N)
100 LET S = N^2
110 PRINT N, R, S
120 NEXT N
130 END
```

19. 12   21. 4

23. 2 REM P IS THE PRINCIPAL BALANCE DUE
        EACH MONTH
     3 REM A IS THE AMOUNT PAID ON THE
        PRINCIPAL EACH MONTH
     10 LET P = 7000
     20 PRINT "NUMBER", "INTEREST",
        "AMOUNT PAID", "BALANCE"
     30 FOR N = 1 TO 36
     40 LET I = P * .01 * 1
     50 LET A = 232.50 - I
     60 LET P = P - A
     70 PRINT N, I, A, P
     80 NEXT N
     90 END

25. a) 10 INPUT A, B, C, D, E
       20 LET M = (A + B + C + D + E) / 5
       30 PRINT M
       40 END

    b) 10 INPUT A, B, C, D, E
       20 LET M = (A + B + C + D + E) / 5
       30 LET X = (A - M)^2 + (B - M)^2
                  + (C - M)^2 + (D - M)^2
                  + (E - M)^2
       40 LET S = SQR (X/5)
       50 PRINT S
       60 END

27. 10 LET F = 1
    20 LET N = 0
    30 LET L = 1
    40 PRINT "FIBONACCI NUMBER"
    50 PRINT F
    60 FOR I = 1 TO 100
    70 LET F = L + N
    80 PRINT F
    90 LET N = L
    95 LET L = F
    100 NEXT I
    110 END

EXERCISES 13.7

The answers for exercises 1 through 9 have been
reduced from the given scale.

1.   3.   5.

7.   9.

11. RT 90 FD 40 LT 90 FD 20 LT 90 FD 20
    RT 90 FD 20
13. RT 90 FD 30 RT 180 FD 60 RT 135 FD 43
    RT 135 FD 30 BK 30 LT 90 FD 30
15. LT 45 FD 40 RT 90 FD 40 RT 90 FD 40
    RT 90 FD 40 RT 135
17. REPEAT 4[FD 20 FD 10 RT 90 FD 10 RT 90
    FD 10 RT 90 FD 10 RT 90 BK 20 RT 90]
19. LT 90 FD 20 RT 90 FD 30 RT 90 FD 40
    RT 90 FD 30 RT 90 FD 20 RT 90

21.   23.   25.

27.   29.

31. SETX 50 SETY 30 SETX -40 SETY -10

33. SETX 40 SETY 40 SETX 0
    SETY 10 SETX 20

35.  **37.**

39. LT 45 FD 20 RT 90 FD 40 RT 90 FD 20
RT 90 FD 20 RT 90 FD 40 LT 90 FD 20
BK 40 LT 90 FD 40 RT 90 FD 40 RT 90
FD 40 HOME
41. REPEAT 4[LT 30 FD 20 RT 120 FD 20
RT 120 FD 20 LT 120]
43. LT 90 PENTAGON REPEAT 2[RT 108 FD 40
LT 72 PENTAGON]

REVIEW EXERCISES

2. *First* — vacuum tubes; *second* — transistors; *third* —
integrated circuits; *fourth* — very large scale
integrated circuits
3. Von Neumann showed how to store programs in a
computer.
4. Input, storage or memory, control, processing,
output
5. $3 + 4^2$ **6.** $4 - (6^2 * 3) + 4$
7. $3 * (SQR(25) + 6)/(4 - 3^2)$ **8.** $5 - 3 * X^2$
9. $2 * X^2 + 4 * X - 6$ **10.** $(3*X*Y - X^2)/(X - 4)$
11. 23   **12.** 10   **13.** 54   **14.** 6   **15.** 7   **16.** 10
17.

```
Start

b, h

A = ½bh

Print A

End
```

18. 10 INPUT B, H
20 LET A = 1/2 * B * H
30 PRINT A
40 END

19. 10 INPUT X
20 LET Y = X^2 + 4 * X - 6
30 PRINT Y
40 END
20. 10 LET X = 1.5
20 LET Y = X^3 - 2 * X + 4
30 PRINT Y
40 END

21. GOOD MORNING
THE SUM OF X AND Y IS 12
THE PRODUCT IS 35

GOOD DAY
22. 10 PRINT "X", "X SQUARED"
20 FOR X = 1 TO 5
30 LET Y = X^2
35 PRINT
40 PRINT X, Y
50 NEXT X
60 END

23. a)

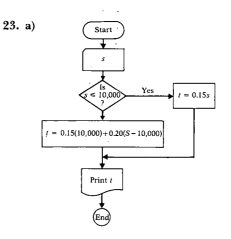

b) 10 INPUT S
20 IF S <= 10000 THEN 60
30 T = .15 * 10000 + .20 *
(S - 10000)
40 PRINT T
50 GO TO 80
60 LET T = .15 * S
70 PRINT T
80 END
24. 10 PRINT "INTEGER", "INTEGER
MULTIPLIED BY 5"
20 FOR I = 1 TO 5
30 PRINT I, 5 * I
40 NEXT I
50 END

25. `1Ø PRINT "X", "Y"`
`2Ø FOR X = 2 TO 5 STEP .1`
`3Ø LET Y = 2 * X^2 - 4 * X + 5`
`4Ø PRINT X,Y`
`5Ø NEXT X`
`6Ø END`

26. `1Ø PRINT "NUMBER", "SUM"`
`2Ø LET S = Ø`
`3Ø FOR N = 1 TO 5`
`4Ø LET S = 2 * N + S`
`5Ø PRINT 2 * N, S`
`6Ø NEXT N`
`7Ø END`

The answers for Exercises 27 through 31 have been reduced from the given scale.

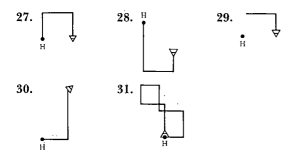

27. **28.** **29.**

30. **31.**

CHAPTER 13 TEST

1. Commodore Grace Hopper
2. First generation used vacuum tubes.
Second generation used transistors.
Third generation used integrated circuits or chips.
Fourth generation use very large-scale integrated circuits and microprocessors.
3. *Input* — to feed information and instructions into computer memory or storage.
Control — opens the proper combinations of gates. Fetches instructions at proper time. Synchronizes different parts of computer.
Processing — performance of calculations and comparisons.
Output — method by which computer gives results to its user.
Storage — stores data and instructions.
4. 24 **5.** 5 **6.** 14 **7.** 5 + (7 * 9), 68

8. (5 + 7) * 9^2, 972 **9.** SQR(64) + 15/3, 13
10. `1Ø LET R = 4`
`2Ø LET C = 6.28*R`
`3Ø PRINT C`
`4Ø END`
11. `1Ø INPUT X, Z`
`2Ø LET R=2*X+Z`
`3Ø PRINT R`
`4Ø END`
12. 9
13. THE VOLUME OF THE BOX IS 336
14. 17
15.

```
Start
  |
R=55
T=3.5
  |
D=R×T
  |
Print D
  |
End
```

16. `1Ø LET R = 55`
`2Ø LET T = 3.5`
`3Ø LET D = R*T`
`4Ø PRINT D`
`5Ø END`

17.

```
Start
  |
N=1
  |
A=N^2
  |
Print A
  |
Is N=10? --Yes--> End
  |No
N=N+1
```

18. `1Ø LET N = 1`
`2Ø LET A = N^2`
`3Ø PRINT A`
`4Ø IF N = 1Ø THEN 7Ø`
`5Ø LET N = N + 1`
`6Ø GO TO 2Ø`
`7Ø END`

The answers for Exercises 19 and 20 have been reduced from the given scales.

19.   **20.**

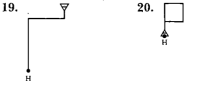

Index

A BRIEF LOOK AT THE HISTORY OF MATHEMATICS

Mathematical Periods	CHAPTER 7 Metric System	CHAPTER 8 Mathematical Systems	CHAPTER 9 Geometry
Egyptian and Babylonian period 3000 B.C.–600 B.C.			**3000 B.C.** Great pyramids constructed. Although the mathematics involved was accurate, the procedures used were trial and error, rather than formal mathematical techniques. **2000 B.C.** Volume of frustrum of pyramid, formulas for land area and granary volumes, rough formulas for area of circle and cylinder in Moscow Papyrus.
Greek period 600 B.C.–A.D. 500			**600 B.C.** Thales, said to be the first mathematician to prove a theorem, is considered the first true mathematician. Pythagoras made numerous contributions to geometry, including the Pythagorean theorem. **300 B.C.** Euclid's *Elements* laid the foundation for present-day high school geometry. The *Elements* emphasized the deductive method. **340 B.C.** Archimedes computed π, conic sections, Archimedes spiral.
Hindu and Arabian period (the dark ages of mathematics) A.D. 500–A.D. 1200			**980** Abul-Wefa—geometric constructions and trigonometric tables. **1100** Omar Khayyam—geometry, solutions of cubic equations, calendar.
Period of transition A.D. 1200–A.D. 1550 Modern period A.D. 1550–present (Century of geniuses 1600–1700)	**1660** Mouton, founding father of the metric system. **1795** Metric system formally adopted in France. **1866** An act passed by Congress made legal the use of the metric system for business transactions. **1875** Treaty of the meter signed by the United States and 19 other countries. Development of the International Bureau of Weights and Measures. **1965** Great Britain adopted the metric system. **1975** Metric Conversion Act signed by President Ford.	**1820** Guass introduced modulo arithmetic. **1825** Abel's work on elliptical functions led to Galois's development (1830) of the theory of groups. **1857** Cayley invented matrices.	**1634** Herigone first used (to represent angle. **1636** Descartes—formula relating edges, faces, and vertices of convex polyhedron. **1640** Desargues introduced projective geometry, conic sections. **1650** Pascal—projective geometry, conic sections, Pascal's triangle. **1733** Saccheri, forerunner of non-Euclidean geometry. **1750** Euler contributed much to geometry and topology, including Euler's formulas. **1805** Legendre reorganized Euclid's *Elements* in his *Elements de Geometrie.* Also contributed to calculus and number theory. **1825** Gauss, Bolyai, and Lobachevski contributed to the development of non-Euclidean geometry. Gauss—point set topology. **1854** Riemann—Riemannian geometry, Riemann surfaces, Riemann integral in calculus. **1895** Poincaré—systematic development of combinatorial topology. **1899** Hilbert—formalization of geometry, contributions to number theory and differential equations.